Lecture Notes in Computer Science 6917

Commenced Publication in 1973
Founding and Former Series Editors:
Gerhard Goos, Juris Hartmanis, and Jan van Leeuwen

Bart Preneel Tsuyoshi Takagi (Eds.)

Cryptographic Hardware and Embedded Systems – CHES 2011

13th International Workshop
Nara, Japan, September 28 – October 1, 2011
Proceedings

Volume Editors

Bart Preneel
Katholieke Universiteit Leuven
3001 Leuven, Belgium
E-mail: bart.preneel@esat.kuleuven.be

Tsuyoshi Takagi
Kyushu University
Institute of Mathematics for Industry
Fukuoka, 819-0395, Japan
E-mail: takagi@imi.kyushu-u.ac.jp

ISSN 0302-9743 e-ISSN 1611-3349
ISBN 978-3-642-23950-2 ISBN 978-3-642-23951-9 (eBook)
DOI 10.1007/978-3-642-23951-9
Springer Heidelberg Dordrecht London New York

Library of Congress Control Number: 2011935857

CR Subject Classification (1998): E.3, D.4.6, K.6.5, E.4, C.2, G.2.1

LNCS Sublibrary: SL 4 – Security and Cryptology

Typesetting: Camera-ready by author, data conversion by Scientific Publishing Services, Chennai, India

Printed on acid-free paper

Springer is part of Springer Science+Business Media (www.springer.com)

Preface

The 13th International Workshop on Cryptographic Hardware and Embedded Systems (CHES 2011) was held at Todai-ji Cultural Center, Nara, Japan, from September 28 to October 1, 2011. The workshop was sponsored by the International Association for Cryptologic Research.

CHES 2011 received 119 submissions from 26 countries all over the world. Each paper was reviewed by at least 4 committee members, for a total of 517 reviews; papers with a committee member as co-author received at least 5 reviews. More than 150 external subreviewers contributed to the review process in their particular areas of expertise. One article was identified as an irregular submission. The Program Committee selected 32 papers for publication in the proceedings. Two of these papers are the result of merging two pairs of closely related submissions. The program was completed with two excellent invited talks given by Ernie Brickell (Intel) and Tetsuya Tominaga (NTT Laboratories). Nominations for the best paper award were solicited among the Program Committee; an ad hoc committee with no conflicts with the shortlisted papers made the final selection. They decided to award the best paper award of CHES 2011 to Michael Hutter and Erich Wenger for their work "Fast Multi-Precision Multiplication for Public-Key Cryptography on Embedded Microprocessors." The runners-up were the papers "To Infinity and Beyond: Combined Attack on ECC Using Points of Low Order" by Junfeng Fan, Benedikt Gierlichs and Frederik Vercauteren, and "Breaking Mifare DESFire MF3ICD40: Power Analysis and Templates in the Real World" by David Oswald and Christof Paar. The authors of these articles were invited to submit an extended version to the *Journal of Cryptology*.

Many people contributed to the success of CHES 2011. First we would like to thank all the authors who submitted their research results. The selection of 32 papers from 119 submissions was a challenging task and we sincerely thank the 42 Program Committee members, as well as the external reviewers, who volunteered to read and discuss the papers over several months. We are greatly indebted to the General Chair, Akashi Satoh, for his relentless efforts that include relocating the conference within short notice because of the earthquake and tsunami in March 2011. We would also like to thank the local Organizing Committee from the Japanese cryptologic community for their continuous support. The submission and review process as well as the editing of the final proceedings were facilitated by the software written by Shai Halevi. The CHES 2011 website was maintained by Jens-Peter Kaps. We would like to thank Shai and Jens-Peter for their excellent support. Finally we want to express our gratitude to our generous sponsors: Cryptographic Research, SASEBO project, Nara Visitors Bureau, NTT, IPA, Mitsubishi Elecric, Morita Tech, NICT, Riscure, ETRI, Tokyo Electron Device, Kayamori Foundation, Technicolor, Telecom ParisTech,

Intrinsic-ID, Hitachi, Oberthur Technologies, IIJ, Toshiba, SPACES project, LG CNS, and Fujitsu.

As embedded systems become ever more pervasive, there is a growing need to develop efficient and secure implementations that help to safeguard our security and privacy. We hope that the papers in this volume prove valuable for your research and professional activities in this area.

September 2011

Bart Preneel
Tsuyoshi Takagi

CHES 2011

Workshop on Cryptographic Hardware and Embedded Systems
Nara, Japan, September 28 – October 1, 2011.

Sponsored by the *International Association for Cryptologic Research.*

General Chair

Akashi Satoh	National Institute of Advanced Industrial Science and Technology, Japan

Program Co-chairs

Bart Preneel	Katholieke Universiteit Leuven, Belgium
Tsuyoshi Takagi	Kyushu University, Japan

Program Committee

Toru Akishita	Sony Corporation, Japan
Paulo Barreto	University of São Paulo, Brazil
Lejla Batina	Radboud University Nijmegen, The Netherlands and Katholieke Universiteit Leuven, Belgium
Daniel J. Bernstein	University of Illinois at Chicago, USA
Guido Bertoni	STMicroelectronics, Italy
Swarup Bhunia	Case Western Reserve University, USA
Chen-Mou Cheng	National Taiwan University, Taiwan
Jean-Sebastien Coron	University of Luxembourg, Luxembourg
Emmanuelle Dottax	Oberthur Technologies, France
Hermann Drexler	Giesecke & Devrient, Germany
Martin Feldhofer	Graz University of Technology, Austria
Pierre-Alain Fouque	ENS, France
Kris Gaj	George Mason University, USA
Benedikt Gierlichs	Katholieke Universiteit Leuven, Belgium
Louis Goubin	Université de Versailles, France
Jorge Guajardo	Robert Bosch LLC, Research and Technology Center, USA
Dong-Guk Han	Kookmin University, Korea
Helena Handschuh	Intrinsic-ID, USA and Katholieke Universiteit Leuven, Belgium
Anwar Hasan	University of Waterloo, Canada
Naofumi Homma	Tohoku University, Japan

Marc Joye	Technicolor, France
Pascal Junod	HEIG-VD, Switzerland
Shinichi Kawamura	AIST, Japan
Paris Kitsos	Hellenic Open University, Greece
Markus Kuhn	Cambridge University, UK
Kerstin Lemke-Rust	University of Applied Sciences Bonn-Rhein-Sieg, Germany
Stefan Mangard	Infineon Technologies, Germany
Mitsuru Matsui	Mitsubishi Electric, Japan
David Naccache	ENS, France
Heike Neumann	NXP, Germany
Elisabeth Oswald	University of Bristol, UK
Christof Paar	Ruhr University of Bochum, Germany
Matt Robshaw	Orange Labs, France
Pankaj Rohatgi	Cryptography Research, USA
Ahmad-Reza Sadeghi	TU Darmstadt and Fraunhofer SIT, Germany
Kazuo Sakiyama	University of Electro Communications, Japan
Erkay Savas	Sabancı University, Turkey
Patrick Schaumont	Virginia Tech, USA
Nigel P. Smart	University of Bristol, UK
Masahiko Takenaka	Fujitsu Laboratories, Japan
Colin Walter	Royal Holloway, University of London, UK

External Reviewers

Diego F. Aranha
Frederik Armknecht
Jean-Philippe Aumasson
Selcuk Baktir
Josep Balasch
Alessandro Barenghi
Claude Barral
Timo Bartkewitz
Georg T. Becker
Thomas Behling
Alexandre Berzati
Markus Bockes
Arnaud Boscher
Murat Cenk
Zhimin Chen
Tung Chou
Christophe Clavier
Jeremy Cooper
Joan Daemen
Elke De Mulder
Fabrizio De Santis
Benedikt Driessen
Orr Dunkelman
Paul Duplys
Ilze Eichhorn
Wieland Fischer
Nicolas Gama
Berndt Gammel
Christophe Giraud
Robert Granger
Johann Großschädl
Eric Guo
Anwar Hasan
Yuichi Hayashi
Olaf Heemskerk
Francisco R. Henriquez
Christoph Herbst
Clemens Heuberger
Stefan Heyse
Markus Hinkelmann
Harunaga Hiwatari
Michael Hutter
Sebastiaan Indesteege
Mawa N. Ismail
Takanori Isobe
Kouichi Itoh
Tetsuya Izu
Josh Jaffe
Dipti Kapadia
Markus Kasper
Michael Kasper
Timo Kasper
Jonathan Katz
Hee Seok Kim
Inyoung Kim
Mario Kirschbaum
Aswin Kishna
Miroslav Knežević
Kazuyuki Kobayashi
Ünal Kocabas

Masanobu Koike
Yuichi Komano
Daniel Krenn
Po-Chun Kuo
Masafumi Kusakawa
Soonhak Kwon
Tanja Lange
Mun-Kyu Lee
Yang Li
Raimondo Luzzi
Gilles Macariot-Rat
Marco Macchetti
Abhranil Maiti
Mark Marson
Ange Martinelli
Pedro Maat C. Massolino
Nicolas Meloni
Filippo Melzani
Atsushi Mitsuda
Hideyuki Miyake
Atsushi Miyamoto
Amir Moradi
Carlos Moreno
Andrew Moss
Bruce Murray
Daisuke Nakatsu
Seetharam Narasimhan
Phong Nguyen
Ruben Niederhagen
Ventzi Nikov
Hanae Nozaki
Katsuyuki Okeya
David Oswald
Dan Page
Jing Pan
Jacques Patarin
Gerardo Pelosi
Geovandro Pereira
Gilles Piret
Thomas Plos
Jerome Plut
Axel Poschmann
Emmanuel Prouff
Jürgen Pulkus
Michaël Quisquater
Francesco Regazzoni
Christof Rempel
Matthieu Rivain
Thomas Roche
Marcin Rogawski
Carsten Rudolph
Koichi Sakumoto
Gokay Saldamli
Jörn-Marc Schmidt
Geert-Jan Schrijen
Steffen Schulz
Peter Schwabe
Michael Scott
Rabia Shahid
Umar Sharif
Kyoji Shibutani
Kouichi Shimizu
Hideo Shimizu
Taizo Shirai
Herve Sibert
Yannick Sierra
Peter Simons
Marcos A. Simplicio Jr.
Dave Singelée
Martijn Stam
Daehyun Strobel
Takeshi Sugawara
Ruggero Susella
Daisuke Suzuki
Robert Szerwinski
Masahiko Takenaka
Shigeki Teramoto
Gilles Van Assche
Vincent van der Leest
Erik van der Sluis
Marten van Hulst
Jasper van Woudenberg
Marc Vauclair
Ingrid Verbauwhede
Frederik Vercauteren
Marion Videau
Christian Wachsmann
Lei Wang
Xinmu Wang
Erich Wenger
Carolyn Whitnall
Jun Yajima
Tolga Yalcin
Panasayya Yalla
Dai Yamamoto
Ralf Zimmermann

Table of Contents

FPGA Implementation

AES

Elliptic Curve Cryptosystems

Lattices

Side Channel Attacks

Invited Talk

Fault Attacks

Lightweight Symmetric Algorithms

PUFs

Public-Key Cryptosystems

Hash Functions

An Exploration of Mechanisms for Dynamic Cryptographic Instruction Set Extension

Philipp Grabher[1], Johann Großschädl[2], Simon Hoerder[1], Kimmo Järvinen[3], Dan Page[1], Stefan Tillich[1], and Marcin Wójcik[1]

[1] University of Bristol, Department of Computer Science, Merchant Venturers Building, Woodland Road, Bristol, BS8 1UB, UK
{grabher,hoerder,page,tillich,wojcik}@cs.bris.ac.uk

[2] University of Luxembourg, FSTC, CSC Research Unit, LACS, 6, rue Richard Coudenhove-Kalergi, L–1359 Luxembourg, Luxembourg
johann.groszschaedl@uni.lu

[3] Aalto University, Department of Information and Computer Science, P.O. Box 15400, FI–00076 Aalto, Finland
kimmo.jarvinen@aalto.fi

Abstract. Instruction Set Extensions (ISEs) supplement a host processor with special-purpose, typically fixed-function hardware components and instructions to utilize them. For cryptographic use-cases, this can be very effective due to the demand for non-standard or niche operations that are not supported by general-purpose architectures. However, one disadvantage of fixed-function ISEs is inflexibility, contradicting a need for "algorithm agility." This paper explores a new approach, namely the provision of re-configurable mechanisms to support dynamic (run-time changeable) ISEs. Our results, obtained using an FPGA-based LEON3 prototype, show that this approach provides a flexible general-purpose platform for cryptographic ISEs with all known advantages of previous work, but relies on careful analysis of the associated security issues.

Keywords: FPGA, embedded processor, instruction set extension.

1 Introduction

Cryptographic kernels could be described as the archetype target for Instruction Set Extensions (ISEs) [31]. Starting with a general-purpose host processor, the idea is to specify an ideally minimal set of (more) special-purpose instructions [25]. By carefully integrating instructions, plus any tightly coupled hardware to support their execution, the goal is more effective implementation of the kernel in question (e.g., with respect to efficiency, memory footprint, or security). This approach can be ideal for cryptography where performance bottlenecks often relate to non-standard or niche operations, and can easily be resolved using a targeted ISE. There exists a wealth of related work to support this premise, see e.g., [10,17,31]. Even when focused on one kernel such as AES, said work spans academic results and evaluation on platforms such as the LEON3, through to commercialisation in workstation-class Intel processors via AES-NI [34].

B. Preneel and T. Takagi (Eds.): CHES 2011, LNCS 6917, pp. 1–16, 2011.

However, at least two valid counterarguments can be considered. First, even though ISEs are often presented theoretically as "non-invasive," their concrete realisation may still be problematic. For example, one can imagine the difficulty of altering incumbent processor designs contributed to the fact that Intel's AES-NI appeared long after suggested by initial work in this area [31]; issues of re-design, re-verification, and re-deployment are scientifically non-trivial and potentially *very* costly. Second, one has to consider the problems of utilisation and flexibility. One aspect is ensuring the cost of design and implementation is worthwhile, another is ensuring ISEs are useful to as many kernels as possible (i.e., making an ISE flexible enough to cater for the future). Cryptography, in particular, has a vested interest in the latter: if an (inflexible) special-purpose ISE for a kernel is deployed and the kernel is then broken, the ISE, associated hardware and sunk design cost subsequently represent useless overhead.

With all these counterarguments in mind, it is interesting to consider how an ISE-based approach, in the most general sense, might be accommodated by next-generation processors. In particular, how might next-generation general-purpose processor designs support dynamic (i.e., changeable at run-time under control of the program) instruction set extension and execution. In both embedded and non-embedded contexts, we already have some answers: focusing on functional units for example, both the Stretch S6000 and ARM-based Triscend A7 include similar concepts which can be dynamically re-configured at run-time, and the new Intel Atom E600C (or "Stellarton") series includes an coarsely integrated (i.e., coprocessor-like) Altera FPGA.

While re-configurable devices such as FPGAs have redefined the traditional roles of hardware and software, re-configurable general-purpose processors are now also a reality; the question is, how does this direction match cryptographic use-cases? This is far from a new topic, but we make progress via four main contributions; we stress that our focus is at the level of micro-architecture and instruction sets, rather than the device level. First we survey strands of related work that support dynamic instruction set extension and execution; second we present PREON (short for Partially Reconfigurable LEON), a novel LEON3-based prototype which includes two such mechanisms; third we evaluate a range of cryptographic primitives on said prototype, demonstrating that it provides a general-purpose platform capable of supporting many existing ISE proposals and specifying some novel additions (e.g., for the two hash functions Skein and JH); and finally, we extend previous security analysis to highlight several issues that require resolution in order to support cryptographic workloads.

2 Background and Analysis

In this section, we present a limited survey of mechanisms that support the concept of dynamic instruction set extension and execution in different ways. Each mechanism has a rich lineage within the field of computer architecture, and is sufficiently mature to exist in production (and in some cases embedded) processor designs [1]. Our overview spans implicit (i.e., invisible to the program) and explicit (i.e., under control of the program) mechanisms; in the latter case we expect that, in addition to the hardware components, there will be generic need for inclusion of management and invocation instruction sets.

Re-Configurable Computation Fabric. The idea is to extend the computational logic (e.g., the ALU) such that, instead of only computing operations which are fixed at design-time, it can be re-configured at run-time. A fairly current and comprehensive overview of the design space is given by Dales [6, Sect. 2.3] and Amano [1], the latter including examples of commercialisation. A more limited list of instances includes

- partly re-configurable functional units, for example CryptoManiac [37] and PipeRench [29],
- tightly integrated run-time re-configurable logic, for example the Triscend A7, Stretch S6000, and Infineon CARMEL, and
- coarsely integrated run-time re-configurable logic, for example Intel's Atom E600C with integrated Altera FPGA.

The choice of fabric can imply extra design constraints whose relevance depends on the context. For example, an FPGA-based fabric could limit the maximum clock frequency, but in an embedded context this may not be of primary concern unless the cost of a mixed technology approach is prohibitive.

Advanced Mechanisms for Instruction Delivery. The idea is to extend the fetch unit so that the mechanism for instruction delivery is (partly) controlled by the program being executed. There is a huge range of related concepts and concrete implementations; a non-exhaustive list includes mechanisms for

- instruction fusion [22], for example load-modify operations in Intel's Core2 micro-architecture and multiply-accumulate in DSP-like (or embedded) processors, allowing composite micro-operations [8] to be specified by a single ISA-level instruction,
- macro-like [32] translation, for example the "sequencer unit" within the IBM RISC Single Chip (RSC) and PowerPC [24] processor, and the ARM Jazelle framework for acceleration of Java programs,
- processor-controlled cache-like structures, for example the trace cache and loop buffer designs within the Core2 and NetBurst micro-architectures, and
- user-controller memory structures such as the register-based buffer of Hines et al. [13], and the now well-studied ideas of scratch-pad memories [2] and non-transparent caches [23].

3 PREON: A LEON3-Based Experimental Prototype

In this section, we introduce PREON, a prototype implementation of selected mechanisms surveyed in Section 2. As a starting point we used the LEON3, an open-source implementation of a 32-bit SPARC V8 compliant processor core developed by Gaisler Research AB. We altered the 7-stage LEON3 pipeline as described in detail below, and equipped it with Harvard-style instruction and data caches (or I-cache and D-cache), each 4 kB in size.

The PREON prototype was synthesised to the SASEBO-GII evaluation platform, which we used to produce the experimental results given in Section 4; the processor core itself required 2338 slices of the Xilinx Virtex-5 FPGA (model XC5VLX50-1FF324). The PREON core is clocked at 24 MHz, although we stress that this is a limit imposed by the SASEBO on-board clock.

3.1 Re-configurable Fabric

The first addition is a re-configurable fabric, tightly integrated with the execution unit. We view the fabric essentially as an FPGA, but clearly less general alternatives are viable and potentially preferable in certain contexts. Inclusion of this mechanism is mainly motivated by the goal of improved computational throughput per-instruction (rather than, say, instruction throughput), without compromising flexibility; the fabric can be re-configured to efficiently compute what the general-purpose processor cannot. At least two further benefits result: first, the mechanism reduces communication latency versus a coarsely integrated alternative (e.g., co-processor), and second, it permits the sharing of resources between the processor and fabric (e.g., storage such as registers).

Design and Programming Interface. Use of the fabric by a program executing on PREON is achieved through a single extra instruction named `fabric`. When executed, the instruction has the semantics

$$GPR[dst] = f(GPR[src_1], GPR[src_2], imm)$$

where f denotes a single-cycle functionality provided by the resident configuration: the 32-bit content of the registers src_1 and src_2 (with an 8-bit immediate value imm) is presented to the fabric as input, and the output (according to the current configuration) is stored in register dst. Essentially, the fabric acts as a replacement for the ALU. One can imagine a few alternative models, but this approach allows the fabric to use imm as a means of first specifying any sub-operation (e.g., to house more than one operation on the re-configurable fabric and select between them), and second supplying any immediate data.

PREON allows a configuration resident in the fabric to maintain short term state, e.g., house a register. On one hand, this is arguably outside the traditional remit of ISEs. On the other hand, it allows a high degree of flexibility since for example, multi-input and/or multi-output operations can be supported. In the former case, the fabric is configured to include an "operand fetch" operation which takes the two operands and stores them internally; the stored operands can then be combined with two more operands in a "normal" operation. This feature could be used to overcome the restrictions of the SPARC V8 3-address instruction format without invasive alteration of the micro- or instruction set architecture. Likewise, multi-cycle operations are possible: the fabric is configured to include a "bubble" or NOP-style operation that can be used to absorb cycles while computing the required output. Some other approaches to increasing the input or output bandwidth are possible, e.g., those proposed by Kluter et al. in [19], but our aim is to limit the amount of extra bespoke hardware required.

Implementation. In practice, one would expect the re-configurable fabric to be implemented using a different technology than the processor core. However, the PREON prototype demands partial re-configuration of a host FPGA representing *both* components. Per [38], this requires partitioning at the top-level: we had to divide our design into a static part (the LEON3 processor) and a dynamic part (the re-configurable fabric). In addition, the SASEBO-GII includes 2 Mbit of on-board SRAM memory which we use to store partial bit-streams. A Xilinx XPS HWICAP core, altered with a wrapper for the LEON3 AMBA bus, is used to interface with the FPGA's Internal Configuration Access Port (ICAP). The size of the SRAM, plus the 700 slices reserved on the FPGA for the dynamic part, set an artificial upper limit on the length of the bit-stream, and hence also limit the complexity of configurations used by PREON.

Integration with the LEON3 core is relatively easy: we simply use the fabric rather than the ALU for `fabric` instructions, abusing the 8-bit Address Space Identifier (ASI) register to supply *imm*. Re-configuration of the fabric is done in software by the processor core; essentially, this boils down to a `memcpy`-style transfer of content from the SRAM into the fabric via the ICAP interface. In theory, it is possible to partition the fabric and allow multiple configurations to be resident at the same time; Dales [6, Section 5] outlines various techniques to manage this, but for simplicity we consider a single configuration only.

3.2 Instruction Register File

Since the LEON3 can already be extended with a scratch-pad for instructions (the so-called ILRAM), it is reasonable to question the novelty of our second addition. Crucially however, the concept of an Instruction Register File (IRF), as described for example by Hines et al. [13], captures program fragments whose form relates to basic blocks rather than functions. In short, we suggest that an appropriate IRF design can provide macro-like cryptographic ISEs: the idea is to record short instruction sequences on-chip, then later replay them from the IRF rather than main memory. Rather than "extended" computational ability during execution, the IS"E" here is an "expansion" from a single instruction to a semantically richer *straight-line* instruction sequence.

Inclusion of the mechanism is motivated by two main goals: it should reduce off-chip memory access (which implies lower power consumption), and provide low-latency and deterministic fetch behaviour (both without the physical overhead of an instruction cache). Our premise is that cryptographic use-cases are ideally suited to take advantage of these features, and also benefit from them as a result, for example, of a need to avoid cache-based attacks.

Design and Programming Interface. Our realisation of the IRF concept uses a few small buffers into which instructions are placed and retrieved. Let $B[i][j]$ denote the j-th entry in the i-th buffer where $0 \leq j < n$ and $0 \leq i < m$, i.e., there are m buffers, each of n elements. Let $C[i]$ denote the number of valid instructions currently held in the i-th buffer, meaning that $0 \leq C[i] < n$ for all i. Three additional instructions are used to control these structures:

```
! AES T-table block #1
! input  : packed AES state in %o0 to %o3
!          packed AES round key in %o4 to %o7
!          T-table base addresses in %i3 to %i6
! output : equivalent of %l4 = T-table0[ ( %o0 >>  0 )& 0xFF ] ^
!                              T-table1[ ( %o1 >>  8 )& 0xFF ] ^
!                              T-table2[ ( %o2 >> 16 )& 0xFF ] ^
!                              T-table3[ ( %o3 >> 24 )& 0xFF ] ^ %o4;
record %g0,   0, %g0
srl    %o0,  22, %l4 ; and %l4, 1020, %l4 ; ld [%l4 + %i3], %l4
srl    %o1,  14, %l5 ; and %l5, 1020, %l5 ; ld [%l5 + %i4], %l5
srl    %o2,   6, %l6 ; and %l6, 1020, %l6 ; ld [%l6 + %i5], %l6
sll    %o3,   2, %l7 ; and %l7, 1020, %l7 ; ld [%l7 + %i6], %l7
xor    %l4, %l5, %l4 ; xor %l4,  %l6, %l4
xor    %l4, %l7, %l4 ; xor %l4,  %o4, %l4
stop   %g0,   0, %g0
! AES T-table block #2
...
! AES T-table block #3
...
! AES T-table block #4
...
! AES round
play   %g0,  0, %g0 ! playback T-table block #1 once
play   %g0, 32, %g0 ! playback T-table block #2 once
play   %g0, 64, %g0 ! playback T-table block #3 once
play   %g0, 96, %g0 ! playback T-table block #4 once
...
```

Fig. 1. A sketched example of IRF use: each "block" of a T-table based AES implementation (including key addition) is recorded as a macro into an IRF buffer, then later played back and expanded to form a (non-final) AES round

- `record` takes an immediate operand i (which specifies a buffer number) and places the fetch unit into recording mode. This acts to redirect instructions into the i-th buffer rather than the pipeline:
  1. initially set $C[i] = 0$,
  2. for each instruction received from the fetch unit, if the instruction is `stop` then act appropriately, otherwise store it to $B[i][C[i]]$, and update $C[i] \leftarrow C[i] + 1 \pmod{n}$.
- `stop` returns the fetch unit to normal mode, redirecting the instruction stream into the pipeline again.
- `play` takes two immediate operands i and c (specifying a buffer number and playback count) and places the fetch unit into playback mode. This acts to inject instructions into the pipeline from the i-th buffer, rather than the fetch unit, c times
  1. freeze the program counter,
  2. inject each j-th instruction from $B[i][j]$, for $0 \leq j < C[i]$, into the pipeline, repeating the process c times then
  3. put the fetch unit back into normal mode, and resume execution from the frozen program counter.

We note that it may be of value to store pre-decoded content in the buffer (like in a trace cache), but defer this topic for further work.

Implementation. Implementation of the IRF in PREON is relatively simple: we follow a conventional (data oriented) register file design, using flip-flops to store content. This structure is controlled by a state machine in the existing LEON3 fetch unit, which applies appropriate operating rules according to the description above. The result contrasts with the more heavy-weight, general-purpose memory approach of an ILRAM in both form and function. More precisely, an ILRAM-based approach implies a larger storage capacity and more involved interface (i.e., memory transaction). Additionally, executing an instruction sequence in an ILRAM demands a branch into and back from said sequence; even if each instruction is retrieved with low latency, the additional branches cause a significant overhead for short sequences. The same is not true of IRF use as there is just a single cycle overhead relating to each `playback`.

Although the description allows general parametrisation (perhaps restricting m and n to powers-of-two), concrete parameters must be selected before use. In theory, the parameters could be selected at run-time using special configuration instructions; this would allow for a high degree of flexibility at relatively marginal cost. However, for simplicity, our PREON prototype currently caters for cases where $m \cdot n = 64$, e.g., $m = 4$, $n = 16$, fixed at design-time after analysis of the associated trade-off. Although larger m or n may improve our results below in theory, our choice tries to balance this against practicality; for example, a total on-chip storage of $64 \cdot 4 = 256$ B matches the capacity of the SSE register file in x86-64 processors.

4 Evaluation of Cryptographic Workloads

4.1 Re-configurable Fabric

Limited evaluations of cryptographic kernels, executed via a similar mechanism with respect to the re-configurable fabric, exist; for example, Dales [6, Section 4.3.2.3] details some experiments with Twofish. In the following, we extend this to include a broader set of modern kernels. Table 1 shows a range of empirical results produced using our prototype PREON implementation. Each result compares a C implementation to a fabric-supported[1] alternative, both with inline assembly statements where appropriate (e.g., to invoke the fabric, or to access SPARC-specific functionality).

Each configuration is designed to match the critical path of the processor; no configuration extends the existing critical path, except the $\mathbb{F}_{3^m}$ multiplier. The execution times (i.e., cycle counts) are averaged over a number of randomised inputs. Although few kernels have data dependent control-flow, this approach takes into account the behaviour of both data and instruction cache. Also note that techniques for automatic identification of configurations, as in [25], seem applicable, but we defer investigation of this topic to future work.

We use the subsections below to discuss each implementation, and conclude with a summary of the results.

[1] To satisfy space restrictions we omit the formal description of each ISE, opting to include a complete description in a full version of this paper.

Table 1. Experimental results comparing the performance of various cryptographic kernels without and with support of ISEs provided by the re-configurable fabric. Note that the static footprint includes instructions *and* any major static data (e.g., T-tables and expanded key schedule), and that initialisation of the fabric is not included in the total number of cycles, rather as a column in the table.

	Without ISE		With ISE			
	Performance (cycles)	Static footprint (bytes)	Performance (cycles)	Static footprint (bytes)	ISE area (slices)	ISE re-config. (µs)
AES						
AES-128 encryption	1281	6068	463	412	115	177
SHA2/SHA3						
SHA-256, 4096-bit message	45241	3304	30528	2492	48	118
JH-256, 4096-bit message	6584962	2052	976372	2116	26	59
Skein-512-256, 4096-bit message	332739	8152	117123	6340	319	470
Grøstl-256, 4096-bit message	258389	16248	152169	1980	112	177
Multiplication in $\mathbb{Z}_N^*$						
1024-bit multiplication	86460	768	25148	428	321	590
Multiplication in $\mathbb{F}_{2^{233}}$						
School-book	30864	548	2290	428	170	295
Width-4 comb	14900	908	12908	724	44	118
Multiplication in $\mathbb{F}_{3^{337}}$						
School-book, bit-sliced	163340	1504	6985	828	690	1062
School-book, bit-serial	445670	1616	11898	420	343	590
Width-4 comb, bit-sliced	82940	4100	56380	3596	70	118
Width-4 comb, bit-serial	247281	8484	40213	2930	54	118

AES. Tillich et al. [31] proposed a set of ISEs that permit efficient implementation of AES on 32-bit architectures, focusing on SPARC V8-based LEON2 in particular. We adopt two ISE classes [31, pp. 275–276], namely `sbox4s` (plus `sbox4r`) and `mixcol4s` (and inverses), and compare them with a T-tables based implementation in software. We note that acceleration of bit-sliced implementations of AES following [10] is viable, but do not investigate this further.

SHA-2/SHA-3. In terms of ISE, SHA-2, focusing on SHA-256 in particular, has been paid relatively little attention. Juliato et al. offer in [17] an exception and explore different hardware/software approaches that include ISEs for rotation and the Ch and Maj functions; we follow their approach fairly directly.

Regarding SHA-3, we stress that we do *not* aim to compare the five finalists directly, but rather evaluate PREON. The finalists can be split into two rough categories: Blake, Keccak and Skein are AXR-based, while Grøstl and JH are AES-based. For the cases of Blake and Keccak, Hoerder et al. [15] highlighted the difficulty of finding appropriate ISEs: their design is already RISC-friendly and potential ISEs therefore fit a more coprocessor-like approach that captures and operates on (most of) the state in each step.

JH-256. We pack eight 4-bit state words into a 32-bit register and utilize three ISEs: one for the S-box and linear transform layer, and two more for the permutation layer; the design of JH means the same ISEs can be used for all parametrisations. To maintain comparability with the reference implementation, we use the ISEs to compute the round constants at run-time. This requires re-packing the round constants at various points, and explains the marginal increase in code footprint.

Skein-512-256. We focus on acceleration of the internal Threefish cipher. In order to match the 32-bit datapath of the LEON3, we specify a four-step ISE to support the MIX function; for comparison, we use the 32-bit oriented reference implementation. To avoid the need for a general-purpose rotation unit, we specify (and supply immediate inputs to select) ISEs for each of 27 rotation distances. The disadvantage of this approach is that the same ISEs can not support a different parametrisation.

Grøstl-256. We pack four 8-bit state words into a 32-bit register and use three ISEs: one for the `SubBytes` step, and two more for the `MixBytes` step. The T-tables based reference implementation is used for comparison.

Multiplication in $\mathbb{Z}_N^*$ *(supporting RSA, ECC).* Großschädl et al. [11] propose an ISE for RSA, or more specifically for Montgomery multiplication; their design is implemented on a SPARC V8-based LEON2. The ISE focuses on Multiply-ACcumulate (MAC) operations (e.g., $S \leftarrow S + a \times b$), and uses three dedicated 32-bit accumulator registers. We replicate this approach fairly directly, housing the accumulators within the fabric configuration itself, and compare our results with a C implementation provided by the authors of [11].

Multiplication in $\mathbb{F}_{2^n}$ *and* $\mathbb{F}_{3^m}$ *(supporting ECC, Pairing-Based Crypto).* Considering the cases $n = 233$ and $m = 337$, arithmetic in $\mathbb{F}_{2^{233}}[X]/X^{233} + X^{73} + 1$ and $\mathbb{F}_{3^{337}}[Y]/Y^{337} + Y^{30} - 1$ underpin specific parametrisations in elliptic curve and pairing-based cryptography, e.g. the former is specified in NIST-B-233. In the characteristic-two case, coefficients of some $x \in \mathbb{F}_{2^n}$ have a natural representation; various published ISEs, including those for SPARC V8-based LEON2 [30] and the Intel CLMUL extension to x86, provide an associated polynomial (or "carry-less") multiplication instruction. A similar concept is possible in the characteristic-three case, but the issue of representation is more complex.

Our implementations accelerate school-book multiplication mainly via a dedicated polynomial multiplication ISE; the width-4 comb based multiplication is accelerated using a "shift with carry" ISE, which is missing in the SPARC V8 instruction set. Particularly for characteristic three, the flexibility of PREON is beneficial: it allows a range of subtle implementation options without changes to the architecture, and resolves some problems with previous work. For example Grabher et al. [10] support bit-slicing, but only using unconventional 6-address instructions; the "operand fetch" idiom in PREON can cope with multi-operand instructions, and still provide significant performance improvements.

Summary and Discussion. It is unsurprising that ISE-based implementations improve either latency (overall cycles) and/or size (memory footprint). For the kernels studied, the latter case implies a hidden side-benefit of reducing memory traffic (primarily loads) and hence reduced reliance on a cache to achieve quoted performance; in a rough sense, one can view the cost of including the fabric as counterbalancing the need for large, efficient layers of memory hierarchy.

Focusing on AES, one can contrast results for dynamic ISE provided by the PREON re-configurable fabric with static ISE (taking AES-NI as an example) and coarsely-integrated FPGA-based coprocessors. As stated in [34], AES-NI achieves a throughput of 3.589 cycles per byte (in CBC mode), i.e., 57.4 cycles per block. However, AES-NI operates on 128-bit SSE registers, which increases the throughput by a factor of 4 in relation to a 32-bit bit datapath. Hence, one can estimate that a 32-bit analogue of AES-NI would require about 230 cycles per block, roughly half that of the PREON-supported implementation. Comparison with an FPGA-based cryptographic coprocessor must consider the interface through which coprocessor and host are connected. For example, Hodjat et al [14] and Schaumont et al. [27] attached an AES coprocessor to the LEON core and achieved a throughput of 704 cycles (via the LEON coprocessor interface) and 1492 cycles (via memory-mapped I/O) per block, even though the AES core itself performs an encryption in only 11 cycles. Using a dedicated high-speed interconnect, such as Xilinx's Fast Simplex Link (FSL), allows improvement to about 202 cycles per block [9]. In summary, even accepting the limited scope and accuracy of this comparison, the PREON prototype offers a very attractive compromise: it is competitive to both static AES extensions *and* coprocessors using metrics of performance and flexibility.

This flexibility is highlighted by the possibility for combination of configurations to support multi-kernel ISEs. Trivial merging of configurations is possible where size permits, but more specific approaches also exist. For example, one may consider multi-field multipliers and, hence, multi-kernel ISEs as described by Vejda et al. [33]. As a more concrete example, AES and Grøstl compute the S-box function in the same way, and therefore it is possible to design a single configuration that supports both kernels. While the performance figures remain the same as above, the total size required is 152 slices, roughly 30% less than a separate implementation.

A less positive issue is that of re-configuration speed. Ideally, one might aim for per-instruction change in an ISE (i.e., the configuration) to maximise the emphasis on flexibility. However, without the facility for multiple resident configurations, the re-configuration speed of our Xilinx Virtex-5 FPGA (100 Mbit/s at 25 MHz, meaning latencies of upto 1000 μs for our more complex cases) is a limiting factor. In a sense this is a property of the technology used to realise the fabric, but even so the re-configuration speed limits "ISE dynamism," which one might reasonably argue is disadvantageous.

4.2 Instruction Register File

Hines et al. [13, Table 2] include a set of security-related benchmarks (e.g., AES and SHA-2), adopting a domain-neutral (and compiler supported) approach to implementation. Since our design differs, we do not offer a direct comparison with these results. Rather, we aim to explore how careful use of our IRF design compares with the natural alternative of cached instruction access.

Again, we use the sub-sections below to discuss each implementation, and conclude with a summary of the results.

Table 2. Experimental results comparing the performance of various cryptographic workloads without and with support from ISEs provided by the IRF (and without and with support from the I-cache in each case). Note that initialisation of the IRF buffers are not included in the total number of cycles.

AES				
	Without ISE		With ISE	
	Performance (cycles)	Fetches (from main memory)	Performance (cycles)	Fetches (from main memory)
Without I-cache	4751	823	2644	309
With I-cache	1281	823	1302	309
Multiplication in $\mathbb{Z}_N^*$				
	Without ISE		With ISE	
	Performance (cycles)	Fetches (from main memory)	Performance (cycles)	Fetches (from main memory)
Without I-cache	189069	34765	67366	4046
With I-cache	51708	34765	48640	4046

AES. Parametrising the IRF with $m = 4$, $n = 16$, we record each "block" of a T-tables based AES implementation into a buffer; these are then replayed (as roughly illustrated in Figure 1) to form each round.

Multiplication in $\mathbb{Z}_N^*$ (supporting RSA, ECC). Parametrising the IRF with $m = 4, n = 16$, we refer directly to the CIOS algorithm in [4, Section 5]. The idea is to record the body of each inner loop into a buffer; we include a final instruction which increments j. Then, by using one `playback` instruction, the buffer can be replayed s times (having set $C = 0$ and $j = 0$ initially); the resulting, expanded instruction sequence implements the entire unrolled loop.

Summary and Discussion. Table 2 outlines our results. Broadly speaking, the conclusion is that use of the IRF can significantly reduce the number of fetches from memory (roughly 2- and 8-fold improvement) without significant negative impact (indeed, in some cases with positive impact) on the performance. The latter result is, naturally, magnified when we switch off the I-cache and (e.g., in the case of Montgomery multiplication) realise benefits of loop unrolling without the associated disadvantage in terms of static footprint.

In an attempt to quantify this in terms of power consumption, we refer to the widely cited figures provided by Segars [28, Slides 34 and 42]. He quotes an ARM9TDMI register file as representing 13% of the datapath power consumption, and ARM920T I- and D-caches as 25% and 19%, respectively, of the total power consumption. Therefore, one can estimate that an ARM920T register file represents about 13% of the quoted 25% total power consumption. As such, one might roughly reason that replacing the I-cache with an IRF reduces the total power consumption by around 20% (per fetch from IRF-resident content).

4.3 Combined Utilisation

A key motivations for our specific selection of mechanisms above is the potential for composing their use: since use of the fabric is via a normal instruction, such instructions can be captured in the IRF like any other. As a final, one-off case study, we produced such a combined implementation of AES. Encryption of a

128-bit block using the T-tables based reference implementation performs 823 instruction fetches from main memory, 208 loads and 4 stores; it takes a total of 1281 cycles and relies on a 6068 B static memory footprint. In contrast, an ISE-based implementation (combining the previous fabric configuration *and* the IRF parametrised with $m = 1, n = 64$) performs 45 instruction fetches from main memory, 48 loads and 4 stores; it takes a total of 434 cycles and relies on a 340 B static memory footprint.

In summary, considered use of the two mechanisms yields a 3-fold improvement in performance, a 10-fold improvement in main memory access (combined fetches, loads and stores), and a 17-fold improvement in memory footprint; this is achieved in a manner which permits similar benefit to other kernels without alteration of the PREON architecture, and largely without the I- and D-caches which underpin the T-tables based approach.

5 Issues Relating to Practical Deployment

Section 4 highlights some practical advantages of the two mechanisms considered. However, these advantages rely on the re-configurable fabric and IRF in PREON housing state: their configuration in both cases, and internal registers in the case of the fabric. As such, one must *also* consider related disadvantages (i.e., the issue of security). In this section, we discuss some (fairly speculative) examples within the context of both PREON and other existing proposals.

Trusted Configuration and Use. Design verification and policy enforcement are well-researched areas in embedded security, and form a key requirement within high-assurance contexts. A review of related techniques is, for example, given by Huffmire et al. [16, Sect. 4.4.1]; they point out that partial re-configuration is rarely used within said contexts due to the increased complexity of design verification. However, the approach of coupling a general-purpose processor to a re-configurable fabric offers a potential solution to this dilemma. Since access control to the re-configurable fabric *must* be integrated into the security model of the processor, this will not cause the same problem as the more general case considered by Huffmire et al. In particular, it is no more difficult to design access policies for the fabric than for the processor itself.

As an example, access to the re-configurable fabric might be limited to the OS kernel via a privilege mode within the processor; this offers a similar protection to that for conventional process state. Likewise, the processor may enforce policies on the configuration bit-stream; a possible approach is to accept only authenticated bit-streams. An effective implementation of such mechanisms is fundamental to mitigation of several problems outlined below.

State "Read Out". Focusing on the fabric, it is obvious that internal registers maintained by one process should not be readable by any other process. More subtle issues are raised by the fabric configuration itself. For example, pushed to an extreme, it is tempting to consider aggressive compile-time or run-time specialisation techniques (cf. Warp processors [21]), e.g., a fabric configuration specialised per key. In this case, it is also vital that the configuration can not be read by another process (or by an external attacker): Kerckhoffs' principle does not apply if secret key material is embedded in the configuration rather than simply used by it. This issue relates vaguely to attacks described in [39].

Information Leakage. When unmanaged, the state of the re-configurable fabric acts as a shared resource between processes. Said resource (or conversely, the lack of appropriate process isolation) represents the potential for various forms of micro-architectural attack; for example, see Wang and Lee [35, Section 4].

A concrete example can be applied to the ProteanARM [6], which requires a process to register the fabric configurations with the OS. When an instruction references a configuration (via an identifier), it is either

1. executed by the fabric (if the configuration is resident),
2. transformed into a call to an equivalent software implementation (where the registration dictates this mode), or
3. causes an exception (whereby the OS can load the configuration if it is not resident).

In a rough sense, this implies that the content of the re-configurable fabric can be "queried" by timing how long a use of the fabric takes to complete. Hence, a heavy-weight version of the so-called "prime + probe" approach to cache-based side-channel attacks seems to apply.

Of course, it is possible to construct mitigating solutions. For example, one can (at least in theory) demand a full context switch of all such resources. In practise, however, it is often very tempting to take short-cuts since the overall cost of switching the context of the re-configurable fabric is extremely high (as illustrated by Section 4). We note that Chan et al. [5] examine a similar issue of processor isolation albeit in a more coprocessor-like context.

Fault Injection. Although one can question whether the statement is still true today, in 2003 Wollinger and Paar [36] mention that "there appears to be no published attempt to perform this kind of [fault] attack against FPGAs." A cursory literature search shows there is (at least) not as much work in this area as one might expect, a notable exception being [3], with more attention paid to ASICs. Some related work includes that of Desmedt et al. [7] and Hadžic et al [12], who introduce the idea of an FPGA "virus." If one accepts mechanisms to support some type of re-configurable fabrics as a viable direction, the examples above suggest issues in terms of security. In particular, does the re-configurable nature of an FPGA mean that "traditional" fault attacks on computation are easier (e.g., by altering logic cells in the same way as RAM)?

Hardware Trojans. In order to reduce the cost of a context switch, the ProteanARM processor [6, Sect. 4] allows a configuration associated with one process to be resident in the fabric while another process is being executed. Imagine the re-configurable fabric is not gated, i.e., that operands are fed to it and computation occurs even if the output is not used. Since the fabric is not forcibly reconfigured for each process, a process may unintentionally provoke computation within the fabric whose configuration is dictated by another process. Or, imagine two (partial) configurations coexisting on a fabric: one might speculate that the behaviour of one (e.g., with respect to thermal properties) could influence the other in some way. These simple examples suggest the potential for a hardware Trojan: the attacker configures the fabric with a high-leakage function which is able to capture (or export) information leaked by some target process.

As above, the issue of isolation is important. We note that the "moats and drawbridges" design concept of Kastner et al. [18] is of special interest in this context: the goal is physical in-fabric isolation of partial configurations, i.e., the separation of Trojan hardware from benign targets.

6 Conclusions

Our results in Section 4 highlight advantages with respect to support for and use of dynamic ISEs in cryptography. Both conceptually simple, relatively non-invasive additions to the LEON3 generalise many existing ISE proposals for this platform and permit high-performance, "algorithm-agile" implementations. In short, such an approach can support ISEs like AES-NI without a need for fixed AES-related functionality. However, to realise said advantages, and following a similar line of reasoning as [20,26], we show in Section 5 that careful analysis and consideration of security is a strict prerequisite.

At least two well-founded criticisms exist. First, one might view speculative attacks against prototype processor designs as moot. Second, and focusing on the re-configurable fabric, one may argue that other design constraints prevent integration of an FPGA into the processor data-path. Recalling Section 2, we again stress that processors of this sort *already* exist; in a sense, commercialised examples offer an interesting vehicle for future work on some questions raised in Section 5. Along similar lines, we stress that it is perfectly viable to instead find a compromise between general- and special-purpose fabric: again, this is an interesting challenge for future work. Partly re-configurable functional units (e.g., those in PipeRench [29] and CryptoManiac [37]) give some direction.

Even though there are some unresolved challenges, our experimental results suggest that the general concept of providing tightly integrated re-configurable components represents an interesting approach for (embedded) processors. We further conclude that provision of exposed, programmer-controlled components rather than automated (cf. the transparent operation of caches) alternatives is an attractive direction for cryptography since they allow at least the potential to avoid classes of existing micro-architectural (e.g., cache-based) attack.

Acknowledgements. The work described in this paper has been supported in part by EPSRC grant EP/H001689/1. The authors would like to thank Atukem Nabina for his general input on FPGA partial re-configuration.

References

1. Amano, H.: A survey on dynamically reconfigurable processors. IEICE Tran. Comm. E89-B(12), 3179–3187 (2006)
2. Banakar, R., Steinke, S., Lee, B.-S., Balakrishnan, M., Marwedel, P.: Scratchpad memory: design alternative for cache on-chip memory in embedded systems. In: CODES, pp. 73–78 (2002)
3. Canivet, G., Maistri, P., Leveugle, R., Clédière, J., Valette, F., Renaudin, M.: Glitch and laser fault attacks onto a secure AES implementation on a SRAM-based FPGA. J. Cryptology 24(2), 247–268 (2011)
4. Koç, Ç.K., Acar, T., Kaliski, B.S.: Analyzing and comparing Montgomery multiplication algorithms. IEEE Micro 16(3), 26–33 (1996)
5. Chan, H., Schaumont, P., Verbauwhede, I.: Process isolation for reconfigurable hardware. In: ERSA, pp. 164–170 (2006)
6. Dales, M.W.: Managing a reconfigurable processor in a general purpose workstation environment. PhD thesis, University of Glasgow (2003)
7. Desmedt, Y.G., Quisquater, J.-J.: Public-key systems based on the difficulty of tampering. In: Odlyzko, A.M. (ed.) CRYPTO 1986. LNCS, vol. 263, pp. 111–117. Springer, Heidelberg (1987)
8. Flynn, M.J., McLaren, M.D.: Microprogramming revisited. In: Proc. of the 22nd ACM National Conference, pp. 457–464 (1967)
9. Gonzalez, I., Gómez-Arribas, F.: Ciphering algorithms in MicroBlaze-based embedded systems. Computers and Digital Techniques 153(2), 87–92 (2006)
10. Grabher, P., Großschädl, J., Page, D.: Light-weight instruction set extensions for bit-sliced cryptography. In: Oswald, E., Rohatgi, P. (eds.) CHES 2008. LNCS, vol. 5154, pp. 331–345. Springer, Heidelberg (2008)
11. Großschädl, J., Tillich, S., Szekely, A.: Performance evaluation of instruction set extensions for long integer modular arithmetic on a SPARC V8 processor. In: DSD, pp. 680–689 (2007)
12. Hadžić, I., Udani, S., Smith, J.M.: FPGA viruses. In: Lysaght, P., Irvine, J., Hartenstein, R.W. (eds.) FPL 1999. LNCS, vol. 1673, pp. 291–300. Springer, Heidelberg (1999)
13. Hines, S.R., Green, J., Tyson, G., Whalley, D.: Improving program efficiency by packing instructions into registers. In: ISCA, pp. 260–271 (2005)
14. Hodjat, A., Verbauwhede, I.: Interfacing a high speed crypto accelerator to an embedded CPU. In: Asilomar Conference on Signals, Systems, and Computers, vol. 1, pp. 488–492 (2004)
15. Hoerder, S., Wójcik, M., Tillich, S., Page, D.: An evaluation of hash functions on a power analysis resistant processor architecture. In: Ardagna, C. (ed.) WISTP 2011. LNCS, vol. 6633, pp. 160–174. Springer, Heidelberg (2011)
16. Huffmire, T., Irvine, C., Nguyen, T.D., Levin, T., Kastner, R., Sherwood, T.: Handbook of FPGA Design Security. Springer, Heidelberg (2010)
17. Juliato, M., Gebotys, C.: Tailoring a reconfigurable platform to SHA-256 and HMAC through custom instructions and peripherals. In: ReConFig, pp. 195–200 (2009)
18. Kastner, R., Levin, T., Nguyen, T., Irvine, C., Brotherton, B., Wang, G., Sherwood, T., Huffmire, T.: Moats and drawbridges: An isolation primitive for reconfigurable hardware based systems. In: IEEE Security and Privacy, pp. 281–295 (2007)
19. Kluter, T., Brisk, P., Ienne, P., Charbon, E.: Way stealing: cache-assisted automatic instruction set extensions. In: DAC, pp. 31–36 (2009)

20. Kocher, P.C., Lee, R.B., McGraw, G., Raghunathan, A.: Security as a new dimension in embedded system design. In: DAC, pp. 753–760 (2004)
21. Lysecky, R., Stitt, G., Vahid, F.: Warp processors. TODAES 11(3), 659–681 (2006)
22. Malik, N., Eickemeyer, R.J., Vassiliadis, S.: Interlock collapsing ALU for increased instruction-level parallelism. SIGMICRO Newsletter 23(1-2), 149–157 (1992)
23. Miller, J.E., Agarwal, A.: Software-based instruction caching for embedded processors. In: ASPLOS, pp. 293–302 (2006)
24. Moore, C.R., Balser, D.M., Muhich, J.S., East, R.E.: IBM single chip RISC processor (RSC). In: ICCD, pp. 200–204 (1991)
25. Pothineni, N., Brisk, P., Ienne, P., Kumar, A., Paul, K.: A high-level synthesis flow for custom instruction set extensions for application-specific processors. In: ASP-DAC, pp. 707–712 (2010)
26. Ravi, S., Raghunathan, A., Kocher, P.C., Hattangady, S.: Security in embedded systems: Design challenges. TECS 3(3), 461–491 (2004)
27. Schaumont, P., Sakiyama, K., Hodjat, A., Verbauwhede, I.: Embedded software integration for coarse-grain reconfigurable systems. In: IPDPS, pp. 137–142 (2004)
28. Segars, S.: Low power design techniques for microprocessors (tutorial session). In: ISSCC (2001)
29. Taylor, R.R., Goldstein, S.C.: A high-performance flexible architecture for cryptography. In: Koç, Ç.K., Paar, C. (eds.) CHES 1999. LNCS, vol. 1717, pp. 231–245. Springer, Heidelberg (1999)
30. Tillich, S., Großschädl, J.: A simple architectural enhancement for fast and flexible elliptic curve cryptography over binary finite fields GF(2^m). In: Yew, P.-C., Xue, J. (eds.) ACSAC 2004. LNCS, vol. 3189, pp. 282–295. Springer, Heidelberg (2004)
31. Tillich, S., Großschädl, J.: Instruction set extensions for efficient AES implementation on 32-bit processors. In: Goubin, L., Matsui, M. (eds.) CHES 2006. LNCS, vol. 4249, pp. 270–284. Springer, Heidelberg (2006)
32. Tucker, A.B., Flynn, M.J.: Dynamic microprogramming: processor organization and programming. CACM 14(4), 240–250 (1971)
33. Vejda, T., Page, D., Großschädl, J.: Instruction set extensions for pairing-based cryptography. In: Takagi, T., Okamoto, T., Okamoto, E., Okamoto, T. (eds.) Pairing 2007. LNCS, vol. 4575, pp. 208–224. Springer, Heidelberg (2007)
34. VeriSign.: An evaluation of new processor instructions for accelerating selected cryptographic algorithms (2010)
35. Wang, Z., Lee, R.B.: Covert and side channels due to processor architecture. In: ACSAC, pp. 473–482 (2006)
36. Wollinger, T., Paar, C.: How secure are FPGAs in cryptographic applications? In *FPL*. In: Y. K. Cheung, P., Constantinides, G.A. (eds.) FPL 2003. LNCS, vol. 2778, pp. 91–100. Springer, Heidelberg (2003)
37. Wu, L., Weaver, C., Austin, T.: CryptoManiac: a fast flexible architecture for secure communication. In: ISCA, pp. 110–119 (2001)
38. Xilinx. Partial reconfiguration user guide (UG702) v12.1 (2010), `http://www.xilinx.com/support/documentation/sw_manuals/xilinx12_1/ug702.pdf`
39. Yang, B., Wu, K., Karri, R.: Scan based side channel attack on dedicated hardware implementations of data encryption standard. In: ITC, pp. 339–344 (2004)

FPGA-Based True Random Number Generation Using Circuit Metastability with Adaptive Feedback Control

Mehrdad Majzoobi[1], Farinaz Koushanfar[1], and Srinivas Devadas[2]

[1] Rice University, ECE
Houston, TX 77005
{mehrdad.majzoobi, farinaz}@rice.edu
[2] Massachusetts Institute of Technology, CSAIL
Cambridge, MA 02139
devadas@mit.edu

Abstract. The paper presents a novel and efficient method to generate true random numbers on FPGAs by inducing metastability in bi-stable circuit elements, e.g. flip-flops. Metastability is achieved by using precise programmable delay lines (PDL) that accurately equalize the signal arrival times to flip-flops. The PDLs are capable of adjusting signal propagation delays with resolutions higher than fractions of a pico second. In addition, a real time monitoring system is utilized to assure a high degree of randomness in the generated output bits, resilience against fluctuations in environmental conditions, as well as robustness against active adversarial attacks. The monitoring system employs a feedback loop that actively monitors the probability of output bits; as soon as any bias is observed in probabilities, it adjusts the delay through PDLs to return to the metastable operation region. Implementation on Xilinx Virtex 5 FPGAs and results of NIST randomness tests show the effectiveness of our approach.

1 Introduction

True Random Number Generators (TRNG) are important security primitives that can be used to generate random numbers for various essential tasks including the generation of (i) secret or public keys, (ii) initialization vectors and seeds for cryptographic primitives and pseudo-random number generators, (iii) padding bits, and (iv) nonces (numbers used once). Since modern cryptographic algorithms often require large key sizes, generating the keys from a smaller sized seed will significantly reduce the entropy of the long keys. In other words, by performing a brute-force attack only on the seed that generated the key, one could break the crypto system. In addition, for applications that demand a constant high-speed and high-quality generation of keys, e.g. secure web servers, algorithmic approaches to pseudo-random number generation are typically inefficient, and hardware accelerated mechanisms are highly desired. True random numbers also find applications in gaming, gambling and lottery drawings.

To date, numerous TRNG designs have been proposed and implemented. Each design uses a different mechanism to extract randomness from some underlying physical phenomena that exhibit uncertainty or unpredictability. Examples of sources of randomness include thermal and shot noise in circuits, secondary effects such as clock

B. Preneel and T. Takagi (Eds.): CHES 2011, LNCS 6917, pp. 17–32, 2011.

jitter and metastability in circuits, Brownian motion, atmospheric noise, nuclear decay, and random photon behavior.

Because of its flexibility and fast time to market, FPGA has become a popular platform for implementing many cryptographic systems that include TRNGs as an essential block. It is important to develop new FPGA TRNG solutions because: (i) not all hardware TRNG methods available for ASICs or other platforms are amenable to FPGA implementation; (ii) the existing FPGA TRNGs have limitations in terms of the throughput-per-unit-area and could be improved; and (iii) active adversarial attacks as well as variations in operational conditions such as fluctuations in temperature and voltage supply may bias and disturb the randomness of TRNGs output bitstream. Since most of the state-of-the-art TRNGs operate in an open-loop fashion, it is important to incorporate a mechanism to constantly provide a feedback signal to adaptively adjust the TRNG system parameters to increase its output bit randomness.

In this work, we propose a novel technique to generate true random numbers on FPGA using the flip-flop metastability as a source of randomness. The introduced TRNG core operates within a closed-loop feedback system that actively monitors the output bit probabilities over windows of bit sequences and generates a proportional feedback signal based on any observed bias in the bit probabilities. The feedback mechanism is made possible by performing fine delay tuning using high precision PDLs with picosecond resolution. The delay tuning ensures that the signals arrive simultaneously at the flip-flop to drive it into a metastable state. Our contributions are as follows.

- We introduce an FPGA-based TRNG system that utilizes flip-flop metastability as the source of randomness.
- A novel feedback mechanism is introduced that performs auto-adjustment on delays in order to make the metastability condition more likely to happen.
- We demonstrate the use of a PDL to perform fine tuning with a precision of higher than a fraction of a pico-second.
- Highly accurate delay measurement results for PDL are demonstrated.
- The proposed TRNG system is implemented on Xilinx Virtex 5 FPGA; the hardware evaluation results demonstrate the high throughput-per-area and the high quality (i.e., true randomness) of the produced output bits.

2 Related Work

The work in [15] uses sampling of phase jitter in oscillator rings to generate a sequence of random bits. The output of a group of identical ring oscillators are fed to a parity generator function (i.e., a multi-input XOR). The output is constantly sampled by a D-flipflop driven using the system clock. In absence of noise and identical phases, the XOR output would be constant (and deterministic). However, in presence of a phase jitter, glitches with varying non-deterministic lengths appear at the output. An implementation of this method on Xilinx Virtex II FPGAs was demonstrated in [12].

Another type of TRNG is introduced in [11] that exploits the arbiter-based Physical Unclonable Function (PUF) structure. PUF provides a mapping from a set of input challenges to a set of output responses based on unique chip-dependent manufacturing process variability. The arbiter-based PUF structure introduced in [3], compares the analog

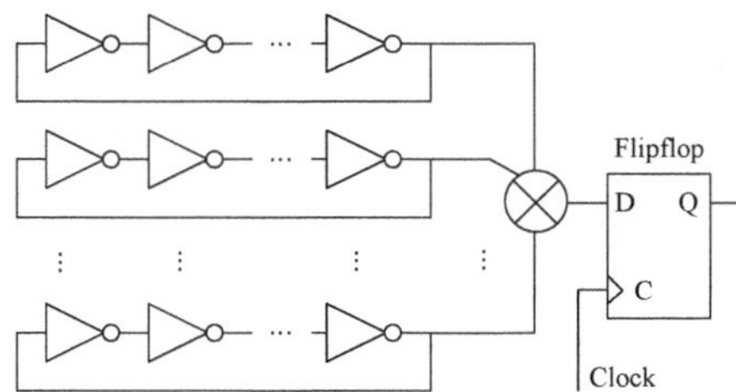

Fig. 1. TRNG based on sampling the ring oscillator phase jitter

delay difference between two parallel timing paths. The paths are built identically, but the physical device imperfections make their timing different. A working implementation of the arbiter-based PUF was demonstrated on both ASICs [5] and FPGA [8,13]. Unlike PUFs where reliable response generation is desired, the PUF-based TRNG goal is to generate unstable responses by driving the arbiter into the metastable state. This is essentially accomplished through violating the arbiter setup/hold time requirements. The PUF-based TRNG in [11] searches for challenges that result in small delay differences at the arbiter input which then cause unreliable response bits.

To improve the quality of the output TRNG bitsteam and increase its randomness, various post-processing techniques are often performed. The work in [15] introduces resilient functions to filter out deterministic bits. The resilient function is implemented by a linear transformation through a generator matrix commonly used in linear codes. The hardware implementation of resilient function is demonstrated in [12] on Xilinx Virtex II FPGAs. The TRNG after post processing achieves a throughput of 2Mbps using 110 ring oscillators with 3 inverters in each. A post-processing may be as simple as von Neumann corrector [10] or may be more complicated such as an extractor function [1] or even a one-way hash function such as SHA-1 [4].

Besides improving the statistical properties of the output bit sequence and removing biases in probabilities, post-processing techniques increase the TRNG resilience against adversarial manipulation and variations in environmental conditions. An active adversary may attempt to bias the output bit probabilities to reduce their entropy. Post-processing techniques typically govern a trade-off between the quality (randomness) of the generated bit versus the throughput. Other online monitoring techniques may be used to assure a higher quality for the generated random bits. For instance, in [11], the generated bit probabilities are constantly monitored; as soon as a bias in the bit sequence is observed, the search for a new challenge vector producing unreliable response bits is initiated. A comprehensive review of hardware TRNGs can be found in [14]. The TRNG system proposed in this paper simultaneously provides randomness, robustness, low area overhead, and high throughput.

3 Programmable Delay Lines

Programmable delay lines (PDLs) alter the signal propagation delay in a controlled fashion. The common mechanisms used to change the delay includes (i) varying the effective load capacitance, (ii) modifying the device current drive (by increasing/decreasing the effective threshold voltage by body biasing), or (iii) incrementally altering the length of the signal propagation path. The first two methods are often employed in either analog fashion and/or in application specific integrated circuits (ASICs) and are not amenable to FPGA implementation.

On reconfigurable digital platforms such as FPGAs, PDLs can be implemented by only changing the signal propagation path length or by altering the circuit fanout that modifies the effective load capacitance. The latter is only feasible if dynamic reconfiguration is available. In other words, changing circuit fanout requires topological changes to the circuit which in turn needs a new configuration. In [2], a technique is proposed to alter the propagation path length by letting the signal bounce a few times inside the switch matrices of FPGA instead of a direct and straight connection. The concept is illustrated in Figure 2. In the switch matrix on the left side, the signal bounces three times off the switch edges before it exits the switch. In the right switch, the signal only bounces once and as a result a shorter propagation path length and a smaller delay is achieved. However, changing the switch connections points and routings require a new configuration, and doing so during the circuit operation is only possible by dynamic reconfigurability.

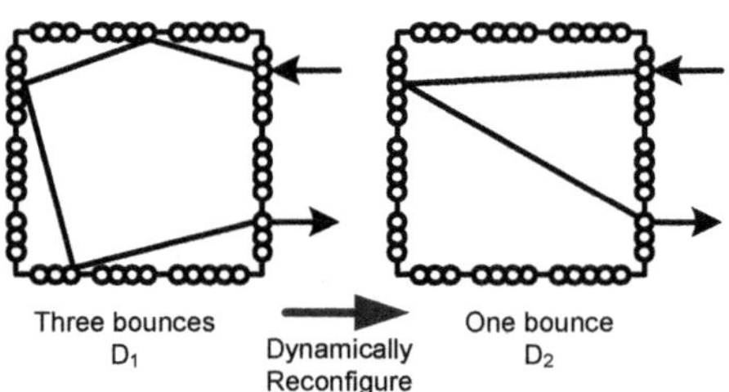

Fig. 2. A PDL implemented by altering the signal routing inside FPGA switch matrix

In this paper, we use a novel technique to vary the signal propagation path length in minute increments/decrements by only using a single lookup table (LUT). The technique changes the propagation path inside the LUT. We use an example to illustrate the concept. Figure 3 shows a 3-input lookup table. The LUT consists of a set of SRAM cells that store the intended functionality and a tree-like structure of multiplexers (MUXs) that enables selection of each individual SRAM cell content. The inputs to the MUXs serve as an address that points to the SRAM cell whose content is selected to appear at the output of LUT. The LUT in Figure 3 is programmed to implement an inverter, where the LUT output is always an inversion of its first input (A_1). The other inputs of LUT, namely A_2 and A_3 are functionally "don't-cares", but their value affect the signal proposition path from A_1 to the output. For instance, as shown in Figure 3, for $A_2A_3 = 00$ and $A_2A_3 = 11$ the signal propagation path length (and thus the propagation delay) from A_1 to O are the shortest and the longest respectively. Xilinx Virtex 5, Virtex 6, and Spartan 6 devices utilize 6-input LUTs. Therefore, by using one single LUT, a programmable delay inverter/buffer with five control inputs can be implemented. The five inputs provide $2^5 = 32$ discrete levels for controlling the delay. The

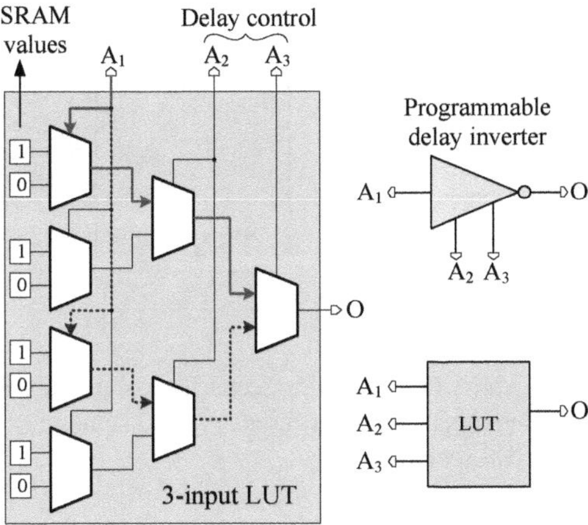

Fig. 3. Precision PDL using a single LUT

measurement data presented in Section 6 obtained from Xilinx Virtex 5 FPGAs suggest that the maximum delay difference from each LUT is approximately 10 pico seconds.

4 Metastability

The proposed TRNG induces metastable conditions in bi-stable logic circuit elements, i.e., flip-flops and latches. The metastable state eventually resolves to a stable state, but the resolution process is extremely sensitive to operational conditions and circuit noise, rendering the result highly unpredictable.

A 'D' flip-flop samples its input at the rising edge of the clock. If sampling takes place within a narrow time window before or after the input signal transitions, a race condition occurs. The race condition takes the flip-flop into a metastable oscillating state. The time window around the sampling moment is typically referred to as setup/hold time. The oscillation eventually settles onto a stable final state of either one or zero. This phenomenon is demonstrated in Figure 4. Note that the probability of settling onto '1' is a monotonic function of the time difference (Δ) between the moment sampling happens and the moment transition occurs at the input. In fact, as shown in [16,9,7], the probability can be accurately modeled by a Gaussian CDF. If the delay difference of the arriving signals is represented by Δ and σ is proportional to the width of the setup/hold time window, then the probability of the output being equal to one can be written as:

$$\text{Prob}\{Out = 1\} = Q(\frac{\Delta}{\sigma}), \tag{1}$$

where $Q(x) = \frac{1}{\sqrt{2\pi}} \int_x^\infty exp(\frac{-u^2}{2})du$. This model can be explained by Central Limit Theorem. Figure 4 demonstrates four scenarios for different signal arrival times. The corresponding probabilities for the scenarios are marked by the scenario number on the probability plot. For instance, in scenarios 1 and 4, since the delay difference is larger than the setup/hold time of the flip-flop, the output is completely deterministic.

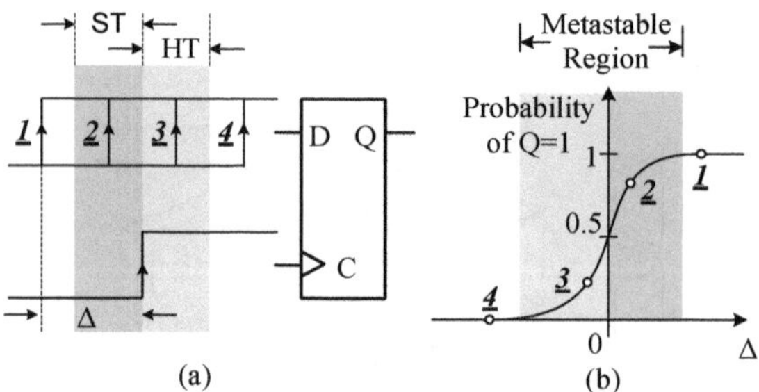

Fig. 4. (a) Flip-flop operation under four sampling scenarios, (b) probability of output being equal to '1' as a function of the input signals delay difference (Δ). The numbers on the probability plot correspond to each signal arrival scenario.

In order to obtain completely non-deterministic and unpredictable output bits with equal probabilities (Prob$\{Output{=}1\}$ = Prob$\{Output{=}0\}$ = 0.5), our method forces the flip-flop into metastability by tuning sampling and signal arrival times so they occur as simultaneously as possible (driving $\Delta \rightarrow 0$) using the PDLs.

5 TRNG System Design

To drive the flip-flop into its metastable state, we use an at-speed monitor-and-control mechanism that establishes a closed loop feedback system. The monitor module keeps track of the output bit probabilities over repeated time intervals. It then passes on the information to the control unit. The control unit based on the received probability information decides to add/subtract the delay to/from top/bottom paths to calibrate the delay difference so that it gets closer to zero. For instance, if the output bits are highly skewed towards 1, then the delay difference (Δ) must be decreased by increasing the top path delay to balance the probabilities. Figure 5 (a) demonstrates this concept.

A straightforward implementation of the monitoring unit can be realized by using a counter. The counter value is incremented every time the flip-flop outputs '1' and is decremented whenever the flip-flop generates a '0'. This is analogous to performing a running sum over the sequence of output bits where zeros are replaced by '-1'. If zeros and ones are equally likely, the value of the counter will stay almost constant.

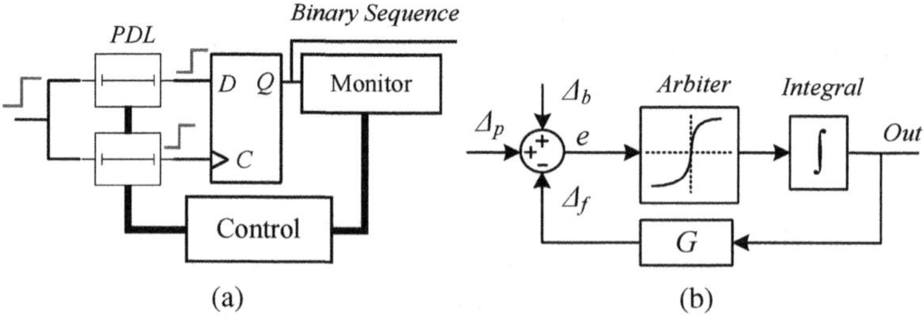

Fig. 5. The TRNG system model

A feedback signal is generated proportional to any deviation from this constant steady state value. The generated error signal is fed back to the signal-to-delay transducer, i.e., the PDL. The delay difference (Δ) is updated/corrected based on the feedback signal.

The described system is in effect a proportional-integral (PI) controller. The system is depicted in Figure 5 (b). In this figure, Δ_b is the constant bias/skew in delays caused by the routing asymmetries. Δ_p is the delay difference induced by changes in environmental and operational conditions such as temperature and supply voltage, and/or delay difference imposed by active adversarial attacks. Δ_f is the correction feedback delay difference injected by the PDL based on the counter value. Equation 2 expresses the total delay difference at the input of the flip-flop. G represents transformation carried out by the PDL from the counter binary value to an analog delay difference. The arbiter and integrator refer to the flip-flop and counter respectively. Therefore, the following relationship holds;

$$\Delta = \Delta_p + \Delta_b - \Delta_f. \tag{2}$$

An example PDL-based implementation of the TRNG system is shown in Figure 6.

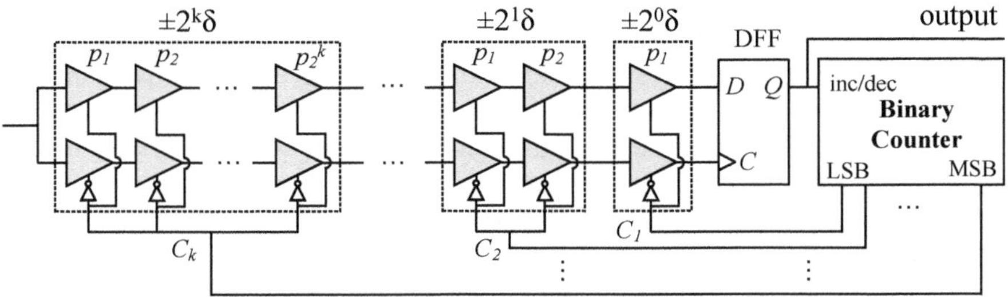

Fig. 6. The TRNG system implementation with a PI controller on FPGA

The PDLs are depicted as gray triangles which provide the finest and most granular level of control over the delays. If the resulting delay difference from one PDL is equal to δ, the effective input/output delay of a PDL, $D(i)$, for the binary input i would be:

$$D(i) = i \times d_c + (1 - i) \times (d_c + \delta). \tag{3}$$

where d_c is a constant delay value. Each programmable delay block consists of two PDLs. The control input of top PDL inside each block is the complement of the bottom PDL control input in order to make a differential programmable delay structure. Based on Equation 4, the differential delay is:

$$D_{diff}(i) = (1 - 2i) \times \delta = (-1)^i \delta, \qquad i = 0 \text{ or } 1. \tag{4}$$

In this example, the programmable delay blocks are packed in groups with sizes of multiples of two to efficiently generate any desirable delay difference using a binary control input. In other words, the first programmable delay block consists of two PDLs, the second one contains 4 PDLs, and so on. With this arrangement, the total incurred delay difference can be written as:

$$\Delta_f = G(\mathbf{C}) = \sum_{i=0}^{K} (-1)^{C_i} 2^i \delta, \tag{5}$$

where $C_i \in \mathbf{C}$ is the ith counter bit with $i = 0$ being the least significant bit (LSB) and $i = K$ being the most significant bit (MSB), and $\mathbf{C}$ represents the counter value. δ is the smallest possible delay difference produced by one PDL.

Let us assume that in the beginning the counter is reset to zero. The resulting feedback delay difference is $\Delta_f = (2^{(K+1)} - 1) \times \delta$ according to Equation 5. This large delay difference skews the output of flip-flop toward '1'. This keeps raising the counter value, lowering the delay difference (Δ). As Δ approaches zero, the flip-flop begins to output '0's more frequently and lowers the rate at which the counter value was previously increasing. At the steady state, the counter value will settle around a constant value with a slight oscillatory behavior. Any outside perturbation on delays will cause transient fluctuations in bit probabilities; however, the automatic adjustment mechanism brings the system back to the equilibrium state.

Counter	I^t	I^b	w
111	1111	0000	+4
110	0111	0000	+3
101	0011	0000	+2
100	0001	0000	+1
000	0000	0001	−1
001	0000	0011	−2
010	0000	0111	−3
011	0000	1111	−4

Fig. 7. Decoding operation

Although the performance of the system in Figure 6 seems ideally flawless, a straightforward hardware implementation was not successful. This is because the design is based on the assumption that δs from PDLs are equal. However, due to manufacturing process variability, the δs slightly vary from one PDL to another. As a result, it is not feasible to generate any desirable delay difference, because the intended weights are not exactly multiples of two anymore. In particular, the input to the largest programmable delay block dominates the system's output behavior.

Instead, we took an alternative approach and used two sets of fine and coarse delay tuning blocks as shown in Figure 8. With n fine tuning delay lines with a resolution of δ_{fn}, and m coarse tuning delay line with resolution of δ_{cs}, any delay difference in the range of $R = [n\delta_{fn} + m\delta_{cs}, -n\delta_{fn} - m\delta_{cs}]$ that satisfies Equation 6 can be produced.

$$\Delta_f = w_{fn}\delta_{fn} + w_{cs}\delta_{cs} \tag{6}$$

where w_{fn} and w_{cs} are integer weights (or levels) such that $-n < w_{fn} < n$ and $-m < w_{cs} < m$. By carefully selecting n,m, δ_{fn}, and δ_{cs}, any delay difference with a resolution of δ_{fn} can be produced within the range R.

The system in Figure 8 is designed such that the weights (or tuning levels) in Equation 6 are a function of the difference in the total number of '1's at PDL inputs on the top and bottom paths;

$$w_{fn} = \sum_{i=1}^{n} I^t[i] - \sum_{i=1}^{n} I^b[i], \quad w_{cs} = \sum_{i=1}^{m} I^t[i] - \sum_{i=1}^{m} I^b[i] \tag{7}$$

where $I^t[i] \in \{0, 1\}$ and $I^b[i] \in \{0, 1\}$ are the input signals to PDLs as demonstrated in Figure 8. Thus, decoder block in Figure 8 needs to perform a mapping from the counter value to the number of '1's at PDL inputs. For example, if $n = 4$, the counter value of '111' corresponds to -4 and '000' corresponds to +4. Table 7 shows an example of decoding operation and corresponding tuning weights for a 3-bit counter. The conversion from the counter value to the effective tuning weight is expressed by Equation 8.

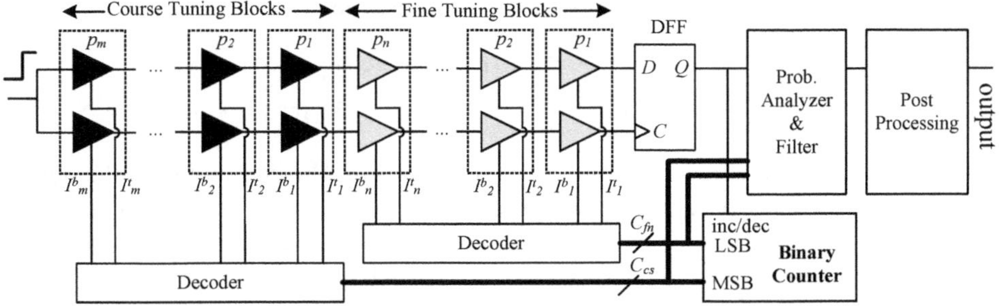

Fig. 8. The complete TRNG system

$$w_{fn} = (-1)^{C_K} \times \left(1 + \sum_{i=0}^{K-1} C_i 2^i\right), \quad K = \lfloor log_2 n \rfloor. \tag{8}$$

The fundamentals of the system's operation shown in Figure 8 are the same as the system in Figure 6 with the only difference lying in how the feedback signal is generated based on the counter states.

Notice that the controller type determines the response time to changes in delays as well as the error in the steady state response. Proportional integral (PI) controllers as opposed to proportional integral derivative (PID) controller due to the lack of derivative function can make the system more stable in the steady state in the case of noisy data. This is because derivative action is more sensitive to higher-frequency terms in the inputs. Additionally, a PI-controlled system is less responsive to inputs (including noise) and so the system will be slower to respond to quick perturbations on the delays than a well-tuned PID system.

The following two observations are important from a security standpoint. First, in the steady state, the counter value oscillates around a constant center value (C_{center}). Let us define the oscillation amplitude as the peak-to-peak range of the oscillations, i.e. the maximum counter value minus the minimum counter value ($C_{max} - C_{min}$). The oscillation is not as periodic as one might think. It is rather a random walk around the center value. Each step in the random walk involves going from one counter value to a one lower or higher value:

$$\text{Step} : C_{current} \rightarrow C_{current} \pm 1$$

The probability of each step (move) is a function of the current location. Intuitively the probability of going outside the range is almost zero:

$$\begin{aligned} \text{Prob}\{C_{max} \rightarrow C_{max} + 1\} &\simeq 0 \\ \text{Prob}\{C_{min} \rightarrow C_{min} - 1\} &\simeq 0 \end{aligned} \tag{9}$$

Also assuming a smooth monotonically increasing probability curve as shown in Figure 4 for the flip-flop, the farther the current counter value is from the center (C_{center}), the lower the probability of moving farther away from the center:

$$\begin{aligned} Prob\{C_i \to C_i + 1\} &< Prob\{C_j \to C_j + 1\} \text{ for } C_j < C_i \\ Prob\{C_i \to C_i - 1\} &< Prob\{C_j \to C_j - 1\} \text{ for } C_j < C_i \end{aligned} \quad (10)$$

Each generated output bit corresponds to a counter value. The probability of the output being to '1' is a function of the feedback counter value. The maximum counter value almost always results in a '0' output, since a '0' value decrements the counter value. Based on Equation 9, transition $C_{max} \to C_{max} + 1$ is unlikely, thus $r(C_{max})$ can almost never be '1'. The following deductions can be explained similarly:

$$\begin{aligned} Prob\{r(C_{center}) = 1\} &\simeq 0.5 \\ Prob\{r(C_{min}) = 1\} &\simeq 1 \\ Prob\{r(C_{max}) = 1\} &\simeq 0 \end{aligned} \quad (11)$$

In other words, during the random walk only those steps that pass close at the center point will result in high entropy and non-deterministic responses. A smaller error in the steady state response means oscillations happen closer to center of the probability transition curve which in turn leads to higher randomness in generated output bits.

In addition, it is desired that the system responds as quickly as possible to external perturbations since the during the recovery time the TRNG generates output bits with highly skewed probabilities.

6 Experimental Results

In this section, we present the LUT-based PDL delay measurement evaluations and TRNG hardware implementation results obtained from Xilinx Virtex 5 LX50T FPGA.

Before moving onto the TRNG system performance evaluation, we shall first discuss the results of our investigation on the maximum achievable resolution of the PDLs. We set up a highly accurate delay measurement system similar to the delay characterization systems presented in [9,7,6].

The circuit under test consists of four PDLs each implemented by a single 6-input LUT. The delay measurement circuit as shown in Figure 9 consists of three flip-flops: launch, sample, and capture flip-flops. At each rising edge of the clock, the launch flip-flop successively sends a low-to-high and high-to-low signal through the PDLs. At the falling edge of the clock, the output from the last PDL is sampled by the sample flip-flop. At the last PDL's output, the sampled signal is compared with the steady state signal. If the signal has already arrived at the sample flip-flop when the sampling takes place, then these two values will be the same; Otherwise they take on different values. In case of inconsistency in sampled and actual values, XOR output becomes high, which indicates a timing error. The capture flip-flop holds the XOR output for one clock cycle.

To measure the absolute delays, the clock frequency is swept from a low frequency to a high target frequency and the rate at which timing errors occur are monitored and recorded. Timing errors start to emerge when the clock half period (T/2) approaches the delay of the circuit under test. Around this point, the timing error rate begins to increase

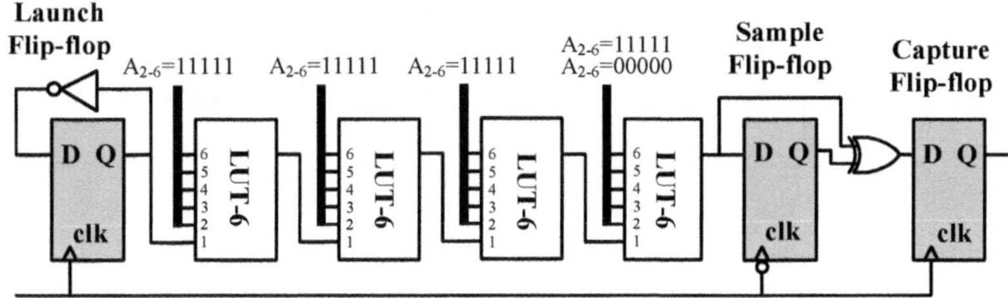

Fig. 9. The delay measurement circuit. The circuit under test consists of four LUTs each implementing a PDL.

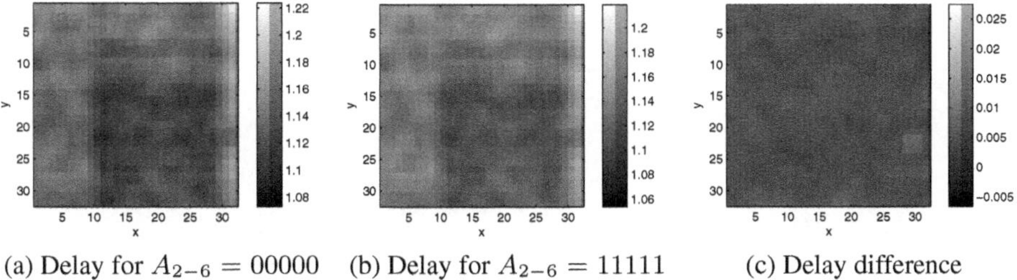

(a) Delay for $A_{2-6} = 00000$ (b) Delay for $A_{2-6} = 11111$ (c) Delay difference

Fig. 10. The measured delay of 32×32 circuit under tests containing a PDL with PDL control inputs being set to (a) $A_{2-6} = 00000$ and (b) $A_{2-6} = 11111$ respectively. The difference between the delays in these two cases is shown in (c).

from 0% and reaches 100%. The center of this transition curve marks the point where the clock half period (T/2) is equal to the effective delay of the circuit under test.

To measure the delay difference incurred by the LUT-based PDL, the measurement is performed twice using different inputs. In the first round of measurement, the inputs to the four PDLs are fixed to $A_{2-6} = 11111$. In the second measurement the inputs to the last PDL are changed to $A_{2-6} = 00000$. In our setup, a 32×32 array of the circuit shown on Figure 9 is implemented on a Xilinx Virtex 5 LX110 FPGA, and the delay from our setup is measured under the two input settings. The clock frequency is swept linearly from 8MHz to 20MHz using a desktop function generator and this frequency is shifted up by 34 times inside the FPGA using the built-in PLL.

The results of the measurement are shown on Figure 10. Each pixel in the image corresponds to one measured delay value across the array. The scale next to the color-map is in nano-seconds. Figure 10 (c) depicts the difference between the measured delays in (a) and (b). As can be seen, the delay values in (b) are on average about 10 pico-seconds larger than the corresponding pixel values in (a). This is in fact equal to the amount of delay difference caused by the coarse PDLs, i.e., δ_{cs}. The delay difference induced by the fine PDL of Figure 11 (a), δ_{fn} is approximately equal to 1/16 of δ_{cs}.

To evaluate the performance of the TRNG system, we implement the system shown in Figure 8 using 32 coarse and fine programmable delay lines ($n = m = 32$). A 12-bit counter performs the running sum operation on the output generated bits. The first six (LSB) bits control the finely tunable PDLs, and the next six (MSB) bits control the

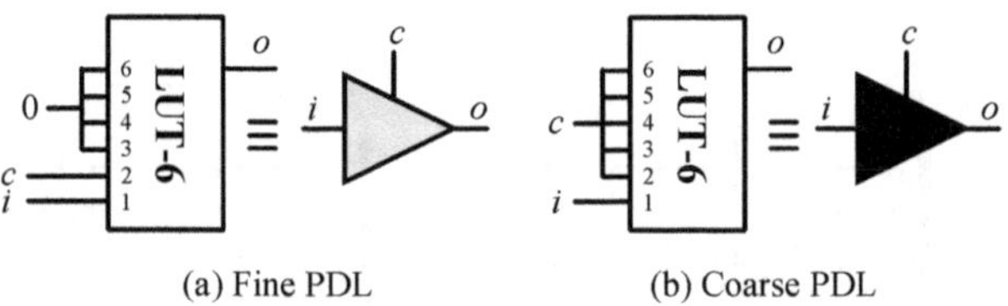

Fig. 11. Coarse and fine PDLs implemented by a single 6-input LUT

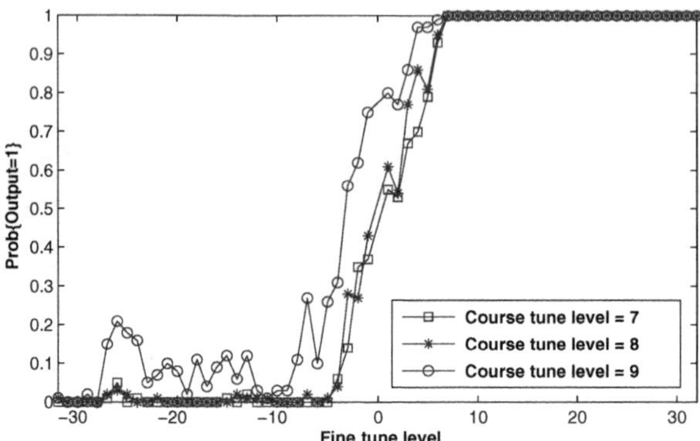

Fig. 12. The probability of flip-flip generating a '1' output as a function of the fine and coarse tuning levels

coarsely tunable PDLs. Both fine and coarse PDLs are implemented by using one LUT as shown in Figure 11. As illustrated in Figure 11, to implement the fine PDL, the LUT inputs A_3 to A_6 are fixed to zero and the only input that controls the delay is A_2. For the coarse PDL, all of the LUT inputs are tied and controlled together.

In the first experiment, we only examine the forward system, which consists of the PDLs, the flip-flop, and the decoders. The tuning weights/levels are swept from the minimum to maximum, and the probability of the flip-flop producing a '1' output is measured at each level. This probability is measured by repeating each experiment over 100 times and counting the number of times the flip-flop outputs a '1'. Since $n = m = 32$, both the fine and coarse tuning levels can go from -32 to 32. Recall that the tuning level represents the difference in the total number of ones at PDL inputs on the top path minus those on the bottom path (see Equation 7). As can be observed from Figure 12, increasing both the coarse and fine tuning levels increase the probability of output being equal to '1'. The non-smoothness of the probability curve is due to variability in the manufacturing process which creates local non-monotonicity. With these observations, we expect the feedback system behavior to stabilize somewhere close to the center of the transition point. Next, we close the feedback loop and initialize the operation. At the beginning, the counter is loaded with all '1's (which results in a decimal value of 2^{12}-1 = 4095). Figure 13 shows the counter value as the operation progresses. The x-axis is the number of clock cycles. Once the operation starts, the

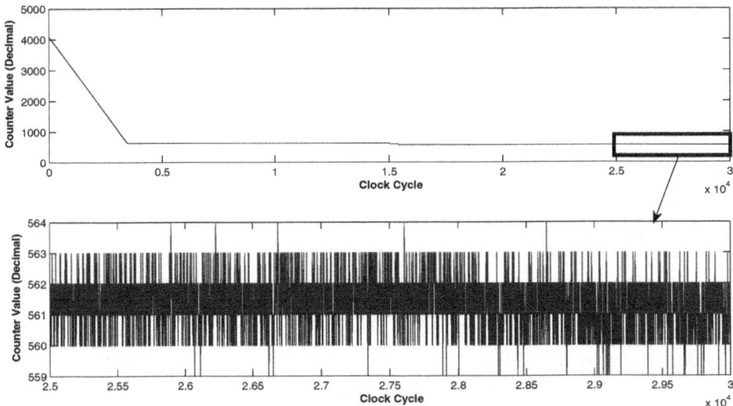

Fig. 13. The transient counter value (decimal) versus the clock cycles

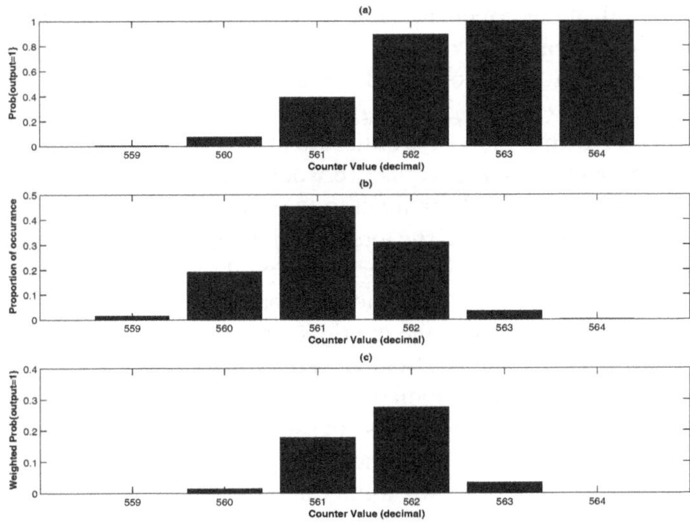

Fig. 14. Distribution of the steady state counter values and associated bit probabilities

counter value keeps decreasing until it reaches the value of approximately 700 after about 3,400 clock cycles. From this point further, the counter value reaches a steady state with a slight oscillatory behavior around a constant value. A close-up of the steady state behavior is depicted in the lower plot of Figure 13. The close-up zooms into the segment between 25,000 to 30,000 clock cycles. As can be observed in the steady state, the counter value oscillates between 559 and 564.

Table 1. NIST Statistical Test Suite results

Statistical Test	Block/Template length	Lowest success ratio
Frequency	-	100%
Frequency within blocks	128	100%
Cumulative sums	-	100%
Runs	-	100%
Longest run within blocks	-	100%
Binary rank	-	100%
FFT	-	100%
Non-overlapping templates	9	90%
Overlapping templates	9	100%
Maurer's universal test	7	100%
Approximate entropy	10	100%
Random excursions	-	100%
Serial	16	100%
Linear complexity	500	90%

Next, we investigate the frequencies at which counter values appear in the steady state. In this experiment, we collect 1,000,000 counter values in the steady state and plot the histogram of the observed values as shown in the middle plot (b) in Figure 14. The normalized histogram suggests that the counter holds the value of 561 more than 40% of the time. Next, it is critical to investigate the probabilities associated with each courter value. In other words, we would like to know for the given counter values – which produce a feedback input to the TRNG core – the probability of the flip-flop output being equal to '1'. The top plot (a) in Figure 14 presents this result. It is interesting to see that most of the counter values produce highly skewed probabilities. Among these counter values, 561 leads to a '1' output slightly more than 40% of the time. We define a metric which is the multiplication of the counter values' frequency of occurrence with the probability of output being equal to one for each counter value. This metric represents the contribution of each counter value to the total number of '1' in the output sequence. The metric values are shown in bottom plot (c) in Figure 14.

To remove the bias in the output sequence in a systematic way as well as to eliminate predictable patterns, we propose a filtering mechanism based on the steady state counter values. The filter unit analyzes the output bit probabilities for each counter value within a window of specific size and flags the counter values that lead to outputs bits with skewed probabilities. Next, it filters out the output bits associated with the flagged counter values. For example, in our implementation, the filter only allows output bits associated with the counter value of 561 to pass through. As a result, the bit-rate is lowered to almost half of the original bit-rate. However, the output bits may still suffer from bias in the bit probabilities. Therefore, a post-processing unit after the filter unit is used to remove any localized biases from the bitstream. In our implementation, we use a von Neumann corrector to perform the post-processing task. The results of the NIST randomness test from running on megabytes of data is shown in Table 1. The comprehensive test results are available online at http://www.ruf.rice.edu/ mm7/trng/.

Table 1 includes the results of the NIST statistical test suite on megabytes of collected data after counter-based filtering and von Neumann correction are performed on the TRNG output bitstream. Due to the large bias in the probabilities, most of the randomness failed when the test was run on the output bitstream before the filtering and correction were carried out.

Finally, according to the ISE Synthesis report, the propagation delay through the TRNG core is equal to 61.06ns which achieves a bit-rate of 16Mbit/sec. The bit-rate drops to 1/8 of the original bit-rate (to 2Mbit/sec) after filtering and von Neumann correction. The TRNG core consumes 128 LUTs that are packed into 16 Virtex 5 CLBs. Note that in practice multiple TRNG cores can run in parallel to offer a higher bit-rate.

7 Conclusion

A novel FPGA-based technique to generate true random numbers through flip-flop metastability was introduced. The presented method took advantage of highly precise programmable delay lines (PDL) to accurately equalize the signal arriving times to flip-flops, thus enforcing a metastable behavior. PDLs as demonstrated in the paper are capable of adjusting signal propagation delays with sub pico-second resolution. With the help of a closed-loop proportional integral (PI) control system, the output bit probabilities are constantly monitored and as soon as any skews in probabilities are observed, feedback signal instantly adjusts the delay taps to revert to the metastable condition. The feedback systems provides resilience against fluctuations in environmental conditions, as well as robustness against active adversarial attacks. Implementation on Xilinx Virtex 5 FPGAs and results of NIST randomness tests show the effectiveness of our true random number generator. The proposed TRNG is capable of producing a throughput of 2 Mbit/sec after post-processing and filtering with a low overhead, using only 5 CLBs.

References

1. Barak, B., Shaltiel, R., Tromer, E.: True random number generators secure in a changing environment. In: Walter, C.D., Koç, Ç.K., Paar, C. (eds.) CHES 2003. LNCS, vol. 2779, pp. 166–180. Springer, Heidelberg (2003)
2. Bergeron, E., Feeley, M., Daigneault, M.A., David, J.: Using dynamic reconfiguration to implement high-resolution programmable delays on an FPGA. In: NEWCAS-TAISA, pp. 265–268 (2008)
3. Gassend, B., Clarke, D., van Dijk, M., Devadas, S.: Silicon physical random functions. In: CCS, pp. 148–160 (2002)
4. Jun, B., Kocher, P.: The Intel random number generator. In: Cryptography Research, Inc. (1999)
5. Lee, J., Lim, D., Gassend, B., Suh, G., van Dijk, M., Devadas, S.: A technique to build a secret key in integrated circuits for identification and authentication applications. In: Symposium on VLSI Circuits, pp. 176–179 (2004)
6. Majzoobi, M., Dyer, E., Elnably, A., Koushanfar, F.: Rapid FPGA characterization using clock synthesis and signal sparsity. In: International Test Conference, ITC (2010)
7. Majzoobi, M., Koushanfar, F.: FPGA time-bounded authentication. IEEE Transactions on Information Forensics and Security PP(99), 1 (2011)

8. Majzoobi, M., Koushanfar, F., Devadas, S.: FPGA PUF using programmable delay lines. In: IEEE Workshop on Information Forensics and Security (WIFS), pp. 1–6 (2010)
9. Majzoobi, M., Elnably, A., Koushanfar, F.: FPGA time-bounded unclonable authentication. In: Böhme, R., Fong, P.W.L., Safavi-Naini, R. (eds.) IH 2010. LNCS, vol. 6387, pp. 1–16. Springer, Heidelberg (2010)
10. von Neumann, J.: Various techniques used in connection with random digits. von Neumann Collected Works 5, 768–770 (1963)
11. O'Donnell, C.W., Suh, G.E., Devadas, S.: PUF-based random number generation. In: MIT CSAIL CSG Technical Memo 481, p. 2004 (2004)
12. Schellekens, D., Preneel, B., Verbauwhede, I.: FPGA vendor agnostic true random number generator. In: Field Programmable Logic and Applications (FPL), pp. 1–6 (2006)
13. Suh, G., Devadas, S.: Physical unclonable functions for device authentication and secret key generation. In: Design Automation Conference (DAC), p. 914 (2007)
14. Sunar, B.: True Random Number Generators for Cryptography. In: Cryptographic Engineering. Springer, Heidelberg (2009)
15. Sunar, B., Martin, W.J., Stinson, D.R.: A provably secure true random number generator with built-in tolerance to active attacks. IEEE Transactions on Computers 58, 109–119 (2007)
16. Wong, J.S.J., Sedcole, P., Cheung, P.Y.K.: Self-Measurement of Combinatorial Circuit Delays in FPGAs. ACM Transactions on Reconfigurable Technology and Systems 2(2), 1–22 (2009)

Generic Side-Channel Countermeasures for Reconfigurable Devices*

Tim Güneysu and Amir Moradi

Horst Görtz Institute for IT Security, Ruhr University Bochum, Germany
{gueneysu,moradi}@crypto.rub.de

Abstract. In this work, we propose and evaluate generic hardware countermeasures against DPA attacks for recent FPGA devices. The proposed set of FPGA-specific countermeasures can be combined to resist a large variety of first-order DPA attacks, even with 100 million recorded power traces. This set includes generic and resource-efficient countermeasures for on-chip noise generation, random-data processing delays and S-box scrambling using dual-ported block memories. In particular, it is possible to build many of these countermeasures into a single IP-core or hard macro that then provides basic protection for any cryptographic implementation just by its inclusion in the design process – what is particularly useful for engineers with no or little background on security and side-channel attacks.

1 Introduction

Since the last fifteen years, side-channel analysis (SCA) [12] attacks have been (publicly) known as a major threat to any unprotected cryptographic implementation in software and hardware. Lots of efforts have already been dedicated towards the development of corresponding countermeasures, in particular against differential power analysis (DPA) [13], such as [5,11,14,15,18,19,20,21,23,24] with this list far from being complete. A particular subject of study has been on algorithmic countermeasures that mask or shuffle security-critical processes of a specific cryptographic system as well as on generic hardware countermeasures, such as noise generators, non-deterministic processors or side-channel resistant logic styles. Based on all these observations it has widely been accepted that a single (and efficient) countermeasure cannot provide complete protection against a large variety of SCA attacks. Hence, a mix of several countermeasures is typically required to provide the security as demanded by the protection profile for a given application (e.g., dictated by an untrusted operation environment and the attacker model). Despite the information theoretic metrics defined by [22], the resistance for such a protection profile, such as against DPA attacks, is typically specified by a number of samples that need to be recorded for a successful attack, as typically done in common criteria evaluations. In other words, if all (known)

* The work described in this paper has been supported by the European Commission through the ICT program under contract ICT-2007-216676 ECRYPT II.

B. Preneel and T. Takagi (Eds.): CHES 2011, LNCS 6917, pp. 33–48, 2011.

SCA attacks with a given number of recorded samples fail, the device is supposed to be sufficiently resistant according to the specified protection profile [16].

In this context, it is primarily the decision of the system designer which combination of countermeasures should be implemented in a device. Unfortunately, choosing a suitable combination of countermeasures is a rather challenging and tedious task in practice. In particular, each device that uses a different technology, architecture or combination of several countermeasures may behave differently. In this context, a hardware developer (in particular one with no or little background on cryptography and/or SCA) would be pleased to have a set of generic, cheap and pre-evaluated hardware-countermeasure implementations at hand that can be easily integrated and combined to achieve a product-specific protection profile.

In this work, we propose and evaluate generic hardware countermeasures against DPA attacks which are suitable for a large variety of recent FPGA devices and cryptographic systems (in particular, devices from Xilinx/Altera). Recently, FPGAs are commonly used for many cryptographic implementations and provide a multitude of different pre-fabricated hardware resources. We will evaluate the most promising reconfigurable resources concerning their usability for generic DPA countermeasures. We finally present resource efficient and generic design approaches for DPA countermeasures that can be combined to resist a large variety of first-order DPA attacks. This includes countermeasures for on-chip noise generation, the insertion of random data processing delays and memory scrambling. In particular, it is possible to combine many of these evaluated countermeasures into a single hard macro to achieve basic protection for any cryptographic implementation – just by its inclusion into the synthesis process. Such an available IP core can be especially valuable for engineers with no or little background on side-channel analysis and/or cryptography.

This work is structured as follows. Section 2 introduces different novel and generic countermeasures which are built from the available resources of recent FPGA devices. In Section 3 we briefly introduce an unprotected AES design as reference implementation for our evaluation. Our measurement setup and the evaluation method used to analyze the impact of each proposed countermeasure are also provided as part of this section. Our results are presented in Section 4 before we conclude in Section 5.

2 Generic Countermeasures for FPGAs

It is well known that generic hardware countermeasures against DPA attacks primarily need to decrease the relation between data processed by a relevant part of the cryptographic implementation and the actual power consumption of the device. There are several options to achieve this goal:

- *Reducing the Signal-to-Noise Ratio*: An attacker attempts to exploit a specific part D_t of a power trace P_t that processes key-dependent data within a (known) cryptographic implementation at a given point in time t. A straightforward countermeasure is thus to bury D_t with lots of additive (Gaussian)

noise N_t so that the overall power trace can be modeled as $P_t = D_t + N_t$. It is evident that the addition of noise is not capable to hide the attackable part D_t completely but it can complicate a practical DPA attack, especially when combined with further countermeasures.
- *Timing Disarrangement*: DPA attacks operate on a high number of (key-dependent) data points that are assumed to be sampled at exactly the same point in time. The attacker usually runs a series of alignment filters to overcome any intrinsic misalignment within the data processing, e.g., due to clock jitter or other operational variations. An effective countermeasure is to further randomize or shuffle the points in time when such attackable operations are processed. Of course, this method can also be overcome [3,5], requiring the attacker to use advanced filtering functions beyond simple peak alignment, such as complex integration and windowing methods.
- *Signal Masking and Scrambling*: An extensive protection against DPA can only be obtained when the attackable part of the signal D_t completely disappears in the power trace. This can be done by applying random masks to D_t. Unfortunately, this strategy is usually very specific, requires expert knowledge and involves significant changes of the cryptographic algorithm at the additional cost of reduced performance and increased resource consumption (see, e.g., for rather costly proposals [7,8]).

In the next subsections we investigate implementation strategies for generic FPGA countermeasures that are widely available for a large number of (symmetric) cryptographic systems. In particular, a primary design goal for this analysis will be to utilize only the (limited amount of) prefabricated resources that are available on recent (Xilinx/Altera) FPGAs.

2.1 Generating Gaussian Noise

Generating white (Gaussian) noise on FPGA devices seems to be simple on the first glance. First, place and route the logic for the main application (that includes particularly a cryptographic component) on the FPGA device. Then, connect all yet unused (but still routable) resources of the device to some random data source and clock them accordingly to the chip enable signal of the cryptographic component. This can even be done automatically, i.e., is certainly possible to create a tool that detects and configures yet unused logic for noise generation in a subsequent development step.

However, our goal in this section is much more specific: we intend to investigate how to configure the available FPGA resources in a way that we achieve a maximum noise level. More precisely, we now analyze three power-consuming strategies that are based on cascading and misusing FPGA resources.

Shift Register LUTs (SRL). Toggling the level of an input signal is known to have the highest impact on a gate's power consumption in CMOS devices – what also holds for SRAM-based FPGA devices. Thus, to generate high noise, we need to toggle as many signals as possible. Taking a closer glance at modern FPGA architectures, these devices consist of large amounts of combined logic

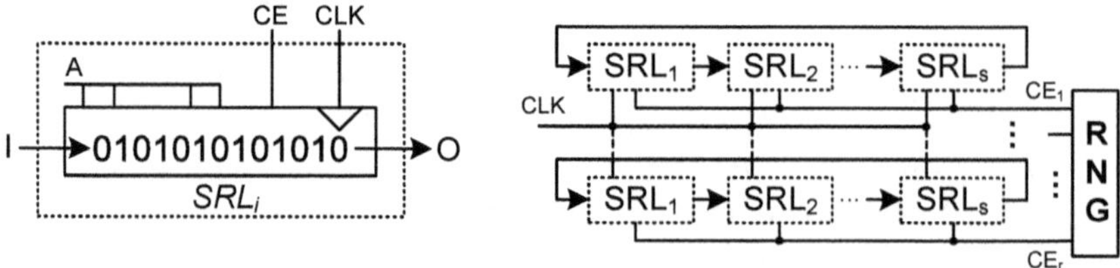

Fig. 1. Noise generation based on shift-register LUT (SRL)

functions made up from flip-flops and lookup tables (LUT). LUTs with n inputs and m outputs can be configured as an n-to-m logic function generator (typically, $n = 4$ or $n = 6$ and $m = 1$). A straightforward approach would then cascade a large number of such LUT/flip-flop pairs (with the LUT being configured as logical NOT) and clock these elements according to a connected random source. However, we certainly could do better. Looking more into the details, a LUT itself consists of 2^n storage bits representing the truth table of its logic function. As a secondary function, the truth table of these LUTs can often be configured as 2^n-bit shift-register LUT (SRL) providing significant savings with respect to conventional shift registers made up from cascades of flip-flops. For an effective noise source, we now configure r cyclic rings of s SRL elements initialized with an alternating (toggle) bit pattern and connect the chip enable signal for each ring to r output bits of a random number generator[1]. The two parameters r and s control the amount of noise variance and noise amplitude, respectively. Figure 1 sketches the noise generating circuit using $r \times s$ SRL elements. Please note that this noise generator does not have an output, hence synthesis tools usually trim such unconnected components. Therefore, the `KEEP` attribute needs to be applied in the HDL description for such constructions to override an undesired optimization/removal by the tools.

BRAM Write Collisions (BWC). Another general observation for hardware devices is that irregular behavior often leads to increased power consumption. A write collision, for example, can occur in the dual-ported block memories (BRAM) of FPGAs when different data is written at the same memory address of a BRAM. For Xilinx FPGAs, for example, the result of such an incident is just undefined [26]. Indeed, it was shown in [9] that the different driving directions lead to data contention on the internal bus lines resulting in metastabilities within the inverter pair of a storage cell. We therefore assume the opposite and conflicting driving directions of the two memory ports to lead to an increased power consumption. We investigated and evaluated this effect using a construction with r BRAMs and s-bit port width as shown in Figure 2. Note that BRAMs are typically a scarce resource in most FPGA applications. But since we only need a single/few empty memory lines to create a write collision, we can also

[1] For test implementations in this work, we used simple PRNGs, however the ideas can easily be combined with available TRNG constructions for FPGAs, e.g., see [25].

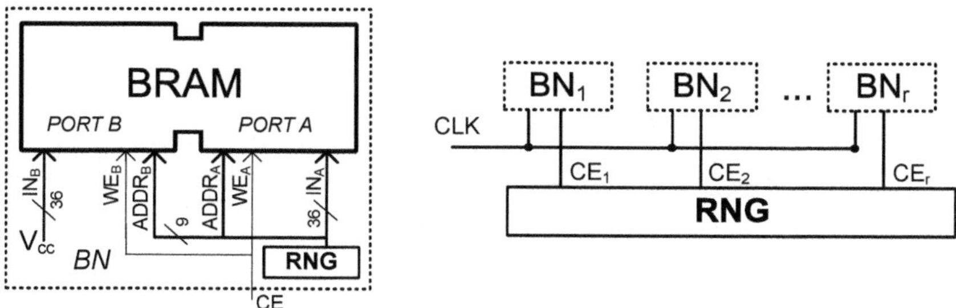

Fig. 2. Noise generator based on memory write collisions

reuse (inactive) BRAMs for noise generation that are actually used otherwise in the main application and whose memory is not entirely used.

Short Circuits in Switch Boxes (SC). Producing a short circuit in a hardware circuit is certainly another strategy to significantly increase the power consumption of a device. Hence, we now try to generate a controlled short circuit (SC) on an FPGA for a very limited amount of time. Note that creating SCs in such a controlled way is not easy with FPGAs, since the design tools run sophisticated design rule checkers that inhibit any misconfiguration. Thus, the development of SC elements need to be done manually and cannot be simplified using HDL tools.

Moreover, we need to consider that SCs have the potential to damage a device. However, FPGA vendors are faced with this issue already by design. Recall that the vendors need to make sure that a configuration file cannot damage a device even if it is corrupted. In case such an illegal configuration is loaded into the FPGA in serial manner, many SCs can happen before the integrity check is finally able to detect the invalid device state. This takes place, in particular, on Xilinx Virtex devices with enabled bitstream encryption when an encrypted configuration is loaded for which no or an incorrect key is present (in this case, the FPGA is getting noticeably hot). Hence, FPGA vendors typically limit the strength of all conflicting drivers to less than 100 μA. Therefore, we can conclude that intentionally constructing short circuits should not have severe consequences on an FPGA's constitution or lifetime.

In order to create an SC in an FPGA configuration based on our Xilinx FPGA setup (cf. Section 3), we refer to the work by Beckhoff et al. [2] which lately demonstrated that LUT input multiplexers of the switch boxes are the most promising source to generate high-power SCs. We used Xilinx low-level tools to create three interconnected LUTs (i.e., two first-level LUTs sourcing a second-level LUT with two input multiplexers). Since it is not possible to directly define the state of routing input multiplexers, we employed Xilinx Design Language (XDL) to convert our placed design into a textual representation that then allows to modify all programmable interconnect points (PIP) freely. We manually reconnected the outgoing PIPs of the first-level LUTs to the identical pinwire of the second-level LUT. This structure (see Figure 3) is then converted into a hard

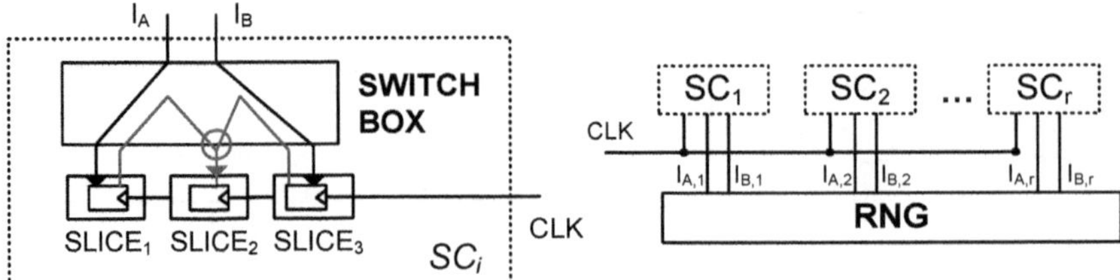

Fig. 3. Short circuits at the input multiplexer to a logic slice (denoted by red wires)

macro that can then be placed multiple times by black-box instantiation inside the FPGA configuration, providing a large number of controllable SC elements. Note that modern FPGAs contain several thousands of LUTs and corresponding input multiplexers. Hence we can easily insert r instances with nearly no resource overhead to scale the amount of noise accordingly to our needs. Note, however, that SC elements should always be spread among the entire chip to distribute the SC load to different power regions to avoid unintended side-effects.

2.2 Clock Randomization (CR)

DPA attacks need to exactly identify the point in time of a power trace when cryptographic data is processed. In order to complicate data alignment, randomized delays or dummy cycles are inserted into the cryptographic operation either by special state machines or non-deterministic processors [3,10]. In this work, we present a novel and very efficient way to randomize data processing by using irregular clock cycle delays and multi-phase shifting obtained from digital clock managers (DCM) in FPGAs. Note that modern FPGAs usually contain a large number of DCMs (often ≥ 4) and clock buffers (≥ 16) of which many remain unused in typical applications. Hence, the following proposal is very appealing to use this type of resource which would be wasted otherwise.

The integrated clock-management functions of many FPGA devices allow jitter correction, clock scaling and phase shifted clock signals. Clock buffers are placed on strategic places of the FPGA to optimize clock distribution. They also enable clock multiplexing (e.g., to drive the design temporarily at reduced clock frequency to implement processor sleep modes) that intrinsically provides a minimum cycle preservation. More precisely, assume we have two different clocks that are multiplexed via a clock buffer to drive a component of the FPGA design. When the clock input is requested to change from one to the other, the clock buffer will wait until the currently selected first clock is low/goes low and remains low until the second input clock has made a transition from high to low. After that, the second clock starts driving the output. In addition, the behavior can be interrupted (resulting in a wait state of undefined length) in case the clock multiplexer is requested to switch clocks again before the first clock change has been completely finished [26].

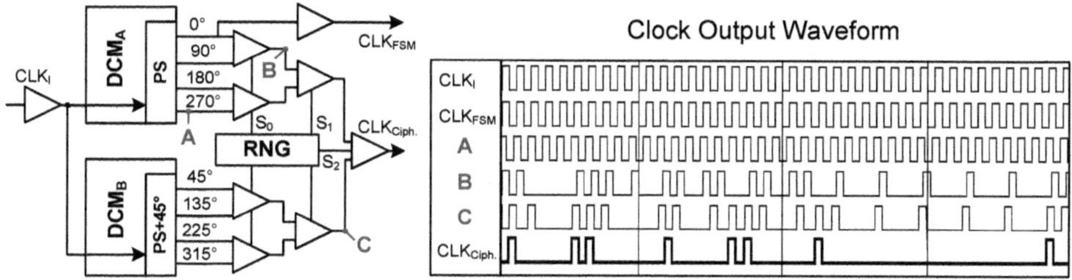

Fig. 4. Clock randomization using DCMs and a tree of clock buffers

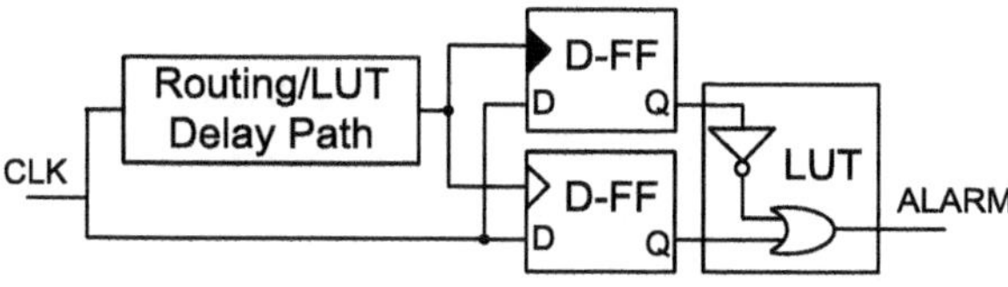

Fig. 5. Circuit to detect external clock manipulations

DCMs in Xilinx FPGAs directly provide outputs for clocks with fixed phase shifts of 0°, 90°, 180°and 270°. Furthermore, the phase of the output clock can also be set to a custom value (which can also be changed dynamically during runtime what is not considered in this work). Our clock randomization countermeasure makes use of a set of l DCMs providing n different output clocks, each phase-shifted by a fixed amount of $360/n$ degrees. A tree of $n-1$ clock multiplexers combines the different clocks to a single clock output that drives the cryptographic core. In addition to that, we need two further clock buffers to sample the input clock and to generate a system clock that is used for the remaining non-cryptographic part of the application (and for noise generating countermeasures). A sample design of this countermeasure with $n = 8$ different phase-shifted clocks is shown in Figure 4.

2.3 Preventing Clock Frequency Manipulations (PCM)

DPA attacks are usually performed at rather low frequencies to easily allow visual peak inspection of the power traces. However, DPA attacks are also at higher clock frequencies possible (e.g., at more than 100 MHz, see [17]), but become more complex due to low-filtering effects of the chip. Hence, to achieve a simple attack setup, an attacker usually desires to reduce the external clock frequency driving the FPGA. This can be prevented as follows. First, a DCM always requires a specific minimum input frequency (specified by the FPGA vendor), otherwise it may not lock (and the main application will not start). Second, a system designer can easily include a detector that triggers an alarm as soon as the clock falls below a specified minimum clock frequency. Figure 5 shows a suitable clock measurement circuit that uses a fixed path delay to shift the phase of a target clock by a fixed amount of $d = 180 + a$ where $a > 0$ denotes an additional phase margin to overcome clock jitter of the input clock. When an attacker attempts to reduce the external input clock frequency (or manipulate the duty cycle) beyond this margin, either one of the flip-flops will sample the alternate part of the clock period, causing finally the alarm to be triggered.

2.4 Block Memory Content Scrambling (BMS)

In this section we present a novel hardware countermeasure for FPGAs based on BRAM-based S-box/T-box scrambling. In many symmetric ciphers, S-boxes are used to introduce a non-linear component in the encryption process and are usually implemented as simple lookup tables. Depending on their size and construction, S-boxes can be realized either using (large amounts of) LUTs or block memories on an FPGA. DPA attacks typically focus on the input and/or outputs of the (known) S-boxes, hence a well-studied countermeasure attempts to mask the S-box data with readily changing, random values. However, it turned out that such a system either significantly reduces the encryption performance and/or requires costly additional operations to pass a random mask through the non-linear S-box [7,8]. Similar to the concept of random permutation tables by Coron [4], we now build a freely running S-box masking scheme specifically for FPGA device. We first assume that we can rewrite the round function $y = R(x)$ of an arbitrary symmetric cipher as composition of a linear part $L(x)$ (including a linear key addition function) and a non-linear part $N(x)$ (implemented as a-on-b-bit S-box), resulting in $y = L(N(x))$ as round function.[2] Assume now N is realized as a lookup table using one port of a dual-ported BRAM of an FPGA device. We further assume that N occupies less than half of the memory available in the BRAM (i.e., 18KBit/36KBit for Xilinx devices). Now we define two memory segments or *contexts* in the BRAM: an active context which contains a recent version of the (masked) S-box currently used for encryption and an inactive context containing a copy which is currently scrambled and remasked by an encryption-independent process. The scrambling itself is a sequential process on the second port of that BRAM that applies a b-bit mask $\mathbf{m}$ to each S-box entry at address i of the active context and stores the result at address $i \oplus \pi(L(\mathbf{m}))$ of the inactive context (here, $\pi(x)$ represents a selection function for the corresponding S-box input bits if $a \neq b$). In other words, it applies an additive mask to the S-box output that is also pushed through the linear part L of the round function to determine the new (permuted) input address index where each updated masked S-box value is finally stored. Note that we refer to this process (that adds a random b-bit mask $\mathbf{m}$ iteratively to all S-box entries and writes each entry back to a permuted address) as *scrambling* rather than *masking*. After all S-box entries have been processed by this scrambling process, the context can be switched (i.e., the inactive context becomes active and vice

[2] Note that this generalization actually holds for a large number of ciphers, though symmetric round constructions often contain several linear operations $L_1, L_2, L_3, \ldots$ involving constructions such as $y = L_1(L_2(N(L_3(\ldots(x)))))$. But assuming that only a single non-linear component N is used in the round function, we can always combine and rearrange subsequent rounds in a way that the linear subcomponents aggregate into a single L as shown above and as could be seen in [4].

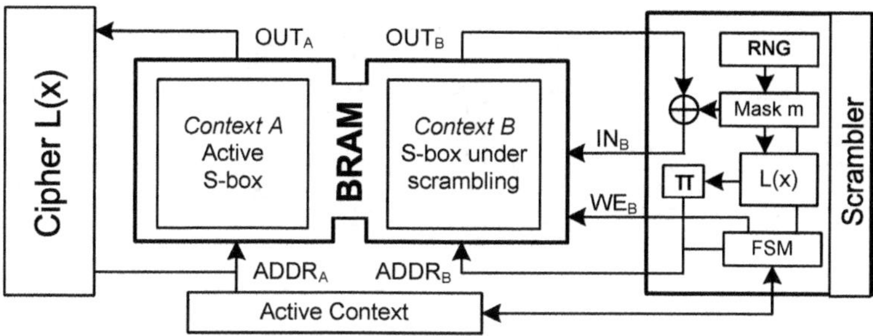

Fig. 6. Construction scrambling S-box entries with random masks

versa) and scrambling restarts on the recently deactivated context. Note that the cipher and scrambler operate concurrently, however, the context switch is never done within a running encryption to avoid data inconsistency issues. This implies that multiple sequential rounds and encryptions are performed using the same scrambled S-box and mask $\mathbf{m}$ (what could be exploited from a theoretical point of view using higher-order attacks). We like to stress at this point that we designed this countermeasure for performance and efficiency, since the concurrent data scrambling process and the instant context switch does not reduce the encryption speed. Note further, that in combination with previous countermeasures such as clock randomization and noise generation, higher-order attacks will also become extremely complex. Figure 6 shows finally the generic construction of the BRAM scrambling circuitry.

3 Case Study

We now investigate how the countermeasures presented above can harden an AES implementation against DPA attacks. We start with an unprotected AES instance which we subsequently augment with our countermeasures to evaluate their effectiveness.

3.1 Reference Architecture

For our experiments we used an unprotected, round-based T-table implementation of the standardized AES block cipher with a 128-bit data path. The round function for such an implementation uses four 8-to-32-bit T-tables T_i to compute a 32-bit share S_j of the 128-bit AES state S according to the following formula:

$$S_j = k_j \oplus T_0[\pi_{0,j}(S)] \oplus T_1[\pi_{1,j}(S)] \oplus T_2[\pi_{2,j}(S)] \oplus T_3[\pi_{3,j}(S)], \tag{1}$$

where $\pi_{i,j}$ represents a static input byte selection function. Note that we can rewrite Equation (1) to

$$R_j(S) = L(N_0(\pi_{0,j}(S)), N_1(\pi_{1,j}(S)), N_2(\pi_{2,j}(S)), N_3(\pi_{3,j}(S)), k_j) \tag{2}$$

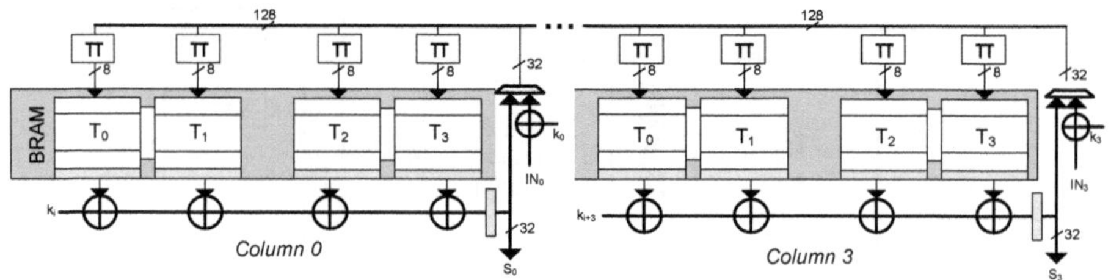

Fig. 7. T-table AES implementation used for this case study

with $N_i = T_i$ and $L(a, b, c, d, e) = a \oplus b \oplus c \oplus d \oplus e$ to comply with the generic round function specification discussed in Section 2.4. Further information about the AES T-table implementations can be found in [6]. Our unprotected AES implementation as shown in Figure 7 requires 21 clock cycles (1 initial clock cycle and 2 for each round) to compute a full AES-128 encryption and consumes 682 slices (1182 LUTs, 397 FF) and 8 BRAMs storing the 16 T-tables on a Xilinx Virtex-II Pro FPGA.

3.2 Measurement Setup and Attack Model

The AES design explained above was implemented on the Xilinx Virtex-II Pro FPGA (xc3vp7) of a SASEBO circuit board which is particularly designed for side-channel attack evaluations [1]. The instantaneous power consumption traces are collected using a LeCroy WP715Zi 1.5 GHz oscilloscope at a sampling rate of 2.5GS/s and by means of a differential probe capturing the voltage drop across a 1Ω resistor placed in the VCCINT (1.6V) path of the target FPGA.

In order to examine the leakage of the implementation and find a suitable power model for the correlation power analysis (CPA) attacks, we started the practical experiments when the target core is clocked at 24MHz which is selected as the reference implementation for further comparisons. Figure 8(a) shows a superposition of 1000 traces of this case indicating the well alignment of the measurements. Since the I/O ports of the BRAMs of the target FPGA have a considerably higher capacitance compared to the other low amount of logic cells of our target design, data transfered through the BRAM buses caused by reading the memory cells while computing the T-table outputs should have noticeable impact on the power consumption. Therefore, we have used the Hamming weight (HW) of the 32-bit T-table output as the hypothetical power model in a CPA attack. The result of such an attack predicting the T-table outputs of the first encryption round using 10 000 traces is shown in Figure 8(b). It shows that – using a *rule of thumb* [17] – around 3000 traces are the minimum number of required measurements. We have examined a couple of other hypothetical models, and the best result was achieved using the aforementioned model.

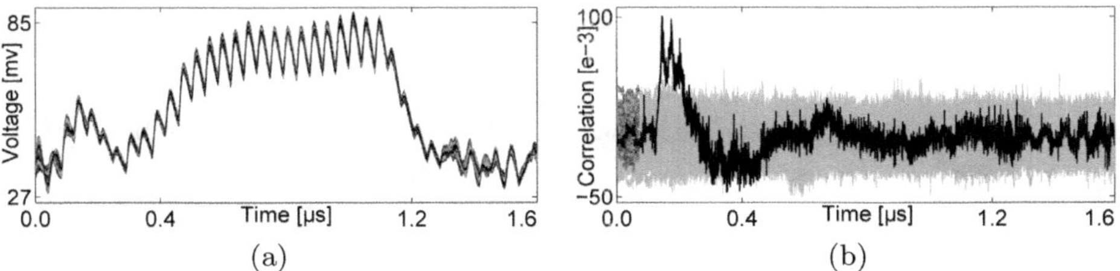

Fig. 8. The AES core in 24MHz (a) superimposition of 1000 traces, (b) CPA attack result using 10 000 traces

4 Evaluation and Results

The results of the attacks evaluating the effect of each method introduced in Section 2 are presented in the following. Since the leakage model of the target device is well known and can be appropriately estimated by a HW model, we limited our evaluations to CPA attacks and considered the number of required measurements as metric for comparisons. The result of each scheme is compared to the results shown above in the reference implementation.

4.1 Noise Generators

Adding each of the noise generation schemes individually as explained in Section 2.1 leads to an increased amount of switching noise and has an effect on the number of required measurements for a successful attack. In order to practically examine their effectiveness, we added the noise generators based on shift-register LUTs (SRL), BRAM write collisions (BWC) and short circuits (SC) individually into the reference implementation and repeated the measurements and corresponding attacks. We used uniform parameters for the individual noise generators, namely $r = 16$ and $s = 36$, taking an additional resource consumption of 576 LUTs (SRL), 16 BRAMs (BWC) and 48 LUTs (SC), respectively. Figure 9 shows the result of the attacks when each of the noise sources is separately enabled in addition to the case when all of them exist in the design. It can be concluded that adding noise sources with quite moderate parameters already increases the number of required measurements slightly, i.e., to around 8000. In this context, the SC noise generator is obviously the most efficient one with respect to the number of consumed resources. However, using solely noise addition (even with much larger parameters for r and s) can certainly not be considered as an optimum way to make any attacks infeasible.

4.2 Clock Randomizing

In Section 2.2 we presented the CR method to randomize the clock source by randomly changing the clock phase to introduce a variable misalignment of the power traces. For our attack, we used a setup based on $l = 2$ DCMs with $n = 8$ phase-shifted clock outputs which are processed and multiplexed by 9

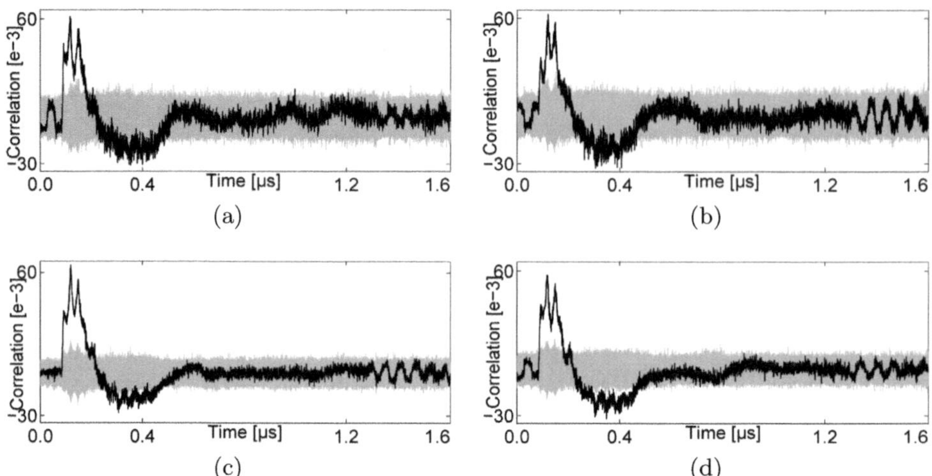

Fig. 9. CPA attack results using 50 000 traces of the AES core in 24MHz including (a) shift-register LUTs, (b) BRAM write collisions, (c) short circuits, and (d) all three noise sources

clock buffers. An encryption clocked by this irregular output takes on average 3.77 times longer than our reference implementation. Embedding this unit into the reference implementation – as expected – led to a variable amount of time required for an encryption. Figure 10(a) shows a superimposition of 1000 traces, already indicating a strong misalignment. Therefore, we performed the attacks using considerably more measurements, i.e., 10 000 000. The results shown in Figure 10(b) signifies the need for around 3 000 000 traces to determine the correct key hypothesis. We like to emphasize that the randomization of processing times is not aligned with the primary clock any longer (unlike in shuffling schemes), rendering a combing technique [24] useless in our case. Since combing is done by adding up the leakage points of consecutive clock cycles of a trace while – as shown in Figure 10(a) – it is here not possible to clearly distinguish the clock cycles. Reducing the input clock frequency to facilitate the attack can be prevented by using a detector for clock manipulations as shown in Section 2.3. However, a windowing approach [3], summing up all points within a defined window, proved to be effective since the operating clock frequency of 24MHz let the power peaks of consecutive clock cycles overlap with each other (i.e., this intrinsic low pass filter has the same effect as windowing). Indeed, we even repeated this type of attack with different window sizes, but with no significant difference to Figure 10(b).

4.3 Block Memory Content Scrambling

In order to integrate the BRAM scrambling technique (BMS) introduced in Section 2.4 into our reference implementation, we need to duplicate the number of BRAMs to provide a separate memory port and space for the scrambling

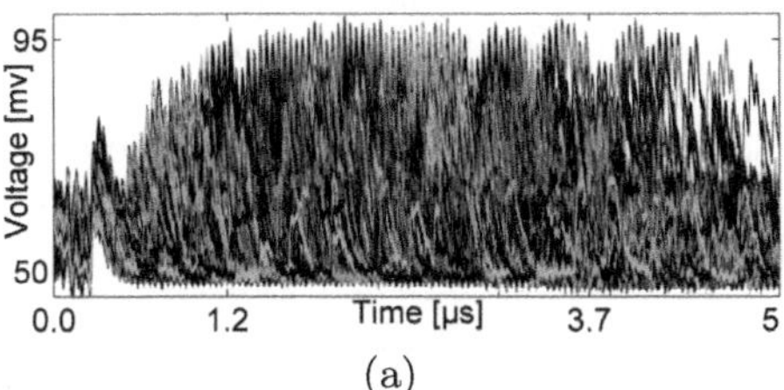

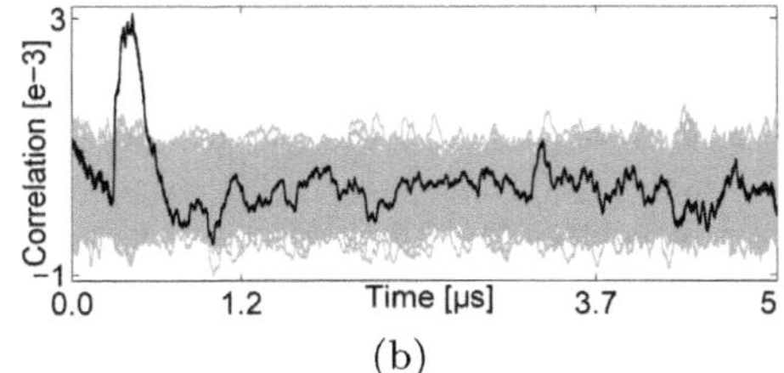

(a) (b)

Fig. 10. The AES core in 24MHz equipped with the clock phase shift unit (a) superimposition of 1000 traces, (b) CPA attack result using 10 000 000 traces

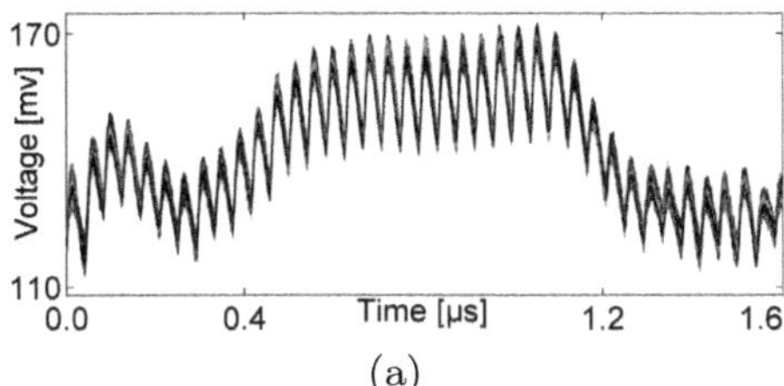

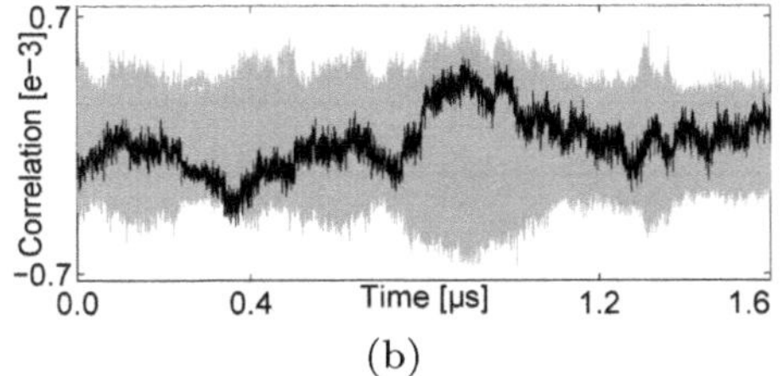

(a) (b)

Fig. 11. The AES core in 24MHz equipped with BRAM scrambling (a) superimposition of 1000 traces, (b) CPA attack result using 100 000 000 traces

process. For the scrambling, we used 16×32 bit masks $\mathbf{m}_i$ to mask the output of all 16 T-tables requiring a total of 512 random bits per scrambling cycle. According to Equation (2), we can apply the linear transformation L of the 32-bit AES round function to the masks by computing $L(\mathbf{m}_i, \mathbf{m}_{i+1}, \mathbf{m}_{i+2}, \mathbf{m}_{i+3}, 0) = \mathbf{M}_i$ for $i = 0 \ldots 3$. From each aggregated 32-bit mask $\mathbf{M}_i$ the corresponding input byte mask is finally selected by $\pi(\mathbf{M}_i)$ and determines the corresponding permutation mask to the next round's T-table input.

Since this scheme does not affect the timing behavior of the design, the variety of power traces should be similar to that of the reference implementation. One difference that may have an influence on the power consumption is the additional scrambler circuit that concurrently modifies the BRAM contents. Therefore, as shown in Figure 11(a) the DC level of the power traces is increased compared to Figure 8(a).

The input and output of the T-tables are now masked by random values which are not shared inside a single round computation, and we expect this design to avoid univariate power analysis attacks. In order to examine this claim, we have measured 100 000 000 traces and have performed the same attack as before. The result is shown in Figure 11(b).

However, a critical point is the reuse of the masks between the subsequent rounds. This means, the 512-bit mask used in the first round is reused in the later rounds since the BRAM sections used during an encryption process stay unaltered. Therefore, a second-order attack is possible combining the leakages, for example, of two consecutive rounds. To do this, the adversary needs to predict

Table 1. Time and resource overhead for each proposed countermeasure for parameters $r = 16$, $s = 36$ and $l = 2$, excluding RNG and output combining function. The plain AES implementation consumes 8 BRAM, 1182 LUT and 397 FF on a Xilinx Virtex-II Pro. Figures with asterisk (*) are not quantified due to implementation-specific characteristics. Abbreviations: LUT=Look-Up-Table, FF=Flip-Flop, DCM=Digital Clock Manager, BRAM=Block Memory, CB=Clock Buffer.

Section/ Method	Generic Implementation		AES T-Table Case Study	
	Logic	*Time*	*Logic*	*Time*
2.1.1/SRL	$r \cdot s$ LUT	*none*	576 LUT	*none*
2.1.2/BWC	r BRAM, $r \cdot s$ LUT	*none*	16 BRAM, 576 LUT	*none*
2.1.3/SC	$3 \cdot r$ LUT	*none*	48 LUT	*none*
2.2/CR	$l-1$ DCM, $4l-1$ CB	*n/a**	1 DCM, 7 CB	3.77×
2.3/PCM	1 LUT, 2 FF, *Delay Path*	*none*	3 LUT, 2 FF	*none*
2.4/BMS	*n/a**	*none*	8 BRAM, 1706 LUT, 1169 FF	*none*

at least 2^{40} bits of the key, i.e., 2^{32} for 4 bytes of the first round key which are at the same column after ShiftRows, and 2^8 for one byte of the second round key. One may also think about combining the leakages of the scrambler module with that of the first encryption round, but this requires to know exactly the instance in time when the scrambler processes the T-table content used in the next encryption. Note further that the scrambler unit operates independently with a separate clock, hence its computations are not synchronized with the encryption process.

4.4 Combining Countermeasures

Although we were already unable to defeat the latter countermeasure with first-order attacks and 100 million traces, we like to stress that we still can strengthen this design by adding the clock randomization and noise generation countermeasures. We have omitted to provide it as a separate result since it is similar to the one depicted in Figure 11(b). The combination of all proposed countermeasures are likely to harden our design against a large number of multivariate attacks which are out of scope of this work. The designer thus can mix and match the proposed countermeasures according to the security requirements of the application and the remaining logic available on the FPGA device. Table 1 shows the overhead of the countermeasures for the generic and specific case considering their impact on the logic resource consumption and execution time of a cryptographic process. Comparing our results with other work, we noticed that countermeasures specifically built for (comparable) FPGA devices are quite rare. We thus only refer to the work [18] implementing an SCA countermeasure based dynamic partial reconfiguration on a Virtex-II Pro. Their countermeasure requires an overhead of 2566 slices (i.e., up to 5132 LUT/FF pairs) and reduces the throughput of a plain AES implementation by a factor of 6.6. Except for BWC, combining all our countermeasures of this work results in a more efficient solution (i.e., a combination of all methods, excluding BWC, consumes 2337 LUT, 1171 FF, 1 DCM, 7 CB and reduces the throughput by a factor of 3.77).

5 Conclusion

In this work we presented several generic hardware countermeasures against DPA attacks that can be efficiently implemented on recent FPGA devices. We could practically demonstrate that a combination of the presented countermeasures (noise generation, clock randomization and memory scrambling) rendered first-order DPA attacks with up to 100 million traces unsuccessful. For independent verification of our results and future work (e.g., second-order SCA), the PROM files with the AES implementation and all countermeasures for the SASEBO are publicly available at `http://www.emsec.rub.de/research/publications`.

References

1. Side-channel Attack Standard Evaluation Board (SASEBO). Further information are available via, `http://www.rcis.aist.go.jp/special/SASEBO/index-en.html`
2. Beckhoff, C., Koch, D., Torresen, J.: Short-Circuits on FPGAs Caused by Partial Runtime Reconfiguration. In: FPL, pp. 596–601. IEEE Computer Society, Los Alamitos (2010)
3. Clavier, C., Coron, J.-S., Dabbous, N.: Differential Power Analysis in the Presence of Hardware Countermeasures. In: Paar, C., Koç, Ç.K. (eds.) CHES 2000. LNCS, vol. 1965, pp. 252–263. Springer, Heidelberg (2000)
4. Coron, J.-S.: A New DPA Countermeasure Based on Permutation Tables. In: Ostrovsky, R., De Prisco, R., Visconti, I. (eds.) SCN 2008. LNCS, vol. 5229, pp. 278–292. Springer, Heidelberg (2008)
5. Coron, J.-S., Kizhvatov, I.: Analysis and Improvement of the Random Delay Countermeasure of CHES 2009. In: Mangard, S., Standaert, F.-X. (eds.) CHES 2010. LNCS, vol. 6225, pp. 95–109. Springer, Heidelberg (2010)
6. Daemen, J., Rijmen, V.: The Design of Rijndael: AES - The Advanced Encryption Standard. Springer, Heidelberg (2002)
7. Golic, J.D., Tymen, C.: Multiplicative Masking and Power Analysis of AES. In: Kaliski Jr., B.S., Koç, Ç.K., Paar, C. (eds.) CHES 2002. LNCS, vol. 2523, pp. 198–212. Springer, Heidelberg (2003)
8. Goubin, L., Patarin, J.: DES and Differential Power Analysis (the "duplication" method). In: Koç, Ç.K., Paar, C. (eds.) CHES 1999. LNCS, vol. 1717, pp. 158–172. Springer, Heidelberg (1999)
9. Güneysu, T.: Using Data Contention in Dual-ported Memories for Security Applications. Journal of Signal Processing Systems, 1–15 (2010)
10. Irwin, J., Page, D., Smart, N.P.: Instruction Stream Mutation for Non-Deterministic Processors. In: ASAP, pp. 286–295. IEEE Computer Society, Los Alamitos (2002)
11. Itoh, K., Yajima, J., Takenaka, M., Torii, N.: DPA Countermeasures by Improving the Window Method. In: Kaliski Jr., B.S., Koç, Ç.K., Paar, C. (eds.) CHES 2002. LNCS, vol. 2523, pp. 303–317. Springer, Heidelberg (2003)

12. Kocher, P.C.: Timing Attacks on Implementations of Diffie-Hellman, RSA, DSS, and other systems. In: Koblitz, N. (ed.) CRYPTO 1996. LNCS, vol. 1109, pp. 104–113. Springer, Heidelberg (1996)
13. Kocher, P.C., Jaffe, J., Jun, B.: Differential Power Analysis. In: Wiener, M. (ed.) CRYPTO 1999. LNCS, vol. 1666, pp. 388–397. Springer, Heidelberg (1999)
14. Macé, F., Standaert, F.-X., Quisquater, J.-J.: Information Theoretic Evaluation of Side-Channel Resistant Logic Styles. In: Paillier, P., Verbauwhede, I. (eds.) CHES 2007. LNCS, vol. 4727, pp. 427–442. Springer, Heidelberg (2007)
15. Mamiya, H., Miyaji, A., Morimoto, H.: Efficient Countermeasures Against RPA, DPA, and SPA. In: Joye, M., Quisquater, J.-J. (eds.) CHES 2004. LNCS, vol. 3156, pp. 343–356. Springer, Heidelberg (2004)
16. Mangard, S.: Hardware Countermeasures Against DPA – A Statistical Analysis of their Effectiveness. In: Okamoto, T. (ed.) CT-RSA 2004. LNCS, vol. 2964, pp. 222–235. Springer, Heidelberg (2004)
17. Mangard, S., Oswald, E., Popp, T.: Power Analysis Attacks: Revealing the Secrets of Smart Cards. Springer, Heidelberg (2007)
18. Mentens, N., Gierlichs, B., Verbauwhede, I.: Power and Fault Analysis Resistance in Hardware Through Dynamic Reconfiguration. In: Oswald, E., Rohatgi, P. (eds.) CHES 2008. LNCS, vol. 5154, pp. 346–362. Springer, Heidelberg (2008)
19. Moradi, A., Poschmann, A.: Lightweight Cryptography and DPA Countermeasures: A survey. In: Sion, R., Curtmola, R., Dietrich, S., Kiayias, A., Miret, J.M., Sako, K., Sebé, F. (eds.) RLCPS, WECSR, and WLC 2010. LNCS, vol. 6054, pp. 68–79. Springer, Heidelberg (2010)
20. Okeya, K., Takagi, T.: A More Flexible Countermeasure Against Side Channel Attacks using Window Method. In: Walter, C.D., Koç, Ç.K., Paar, C. (eds.) CHES 2003. LNCS, vol. 2779, pp. 397–410. Springer, Heidelberg (2003)
21. Prouff, E., McEvoy, R.: First-order Side-Channel Attacks on the Permutation Tables Countermeasure. In: Clavier, C., Gaj, K. (eds.) CHES 2009. LNCS, vol. 5747, pp. 81–96. Springer, Heidelberg (2009)
22. Standaert, F.-X., Malkin, T.G., Yung, M.: A Unified Framework for the Analysis of Side-Channel Key Recovery Attacks. In: Joux, A. (ed.) EUROCRYPT 2009. LNCS, vol. 5479, pp. 443–461. Springer, Heidelberg (2009)
23. Standaert, F.-X., Örs, S.B., Preneel, B.: Power Analysis of an FPGA: Implementation of Rijndael: Is Pipelining a DPA Countermeasure? In: Joye, M., Quisquater, J.-J. (eds.) CHES 2004. LNCS, vol. 3156, pp. 30–44. Springer, Heidelberg (2004)
24. Tillich, S., Herbst, C.: Attacking State-of-the-art Software Countermeasures—A Case Study for AES. In: Oswald, E., Rohatgi, P. (eds.) CHES 2008. LNCS, vol. 5154, pp. 228–243. Springer, Heidelberg (2008)
25. Varchola, M.: FPGA Based True Random Number Generators for Embedded Cryptographic Applications. PhD thesis, Technical University of Kosice (2008)
26. Xilinx Inc. User Guides for Xilinx FPGA devices (April 2011), `http://www.xilinx.com`

Improved Collision-Correlation Power Analysis on First Order Protected AES

Christophe Clavier[1], Benoit Feix[1,2], Georges Gagnerot[1,2], Mylène Roussellet[2], and Vincent Verneuil[2,3]

[1] XLIM-CNRS, Université de Limoges, France
firstname.familyname@unilim.fr
[2] INSIDE Secure, Aix-en-Provence, France
firstname-first-letterfamilyname@insidefr.com
[3] Institut de Mathématiques de Bordeaux, France

Abstract. The recent results presented by Moradi et al. on AES at CHES 2010 and Witteman et al. on square-and-multiply always RSA exponentiation at CT-RSA 2011 have shown that collision-correlation power analysis is able to recover the secret keys on embedded implementations. However, we noticed that the attack published last year by Moradi et al. is not efficient on correctly first-order protected implementations. We propose in this paper improvements on collision-correlation attacks which require less power traces than classical second-order power analysis techniques. We present here two new methods and show in practice their real efficiency on two first-order protected AES implementations. We also mention that other symmetric embedded algorithms can be targeted by our new techniques.

Keywords: Advanced Encryption Standard, Side Channel Analysis, Collision, Correlation, DPA, Masking.

1 Introduction

Side-channel analysis was introduced by Kocher et al. in 1998 [9] and marks the outbreak of this new research field in the applied cryptography area. Meanwhile, many side-channel techniques have been published. For example Brier et al. proposed the Correlation Power Analysis (CPA) [4] which has shown to be very efficient as it significantly reduces the number of curves needed for recovering a secret key, and more recently the Mutual Information Analysis from Gierlichs et al. [7] has generated a lot of interest.

Since side-channel attacks potentially concern any kind of embedded implementations of symmetric or asymmetric algorithms, it is recommended to apply various masking countermeasures (among others) in sensitive products [1,14]. Second-order or higher-order side-channel analysis can however defeat such countermeasures by combining leakages from different instants of the execution of an algorithm and canceling the effect of a mask [12,13]. Such attacks are considered very difficult to implement and generally require an important number of power curves.

B. Preneel and T. Takagi (Eds.): CHES 2011, LNCS 6917, pp. 49–62, 2011.

A specific approach for side-channel analysis is using information leakages to detect collisions between data manipulated in algorithms. Side-channel collision attacks against a block cipher were first proposed by Schramm et al. in 2003 [19]. Their attack uses differential analysis to exploit collisions in adjacent S-Boxes of the DES algorithm. In [18] an attack against the AES is proposed to detect collisions in the output of the first round `MixColumns`. Later, Bogdanov [2] improved this attack by looking for equal S-Boxes inputs in several AES executions. He then studied in [3] statistical techniques to detect collisions between power curves. Two recent papers have updated the state-of-the-art by introducing correlation based collision detection: Moradi et al. [15] proposed a collision attack to defeat an AES implementation using masked S-Boxes, while Witteman et al. [22] applied a cross-correlation analysis to an RSA implementation using message blinding.

In this paper, we present two collision-correlation attacks on software AES implementations protected against first-order power analysis using masked S-Boxes and practical results on both simulated and real power curves. Our attacks are much more efficient and generic compared to the one presented in [15]. Moreover we believe our techniques to be applicable to other embedded implementations of symmetric block ciphers.

The remainder of the paper is organized as follows: Section 2 presents the two AES first-order protected implementations targeted by our study. Then in Section 3 we present our attacks and practical results on simulated power curves and on a physical integrated circuit. In Section 4 we compare our technique with second-order power analysis. Section 5 deals with the possible countermeasures and finally we conclude this paper in Section 6.

2 Targeted Implementations

The AES Algorithm. For the sake of simplicity in this paper we focus on the AES-128 which includes 10 rounds, each one decomposed into four functions: `AddRoundKey`, `SubBytes`, `ShiftRows` and `MixColumns`. It encrypts a 128-bit message $M = (m_0, \ldots, m_{15})$ using a 128-bit secret key $K = (k_0, \ldots, k_{15})$ and produces a 128-bit ciphertext $C = (c_0, \ldots, c_{15})$. Note however that the techniques presented in this paper are easily applicable to AES-192 and AES-256.

The only non-linear function of the AES is `SubBytes` (also referred to as the S-Boxes S in the following) which is a substitution function defined by the pseudo-inversion I in $\mathrm{GF}(2^8)$ and an affine transformation. In this paper, we consider the two following solutions that have been proposed to protect this function against first-order attacks.

2.1 Blinded Lookup Table

The first targeted implementation uses a *masked* substitution table as proposed by Kocher et al. [10] and Akkar et al. [1]. This masked table S' is defined by $S'(x_i \oplus u_i) = S(x_i) \oplus v_i$, with u_i (resp. v_i) the mask of the i-th input byte x_i (resp. output byte) of function `SubBytes`, $x_i, y_i, u_i, v_i \in \mathrm{GF}(2^8)$, $0 \leq i \leq 15$.

This table is usually computed before the AES execution and stored in volatile memory.

We further consider that the same masks u and v are applied on all S-Boxes during one execution (or a round at least) of the algorithm, i.e. $u_i = u$ and $v_i = v$ for $0 \leq i \leq 15$. We believe that this hypothesis is realistic for embedded security products considering that an expensive recomputation of the 256-byte substitution table S' is necessary for each new pair (u, v) and that the storage of many masked tables is not conceivable in memory constrained devices.

2.2 Blinded Inversion Calculation

An alternative solution has been proposed by Oswald et al. [16] and improved on by Canright et al. [5]. It consists in computing the inversion in GF(2^8) using a multiplicative mask. To do this efficiently it is proposed to decompose the computation using inversions in the subfield GF(2^4) (and possibly in GF(2^2)). Such masking method is well suited for hardware implementations.

We recall some properties of the masked inversion. Let I' denote the masked pseudo-inversion such that $I'(x_i \oplus u_i) = I(x_i) \oplus u_i$. The element $x_i \oplus u_i$ in GF(2^8) is mapped to a couple $(x_{i,h} \oplus u_{i,h}, x_{i,l} \oplus u_{i,l})$ of GF(2^4) such that $x_i \oplus u_i \cong (x_{i,h} \oplus u_{i,h})X + (x_{i,l} \oplus u_{i,l})$. As detailed in [16] many calculations occur on these subfield elements to compute the masked inversion of $x_i \oplus u_i$. The exact details of these computations can be found in [16]. Note that in these formulas neither $x_{i,h}$ nor $x_{i,l}$ is directly inversed in GF(2^4) but the following value:

$$d_i \oplus u_{i,h} = x_{i,h}^2 \times 14 \oplus (x_{i,h} \times x_{i,l}) \oplus x_{i,l}^2 \oplus u_{i,h} \ .$$

Then the masked inversion in GF(2^4) of $d_i \oplus u_{i,h}$ gives $d_i^{-1} \oplus u_{i,h}$ and is used to compute $I'(x_i \oplus u_i)$.

The 16 input bytes of `SubBytes` are blinded using different masks u_i, but one can notice that input and output masks of the inversion stage are identical. Therefore another threat to take into consideration is the zero value power analysis. This technique has been introduced in [8] and [11], and recently implemented on the masked inversion in [15]. Finally, note that the technique presented in this paper also applies to the improved version of Canright et al. [5] when input and output of the inversion are masked with the same value.

2.3 Measurements and Validation of Implementations

Curve Acquisition. We have developed software implementations on a contact smart card using a 16-bit RISC CPU with low power consumption. Two different methods were used to validate our attacks.

First, we used *simulated curves*: a proprietary tool was used to simulate power curves based on the chip architecture and the code executed. This tool generates ideal power consumption curves without any noise which enables to validate in practice the resistance of an implementation to a set of side-channel attacks leaving aside the acquisition and signal processing problems.

Second, we used *real curves*: we made physical measurements on the chip itself using a MicroPross MP100 reader and a Lecroy WavePro numerical oscilloscope.

First-Order Resistance Validation. Since our aim was to present techniques able to defeat first-order protected devices, we performed the classical first-order differential and correlation analysis on the two implementations presented above, before testing our collision attacks.

To do so, we applied DPA and CPA on the `AddRoundKey`, `SubBytes` and `MixColumns` functions at the first and the last rounds of our implementations. We also performed detailed SPA for each input byte value using many average curves to detect any noticeable (biased) power traces that would reveal a potential leakage. In any case no leakage were observed. We also verified that both implementations were immune to zero value power analysis and to the attack presented by Moradi et al.

We have thus verified that to the best of our knowledge both considered AES implementations are resistant to known first-order attacks. Nevertheless we present in the next section two new collision-correlation techniques which jeopardize these implementations.

3 Description of Our Attacks

In this section, we present the general principle of collision-correlation attacks and then detail how it can be applied on the two considered AES implementations.

3.1 The Collision-Correlation Method

The principle of the attacks presented in this paper is to detect internal collisions between data processed in blinded S-Boxes on the first round of an AES execution. We demonstrate in the following that if i) we are able to detect that the same data is processed at instants t_0 and t_1, and ii) the S-Boxes are blinded such that either the same mask is applied to all message bytes or the mask is identical at the input and the output of each S-Box, then it is possible to infer information on the secret key with very few curves.

In the following, we will denote $(T^n)_{0 \leq n \leq N-1}$ a set of N power traces captured from a device processing N encryptions of the same message M. Then we consider two instructions[1] whose processing starts at times t_0 and t_1 and denote l the number of points acquired per instruction processing. As depicted in Fig. 1 we finally consider $\Theta_0 = (T^n_{t_0})_n$ and $\Theta_1 = (T^n_{t_1})_n$ the two series of power consumptions segments at instants t_0 and t_1.

Note that in practice the N power curves should start at the same instant of the encryption and be perfectly aligned. Such conditions generally require signal processing to be performed first. Note also that as the sampling rate is usually such that $l > 1$ points are acquired per instruction, we can generalize the definition of Θ_0 and Θ_1 as being series of l-sample curve segments instead of series of single power consumption samples.

[1] In our attacks we only consider the correlation between two identical instructions, but it may even be possible to detect that two different instructions manipulate identical data, e.g. by spotting a data bus using EMA.

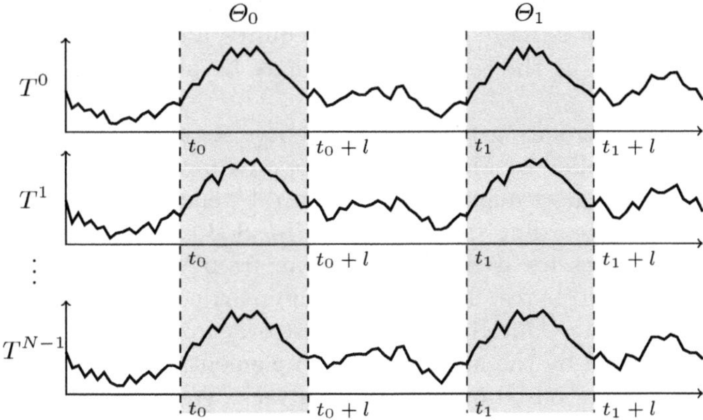

Fig. 1. General description of the collision-correlation attack

The final stage of the attack consists in applying a statistical treatment to (Θ_0, Θ_1) in order to identify if the same data was involved in $T^n_{t_0}$ and $T^n_{t_1}$ for $0 \leq n \leq N-1$. Let `Collision`(Θ_0, Θ_1) denote a decision function returning `true` or `false` depending on whether this property is presumed to be fulfilled or not. Such a decision function would usually compare the value of a synthetic criterion with a practically determined threshold. Possible examples of such a criterion include the mean[2] squared difference, the least squared difference with binary or ternary voting [3], and the maximum Pearson correlation factor. As we used this latter criterion in our study, we recall that an estimation of the Pearson correlation factor between series of curve segments Θ_0 and Θ_1 at time offset t $(0 \leq t \leq l-1)$ expressed as

$$\begin{aligned}\hat{\rho}_{\Theta_0,\Theta_1}(t) &= \frac{\mathrm{Cov}(\Theta_0(t), \Theta_1(t))}{\sigma_{\Theta_0(t)}\sigma_{\Theta_1(t)}} \\ &= \frac{N\sum(T^n_{t_0+t}T^n_{t_1+t}) - \sum T^n_{t_0+t}\sum T^n_{t_1+t}}{\sqrt{N\sum(T^n_{t_0+t})^2 - (\sum T^n_{t_0+t})^2}\sqrt{N\sum(T^n_{t_1+t})^2 - (\sum T^n_{t_1+t})^2}}\end{aligned}$$

where summations are taken over $0 \leq n \leq N-1$, and $\Theta_i(t) = (T^n_{t_i+t})_n$ for $i \in \{0, 1\}$.

`Collision`(Θ_0, Θ_1) thus consists in comparing $\max_{0 \leq t \leq l-1}(\hat{\rho}_{\Theta_0,\Theta_1}(t))$ to a given threshold. In our experiments a preliminary characterization of the targeted device enabled us to find proper values for l and the threshold.

Note that in this collision-correlation technique we compute the correlation factor between a set of real power consumptions Θ_0 with another set of real power consumptions Θ_1, rather than with model dependent estimations. As Bogdanov already described in [3] about binary and ternary voting techniques, an interesting property of this method is that, unlike Hamming weight based CPA, our

[2] The mean being taken over the N traces as well as over the l samples.

criterion does not rely on a particular leakage model. The consequences of this are that i) the attack is more generic and requires much less knowledge of the targeted device, and ii) the secret S-Boxes may be attacked as well as known ones.

As said above, correlating two instants (curve segments) on different traces has already been applied by Moradi et al. [15] on a particular AES implementation. However they collect many traces obtained by encrypting random messages and average them according to the value of an S-Box input byte. This results in 2^8 averaged curves for each byte position, from which they try to detect collisions between two bytes. They successfully carried out this attack on their implementation of the Canright et al. [5] first-order protected implementation. However as indicated by the authors their implementation presented a remaining first-order leakage based on zero-value attack. We applied Moradi's attack to the first-order protected implementations considered in this study without success. We thus consider that this attack is not applicable to most first-order protected implementations. Indeed averaging different traces implies the use of new random mask values which should spoil the influence of the unmasked data and make the collision of intermediate values undetectable. The technique we develop in this paper improves on Moradi's attack in order to detect data collisions by comparing two instants on a same trace and repeating it on many executions without the destructive averaging process. In the following we detail two applications of our attack on two different implementations.

Remark. Collision based analyses are also known as *cross-correlation attacks* in [22] and *multiple-differential collision attacks* in [3]. We prefer the term *collision-correlation attacks* since cross-correlation may be ambiguous depending on the context, and multiple-differential collision attacks seems us too generic for our method.

3.2 Attack on the Blinded Lookup Table Implementation

First, we present an application using principle presented above on the implementation described in Section 2.1. This attack targets the execution of the first round `SubBytes` function. Each 16 masked input byte $x'_i = x_i \oplus u$ is substituted by a masked output byte $y'_i = y_i \oplus v$ where $y'_i = S'(x'_i)$. We try to detect when two `SubBytes` inputs (and outputs) are equal within the first AES round as depicted on Fig. 2.

Detecting a collision in the first AES round between bytes i_1 and i_2 yields that $x_{i_1} \oplus u = x_{i_2} \oplus u$ and considering that $x_i = m_i \oplus k_i \oplus u$ implies the following relation of the two involved key bytes:

$$k_{i_1} \oplus k_{i_2} = m_{i_1} \oplus m_{i_2} \ . \tag{1}$$

Description. Practically, we encrypted N times the same message M and collected the N traces corresponding to the first AES round. For each of the N

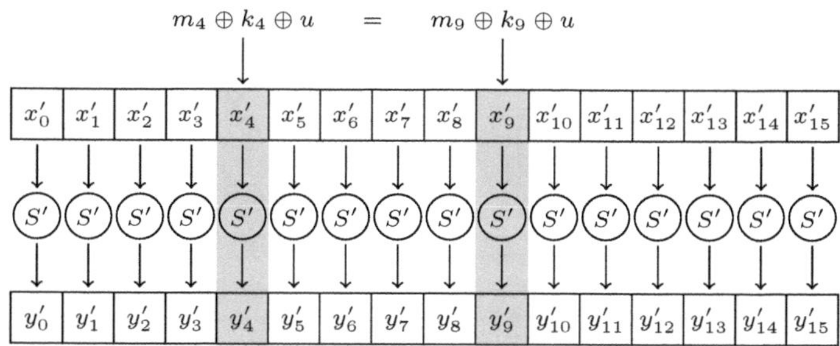

Fig. 2. Collision between the computation of two S-Boxes on bytes 4 and 9 on the blinded lookup table implementation

traces we identified the 16 instants t_i corresponding to the beginning of the computation $S'(x_i \oplus u)$. This allowed us to extract 16 segments from each trace and construct the series Θ_i used for collision-correlation as explained in Section 3.1.

Performing $\texttt{Collision}(\Theta_{i_1}, \Theta_{i_2})$ for all the 120 possible pairs (i_1, i_2) yields a set of relations $(i_1, i_2, m_{i_1} \oplus m_{i_2})$ given by Eq. (1). By repeating this process for several random messages M one can accumulate enough relations so that the secret key is recovered up to a guess on one key byte.

Based on 10 000 simulations we observed that on average 59 random messages (each one being encrypted N times) provide enough relations to retrieve the key up to an unknown byte.

Practical Results. We present hereafter our results on both simulated and real curves.

On simulated curves. The threshold of `Collision` was fixed to having at least one point among the l points correlation curve equal to 1. Under this condition our attack was successful for $N = 16$. Since a mean of 59 different messages are required, then $16 \times 59 = 944$ traces are sufficient on the average for the attack to succeed on simulated curves.

Figures 3 and 4 show the correlation curves obtained for two different messages. Both figures present the 120 outputs of $\hat{\rho}_{\Theta_{i_1}, \Theta_{i_2}}(t)$, $i_1 < i_2$ for each message. The black curve on Fig. 3 corresponds to a collision found for the first message, whereas the second message yields no collision.

On real curves. The attack was successful using $N = 25$ so that less than 1 500 traces allow to recover the key. Notice how few traces are needed to detect a collision by correlation. This confirms that the collision-correlation technique is much more efficient than classical model-based CPA which would not obtain high correlation levels with only 25 traces. Figure 5 shows an example of a correlation peak when an equality between two S-Box outputs occurs, while Fig. 6 shows the correlation curve when all S-Box outputs are different.

Note that in the case of real curves the threshold is slightly different. To identify a clear relation between two S-Box outputs the correlation curve must

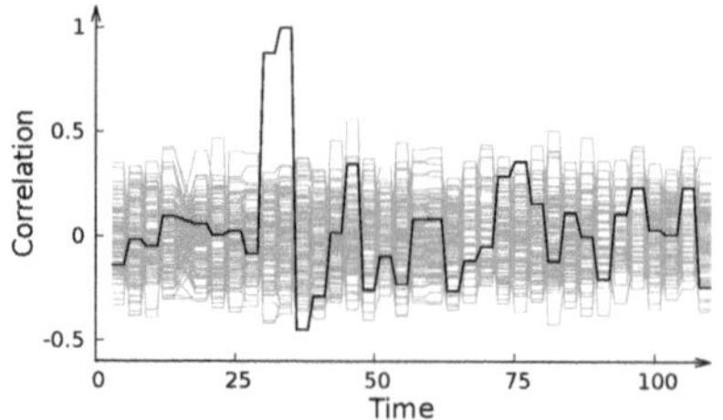

Fig. 3. Correlation curves obtained for a message giving one collision (black curve)

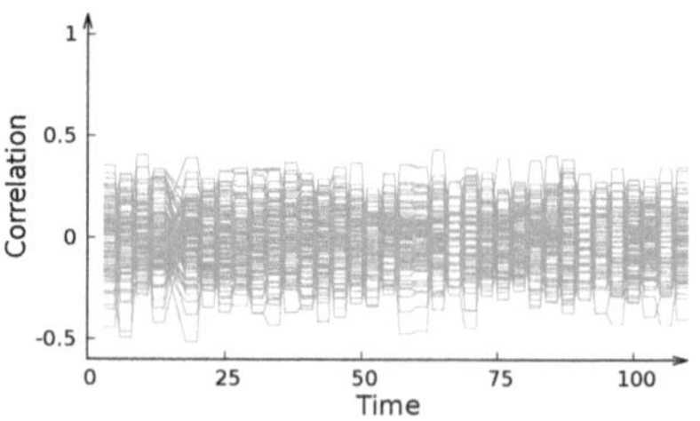

Fig. 4. Correlation curves obtained for a message giving no collision

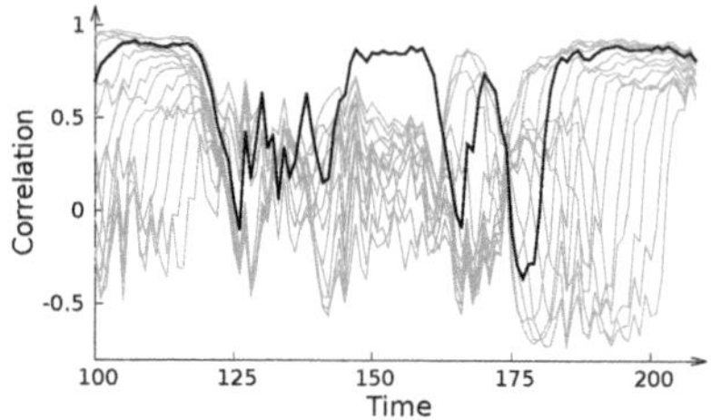

Fig. 5. Correlation peak on real curves when a collision occurs (black curve)

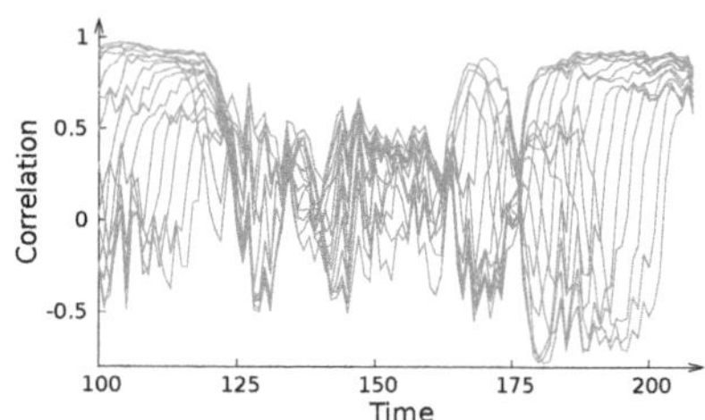

Fig. 6. No correlation peak occurs on real curves when intermediate data differ

be greater than 0.8 in the interval [130, 160]. So only these $l = 30$ points must be considered when computing $\texttt{Collision}(\Theta_0, \Theta_1)$.

Attack Improvement. The method for obtaining information about the key as described above basically exploits collision events where a pair (i_1, i_2) of indices gives a high correlation between Θ_{i_1} and Θ_{i_2} revealing the value of $k_{i_1} \oplus k_{i_2}$. While very informative, such collision events occur much less frequently than non-collision ones, that is when Θ_{i_1} and Θ_{i_2} show no significant correlation between each other. Non-collision events individually bring quite few information – namely that $k_{i_1} \oplus k_{i_2}$ is different from $m_{i_1} \oplus m_{i_2}$ – but they are so numerous that it appears worth trying to exploit them also.

As was already noted in [6,2], the problem of solving a set of equations involving sub-parts of the key can be formulated in terms of a labelled undirected graph. Each vertex i represents a key byte index and the knowledge of the XOR between two key bytes is represented by an edge (i_1, i_2) labelled with $k_{i_1} \oplus k_{i_2}$. At the beginning the graph does not include any edges. Each time a collision occurs between two unrelated key bytes a new edge is put on the graph and results in the merge of two connected components into a single larger one. All key byte values belonging to the same connected component can be derived from each other, and the goal of the attacker is to end up with a fully connected graph.

For a given message, only 0, 1, or 2 from the 120 pairs (i_1, i_2) lead to collisions in most cases. All other pairs reveal some impossible value for each $k_{i_1} \oplus k_{i_2}$.

Gathering all the information provided by these non-collisions, for each (i_1, i_2) we maintain a blacklist of impossible values for the XOR of the two key bytes[3].

Given the information provided by previous messages to the current graph and blacklists, we adaptively choose the next message in order to maximize its usefulness which we define as the number of pairs (i_1, i_2) where one can expect new information (either positive or negative) to be obtained. As a first idea we could define the penalty of a candidate message as the number of pairs (i_1, i_2) for which $m_{i_1} \oplus m_{i_2}$ is already blacklisted. Obviously the chosen message should minimize the penalty. Actually this is slightly more complex and the definition of the penalty of a message should be refined. Indeed we must also consider cases where the message is useful for (i_1, i_2) and (i_1, i'_2) – that neither $m_{i_1} \oplus m_{i_2}$ nor $m_{i_1} \oplus m_{i'_2}$ are blacklisted – but the value of $k_{i_2} \oplus k_{i'_2}$ is known to be precisely equal to $m_{i_2} \oplus m_{i'_2}$. In such a case the two usefulness opportunities brought by the message on pairs (i_1, i_2) and (i_1, i'_2) would bring the same information so that they should count for a single one and the penalty of that message must be increased by one.

In order to find a message with minimal penalty we devised a heuristic which works in two steps. In the first step we consider some random messages (say a few hundred) and select the one with the lowest penalty. This first step ends with a somewhat good candidate. Then in a second step we repeatedly attempt to decrease further the penalty by trying small modifications on this candidate until no more improvements occur by small modifications.

We simulated our method for adaptively choosing the messages. In these simulations we assumed that the attacker is always able to correctly distinguish between collision and non-collision events. Based on 1 000 simulations with random keys, we show that the key is fully recovered (up to the knowledge of one of its bytes) with as few as 27.5 messages instead of 59 messages with the basic method. As distinguishing between a collision and a non-collision necessitates only 25 traces per message, a mere 700 executions would suffice to recover the key by analysing real curves.

3.3 Attack on the Blinded Inversion Implementation

The previous attack cannot be applied to the blinded inversion implementation described in Section 2.2 since the different S-Box input and output bytes are masked with different values u_i. However there may exist a possible leakage leading to what we may call a *Zero & One value attack.*

One can notice that values 0 and 1 produce a collision between the input and the output of the masked pseudo-inversion stage I' as depicted on Fig. 7. This is due to the following properties of the pseudo-inversion:

$$I(0) = 0 \quad \Rightarrow \quad I'(0 \oplus u_i) = 0 \oplus u_i$$
$$I(1) = 1 \quad \Rightarrow \quad I'(1 \oplus u_i) = 1 \oplus u_i$$

[3] Some of these blacklists must also be updated when two connected components are merged.

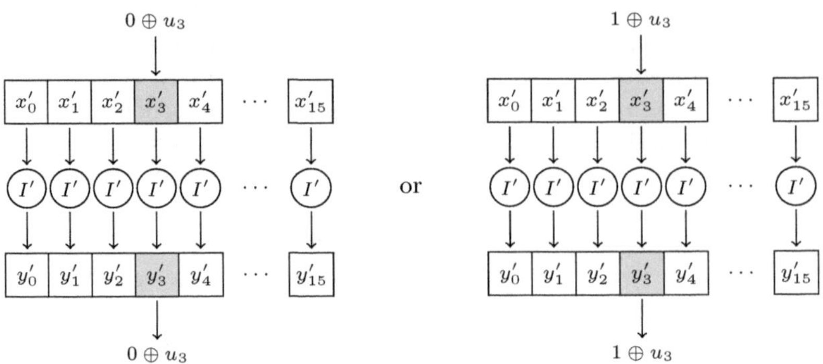

Fig. 7. Collision between the input and the output on byte 3 of the blinded inversion I' (values 0 and 1 lead to a collision)

The two cases leading to a collision are indistinguishable from one another. Detecting a collision between the input and the output of a blinded inversion gives either $x'_i = 0 \oplus u_i$ or $x'_i = 1 \oplus u_i$ which reveals a key byte except one bit:

$$k_i = m_i \quad \text{or} \quad k_i = m_i \oplus 1 \, .$$

Description. Assume we want to recover the 7 most significant bits of k_0. For every even byte value g we encrypt N times a single message M with $m_0 = g$ and collect the corresponding power consumption traces $T^{n,g}$, $0 \leq n \leq N-1$. Note that in this attack we only need to guess the 7 most significant bits because the least significant one is indistinguishable. Let's denote t_0 and t_1 the instants when $x_0 \oplus u_0$ is loaded before the pseudo-inversion I, and when the result is stored respectively. For each of the N traces we extract the two segments $T^{n,g}_{[t_0,t_0+l-1]}$ and $T^{n,g}_{[t_1,t_1+l-1]}$ and construct the series $\Theta^g_0 = (T^{n,g}_{[t_0,t_0+l-1]})_n$ and $\Theta^g_1 = (T^{n,g}_{[t_1,t_1+l-1]})_n$. For this step of our attack it is helpful to have some experience on the targeted implementation identify exactly where these two segments are located.

Applying the decision function $\texttt{Collision}(\Theta^g_0, \Theta^g_1)$ for all the 128 possible values g will reveal two possibilities for k_0. Repeating this step for all key bytes allows the key space to be reduced to 2^{16} values only. Note that a trick which allows to considerably reduce the number of traces is to encrypt the messages $M^g = (g, g, \ldots, g)$ with all bytes equal.

Results on Simulated Curves. As for previous attack on simulated curves, a relation is established when at least one point among the l points correlation curve is equal to 1. The attack is successful using $N = 16$ curves for each key guess. Figure 8 shows the 128 correlation curves for all possible guesses on k_0. The black curve corresponds to the correct guess for k_0.

The attack on this second implementation has thus been validated on simulated curves. We did not acquire real curves for this implementation. Based on

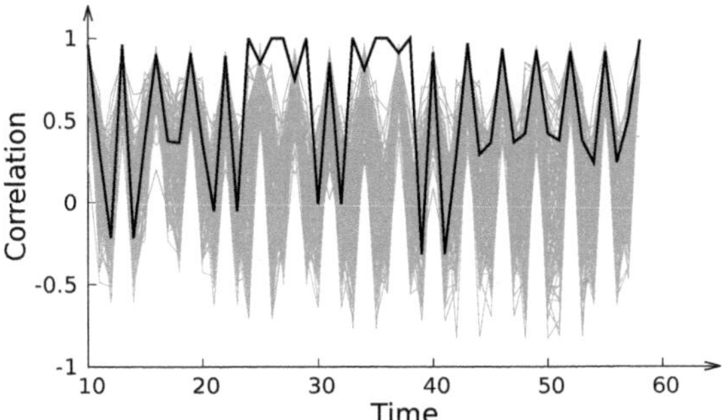

Fig. 8. Collision-correlation curves in the pseudo-inversion of the first byte in $\mathrm{GF}(2^8)$

what has been observed on the previous attack (successful results obtained using simulations have led to successful results on the chip in practice), we believe that the attack would be successful on the real chip too, using a value for N of the same order to what was necessary for the first attack.

4 Comparison with Second Order Analysis

In this section, we present a brief comparison between the collision-correlation method and some known second-order attacks. Our analysis was inspired from the recent framework introduced by Standaert et al. in [20] and refined later in [21]. This comparison gives an overview on the efficiency of these different second-order techniques, and highlights how much the collision-correlation analysis improves on second-order attacks.

Our analysis targets the first implementation only. We compared the collision-correlation analysis with the second-order analysis involving the absolute difference combining function f_1, the squared absolute difference combining function f_2 and the normalized product combining function f_3, when using as distinguisher the Pearson linear correlation factor $\hat{\rho}$. Note that we did not used Mutual Information Analysis, whose results remain less efficient than the classical CPA in practice.

For sake of simplicity, we consider that the power consumption at instant t is the Hamming weight of the intermediate data involved in the computation plus a centered Gaussian noise ω_σ with standard deviation σ. Therefore $\mathrm{HW}_n(z)$ corresponds to the handling of the value z for the n-th encryption. We now define θ_0 and θ_1 as:

$$\theta_0 = (\mathrm{HW}_n(S(m_i \oplus k_i \oplus u) \oplus v) + \omega_\sigma)_{0 \leq n \leq N-1}$$
$$\theta_1 = (\mathrm{HW}_n(S(m_j \oplus k_j \oplus u) \oplus v) + \omega_\sigma)_{0 \leq n \leq N-1}$$

Let g_i (resp. g_j) denote a guess on k_i (resp. k_j). We compute the estimated values $w_{g_i,g_j} = \mathrm{HW}(S(m_i \oplus g_i) \oplus S(m_j \oplus g_j))$. Considering the N messages we obtain

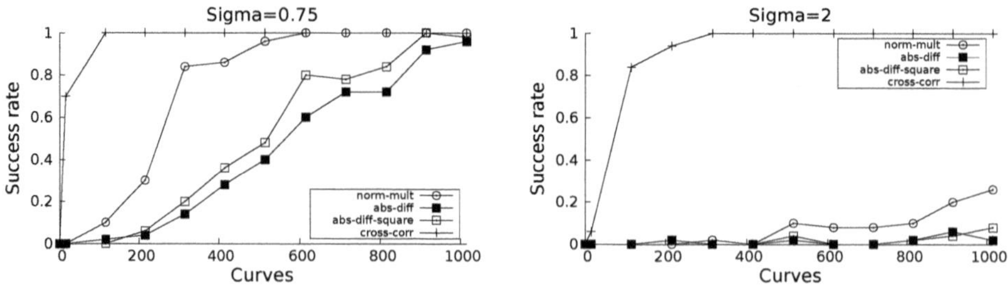

Fig. 9. Success rates of different simulated second-order attacks

the series $W_{g_i,g_j} = (w^n_{g_i,g_j})_{0 \leq n \leq N-1}$. Using the combining function f_j, the right key bytes are obtained for the highest correlation value $\hat{\rho}(f_j(\theta_0, \theta_1), W_{g_i,g_j})$.

Then as in [21] we execute many times the attack with the different combining functions and calculate the success rate of each one. Figure 9 shows two comparison graphs, one for $\sigma = 0.75$ and the other for $\sigma = 2$. Both graphs plot the success rates on 50 runs with respect to the number of curves used.

We emphasise that in this comparison the second-order attacks are shown in a very favorable light. Indeed the correlation model used here is exactly the one applied to simulate the curves. In practice an attacker would not have such good properties.

5 Countermeasures

The attacks presented in this paper defeat first-order protected implementations. Therefore, an obvious countermeasure would be to apply second-order masking. To the best of our knowledge, the best solution should be the countermeasure presented by Rivain et al. [17]. It allows the implementation of proven d-order DPA resistant AES for any $d \geq 1$.

Another countermeasure against our first attack may simply consist in executing the `SubBytes` function in a random order. Even if this method is not theoretically perfect, it may be sufficient to practically resist to second-order attacks. Considering the second implementation, we think that its main weakness is the use of a same mask before and after each byte pseudo-inversion. If the result is masked with a different value then the collision-correlation attack is no longer feasible.

It is also necessary to consider that depending on the quality of the hardware countermeasures provided by the device, these attacks can become much more complicated in practice.

6 Conclusion

We have presented a new collision-correlation analysis method on first-order secured AES implementations. We highlighted the fact that this kind of attack is more powerful and practicable than previous second-order power analyses, and increases the risk of these implementations being broken in practice. This confirms the necessity for developers to take into account how collisions of masked data may be unsafe in cryptographic implementations. A possible countermeasure could be the use of second (or higher) order resistant schemes.

Though we presented practical results on software implementations, we believe that this technique may also be a threat for hardware coprocessors. Therefore the collision-correlation threat should be taken into consideration by developers and designers during their embedded cryptographic design.

Acknowledgments. The authors would like to thank Sean Commercial for his valuable comments and advice on this manuscript. We would also like to thank the anonymous reviewers of this paper for their fruitful comments and advice.

References

1. Akkar, M.-L., Giraud, C.: An Implementation of DES and AES, Secure against Some Attacks. In: Koç, Ç.K., Naccache, D., Paar, C. (eds.) CHES 2001. LNCS, vol. 2162, pp. 309–318. Springer, Heidelberg (2001)
2. Bogdanov, A.: Improved side-channel collision attacks on AES. In: Adams, C., Miri, A., Wiener, M. (eds.) SAC 2007. LNCS, vol. 4876, pp. 84–95. Springer, Heidelberg (2007)
3. Bogdanov, A.: Multiple-differential side-channel collision attacks on AES. In: Oswald, E., Rohatgi, P. (eds.) CHES 2008. LNCS, vol. 5154, pp. 30–44. Springer, Heidelberg (2008)
4. Brier, E., Clavier, C., Olivier, F.: Correlation Power Analysis with a Leakage Model. In: Joye, M., Quisquater, J.-J. (eds.) CHES 2004. LNCS, vol. 3156, pp. 16–29. Springer, Heidelberg (2004)
5. Canright, D., Batina, L.: A Very Compact "Perfectly Masked" S-Box for AES. In: Bellovin, S.M., Gennaro, R., Keromytis, A.D., Yung, M. (eds.) ACNS 2008. LNCS, vol. 5037, pp. 446–459. Springer, Heidelberg (2008)
6. Clavier, C.: An improved SCARE cryptanalysis against a secret A3/A8 GSM algorithm. In: McDaniel, P., Gupta, S.K. (eds.) ICISS 2007. LNCS, vol. 4812, pp. 143–155. Springer, Heidelberg (2007)
7. Gierlichs, B., Batina, L., Tuyls, P., Preneel, B.: Mutual Information Analysis. In: Oswald, E., Rohatgi, P. (eds.) CHES 2008. LNCS, vol. 5154, pp. 426–442. Springer, Heidelberg (2008)
8. Di Golic, J., Tymen, C.: Multiplicative masking and power analysis of AES. In: Kaliski Jr., B.S., Koç, Ç.K., Paar, C. (eds.) CHES 2002. LNCS, vol. 2523, pp. 198–212. Springer, Heidelberg (2003)
9. Kocher, P.C., Jaffe, J., Jun, B.: Differential Power Analysis. In: Wiener, M. (ed.) CRYPTO 1999. LNCS, vol. 1666, pp. 388–397. Springer, Heidelberg (1999)
10. Kocher, P.C., Jaffe, J.M., June, B.C.: DES and Other Cryptographic Processes with Leak Minimization for Smartcards and other CryptoSystems, Journal = US Patent 6,278,783 (1998)
11. Mangard, S., Oswald, E., Popp, T.: Power Analysis Attacks: Revealing the Secrets of Smart Cards. Springer, Heidelberg (2007)

12. Mangard, S., Pramstaller, N., Oswald, E.: Successfully Attacking Masked AES Hardware Implementations. In: Rao, J.R., Sunar, B. (eds.) CHES 2005. LNCS, vol. 3659, pp. 157–171. Springer, Heidelberg (2005)
13. Messerges, T.S.: Using Second-Order Power Analysis to Attack DPA Resistant Software. In: Paar, C., Koç, Ç.K. (eds.) CHES 2000. LNCS, vol. 1965, pp. 238–251. Springer, Heidelberg (2000)
14. Messerges, T.S.: Securing the AES Finalists Against Power Analysis Attacks. In: Schneier, B. (ed.) FSE 2000. LNCS, vol. 1978, pp. 150–164. Springer, Heidelberg (2001)
15. Moradi, A., Mischke, O., Eisenbarth, T.: Correlation-enhanced power analysis collision attack. In: Mangard, S., Standaert, F.-X. (eds.) CHES 2010. LNCS, vol. 6225, pp. 125–139. Springer, Heidelberg (2010)
16. Oswald, E., Mangard, S., Pramstaller, N., Rijmen, V.: A Side-Channel Analysis Resistant Description of the AES S-Box. In: Gilbert, H., Handschuh, H. (eds.) FSE 2005. LNCS, vol. 3557, pp. 413–423. Springer, Heidelberg (2005)
17. Rivain, M., Prouff, E.: Provably Secure Higher-Order Masking of AES. In: Mangard, S., Standaert, F.-X. (eds.) CHES 2010. LNCS, vol. 6225, pp. 413–427. Springer, Heidelberg (2010)
18. Schramm, K., Leander, G., Felke, P., Paar, C.: A Collision-Attack on AES: Combining Side Channel- and Differential-Attack. In: Joye, M., Quisquater, J.-J. (eds.) CHES 2004. LNCS, vol. 3156, pp. 163–175. Springer, Heidelberg (2004)
19. Schramm, K., Wollinger, T., Paar, C.: A New Class of Collision Attacks and Its Application to DES. In: Johansson, T. (ed.) FSE 2003. LNCS, vol. 2887, pp. 206–222. Springer, Heidelberg (2003)
20. Standaert, F.-X., Malkin, T.G., Yung, M.: A Unified Framework for the Analysis of Side-Channel Key Recovery Attacks. In: Joux, A. (ed.) EUROCRYPT 2009. LNCS, vol. 5479, pp. 443–461. Springer, Heidelberg (2009)
21. Standaert, F.-X., Veyrat-Charvillon, N., Oswald, E., Gierlichs, B., Medwed, M., Kasper, M., Mangard, S.: The World Is Not Enough: Another Look on Second-Order DPA. In: Abe, M. (ed.) ASIACRYPT 2010. LNCS, vol. 6477, pp. 112–129. Springer, Heidelberg (2010)
22. Witteman, M.F., van Woudenberg, J.G.J., Menarini, F.: Defeating RSA multiply-always and message blinding countermeasures. In: Kiayias, A. (ed.) CT-RSA 2011. LNCS, vol. 6558, pp. 77–88. Springer, Heidelberg (2011)

Higher-Order Glitches Free Implementation of the AES Using Secure Multi-party Computation Protocols

Emmanuel Prouff[1] and Thomas Roche[2,*]

[1] Oberthur Technologies, 71-73, rue des Hautes Pâtures 92726 Nanterre, France
e.prouff@oberthur.com
[2] ANSSI, 51 boulevard de La Tour-Maubourg 75700 Paris 07 SP, France
thomas.roche@ssi.gouv.fr

Abstract. Higher-order side channel attacks (HO-SCA) is a powerful technique against cryptographic implementations and the design of appropriate countermeasures is nowadays an important topic. In parallel, another class of attacks, called *glitches attacks*, have been investigated which exploit the hardware glitches phenomena occurring during the physical execution of algorithms. We introduce in this paper a circuit model that encompasses sufficient conditions to resist glitches effects. This allows us to construct the first countermeasure thwarting both glitches and HO-SCA attacks. Our new construction requires Secure Multi-Party Computation protocols and we propose to apply the one introduced by Ben'Or *et al.* at STOC in 1988. The adaptation of the latter protocol to the context of side channel analysis results in a completely new higher-order masking scheme, particularly interesting when addressing resistance in the presence of glitches. An application of our scheme to the AES block cipher is detailed.

1 Introduction

Higher-Order Side-Channel Analysis (HO-SCA for short) is a class of *physical cryptanalyses* against cryptosystems. They generalize the seminal side-channel attacks introduced in the late nineties by Kocher *et al.* [11]. Contrary to the latter attacks that only exploit instantaneous leakages, HO-SCA attacks mix observations of several leakages to recover information about the secret parameters of the targeted algorithm. The number of different leakages (*e.g.* related to different times during the processing or to different locations in a circuit) defines the attack order.

A very common countermeasure to protect block cipher implementations against HO-SCA is to randomize the key-dependent variables by *masking* (*a.k.a.*

* This work has been carried out when the author was Post-doc at the University of Paris 8, Département de mathématiques, 2, rue de la Liberté; 93526 Saint-Denis, France.

B. Preneel and T. Takagi (Eds.): CHES 2011, LNCS 6917, pp. 63–78, 2011.

sharing) techniques [3, 8]. The masking can be characterized by both the number n of random shares and the smallest number $d+1$ of them that are required to re-construct the variable. In this case, it is called a *(n,d)-sharing.* A scheme specifying how to apply (n,d)-sharing to protect an algorithm implementation, is called *d^{th}-order masking scheme.* It aims at defining a *modus operandi* that thwarts any SCA attack of order lower than or equal to d. Even though a $(d+1)^{\text{th}}$-order SCA exploiting the leakage of $d+1$ shares can always theoretically be successfully performed, it has been shown in [3] that the complexity of such an attack increases exponentially with the order due to noises effects. Hence a dth-order masking scheme is a sound countermeasure against HO-SCA attacks when d is high enough w.r.t. the noise. Some d^{th}-order masking schemes have been proposed with formal security proof [9, 23–25]. However in 2004 Mangard *et al.* [12] pointed out a weakness in the adversary model involved to construct these masking schemes: when an implementation is processed, so-called *hardware glitches* — common in CMOS technology — occur that deteriorate the effect of masking and leak more information than the simple value of the variables specified by the implementation. Since the seminal work of Mangard *et al.* , several papers have successfully applied so-called *glitches attacks* against implementations that were secure in the classical HO-SCA adversary model (see *e.g.* [12]). Up to now, a unique masking scheme has been presented as secure in presence of glitches [16–19]. This masking scheme is only resistant against 1st-order SCA, which leaves open the problem of specifying cryptographic implementations secure against higher-order SCA attacks in presence of glitches.

1.1 Related Works

The problematic of specifying cryptosystem implementations thwarting d^{th}-order SCA attacks in presence of glitches is recent. Actually, to the best of our knowledge, most of the work on this subject have been done by Nikova *et al.* [16–19]. In those papers, a masking scheme is proposed that is claimed to resist 1st-order SCA in presence of glitches. Unfortunately, adapting the proposed schemes to also counteract SCA attacks of order greater than 1 seems difficult while keeping the efficiency and performance on acceptable level. On the other hand, two implementations proven secure against d^{th}-order SCA attacks for any d have been proposed by Ishai *et al.* [9] and Rivain *et al.* [24]. Ishai *et al.* 's solution is dedicated to hardware implementations. It clearly does not thwart glitches attacks and modifying it to achieve resistance against the latter ones is an (open) issue. Rivain *et al.* 's solution is dedicated to software context. When embedded in a physical device there is no guaranty that no glitches effects will occur during the processing. Providing such a guaranty may be possible by ensuring that all the elementary operations are performed sequentially on a circuit which is re-initialized between each operation. However, such a process, if possible, would induce a prohibitive computational overhead.

1.2 Our Contribution

In this paper, we introduce a generic framework for the design of Hardware or Software implementations that counteract HO-SCA in presence of glitches (Sect. 2). This framework is built around the notion of *multi-party circuit* which is defined as the composition of several sub-circuits whose respective side-channel leakages are strictly independent. Secondly, we establish a parallel between the construction of a HO-SCA resistant implementation inside our new framework and the classical problem of building Secure Multi-Party Computation (SMC) protocols in the context of *semi-honest* adversaries. As a matter of fact, starting from the seminal work of Ben-Or *et al.* on *non-cryptographic* SMC protocols [1], we show how to design a multi-party circuit that can implement any function with a minimum number of sub-circuits (Sect. 3). As an example, we specify such a circuit for the AES-128 algorithm (Sect. 4 and [21] for a description of the scheme for the full AES-128). To the best of our knowledge, our proposal is the first implementation that claims to thwart d^{th}-order SCA attacks in presence of glitches.

2 Preliminaries and Multi-party Circuits

In this section, we introduce a framework in which the resistance of a (hardware or software) implementation against HO-SCA in presence of glitches attacks can be stated. First, we give a formal definition of the attacks. Then, in Sect(s). 2.2 and 2.3 we exhibit sufficient conditions and general principles to thwart the attacks.

2.1 Computation and Adversary Models

SCA attacks exploit a dependency between a subpart of the secret parameter and the variations of a physical leakage as function of a known input. This dependency results from the manipulation of some variables, called *sensitive*, by the implementation. We say that a variable $\mathbf{Z}$ is *sensitive* if it depends on both a known variable and a secret parameter. The physical leakage on such a variable will be viewed as a *noisy leakage function* $L(\mathbf{Z})$ which returns a noisy information about it. The noisy property of $L(\cdot)$ is captured by assuming that the bias introduced in the distribution of $\mathbf{Z}$ given the leakage $L(\mathbf{Z})$ is bounded.

In this paper, we shall see the implementation targeted by a SCA attack as a *circuit*, whose formal definition is given hereafter:

Definition 1 (Circuit $\mathcal{C}_f$). *Let f be a function and let $\mathcal{O}$ be a set of elementary operations. A* circuit $\mathcal{C}_f$ implementing f thanks to operations in $\mathcal{O}$ *is an oriented graph were each cell c_i defines an elementary operation and each edge bears an intermediate variable V_{ij} which is an output of the operation c_i and an input of the operation c_j.*

Remark 1. The above notion of circuit encompasses both hardware and software implementations. In the hardware context, the set $\mathcal{O}$ may only contain the logical binary operations XOR and AND. In the software context, the set $\mathcal{O}$ may only contain field operations $\oplus$ and $\otimes$ in $\mathrm{GF}(2^m)$, where m is the architecture size.

Several adversary models have been proposed in the literature to capture a practical SCA attacker against a circuit $\mathcal{C}_f$. Among them, the *probing adversary model* is the most popular one. We give hereafter a formal definition of this adversary in our framework.

Definition 2 (d^{th}-order Probing Adversary Model). *Let $\mathcal{C}_f$ be a circuit with $(V_{ij})_{(i,j)\in I}$ as edges and let d be a positive integer. Let $\mathcal{L}$ be a set of noisy leakage functions. A* d^{th}-order Probing Adversary *against $\mathcal{C}_f$ is an adversary that can choose a subset J of I with $\#J = d$ and can observe the random variable $(L_{ij}(V_{ij}))_{(i,j)\in J}$, where $(L_{ij}(\cdot))_{ij}$ is a d-tuple of functions in $\mathcal{L}$.*

Remark 2. In the hardware context (when the circuit $\mathcal{C}_f$ is defined with respect to operations XOR and AND), the Probing Adversary Model with $\mathcal{L}$ reduced to the identity function exactly fits with the definition given by Ishai *et al.* in [9]. In the software context (when the circuit $\mathcal{C}_f$ is defined with respect to operations on m-bit words with m being the architecture size), our definition corresponds to the classical notion of dth-order SCA [2, 10, 14]. In this case, $\mathcal{L}$ is usually defined as the set of noisy leakage functions $L(\cdot) = \mathrm{HW}(\cdot) + \mathcal{B}(\mu, \sigma)$, where $\mathcal{B}$ is a gaussian independent noise with mean μ and standard deviation σ and where HW denotes the Hamming weight function.

Notation. An attack performed by the adversary in Definition 2 is called *d^{th}-order probing attack*. A circuit secure against those attacks is said to be *d-probing secure.*

The two works of Ishai *et al.* [9] and Rivain and Prouff [24] show that achieving perfect security in the Probing Adversary Model is possible. In parallel however, several publications have shown that schemes secure in this model can still be broken in practice (see *e.g.* [12] and [13]). The reason for this is that in physical implementations (*e.g.* in CMOS), a lot of unintended switching activities occur. These unintended switching activities are usually referred to as dynamic hazards or *glitches.*

The effect of glitches on the side-channel resistance of masked circuits has first been analyzed in [12]. The same year, a technique to model the effect of glitches on the side-channel resistance of circuits has been published in [27]. Hereafter we model an adversary which performs glitches attacks against a circuit. For such a purpose, and following the same approach as in [27], we transform our *static* definition of a circuit into a *dynamic* one. Actually, this simply amounts to assume that the random variable V_{ij} not only depends on the pair (i, j) but also on a third time parameter t. We denote by $V_{ij}(t)$ the dynamic version of V_{ij} and call *dynamic* the corresponding circuit. The main difference between the two models is that, in the probing model, the value taken by V_{ij} is assumed to be constant, whereas its dynamic version $V_{ij}(t)$ can change over the time, even when the circuit input is fixed. The *internal state transition at time t' of a circuit $\mathcal{C}_f$* with $(V_{ij})_{(i,j)\in I}$ as edges refers to all the non-zero transitions of the value taken by the $V_{ij}(t)$ at time $t = t'$. It is denoted by $\mathcal{C}_f(t')$.

Definition 3 (d^{th}-order Glitches Adversary Model). *Let $\mathcal{C}_f$ be a circuit and let d be a positive integer. Let $\mathcal{L}$ be a set of leakage functions. A dth-*order Glitches Adversary against *$\mathcal{C}_f$ is an adversary that can choose d times t_1, t_2, ..., t_d and can observe the internal state transition at the d selected times $(L_i(\mathcal{C}_f(t_i))_{i\leq d}$, where, for each $i \leq d$, $L_i(\cdot)$ is a function in $\mathcal{L}$.*

Notation. An attack performed by the adversary defined in Definition 3 is called *d^{th}-order glitches attack.* A circuit secure against those attacks is said to be *d-glitches free.*

Security in the glitches adversary model implies security in the probing adversary model whereas the converse is false. In the following sections, we introduce the notions of *d-probing security* and of *d-glitches freeness.* In both cases, we exhibit generic constructions principles that enable to design circuits achieving them.

2.2 Security in the Probing Adversary Model

The following definition formalizes the notion of security w.r.t. d^{th}-order probing adversaries. Note that this definition corresponds to that involved in numerous papers (*e.g.* [2,4,10,14,20]).

Definition 4 (d-probing Security). *Let d be a positive integer. A circuit $\mathcal{C}_f$ with family of edges $\mathcal{V}$ is d-*probing secure *iff no family of at most d elements in $\mathcal{V}$ is sensitive.*

To achieve d-probing security, the most widely used approach is to split each sensitive variable appearing in the algorithmic description of the cryptosystem into several shares and to replace operations on the sensitive data by operations on the shares. We give hereafter a formal description of this technique called *(n,d)-sharing* in the sequel.

Definition 5 ((n,d)-sharing). *Let n and d be two positive integers such that $n > d$. A (n,d)-sharing of a variable $Z \in \mathrm{GF}(2^m)$ is a family of n variables $(Z_i)_{1\leq i\leq n}$ such that:*

1. *there exists F from $\mathrm{GF}(2^m)^n$ into $\mathrm{GF}(2^m)$ s.t. $F(Z_1,\cdots,Z_n) = Z$,*
2. *for every $I \subset [1;n]$ s.t. $\#I \leq d$, we have $\Pr(Z \mid (Z_i)_{i\in I}) = \Pr(Z)$.*

In relation with the notion of (n,d)-sharing we will sometimes use the notion of independent sharings. A formal definition is given hereafter.

Definition 6 (Independence of (n,d)-sharings). *Let n and d be two positive integers such that $n > d$. Two (n,d)-sharings $(Z_i)_{i\leq n}$ and $(Z'_i)_{i\leq n}$ of two (not necessarily distinct) variables Z and Z' are said to be* independent *if for every pair of subsets (I,I') in $[1;n]$, each of cardinality lower than or equal to d, we have $\Pr((Z_i)_{i\in I} \mid (Z'_i)_{i\in I'}) = \Pr((Z_i)_{i\in I})$.*

Remark 3. Two (n, d)-sharings are independent if they involve independent masking materials (*i.e.* different random values). The replacement of a (n, d)-sharing of Z by a new independent one is sometimes called *re-randomization* or *masks refreshing* in the literature [1, 24].

In the SCA literature, the family $(Z_i)_{1 \le i \le n}$ is usually called a *masked representation at order* d of Z. The function F is usually simply defined as the sum of the Z_i and n is chosen equal to $(d + 1)$. Several ways have been proposed to apply such a d-sharing to protect a circuit against probing attacks. To the best of our knowledge, only the circuit implementations proposed by Ishai *et al.* [9] and Rivain and Prouff [24] are d-probing secure for any fixed value d. Unfortunately, those schemes are not by construction secure in the Glitches Adversary Model. In the following, we introduce a new strategy to directly obtain circuits secure in the Glitches Adversary Model.

2.3 Security in the Gliches Adversary Model

To achieve security in the Glitches Adversary Model, we develop hereafter a strategy that consists in *hermetically* separating some parts of the computation. The idea is to split a circuit $\mathcal{C}_f$ implementing a function f into several sub-circuits $\mathcal{C}_{f_i}$ such that the observation of d or fewer sub-circuits gives no information on the original circuit input. The security/pertinence of this approach is essentially based on the following simple observation: if n sub-circuits $\mathcal{C}_{f_i}$ operate each on a single element of a (n, d)-sharing and if the processings leak independently, then the circuit composed of the n sub-circuits is d-glitches free. Of course, this observation alone does not directly permit to design a d-glitches free circuit implementing any function f. Indeed, the above construction implies that each sub-circuit is input with a single share of a sharing and cannot access the other shares. By consequence, only a certain type of function f (homomorphic with respect to the sharing) can be split into n independent computations, each operating only a single share of f's input and returning a sharing of $\mathcal{C}_f$'s output.

Secure Multi-Party Computation protocols, and in particular the protocol in [1] recalled in next section, are methods that extend the above idea to any function f (*i.e.* not only homomorphic function for which, as we have seen, the solution is straightforward). This extension however requires to provide each sub-circuit $\mathcal{C}_{f_i}$ with the ability to send to the other ones information about some of its intermediate results. To ensure that this new ability does not impact on the d-glitches freeness of the circuit composed of the n sub-circuits, this sent information must itself be shared between all the other n sub-circuits. This implies that each sub-circuit must only be able to access a single share of a (n, d)-sharing of the intermediate result of another sub-circuit. To ensure that a sub-circuit cannot receive several shares of a same (n, d)-sharing or cannot access several shares related to dependent (n, d)-sharings, we set the condition that all the shares accessed by a sub-circuit come from distinct and independent (n, d)-sharings. To formalize our construction, each sub-circuit $\mathcal{C}_{f_i}$ is extended with a family of $n-1$ channels $(\$_{ij})_{j\neq i}$, the channel $\$_{ij}$ being dedicated to the communication between $\mathcal{C}_{f_i}$ and $\mathcal{C}_{f_j}$ and being not accessible by another sub-circuit. Through those channels, each $\mathcal{C}_{f_i}$ can access a share of any intermediate result of another sub-circuit, each new accessed share being related to a (n,d)-sharing independent of the previous ones. The family of extended circuits $(\mathcal{C}_{f_i}, (\$_{ij})_{i\neq j})_i$ is called a (n,d)-multi-party circuit. We give a formal definition of it hereafter.

Definition 7. *Let f be a function and let Z denote its input. Let $(Z_i)_i$ be a (n,d)-sharing of Z. A circuit $\mathcal{C}_f$ composed of n extended sub-circuits $(\mathcal{C}_{f_1}, (\$_{1j})_{j\neq 1})$, ..., $(\mathcal{C}_{f_n}, (\$_{nj})_{j\neq n})$ is* a (n,d)-multi-party circuit *iff:*

- *every $\mathcal{C}_{f_i}$ is input with a share Z_i only,*
- *each $\mathcal{C}_{f_i}$ can access, through $\$_{ij}$, to a share of an (n,d)-sharing of an intermediate result of the sub-circuit $\mathcal{C}_{f_j}$,*
- *all the shares accessed by a sub-circuit $\mathcal{C}_{f_i}$ relate to mutually independent (n,d)-sharings,*
- *the $\mathcal{C}_{f_i}'s$ outputs form a (n,d)-sharing of $f(Z)$.*
- *for $i\neq j$, $\mathcal{C}_{f_i}$ leaks independently to $\mathcal{C}_{f_j}$.*

The following proposition states on the security of our construction against glitches attacks[1].

Proposition 1. *A* (n,d)-multi-party circuit *is secure in the dth-order Glitches Adversary Model.*

In the next section we recall the basics about Secure Multi-party Computation and, in particular, a SMC protocol introduced by Ben Or *et al.* in [1]. Then, we argue that this protocol can be adapted to our context in order to design (n,d)-multi-party circuits as long as n is greater than $2d$.

3 Secure Multi-party Computation

Secure Multi-Party Computation represents a rich area of research initiated by the seminal work of Yao in 1986 [28]. For a n-ary function f and a family of n players $(I_i)_{i\leq n}$, each holding a private value Z_i, a *secure multi-party computation* is a joint protocol enabling the players I_i to compute $f(Z_1,\cdots,Z_n)$ while under attack by an external adversary and/or by a subset of malicious players (also called the *colluding players*). The purpose of the attack is to learn the private information of the – non-colluding – honest players or to cause the computation to be incorrect. As a result, there are two important requirements of a multi-party computation protocol: correctness and privacy. Those two requirements relate to two different kinds of adversaries. The first one, usually called *active*, is allowed to let the malicious parties deviate from the protocol in arbitrary ways. It is out of the scope of this paper. The second adversary kind, called *passive* or *semi-honest*, is only allowed to create collusion of players to gain

[1] A sketch of proof can be found in the extended version of this paper [21].

information about the secret. The corrupted players still follow the protocol and never forge wrong data. A security threshold parameter $d \leq n$ is used to indicate the maximum number of players the adversary is allowed to corrupt. A SMC protocol secure against a passive or active adversary with threshold d is called a *d-private protocol.* We show in this section how the problem of designing SMC protocols secure for this adversary model is related to the problem of designing multi-party circuits secure in the Glitches Adversary Model. Before that, we recall in the next section the main aspects of the SMC protocol introduced by Ben Or *et al.* in [1] on the basis of an idea proposed by Shamir in 1988.

3.1 Shamir's Secret Sharing Scheme and BGW's Protocol

In a seminal paper [26], Shamir has introduced a simple and elegant way to share a secret Z (considered here in $\mathrm{GF}(2^m)$) between $n < 2^m$ players such that no collusion of $d < n$ players can retrieve information about Z. In Shamir's protocol, an entity called *the Dealer* generates a degree-d polynomial $P_Z(X) \in \mathrm{GF}(2^m)[X]$ with constant term Z and secret coefficients a_i (*i.e.* $P_Z(X) = Z + \sum_{i=1}^{d} a_i X^i$). Then, he chooses n distinct non-zero elements $\alpha_1, \cdots, \alpha_n$ in $\mathrm{GF}(2^m)$, makes them publicly available and *distributes* to each player I_i the value $Z_i = P_Z(\alpha_i)$. To *re-construct* the secret Z, the players publish their private values Z_i, reconstruct P_Z by polynomial interpolation (always possible since $n > d$) and evaluate $P_Z(X)$ in 0 (we have $Z = P_Z(0)$). It can be easily checked that Shamir's sharing fits with the notion of (n, d)-sharing given in Definition 5 with reconstruction function F defined s.t. $F(Z_1, \cdots, Z_n) = \sum_{i=1}^{n} Z_i \prod_{k=1,k\neq i}^{n} -\alpha_k(\alpha_i - \alpha_k)^{-1}$ (due to Lagrange's Interpolation, $F(Z_1, \cdots, Z_n)$ equals $P_Z(0)$ that is Z).

Remark 4. The products $\prod_{k=1,k\neq i}^{n} -\alpha_k(\alpha_i - \alpha_k)^{-1}$ for all i can be precomputed once for all. They actually correspond to the first row $(\lambda_1, \cdots, \lambda_n)$ of the inverse of the Vandermonde $(n \times n)$-matrix $(\alpha_i^j)_{1\leq i,j\leq n}$. We hence have $F(Z_1, \cdots, Z_n) = \sum_{i=1}^{n} \lambda_i Z_i$.

Starting from Shamir's secret sharing, Ben Or *et al.* have defined in [1] a d-private SMC protocol in the case where the number of players n satisfies $n > 2d$. This construction, called *BGW's protocol* in the following, is in fact a constructive proof of Theorem 1 (see [1]):

Theorem 1. *For every (probabilistic) function f and $n > 2d$, there exists a d-private protocol.*

In BGW's protocol, the input $(Z_1, \cdots, Z_n)$ of the function f whose computation must be made d-private is assumed to correspond to Shamir's sharing of a secret variable Z. Namely, they correspond to the evaluation of a degree-d secret polynomial $P_Z(X)$ in n distinct non-zero public points α_1, ..., α_n. It is moreover assumed that each player I_i has been initially provided with a share Z_i which is unknown to the others. Then, the function f is modeled as a sequence of computations operating either an affine transformation on an intermediate state V or additions/multiplications between two intermediate states V and V'.

Let us denote by C such a (univariate or bivariate) computation. BGW's protocol ensures that the intermediate states V and V' at input of a bivariate operation C have been shared w.r.t. two random polynomials P_V and $P_{V'}$ which are independent but evaluated in the same public points α_1, ..., α_n (*i.e.* V_i and V_i' satisfy $V_i = P_V(\alpha_i)$ and $V_i' = P_{V'}'(\alpha_i)$ respectively). Moreover, for each C to process, BGW's protocol is designed such that each player I_i has either a single share V_i (if C is univariate) or a single pair of shares (V_i, V_i') (if C is bivariate). Eventually, BGW's protocol describes a d-private multi-party computation for each kind of operation C depending on its nature. We recall them hereafter.

If C is an affine transformation applied on a shared variable V, then the protocol simply consists in asking each player I_i to apply C on its private share V_i. After this step, each player owns a new share $\mathsf{C}(V_i)$ and the family $(\mathsf{C}(V_i))_i$ is a (n, d)-sharing of $\mathsf{C}(V)$. Indeed, since C is affine, $\mathsf{C}(P_V(X))$ is a degree-d polynomial such that $\mathsf{C}(P_V(0)) = \mathsf{C}(V)$ and each $\mathsf{C}(V_i)$ corresponds to the evaluation of P_V in α_i (*i.e.* $\mathsf{C}(V_i) = \mathsf{C}(P_V(\alpha_i))$).

If C is the addition operation $\oplus$ applied on two shared intermediate states V and V', then the protocol consists in asking each player I_i to compute $\mathsf{C}(V_i, V_i') = V_i \oplus V_i'$. After this step, each player owns a new share $\mathsf{C}(V_i, V_i')$ and the family of shares $(\mathsf{C}(V_i, V_i'))_i$ is a (n, d)-sharing of $\mathsf{C}(V, V')$. Indeed, by construction of the V_i and V_i', we have $\mathsf{C}(V_i, V_i') = P_V(\alpha_i) \oplus P_{V'}(\alpha_i)$ which implies that $(\mathsf{C}(V_i, V_i'))_i$ corresponds to the evaluation of the polynomial $(P_V(X) \oplus P_{V'}(X))$ in $(\alpha_i)_i$. This polynomial is of degree at most d and satisfies $(P_V(0) \oplus P_{V'}(0)) = V \oplus V' = \mathsf{C}(V, V')$.

If C is the multiplication operation $\otimes$ applied on two shared intermediate states V and V', then the protocol is more complex than the previous ones. It involves the first row $(\lambda_1, \cdots, \lambda_n)$ of the inverse of the Vandermonde $(n \times n)$-matrix $(\alpha_i^j)_{1 \leq i,j \leq n}$ and it is composed of three steps[2]. Each player I_i:

1. computes $\mathsf{C}(V_i, V_i') = V_i \otimes V_i' = P_V(\alpha_i) \otimes P_{V'}(\alpha_i)$.
2. randomly generates a degree-d polynomial Q_i such that $Q_i(0) = \mathsf{C}(V_i, V_i')$ and for every $j \neq i$, sends the value $Q_i(\alpha_j)$ to player I_j.
3. computes the linear combination $\mathbf{Q}(\alpha_i) = \sum_{j=1}^{n} \lambda_j Q_j(\alpha_i)$.

The shares $\mathsf{C}(V_i, V_i')$ computed by the players at Step 1 correspond to the degree-$2d$ polynomial $P_V(X) \times P_{V'}(X)$. As desired, the constant term of this polynomial is $\mathsf{C}(V, V')$. However, the family $(\mathsf{C}(V_i, V_i'))_i$ built by Step 1 is not a (n, d)-sharing since the corresponding polynomial is firstly not of degree d and secondly, is not a random polynomial (its distribution over the set of degree-$2d$ polynomials with constant term $\mathsf{C}(V, V)$ is not uniform). To overcome this issue, Steps 2 and 3 perform both a degree reduction and a re-randomization of the shares. More precisely, Step 2 allows player I_i to compute the (n, d)-sharing $(Q_i(\alpha_j))_j$ of its share $\mathsf{C}(V_i, V_i')$ thanks to a random polynomial Q_i, and to send those shares to the other players. Then, in Step 3 each player I_i computes $\mathbf{Q}(\alpha_i) = \sum_{j=1}^{n} \lambda_j Q_j(\alpha_i)$. The family $(\mathbf{Q}(\alpha_i))_i$ corresponds to the evaluation in $(\alpha_i)_i$ of

[2] The protocol described in this paper is an improved version of the protocol originally proposed by Ben-Or *et al.* [1]. It has been introduced by Gennero *et al.* in [6].

the polynomial $\mathbf{Q}(X) = \sum_j \lambda_j Q_j(X)$, which, by construction, is of degree d and admits $\mathtt{C}(V, V')$ as constant term. It is therefore a (n, d)-sharing of $\mathtt{C}(V, V')$ (see [6] for more details).

3.2 SMC Protocol and Multi-party Circuits

The design of a (n, d)-multi-party circuit from BGW's protocol is merely based on the following remark: the d-privacy for a set of n semi-honest players evaluating a function f coincides with the d-probing security for a set of n circuits implementing f. Hence, if each player I_i in BGW's protocol is replaced by an extended sub-circuit $(\mathcal{C}_{f_i}, (\$_{ij})_{j \neq i})$, then the previous description specifies a (n, d)-multi-party circuit. Moreover, such a design can be specified for any function f as long as n and d satisfy $n > 2d$. If f is defined over the finite field $\mathrm{GF}(2^m)$ with addition and multiplication laws $\oplus$ and $\otimes$, the sub-circuits $\mathcal{C}_{f_i}$ are defined with respect to the elementary operations $\{\mathcal{A}(m), \oplus, \otimes\}$, where $\mathcal{A}(m)$ denotes the set of affine functions over $\mathrm{GF}(2^m)$. By construction, each extended sub-circuit $(\mathcal{C}_{f_i}, (\$_{ij})_{j \neq i})$ always operates on a single share of a sensitive variable and has never access to the other shares nor a function of them. Consequently, the observation of an extended sub-circuit cannot give more information than a single share of a (n, d)-sharing. Hence, since the sub-circuit executions do not overlap, and by definition of a (n, d)-sharing, a glitches adversary must observe the behavior of at least $d + 1$ extended sub-circuits $(\mathcal{C}_{f_i}, (\$_{ij})_{j \neq i})$ to recover sensitive information.

Eventually, to fully specify how to put BGW's protocol into practice for a (n, d)-multi-party circuit, it just remains to clarify the following practical points:

(a) Messages Exchange between Sub-circuits (a.k.a. players) during Step 2 of the Secure Processing of $\otimes$. The exchange of messages is done thanks to the channels $\$_{ij}$. In software, each channel $\$_{ij}$ between a pair of sub-circuits $(\mathcal{C}_{f_i}, \mathcal{C}_{f_j})$ may simply consist in a RAM space which is not accessed by another circuits $\mathcal{C}_{f_k}$ with $k \neq i, j$. In hardware, the designer can code a unique communication channel for each pair of circuits running sequentially. Another solution could be to run each sub-circuit in a different environment (*e.g.* different platforms) and to implement a channel between each pair of them.

(b) The Initial Shares Distribution by a Honest Entity (the Dealer). In our context, the role of the Dealer is played by a special procedure run before processing the multi-party circuit. This procedure shares the sensitive variable as usually done in the literature to counteract d^{th}-order probing attacks. To also achieve security in the dth-order Glitches Adversary Model, the computation is split into elementary operations that are processed sequentially. Although expensive this strategy can always be followed to go from probing attack security to glitches attacks security.

Because it is the most tricky part in BGW's protocol, we develop hereafter the algorithm processed by each sub-circuit $\mathcal{C}_{f_i}$ when computing the (n, d)-sharing of the product of two shared values over a field $\mathrm{GF}(2^m)$.

Notation. Instruction **read**$(X, \$_{ij})$ reads the content of $\$_{ij}$ (viewed as a channel or a memory address) and update X accordingly. Instruction **write**$(X, \$_{ij})$ writes the value X on $\$_{ij}$.

Algorithm1. Secure Multiplication Part Dedicated to a Sub-Circuit $\mathcal{C}_{f_i}, (\$_{ij})$

INPUT: the ith element $P_V(\alpha_i)$ of a (n,d)-sharing of V, the i^{th}element $P_{V'}(\alpha_i)$ of a (n,d)-sharing of V' and a set of channels $(\$_{ij})_{j \neq i}$.
OUTPUT: the ith element $\mathbf{Q}_{V \otimes V'}(\alpha_i)$ of a (n,d)-sharing of $V \otimes V'$.
PUBLIC: the points α_i, the first row $(\lambda_1, \cdots, \lambda_n)$ of the inverse of the matrix (α_i^j) with i and j lower than n.

1. **do** $W_i \leftarrow P_V(\alpha_i) \otimes P_{V'}(\alpha_i)$
**** Randomly generate a d-tuple* (a_j) *of coefficients in* $\mathrm{GF}(2^m)$
2. **for** $j = 1$ **to** d **do** $a_j \leftarrow \texttt{rand}(\mathrm{GF}(2^m))$
**** Compute a* (n,d)*-sharing* $(Q_i(\alpha_1), \cdots, Q_i(\alpha_n))$ *of* W_i.
3. **for** $j = 1$ **to** n **do** $Q_i(\alpha_j) \leftarrow W_i \oplus \bigoplus_{k=1}^{d} a_k \alpha_i^k$
**** Send the shares of* W_i *to the other sub-circuits* $\mathcal{C}_{f_j}$ *through* $\$_{ij}$.
4. **for** $j = 1$ **to** n, $j \neq i$ **write**$(Q_i(\alpha_j), \$_{ij})$
**** Receive a share* $Q_j(\alpha_i)$ *from each sub-circuit* C_j *through* $\$_{ji}$.
5. **for** $j = 1$ **to** n, $j \neq i$ **read**$(Q_j(\alpha_i), \$_{ij})$
**** Compute the share* $\mathbf{Q}_{V \otimes V'}(\alpha_i)$
6. **do** $\mathbf{Q}_{V \otimes V'}(\alpha_i) \leftarrow \bigoplus_{j=1}^{n} \lambda_j Q_j(\alpha_i)$.

Steps 2 enables to randomly generate a degree-d polynomial $Q_i(X) = W_i + \bigoplus_{j=1}^{d} a_j X^j$. Step 3 evaluates $Q_i(X)$ in each public point α_j to construct a (n,d)-sharing $(Q_i(\alpha_j))_j$ of W_i. Step 4 sends those shares to the other sub-circuits and Step 5 enables $\mathcal{C}_{f_i}$ to receive the shares $Q_j(\alpha_i)$ computed by the other sub-circuits $\mathcal{C}_{f_j}$. Eventually, Step 6 computes a share of $V \otimes V'$ for sub-circuit $\mathcal{C}_{f_i}$.

3.3 Complexity of the Scheme and Comparison

Complexity Evaluation. Except the multiplication, the proposed scheme replaces each operation of a given function by n similar operations. Concerning the multiplication $\otimes$ over $\mathrm{GF}(2^m)$, it is performed by asking each of the sub-circuits to run Algorithm 1. As a consequence, the multiplication $\otimes$ is replaced by $n^2(d+1)+n$ multiplications, $n^2(d+1)-n$ additions and $2(n-1)$ **read**/**write** operations. In the following table, we develop this complexity for $n = 2d+1$ (which is the smallest value n allowed in BGW's protocol). For comparison purpose, we also give the complexity of the secure multiplication proposed in [24]. We recall that when applied with $n = 2d + 1$, the multiplication algorithm proposed in Algorithm 1 offers the same (perfect) resistance against d-probing attacks than the method proposed in [24].

Comparison with Other State of the Art Solutions. In [18], Nikova *et al.* have already attempted to apply the multi-party computation theory in the context of hardware implementations, with 1-glitches freeness in mind. Contrary

Table 1. Complexity of the secure processing of a field multiplication

Method	multiplications	additions	random bytes
This paper	$4d^3 + 8d^2 + 3d$	$4d^3 + 8d^2 + 7d + 2$	$d(2d + 1)$
[24]	$2d^2 + 2d$	$d^2 + d + 1$	$d(d + 1)/2$

to our proposal where data are shared thanks to Shamir's scheme, Nikova *et al.* 's construction relies on the classical additive sharing (namely the circuit's input is additively masked with several, say $n - 1$, random variables). To secure the processing on the masked input and the masks, they propose to split the computation according to a set of security rules. The obtained circuit sharing differs from ours in the two following main points. First, the security is only proven against first-order attacks, which implies that Nikova *et al.* 's construction can not be used to design (n, d)-multi-party circuits for $d > 1$. Secondly, the sharing is not explicit and involves an exhaustive search that becomes impossible when the size m of the circuit input is greater than 5. Moreover, there is no guaranty that the approach works for any circuit. In particular, Moradi *et al.* [15] discuss the difficulty of applying Nikova *et al.* 's scheme to the AES s-box. In [19], the scheme has been applied to Noekeon. Instead of taking the direction proposed in the present paper where the circuit is divided into several sub-circuits leaking independently, they take an opposite position where the different shares of a variable are manipulated *simultaneously* by the same circuit. Our scheme could also be implemented in such a way but the resulting security against d^{th}-order attacks would be significantly reduced.

On the opposite side, the constructions proposed in [9] (operations in GF(2)) and in its extension [24] (operations in $\mathrm{GF}(2^m)$) are d-probing secure but not 1-glitches free. The cost of the secure multiplication in BGW's protocol is greater than that of the d-probing secure multiplication proposed in [24] (see Tab. 1). The overhead between the two methods is essentially explained by the fact that BGW's multiplication is designed to achieve d-glitches freeness whereas Rivain-Prouff's one is not. As a matter of fact and as far as we know, the only sound way to induce glitches freeness in Rivain-Prouff's multiplication would be to implement it on a multi-party circuit such that each elementary operation is processed on a separate sub-circuit. This implies the use of $O(d^2)$ sub-circuits when BGW's protocol, and then our scheme, was designed to minimize the number of players (thus the number of sub-circuits here) to $2d + 1$. Hence, even though the overall bit-complexity of our scheme is one order of magnitude more expensive than Rivain-Prouff's scheme, its limited cost in sub-circuits number makes it competitive when the design of sub-circuits is prohibitive.

4 Glitches Free HO-Masking of the AES

We apply here the construction proposed in Section 3.2 to design a multi-party circuit implementing the AES-128 nonlinear layer `SubBytes` which applies the same *substitution-box* (s-box) to every byte of the internal state. The s-box S

is defined as the left-composition of an affine transformation Γ_A over GF(256) with the power function $x \mapsto x^{254}$ over the field GF(256). In the following, we propose a (n,d)-multi-party circuit $(\mathcal{C}_{f_1},\cdots,\mathcal{C}_{f_n})$ implementing the power function. The secure implementation of the full AES-128 is not detailed here, but it can be straightforwardly deduced from the algorithms presented in this section and in the previous section.

As shown in [24], the exponentiation $x \mapsto x^{254}$ can be processed thanks to a chain of operations composed of raisings to powers in the form 2^j (which are linear over GF(256)) and 4 field multiplications. For any j, let us denote by η_j the power function $x \mapsto x^{2^j}$, the exponentiation algorithm proposed in [24] is recalled hereafter:

Algorithm2. Exponentiation to the 254

INPUT: V

OUTPUT: $Y = V^{254}$

1. $Z \leftarrow \eta_1(V)$ $\quad [Z = V^2]$
2. $Y \leftarrow Z \otimes V$ $\quad [Y = V^2V = V^3]$
3. $W \leftarrow \eta_2(Y)$ $\quad [W = (V^3)^4 = V^{12}]$
4. $Y \leftarrow Y \otimes W$ $\quad [Y = V^3V^{12} = V^{15}]$
5. $Y \leftarrow \eta_4(Y)$ $\quad [Y = (V^{15})^{16} = V^{240}]$
6. $Y \leftarrow Y \otimes W$ $\quad [Y = V^{240}V^{12} = V^{252}]$
7. $Y \leftarrow Y \otimes Z$ $\quad [Y = V^{252}V^2 = V^{254}]$

Starting from Algorithm 2 and applying Algorithm 1 to securely process the multiplications $\otimes$, we develop hereafter the s-box computation routine processed by each extended sub-circuit $(\mathcal{C}_{f_i},(\$_{ij})_{j\neq i})$.

Algorithm3. Secure S-box Processing Routine Dedicated to an Extended Circuit $(\mathcal{C}_{f_i},(\$_{ij})_{j\neq i})$

INPUT: the ith element $P_V(\alpha_i)$ of a (n,d)-sharing of V and a family of channels $(\$_{ij})_{j\neq i}$.

OUTPUT: the ith element $P_V(\alpha_i)$ of a (n,d)-sharing of $S(V)$.

PUBLIC: the n distinct points α_i, the first row $(\lambda_1,\cdots,\lambda_n)$ of the inverse of the matrix (α_i^j) with i and j lower than n.

1. **do** $P_Z(\alpha_i) \leftarrow \eta_1(P_V(\alpha_i))$ $\quad [Z = \eta_1(V)]$
2. **do** $P_Y(\alpha_i) \leftarrow$ Algorithm $1(P_V(\alpha_i), P_Z(\alpha_i), (\$_{ij})_{j\neq i})$ $\quad [Y = V \otimes Z]$
3. **do** $P_W(\alpha_i) \leftarrow \eta_2(P_Y(\alpha_i))$ $\quad [W = \eta_2(Y)]$
4. **do** $P_Y(\alpha_i) \leftarrow$ Algorithm $1(P_Y(\alpha_i), P_W(\alpha_i), (\$_{ij})_{j\neq i})$ $\quad [Y = Y \otimes W]$
5. **do** $P_Y(\alpha_i) \leftarrow \eta_4(P_Y(\alpha_i))$ $\quad [Y = \eta_4(Y)]$
6. **do** $P_Y(\alpha_i) \leftarrow$ Algorithm $1(P_Y(\alpha_i), P_W(\alpha_i), (\$_{ij})_{j\neq i})$ $\quad [Y = Y \otimes W]$
7. **do** $P_Y(\alpha_i) \leftarrow$ Algorithm $1(P_Y(\alpha_i), P_Z(\alpha_i), (\$_{ij})_{j\neq i})$ $\quad [Y = Y \otimes Z]$
8. **do** $P_V(\alpha_i) \leftarrow \Gamma_A(P_Y(\alpha_i))$ $\quad [Y = \Gamma_A(Y)]$

5 Conclusion

Thanks to the notion of multi-party circuit, we have shown in this paper that it is possible to prove, under realistic assumptions, the resistance of a dth-order masking scheme in the presence of glitches. This new framework enables to convert any classical d^{th}-order secure scheme into an implementation immune to glitches effects. The complexity of the new implementation greatly depends on the number of sub-circuits in which the initial scheme has been shared and the latter scheme must therefore be carefully chosen. Here, we have proposed to adapt the SMC protocol proposed in [1] to define a circuit sharing that is particularly well suited to our problematic. We have applied it to build a d-glitches free AES-128 implementation. As a side effect of basing our security on SMC scheme, the protocol is intrinsically immune against fault injection attacks when fewer than 1/3 of the sub-circuits are corrupted. This is a real asset of the proposed scheme when both active and passive attacks must be thwarted by the implementation. In addition, our work, together with the recent analysis [7] of Shamir's secret sharing scheme conducted in the context of SCA, shows that this sharing is a valuable alternative to the classical Boolean masking. It indeed not only enables to define glitches-free implementations, but it is also intrinsically more resistant against higher-order SCA (see [7,21] for an argumentation of this point). We based our study on very strong hypothesis on the attacker power. Even if such brutal approach allows us to develop sound proofs of security, the resulted secure implementation is costly. Future works could investigate more realistic (weaker) adversary models in order to build lighter secure implementations, or, to the same purpose, study alternative SMC protocols, less generic than BGW's protocol but more efficient. Another avenue could be to study some existing optimizations of BGW's protocol (*e.g.* the optimization based on Franklin and Yung's trick [5] that is based on efficient parallel computations.

References

1. Ben-Or, M., Goldwasser, S., Wigderson, A.: Completeness theorems for non-cryptographic fault-tolerant distributed computation. In: STOC 1988: Proceedings of the Twentieth Annual ACM Symposium on Theory of Computing, pp. 1–10. ACM, New York (1988)
2. Blömer, J., Guajardo, J., Krummel, V.: Provably Secure Masking of AES. In: Handschuh, H., Hasan, M.A. (eds.) SAC 2004. LNCS, vol. 3357, pp. 69–83. Springer, Heidelberg (2004)
3. Chari, S., Jutla, C.S., Rao, J.R., Rohatgi, P.: Towards Sound Approaches to Counteract Power-Analysis Attacks. In: Wiener, M. (ed.) CRYPTO 1999. LNCS, vol. 1666, pp. 398–412. Springer, Heidelberg (1999)
4. Coron, J.-S.: A New DPA Countermeasure Based on Permutation Tables. In: Ostrovsky, R., De Prisco, R., Visconti, I. (eds.) SCN 2008. LNCS, vol. 5229, pp. 278–292. Springer, Heidelberg (2008)
5. Franklin, M., Yung, M.: Communication complexity of secure computation (extended abstract). In: STOC 1992: Proceedings of the Twenty-Fourth Annual ACM Symposium on Theory of Computing, pp. 699–710. ACM, New York (1992)

6. Gennaro, R., Rabin, M.O., Rabin, T.: Simplified vss and fact-track multiparty computations with applications to threshold cryptography. In: PODC, pp. 101–111 (1998)
7. Goubin, L., Martinelli, A.: Protecting AES with Shamir's Secret Sharing Scheme. In: Preneel, B., Takagi, T. (eds.) CHES 2011. LNCS, vol. 6917, pp. 63–78. Springer, Heidelberg (2011)
8. Goubin, L., Patarin, J.: DES and Differential Power Analysis – The Duplication Method. In: Koç, Ç.K., Paar, C. (eds.) CHES 1999. LNCS, vol. 1717, pp. 158–172. Springer, Heidelberg (1999)
9. Ishai, Y., Sahai, A., Wagner, D.: Private circuits: Securing hardware against probing attacks. In: Boneh, D. (ed.) CRYPTO 2003. LNCS, vol. 2729, pp. 463–481. Springer, Heidelberg (2003)
10. Joye, M., Paillier, P., Schoenmakers, B.: On Second-order Differential Power Analysis. In: Rao and Sunar [22], pp. 293–308
11. Kocher, P., Jaffe, J., Jun, B.: Introduction to Differential Power Analysis and Related Attacks. Technical report, Cryptography Research Inc. (1998)
12. Mangard, S., Popp, T., Gammel, B.M.: Side-Channel Leakage of Masked CMOS Gates. In: Menezes, A. (ed.) CT-RSA 2005. LNCS, vol. 3376, pp. 351–365. Springer, Heidelberg (2005)
13. Mangard, S., Schramm, K.: Pinpointing the Side-Channel Leakage of Masked AES Hardware Implementations. In: Goubin, L., Matsui, M. (eds.) CHES 2006. LNCS, vol. 4249, pp. 76–90. Springer, Heidelberg (2006)
14. Messerges, T.S.: Using Second-Order Power Analysis to Attack DPA Resistant Software. In: Paar, C., Koç, Ç.K. (eds.) CHES 2000. LNCS, vol. 1965, pp. 238–251. Springer, Heidelberg (2000)
15. Moradi, A., Poschmann, A., Ling, S., Paar, C., Wang, H.: Pushing the limits: A very compact and a threshold implementation of AES. In: Paterson, K.G. (ed.) EUROCRYPT 2011. LNCS, vol. 6632, pp. 69–88. Springer, Heidelberg (2011)
16. Nikova, S., Rechberger, C., Rijmen, V.: Threshold implementations against side-channel attacks and glitches. In: Ning, P., Qing, S., Li, N. (eds.) ICICS 2006. LNCS, vol. 4307, pp. 529–545. Springer, Heidelberg (2006)
17. Nikova, S., Rijmen, V., Schläffer, M.: Secure Hardware Implementation of Nonlinear Functions in the Presence of Glitches. In: Lee, P.J., Cheon, J.H. (eds.) ICISC 2008. LNCS, vol. 5461, pp. 218–234. Springer, Heidelberg (2009)
18. Nikova, S., Rijmen, V., Schläffer, M.: Secure hardware implementation of non-linear functions in the presence of glitches. Technical report, VAMPIRE II (2010)
19. Nikova, S., Rijmen, V., Schläffer, M.: Secure hardware implementation of nonlinear functions in the presence of glitches. J. Cryptology 24(2), 292–321 (2011)
20. Prouff, E., Roche, T.: Attack on a higher-order masking of the AES based on homographic functions. In: Gong, G., Gupta, K.C. (eds.) INDOCRYPT 2010. LNCS, vol. 6498, pp. 262–281. Springer, Heidelberg (2010)
21. Prouff, E., Roche, T.: Higher-Order Glitches Free Implementation of the AES using Secure Multi-Party Computation Protocols. Cryptology ePrint Archive (to appear, 2011), `http://eprint.iacr.org/`
22. Rao, J.R., Sunar, B. (eds.): CHES 2005. LNCS, vol. 3659. Springer, Heidelberg (2005)
23. Rivain, M., Dottax, E., Prouff, E.: Block Ciphers Implementations Provably Secure Against Second Order Side Channel Analysis. Cryptology ePrint Archive, Report 2008/021 (2008), `http://eprint.iacr.org/`

24. Rivain, M., Prouff, E.: Provably secure higher-order masking of AES. In: Mangard, S., Standaert, F.-X. (eds.) CHES 2010. LNCS, vol. 6225, pp. 413–427. Springer, Heidelberg (2010)
25. Schramm, K., Paar, C.: Higher Order Masking of the AES. In: Pointcheval, D. (ed.) CT-RSA 2006. LNCS, vol. 3860, pp. 208–225. Springer, Heidelberg (2006)
26. Shamir, A.: How to Share a Secret. Commun. ACM 22(11), 612–613 (1979)
27. Suzuki, D., Saeki, M., Ichikawa, T.: DPA Leakage Models for CMOS Logic Circuits. In: Rao and Sunar [22], pp. 366–382
28. Yao, A.C.-C.: How to generate and exchange secrets. In: Proceedings of the 27th Annual Symposium on Foundations of Computer Science, pp. 162–167. IEEE Computer Society, Washington, DC, USA (1986)

Protecting AES with Shamir's Secret Sharing Scheme

Louis Goubin[1] and Ange Martinelli[1,2]

[1] Versailles Saint-Quentin-en-Yvelines University
Louis.Goubin@prism.uvsq.fr
[2] Thales Communications
jean.martinelli@fr.thalesgroup.com

Abstract. Cryptographic algorithms embedded on physical devices are particularly vulnerable to Side Channel Analysis (SCA). The most common countermeasure for block cipher implementations is masking, which randomizes the variables to be protected by combining them with one or several random values. In this paper, we propose an original masking scheme based on Shamir's Secret Sharing scheme [22] as an alternative to Boolean masking. We detail its implementation for the AES using the same tool than Rivain and Prouff in CHES 2010 [16]: multi-party computation. We then conduct a security analysis of our scheme in order to compare it to Boolean masking. Our results show that for a given amount of noise the proposed scheme - implemented to the first order - provides the same security level as 3rd up to 4th order boolean masking, together with a better efficiency.

Keywords: Side Channel Analysis (SCA), Masking, AES Implementation, Shamir's Secret Sharing, Multi-party computation.

1 Introduction

Side Channel Analysis is a cryptanalytic method in which an attacker analyzes the *side channel leakage* (*e.g.* the power consumption, ...) produced during the execution of a cryptographic algorithm embedded on a physical device. SCA exploits the fact that this leakage is statistically dependent on the intermediate variables that are involved in the computation. Some of these variables are called *sensitive* in that they are related to a secret data (*e.g.* the key) and a known data (*e.g.* the plain text), and recovering information on them therefore enables efficient key recovery attacks [11,3,8].
The most common countermeasure to protect implementations of block ciphers against SCA is to use masking techniques [4,9] to randomize the sensitive variables. The principle is to combine one or several random values, called *masks*, with every processed sensitive variable. Masks and masked variables propagate throughout the cipher in such a way that any intermediate variable is independent of any sensitive variable. This method ensures that the leakage at an instant t is independent of any sensitive variable, thus rendering SCA difficult to perform. The masking can be improved by increasing the number of random masks

B. Preneel and T. Takagi (Eds.): CHES 2011, LNCS 6917, pp. 79–94, 2011.

that are used per sensitive variable. A masking that involves d random masks is called a dth*-order masking* and can always be theoretically broken by a $(d+1)^{th}$-*order SCA*, namely an SCA that targets $d+1$ intermediate variables at the same time [13,21,18]. However, the noise effects imply that the complexity of a dth-order SCA increases exponentially with d in practice [4]. The dth*-order SCA resistance* (for a given d) is thus a good security criterion for implementations of block ciphers. In [17] Rivain and Prouff give a general method to implement a dth-order masking scheme to the AES using secure Multi-Party Computation. Instead of looking for perfect theoretical security against d^{th}-order SCA as done in [17], an alternative approach consists in looking for practical resistance to these attacks. It may for instance be observed that the efficiency of higher-order SCA is related to the way the masks are introduced to randomize sensitive variables. The most widely studied masking schemes are based on *Boolean masking* where masks are introduced by exclusive-or (XOR). First order boolean masking enables securing implementations against first-order SCA quite efficiently[1,16]. It is however especially vulnerable to higher-order SCA [13] due to the intrinsic physical properties of electronic devices. Other masking schemes may provide better resistance against these attacks using various operations to randomize sensitive variables. This approach will be further investigated in this paper.

Related work. In [25,6], the authors propose to use an affine function instead of just XOR to mask sensitive variables, thus improving the security of the scheme for a low complexity overhead. However, this countermeasure is developed only to the 1th order and it is not clear how it can be extended to higher orders. In [10,16] the authors explain how to use secure Multi-Party Computation to process the cipher on shared variables. They use a sharing scheme based on XOR, implementing boolean masking to any order to secure the AES block cipher. At last, in [19], Prouff and Roche give a hardware oriented glitch free way to implement block ciphers using Shamir's Secret Sharing scheme and Ben-Or *et al.* secure multi-party computation [2] protocol operating on $2d+1$ shares to thwart d-th order SCA.

Our contribution. In this paper, we propose to combine both approaches in implementing a masking scheme based upon Shamir's Secret Sharing scheme [22], called *SSS masking* and processed using Multi-party Computation methods. Namely, we present an implementation of the block cipher such that every 8-bit intermediate result $z \in \text{GF}(256)$ is manipulated under the form $(x_i, P(x_i))_{i=0..d}$, where $x_i \in \text{GF}(256)^*$ is a random value generated before each new execution of the algorithm and $P(X) \in \text{GF}(256)[X]$ is a polynomial of degree d such that $P(0) = z$. Our scheme maintains the same compatibility as Boolean masking with the linear transformations of the algorithm. Moreover, the fact that the masks are never processed alone prevents them to be targeted by a higher-order SCA, thus greatly improves the resistance of the scheme to such attacks.

Organization of the paper. We fist recall Shamir's secret sharing scheme in Sect. 2. In Sect. 3, we show how SSS masking can be applied to the AES and give some implementation results. Sect. 4 analyzes the resistance of our method to high-order SCA and Sect. 5 concludes the paper.

2 Shamir's Secret Sharing Scheme

In some cryptographic context ones may need to share a secret between (at least) d users without any $k < d$ users being able to recover the secret alone. In [22] Shamir exposes the problematic and gives a secret sharing scheme using polynomial interpolation as a recovery mean. Namely, every user has a pair $(x_i, P(x_i))_{x_i \neq 0}$, where P is a polynomial of degree k, and the secret is given by $P(0)$. In this configuration, one needs at least $k+1$ shares to recover P, then $P(0)$. We recall hereafter the sharing and reconstruction algorithms for a value of $k = d-1$, operating on n-bits words. With these parameters, in order to share a secret a_0 into d shares, one needs to choose $d-1$ random numbers $(a_{d-1}, \ldots, a_1)$ to construct the polynomial $P(x) = a_{d-1} \cdot x^{d-1} + a_{d-2} \cdot x^{d-2} + \cdots + a_1 \cdot x + a_0$. Every share i is then given by (x_i, y_i) where $y_i = P(x_i)$, and the x_i's are all distinct and non-zero. Formally we have Algorithm 1.

Algorithm 1. Shamir's Secret Sharing scheme

INPUT: A secret a_0, random values $(x_i)_{i=0..d-1}$

OUTPUT: Shares $(x_i, y_i)_{i=0..d-1}$

1. $(a_i)_{i=1..d-1} \leftarrow \mathtt{Rand}(n)$
2. **for** $i = 0$ **to** d **do**
3. $\quad y_i \leftarrow a_{d-1} \cdot x_i^{d-1} + a_{d-2} \cdot x_i^{d-2} + \cdots + a_1 \cdot x_i + a_0$
4. **return** $(x_i, y_i)_{i=0..d-1}$

The reconstruction step is directly derived from polynomial interpolation and proceeds as follows:

$$a_0 = \sum_{0}^{d} y_i \cdot \beta_i \tag{1}$$

where each β_i is a precomputed value such that $\beta_i = \prod_{j=0, j\neq i}^{d} \frac{-x_j}{x_i - x_j}$.

3 Higher Order Masking of AES

The AES block cipher iterates a round transform composed of four stages: `AddRoundKey`, `ShiftRows`, `MixColumn` and `SubByte`. In this section we will show how to securely mask the different layers at order d using *SSS masking*.

3.1 Masking Field Operations

In order to secure the AES there are five field operations that must be protected: the addition with an unmasked constant, the addition with a masked variable, the multiplication by a scalar, the square, and the multiplication between two shared variables. Moreover, as the AES Sbox is the composition of the inversion in GF(256) and an affine function modulus the polynomial $X^8 + 1$, this affine transform also has to be secured.

Let b be a sensitive variable shared as $(x_i, y_i)_{i=0..d}$ following Shamir's secret sharing scheme. Addition with an unmasked constant u can be directly computed by XORing u to the second component of the shares, such as:

$$(x'_i, y'_i) \leftarrow (x_i, y_i \oplus u).$$

Let $(x_i, u_i)_{i=0..d}$ be the shared representation of a variable u. The shared representation of the addition $b \oplus u$ is obtained as: $(x'_i, y'_i) \leftarrow (x_i, y_i \oplus u_i)$. Similarly the multiplication by a scalar p is computed as: $(x'_i, y'_i) \leftarrow (x_i, y_i \cdot p)$.

While working in a field of characteristic 2 squaring is GF(256)-linear: $(x'_i, y'_i) \leftarrow (x_i^2, y_i^2)$. Remark that the coefficient $x'_i \neq x_i$. This matter must be taken in account in the computation as shown in Algorithm 4.

The product of two variables protected by secret sharing cannot be solved with the linear property of the transformation, as multiplying two polynomials with the same degree gives a polynomial with degree double of the original polynomial. Two different approaches can be studied. The first, developed in [19], is to use the proven secure multi-party computation scheme of [2] operating on $2d + 1$ shares to process the product. However this approach has a very high complexity. The second possibility is to exploit the context of side channel countermeasure that allows us to compute values unknown in classical multi-party computation in order to improve the complexity at the loss of the security proof. We give Algorithm 2 to compute secure shared field multiplication.

Algorithm 2. Share multiplication `SecMult`

INPUT: Shared representation of b, $(x_i, y_i)_{i=0..d}$ and u, $(x_i, w_i)_{i=0..d}$

OUTPUT: Shares $(x_i, y'_i)_{i=0..d}$ representing the product of b and u

1. **for** $j = 0$ **to** d **do**
2.   **for** $k = 0$ **to** d **do**
3. $z_{j,k} \leftarrow y_j \cdot w_k$
4. **for** $i = 0$ **to** d **do**
5. $(x_i, y'_i) \leftarrow \left(x_i, \left(\sum_{j=1}^{d} \sum_{0 \leq k < j} (z_{j,k} \oplus z_{k,j}) \cdot \beta_{j,k}(x_i) \right) + \sum_{j=0}^{d} z_{j,j} \cdot \beta_{j,j}(x_i) \right)$
6. **return** $(x_i, y'_i)_{i=0..d}$

where the $\beta_{j,k}(x_i)$ are precomputed values defined as follows.

Recall that $\beta_j(x) = \prod_{l=0, l \neq j}^{d} \frac{x - x_l}{x_j - x_l}$. We have

$$\begin{aligned}\beta_j(x)\cdot\beta_k(x) &= \prod_{l=0,l\neq j}^{d}\frac{x-x_l}{x_j-x_l}\cdot\prod_{m=0,m\neq k}^{d}\frac{x-x_m}{x_k-x_m} \\ &= \alpha_{2d}x^{2d}+\cdots+\alpha_d x^d+\cdots+\alpha_1 x+\alpha_0\end{aligned} \tag{2}$$

We then define $\beta_{j,k}(x)=\beta_{k,j}(x)=\alpha_d x^d+\cdots+\alpha_1 x+\alpha_0$.

Proposition 1. *Algorithm 2 holds because the polynomial* $P(x)=\sum_{j=0}^{d}\sum_{k=0}^{d} y_j\cdot w_k\cdot\beta_{j,k}(x)$ *is such that:* $\begin{cases} degree(P)=d \\ P(0)=b\cdot u \\ \forall x\in\{x_i\}_{i=0..d},\ P(x_i)=y_i' \end{cases}$

Proof. – By construction of the $\beta_{k,j}(x)$, $degree(P)=d$.
- Let b,u be shared respectively in $(x_i, y_i=R(x_i))$ and $(x_i, w_i=Q(x_i))$. $R(x)=\sum_0^d y_i\cdot\beta_i(x)$ and $b=R(0)$ and $Q(x)=\sum_0^d w_i\cdot\beta_i(x)$ and $u=Q(0)$. As the truncation does not modify the constant term of the polynomial,

$$P(0)=R(0)\cdot Q(0)=b\cdot u.$$

- At last, by construction $\forall x\in\{x_i\}_{i=0..d}$,

$$y_i'=\sum_{j=0}^{d}\sum_{k=0}^{d} y_j\cdot w_k\cdot\beta_{j,k}(x_i)=P(x_i)$$

□

Intuitivley, the security of the scheme against k-th order SCA ($k\leq d$) is based on the following points:

- according to polynomial interpolation, one needs at least $d+1$ shares to define a polynomial of degree d,
- the computation of the $\beta_{j,k}(x_i)$ is independent of any secret,
- the knowledge of $y_j\cdot w_k$ does not leak more information on b (resp. u) than the knowledge of y_j (resp. w_k),

However the security proof of Algorithm 2 does not seems to be an easy matter and is still an open work.

Finally the affine function A involved in the AES Sbox, as if non linear with respect to the polynomial mask, can nevertheless be implemented using straightforwardly as : $(x_i', y_i')\leftarrow(x_i, A(y_i))$. Indeed, if $y_i=P(x_i)$, since A s affine $A(P)$ is a polynomial of degree d with $A(P(0))=A(b)$ and every $A(y_i)$ is the polynomial value of $A(P)(x)$ in x_i.

3.2 Complexity of the Operations

In order to evaluate the complexity overhead of SSS masking with respect to boolean masking, we compare the complexity of each operation involved in the AES computation for both kind of masking. As shown in the previous section, the multiplication between two shared variables is the most consuming operation, but this is also the case for boolean masking (see [16]). Table 1 resumes the complexities of both schemes.

Table 1. Complexity of masked operations

Operation \ Scheme	Boolean Masking [16]	SSS Masking
XOR with a constant	1 XOR	$d+1$ XORs
Shared XOR	$d+1$ XORs	$d+1$ XORs
Scalar Multiplication	$d+1$ multiplications	$d+1$ multiplications
Squaring	$d+1$ squaring	$2(d+1)$ squaring
Shared Field Mult.	$2d(d+1)$ XORs $d(d+1)/2$ random numbers $(d+1)^2$ field products	$d(d+1)(d+2)$ XORs $(d+1)^2(2+\frac{d}{2})$ field products
Sbox Affine transform.	1 XOR d ring products	d XORs d ring products

3.3 Masking the Full S-Box

We have defined secure squaring and multiplication in Section 3.1, we then use the exponentiation algorithm given in [16], and resumed afterward (Algo 3), to implement the power function involved in the AES S-box.

Algorithm 3. Secure Exponentiation to the power 254 over GF(2^8)

INPUT: Shared representation of b, $(x_i, y_i)_{i=0..d}$

OUTPUT: Shares $(x_i, y_i')_{i=0..d}$ of the value b^{254}

1. **for** $i = 0$ **to** d **do**$(\alpha_i, \zeta_i) \leftarrow (x_i^2, y_i^2)$
2. $(x_i, \zeta_i)_i \leftarrow$ `RefreshMasks`$((\alpha_i, \zeta_i)_i, 2)$
3. $(x_i, \gamma_i)_i \leftarrow$ `SecMult`$((x_i, \zeta_i), (x_i, y_i))$
4. **for** $i = 0$ **to** d **do**$(\alpha_i, \delta_i) \leftarrow (x_i^4, \gamma_i^4)$
5. $(x_i, \delta_i)_i \leftarrow$ `RefreshMasks`$((\alpha_i, \delta_i)_i, 4)$
6. $(x_i, \gamma_i)_i \leftarrow$ `SecMult`$((x_i, \gamma_i), (x_i, \delta_i))$
7. **for** $i = 0$ **to** d **do**$(\alpha_i, \gamma_i) \leftarrow (x_i^{16}, \gamma_i^{16})$
8. $(x_i, \gamma_i)_i \leftarrow$ `RefreshMasks`$((\alpha_i, \gamma_i)_i, 16)$
9. $(x_i, \gamma_i)_i \leftarrow$ `SecMult`$((x_i, \gamma_i), (x_i, \delta_i))$
10. $(x_i, y_i')_i \leftarrow$ `SecMult`$((x_i, \gamma_i), (x_i, \zeta_i))$
11. `return` $(x_i, y_i')_{i=0..d}$

Here the `RefreshMasks` operation is needed to ensure the conservation of the x_i's during the computation and the independence of the coefficients of the polynomials before `SecMult` operation. Formally it follows Algorithm 4.

Algorithm 4. `RefreshMasks`

INPUT: Shared representation of b, $(\alpha_i, y_i)_{i=0..d}$, chosen $(x_i)_{i=0..d}$, t such that $\alpha_i = x_i^{2^t}$

OUTPUT: Shared representation of b, $(x_i, y_i')_{i=0..d}$

1. **for** $i = 0$ **to** d **do**
2. $\quad \beta_i' \leftarrow \beta_i^{2^t}$
3. $\quad$ Share y_i in (x_j, z_{i_j}) using Algo 1
4. **for** $i = 0$ **to** d **do**
5. $\quad (x_i, y_i') \leftarrow \left(x_i, \sum_{j=0}^{d} \beta_j' \cdot z_{j_i}\right)$
6. **return** $(x_i, y_i')_{i=0..d}$

Algorithm 4 consists in re-sharing each shares separately using a new random polynomial, then to reconstruct the original shares to obtain $d+1$ shares corresponding to this new polynomial. Eventually the complexity of Algorithm 3 is resumed in Table 2. As a matter of comparison, we recall hereafter the complexity of Boolean masking as given in [16].

Table 2. Complexity of inversion algorithms

order	XORs	multiplications	^2^j	Rand. bytes	RAM (bytes)
$O1$-SSS	36	54	14	6	20
$O2$-SSS	150	165	21	18	33
Od-SSS	$7d^3 + 18d^2 + 11d$	$5d^3 + 18d^2 + 22d + 9$	$7(d+1)$	$3d^2 + 3d$	$d^2 + 10d + 9$
$O1$-Bool.	20	16	6	6	7
$O2$-Bool.	56	36	9	16	12
$O3$-Bool.	108	64	12	20	18
$O4$-Bool.	176	100	15	48	25
Od-Bool.	$7d^2 + 12d$	$4d^2 + 8d + 4$	$3(d+1)$	$2d^2 + 4d$	$\frac{1}{2}d^2 + \frac{7}{2}d + 3$

As a matter of fact, the number of operations involved in SSS masking is larger than that of boolean masking for a given order d, as the number of field multiplications and XOR operations are cubic in the order instead of quadratic for boolean masking. We can ask ourselves if this observation remains true for a given security level. This question will be studied in section 4.

3.4 Masking the Whole AES

In the following, we describe how to mask an AES computation at the dth order using SSS masking. We will assume that the secret key has been previously masked and that its $d+1$ shares are provided as input to the algorithm (otherwise a straightforward 1st-order attack would be possible). At the beginning of the computation, the state s (holding the plaintext) is split into $d+1$ shares (x_0, y_0), (x_1, y_1), ... , (x_d, y_d) with respect to Shamir's secret sharing scheme. In the next sections, we describe how to perform the different AES transformations on the state shares in order to guarantee the completeness as well as the dth-order security.

Masking AddRoundKey. The AddRoundKey stage at round r consists in XORing the rth round key k_r to the state. The masked key schedule provides $d+1$ shares $(x_i, k_{r,i})_i$ for every round key k_r. The XOR operation is then processed as described in section 3.1: $M(\mathbf{s} \oplus k_r) \rightarrow (x_i, y_i \oplus k_{r,i})_{i=0..d}$

Masking ShiftRows. As the ShiftRows layer operates on each byte separately and does not change their value, we have: $M(\mathsf{ShiftRows}(\mathbf{s})) = \mathsf{ShiftRows}(M(\mathbf{s}))$

Masking MixColumn. Since each output byte of $\mathsf{MixColumns}_c$ can be expressed as a linear function of the bytes of the input state over GF(256) we have:

$$\mathsf{MixColumns}_c(M(s_0), M(s_1), M(s_2), M(s_3)) = (M(s'_0), M(s'_1), M(s'_2), M(s'_3)).$$

This suggests to perform the following steps to securely process $\mathsf{MixColumns}_c$ on the masked representation of the state columns.

$$\begin{cases} M(s'_0) = (x_i, y'_{0,i}) \leftarrow (x_i,\ \mathsf{xtimes}(y_{0,i} \oplus y_{1,i}) \oplus tmp_i \oplus y_{0,i}) \\ M(s'_1) = (x_i, y'_{1,i}) \leftarrow (x_i,\ \mathsf{xtimes}(y_{1,i} \oplus y_{2,i}) \oplus tmp_i \oplus y_{1,i}) \\ M(s'_2) = (x_i, y'_{2,i}) \leftarrow (x_i,\ \mathsf{xtimes}(y_{2,i} \oplus y_{3,i}) \oplus tmp_i \oplus y_{2,i}) \\ M(s'_3) = (x_i, y'_{2,i}) \leftarrow (x_i,\ y'_{0,i} \oplus y'_{1,i} \oplus y'_{2,i} \oplus tmp_i). \end{cases} \tag{3}$$

where $tmp_i = y_{0,i} \oplus y_{1,i} \oplus y_{2,i} \oplus y_{3,i}$ and where xtimes denotes a look-up table for the function $x \mapsto \mathsf{02} \cdot x$. The completeness holds because the single operation that modify the random factors $(a_i)_{i=1..d}$ is the xtimes one, and is applied similarly to each share.

Masking SubByte. The SubBytes transformation consists in applying the AES S-box S to each byte of the state. In order to mask this transformation, we apply the secure S-box computation described in Section 3.3 to the $(d+1)$-tuple of every byte shares of the state.

KeySchedule. Finally, since the round key derivation is a composition of the previous transformations, it can be protected using the exact same methods as previously described.

Overall Complexity. In order to give an idea of the global complexity of the scheme, and to compare it to the boolean masking, we give in Table 3 the overall number of operations involved in the ciphering.

Table 3. Complexity of cipher implementations

Masking	XORs/ANDs	Table look-ups	Random bits	RAM (bits)	ROM (bits)
1O boolean	17640	16144	16896	312	6128
2O boolean	37800	32272	46080	352	6128
3O boolean	65640	54160	87552	400	6128
1O SSS	31760	37296	16240	400	6128

4 Security Analysis

In what follows, we shall consider that an intermediate variable U_i is associated with a leakage variable L_i representing the information leaking about U_i through side channel. We will assume that the leakage can be expressed as a deterministic *leakage function* φ of the intermediate variable U_i with an independent additive noise B_i. Namely, we will assume that the leakage variable L_i satisfies:

$$L_i = \varphi(U_i) + B_i \ . \tag{4}$$

In the following, we call *d^{th}-order leakage* a tuple of d leakage variables L_i corresponding to d different intermediate variables U_i that jointly depend on some sensitive variable. As already argued in Sect. 3.4, when an implementation is correctly protected by SSS masking, no first-order leakage of sensitive information occurs. This directly comes from Shamir's secret sharing scheme security. In the following we will focus on higher orders attacks against protected implementations, secured by boolean or SSS masking.

4.1 Information Theoretic Analysis

In order to evaluate the information leaked by 1O-SSS masking and to compare it to that of various orders Boolean masking, we compute, as suggested in [23], the theoretical mutual information $I(S|L_d)$ for a class discrete variable S of the secret, and a d-order leakage L_d, with respect to increasing noise standard deviation σ. Namely we consider the three following leakages:

- 2nd-order leakage of 1O-Boolean masking with targeted variables $(Z \oplus m_1, m_1)$
- 3rd-order leakage of 2O-Boolean masking with targeted variables $(Z \oplus m_1 \oplus m_2, m_1, m_2)$
- 2nd-order leakage of 1O-SSS masking with targeted variables $((x_1, a \cdot x_1 \oplus Z), (x_2, a \cdot x_2 \oplus Z))$

The variables Z, m_1, m_2 and a are assumed uniformly distributed over GF(256) and mutually independent, and x_1, x_2 are assumed uniformly distributed over GF(256)* with $x_1 \neq x_2$. For each kind of leakage, we compute the mutual information between Z and the tuple of leakages in the Hamming weight (HW) model with Gaussian noise: the leakage L_i related to a variable U_i is distributed according to equation (4) with $\varphi =$ HW and $B_i \sim \mathcal{N}(0, \sigma^2)$ (the different B_i's are also assumed independent). In this context, the signal-to-noise ratio (SNR) of the leakage is defined as $\mathrm{Var}\left[\varphi(U_i)\right] / \mathrm{Var}\left[B_i\right] = 2/\sigma^2$.

Fig. 1 shows the mutual information values obtained for each kind of leakage with respect to an increasing noise standard deviation. These results demonstrate the information leakage reduction implied by the use of SSS masking. As expected, SSS masking leaks less information than first and second order Boolean masking for the considered Signal to Noise ratios (SNRs). We will now see to which extent this reduction also applies to the efficiency of SCA on SSS masking.

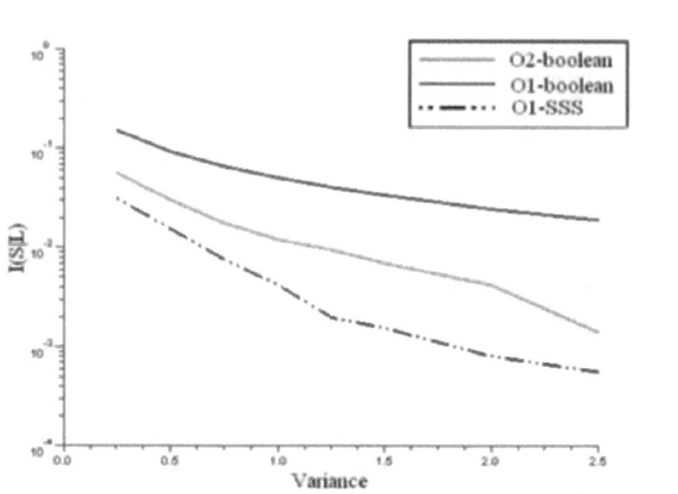

Fig. 1. Mutual Information values with respect to σ^2 (logarithmic scale)

4.2 Higher-Order DPA Evaluation

Let us assume that Z depends on the plaintext and of a subkey $k^\star$, and let us denote by $Z(k)$ the hypothetic value of Z for a guess k on $k^\star$. In a *higher-order DPA* (HO-DPA) [13,18], the attacker tests the guess k by estimating the correlation coefficient $\rho\left[\hat{\varphi}(Z(k)), \mathcal{C}(\mathbf{L})\right]$, where $\mathcal{C}$ is a *combining function* that converts the multivariate leakage $\mathbf{L}$ into a univariate signal and where $\hat{\varphi}$ is a *prediction function* chosen such that $\hat{\varphi}(Z)$ is correlated as much as possible to $\mathcal{C}(\mathbf{L})$. The guess k leading to the greatest correlation (in absolute value) is selected as key-candidate. In [12], the authors show that the number of traces required to mount a successful DPA attack is roughly quadratic in ρ^{-1} where ρ is the correlation coefficient $\rho\left[\hat{\varphi}(Z), \mathcal{C}(\mathbf{L})\right]$ (that is the expected correlation for the correct key guess). The latter can therefore be used as a metric for the efficiency of a (HO-)DPA attack.

The analysis conducted in [18] states that a good choice for $\mathcal{C}$ is the *normalized product combining*:

$$\mathcal{C} : \mathbf{L} \mapsto \prod_i (L_i - \mathrm{E}\left[L_i\right]). \tag{5}$$

Although the effectiveness of the normalized product combining has been only studied in [18] in the context of Boolean masking, we can argue that this combining function stays a natural choice against any kind of masking since $\rho\left[\hat{\varphi}(Z(k)), \mathcal{C}(\mathbf{L})\right]$ is related to the *multivariate correlation*[1] between $\hat{\varphi}(Z(k))$

[1] What we call multivariate correlation here is the straightforward generalization of the correlation coefficient to more than two variables (see [24]).

and every coordinate of $\mathbf{L}$ [24]. Besides, in the presence of (even little) noise in the side-channel leakage, the HO-DPA with normalized product combining is nowadays the most efficient unprofiled attack against Boolean masking in the literature (see for instance [18,24,17]).

From Corollary 8 in [18], the optimal correlation ρ_{SSS} for the correct key hypothesis can be obtained as:

$$\rho_{\text{SSS}} = \sqrt{\frac{\text{Var}\left[\text{E}\left[\overline{L}_1 \times \overline{L}_2 | Z = z\right]\right]}{\text{Var}\left[\overline{L}_1 \times \overline{L}_2\right]}} \tag{6}$$

Formally, when the leakage satisfies (4) with $\varphi = \text{HW}$ and $B_i \sim \mathcal{N}(0, \sigma^2)$, the coefficient ρ_{SSS} obtained for the 2-nd order leakage of 1-st order SSS masking satisfies:

$$\rho_{\text{SSS}} = \sqrt{\frac{n^3 \cdot (2^{n+1} - 4^n - 1)}{\alpha_2 \cdot \sigma^4 + \alpha_1 \cdot \sigma^2 + \alpha_0}}, \tag{7}$$

where n is the bit-size of Z, and

$$\begin{aligned}
\alpha_2 &= 192 \cdot 2^n - 2^{4n+4} - 64 - 208 \cdot 4^n + 96 \cdot 8^n \\
\alpha_1 &= (40 \cdot 8^n - 64 \cdot 4^n - 8 \cdot 16^n + 32 \cdot 2^n)n^2 \\
&\quad +(88 \cdot 8^n + 128 \cdot 2^n - 2^{4n+4} - 168 \cdot 4^n - 32)n \\
\alpha_0 &= (8^n - 3 \cdot 4^n + 6 \cdot 2^n - 4)n^4 + (-4 \cdot 16^n + 14 \cdot 8^n - 16 \cdot 4^n + 2 \cdot 2^n + 4)n^3 \\
&\quad +(23 \cdot 8^n - 4 \cdot 16^n - 44 \cdot 4^n + 34 \cdot 2^n - 8)n^2 + (10 \cdot 4^n - 3 \cdot 8^n - 9 \cdot 2^n + 2)n
\end{aligned} \tag{8}$$

Remark 1. In order to endorse our choice of targeted variables, we also computed the correlation coefficient corresponding to another 2rd-order leakage of SSS masking targeting the pair $(a, a \cdot x + Z)$ with the corresponding pair of prediction functions: the Dirac function δ_0 ($\delta_0(x) = 0 \Leftrightarrow x \neq 0$) and the Hamming weight. We observed for several values of n and σ that the correlation coefficient was always lower than ρ_{SSS}.

Regarding Boolean masking, it has been shown in [20] that the correlation ρ_{bool} corresponding to HO-DPA with normalized product combining against dth-order Boolean masking satisfies (in the Hamming weight model):

$$\rho_{\text{bool}} = (-1)^d \frac{\sqrt{n}}{(n + 4\sigma^2)^{\frac{d+1}{2}}} . \tag{9}$$

From (7) and (9), we define the ratio ν as: $\nu = \frac{\rho_{\text{SSS}}}{\rho_{\text{bool}}}$.

Let us denote by N_{SSS} (resp. N_{bool}) the number of leakage measurements for a successful attack on SSS masking (resp. Boolean masking). According to [12], N_{SSS} and N_{bool} are roughly quadratic in the values of the correlation values. Hence the ratio $\frac{N_{\text{SSS}}}{N_{\text{bool}}}$ satisfies:

$$\frac{N_{\text{SSS}}}{N_{\text{bool}}} \approx \frac{1}{\nu^2} . \tag{10}$$

Table 4. Ratio $1/\nu^2$ for several Boolean masking orders with respect to 1*O*-order SSS masking

Attack \ SNR	$+\infty$	1	1/2	1/5	1/10
3O-DPA on 2O Boolean Masking	4544.83	899.99	374.33	94.17	22.70
4O-DPA on 3O Boolean Masking	568.10	56.25	15.60	1.96	0.21
5O-DPA on 4O Boolean Masking	71.01	3.52	0.65	0.04	0.002

Table 4 illustrate this relation giving the value $1/\nu^2$ for different values of SNRs. Values greater (resp. lower) than 1 indicate that SSS masking is more (resp. less) secure than the considered attack.

Due to (10), SSS masking is more secure than dth-order Boolean masking if and only if $|\nu| \leq 1$. Comparing the resistance of the Boolean masking and SSS masking against HO-DPA thus amounts to study when $|\nu| \leq 1$ is satisfied. We can note that $|\nu|$ is an increasing function of σ and a decreasing function of n. Let us denote by ϑ the maximal variance of the noise such that $|\nu| \leq 1$ is satisfied. For the first values of d, we have:

$$\vartheta = \begin{cases} +\infty & \text{if } d = 1, \\ 282.2683 & \text{if } d = 2, \\ 13.2072 & \text{if } d = 3, \\ 3.4036 & \text{if } d = 4. \end{cases} \tag{11}$$

Eventually Fig. 2 sums up our main theoretical results. In particular, it illustrates the fact that the coefficient ρ_{SSS} is lower than ρ_{bool} (computed for $d = 3$) only when the noise variance σ^2 is lower than 13.2072.

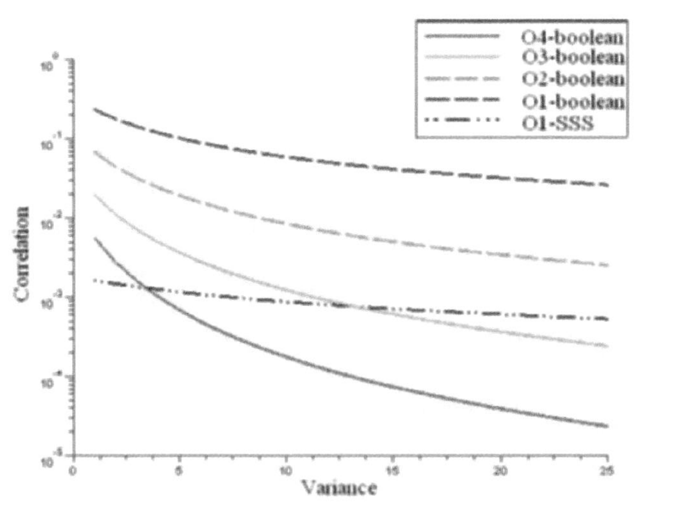

Fig. 2. Correlation values with respect to σ^2 (logarithmic scale)

Table 5. Number of leakage measurements for a 90% success rate

Attack \ SNR	$+\infty$	1	1/2	1/5	1/10
Attacks against Boolean Masking					
2O-DPA on 1O Boolean Masking	150	500	1500	6000	20 000
2O-MIA on 1O Boolean Masking	100	5000	15 000	50 000	160 000
3O-DPA on 2O Boolean Masking	1500	9000	35 000	280 000	$> 10^6$
3O-MIA on 2O Boolean Masking	160	160 000	650 000	$> 10^6$	$> 10^6$
Attacks against SSS Masking					
2O-DPA on 1O SSS Masking	$> 10^6$	$> 10^6$	$> 10^6$	$> 10^6$	$> 10^6$
2O-MIA on 1O SSS Masking	500 000	$> 10^6$	$> 10^6$	$> 10^6$	$> 10^6$
3O-DPA on 2O SSS Masking	$> 10^6$	$> 10^6$	$> 10^6$	$> 10^6$	$> 10^6$
3O-MIA on 2O SSS Masking	$> 10^6$	$> 10^6$	$> 10^6$	$> 10^6$	$> 10^6$

4.3 Attack Simulations

In order to confront our theoretical analyses to practical evaluation, we performed several attacks simulations. We then applied several side-channel distinguishers to leakage measurements simulated in the Hamming weight model with Gaussian noise. The leakage measurements have been simulated as samples of the random variables L_i defined according to equation (4) with $\varphi = \mathrm{HW}$ and $B_i \sim \mathcal{N}(0, \sigma^2)$ (the different B_i's are also assumed independent). For all the attacks, the sensitive variable Z was chosen to be an AES S-box output of the form $\mathsf{S}(M \oplus k^\star)$ where M represents a varying plaintext byte and $k^\star$ represents the key byte to recover.

Side-Channel Distinguishers. We applied two kind of side-channel distinguishers: higher-order DPA such as described in Sect. 4.2 and higher-order MIA. In a HO-MIA [15,7], the distinguisher is the mutual information: the guess k is tested by estimating $\mathrm{I}(\hat{\varphi}(Z(k)); \mathbf{L})$. As mutual information is a multivariate operator, this approach does not involve a combining function.

Targeted Variables. Each attack was applied against the leakages of SSS masking, and Boolean masking. The target variables are those listed in Sect. 4.1 for Z being $\mathsf{S}(X \oplus k^\star)$.

Prediction Functions. For each DPA, we choose $\hat{\varphi}$ to be the optimal prediction function :

$$\hat{\varphi} : z \mapsto \mathrm{E}\left[\mathcal{C}(\mathbf{L}) | Z = z\right]. \tag{12}$$

This leads us to select the Hamming weight function in the attacks against both SSS and Boolean masking of any order.

For the MIA attacks, we choose $\hat{\varphi}$ such that it maximizes the mutual information $\mathrm{I}(\hat{\varphi}(Z(k)); \mathbf{L})$ for $k = k^\star$ while ensuring that the mutual information is lower for $k \neq k^\star$. In our case, every HO-MIA against both SSS and Boolean masking is performed with $\hat{\varphi} = \mathrm{HW}$ since the distribution of $(\mathrm{HW}(Z \oplus m_0), \mathrm{HW}(m_0))$ (resp. $(\mathrm{HW}(Z \oplus a_0 \cdot x_0, x_0), \mathrm{HW}(Z \oplus a_0 \cdot x_1, x_1)))$ only depends on $\mathrm{HW}(Z)$. Therefore

$$\mathrm{I}\big(Z; (\mathrm{HW}(Z \oplus m_0), \mathrm{HW}(m_0))\big) = \mathrm{I}\big(\mathrm{HW}(Z); (\mathrm{HW}(Z \oplus m_0), \mathrm{HW}(m_0))\big).$$

Note that the same relation holds at every masking order.

Pdf Estimation Method. For the (HO-)MIA attacks, we use the histogram estimation method with rule of [8] for the *bin-widths* selection.

Attack Simulation Results. Each attack simulation is performed 100 times for various SNR values ($+\infty$, 1, 1/2, 1/5 and 1/10). Table 5 summarizes the number of leakage measurements required to observe a success rate of 90% in retrieving $k^\star$ for the different attacks.These results shows the security improvement provided by SSS masking with respect to boolean masking. This gain of security can be explained by the fact that an attacker does not have direct access to the mask $a_1 \cdot x_i$, hence the relation between the key and the targeted variables is much more noisy than for boolean masking.

5 Conclusion

In this paper we propose a new alternative to boolean masking to secure implementations of AES against side channel attacks using Shamir's Secret Sharing scheme to share sensitive variables. We give implementation results and conduct a security analysis that clearly show that our scheme can provide a good complexity-security trade-off compared to boolean masking. In particular, on smart card implementation, where SNR value is around 1/2, $1O$ SSS masking provides both a better security and complexity than $3O$ boolean masking. On hardware implementations where the noise can be drastically reduced, $1O$ SSS masking is to be compared to 4th order boolean masking, which increase the advantage of SSS masking. These results show that the opening to secret sharing and secure multi-party computation can provide a good alternative to boolean masking. This may be an interesting way to thwart HO-SCA. It is an open research topic to try the security and complexity of such a masking using other kinds of secret sharing scheme.

References

1. Akkar, M.-L., Giraud, C.: An implementation of DES and AES, secure against some attacks. In: Koç, Ç.K., Naccache, D., Paar, C. (eds.) CHES 2001. LNCS, vol. 2162, pp. 309–318. Springer, Heidelberg (2001)
2. Ben-Or, M., Goldwasser, S., Wigderson, A.: Completeness theorems for non-cryptographic fault-tolerant distributed computation (extended abstract). In: STOC, pp. 1–10. ACM, New York (1988)
3. Brier, É., Clavier, C., Olivier, F.: Correlation Power Analysis with a Leakage Model. In: Joye, M., Quisquater, J.-J. (eds.) CHES 2004. LNCS, vol. 3156, pp. 16–29. Springer, Heidelberg (2004)
4. Chari, S., Jutla, C.S., Rao, J.R., Rohatgi, P.: Towards Sound Approaches to Counteract Power-Analysis Attacks. In: Wiener [26], pp. 398–412
5. Clavier, C., Gaj, K. (eds.): CHES 2009. LNCS, vol. 5747. Springer, Heidelberg (2009)

6. Fumaroli, G., Martinelli, A., Prouff, E., Rivain, M.: Affine masking against higher-order side channel analysis. In: Biryukov, A., Gong, G., Stinson, D.R. (eds.) SAC 2010. LNCS, vol. 6544, pp. 262–280. Springer, Heidelberg (2011)
7. Gierlichs, B., Batina, L., Preneel, B., Verbauwhede, I.: Revisiting Higher-Order DPA Attacks: Multivariate Mutual Information Analysis. Cryptology ePrint Archive, Report 2009/228 (2009), `http://eprint.iacr.org/`
8. Gierlichs, B., Batina, L., Tuyls, P., Preneel, B.: Mutual Information Analysis. In: Oswald, E., Rohatgi, P. (eds.) CHES 2008. LNCS, vol. 5154, pp. 426–442. Springer, Heidelberg (2008)
9. Goubin, L., Patarin, J.: DES and Differential Power Analysis – The Duplication Method. In: Koç, Ç.K., Paar, C. (eds.) CHES 1999. LNCS, vol. 1717, pp. 158–172. Springer, Heidelberg (1999)
10. Ishai, Y., Sahai, A., Wagner, D.: Private Circuits: Securing Hardware against Probing Attacks. In: Boneh, D. (ed.) CRYPTO 2003. LNCS, vol. 2729, pp. 463–481. Springer, Heidelberg (2003)
11. Kocher, P., Jaffe, J., Jun, B.: Differential Power Analysis. In: Wiener [26], pp. 388–397
12. Mangard, S., Oswald, E., Popp, T.: Power Analysis Attacks – Revealing the Secrets of Smartcards. Springer, Heidelberg (2007)
13. Messerges, T.S.: Using Second-Order Power Analysis to Attack DPA Resistant Software. In: Paar, C., Koç, Ç.K. (eds.) CHES 2000. LNCS, vol. 1965, pp. 238–251. Springer, Heidelberg (2000)
14. Pointcheval, D. (ed.): CT-RSA 2006. LNCS, vol. 3860. Springer, Heidelberg (2006)
15. Prouff, E., Rivain, M.: Theoretical and Practical Aspects of Mutual Information Based Side Channel Analysis. In: Abdalla, M., Pointcheval, D., Fouque, P.-A., Vergnaud, D. (eds.) ACNS 2009. LNCS, vol. 5536, pp. 499–518. Springer, Heidelberg (2009)
16. Rivain, M., Prouff, E.: Provably Secure Higher-Order Masking of AES. In: Mangard, S., Standaert, F.-X. (eds.) CHES 2010. LNCS, vol. 6225, pp. 413–427. Springer, Heidelberg (2010)
17. Prouff, E., Rivain, M.: Theoretical and Practical Aspects of Mutual Information Based Side Channel Analysis (Extended Version). Int. Journal of Applied Cryptography, IJACT 2(2) (2010)
18. Prouff, E., Rivain, M., Bévan, R.: Statistical Analysis of Second Order Differential Power Analysis. IEEE Trans. Comput. 58(6), 799–811 (2009)
19. Prouff, E., Roche, T.: Higher-order glitches free implementation of the aes using secure multi-party computation protocols. In: Preneel, B., Takagi, T. (eds.) CHES 2011. LNCS, vol. 6917, pp. 79–94. Springer, Heidelberg (2011)
20. Rivain, M., Prouff, E., Doget, J.: Higher-Order Masking and Shuffling for Software Implementations of Block Ciphers. In: Clavier, C., Gaj, K. (eds.) CHES 2009. LNCS, vol. 5747, pp. 171–188. Springer, Heidelberg (2009)
21. Schramm, K., Paar, C.: Higher Order Masking of the AES. In: Pointcheval, D. (ed.) CT-RSA 2006. LNCS, vol. 3860, pp. 208–225. Springer, Heidelberg (2006)
22. Shamir, A.: How to Share a Secret. Communications of the ACM 22(11), 612–613 (1979)
23. Standaert, F.-X., Malkin, T.G., Yung, M.: A Unified Framework for the Analysis of Side-Channel Key Recovery Attacks. In: Joux, A. (ed.) EUROCRYPT 2009. LNCS, vol. 5479, pp. 443–461. Springer, Heidelberg (2009)

24. Standaert, F.-X., Veyrat-Charvillon, N., Oswald, E., Gierlichs, B., Medwed, M., Kasper, M., Mangard, S.: The world is not enough: Another look on second-order dpa. Cryptology ePrint Archive, Report 2010/180 (2010), http://eprint.iacr.org/
25. von Willich, M.: A technique with an information-theoretic basis for protecting secret data from differential power attacks. In: Honary, B. (ed.) Cryptography and Coding 2001. LNCS, vol. 2260, pp. 44–62. Springer, Heidelberg (2001)
26. Wiener, M. (ed.): CRYPTO 1999. LNCS, vol. 1666. Springer, Heidelberg (1999)

A Fast and Provably Secure Higher-Order Masking of AES S-Box*

HeeSeok Kim, Seokhie Hong, and Jongin Lim

Center for Information Security Technologies, Korea University, Korea
{80khs,shhong,jilim}@korea.ac.kr
http://cist.korea.ac.kr

Abstract. This paper proposes an efficient and secure higher-order masking algorithm for AES S-box that consumes the most computation time of the higher-order masked AES. During the past few years, much of the research has focused on finding higher-order masking schemes for this AES S-box, but these are still slow for embedded processors use. Our proposed higher-order masking of AES S-box is constructed based on the inversion operation over the composite field. We replace the subfield operations over the composite field into the table lookup operation, but these precomputation tables do not require much ROM space because these are the operations over $GF(2^4)$. In the implementation results, we show that the higher-order masking scheme using our masked S-box is about 2.54 (second-order masking) and 3.03 (third-order masking) times faster than the fastest method among the existing higher-order masking schemes of AES.

Keywords: AES, side channel attack, higher-order masking, higher-order DPA, differential power analysis.

1 Introduction

Since Kocher introduced the concept of differential power analysis (DPA) [13], the security of block ciphers has received considerable attention, and it is now obvious that the unprotected implementations of block ciphers in embedded processors can be broken by DPA. During the past few years, much of the research on DPA attacks has focused on finding secure countermeasures. Among these countermeasures, a masking method based on algorithmic techniques is known to be inexpensive and secure against a first-order DPA (FODPA) [3,5,9,11,15,16].

Recently, the important effort has been carried out to find a masking method that is secure against the higher-order DPA (HODPA) [14,22] as well as FODPA [22,17,18]. These masking schemes are called the higher-order masking schemes. Also, the higher-order masking scheme to counteract d-th order DPA [22] is

* "This research was supported by the MKE(The Ministry of Knowledge Economy), Korea, under the "itrc" support program supervised by the NIPA(National IT Industry Promotion Agency)" (NIPA-2011-C1090-1001-0004).

B. Preneel and T. Takagi (Eds.): CHES 2011, LNCS 6917, pp. 95–107, 2011.

called the d-th order masking scheme[1]. In this d-th order masking scheme, every intermediate value I of an original cipher is randomly split into $(d+1)$-tuple (I_0, I_1, ..., I_d). Here, for any intermediate value I, there exists a group operation $\perp$ such that $\perp(I_0, I_1, ..., I_d)$ equals I. This randomly split $(d+1)$-tuple can block that the combination of d or less elements in this tuple is dependent on the intermediate value I. Thus, security against d-th order DPA can be provided.

In the higher-order masking scheme of standard block cipher AES [1], the most important part is the S-box operation, which is the only non-linear operation of AES. Actually, most of the cost for higher-order masked AES is required by this non-linear part. Thus, to construct the higher-order masking scheme of AES in all previous works, the most important consideration has been to mask this S-box operation. To mask this operation, the initial works of [22] and [17] carry out table re-computation before all S-box operation. However, these methods require much computation time. M. Rivain and E. Prouff introduced a higher-order masked S-box operation based on the exponentiation operation to solve this problem [18]. This method can considerably reduce computation time compared with the existing methods. However, this scheme is still about 60 (second-order masking) and 130 (third-order masking) times slower than the straightforward AES.

In this paper, we propose a new higher-order masking of AES S-box based on the inversion operation over the composite field [20,21]. This method uses the precomputation tables for subfield operations such as the multiplication, square, and scalar multiplication over $GF(2^4)$. These tables can considerably reduce the time required. Also, because these tables are used for the operations over $GF(2^4)$, our method does not require much ROM space. The security of this new algorithm can be easily proved via the proofs in [18]. In the implementation results, our method is about 2.54 (second-order masking) and 3.03 (third-order masking) times faster than the method in [18]. Also, to use the higher-order masking scheme in embedded processors, we show the implementation results that apply the higher-order masking scheme to the first two and the last two rounds only, and the first-order masking to the other rounds and key-schedule. This is because HODPA generally attacks the first and last few rounds. Implementation results for this reduced masking are just 8.6 (second-order masking) and 13.8 (third-order masking) times slower than the straightforward AES. These numerical values mean that the reduced masking using our higher-order masked AES S-box can be sufficiently used in embedded processors.

The remainder of this paper is organized as follows. Section 2 describes the higher-order masking of AES and the inversion operation over the composite field. In Section 3, we introduce the new higher-order masking of AES S-box. Section 4 simply demonstrates the security of our method based on the proofs in [18] and Section 5 shows its efficiency. Finally, in Section 6, we offer the conclusion.

---

[1] Since the method of [22] has been known insecure for $d \geq 3$ [7], the method of [18] is currently the only higher-order masking scheme for $d \geq 3$.

2 Preliminaries

2.1 Advanced Encryption Standard (AES)

The Advanced Encryption Standard (AES), also known as Rijndael, is a block cipher adopted as an encryption standard by the US government [1]. This block cipher is composed of an SPN structure [4,6]. For an N-bit SPN-type block cipher of r rounds, each round consists of three layers. These layers are the key mixing layer, substitution layer, and linear transformation layer. In the key mixing layer, the round input is bitwise exclusive-ORed with the subkey for each round. In the substitution layer, the value resulting from the key mixing layer is partitioned into N/n blocks of n bits and each block of n bits then outputs other n bits through a non-linear bijective mapping $\pi : \mathbb{F}_{2^n} \rightarrow \mathbb{F}_{2^n}$. In the case of AES, an S-box fulfills the role of this bijective mapping π. AES S-box is defined by a multiplicative inverse $b = a^{-1}$ in $GF(2^8)$ (except if $a = 0$ then $b = 0$) and an affine transformation as in the following equations:

$$\begin{array}{rcccc} S & : & GF(2^8) & \rightarrow & GF(2^8) \\ & & x & \rightarrow & Mx^{254} \oplus v \end{array}$$

where the value of x^{254} is regarded as a $GF(2)$-vector of dimension 8, M is an $8{\times}8$ $GF(2)$-matrix, and v is an $8{\times}1$ $GF(2)$-vector. The resulting value of the substitution layer becomes the N bit input $(i_0, i_1, , i_{N-1})$ of the linear transformation layer. The linear transformation layer consists of multiplication by the $N{\times}N$ matrix. That is, the resulting value of this layer is $O = LI$, where L is an $N{\times}N$ matrix and I is $(i_0, i_1, , i_{N-1})^T$. In AES, ShiftRows and MixColumns play this role. The linear transformation layer is omitted from the last round since it is easily shown that its inclusion adds no cryptographic strength.

2.2 Higher-Order Masking of AES

A d-th order masking method e' for an encryption algorithm $c \leftarrow e(m, k)$ is defined as follows, where m, k and c are plaintext, key and ciphertext, respectively [18]:

$$(c_0, c_1, ..., c_d) \leftarrow e'((m_0, m_1, ..., m_d), (k_0, k_1, ..., k_d))$$

In the equation above, d-th order masking method e' must satisfy the equations of $c = \perp_{i=0}^{d} c_i$, $m = \perp_{i=0}^{d} m_i$ and $k = \perp_{i=0}^{d} k_i$, where $\perp$ is the specific group operation. In this paper, we consider that this group operation is the exclusive-or (XOR, $\oplus$).

To guarantee security against d-th order DPA which exploits the leakages related to d intermediate values [22], the intermediate value I of the encryption algorithm e must also be replaced into (I_0, I_1, ..., I_d). Here, the sum of d or less intermediate values $\bigoplus_{i \in S} I_i$ is independent of the sensitive data value (the intermediate value of the original cipher) I where S is a subset of $\{0, 1, ..., d\}$ and $1 \leq size(S) \leq d$.

M. Rivain and E. Prouff introduced the d-th order masking scheme for AES satisfying the conditions above. In this scheme, the higher-order masking method can be easily applied to every operation, except S-box, because these operations are linear operations. Namely, the linear operation $O \leftarrow L(I)$ can be replaced into $(O_0, O_1, ..., O_d) \leftarrow L'((I_0, I_1, ..., I_d))$ where $O_i = L(I_i)$, because the operation L' satisfies the equation of $\bigoplus_{i=0}^{d} O_i = L(\bigoplus_{i=0}^{d} I_i)$.

However, it is not easy to apply the higher-order masking to the S-box operation. Actually, the higher-order masking scheme spends most of the time for computing this non-linear operation. In the initial works on the higher-order masking method [22,17], it is general to carry out table re-computation before S-box operation, but this operation consumes a lot of time.

To reduce the time required, M. Rivain and E. Prouff introduced the d-th order secure **SecSbox** operation based on the exponentiation operation [18]. They found the following addition chain that can minimize not the number of squares but the number of multiplications because the square is a linear operation.

$$x \xrightarrow{S} x^2 \xrightarrow{M} x^3 \xrightarrow{2S} x^{12} \xrightarrow{M} x^{15} \xrightarrow{4S} x^{240} \xrightarrow{M} x^{252} \xrightarrow{M} x^{254}$$

In the chain above, S, M, $2S$ and $4S$ mean the square, multiplication, two squares and four squares, respectively. Here, the square, two squares and four squares are the linear operations. Thus, these operations are easily implemented using the look-up tables (LUTs). However, the other four multiplications must be carefully constructed because these are non-linear operations. They designed the d-th order secure multiplication algorithm **SecMult** using the Ishai-Sahai-Wagner (ISW) scheme [10]. This algorithm is described in Algorithm 1.

Algorithm 1. SecMult function [18]

Input: two $(d+1)$-tuples $(a_0, a_1, ..., a_d)$, $(b_0, b_1, ..., b_d)$ where $\bigoplus_{i=0}^{d} a_i = a$, $\bigoplus_{i=0}^{d} b_i = b$
Output: $(d+1)$-tuple $(c_0, c_1, ..., c_d)$ satisfying $\bigoplus_{i=0}^{d} c_i = ab$

1. For $i = 0$ to d do
   (a) For $j = i + 1$ to d do
      i. $r_{i,j} \leftarrow \mathbf{rand}(8)$
      ii. $r_{j,i} \leftarrow (r_{i,j} \oplus a_i b_j) \oplus a_j b_i$
2. For $i = 0$ to d do
   (a) $c_i \leftarrow a_i b_i$
   (b) For $j = 0$ to d, $j \neq i$ do, $c_i \leftarrow c_i \oplus r_{i,j}$

2.3 The Inversion Operation over a Composite Field

In order to reduce the cost of AES S-box, inversion methods over a composite field have been proposed [20,21]. These methods transform an element over $GF(2^8)$ into an element over the composite field having low inversion cost by the isomorphism function δ, and the inversion operation is actually carried out over this composite field. Then, the inversion operation over $GF(2^8)$ is completed by carrying out the inverse mapping δ^{-1} into the element over $GF(2^8)$.

In [21], the composite field is built by repeating degree-2 extensions with the following irreducible polynomials:

$$\begin{array}{rcl} GF(2^2) & : & P_0(x) = x^2 + x + 1, \text{ where } P_0(\alpha) = 0, \\ GF((2^2)^2) & : & P_1(x) = x^2 + x + \alpha, \text{ where } P_1(\beta) = 0, \\ GF(((2^2)^2)^2) & : & P_2(x) = x^2 + x + \lambda, \text{ where } \lambda = (\alpha + 1)\beta, \; P_2(\gamma) = 0. \end{array}$$

Two isomorphism functions δ and δ^{-1} according to the above irreducible polynomials are as follows:

$$\begin{array}{rccc} \delta : & GF(2^8) & \rightarrow & GF(((2^2)^2)^2) \\ \delta^{-1} : & GF(((2^2)^2)^2) & \rightarrow & GF(2^8) \end{array}$$

$$\delta : \begin{pmatrix} 1\,1\,0\,0\,0\,0\,1\,0 \\ 0\,1\,0\,0\,1\,0\,1\,0 \\ 0\,1\,1\,1\,1\,0\,0\,1 \\ 0\,1\,1\,0\,0\,0\,1\,1 \\ 0\,1\,1\,1\,0\,1\,0\,1 \\ 0\,0\,1\,1\,0\,1\,0\,1 \\ 0\,1\,1\,1\,1\,0\,1\,1 \\ 0\,0\,0\,0\,0\,1\,0\,1 \end{pmatrix} \qquad \delta^{-1} : \begin{pmatrix} 1\,0\,1\,0\,1\,1\,1\,0 \\ 0\,0\,0\,0\,1\,1\,0\,0 \\ 0\,1\,1\,1\,1\,0\,0\,1 \\ 0\,1\,1\,1\,1\,1\,0\,0 \\ 0\,1\,1\,0\,1\,1\,1\,0 \\ 0\,1\,0\,0\,0\,1\,1\,0 \\ 0\,0\,1\,0\,0\,0\,1\,0 \\ 0\,1\,0\,0\,0\,1\,1\,1 \end{pmatrix}$$

The inversion operation of an input value $A = a_h\gamma + a_l \in GF(((2^2)^2)^2)$, where a_h and a_l are elements in $GF((2^2)^2)$, is performed as follows in this composite field.

First, $A^{-1} \in GF(((2^2)^2)^2)$ is computed by $C^{-1}A^{16}$ ($C = A^{17} \in GF((2^2)^2)$). This computation method unavoidably requires the operations of A^{16} and A^{17}, but A^{16} can be computed simply with only 4 bitwise XOR operations of $a_h\gamma + (a_h + a_l)$. Furthermore, A^{17} is computed simply by $\lambda a_h^2 + (a_h + a_l)a_l$ due to $\gamma^2 + \gamma = \lambda$. After computing the inverse of A^{17} over $GF((2^2)^2)$, the computation of $A^{-1} = C^{-1}A^{16}$ can be completed. Figure 1 represents the AES S-box operation including the inversion operation over the composite field $GF(((2^2)^2)^2)$ where A_f is an affine transformation. For additional operations over both fields $GF((2^2)^2)$ and $GF(2^2)$ refer to [21].

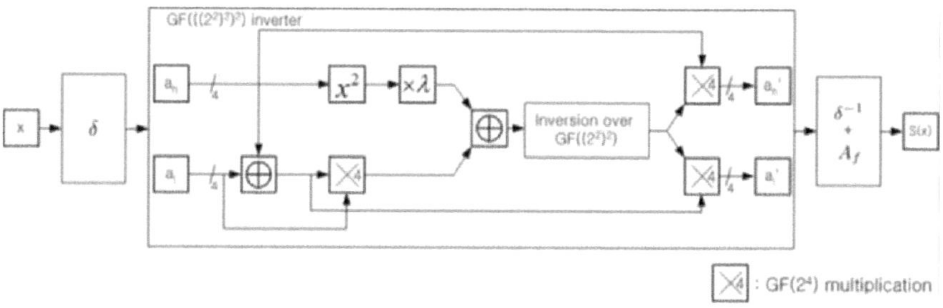

Fig. 1. S-box operation of AES

In Fig. 1, an equation for computing the output value $S(x)$ of x is performed as follows:

Step 1. $a_h\gamma + a_l = \delta(x)$ where a_h and a_l are elements in $GF((2^2)^2)$
Step 2. $d = \lambda a_h^2 + a_l(a_h + a_l) \in GF((2^2)^2)$
Step 3. $d' = d^{-1} \in GF((2^2)^2)$
Step 4. $a_h' = d'a_h \in GF((2^2)^2)$
Step 5. $a_l' = d'(a_h + a_l) \in GF((2^2)^2)$
Step 6. $S(x) = A_f(\delta^{-1}(a_h'\gamma + a_l')) \in GF(((2^2)^2)^2)$

3 A Fast and Provably Secure Higher-Order Masking of AES S-Box

In this section, we propose a fast and provably secure higher-order masking of AES S-box. As aforementioned, the higher-order masking scheme of AES spends the most time computing the masked S-box operation. The main purpose of this paper is to reduce running time of the higher-order masking algorithm. Thus, our masked S-box uses the six precomputation tables. Most of the elements of these tables are the 4-bit data, but we allocate one byte for each element to simplify accessing the table elements. These precomputation tables are as follows with the notations defined in Section 2.3:

1. Squaring table $T1$ (The requirement for 16 bytes of ROM)
   - **Input :** $X \in GF(2^4)$
   - **Output :** $T1[X] = X^2 \in GF(2^4)$
2. Two squaring table $T2$ (The requirement for 16 bytes of ROM)
   - **Input :** $X \in GF(2^4)$
   - **Output :** $T2[X] = X^4 \in GF(2^4)$
3. Squaring-scalar multiplication table $T3$ (The requirement for 16 bytes of ROM)
   - **Input :** $X \in GF(2^4)$
   - **Output :** $T3[X] = \lambda X^2 \in GF(2^4)$
4. Multiplication table $T4$ (The requirement for 256 bytes of ROM)
   - **Input :** $X, Y \in GF(2^4)$
   - **Output :** $T4[X][Y] = XY \in GF(2^4)$
5. Isomorphism table $T5$ (The requirement for 256 bytes of ROM)
   - **Input :** $X \in GF(2^8)$
   - **Output :** $T5[X] = \delta(X) \in GF(((2^2)^2)^2)$
6. Inverse isomorphism-Affine transformation table $T6$ (The requirement for 256 bytes of ROM)
   - **Input :** $X \in GF(((2^2)^2)^2))$
   - **Output :** $T6[X] = A_f(\delta^{-1}(X)) \in GF(2^8)$

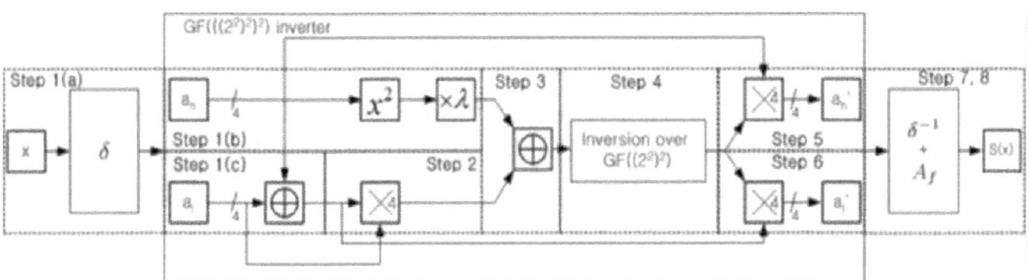

Fig. 2. The eight steps of the proposed masked S-box

As aforementioned, the d-th order masking scheme must compute every intermediate value I of the original encryption algorithm e with the form of $(I_0, I_1, ..., I_d)$ where $\bigoplus_{i=0}^{d} I_i = I$. We design a new higher-order masked S-box satisfying this condition in each step over the composite field. We first classify the S-box operation of Fig. 1 into the nine steps of Fig. 2.

In Fig. 2, the operations of Step 1(a), 1(b), 7 and 8 are the linear operations or affine transformation. Here, the linear operation $O \leftarrow L(I)$ can be easily replaced into $(O_0, O_1, ..., O_d) \leftarrow L'((I_0, I_1, ..., I_d))$, where $O_i = L(I_i)$, because the operation L' satisfies the equation of $\bigoplus_{i=0}^{d} O_i = L(\bigoplus_{i=0}^{d} I_i)$. Affine transformation $O \leftarrow A_f(I)$ can also be replaced into $(O_0, O_1, ..., O_d) \leftarrow A'_f((I_0, I_1, ..., I_d))$, where $O_0 = A_f(I_0) \oplus (0x63 \times (d \ mod \ 2))$ and $O_i = A_f(I_i)(i \neq 0)$, because the operation A'_f satisfies the equation of $\bigoplus_{i=0}^{d} O_i = A_f(\bigoplus_{i=0}^{d} I_i)$. We show hereafter some methods to apply the d-th order masking to the remaining non-linear operations.

- **Masking XOR operation:** The masked XOR operation can be easily constructed. This operation outputs $(x_0 \oplus y_0, x_1 \oplus y_1, ..., x_d \oplus y_d)$ from two input $(d+1)$-tuples $(x_0, x_1, ..., x_d)$, $(y_0, y_1, ..., y_d)$. Here, two input $(d+1)$-tuples must be independent of each other as mentioned in [18].

- **Masking $GF(2^4)$ multiplication:** We construct the masked $GF(2^4)$ multiplication using Algorithm 1 of [18]. This algorithm is described in Algorithm 2. The only difference between two algorithms is whether or not the multiplication table is used. Our masked S-box computes the inversion by using the subfield operation over $GF(2^4)$. Thus, most operations, including $GF(2^4)$ multiplication, can be computed by using the precomputation tables. These require ROM space, but the size required is very small.

- **Masking $GF(2^4)$ inversion:** The masking method for $GF(2^4)$ inversion can be designed variously. One way is to use the composite field operation over $GF((2^2)^2)$ similarly to the masked operation over $GF(((2^2)^2)^2)$. However, this method requires as many table lookup operations as that over $GF(((2^2)^2)^2)$. Therefore, we use the operation of x^{14} over $GF(2^4)$. The addition chain of x^{14} to minimize the number of multiplications can be constructed as follows:

$$x \xrightarrow{S} x^2 \xrightarrow{M} x^3 \xrightarrow{2S} x^{12} \xrightarrow{M} x^{14}$$

The masked inversion algorithm over $GF(2^4)$ using this addition chain is shown in Algorithm 3. In Algorithm 3, for **RefreshMasks** algorithm to eliminate the dependence between the two input tuples of **SecMult4** function refer to [18].

Algorithm 2. $GF(2^4)$ **SecMult4** function using the multiplication table $T4$

Input: two $(d+1)$-tuples $(a_0, a_1, ..., a_d)$, $(b_0, b_1, ..., b_d)$ where $\bigoplus_{i=0}^{d} a_i = a$, $\bigoplus_{i=0}^{d} b_i = b$
Output: $(d+1)$-tuple $(c_0, c_1, ..., c_d)$ satisfying $\bigoplus_{i=0}^{d} c_i = ab \in GF(2^4)$

1. For $i = 0$ to d do
   (a) For $j = i + 1$ to d do
      i. $r_{i,j} \leftarrow$ **rand**(4)
      ii. $r_{j,i} \leftarrow (r_{i,j} \oplus T4[a_i][b_j]) \oplus T4[a_j][b_i]$
2. For $i = 0$ to d do
   (a) $c_i \leftarrow T4[a_i][b_i]$
   (b) For $j = 0$ to d, $j \neq i$ do, $c_i \leftarrow c_i \oplus r_{i,j}$

Algorithm 3. $GF(2^4)$ **SecInv** function

Input: $(d+1)$-tuple $(x_0, x_1, ..., x_d)$ satisfying $\bigoplus_{i=0}^{d} x_i = x \in GF(2^4)$
Output: $(d+1)$-tuple $(y_0, y_1, ..., y_d)$ satisfying $\bigoplus_{i=0}^{d} y_i = x^{-1} = x^{14} \in GF(2^4)$

1. For $i = 0$ to d do
   (a) $w_i = T1[x_i]$
2. **RefreshMasks**$((w_0, w_1, ..., w_d))$
3. $(z_0, z_1, ..., z_d)$=**SecMult4**$((w_0, w_1, ..., w_d), (x_0, x_1, ..., x_d))$
4. For $i = 0$ to d do
   (a) $z_i = T2[z_i]$
5. $(y_0, y_1, ..., y_d)$=**SecMult4**$((z_0, z_1, ..., z_d), (w_0, w_1, ..., w_d))$

Algorithm 4 presents the entire operation of the proposed d-th order masking for AES S-box. The meaning of the operations carried out in each step is described in Fig. 2.

4 Security Analysis

The security of the proposed algorithm can be easily proved by the proofs in [18]. It is straightforward to prove the security for all operations, except the **SecMult4** algorithm because each element of the input tuple is independently operated in these operations. Also, the security of **SecMult4** algorithm can be proved by Theorem 1 of [18]. The remaining consideration is the independence between two input $(d+1)$-tuples of the **SecMult4** algorithm. In Algorithm 4, two input tuples of **SecMult4** algorithm are independent of each other. However, in Step 3 of Algorithm 3, the input tuples of **SecMult4** function (x, x^2) have the dependency. However, **RefreshMasks** in Step 2 can eliminate this dependency as mentioned in Section 3. Thus, the proposed algorithm provides security against the d-th order DPA, that is, it can guarantee that every combination of d or less intermediate values is independent of any sensitive data value.

Algorithm 4. d-th order masking of AES S-box

Input: $(d+1)$-tuple $(x_0, x_1, ..., x_d)$ satisfying $\bigoplus_{i=0}^{d} x_i = x \in GF(2^8)$
Output: $(d+1)$-tuple $(y_0, y_1, ..., y_d)$ satisfying $\bigoplus_{i=0}^{d} y_i = Sbox(x) \in GF(2^8)$

1. For $i = 0$ to d do
   (a) $(H_i||L_i) = T5[x_i]$ /* H_i, $L_i \in GF(2^4)$ */
   (b) $w_i = T3[H_i]$
   (c) $t_i = H_i \oplus L_i$
2. $(L_0, L_1, ..., L_d)$=**SecMult4**$((t_0, t_1, ..., t_d), (L_0, L_1, ..., L_d))$
3. For $i = 0$ to d do
   (a) $w_i = w_i \oplus L_i$
4. $(w_0, w_1, ..., w_d)$=**SecInv**$((w_0, w_1, ..., w_d))$
5. $(H_0, H_1, ..., H_d)$=**SecMult4**$((w_0, w_1, ..., w_d), (H_0, H_1, ..., H_d))$
6. $(L_0, L_1, ..., L_d)$=**SecMult4**$((w_0, w_1, ..., w_d), (t_0, t_1, ..., t_d))$
7. For $i = 0$ to d do
   (a) $y_i = T6[H_i||L_i]$
8. If d is odd, $y_0 = y_0 \oplus 0x63$
9. Return $(y_0, y_1, ..., y_d)$.

5 Performance Analysis and Implementation Results

In our d-th order masked S-box, **SecMult4** function requires $(d+1)^2$ table lookup operations, $2d(d+1)$ XOR operations, and the generation of $2d(d+1)$ random bits. Considering 5 **SecMult4** function calls and other minor operations (**RefreshMasks**, table lookup operations, and 4-bit shift operations[2]), our masked S-box requires totally $(5d^2 + 13d + 8)$ table lookup operations, $(10d^2 + 16d + 5)$ XOR operations, $(\frac{10d^2+14d}{8} + 2(d+1))$ 4-bit shift operations, $(\frac{10d^2+14d}{8} + 2(d+1))$ bitwise AND operations and the generation of $(10d^2 + 14d)$ random bits[3].

[2] 4-bit shift operation may require 4 instruction calls unless the single instruction carrying out 4-bit shift is supported. However, some microcontrollers like 8051 and AVR family support a single SWAP operation, which swaps high and low nibbles in a register.

[3] To get the random nibbles from $(10d^2 + 14d)$ random bits, we split 1 random byte into two nibbles. This method requires one 4-bit shift operation and one bitwise AND operation. Therefore, the generation of $(10d^2 + 14d)$ random bits involves $(\frac{10d^2+14d}{8})$ 4-bit shift and bitwise AND operations.

Table 1. Comparison of two d-th order masked S-box schemes in terms of the total number of operations

	Ours	[18]
Table Lookup	$5d^2 + 13d + 8$	$12d^2 + 31d + 19$
XOR	$10d^2 + 16d + 5$	$8d^2 + 12d$
Random Bits	$10d^2 + 14d$	$16d^2 + 32d$
etc	4-bit logical shift : $\frac{5}{4}d^2 + \frac{15}{4}d + 2$, 8-bit bitwise AND : $\frac{5}{4}d^2 + \frac{15}{4}d + 2$	8-bit Addition : $8(d+1)^2$, 8-bit logical AND : $4(d+1)^2$

On the other hand, the d-th order masked S-box in [18] involves 4 **SecMult** function calls, i.e., $4(d+1)^2$ multiplications over $GF(2^8)$. The multiplications over $GF(2^8)$ can be efficiently implemented with $log/alog$ tables (see Appendix A). Here, we remove the conditional branching operation because this operation leaks some information that can be exploited by simple power analysis (SPA) [13]. Also, we remove the reduction operation modulo 255 to improve the computation speed. 4 **SecMult** function calls require $12(d+1)^2$ table lookup operations because one multiplication over $GF(2^8)$ involves 3 table lookup operations. Table 1 compares two d-th order masked S-box schemes in terms of the total number of operations.

To compare the performance of our masked S-box with the existing countermeasures in embedded processors, we use Algorithms 6 and 7 in [19]. Here, we replace the masked S-box operation of these algorithms into our algorithm.

We implement AES-128 in C-language for ATmega128 8-bit architecture [2]. First, the straightforward AES requires 11,170 clock cycles. We implement the first-order masking of AES using the method in [9], but we do not apply the dummy operation and the shuffling method, which provide partial security against the second-order DPA. This requires 19,525 clock cycles.

In the implementation of the second-order masking methods, we compare our method with the methods in [17] and [18]; [18] is the most recent work on the higher-order masking of AES. Our method requires slightly more ROM size than the method in [18], but is 2.54 times faster and also 4.51 times faster than the method in [17].

However, our method is still 23 times slower than the straightforward AES. Thus, we consider the additional case for applying the second-order masking to only the first two and the last two rounds. This is because the second-order DPA generally attacks the first and last few rounds. Also, we apply the first-order masking to the key-schedule and the rest of the rounds (3∼8 rounds). After finishing the key-schedule operation, we change the first two and the last two round keys into the form of $(d+1)$-tuple by using d random numbers. To provide security against the analysis such as [8] and [12], we apply the first-order masking to 3∼8 rounds. The implementation result is just 8.6 times slower than the straightforward AES. These numerical values mean that this algorithm can be used practically in the embedded processors.

Method	Cycles (x 10^3 cc)			Table Size (Bytes)	
	KeyExpand	Encryption	Total	RAM	ROM
Original AES					
No Masking (Straightforward AES)	2.2	9.0	11.2	0	256
First Order Masking					
ACNS'06 [9] (No dummy, No Shuffling)	4.6	14.9	19.5	256	256
Second Order Masking					
FSE'08 [17] (Complete second-order masking)	247.4	950.0	1197.4	256	256
CHES'10 [18] (Complete second-order masking)	144.1	531.2	675.4	0	768
Ours (Complete second-order masking)	66.2	199.3	265.5	0	816
Ours (KeyExpand (first) Enc. (1,2,9,10:second, 3~8:first))	5.2	90.6	95.8	256	1062
Third Order Masking					
CHES'10 [18] (Complete third-order masking)	293.4	1102.9	1396.3	0	768
Ours (Complete third-order masking)	114.6	346.8	461.3	0	816
Ours (KeyExpand (first) Enc. (1,2,9,10:third, 3~8:first))	5.5	149.6	155.1	256	1062

We also implement the third-order masking, and compare it with the method of [18]. Here, our method is 3.03 times faster than the existing countermeasure and 41.03 times slower than the straightforward AES. Also, the reduced masking is just 13.8 times slower than the straightforward AES.

6 Conclusion

In this paper, we proposed a new higher-order masking method for AES S-box. Our method could considerably reduce the computation time of the higher-order masked AES. Our method was 2.54 times faster than the most recent method of the second-order masking, and it was 3.03 times faster than that of the third-order masking. Also, in order to use the second-order masking algorithm in embedded processors, we only applied the second (third) order masking to the first two and the last two rounds in the encryption of AES. The results for these implementations were just 8.6 (second) and 13.8 (third) times slower than the straightforward AES. These numerical values mean that our higher-order masked S-box can achieve practical use of the higher-order masked AES in embedded processors.

References

1. NIST, FIPS 197: Advanced Encryption Standard (2001)
2. Atmel Corporation: Datasheet: ATmega128(L), `http://www.atmel.com/products/avr/`
3. Akkar, M.-L., Giraud, C.: An Implementation of DES and AES, Secure against Some Attacks. In: Koç, Ç.K., Naccache, D., Paar, C. (eds.) CHES 2001. LNCS, vol. 2162, pp. 309–318. Springer, Heidelberg (2001)
4. Adams, C., Tavares, S.: The Structured Design of Cryptographically Good SBoxes. Journal of Cryptology 3(1), 27–42 (1990)
5. Blömer, J., Guajardo, J., Krummel, V.: Provably Secure Masking of AES. In: Handschuh, H., Hasan, M.A. (eds.) SAC 2004. LNCS, vol. 3357, pp. 69–83. Springer, Heidelberg (2004)
6. O'Connor, L.: On the Distribution of Characteristics in Bijective Mappings. In: Helleseth, T. (ed.) EUROCRYPT 1993. LNCS, vol. 765, pp. 360–370. Springer, Heidelberg (1994)
7. Coron, J.-S., Prouff, E., Rivain, M.: Side Channel Cryptanalysis of a Higher Order Masking Scheme. In: Paillier, P., Verbauwhede, I. (eds.) CHES 2007. LNCS, vol. 4727, pp. 28–44. Springer, Heidelberg (2007)
8. Handschuh, H., Preneel, B.: Blind Differential Cryptanalysis for Enhanced Power Attacks. In: Biham, E., Youssef, A.M. (eds.) SAC 2006. LNCS, vol. 4356, pp. 163–173. Springer, Heidelberg (2007)
9. Herbst, C., Oswald, E., Mangard, S.: An AES Smart Card Implementation Resistant to Power Analysis Attacks. In: Zhou, J., Yung, M., Bao, F. (eds.) ACNS 2006. LNCS, vol. 3989, pp. 239–252. Springer, Heidelberg (2006)
10. Ishai, Y., Sahai, A., Wagner, D.: Private Circuits: Securing Hardware against Probing Attacks. In: Boneh, D. (ed.) CRYPTO 2003. LNCS, vol. 2729, pp. 463–481. Springer, Heidelberg (2003)
11. Kim, H., Kim, T., Han, D., Hong, S.: Efficient Masking Methods Appropriate for the Block Ciphers ARIA and AES. ETRI Journal 32(3), 370–379 (2010)
12. Kim, J., Hong, S., Han, D., Lee, S.: Improved Side-Channel Attack on DES with the First Four Rounds Masked. ETRI Journal 31(5), 625–627 (2009)
13. Kocher, P.C., Jaffe, J., Jun, B.: Differential Power Analysis. In: Wiener, M. (ed.) CRYPTO 1999. LNCS, vol. 1666, pp. 388–397. Springer, Heidelberg (1999)
14. Messerges, T.S.: Using Second-Order Power Analysis to Attack DPA Resistant Software. In: Paar, C., Koç, Ç.K. (eds.) CHES 2000. LNCS, vol. 1965, pp. 238–251. Springer, Heidelberg (2000)
15. Oswald, E., Mangard, S., Pramstaller, N., Rijmen, V.: A Side-Channel Analysis Resistant Description of the AES S-Box. In: Gilbert, H., Handschuh, H. (eds.) FSE 2005. LNCS, vol. 3557, pp. 413–423. Springer, Heidelberg (2005)
16. Oswald, E., Schramm, K.: An Efficient Masking Scheme for AES Software Implementations. In: Song, J.-S., Kwon, T., Yung, M. (eds.) WISA 2005. LNCS, vol. 3786, pp. 292–305. Springer, Heidelberg (2006)
17. Rivain, M., Dottax, E., Prouff, E.: Block Ciphers Implementations Provably Secure Against Second Order Side Channel Analysis. In: Nyberg, K. (ed.) FSE 2008. LNCS, vol. 5086, pp. 127–143. Springer, Heidelberg (2008)
18. Rivain, M., Prouff, E.: Provably Secure Higher-Order Masking of AES. In: Mangard, S., Standaert, F.-X. (eds.) CHES 2010. LNCS, vol. 6225, pp. 413–427. Springer, Heidelberg (2010)

19. Rivain, M., Prouff, E.: Provably Secure Higher-Order Masking of AES, Cryptology ePrint Archive (2010), http://eprint.iacr.org/
20. Rudra, A., Dubey, P.K., Jutla, C.S., Kumar, V., Rao, J.R., Rohatgi, P.: Efficient Rijndael Encryption Implementation with Composite Field Arithmetic. In: Koç, Ç.K., Naccache, D., Paar, C. (eds.) CHES 2001. LNCS, vol. 2162, pp. 171–188. Springer, Heidelberg (2001)
21. Satoh, A., Morioka, S., Takano, K., Munetoh, S.: A Compact Rijndael Hardware Architecture with S-Box Optimization. In: Boyd, C. (ed.) ASIACRYPT 2001. LNCS, vol. 2248, pp. 239–254. Springer, Heidelberg (2001)
22. Schramm, K., Paar, C.: Higher Order Masking of the AES. In: Pointcheval, D. (ed.) CT-RSA 2006. LNCS, vol. 3860, pp. 208–225. Springer, Heidelberg (2006)

A Multiplication in GF(2^8) without SPA Leakage

In [18], the higher-order masked S-box computes AES multiplications over $GF(2^8)$. The most efficient way for this operation is to use *log* and *alog* tables where $log[\alpha^i] = i$ and $alog[i] = \alpha^i$ for a generator α of $GF(256)^*$ and $0 \leqslant i < 255$. Multiplication using these two tables is computed according to the following equation:

$$ab = \begin{cases} alog[(log[a] + log[b]) \ mod \ 255] & \text{if both } a \text{ and } b \text{ are not zero} \\ 0 & \text{otherwise} \end{cases}$$

However, the reduction modulo 255 requires heavy computational cost and the conditional branching operation has to be removed to eliminate the possibility of SPA. We compute the reduction modulo 255 by using the reduction modulo 256 and remove the conditional branching operation. This algorithm is described in Algorithm 5. To reduce the number of operations, we also use $\overline{alog}$ table as in the following equation, which requires 1 byte more ROM size than *alog* table.

$$\overline{alog}[x] = \begin{cases} alog[x] & \text{if } 0 \leqslant x < 255 \\ alog[0] & \text{if } x = 255 \end{cases}$$

Algorithm 5. Multiplication in $GF(2^8)$ without SPA Leakage

Input: a, $b \in GF(2^8)$, f is an irreducible polynomial over $GF(2^8)$
Output: $ab \ mod \ f$

1. $t = log[a]$
2. $s = (t + log[b]) \ mod \ 2^8$
3. $r = \overline{alog}[(s < t) + s]$ /* $s < t$: the carry associated with Step 2 */
4. Return $(a \&\& b) * r$ /* &&: the logical AND operation */

Software Implementation of Binary Elliptic Curves: Impact of the Carry-Less Multiplier on Scalar Multiplication

Jonathan Taverne[1,⋆], Armando Faz-Hernández[2], Diego F. Aranha[3,⋆⋆], Francisco Rodríguez-Henríquez[2], Darrel Hankerson[4], and Julio López[3]

[1] Université de Lyon, Université Lyon1, ISFA, France
jonathan.taverne@etu.univ-lyon1.fr
[2] Computer Science Department, CINVESTAV-IPN, México
armfaz@computacion.cs.cinvestav.mx, francisco@cs.cinvestav.mx
[3] Institute of Computing, University of Campinas, Brazil
dfaranha@ic.unicamp.br, jlopez@ic.unicamp.br
[4] Auburn University, USA
hankedr@auburn.edu

Abstract. The availability of a new carry-less multiplication instruction in the latest Intel desktop processors significantly accelerates multiplication in binary fields and hence presents the opportunity for reevaluating algorithms for binary field arithmetic and scalar multiplication over elliptic curves. We describe how to best employ this instruction in field multiplication and the effect on performance of doubling and halving operations. Alternate strategies for implementing inversion and half-trace are examined to restore most of their competitiveness relative to the new multiplier. These improvements in field arithmetic are complemented by a study on serial and parallel approaches for Koblitz and random curves, where parallelization strategies are implemented and compared. The contributions are illustrated with experimental results improving the state-of-the-art performance of halving and doubling-based scalar multiplication on NIST curves at the 112- and 192-bit security levels, and a new speed record for side-channel resistant scalar multiplication in a random curve at the 128-bit security level.

Keywords: Elliptic curve cryptography, finite field arithmetic, parallel algorithm, efficient software implementation.

1 Introduction

Improvements in the fabrication process of microprocessors allow the resulting higher transistor density to be converted into architectural features such as inclusion of new instructions or faster execution of the current instruction set. Limits on the conventional ways of increasing a processor's performance such as incrementing the clock rate, scaling the memory hierarchy [38] or improving support for instruction-level parallelism [37] have pushed manufacturers to embrace parallel processing as the

⋆ This work was performed while the author was visiting CINVESTAV-IPN.
⋆⋆ A portion of this work was performed while the author was visiting University of Waterloo.

B. Preneel and T. Takagi (Eds.): CHES 2011, LNCS 6917, pp. 108–123, 2011.

mainstream computing paradigm and consequently amplify support for resources such as multiprocessing and vectorization. Examples of the latter are the recent inclusions of the SSE4 [22], AES [19] and AVX [14] instruction sets in the latest Intel microarchitectures.

Since the dawn of elliptic curve cryptography in 1985, several field arithmetic assumptions have been made by researchers and designers regarding its efficient implementation in software platforms. Some analysis (supported by experiments) assumed that inversion to multiplication ratios (I/M) were sufficiently small (e.g., $I/M \approx 3$) that point operations would be done in affine coordinates, favoring certain techniques. However, the small ratios were a mix of old hardware designs, slower multiplication algorithms compared with [32], and composite extension degree. It seems clear that sufficient progress was made in multiplication so there is incentive to use projective coordinates. Our interest in the face of much faster multiplication is at the other end—is I/M large enough to affect methods that commonly assumed this ratio is modest?

On the other hand, authors in [16] considered that the cost of a point halving computation was roughly equivalent to 2 field multiplications. The expensive computations in halving are a field multiplication, solving a quadratic $z^2 + z = c$, and finding a square root over $\mathbb{F}_{2^m}$. However, quadratic solvers presented in [21] are multiplication-free and hence, provided that a fast binary field multiplier is available, there would be concern that the ratio of point halving to multiplication may be much larger than 2. Having a particularly fast multiplier would also push for computing square roots in $\mathbb{F}_{2^m}$ as efficiently as possible. Similarly, the common software design assumption that field squaring is essentially free (relative to multiplication) may no longer be valid.

A prevalent assumption is that large-characteristic fields are faster than binary field counterparts for software implementations of elliptic curve cryptography.[1] In spite of simpler arithmetic, binary field realizations could not be faster than large-characteristic analogues mostly due to the absence of a native *carry-less multiplier* in contemporary high-performance processors. However, using a bit-slicing technique, Bernstein [6] was able to compute a batch of 251-bit scalar multiplications on a binary Edwards curve, employing 314,323 clock cycles per scalar multiplication, which, before the results presented in this work and to the best of our knowledge, was the fastest reported time for a software implementation of binary elliptic point multiplication.

In this work, we evaluate the impact of the recently introduced carry-less multiplication instruction [20] in the performance of binary field arithmetic and scalar multiplication over elliptic curves. We also consider parallel strategies in order to speed scalar multiplication when working on multi-core architectures. In contrast to parallelization applied to a batch of operations, the approach considered here applies to a single point multiplication. These approaches target different environments: batching makes sense when throughput is the measurement of interest, while the lower level parallelization is of interest when latency matters and the device is perhaps weak but has multiple processing units. Furthermore, throughout this paper we will assume that we are working in the unknown point scenario, i.e., where the elliptic curve point to be processed is not known in advance, thus precluding off-line precomputation. We will assume that there

[1] In hardware realizations, the opposite thesis is widely accepted: elliptic curve scalar point multiplication can be computed (much) faster using binary extension fields.

is sufficient memory space for storing a few multiples of the point to be processed and look-up tables for accelerating the computation of the underlying field arithmetic.

As the experimental results will show, our implementation of multiplication via this native support was significantly faster than previous timings reported in the literature. This motivated a study on alternative implementations of binary field arithmetic in hope of restoring the performance ratios among different operations in which the literature is traditionally based [21]. A direct consequence of this study is that performance analysis based on these conventional ratios [5] will remain valid in the new platform. Our main contributions are:

- A strategy to efficiently employ the native carry-less multiplier in binary field multiplication.
- Branchless and/or vectorized approaches for implementing half-trace computation, integer recoding and inversion. These approaches allow the halving operation to become again competitive with doubling in the face of a significantly faster multiplier, and help to reduce the impact of integer recoding and inversion in the overall speed of scalar multiplication, even when projective coordinates are used.
- Parallelization strategies for dual core execution of scalar multiplication algorithms in random and Koblitz binary elliptic curves.

We obtain a new state-of-the-art implementation of arithmetic in binary elliptic curves, including improved performance for NIST-standardized Koblitz curves and random curves suitable for halving and a new speed record for side-channel resistant point multiplication in a random curve at the 128-bit security level.

The remainder of the paper progresses as follows. Section 2 elaborates on exploiting carry-less multiplication for high-performance field multiplication along with implementation strategies for half-trace and inversion. Sections 3 and 4 discuss serial and parallel approaches for scalar multiplication. Section 5 presents extensive experimental results and comparison with related work. Section 6 concludes the paper with perspectives on the interplay between the proposed implementation strategies and future enhancements in the architecture under consideration.

2 Binary Field Arithmetic

A binary extension field $\mathbb{F}_{2^m}$ can be constructed by means of a degree-m polynomial f irreducible over $\mathbb{F}_2$ as $\mathbb{F}_{2^m} \cong \mathbb{F}_2[z]/\left(f(z)\right)$. In the case of software implementations in modern desktop platforms, field elements $a \in \mathbb{F}_{2^m}$ can be represented as polynomials of degree at most $m-1$ with binary coefficients a_i packed in $n_{64} = \lceil \frac{m}{64} \rceil$ 64-bit processor words. In this context, the recently introduced carry-less multiplication instruction can play a significant role in order to efficiently implement a multiplier in $\mathbb{F}_{2^m}$. Along with field multiplication, other relevant field arithmetic operations such as squaring, square root, and half-trace, will be discussed in the rest of this section.

2.1 Multiplication

Field multiplication is the performance-critical operation for implementing several cryptographic primitives relying on binary fields, including arithmetic over elliptic curves and the Galois Counter Mode of operation (GCM). For accelerating the latter when used in combination with the AES block cipher [19], Intel introduced the carry-less multiplier in the Westmere microarchitecture as an instruction operating on 64-bit words stored in 128-bit vector registers with opcode *pclmulqdq* [20]. The instruction latency currently peaks at 15 cycles while reciprocal throughput ranks at 10 cycles. In other words, when operands are not in a dependency chain, effective latency is 10 cycles [15].

The instruction certainly looks expensive when compared to the 3-cycle 64-bit integer multiplier present in the same platform, which raises speculation whether Intel aimed for an area/performance trade-off or simply balanced the latency to the point where the carry-less multiplier did not interfere with the throughput of the hardware AES implementation. Either way, the instruction features suggest the following empirical guidelines for organizing the field multiplication code: (i) as memory access by vector instructions continues to be expensive [6], the maximum amount of work should be done in registers, for example through a Comba organization [12]; (ii) as the number of registers employed in multiplication should be minimized for avoiding false dependencies and maximize throughput, the multiplier should have 128-bit granularity; (iii) as the instruction latency allows, each 128-bit multiplication should be implemented with three carry-less multiplications in a Karatsuba fashion [25].

In fact, the overhead of Karatsuba multiplication is minimal in binary fields and the Karatsuba formula with the smaller number of multiplications for multiplying $\lceil \frac{n_{64}}{2} \rceil$ 128-bit digits proved to be optimal in all the considered field sizes. This observation comes in direct contrast to previous vectorized implementations of the *comb* method for binary field multiplication due to López and Dahab [32, Algorithm 5], where the memory-bound precomputation step severely limits the number of Karatsuba steps which can be employed, fixing the cutoff point to large fields [2] such as $\mathbb{F}_{2^{1223}}$. To summarize, multiplication was implemented as a 128-bit granular Karatsuba multiplier with each 128-digit multiplication solved by another Karatsuba instance requiring three carry-less multiplications, cheap additions and efficient shifts by multiples of 8 bits. A single 128-digit level of Karatsuba was used for fields $\mathbb{F}_{2^{233}}$ and $\mathbb{F}_{2^{251}}$ where $\lceil \frac{n_{64}}{2} \rceil = 2$, while two instances were used for field $\mathbb{F}_{2^{409}}$ where $\lceil \frac{n_{64}}{2} \rceil = 4$. Particular approaches which led to lower performance in our experiments were organizations based on optimal Toom-Cook [10] due to the higher overhead brought by minor operations; and on a lower 64-bit granularity combined with alternative multiple-term Karatsuba formulas [33] due to register exhaustion to store all the intermediate values, causing a reduction in overall throughput.

2.2 Squaring, Square-Root and Multi-squaring

Squaring and square-root are considered cheap operations in a binary field, especially when $\mathbb{F}_{2^m}$ is defined by a square-root friendly polynomial [3,1], because they require only linear manipulation of individual coefficients [21]. These operations are traditionally implemented with the help of large precomputed tables, but vectorized implementations are possible with simultaneous table lookups through byte shuffling instructions [2]. This approach is enough to keep square and square-root efficient relative to multiplication even with a dramatic acceleration of field multiplication. For illustration, [2] reports multiplication-to-squaring ratios as high as 34 without a native multiplier, far from the conventional ratios of 5 [5] or 7 [21] and with a large room for future improvement.

Multi-squaring, or exponentiation to 2^k, can be efficiently implemented with a time-memory trade-off proposed as m-squaring in [1,11] and here referred as *multi-squaring*. For a fixed k, a table T of $16\lceil \frac{m}{4} \rceil$ field elements can be precomputed such that $T[j, i_0 + 2i_1 + 4i_2 + 8i_3] = (i_0 z^{4j} + i_1 z^{4j+1} + i_2 z^{4j+2} + i_3 z^{4j+3})^{2^k}$ and $a^{2^k} = \sum_{j=0}^{\lceil \frac{m}{4} \rceil} T[j, \lfloor a/2^{4j} \rfloor \bmod 2^4]$. The threshold where multi-squaring became faster than simple consecutive squaring observed in our implementation was around $k \geq 6$ for $\mathbb{F}_{2^{233}}$ and $k \geq 10$ for $\mathbb{F}_{2^{409}}$.

2.3 Inversion

Inversion modulo $f(z)$ can be implemented via the polynomial version of the Extended Euclidean Algorithm (EEA), but the frequent branching and recurrent shifts by arbitrary amounts present a performance obstacle for vectorized implementations, which makes it difficult to write consistently fast EEA codes across different platforms. A branchless approach can be implemented through Itoh-Tsuji inversion [23] by computing $a^{-1} = a^{(2^{m-1}-1)2}$, as proposed in [18]. In contrast to the EEA method, the Itoh-Tsujii approach has the additional merit of being similarly fast (relative to multiplication) across common processors.

The overall cost of the method is $m - 1$ squarings and a number of multiplications dictated by the length of an addition chain for $m - 1$. The cost of squarings can be reduced by computing each required 2^i-power as a multi-squaring [11]. The choice of an addition chain allows the implementer to control the amount of required multiplications and the precomputed storage for multi-squaring, since the number of 2^i-powers involved can be balanced.

Previous work obtained inversion-to-multiplication ratios between 22 and 41 by implementing EEA in 64-bit mode [2], while the conventional ratios are between 5 and 10 [21,5]. While we cannot reach the small ratios with Itoh-Tsujii for the parameters considered here, we can hope to do better than applying the method from [2] which will give significantly larger ratios with the carry-less multiplier. Hence the cost of squarings and multi-squarings should be minimized to the lowest possible allowed by storage capacity.

To summarize, we use addition chains of 10, 10 and 11 steps for computing field inversion over the fields $\mathbb{F}_{2^{233}}$, $\mathbb{F}_{2^{251}}$ and $\mathbb{F}_{2^{409}}$, respectively.[2] We extensively used the multi-squaring approach described in the preceding section. For example, in the case of $\mathbb{F}_{2^{233}}$, we selected the addition chain $1 \rightarrow 2 \rightarrow 3 \rightarrow 6 \rightarrow 7 \rightarrow 14 \rightarrow 28 \rightarrow 29 \rightarrow 58 \rightarrow 116 \rightarrow 232$, and used 3 pre-computed tables for computing the iterated squarings $a^{2^{29}}$, $a^{2^{58}}$ and $a^{2^{116}}$. The rest of the field squaring operations were computed by executing consecutive squarings. We recall that each table stores a total of $16\lceil \frac{m}{4} \rceil$ field elements.

[2] In the case of inversion over $\mathbb{F}_{2^{409}}$, the minimal length addition chain to reach $m - 1 = 408$ has 10 steps. However, we preferred to use an 11-step chain to save one look-up table.

2.4 Half-Trace

Half-trace plays a central role in point halving and its performance is essential if halving is to be competitive against doubling. For an odd integer m, the half-trace function $H : \mathbb{F}_{2^m} \rightarrow \mathbb{F}_{2^m}$ is defined by $H(c) = \sum_{i=0}^{(m-1)/2} c^{2^{2i}}$ and satisfies the equation $\lambda^2 + \lambda = c + \mathrm{Tr}(c)$ required for point halving. One efficient desktop-targeted implementation of the half-trace is described in [3] and presented as Algorithm 1, making extensive use of precomputations. This implementation is based on two main steps: the elimination of even power coefficients and the accumulation of half-trace precomputed values.

Step 5 in Algorithm 1, as shown in [21], consists in reducing the number of non-zero coefficients of c by removing the coefficients of even powers i via $H(z^i) = H(z^{i/2}) + z^{i/2} + \mathrm{Tr}(z^i)$. That will lead to memory and time savings during the last step of the half-trace computation, the accumulation (step 6). This is done by extraction of the odd and even bits and can benefit from vectorization in the same way as square-root in [2]. However, in the case of half-trace there is a bottleneck caused by data dependencies. For efficiency, the bank of 128-bit registers is used as much as possible, but at one point in the algorithm execution the number of available bits to process decreases. For 64-bit and 32-bit digits, the use of 128-bit registers is still beneficial, but for a smaller size, the conventional approach (not vectorized) becomes again competitive.

Unlike the direction taken in [21], the approach in [3] does not attempt to minimize memory requirements but rather it greedily strives to speed the accumulation part (step 6). Precomputation is extended so as to reduce the number of accesses to the lookup table. The following values of the half-trace are stored: $H(l_0c^{8i+1} + l_1c^{8i+3} + l_2c^{8i+5} + l_3c^{8i+7})$ for all $i \geq 0$ such that $8i < m - 3$ and $l_j \in \mathbb{F}_2$. The memory size in bytes taken by the precomputations follows the formula $16 \times n_{64} \times 8 \times \lceil \frac{m}{8} \rceil$.

Algorithm 1. Solve $x^2 + x = c$

Input: $c = \sum_{i=0}^{m-1} c_i z^i \in \mathbb{F}_{2^m}$ where m is odd and $\mathrm{Tr}(c) = 0$
Output: a solution s of $x^2 + x = c$
1: compute $H(l_0c^{8i+1}+l_1c^{8i+3}+l_2c^{8i+5}+l_3c^{8i+7})$ for $i \in I = \{0, \ldots, \lfloor \frac{m-3}{8} \rfloor\}$ and $l_j \in \mathbb{F}_2$
2: $s \leftarrow 0$
3: **for** $i = (m-1)/2$ **downto** 1 **do**
4:     **if** $c_{2i} = 1$ **then**
5:         $c \leftarrow c + z^i$, $s \leftarrow s + z^i$
6: **return** $s \leftarrow s + \sum_{i \in I} c^{8i+1}H(z^{8i+1}) + c^{8i+3}H(z^{8i+3}) + c^{8i+5}H(z^{8i+5}) + c^{8i+7}H(z^{8i+7})$

While considering different organizations of the half-trace code, we made the following serendipitous observation: inserting as many `xor` operations as the data dependencies permitted from the accumulation stage (step 6) into step 5 gave a substantial speed-up of 20% to 25% compared with code written in the order as described in Algorithm 1. Plausible explanations are compiler optimization and processor pipelining characteristics. The result is a half-trace-to-multiplication ratio near 1, and this ratio can be reduced if memory can be consumed more aggressively.

3 Random Binary Elliptic Curves

Given a finite field $\mathbb{F}_q$ for $q = 2^m$, a non-supersingular elliptic curve $E(\mathbb{F}_q)$ is defined to be the set of points $(x, y) \in \mathbb{F}_q \times \mathbb{F}_q$ that satisfy the affine equation

$$y^2 + xy = x^3 + ax^2 + b, \quad (1)$$

where a and $0 \neq b \in \mathbb{F}_q$, together with the point at infinity denoted by $\mathcal{O}$. It is known that $E(\mathbb{F}_q)$ forms an additive Abelian group with respect to the elliptic point addition operation.

Let k be a positive integer and P a point on an elliptic curve. Then *elliptic curve scalar multiplication* is the operation that computes the multiple $Q = kP$, defined as the point resulting of adding P to itself $k - 1$ times. One of the most basic methods for computing a scalar multiplication is based on a double-and-add variant of Horner's rule. As the name suggests, the two most prominent building blocks of this method are the *point doubling* and *point addition* primitives. By using the non-adjacent form (NAF) representation of the scalar k, the addition-subtraction method computes a scalar multiplication in about m doubles and $m/3$ additions [21]. The method can be extended to a *width-ω NAF* $k = \sum_{i=0}^{t-1} k_i 2^i$ where $k_i \in \{0, \pm 1, \ldots, \pm 2^m - 1\}$, $k_{t-1} \neq 0$, and at most one of any ω consecutive digits is nonzero. The length t is at most one larger than the bitsize of k, and the density is approximately $1/(\omega + 1)$; for $\omega = 2$, this is the same as NAF.

3.1 Sequential Algorithms for Random Binary Curves

The traditional left-to-right double-and-add method is illustrated in Algorithm 2 where $n = 0$ (that is, the computation corresponds to the left column) and the *width-ω NAF* $k = \sum_{i=0}^{t-1} k_i 2^i$ expression is computed from left to right, i.e., it starts processing k_{t-1} first, then k_{t-2} until it ends with the coefficient k_0. Step 1 computes $2^{\omega-2} - 1$ multiples of the point P. Based on the Montgomery trick, authors in [13] suggested a method to precompute the affine points in large-characteristic fields $\mathbb{F}_p$, employing only one inversion. Exporting that approach to $\mathbb{F}_{2^m}$, we obtained formulae that offer a saving of 4 multiplications and 15 squarings for $\omega = 4$ when compared with a naive method that would make use of the Montgomery trick in a trivial way (see Table 1 for a summary of the computational effort associated to this phase).

For a given ω, the evaluation stage of the algorithm has approximately $m/(\omega + 1)$ point additions, and hence increasing ω has diminishing returns. For the curves given by NIST [34] and with on-line precomputation, $\omega \leq 6$ is optimal in the sense that total point additions are minimized. In many cases, the recoding in ωNAF(k) is performed on-line and can be considered as part of the precomputation step.

The most popular way to represent points in binary curves is López-Dahab projective coordinates that yield an effective cost for a mixed point addition and point doubling operation of about $8M + 5S \approx 9M$ and $4M + 5S \approx 5M$, respectively (see Tables 2 and 3). Kim and Kim [26] report alternate formulas for point doubling requiring four multiplications and five squarings, but two of the four multiplications are by the constant b, and these have the same cost as general multiplication with the native carry-less multiplier. For mixed addition, Kim and Kim require eight multiplications but save

---

Algorithm 2. Double-and-add, halve-and-add scalar multiplication: parallel

---

Input: ω, scalar k, $P \in E(\mathbb{F}_{2^m})$ of odd order r, constant n (e.g., from Table 1(b))
Output: kP

Compute $P_i = iP$ for $i \in I = \{1, 3, \ldots, 2^{\omega-1} - 1\}$
$Q_0 \leftarrow \mathcal{O}$
Recode: $k' = 2^n k \bmod r$ and obtain rep $\omega\text{NAF}(k')/2^n = \sum_{i=0}^{t} k'_i 2^{i-n}$
Initialize $Q_i \leftarrow \mathcal{O}$ for $i \in I$
{Barrier}
for $i = t$ **downto** n **do**
  $Q_0 \leftarrow 2Q_0$
  if $k'_i > 0$ **then**
    $Q_0 \leftarrow Q_0 + P_{k'_i}$
  else if $k'_i < 0$ **then**
    $Q_0 \leftarrow Q_0 - P_{-k'_i}$

for $i = n - 1$ **downto** 0 **do**
  $P \leftarrow P/2$
  if $k'_i > 0$ **then**
    $Q_{k'_i} \leftarrow Q_{k'_i} + P$
  else if $k'_i < 0$ **then**
    $Q_{-k'_i} \leftarrow Q_{-k'_i} - P$
{Barrier}
return $Q \leftarrow Q_0 + \sum_{i \in I} iQ_i$

---

two field reductions when compared with López-Dahab, giving their method the edge. Hence, in this work we use López-Dahab for point doubling and Kim and Kim for point addition.

Right-to-Left Halve-and-Add. Scalar multiplication based on point halving replaces point doubling by a potentially faster *halving* operation that produces Q from P with $P = 2Q$. The method was proposed independently by Knudsen [28] and Schroeppel [35] for curves $y^2 + xy = x^3 + ax^2 + b$ over $\mathbb{F}_{2^m}$. The method is simpler if the trace of a is 1, and this is the only case we consider. The expensive computations in halving are a field multiplication, solving a quadratic $z^2 + z = c$, and finding a square root. On the NIST random curves studied in this work, we found that the cost of halving is approximately $3M$, where M denotes the cost of a field multiplication.

Let the base point P have odd order r, and let t be the number of bits to represent r. For $0 < n \leq t$, let $\sum_{i=0}^{t} k'_i 2^i$ be given by the width-ω NAF of $2^n k \bmod r$. Then $k \equiv k'/2^n \equiv \sum_{i=0}^{t} k'_i 2^{i-n} \pmod{r}$ and the scalar multiplication can be split as

$$kP = (k'_t 2^{t-n} + \cdots + k'_n)P + (k'_{n-1} 2^{-1} + \cdots + k'_0 2^{-n})P. \quad (2)$$

When $n = t$, this gives the usual representation for point multiplication via halving, illustrated in Algorithm 2 (that is, the computation is essentially the right column). The cost for postcomputation appears in Table 1.

3.2 Parallel Scalar Multiplication on Random Binary Curves

For parallelization, choose $n < t$ in (2) and process the first portion by a double-and-add method and the second portion by a method based on halve-and-add. Algorithm 2 illustrates a parallel approach suitable for two processors. Recommended values for n to balance cost between processors appear in Table 1.

Table 1. Costs and parameter recommendations for $\omega \in \{3,4,5\}$

ω	Algorithm 2 Precomp	Algorithm 2 Postcomp	[21, Alg 3.70] Precomp	[21, Alg 3.70]′ Postcomp
3	$14M$,$11S$,I	$43M$,$26S$	$2M$,$3S$,I	$26M$,$13S$
4	$38M$,$15S$,I	$116M$,$79S$	$9M$,$9S$,I	$79M$,$45S$
5	N/A	N/A	$23M$,$19S$,$2I$	$200M$,$117S$

(a) Pre- and post-computation costs.

ω	Algorithm 2 B-233	Algorithm 2 B-409	Algorithm 3 K-233	Algorithm 3 K-409
3	128	242	131	207
4	132	240	135	210
5	N/A	N/A	136	213

(b) Recommended value for n.

3.3 Side-Channel Resistant Multiplication on Random Binary Curves

Another approach for scalar multiplication offering some resistance to side-channel attacks was proposed by López and Dahab [31] based on the Montgomery laddering technique. This approach requires $6M + 5S$ in $\mathbb{F}_{2^m}$ per iteration independently of the bit pattern in the scalar, and one of these multiplications is by the curve coefficient b. The curve being lately used for benchmarking purposes [7] at the 128-bit security level is an Edwards curve (CURVE2251) corresponding to the Weierstraß curve $y^2 + xy = x^3+(z^{13}+z^9+z^8+z^7+z^2+z+1)$. It is clear that this curve is especially tailored for this method due to the short length of b, reducing the cost of the algorithm to approximately $5.25M + 5S$ per iteration. At the same time, halving-based approaches are non-optimal for this curve due to the penalties introduced by the 4-cofactor [27]. Considering this and to partially satisfy the side-channel resistance offered by a bitsliced implementation such as [6], we restricted the choices of scalar multiplication at this security level to the Montgomery laddering approach.

4 Koblitz Elliptic Curves

A Koblitz curve $E_a(\mathbb{F}_q)$, also known as an Anomalous Binary Curve [29], is a special case of (1) where $b = 1$ and $a \in \{0,1\}$. In a binary field, the map taking x to x^2 is an automorphism known as the Frobenius map. Since Koblitz curves are defined over the binary field $\mathbb{F}_2$, the Frobenius map and its inverse naturally extend to automorphisms of the curve denoted τ and τ^{-1}, respectively, where $\tau(x,y) = (x^2,y^2)$. Moreover, $(x^4,y^4) + 2(x,y) = \mu(x^2,y^2)$ for every (x,y) on E_a, where $\mu = (-1)^{1-a}$; that is, τ satisfies $\tau^2 + 2 = \mu\tau$ and we can associate τ with the complex number $\tau = \frac{\mu+\sqrt{-7}}{2}$.

Solinas [36] presents a τ-adic analogue of the usual NAF as follows. Since short representations are desirable, an element $\rho \in \mathbb{Z}[\tau]$ is found with $\rho \equiv k \pmod{\delta}$ of as small norm as possible, where $\delta = (\tau^m - 1)/(\tau - 1)$. Then for the subgroup of interest, $kP = \rho P$ and a width-ω τ-adic NAF ($\omega\tau$NAF) for ρ is obtained in a fashion that parallels the usual ωNAF. As in [36], define $\alpha_i = i \bmod \tau^\omega$ for $i \in \{1,3,\ldots,2^{\omega-1}-1\}$. A $\omega\tau$NAF of a nonzero element ρ is an expression $\rho = \sum_{i=0}^{l-1} u_i\tau^i$ where each $u_i \in \{0,\pm\alpha_1,\pm\alpha_3,\ldots,\pm\alpha_{2^{\omega-1}-1}\}$, $u_{l-1} \neq 0$, and at most one of any consecutive ω coefficients is nonzero. Scalar multiplication kP can be performed with the $\omega\tau$NAF expansion of ρ as

$$u_{l-1}\tau^{l-1}P + \cdots + u_2\tau^2P + u_1\tau P + u_0P \tag{3}$$

with $l-1$ applications of τ and approximately $l/(\omega+1)$ additions.

The length of the representation is at most $m + a$, and Solinas presents an efficient technique to find an estimate for ρ, denoted $\rho' = k$ partmod δ with $\rho' \equiv \rho \pmod{\delta}$, having expansion of length at most $m+a+3$ [36,9]. Under reasonable assumptions, the algorithm will usually produce an estimate giving length at most $m + 1$. For simplicity, we will assume that the recodings obtained have this as an upper bound on length; small adjustments are necessary to process longer representations. Under these assumptions and properties of τ, scalars may be written $k = \sum_{i=0}^{m} u_i\tau^i = \sum_{i=0}^{m} u_i\tau^{-(m-i)}$ since $\tau^{-i} = \tau^{m-i}$ for all i.

4.1 Sequential Algorithms for Koblitz Curves

A traditional left-to-right τ-and-add method for (3) appears as [21, Alg 3.70], and is essentially the left-hand portion of Algorithm 3. Precomputation consists of $2^{\omega-2} - 1$ multiples of the point P, each at a cost of approximately one point addition (see Table 1 for a summary of the computational effort associated to this phase).

Alternatively, we can process bits right-to-left and obtain a variant we shall denote as [21, Alg 3.70]′ (an analogue of [21, Alg 3.91]). The multiple points of precomputation P_u are exchanged for the same number of accumulators Q_u along with postcomputation of form $\sum \alpha_u Q_u$. The cost of postcomputation is likely more than the precomputation of the left-to-right variant; see Table 1 for a summary in the case where postcomputation uses projective additions. However, if the accumulator in Algorithm 3 is in projective coordinates, then the right-to-left variant has a less expensive evaluation phase since τ is applied to points in affine coordinates.

4.2 Parallel Algorithm for Koblitz Curves

The basic strategy in our parallel algorithm is to reformulate the scalar multiplication in terms of both the τ and the τ^{-1} operators as $k = \sum_{i=0}^{m} u_i\tau^i = u_0 + u_1\tau^1 + \cdots + u_n\tau^n + u_{n+1}\tau^{-(m-n-1)} + \cdots + u_m = \sum_{i=0}^{n} u_i\tau^i + \sum_{i=n+1}^{m} u_i\tau^{-(m-i)}$ where $0 < n < m$. Algorithm 3 illustrates a parallel approach suitable for two processors. Although similar in structure to Algorithm 2, a significant difference is the shared precomputation rather than the pre and postcomputation required in Algorithm 2.

The scalar representation is given by Solinas [36] and hence has an expected $m/(\omega + 1)$ point additions in the evaluation-stage, and an extra point addition at the end. There are also approximately m applications of τ or its inverse. If the field representation is such that these operators have similar cost or are sufficiently inexpensive relative to field multiplication, then the evaluation stage can be a factor 2 faster than a corresponding non-parallel algorithm.

As discussed before, unlike the ordinary width-ω NAF, the τ-adic version requires a relatively expensive calculation to find a short ρ with $\rho \equiv k \pmod{\delta}$. Hence, (a portion of) the precomputation is "free" in the sense that it occurs during scalar recoding. This can encourage the use of a larger window size ω. The essential features exploited by Algorithm 3 are that the scalar can be efficiently represented in terms of the Frobenius map and that the map and its inverse can be efficiently computed, and hence the algorithm adapts to curves defined over small fields.

Algorithm 3 is attractive in the sense that two processors are directly supported without "extra" computations. However, if multiple applications of the "doubling step" are

Algorithm 3. $\omega\tau$NAF scalar multiplication: parallel

Input: $\omega, k \in [1, r-1]$, $P \in E_a(\mathbb{F}_{2^m})$ of order r, constant n (e.g., from Table 1(b))
Output: kP

$\rho \leftarrow k$ partmod δ
$\sum_{i=0}^{l-1} u_i \tau^i \leftarrow \omega\tau\text{NAF}(\rho)$
$P_u = \alpha_u P$, for $u \in \{1, 3, 5, \ldots, 2^{\omega-1} - 1\}$
{Barrier}

$Q_0 \leftarrow \mathcal{O}$
for $i = n$ **downto** 0 **do**
  $Q_0 \leftarrow \tau Q_0$
  if $u_i = \alpha_j$ **then**
    $Q_0 \leftarrow Q_0 + P_j$
  else if $u_i = -\alpha_j$ **then**
    $Q_0 \leftarrow Q_0 - P_j$

$Q_1 \leftarrow \mathcal{O}$
for $i = n+1$ **to** m **do**
  $Q_1 \leftarrow \tau^{-1} Q_1$
  if $u_i = \alpha_j$ **then**
    $Q_1 \leftarrow Q_1 + P_j$
  else if $u_i = -\alpha_j$ **then**
    $Q_1 \leftarrow Q_1 - P_j$

{Barrier}
return $Q \leftarrow Q_0 + Q_1$

sufficiently inexpensive, then more processors and additional curves can be accommodated in a straightforward fashion without sacrificing the high-level parallelism of Algorithm 3. As an example for Koblitz curves, a variant on Algorithm 3 discards the applications of τ^{-1} (which may be more expensive than τ) and finds $kP = k^1(\tau^j P) + k^0 P = \tau^j(k^1 P) + k^0 P$ for suitable k^i and $j \approx m/2$ with traditional methods to calculate $k^i P$. The application of τ^j is low cost if there is storage for a per-field matrix as it was first discussed in [1].

5 Experimental Results

We consider example fields $\mathbb{F}_{2^m}$ for $m \in \{233, 251, 409\}$. These were chosen to address 112-bit and 192-bit security levels, according to the NIST recommendation, and the 251-bit binary Edwards elliptic curve presented in [6]. The field $\mathbb{F}_{2^{233}}$ was also chosen as more likely to expose any overhead penalty in the parallelization compared with larger fields from NIST. Our C library coded all the algorithms using the GNU C 4.6 (GCC) and Intel 12 (ICC) compilers, and the timings were obtained on a 3.326 GHz 32nm Intel *Westmere* processor i5 660.

Obtaining times useful for comparison across similar systems can be problematic. Intel, for example, introduced "Pentium 4" processors that were fundamentally different than earlier designs with the same name. The common method via time stamp counter (TSC) requires care on recent processors having "turbo" modes that increase the clock (on perhaps 1 of 2 cores) over the nominal clock implicit in TSC, giving an underestimate of actual cycles consumed. Benchmarking guidelines on eBACS [7], for example, recommend disabling such modes, and this is the method followed in this paper.

Timings for field arithmetic appear in Table 2. The López-Dahab multiplier described in [2] was implemented as a baseline to quantify the speedup due to the native multiplier. For the most part, timings for GCC and ICC are similar, although López-Dahab multiplication is an exception. The difference in multiplication times between $\mathbb{F}_{2^{233}} = \mathbb{F}_2[z]/(z^{233}+z^{74}+1)$ and $\mathbb{F}_{2^{251}} = \mathbb{F}_2[z]/(z^{251}+z^7+z^4+z^2+1)$ is in reduction. The relatively expensive square root in $\mathbb{F}_{2^{251}}$ is due to the representation chosen;

Table 2. Timings in clock cycles for field arithmetic operations. "op/M" denotes ratio to multiplication obtained from ICC.

Base field operation	$\mathbb{F}_{2^{233}}$			$\mathbb{F}_{2^{251}}$			$\mathbb{F}_{2^{409}}$		
	GCC	ICC	op/M	GCC	ICC	op/M	GCC	ICC	op/M
Multiplication	128	128	1.00	161	159	1.00	345	348	1.00
López-Dahab Mult.	256	367	2.87	338	429	2.70	637	761	2.19
Square root	67	60	0.47	155	144	0.91	59	56	0.16
Squaring	30	35	0.27	56	59	0.37	44	49	0.14
Half trace	167	150	1.17	219	212	1.33	322	320	0.92
Multi-Squaring	191	184	1.44	195	209	1.31	460	475	1.36
Inversion	2,951	2,914	22.77	3,710	3,878	24.39	9,241	9,350	26.87
4-τNAF	9,074	11,249	87.88	-	-	-	23,783	26,633	76.53
3-NAF	5,088	5,059	39.52	-	-	-	13,329	14,373	41.30
4-NAF	4,280	4,198	32.80	-	-	-	11,406	12,128	34.85
Recoding (halving)	1,543	1,509	11.79	-	-	-	3,382	3,087	8.87
Recoding (parallel)	999	1,043	8.15	-	-	-	2,272	2,188	6.29

Table 3. Timings in clock cycles for curve arithmetic operations. "op/M" denotes ratio to multiplication obtained from ICC.

Elliptic curve operations	B-233			B-409		
	GCC	ICC	op/M	GCC	ICC	op/M
Doubling (LD)	690	710	5.55	1,641	1,655	4.76
Addition (KIM Mixed)	1,194	1,171	9.15	2,987	3,000	8.62
Addition (LD Mixed)	1,243	1,233	9.63	3,072	3,079	8.85
Addition (LD General)	1,954	1,961	15.32	4,893	4,922	14.14
Halving	439	417	3.26	894	878	2.52

if square roots are of interest, then there are reduction polynomials giving faster square root and similar numbers for other operations. Inversion via exponentiation (§2) gives I/M similar to that in [2] where an Euclidean algorithm variant was used with similar hardware but without the carry-less multiplier.

Table 4 shows timings obtained for different variants of sequential and parallel scalar multiplication. We observe that for ωNAF recoding with $\omega = 3, 4$, the halve-and-add algorithm is always faster than its double-and-add counterpart. This performance is a direct consequence of the timings reported in Table 3, where the cost of one point doubling is roughly 5.5 and 4.8 multiplications whereas the cost of a point halving is of only 3.3 and 2.5 multiplications in the fields $\mathbb{F}_{2^{233}}$ and $\mathbb{F}_{2^{409}}$, respectively. The parallel version that concurrently executes these algorithms in two threads computes one scalar multiplication with a latency that is roughly 37.7% and 37.0% smaller than that of the halve-and-add algorithm for the curves B-233 and B-409, respectively.

The bold entries for Koblitz curves identify fastest timings per category (i.e., considering the compiler, curve, and the specific value of ω used in the ωNAF recoding). For smaller ω, [21, Alg 3.70]$'$ has an edge over [21, Alg 3.70] because τ is applied to points in affine coordinates; this advantage diminishes with increasing ω due to post-computation cost. "(τ, τ)-and-add" denotes the parallel variant described in §4.2. There

Table 4. Timings in 10^3 clock cycles for scalar multiplication in the unknown-point scenario

ω	Scalar mult random curves	B-233 GCC	B-233 ICC	B-409 GCC	B-409 ICC
3	Double-and-add	240	238	984	989
	Halve-and-add	196	192	755	756
	(Dbl,Halve)-and-add	122	118	465	466
4	Double-and-add	231	229	941	944
	Halve-and-add	188	182	706	705
	(Dbl,Halve)-and-add	122	116	444	445

Side-channel resistant scalar multiplication	CURVE2251 GCC	CURVE2251 ICC
Montgomery laddering	296	282

ω	Scalar mult Koblitz curves	K-233 GCC	K-233 ICC	K-409 GCC	K-409 ICC
3	[21, Alg 3.70]	111	110	413	416
	[21, Alg 3.70]′	98	**98**	**381**	389
	(τ,τ)-and-add	**73**	74	**248**	248
	Alg. 3	80	78	253	248
4	[21, Alg 3.70]	97	95	353	355
	[21, Alg 3.70]′	90	**89**	**332**	339
	(τ,τ)-and-add	68	**65**	216	214
	Alg. 3	73	69	218	**214**
5	[21, Alg 3.70]	92	**90**	326	328
	[21, Alg 3.70]′	95	93	**321**	332
	(τ,τ)-and-add	63	**58**	197	**191**
	Alg. 3	68	63	197	194

is a storage penalty for a linear map, but applications of τ^{-1} are eliminated (of interest when τ is significantly less expensive). Given the modest cost of the multi-squaring operation (with an equivalent cost of less than 1.44 field multiplications, see Table 2), the (τ,τ)-and-add parallel variant is usually faster than Algorithm 3. When using $\omega = 5$, the parallel (τ,τ)-and-add algorithm computes one scalar multiplication with a latency that is roughly 35.5% and 40.5% smaller than that of the best sequential algorithm for the curves K-233 and K-409, respectively.

Per-field storage and coding techniques compute half-trace at cost comparable to field multiplication, and methods based on halving continue to be fastest for suitable random curves. However, the hardware multiplier and squaring (via shuffle) give a factor 2 advantage to Koblitz curves in the examples from NIST. This is larger than in [16,21], where a 32-bit processor in the same general family as the i5 has half-trace at approximately half the cost of a field multiplication for B-233 and a factor 1.7 advantage to K-163 over B-163 (and the factor would have been smaller for K-233 and B-233). It is worth remarking that the parallel scalar multiplications versions shown in Table 4 look best for bigger curves and larger ω.

6 Conclusion and Future Work

In this work we achieve the fastest timings reported in the open literature for software computation of scalar multiplication in NIST and Edwards binary elliptic curves defined at the 112-bit, 128-bit and 192-bit security levels. The fastest curve implemented, namely NIST K-233, can compute one scalar multiplication in less than 17.5μs, a result that is not only much faster than previous software implementations of that curve, but is also quite competitive with the computation time achieved by state-of-the-art hardware accelerators working on similar or smaller curves [24,1].

These fast timings were obtained through the usage of the native carry-less multiplier available in the newest Intel processors. At the same time, we strive to use the best algorithmic techniques, and the most efficient elliptic curve and finite field arithmetic formulae. Further, we proposed effective parallel formulations of scalar multiplication algorithms suitable for deployment in multi-core platforms.

The curves over binary fields permit relatively elegant parallelization with low synchronization cost, mainly due to the efficient halving or τ^{-1} operations. Parallelizing at lower levels in the arithmetic would be desirable, especially for curves over prime fields. Grabher *et al.* [17] apply parallelization for extension field multiplication, but times for a base field multiplication in a 256-bit prime field are relatively slow compared with Beuchat *et al.* [8]. On the other hand, a strategy that applies to all curves performs point doubles in one thread and point additions in another. The doubling thread stores intermediate values corresponding to nonzero digits of the NAF; the addition thread processes these points as they become available. Experimentally, synchronization cost is low, but so is the expected acceleration. Against the fastest times in Longa and Gebotys [30] for a curve over a 256-bit prime field, the technique would offer roughly 17% improvement, a disappointing return on processor investment.

The new native support for binary field multiplication allowed our implementation to improve by 10% the previous speed record for side-channel resistant scalar multiplication in random elliptic curves. It is hard to predict what will be the superior strategy between a conventional non-bitsliced or a bitsliced implementation on future revisions of the target platform: the latency of the carry-less multiplier instruction has clear room for improvement, while the new AVX instruction set has 256-bit registers. An issue with the current Sandy Bridge version of AVX is that `xor` throughput for operations with register operands was decreased significantly from 3 operations per cycle in SSE to 1 operation per cycle in AVX. The resulting performance of a bitsliced implementation will ultimately rely on the amount of work which can be scheduled to be done mostly in registers.

Acknowledgments. We wish to thank the University of Waterloo and especially Professor Alfred Menezes for useful discussions related to this work during a visit by three of the authors, where the idea of this project was discussed, planned and a portion of the development phase was done. Diego F. Aranha and Julio López thank CNPq, CAPES and FAPESP for financial support.

References

1. Ahmadi, O., Hankerson, D., Rodríguez-Henríquez, F.: Parallel formulations of scalar multiplication on Koblitz curves. J. UCS 14(3), 481–504 (2008)
2. Aranha, D.F., López, J., Hankerson, D.: Efficient Software Implementation of Binary Field Arithmetic Using Vector Instruction Sets. In: Abdalla, M., Barreto, P.S.L.M. (eds.) LATINCRYPT 2010. LNCS, vol. 6212, pp. 144–161. Springer, Heidelberg (2010)
3. Avanzi, R.M.: Another Look at Square Roots (and Other Less Common Operations) in Fields of Even Characteristic. In: Adams, C., Miri, A., Wiener, M. (eds.) SAC 2007. LNCS, vol. 4876, pp. 138–154. Springer, Heidelberg (2007)
4. Bellare, M. (ed.): CRYPTO 2000. LNCS, vol. 1880. Springer, Heidelberg (2000)
5. Bernstein, D., Lange, T.: Analysis and optimization of elliptic-curve single-scalar multiplication. In: Proceedings 8th International Conference on Finite Fields and Applications (Fq8), vol. 461, pp. 1–20. AMS, Providence (2008)

6. Bernstein, D.J.: Batch Binary Edwards. In: Halevi, S. (ed.) CRYPTO 2009. LNCS, vol. 5677, pp. 317–336. Springer, Heidelberg (2009)
7. Bernstein, D.J., Lange, T. (eds.): eBACS: ECRYPT Benchmarking of Cryptographic Systems, http://bench.cr.yp.to (accessed March 30, 2011)
8. Beuchat, J.-L., González-Díaz, J.E., Mitsunari, S., Okamoto, E., Rodríguez-Henríquez, F., Teruya, T.: High-speed software implementation of the optimal ate pairing over barreto–naehrig curves. In: Joye, M., Miyaji, A., Otsuka, A. (eds.) Pairing 2010. LNCS, vol. 6487, pp. 21–39. Springer, Heidelberg (2010)
9. Blake, I.F., Murty, V.K., Xu, G.: A note on window τ-NAF algorithm. Inf. Process. Lett. 95(5), 496–502 (2005)
10. Bodrato, M.: Towards Optimal Toom-Cook Multiplication for Univariate and Multivariate Polynomials in Characteristic 2 and 0. In: Carlet, C., Sunar, B. (eds.) WAIFI 2007. LNCS, vol. 4547, pp. 116–133. Springer, Heidelberg (2007)
11. Bos, J.W., Kleinjung, T., Niederhagen, R., Schwabe, P.: ECC2K-130 on Cell CPUs. In: Bernstein, D.J., Lange, T. (eds.) AFRICACRYPT 2010. LNCS, vol. 6055, pp. 225–242. Springer, Heidelberg (2010)
12. Comba, P.G.: Exponentiation Cryptosystems on the IBM PC. IBM Systems Journal 29(4), 526–538 (1990)
13. Dahmen, E., Okeya, K., Schepers, D.: Affine Precomputation with Sole Inversion in Elliptic Curve Cryptography. In: Pieprzyk, J., Ghodosi, H., Dawson, E. (eds.) ACISP 2007. LNCS, vol. 4586, pp. 245–258. Springer, Heidelberg (2007)
14. Firasta, N., Buxton, M., Jinbo, P., Nasri, K., Kuo, S.: Intel AVX: New frontiers in performance improvement and energy efficiency. White paper, http://software.intel.com/
15. Fog, A.: Instruction tables: List of instruction latencies, throughputs and micro-operation breakdowns for Intel, AMD and VIA CPUs, http://www.agner.org/optimize/instruction_tables.pdf (accessed March 01, 2011)
16. Fong, K., Hankerson, D., López, J., Menezes, A.: Field inversion and point halving revisited. IEEE Transactions on Computers 53(8), 1047–1059 (2004)
17. Grabher, P., Großschädl, J., Page, D.: On software parallel implementation of cryptographic pairings. Cryptology ePrint Archive, Report 2008/205 (2008), http://eprint.iacr.org/
18. Guajardo, J., Paar, C.: Itoh-Tsujii inversion in standard basis and its application in cryptography and codes. Designs, Codes and Cryptography 25(2), 207–216 (2002)
19. Gueron, S.: Intel Advanced Encryption Standard (AES) Instructions Set. White paper, http://software.intel.com/
20. Gueron, S., Kounavis, M. E.: Carry-less multiplication and its usage for computing the GCM mode. White paper, http://software.intel.com/
21. Hankerson, D., Menezes, A.J., Vanstone, S.: Guide to Elliptic Curve Cryptography. Springer, Secaucus (2004)
22. Intel. Intel SSE4 Programming Reference. Technical Report, http://software.intel.com/
23. Itoh, T., Tsujii, S.: A fast algorithm for computing multiplicative inverses in GF(2^m) using normal bases. Inf. Comput. 78(3), 171–177 (1988)
24. Järvinen, K.: Optimized FPGA-based elliptic curve cryptography processor for high-speed applications. Integration, the VLSI Journal (to appear)
25. Karatsuba, A., Ofman, Y.: Multiplication of many-digital numbers by automatic computers. Doklady Akad. Nauk SSSR 145, 293–294 (1962); Translation in Physics-Doklady 7, 595–596 (1963)
26. Kim, K.H., Kim, S.I.: A new method for speeding up arithmetic on elliptic curves over binary fields. Cryptology ePrint Archive, Report 2007/181 (2007), http://eprint.iacr.org/
27. King, B., Rubin, B.: Improvements to the Point Halving Algorithm. In: Wang, H., Pieprzyk, J., Varadharajan, V. (eds.) ACISP 2004. LNCS, vol. 3108, pp. 262–276. Springer, Heidelberg (2004)

28. Knudsen, E.: Elliptic Scalar Multiplication Using Point Halving. In: Lam, K.-Y., Okamoto, E., Xing, C. (eds.) ASIACRYPT 1999. LNCS, vol. 1716, pp. 135–149. Springer, Heidelberg (1999)
29. Koblitz, N.: CM-Curves with Good Cryptographic Properties. In: Feigenbaum, J. (ed.) CRYPTO 1991. LNCS, vol. 576, pp. 279–287. Springer, Heidelberg (1992)
30. Longa, P., Gebotys, C.H.: Efficient techniques for high-speed elliptic curve cryptography. In: Mangard, S., Standaert, F.-X. (eds.) CHES 2010. LNCS, vol. 6225, pp. 80–94. Springer, Heidelberg (2010)
31. López, J., Dahab, R.: Fast Multiplication on Elliptic Curves over GF(2^m) without Precomputation. In: Koç, Ç.K., Paar, C. (eds.) CHES 1999. LNCS, vol. 1717, pp. 316–327. Springer, Heidelberg (1999)
32. López, J., Dahab, R.: High-Speed Software Multiplication in GF(2^m). In: Roy, B., Okamoto, E. (eds.) INDOCRYPT 2000. LNCS, vol. 1977, pp. 203–212. Springer, Heidelberg (2000)
33. Montgomery, P.L.: Five, six, and seven-term Karatsuba-like formulae. IEEE Transactions on Computers 54(3), 362–369 (2005)
34. National Institute of Standards and Technology (NIST). Recommended Elliptic Curves for Federal Government Use. NIST Special Publication (July 1999), http://csrc.nist.gov/csrc/fedstandards.html
35. Schroeppel, R.: Elliptic curves: Twice as fast! Presentation at the CRYPTO 2000 [4] Rump Session (2000)
36. Solinas, J.A.: Efficient arithmetic on Koblitz curves. Designs, Codes and Cryptography 19(2-3), 195–249 (2000)
37. Wall, D.W.: Limits of instruction-level parallelism. In: 4th International Conference on Architectural Support for Programming Languages and Operating System (ASPLOS 1991), pp. 176–188. ACM, New York (1991)
38. Wulf, W.A., McKee, S.A.: Hitting the Memory Wall: Implications of the Obvious. SIGARCH Computer Architecture News 23(1), 20–24 (1995)

High-Speed High-Security Signatures

Daniel J. Bernstein[1], Niels Duif[2], Tanja Lange[2], Peter Schwabe[3], and Bo-Yin Yang[4]

[1] Department of Computer Science
University of Illinois at Chicago, Chicago, IL 60607–7045, USA
djb@cr.yp.to

[2] Department of Mathematics and Computer Science
Technische Universiteit Eindhoven, P.O. Box 513, 5600 MB Eindhoven, Netherlands
nielsduif@hotmail.com, tanja@hyperelliptic.org

[3] Department of Electrical Engineering
National Taiwan University
1, Section 4, Roosevelt Road, Taipei 10617, Taiwan
peter@cryptojedi.org

[4] Institute of Information Science
Academia Sinica, 128 Section 2 Academia Road, Taipei 115-29, Taiwan
by@crypto.tw

Abstract. This paper shows that a $390 mass-market quad-core 2.4GHz Intel Westmere (Xeon E5620) CPU can create 108000 signatures per second and verify 71000 signatures per second on an elliptic curve at a 2^{128} security level. Public keys are 32 bytes, and signatures are 64 bytes. These performance figures include strong defenses against software side-channel attacks: there is no data flow from secret keys to array indices, and there is no data flow from secret keys to branch conditions.

Keywords: Elliptic curves, Edwards curves, signatures, speed, software side channels, foolproof session keys.

1 Introduction

This paper introduces software for public-key signatures with several attractive features:

– **Fast single-signature verification.** The software takes only 280880 cycles to verify a signature on Intel's widely deployed Nehalem/Westmere lines of CPUs. (This performance measurement is for short messages; for very long messages, verification time is dominated by hashing time.) Nehalem and

This work was supported by the National Science Foundation under grant 1018836, by the European Commission under Contract ICT-2007-216676 ECRYPT II, and by the National Science Council, National Taiwan University and Intel Corporation under Grant NSC99-2911-I-002-001 and 99-2218-E-001-007, and the Academia Sinica Career Award. Part of this work was carried out when Peter Schwabe was employed by Academia Sinica, Taiwan. Part of this work was carried out when Niels Duif was employed by Compumatica secure networks BV, the Netherlands. Permanent ID of this document: a1a62a2f76d23f65d622484ddd09caf8. Date: 2011.07.04.

B. Preneel and T. Takagi (Eds.): CHES 2011, LNCS 6917, pp. 124–142, 2011.

Westmere include all Core i7, i5, and i3 CPUs released between 2008 and 2010, and most Xeon CPUs released in the same period.

- **Even faster batch verification.** The software performs a batch of 64 separate signature verifications (verifying 64 signatures of 64 messages under 64 public keys) in only 8.55 million cycles, i.e., under 134000 cycles per signature. The software fits easily into L1 cache, so contention between cores is negligible: a quad-core 2.4GHz Westmere verifies 71000 signatures per second, while keeping the maximum verification latency below 4 milliseconds.
- **Very fast signing.** The software takes only 88328 cycles to sign a message. A quad-core 2.4GHz Westmere signs 108000 messages per second.
- **Fast key generation.** Key generation is almost as fast as signing. There is a slight penalty for key generation to obtain a secure random number from the operating system; `/dev/urandom` under Linux costs about 6000 cycles.
- **High security level.** All known attacks take at least 2^{128} operations. This is the security level achieved by AES-128, NIST P-256, RSA with $\approx$ 3000-bit keys, etc. The same techniques would also produce speed improvements at other security levels.
- **Foolproof session keys.** Signatures in this paper are generated deterministically; key generation consumes new randomness but new signatures do not. This is not only a speed feature but also a security feature, directly relevant to the recent collapse of the Sony PlayStation 3 security system. See Section 2 for further discussion.
- **Collision resilience.** Hash-function collisions do not break this system. This adds a layer of defense against the possibility of weakness in the selected hash function.
- **No secret array indices.** The software never reads or writes data from secret addresses in RAM; the pattern of addresses is completely predictable. The software is therefore immune to cache-timing attacks, hyperthreading attacks, and other side-channel attacks that rely on leakage of addresses through the CPU cache.
- **No secret branch conditions.** The software never performs conditional branches based on secret data; the pattern of jumps is completely predictable. The software is therefore immune to side-channel attacks that rely on leakage of information through the branch-prediction unit.
- **Small signatures.** Signatures fit into 64 bytes. These signatures are actually compressed versions of longer signatures; the times for compression and decompression are included in the cycle counts reported above.
- **Small keys.** Public keys consume only 32 bytes. The times for compression and decompression are again included.

We have submitted our software to the eBATS project [9] for public benchmarking, and placed the software into the public domain to maximize reusability. The numbers 88328 and 280880 shown above are from the eBATS reports for our software on a Westmere CPU (Intel Xeon E5620, `hydra2`).

Our signatures are elliptic-curve signatures, carefully engineered at several levels of design and implementation to achieve very high speeds without compromising security. Section 2 specifies the signature system; Section 3 explains the techniques we use for finite-field arithmetic; Section 4 discusses fast signatures; Section 5 discusses fast verification.

Comparison to Previous ECC Work. Carrying out high-security elliptic-curve signature verification in only 134000 cycles on a single core of a typical Intel CPU is unprecedented. The following paragraphs discuss previous work.

Readers should be aware of several difficulties in comparing ECC performance results. First, most papers on fast ECC have been limited to ECDH (variable-base-point single-scalar multiplication) and have not implemented ECC signature verification, although there are certainly some exceptions — for example, [12] reported verification 1.33× slower than ECDH, and [22] reported verification 1.36× slower than ECDH. Second, most implementations use secret array indices and secret branch conditions and therefore must be assumed to be breakable by side-channel attacks, as illustrated by the successful OpenSSL attack in [14]; this is not an issue for public-key signature verification but it is an issue for signing and for ECDH. Third, most papers report results for only a few CPUs, so anyone without access to the same CPUs must engage in error-prone extrapolation from one CPU to another; this is not an issue for systems included in the eBATS benchmarks, but we are aware of two recent ECC implementations (discussed below) that are not included in eBATS.

Before this paper, the closest system to ours in eBATS was `ecdonaldp256`: ECDSA signatures using the NIST P-256 elliptic curve. On `hydra2` this system takes 1690936 cycles for key generation, 1790936 cycles for signing, and 2087500 cycles for verification. Better speeds were reported for ECDH: third place was `curve25519`, an implementation by Gaudry and Thomé [23] of Bernstein's Curve25519 [6]; second place was 307180 cycles for `ecfp256e`, an implementation by Hisil [27] of ECDH on an Edwards curve with similar security properties to Curve25519; and first place was 278256 cycles for `gls1271`, an implementation by Galbraith, Lin, and Scott [22] of ECDH on an Edwards curve with an endomorphism. The recent papers [26] and [29] point out security problems with endomorphisms in some ECC-based protocols, but as far as we can tell those security issues are not relevant to ECDH with standard hashing of the ECDH output, and are not relevant to ECC signatures.

Longa and Gebotys in [34] reported 281000 cycles on a Core 2 Duo E6750 for ECDH on a curve similar to `ecfp256e`, and 229000 cycles for ECDH on a curve similar to `gls1271`. The software in [34] is not included in the eBATS benchmarks and apparently is not publicly available, so we are unable to benchmark it on a Westmere. More recently Käsper in [30] reported 457813 cycles for side-channel-protected ECDH on the NIST P-224 curve on a Core 2 Duo E8400; this software is not in eBATS but has been integrated into OpenSSL.

To aid comparisons we also implemented ECDH, specifically `curve25519`, with the same side-channel defenses as our signature software (no secret array indices, and no secret branch conditions). We submitted our ECDH software to eBATS, which reports that the software uses 226872 cycles on `hydra2` for variable-base-point single-scalar multiplication. This is a new speed record for public ECDH software, a new speed record for side-channel-protected ECDH (out of all the papers mentioned above, the only ones that report side-channel protection are [6] and [30]), and a new speed record for ECDH without endomorphisms. It is even slightly better than the speed in [34] for non-side-channel-protected ECDH with endomorphisms.

Given this ECDH speed, given the ECDH-to-verification slowdowns reported in [12] and [22], and given the extra costs that we incur for decompressing keys and signatures, one would expect a verification speed close to 400000 cycles. We do better than this for several reasons, the most important reason being our use of batching. This requires careful design of the signature system, as discussed later in this paper: ECDSA, like DSA and most other signature systems, is incompatible with fast batch verification.

Comparison to Other Signature Systems. The eBATS benchmarks cover 42 different signature systems, including various sizes of RSA, DSA, ECDSA, hyperelliptic-curve signatures, and multivariate-quadratic signatures. This paper beats almost all of the signature times and verification times (and key-generation times, which are an issue for some applications) by more than a factor of 2. The only exceptions are as follows:

- Multivariate-quadratic signatures are competitive in speed. For example, `sflashv2` takes 124740 cycles to sign and 165884 cycles to verify; `mqqsig256` takes 4216 cycles to sign and 134920 cycles to verify; smaller `mqqsig` versions are even faster. However, `sflashv2` was broken by Dubois, Fouque, Shamir, and Stern in [19]. We are not aware of any security evaluation of `mqqsig`, which was introduced last year in [24], but we disregard `mqqsig256` for the simple reason that it has a 789552-byte public key.
- `donald512` (512-bit DSA) takes 337084 cycles to verify. This is comparable to our single-signature verification speed but much slower than our batch verification speed. This is also at a far lower security level, breakable in about 2^{60} operations rather than 2^{128}.
- Some RSA-type systems provide faster verification — but this advantage decreases as the security level increases, and for many applications the advantage is outweighed by much slower signatures and much larger keys. For example, `rwb0fuz1024` (1024-bit Rabin–Williams) uses 12304 cycles to verify but 1751284 cycles to sign and 128 bytes for a public key; `ronald1024` (1024-bit RSA) uses 60628 cycles to verify but 2176212 cycles to sign and 128 bytes for a public key; `ronald3072` (3072-bit RSA) uses 230260 cycles to verify but an astonishing 31469536 cycles to sign and 384 bytes for a public key. This paper uses 134000 cycles to verify (in batches), 89416 cycles to sign, and 32 bytes for a public key.

The conventional wisdom is that RSA signatures are much better than ECC signatures in applications where each signature is verified many times, since RSA verification is much faster than ECC verification. Our ECC speed results call this conventional wisdom into question. We do not claim that our verification speeds cannot be beaten by RSA at the same security level, but we do claim that they are fast enough to make ECC an attractive option even for verification-intensive applications such as [43].

2 The Signature System

This section specifies the signature system used in this paper, and a generalized signature system EdDSA that can be used with other choices of elliptic curves.

There is an extensive literature on variants of the classic signature system introduced by ElGamal in [21]; notable variants include Schnorr's signature system [44], DSA, and ECDSA. Our generalized system is another of these variants. We do not claim novelty for any of the individual modifications that we use, but we emphasize that selecting a good combination of modifications is critical for top performance. The most obvious modification is that we use twisted Edwards curves rather than Weierstrass curves; this explains our choice of the name EdDSA (Edwards-curve Digital Signature Algorithm).

EdDSA Parameters. EdDSA has six parameters: an integer $b \geq 10$; a cryptographic hash function H producing $2b$-bit output; a prime power q congruent to 1 modulo 4; a $(b-1)$-bit encoding of elements of the finite field $\mathbf{F}_q$; a non-square element d of $\mathbf{F}_q$; a prime ℓ between 2^{b-4} and 2^{b-3} satisfying an extra constraint described below; and an element $B \neq (0,1)$ of the set

$$E = \{(x,y) \in \mathbf{F}_q \times \mathbf{F}_q : -x^2 + y^2 = 1 + dx^2y^2\}.$$

The condition that d is not a square implies that $d \notin \{0,-1\}$, so this set E forms a group with neutral element $0 = (0,1)$ under the twisted Edwards addition law

$$(x_1,y_1) + (x_2,y_2) = \left(\frac{x_1y_2 + x_2y_1}{1 + dx_1x_2y_1y_2}, \frac{y_1y_2 + x_1x_2}{1 - dx_1x_2y_1y_2}\right)$$

introduced by Bernstein, Birkner, Joye, Lange, and Peters in [7]. Completeness of the addition law — the fact that the denominators $1 \pm dx_1x_2y_1y_2$ are nonzero — follows as explained in [7, Section 6]: -1 is a square in $\mathbf{F}_q$ (since q is congruent to 1 modulo 4), so this addition law on E is $\mathbf{F}_q$-isomorphic to the Edwards addition law on the Edwards curve $x^2 + y^2 = 1 - dx^2y^2$, which is complete by [8, Theorem 3.3] since $-d$ is not a square in $\mathbf{F}_q$. The latter follows from d being a non-square and -1 being a square in $\mathbf{F}_q$. The extra constraint mentioned above is that $\ell B = 0$, where nB means the nth multiple of B in this group.

We use the encoding of $\mathbf{F}_q$ to define some field elements as being negative: specifically, x is negative if the $(b-1)$-bit encoding of x is lexicographically larger than the $(b-1)$-bit encoding of $-x$. If q is an odd prime and the encoding is the little-endian representation of $\{0,1,\ldots,q-1\}$ then the negative elements of $\mathbf{F}_q$ are $\{1,3,5,\ldots,q-2\}$.

An element $(x,y) \in E$ is encoded as a b-bit string $\underline{(x,y)}$, namely the $(b-1)$-bit encoding of y followed by a sign bit; the sign bit is 1 iff x is negative. This encoding immediately determines y, and it determines x via the equation $x = \pm\sqrt{(y^2-1)/(dy^2+1)}$.

EdDSA Keys and Signatures. An EdDSA secret key is a b-bit string k. The hash $H(k) = (h_0, h_1, \ldots, h_{2b-1})$ determines an integer

$$a = 2^{b-2} + \sum_{3 \leq i \leq b-3} 2^i h_i \in \{2^{b-2}, 2^{b-2}+8, \ldots, 2^{b-1}-8\},$$

which in turn determines the multiple $A = aB$. The corresponding EdDSA public key is $\underline{A}$. Bits $h_b, \ldots, h_{2b-1}$ of the hash are used as part of signing, as discussed in a moment.

The signature of a message M under this secret key k is defined as follows. Define $r = H(h_b, \ldots, h_{2b-1}, M) \in \{0, 1, \ldots, 2^{2b} - 1\}$; here we interpret $2b$-bit strings in little-endian form as integers in $\{0, 1, \ldots, 2^{2b} - 1\}$. Define $R = rB$. Define $S = (r + H(\underline{R}, \underline{A}, M)a) \bmod \ell$. The signature of M under k is then the $2b$-bit string $(\underline{R}, \underline{S})$, where $\underline{S}$ is the b-bit little-endian encoding of S. Applications wishing to pack data into every last nook and cranny should note that the last three bits of signatures are always 0 because ℓ fits into $b - 3$ bits.

Verification of an alleged signature on a message M under a public key works as follows. The verifier parses the key as $\underline{A}$ for some $A \in E$, and parses the alleged signature as $(\underline{R}, \underline{S})$ for some $R \in E$ and $S \in \{0, 1, \ldots, \ell - 1\}$. The verifier computes $H(\underline{R}, \underline{A}, M)$ and then checks the group equation $8SB = 8R + 8H(\underline{R}, \underline{A}, M)A$ in E. The verifier rejects the alleged signature if the parsing fails or if the group equation does not hold.

To see that signatures pass verification, simply multiply B by the equation $S = (r + H(\underline{R}, \underline{A}, M)a) \bmod \ell$, and use the fact that $\ell B = 0$, to see that $SB = rB + H(\underline{R}, \underline{A}, M)aB = R + H(\underline{R}, \underline{A}, M)A$. The verifier is *permitted* to check this stronger equation and to reject alleged signatures where the stronger equation does not hold. However, this is not *required*; checking that $8SB = 8R + 8H(\underline{R}, \underline{A}, M)A$ is enough for security.

Weak Keys. Forgeries are trivial if A is a known multiple of B. For example, an attacker who knows that $A = 37B$ can choose r and compute $S = (r + 37H(\underline{R}, \underline{A}, M)) \bmod \ell$. As an even more extreme example, an attacker who knows that $A = 0B$ can choose r and compute $S = r \bmod \ell$, independently of M. We could declare that $0B$ and $37B$ are "broken" by these two "attacks" and that users must check for, and reject, these "weak keys"; but the same confused logic would require rejecting *all* keys in *all* cryptosystems, and would have no relevance to the standard definition of signature security.

Legitimate users choose $A = aB$, where a is a random secret; the derivation of a from $H(k)$ ensures adequate randomness. These users have negligible chance of generating any particular multiple of B targeted by the attacker (and no chance of generating $0B$). The chance of the attacker randomly guessing a is far smaller than the chance of the attacker computing a by known discrete-logarithm algorithms; standard elliptic-curve security criteria are designed so that the latter algorithms have negligible chance of succeeding in any reasonable amount of time.

Malleability. We also see no relevance of "malleability" to the standard definition of signature security. For example, if we slightly modified the system then replacing S by $-S$ and replacing A by $-A$ (a slight variant of the "attack" of [45]) would convert one valid signature into another valid signature of the same message under a new public key; but it would still not accomplish the attacker's goal, namely to forge a signature on a new message under a target public key. One such modification would be to omit $\underline{A}$ from the hashing; another such modification would be to have $\underline{A}$ encode only $|A|$, rather than A.

Choice of Curve. Our recommended curve for EdDSA is a twisted Edwards curve birationally equivalent to the curve Curve25519 from [6]. Any efficiently computable birational equivalence preserves ECDLP difficulty, so the well-known difficulty of computing ECDLP for Curve25519 immediately implies the difficulty of computing ECDLP for our curve. We use the name Ed25519 for EdDSA with this particular choice of curve.

Specifically, Ed25519 is EdDSA with the following parameters: $b = 256$; H is SHA-512; q is the prime $2^{255} - 19$; the 255-bit encoding of $\mathbf{F}_{2^{255}-19}$ is the usual little-endian encoding of $\{0, 1, \ldots, 2^{255} - 20\}$; ℓ is the prime $2^{252} + 27742317777372353535851937790883648493$ from [6]; $d = -121665/121666 \in \mathbf{F}_q$; and B is the unique point $(x, 4/5) \in E$ for which x is positive.

Curve25519 from [6] is the Montgomery curve $v^2 = u^3 + 486662u^2 + u$ over the same field $\mathbf{F}_q$. Bernstein and Lange pointed out in [8, Section 2] that Curve25519 is birationally equivalent to an Edwards curve, specifically $x^2 + y^2 = 1 + (121665/121666)x^2y^2$; the equivalence is $x = \sqrt{486664}u/v$ and $y = (u-1)/(u+1)$. As above this Edwards curve is isomorphic to $-x^2 + y^2 = 1 - (121665/121666)x^2y^2$ since -1 is a square in $\mathbf{F}_q$. Our choice of base point B corresponds to the choice $u = 9$ made in [6].

Pseudorandom Generation of *r*. ECDSA, like many other signature systems, asks users to generate not merely a random long-term secret key, but also a new random secret session key r for each message to be signed. If r becomes public then, assuming $H(\underline{R}, \underline{A}, M) \bmod \ell \neq 0$, the long-term secret key a can be simply computed as $a = (S - r)/H(\underline{R}, \underline{A}, M) \bmod \ell$. If the same value r is ever used for 2 different messages the secret key can be computed as well, as ElGamal noted in [21]. It was reported in [15] that the latter failure had occurred in Sony's ECDSA implementation for code-signing for the PlayStation3, immediately revealing Sony's long-term secret key.

Furthermore, it is well known that ECDSA's session keys are much less tolerant than the long-term key of slight deviations from randomness, even if the session keys are not revealed or reused. For example, Nguyen and Shparlinski in [40] presented an algorithm using lattice methods to compute the long-term ECDSA key from the knowledge of as few as 3 bits of r for hundreds of signatures, whether this knowledge is gained from side-channel attacks or from non-uniformity of the distribution from which r is taken.

EdDSA avoids these issues by generating $r = H(h_b, \ldots, h_{2b-1}, M)$, so that different messages will lead to different, hard-to-predict values of r. No per-message randomness is consumed. Standard PRF hypotheses imply that this session key r is indistinguishable from a truly random string generated independently for each M, so there is no loss of security. This idea of generating random signatures in a secretly deterministic way, in particular obtaining pseudorandomness by hashing a long-term secret key together with the input message, was proposed by Barwood in [3]; independently by Wigley in [47]; a few months later in a patent application [36] by Naccache, M'Raïhi, and Levy-dit-Vehel; later by

M'Raïhi, Naccache, Pointcheval, and Vaudenay in [35]; and much later by Katz and Wang in [31]. The patent application was abandoned in 2003.

EdDSA samples r from the interval $[0, 2^{2b} - 1]$, ensuring almost uniformity of the distribution modulo ℓ. The guideline [1, Section 4.1.1, Algorithm 2] specifies that the interval should be of size at least $[0, 2^{b+61} - 1]$, i.e., 64 bits more than ℓ; for Ed25519 there are 259 extra bits.

Comparison to Previous ElGamal Variants. The ElGamal signature system works as follows: generate a random rB for each message to be signed, and compute the signature (X, S), where X is the x-coordinate of $R = rB$ and $S = r^{-1}(H(M) + Xa) \bmod \ell$. The verifier can compute $R = S^{-1}H(M)B + S^{-1}XA$ using the public key $A = aB$ and can then verify that $X = x(R)$. (We disregard the possibility $S = 0$, which has negligible chance of occurring even under adversarial input; ECDSA is defined to check for this possibility and generate a new r, but sensible implementations will skip that check.) ElGamal's system actually uses the multiplicative group $\mathbf{F}_q^*$ with non-prime $\ell = q - 1$; ECDSA uses an elliptic-curve group with prime ℓ.

Schnorr in [44] replaced ElGamal's equation $S = r^{-1}(H(M) + x(R)a) \bmod \ell$ with the equation $S = (r + H(\underline{R}, M)a) \bmod \ell$. Schnorr's system has two attractive features:

- No inversions. This is an obvious advantage, saving time and reducing code size both for the signer and for the verifier.
- Collision resilience. The presence of $\underline{R}$ in the hash means that the attacker cannot break Schnorr's system by merely finding hash collisions.

Practical use of Schnorr's system was hampered by a patent (which expired in 2008), but the system became well known to theoreticians: the hashing of $\underline{R}$ allowed a proof (using the "forking lemma") that breaking Schnorr's system is as difficult "in the random-oracle model" as breaking DLP. See, for example, [42], [5], and [39]. We do not mean to exaggerate the real-world relevance of "provable security", but we find it obvious that Schnorr's system is a conservative, well-studied signature system.

Schnorr's signatures were not exactly (R, S): Schnorr, like ElGamal, compressed R to the hash $H(\underline{R}, M)$. The verifier can undo this compression by computing R as $SB - H(\underline{R}, M)A$. Note that this compression is public, so it cannot affect security. Neven, Smart, and Warinschi in [39] proposed taking advantage of collision resilience by choosing H to output only $b/2$ bits, reducing the size of compressed signatures by 25%; but the same proposal had actually appeared twenty years earlier in Schnorr's original paper. See [44, Section 2]. Compression of R to a hash had a much larger effect in ElGamal's original system: the system used b bits of output from H (and could not use fewer, because it was not collision-resilient), but the system used multiplicative groups rather than elliptic curves, so R needed many more than b bits. The same compression also appears in ECDSA but has no benefit there: ECDSA's hash is the same size as $\underline{R}$.

Our verification equation is the same as Schnorr's verification equation with double-size hashing instead of half-size hashing, with $\underline{A}$ inserted as an extra hash input, and *without* the compression described in the previous paragraph. These modifications obviously do not compromise security. The use of double-size hashing helps alleviate concerns regarding hash-function security; the use of $\underline{A}$ is an inexpensive way to alleviate concerns that several public keys could be attacked simultaneously; and the avoidance of compression allows an important verification speedup, as discussed in Section 5. We also reuse the double-size hash to alleviate concerns regarding nonce randomness, as discussed above.

3 Fast Arithmetic Modulo $2^{255} - 19$

This section explains how our software represents elements of the field $\mathbf{F}_{2^{255}-19}$, and how our software performs efficient field arithmetic. The machine instructions used in the software are available on all 64-bit Intel and AMD CPUs, but we target Intel's Nehalem/Westmere CPUs.

Multipliers on Nehalem CPUs. Field multiplications (and squarings) are the main bottlenecks in elliptic-curve performance on most CPUs. The most important tool for fast field multiplication is a fast CPU multiplication instruction. Nehalem CPUs offer three different multiplication instructions that can be used to implement high-speed field arithmetic:

- The `mulpd` instruction can perform two double-precision floating-point multiplications in SIMD fashion every cycle. Newer Sandy Bridge CPUs include a `vmulpd` instruction that can perform up to 4 double-precision floating-point multiplications per cycle, but this instruction is not available on our target CPUs.
- The `mul` instruction can multiply two 64-bit unsigned integers, producing a 128-bit result, every two cycles.
- The `pmuldq`/`pmuludq` instructions can perform two multiplications of 32-bit integers, producing 64-bit results, every cycle. The `pmuldq` instruction performs signed multiplication; the `pmuludq` instruction performs unsigned multiplication.

Multiplication and Edwards-curve arithmetic involve data-level parallelism that we could exploit with `mulpd` and `pmuldq`, but this approach would incur a serious overhead of shuffle instructions needed to arrange data in registers as described in, e.g., [17] and [38]. This overhead is eliminated when several independent computations are run in parallel, but two 64-bit results every cycle are not fundamentally better than one 128-bit result every two cycles. We therefore decompose field multiplication into multiplications of 64-bit unsigned integers.

Radix-2^{64} Representation. The standard way to split 255-bit values into 64-bit limbs is a 4-limb, radix-2^{64} representation. Each element x of the field is represented as (x_0, x_1, x_2, x_3) with $x = \sum_{i=0}^{3} x_i 2^{64i}$. The multiplication of two elements x and y is decomposed into 16 multiplications of 64-bit unsigned integers; the 128-bit results are added up to produce the result in 8 limbs $r_0, \ldots, r_7$.

Reduction modulo $2^{255} - 19$ exploits the fact that $2^{256} \equiv 38$, so $38 \cdot r_4$ is added to r_0, $38 \cdot r_5$ to r_1 and so on.

A detail worth noting of this representation is that it uses 256 bits to represent 255-bit field elements. We use this one extra bit and do not always reduce modulo $2^{255} - 19$ but modulo $2^{256} - 38$. For a similar representation this has been shown to be useful for example in [10].

Our implementation of the signature scheme based on this representation of field elements yields high performance on many microprocessors such as AMD K10 or 65-nm Intel Core 2 processors. However, on our target platform, the Intel Nehalem/Westmere CPUs, this representation triggers a serious bottleneck. Every 128-bit result of the `mul` instruction is produced in two 64-bit registers. Adding two of these results requires two addition instructions. In the field multiplication most of these additions produce carries; the carry bits need to be handled by subsequent additions. The Intel Nehalem and Westmere CPUs can perform three additions per cycle, but only if these additions do not have to handle a carry bit from a previous addition (`add` instruction). An add with carry (`adc` instruction) can only be done once every two cycles; i.e., carry bits decrease addition throughput by a factor of 6. This bottleneck is triggered not only inside field multiplication and squaring but also inside additions.

Radix-2^{51} Representation. To reduce the number of expensive `adc`/`subc` instructions, we instead represent an element x of $\mathbf{F}_{2^{255}-19}$ as $(x_0, x_1, x_2, x_3, x_4)$ with $x = \sum_{i=0}^{4} x_i 2^{51i}$.

The 5 limbs are unsigned integers. We can represent each element of the field $\mathbf{F}_{2^{255}-19}$ with each $x_i \in [0, \ldots, 2^{51} - 1]$. In fact our implementation does not enforce these bounds except for comparisons. Multiplication accepts inputs with each limb having up to 54 bits and produces results of which each limb can be only slightly larger than 2^{51}.

Multiplication and Squaring. Schoolbook multiplication of two field elements x and y, each represented in 5 unsigned integers, takes 25 `mul` instructions. The results are again produced in two 64-bit integer registers, but as both inputs have only up to 54 bits, the value in the upper result register has only up to 44 bits. Adding two multiplication results now takes only one `adc` and one `add` instruction. Furthermore reduction can be carried out simultaneously to multiplication. For example, we do not compute a coefficient r_5. Whenever the result of a `mul` instruction belongs to r_5, for example in the multiplication of $x_2 \cdot y_3$, we multiply one of the inputs by 19 and add the result to r_0. Similarly we do not compute r_6, r_7, r_8 and r_9 but directly add into $r_1, \ldots, r_4$. Multiplying one input by 19 yields a result with less than 64 bits so we can use the faster `imul` instruction for these multiplications. The 5 result coefficients require 10 64-bit registers; the AMD64 architecture has 15 such registers, so we can keep the result coefficients inside registers throughout the computation.

After the multiplication we need to reduce (carry) the 5 coefficients to obtain a result with coefficients that are at most slightly larger than 2^{51}. Denote the two registers holding coefficient r_0 as r_{00} and r_{01} with $r_0 = 2^{64} r_{01} + r_{00}$. Similarly denote the two registers holding coefficient r_1 as r_{10} and r_{11}. We first shift r_{01}

left by 13, while shifting in the most significant bits of r_{00} (`shld` instruction) and then compute the logical and of r_{00} with $2^{51} - 1$. We do the same with r_{10} and r_{11} and add r_{01} into r_{10} after the logical and with $2^{51} - 1$. We proceed this way for coefficients $r_2, \ldots, r_4$; register r_{41} is multiplied by 19 before adding it to r_{00}. Now all 5 coefficients fit into 64-bit registers but are still too large to be used as input to another multiplication. We therefore carry from r_0 to r_1, from r_1 to r_2, from r_2 to r_3, from r_3 to r_4, and finally from r_4 to r_0. Each of these carries is done as one copy, one right shift by 51, one logical and with $2^{51} - 1$, and one addition.

Squaring needs only 15 `mul` instructions. Some inputs are multiplied by 2; this is combined with multiplication by 19 where possible. The coefficient reduction after squaring is the same as for multiplication.

Multiplication and squaring are implemented as separate functions, but calls to these functions are used only for inversion (see below). Edwards-curve arithmetic uses inlined functions for point addition and doubling.

Addition, Subtraction, and Inversion. The results of additions do not have to be reduced if they are used as input to a multiplication. Long sequences of additions that let coefficients grow larger than 54 bits would be a problem but we do not have such sequences of additions. Field addition is therefore nothing but 5 integer additions without carries (`add` instruction). Subtraction is slightly more expensive because we use unsigned coefficients. Therefore we first add a multiple of q and then perform subtraction. This costs 5 `add` and 5 `sub` instructions.

Inversion is implemented as exponentiation with exponent $q - 2$. It uses the same sequence of 255 squarings and 11 multiplications as [6].

4 Signing Messages

Signature generation has three steps: (1) computing $r = H(h_b, \ldots, h_{2b-1}, M)$; (2) computing $R = rB$; (3) computing $S = (r + H(\underline{R}, \underline{A}, M)a) \bmod \ell$.

Our primary concern is with short messages M, obviously the top concern for a server trying to keep up with a given volume of data; longer messages take more cycles per signature but far fewer cycles per byte. The computations of H take negligible time for short messages. The reduction modulo ℓ also takes negligible time with standard branchless techniques. For the rest of this section we focus on the main signing bottleneck, namely computing rB given r.

High-Level Strategy. We begin by computing the 253-bit integer $r \bmod \ell$. We then write $r \bmod \ell$ as $r_0 + 16r_1 + \cdots + 16^{63}r_{63}$ with

$$r_i \in \{-8, -7, -6, -5, -4, -3, -2, -1, 0, 1, 2, 3, 4, 5, 6, 7\}.$$

For each i we look up $16^i|r_i|B$ in a precomputed table, and then conditionally negate $16^i|r_i|B$ to obtain $16^i r_i B$. Finally we compute rB as $\sum_i 16^i r_i B$.

There is nothing new in our computation at this level. Computing rB as a sum of precomputed pieces is a special case of a standard scalar-multiplication algorithm published by Pippenger in [41] (subsequently reinvented in [11] and

[33]); allowing negative coefficients is a standard tweak. The devil lies in the lower-level details — choosing the optimal radix 16, and computing $16^i r_i B$ and $\sum_i 16^i r_i B$ as efficiently as possible. These details are discussed below.

Low Level, Part 1: Table Lookups. Recall that, as a side-channel defense, we prohibit secret array indices. In particular, we cannot use $|r_i|$ as an array index. We instead load all table entries $0B, 16^i B, 2 \cdot 16^i B, 3 \cdot 16^i B, 4 \cdot 16^i B, 5 \cdot 16^i B, 6 \cdot 16^i B, 7 \cdot 16^i B, 8 \cdot 16^i B$ and use arithmetic operations, without branching, to combine the table entries into $16^i |r_i| B$. We similarly use arithmetic operations to compute $16^i r_i B$ from $16^i |r_i| B$ and $-16^i |r_i| B$.

We actually store table entries only for $i \in \{0, 2, 4, \ldots, 62\}$, at the expense of 4 elliptic-curve doublings. The table then contains $8 \cdot 32 = 256$ curve points (aside from $0B$, which is not stored). Each point is represented as three integers (see below) modulo $2^{255} - 19$. Each integer in turn is represented as five 8-byte words. Overall the table consumes 30 kilobytes of RAM.

We could instead use radix 32 or larger. Radix 32 would involve twice as many table loads (since we load all table entries), and twice as much arithmetic to combine table entries, but these costs would be outweighed by the benefit of fewer elliptic-curve additions. A more serious concern is that the table would be twice as large, consuming 60KB instead of 30KB. This is only a minor issue for a typical cryptographic speed test on our target CPUs (each Nehalem/Westmere core has its own fast 256KB L2 cache efficiently handling our sequential loads), but 30KB is clearly more attractive inside a larger application that needs to fit several different subroutines into L2 cache.

In the opposite direction, we could chop the table in half again at the expense of 8 more doublings; we could also switch to radix 8, 4, or 2. These changes would also allow reasonably fast signing on much smaller CPUs.

Low Level, Part 2: Elliptic-Curve Addition. We use extended coordinates for the twisted Edwards curve $-x^2 + y^2 = 1 + dx^2y^2$, as proposed by Hisil, Wong, Carter, and Dawson in [28]. These coordinates are $(X : Y : Z : T)$ with $XY = ZT$ representing $x = X/Z$ and $y = Y/Z$. The addition formulas from [28, Section 3.1] are complete for our curve and use just 9 field multiplications to add a table entry (x_0, y_0) into $(X : Y : Z : T)$. Note that these formulas rely on the -1 in $-x^2$; this is why EdDSA uses the -1 twist.

One of the field multiplications is a multiplication by $d = -121665/121666$. We could replace this with a small number of multiplications by 121665 and 121666, as in [7, Section 6], but our current software treats d as a generic field element to save code size. We considered switching to a new curve using a small integer d (such as 646, which has a near-prime group order; note that we do not need the twist security of Curve25519), but decided that the resulting speedup was too small to justify departing from an established curve.

A different way to save a multiplication is to use the dual addition formulas from [28, Section 3.2]. However, those formulas are not complete; they would require a detailed analysis of intermediate results in our computation to see whether any of the intermediate additions could trigger any of the exceptional cases in the formulas.

Instead we represent a precomputed point (x_0, y_0) as $(y_0 - x_0, y_0 + x_0, 2dx_0y_0)$. These values depend only on x_0 and y_0 and are usually computed in the first part of addition in extended coordinates; providing them as part of the precomputation saves the multiplication by d, the multiplication x_0y_0, and 2 field additions, at the expense of increasing the storage requirements by a factor of 1.5. We comment that for hardware implementations this approach reduces the information exposed to template attacks trying to link multiple uses of the same precomputed point: all operations involving the precomputed point also involve the intermediate point. For details see [20, Section 5.1.2].

Results. Overall we spend a bit less than 1000 cycles for each iteration of our main signing loop, i.e., for one table lookup and one elliptic-curve mixed addition. We also spend about 21000 cycles to invert Z at the end of the computation. The complete signing procedure for a short message takes 88328 cycles.

5 Verifying Signatures

Fast signature verification seems considerably more difficult than fast signature generation, for two reasons. First, the verifier has to recover the elliptic-curve points A and R from the compressed points $\underline{A}$ and $\underline{R}$. Second, checking $SB = R + H(\underline{R}, \underline{A}, M)A$ seems to require not merely a fixed-base scalar multiplication SB but also a much more expensive variable-base scalar multiplication $H(\underline{R}, \underline{A}, M)A$. This section explains several techniques that we use to address these problems.

Fast Decompression. Recall that the encoding $\underline{R}$ of a point $R = (x, y)$ contains a straightforward encoding of y but contains only a sign bit for x. One must therefore recover x via the equation $x = \pm\sqrt{(y^2-1)/(dy^2+1)}$; note that $dy^2 + 1 \neq 0$ since $-d$ is not a square. The division and square root here seem to involve two exponentiations, about twice as expensive as the usual Weierstrass-curve decompression.

Of course, we could use Montgomery's trick to merge the two divisions involved in decompressing two points, but two square roots and a division are still more expensive than two Weierstrass-curve decompressions. We could also skip the compression and decompression for applications willing to use 64-byte keys and 96-byte signatures; but we think that 32-byte keys and 64-byte signatures are considerably more attractive.

To save time we look more closely at the standard computation of square roots in $\mathbf{F}_q$. The prime $q = 2^{255} - 19$ is congruent to 5 modulo 8, so any square $\alpha \in \mathbf{F}_q$ satisfies $\alpha^2 = \beta^4$ where $\beta = \alpha^{(q+3)/8}$, i.e., $\pm\alpha = \beta^2$. The standard computation is a single exponentiation to compute β, followed by a quick multiplication of β by $\sqrt{-1}$ if $\beta^2 = -\alpha$.

In the decompression context we are given α as a fraction u/v, where $u = y^2 - 1$ and $v = dy^2 + 1$. Instead of computing α we merge the division with the square-root computation:

$$\beta = (u/v)^{(q+3)/8} = u^{(q+3)/8}v^{q-1-(q+3)/8} = u^{(q+3)/8}v^{(7q-11)/8} = uv^3(uv^7)^{(q-5)/8}.$$

We check whether $\beta^2 = -\alpha$ by checking whether $v\beta^2 = -u$, and if so we multiply β by $\sqrt{-1}$. The entire computation of $\sqrt{u/v}$, starting from u and v, takes just a few multiplications more than a single exponentiation. In other words, Edwards-curve decompression is as inexpensive as Weierstrass-curve decompression.

Fast Single-Signature Verification. To verify a single signature we use standard techniques for double-scalar multiplication to compute $SB - H(\underline{R}, \underline{A}, M)A$, and we then check whether the result is the same as R. (We actually check whether the encoding of the result is the same as the encoding of R, so that we can skip decompression of R.) The speed of Edwards-curve addition, especially with the -1 twist, makes these techniques particularly efficient; using the tables discussed in Section 4 does not seem to offer any advantage. This computation fits in very little space.

We have also considered the verification method suggested by Antipa, Brown, Gallant, Lambert, Struik, and Vanstone in [2], but our very efficient elliptic-curve arithmetic makes the overheads in this method — extra decompression and a Euclidean computation — much more troublesome. In the batch context discussed below, the only extra overhead of the method of [2] would be the Euclidean computation, but the benefit would also be much smaller.

Fast Batch Verification. For any system bottlenecked by signature verification, the problem is not to verify *one* signature at a time, but to verify many signatures as quickly as possible.

Naccache, M'Raïhi, Vaudenay, and Raphaeli in [37, Section 2.2] proposed verifying a batch of linear signature equations by verifying a random linear combination of the equations. This proposal is not directly applicable to ElGamal, DSA, Schnorr, ECDSA, et al., because all of those systems require *computing* linear functions (to compute R) rather than merely *verifying* linear functions; but if R is transmitted instead of $H(\cdots)$, as suggested in [37], then this problem disappears.

Unfortunately, the verification algorithm in [37] was quite slow: [37, Table 1] reported "$29n$" multiplications to verify n signatures from the same signer at a highly questionable 2^{20} security level. If the same technique were adapted to ECDSA and increased to a 2^{128} security level then it would require nearly 200 elliptic-curve additions for each signature from the same signer — somewhat faster than verifying each signature separately, but not much.

The followup paper [4] by Bellare, Garay, and Rabin proposed a more complicated verification technique using, e.g., 3200 multiplications to verify 100 exponentiations, or 6480 multiplications to verify 100 DSA signatures, in both cases at a substandard 2^{60} security level. See [4, Appendix A.1]. The number of multiplications per signature begins to drop as the batch size grows towards 1000 — see [4, Figure 3] — but such large batches do not fit into cache on typical CPUs.

The unimpressive theoretical performance of these batch-verification techniques can be traced directly to the naive exponentiation algorithms used in [37] and [4]. We do much better by using random linear combinations, as in [37], together with state-of-the-art scalar-multiplication techniques.

Specifically, we start from a batch of (M_i, A_i, R_i, S_i) where $(\underline{R_i}, \underline{S_i})$ is an alleged signature of M_i under key $\underline{A_i}$. We choose independent uniform random 128-bit integers z_i, compute $H_i = H(\underline{R_i}, \underline{A_i}, M_i)$, and verify the equation

$$\left(-\sum_i z_i S_i \bmod \ell\right) B + \sum_i z_i R_i + \sum_i (z_i H_i \bmod \ell) A_i = 0$$

by a multi-scalar multiplication. There are two reasonable choices of scalar-multiplication methods here, namely Pippenger's method in [41] and the Bos–Coster method reported in [18, Section 4]. We use the Bos–Coster method because it fits into less storage; see below for details. Note that z_i is not secret, so side-channel protection is not required.

The number of scalars here is $2n + 1$. Half of the scalars are 253-bit and half are 128-bit. If public keys appear repeatedly, the situation considered in [37] and [4], then we could save some time by merging the 253-bit scalars; this merging also explains why we do not use the similar signature equation $SB = A + H(\underline{R}, \underline{A}, M)R$, which would allow only merging 128-bit scalars. Our software focuses on general-purpose verification with arbitrary keys.

If verification succeeds then we are confident that $8S_iB = 8R_i + 8H_iA_i$ for each i, i.e., that each signature is valid. The logic is simple: the differences $P_i = 8R_i + 8H_iA_i - 8S_iB$ are elements of a cyclic group of prime order ℓ, and have been verified to satisfy $\sum_i z_iP_i = 0$; but this equation cannot hold with probability more than 2^{-128} unless all $P_i = 0$. For example, if P_4 is nonzero then the choices of $z_1, z_2, z_3, z_5, z_6, \ldots$ determine exactly one choice of z_4 satisfying $\sum_i z_iP_i = 0$, and z_4 has chance at most 2^{-128} of matching that choice.

If verification fails then there must be at least one invalid signature. We then fall back to verifying each signature separately. There are several techniques to identify a *small* number of invalid signatures in a batch, but all known techniques become slower than separate verification as the number of invalid signatures increases; separate verification provides the best defense against denial-of-service attacks.

Fast Multi-scalar Multiplication. The Bos–Coster method mentioned above is as follows: to compute $n_1P_1 + n_2P_2 + \cdots$, where $n_1 \geq n_2 \geq \cdots$, we recursively compute $(n_1 - n_2)P_1 + n_2(P_1 + P_2) + \cdots$. For n_1 much larger than n_2, say $2^{k+1}n_2 > n_1 \geq 2^kn_2$, we could gain speed by instead recursively computing $(n_1 - 2^kn_2)P_1 + n_2(2^kP_1 + P_2) + \cdots$, but we have found this to occur so rarely that checking for it is not worthwhile.

We keep the scalars n_i in a heap so that identifying the two largest scalars is easy. The usual method to insert a new element into a heap is top-down, starting at the root and swapping down for a variable number of steps. We instead use Floyd's 1964 bottom-up algorithm discussed in [32, Exercise 5.2.3–18] (often miscredited to [16] and [46]): start at the root, swap down to the bottom, and then swap up for a variable number of steps. This has the advantage of somewhat reducing the number of comparisons, and the not-so-well-known advantage of drastically reducing the number of branches, especially for balanced heaps.

Results. The complete verification procedure takes under 134000 cycles per signature for batch size 64. Our batch-verification software is included in, although not yet benchmarked by, the public eBATS benchmarking framework.

Doubling the batch size to 128 no longer fits into L1 cache but still improves performance on our target CPU, taking under 125000 cycles per signature. Larger batches take under 114000 cycles per signature while still fitting into L2 cache. Our software spends about 44000 cycles on decompression, so verification of uncompressed signatures (32 extra bytes) using uncompressed public keys (another 32 extra bytes) would take only about 81000 cycles for batch size 128, even faster than signing. However, in this paper we have emphasized the performance that we obtain without using so much space.

References

[1] — (no editor), Technical guideline TR-03111, elliptic curve cryptography (2009), Citations in this document: §2

[2] Antipa, A., Brown, D.R.L., Gallant, R.P., Lambert, R., Struik, R., Vanstone, S.A.: Accelerated verification of ECDSA signatures. In: Preneel, B., Tavares, S. (eds.) SAC 2005. LNCS, vol. 3897, pp. 307–318. Springer, Heidelberg (2006), Citations in this document: §5, §5

[3] Barwood, G.: Digital signatures using elliptic curves, message `32f519ad.19609226@news.dial.pipex.com` posted to `sci.crypt` (1997), `http://groups.google.com/group/sci.crypt/msg/b28aba37180dd6c6`, Citations in this document: §2

[4] Bellare, M., Garay, J.A., Rabin, T.: Fast batch verification for modular exponentiation and digital signatures. In: Nyberg, K. (ed.) Eurocrypt '98. LNCS, vol. 1403, pp. 236–250. Springer, Heidelberg (1998), Citations in this document: §5, §5, §5, §5, §5

[5] Bellare, M., Neven, G.: Multi-signatures in the plain public-key model and a general forking lemma. In: CCS 2006, pp. 390–399 (2006), Citations in this document: §2

[6] Bernstein, D.J.: Curve25519: new Diffie-Hellman speed records. In: Yung, M., Dodis, Y., Kiayias, A., Malkin, T. (eds.) PKC 2006. LNCS, vol. 3958, pp. 207–228. Springer, Heidelberg (2006), Citations in this document: §1, §1, §2, §2, §2, §2, §3

[7] Bernstein, D.J., Birkner, P., Joye, M., Lange, T., Peters, C.: Twisted Edwards curves. In: Vaudenay, S. (ed.) Africacrypt 2008. LNCS, vol. 5023, pp. 389–405. Springer, Heidelberg (2008), Citations in this document: §2, §2, §4

[8] Bernstein, D.J., Lange, T.: Faster addition and doubling on elliptic curves. In: Kurosawa, K. (ed.) Asiacrypt 2007. LNCS, vol. 4833, pp. 29–50. Springer, Heidelberg (2007), Citations in this document: §2, §2

[9] Bernstein, D.J., Lange, T. (eds.): eBACS: ECRYPT Benchmarking of Cryptographic Systems (2011), `http://bench.cr.yp.to/ebats.html` (accessed July 4, 2011), Citations in this document: §1

[10] Bos, J.W.: High-performance modular multiplication on the Cell processor. In: Hasan, M.A., Helleseth, T. (eds.) WAIFI 2010. LNCS, vol. 6087, pp. 7–24. Springer, Heidelberg (2010), Citations in this document: §3

[11] Brickell, E.F., Gordon, D.M., McCurley, K.S., Wilson, D.B.: Fast exponentiation with precomputation (extended abstract). In: Rueppel, R.A. (ed.) Eurocrypt '92. LNCS, vol. 658, pp. 200–207. Springer, Heidelberg (1993), Citations in this document: §4

[12] Brown, M., Hankerson, D., López, J., Menezes, A.: Software implementation of the NIST elliptic curves over prime fields (2000); see also newer version [13], `http://www.cacr.math.uwaterloo.ca/techreports/2000/corr2000-56.ps`, Citations in this document: §1, §1

[13] Brown, M., Hankerson, D., López, J., Menezes, A.: Software implementation of the NIST elliptic curves over prime fields. In: Naccache, D. (ed.) CT-RSA 2001. LNCS, vol. 2020, pp. 250–265. Springer, Heidelberg (2001); see also older version [12]. MR 1907102

[14] Brumley, B.B., Hakala, R.M.: Cache-timing template attacks. In: Matsui, M. (ed.) Asiacrypt 2009. LNCS, vol. 5912, pp. 667–684. Springer, Heidelberg (2009), Citations in this document: §1

[15] "Bushing", "marcan" Cantero, H.M., Boessenkool, S., Peter, S.: PS3 epic fail (2010), `http://events.ccc.de/congress/2010/Fahrplan/attachments/1780_27c3_console_hacking_2010.pdf`, Citations in this document: §2

[16] Carlsson, S.: Average-case results on heapsort. BIT 27, 2–17 (1987), Citations in this document: §5

[17] Costigan, N., Schwabe, P.: Fast elliptic-curve cryptography on the Cell Broadband Engine. In: Preneel, B. (ed.) Africacrypt 2009. LNCS, vol. 5580, pp. 368–385. Springer, Heidelberg (2009), Citations in this document: §3

[18] de Rooij, P.: Efficient exponentiation using precomputation and vector addition chains. In: De Santis, A. (ed.) Eurocrypt '94. LNCS, vol. 950, pp. 389–399. Springer, Heidelberg (1995), Citations in this document: §5

[19] Dubois, V., Fouque, P.-A., Shamir, A., Stern, J.: Practical cryptanalysis of SFLASH. In: Menezes, A. (ed.) Crypto 2007. LNCS, vol. 4622, pp. 1–12. Springer, Heidelberg (2007), Citations in this document: §1

[20] Duif, N.: Smart card implementation of a digital signature scheme for Twisted Edwards curves, M.A. thesis, Technische Universiteit Eindhoven (2011), Citations in this document: §4

[21] ElGamal, T.: A public key cryptosystem and a signature scheme based on discrete logarithms. IEEE Transactions on Information Theory 31, 469–472 (1985), Citations in this document: §2, §2

[22] Galbraith, S.D., Lin, X., Scott, M.: Endomorphisms for faster elliptic curve cryptography on a large class of curves. In: Joux, A. (ed.) Eurocrypt 2009. LNCS, vol. 5479, pp. 518–535. Springer, Heidelberg (2009), Citations in this document: §1, §1, §1

[23] Gaudry, P., Thomé, E.: The mpFq library and implementing curve-based key exchanges. In: SPEED 2007, pp. 49–64 (2007), Citations in this document: §1

[24] Gligoroski, D., Odegøard, R.S., Jensen, R.E., Perret, L., Faugère, J.-C., Knapskog, S.J., Markovski, S.: The digital signature scheme MQQ-SIG (2010), Citations in this document: §1

[25] Goh, E.-J., Jarecki, S., Katz, J., Wang, N.: Efficient signature schemes with tight reductions to the Diffie-Hellman problems. Journal of Cryptology 20, 493–514 (2007), See [31]

[26] Granger, R.: On the static Diffie-Hellman problem on elliptic curves over extension fields. In: Abe, M. (ed.) Asiacrypt 2010. LNCS, vol. 6477, pp. 283–302. Springer, Heidelberg (2010), Citations in this document: §1

[27] Hisil, H.: Elliptic curves, group law, and efficient computation, Ph.D. thesis, Queensland University of Technology (2010), Citations in this document: §1

[28] Hisil, H., Wong, K.K.-H., Carter, G., Dawson, E.: Twisted Edwards curves revisited. In: Pieprzyk, J. (ed.) Asiacrypt 2008. LNCS, vol. 5350, pp. 326–343. Springer, Heidelberg (2008), Citations in this document: §4, §4, §4

[29] Joux, A., Vitse, V.: Elliptic curve discrete logarithm problem over small degree extension fields. Application to the static Diffie-Hellman problem on $E(\mathbf{F}_{q^5})$ (2010), Citations in this document: §1

[30] Käsper, E.: Fast elliptic curve cryptography in OpenSSL. In: RLCPS 2011 (to appear, 2011), Citations in this document: §1, §1

[31] Katz, J., Wang, N.: Efficiency improvements for signature schemes with tight security reductions. In: CCS 2003, pp. 155–164 (2003); portions incorporated into [25], Citations in this document: §2

[32] Knuth, D.E.: The art of computer programming, volume 3: sorting and searching, 2nd edn. Addison-Wesley, Reading (1998), Citations in this document: §5

[33] Lim, C.H., Lee, P.J.: More flexible exponentiation with precomputation. In: Desmedt, Y.G. (ed.) CRYPTO 1994. LNCS, vol. 839, pp. 95–107. Springer, Heidelberg (1994), Citations in this document: §4

[34] Longa, P., Gebotys, C.: Efficient techniques for high-speed elliptic curve cryptography. In: Mangard, S., Standaert, F.-X. (eds.) CHES 2010. LNCS, vol. 6225, pp. 80–94. Springer, Heidelberg (2010), Citations in this document: §1, §1, §1

[35] M'Raïhi, D., Naccache, D., Pointcheval, D., Vaudenay, S.: Computational alternatives to random number generators. In: Tavares, S., Meijer, H. (eds.) SAC '98. LNCS, vol. 1556, pp. 72–80. Springer, Heidelberg (1999), Citations in this document: §2

[36] Naccache, D., M'Raïhi, D., Levy-dit-Vehel, F.: Patent application WO/1998/051038: pseudo-random generator based on a hash coding function for cryptographic systems requiring random drawing (1997), Citations in this document: §2

[37] Naccache, D., M'Raïhi, D., Vaudenay, S., Raphaeli, D.: Can D.S.A. be improved? Complexity trade-offs with the digital signature standard. In: De Santis, A. (ed.) Eurocrypt '94. LNCS, vol. 950, pp. 77–85. Springer, Heidelberg (1995), Citations in this document: §5, §5, §5, §5, §5, §5, §5

[38] Naehrig, M., Niederhagen, R., Schwabe, P.: New software speed records for cryptographic pairings. In: Abdalla, M., Barreto, P.S.L.M. (eds.) Latincrypt 2010. LNCS, vol. 6212, pp. 109–123. Springer, Heidelberg (2010), Citations in this document: §3

[39] Neven, G., Smart, N.P., Warinschi, B.: Hash function requirements for Schnorr signatures. Journal of Mathematical Cryptology 3, 69–87 (2009), Citations in this document: §2, §2

[40] Nguyen, P.Q., Shparlinski, I.: The insecurity of the elliptic curve digital signature algorithm with partially known nonces. Designs, Codes and Cryptography 30, 201–217 (2003), Citations in this document: §2

[41] Pippenger, N.: On the evaluation of powers and related problems (preliminary version). In: FOCS '76, pp. 258–263 (1976), Citations in this document: §4, §5

[42] Pointcheval, D., Stern, J.: Security arguments for digital signatures and blind signatures. Journal of Cryptology 13, 361–396 (2000), Citations in this document: §2

[43] Rangasamy, J., Stebila, D., Boyd, C., González Nieto, J.: An integrated approach to cryptographic mitigation of denial-of-service attacks. In: ASIACCS 2011 (2011), Citations in this document: §1

[44] Schnorr, C.-P.: Efficient identification and signatures for smart cards. In: Brassard, G. (ed.) Crypto '89. LNCS, vol. 435, pp. 239–252. Springer, Heidelberg (1990), Citations in this document: §2, §2, §2

[45] Stern, J., Pointcheval, D., Malone-Lee, J., Smart, N.P.: Flaws in applying proof methodologies to signature schemes. In: Yung, M. (ed.) Crypto 2002. LNCS, vol. 2442, pp. 93–110. Springer, Heidelberg (2002), Citations in this document: §2

[46] Wegener, I.: Bottom-up-heapsort, a new variant of heapsort, beating, on average, quicksort (if n is not very small). Theoretical Computer Science 118, 81–98 (1993), Citations in this document: §5

[47] Wigley, J.: Removing need for rng in signatures, message `5gov5d$pad@wapping.ecs.soton.ac.uk` posted to `sci.crypt` (1997), `http://groups.google.com/group/sci.crypt/msg/a6da45bcc8939a89`, Citations in this document: §2

To Infinity and Beyond: Combined Attack on ECC Using Points of Low Order*

Junfeng Fan, Benedikt Gierlichs, and Frederik Vercauteren

Katholieke Universiteit Leuven, COSIC & IBBT
Kasteelpark Arenberg 10, B-3001 Leuven-Heverlee, Belgium
firstname.lastname@esat.kuleuven.be

Abstract. We present a novel combined attack against ECC implementations that exploits specially crafted, but valid input points. The core idea is that after fault injection, these points turn into points of very low order. Using side channel information we deduce when the point at infinity occurs during the scalar multiplication, which leaks information about the secret key. In the best case, our attack breaks a simple and differential side channel analysis resistant implementation with input/output point validity and curve parameter checks using a single query.

Keywords: Fault attack, side channel attack, elliptic curve cryptography.

1 Introduction

Elliptic curve cryptography (ECC) is a public-key cryptosystem that was independently proposed by Miller [33] and Koblitz [29]. In the context of embedded implementations, ECC is an interesting alternative to systems like RSA [37] because it allows for more compact and more efficient implementations.

The ubiquity of embedded cryptography in applications such as smart cards, RFID tags, access control, etc. leads to a new security threat that does not target the mathematical strength of the cryptographic algorithms but the physical strength of concrete implementations using side channel and fault attacks. Side channel attacks (SCAs) were first described by Kocher in [30] and use the fact that physical devices leak information through measurable quantities such as power consumption [31], timing behavior [30], electromagnetic radiation [24,36], etc. Fault attacks (FAs) were introduced by Boneh et al. [10], and rely on the fact that an adversary can actively inject faults into a device which typically leads the device to compute an incorrect result. Ways to inject faults include clock and power glitches [4,6], lasers [38], etc.

* This work was supported in part by the European Commission's ECRYPT II NoE (ICT-2007-216676), by the Belgian State's IAP program P6/26 BCRYPT, by the K.U. Leuven-BOF (OT/06/40) and by the Research Council K.U. Leuven: GOA TENSE (GOA/11/007). Benedikt Gierlichs and Frederik Vercauteren are Postdoctoral Fellows of the Fund for Scientific Research - Flanders (FWO).

B. Preneel and T. Takagi (Eds.): CHES 2011, LNCS 6917, pp. 143–159, 2011.

Straightforward implementations of ECC can be easily broken by a range of well known attacks, including simple and differential side channel analysis (SSCA, DSCA) as shown by Coron [20] and differential fault analysis as demonstrated by Biehl et al. [9] and later generalized by Ciet and Joye [16]. We refer to Fan et al. [23] for a comprehensive overview of the existing countermeasures to thwart these attacks and simply focus on the main ideas.

Resistance against SSCA can be achieved by regular scalar multiplication algorithms [20,28], unified addition and doubling formulae [12,19] or side channel atomicity [14]. Basically any solution that ensures a constant sequence of operations in the scalar multiplication algorithm, identical or indistinguishable point operations, is viable.

DSCA can be thwarted by ensuring that the scalar multiplication algorithm processes strictly unpredictable, e.g. randomized, data. Typical randomization techniques include base point blinding [20], randomized projective coordinates [20], curve isomorphisms [27] and field isomorphisms [27]. Alternative approaches include key randomization [20] and random key splitting [15] before each scalar multiplication, but they require that an adversary cannot extract any information from a single trace [18]. However, as shown by Goubin [25] most of these countermeasures can be broken in the chosen message scenario when the curve admits "special points", i.e. where one of the coordinates is zero. Smart [40] provides several easy countermeasures preventing Goubin's attack: for special points of low order, cofactor multiplication is proposed and to avoid special points of large order, all points are first mapped to an isogenous curve, before scalar multiplication is executed. Note that all NIST curves over large prime fields have cofactor equal to one.

Due to ECC's group structure, an elegant and efficient way to detect faults is to check if the input to and the output of the scalar multiplication algorithm are valid points on the curve as explained by Biehl et al. [9]. Ciet and Joye point out that one must additionally check the curve parameters for faults [16], which in the remainder of the paper we consider to be part of the initial validity check.

In this paper we present a novel attack that combines fault injection with SSCA (cf. combined attack [3]) and specially crafted, but valid input points P. The core idea is that, after a single fault injection, P turns into a point P' of very low order ℓ (e.g. $\ell = 2, 3, \dots, 200$) with practical probability. Since the point P' has low order, the point at infinity will appear during the computation of $k \cdot P'$. This event can be detected via side channels and leaks information about the key k. Our attack cannot be prevented by most of the countermeasures mentioned above such as input and output validity checks, cofactor multiplication and isogeny defence (which foil Goubin's attack), SSCA countermeasures and it bypasses many DSCA countermeasures.

The paper is organized as follows. In Section 2, we recall the necessary background on elliptic curves and in Section 3, we describe an effective algorithm to compute valid points on an elliptic curve that, after a bit-flip in one of their coordinates turn into points of a given small order. In Section 4, we exploit these points to derive our new attack and illustrate it on a very basic implementation.

In Section 5, we discuss the assumptions underlying our attack and analyze its applicability when the basic implementation is enhanced with common countermeasures. Finally, Section 6 concludes the paper.

2 Background on Elliptic Curves

In this section we briefly review the necessary background on elliptic curves over $\mathbb{F}_p$. An elliptic curve E over $\mathbb{F}_p$ with $p > 3$ can always be given by a short Weierstrass equation $y^2 = x^3 + ax + b$, with $a, b \in \mathbb{F}_p$ and $4a^3 + 27b^2 \neq 0$. For every finite field K containing $\mathbb{F}_p$ one now considers the set of K-rational points

$$E(K) := \{(x, y) \in K \times K \mid y^2 = x^3 + ax + b\} \cup \{\mathcal{O}\}$$

where $\mathcal{O}$ denotes the point at infinity.

2.1 Group Law

The use of elliptic curves in cryptography stems from the fact that $E(K)$ naturally possesses the structure of an abelian group. It is common practice to denote the group operations in an additive way (i.e. using + and – symbols), as opposed to the multiplicative notation when dealing with groups like $\mathbb{F}_p^*$. The group law is defined by the following general rules: $\mathcal{O}$ is the zero element, and any three points that lie on a line add up to zero.

Group Law Formulae. Working this out yields the following explicit rules for adding two points $P = (x_P, y_P)$ and $Q = (x_Q, y_Q)$. If $Q = -P$, i.e. if $x_P = x_Q$ and $y_P = -y_Q$, then $P + Q = \mathcal{O}$. If $P \neq \pm Q$, we obtain the following addition formula: $R = (x_R, y_R) = P + Q$ with

$$x_R = \left(\frac{y_Q - y_P}{x_Q - x_P}\right)^2 - x_P - x_Q \quad \text{and} \quad y_R = \left(\frac{y_Q - y_P}{x_Q - x_P}\right)(x_P - x_R) - y_P \,. \quad (1)$$

If $P = Q$, we obtain the doubling formula: $R = (x_R, y_R) = 2 \cdot P$ with

$$x_R = \left(\frac{3x_P^2 + a}{2y_P}\right)^2 - x_P - x_Q \quad \text{and} \quad y_R = \left(\frac{3x_P^2 + a}{2y_P}\right)(x_P - x_R) - y_P \,. \quad (2)$$

Note that the above formula for addition does not depend on the curve equation at all and that the formula for doubling only involves the parameter a. This simple fact has been exploited in several attacks before [9] and will also be crucial in our attack.

Since inversions are typically much more expensive than multiplications, several types of projective coordinate systems have been developed. Standard projective coordinates [5] represent an elliptic curve point $P = (x, y)$ by (X, Y, Z) where $x = X/Z$ and $y = Y/Z$, whereas Jacobian projective coordinates [5] use $x = X/Z^2$ and $y = Y/Z^3$. The above addition/doubling formulae can easily be reformulated using projective coordinates, but the resulting formulae will also depend on a only.

Group Law Implementation. An implementer of an elliptic curve system is not only faced with the choice of the elliptic curve model to use, such as short Weierstrass, Montgomery [34], Edwards [7,8], Hessian [39], etc., and the choice of an appropriate coordinate system like projective [5] or Jacobian [5], but also with the handling of borderline cases. Indeed, the above addition formula (1) can only handle the cases where $P \neq \pm Q$, $P \neq \mathcal{O}$ and $Q \neq \mathcal{O}$. Similarly, the doubling formula (2) will fail when P is a point of order two or $P = \mathcal{O}$.

The way in which the implementation handles these borderline cases leads to the following classification: *full and partial domain correctness*. In the full domain correctness case, the implementation computes $P+Q$ and $2 \cdot P$ correctly for all P, Q. In the partial domain correctness case the implementation either stops working (e.g. division by zero occurs), computes on invalid points or ends in a fixed point (both cases occur when using the above formulae in projective or Jacobian coordinates, see Table 1).

Table 1. Borderline cases for projective and Jacobian coordinates

$E(\mathbb{F}_p)$: $y^2 = x^3 + ax + b$					
Coordinate System	Operation	Using a	Using b	Input	Output
Projective	PA($\boldsymbol{P_1}, \boldsymbol{P_2}$)	-	-	$\boldsymbol{P_1}$=$\boldsymbol{P_2}$	(0,0,0)
				$\boldsymbol{P_1}$=-$\boldsymbol{P_2}$	(0,*,0)
				$\boldsymbol{P_1}$=(0,*,0)	(0,0,0)
	PD($\boldsymbol{P_1}$)	+	-	Order($\boldsymbol{P_1}$)=2	(0,*,0)
				$\boldsymbol{P_1}$=(0,*,0)	(0,0,0)
Jacobian	PA($\boldsymbol{P_1}, \boldsymbol{P_2}$)	-	-	$\boldsymbol{P_1}$=$\boldsymbol{P_2}$	(0,0,0)
				$\boldsymbol{P_1}$=-$\boldsymbol{P_2}$	(*,*,0)
				$\boldsymbol{P_1}$=(*,*,0)	(*,*,0)
				$\boldsymbol{P_1}$=(0,0,0)	(0,0,0)
	PD($\boldsymbol{P_1}$)	+	-	Order($\boldsymbol{P_1}$)=2	(*,*,0)
				$\boldsymbol{P_1}$=(*,*,0)	(*,*,0)
				$\boldsymbol{P_1}$=(0,0,0)	(0,0,0)

2.2 Scalar Multiplication

The basic operation in classical cryptosystems such as RSA and ECC is exponentiation in the underlying group. For elliptic curves, this exponentiation is called scalar multiplication since given a point P and a scalar k, it computes $k \cdot P$ by repeatedly using the double/add operations.

The most basic scalar multiplication algorithm is the binary double-and-add algorithm, which computes $k \cdot P$ according to the binary expansion of $k = \sum_{i=0}^{n-1} k_i 2^i$. Depending on the direction in which the bits of k are scanned, we obtain a left-to-right or right-to-left variant.

The left-to-right variant is described in Algorithm 1 and will be used to illustrate our attack. The applicability of our attack to SSCA and DSCA resistant scalar multiplication algorithms will be discussed in Section 5.

Algorithm 1. Double and Add Left-to-Right

Input: $\boldsymbol{P}$, $k = (k_{n-1}, k_{n-2}, \dots, k_0)_2$
Output: $\boldsymbol{Q} = k \cdot \boldsymbol{P}$

```
R ← P ;
for i ← n − 2 down to 0 do
    R ← 2 · R ;
    if (k_i = 1) then R ← R + P ;
end
return R
```

3 Elliptic Curve Points with Low Order Neighbours

In this section, we consider the following problem, the solution of which is crucial for our attack: given an elliptic curve $E : y^2 = x^3 + ax + b$ over $\mathbb{F}_p$, two integers ℓ and Δ, is it possible to construct a point $P := (x_P, y_P)$ in $E(\mathbb{F}_p)$ with the following properties:

- there exists a curve $E' : y^2 = x^3 + ax + b'$ over $\mathbb{F}_p$
- with a point $P' = (x_{P'}, y_{P'}) \in E'(\mathbb{F}_p)$ of order ℓ
- such that the Hamming distance of the bit-representations $x_P||y_P$ and $x_{P'}||y_{P'}$ equals Δ.

When $\Delta = 1$, i.e. the coordinates differ in a single bit, we call the points P and P' *neighbours*. We will describe an effective construction of points P with neighbours P' of a given order ℓ and with $x_P = x_{P'}$, i.e. the bit-flip occurred in the y-coordinate only. The construction can be easily extended to encompass bit-flips in x_P, and indirect neighbours, i.e. $\Delta > 1$.

In Section 3.1 we first show how to construct points of given order and in Section 3.2 we adapt this method to find points with low order neighbours.

3.1 Constructing Points of Given Order

Given an elliptic curve E over $\mathbb{F}_p$, we can consider the points on E of order dividing n, i.e. points $P \in E(\overline{\mathbb{F}}_p)$ with $n \cdot P = \mathcal{O}$, where the coordinates of P can lie in any extension field of $\mathbb{F}_p$. These points can be characterized explicitly using the so called division polynomials [5]. For $n \in \mathbb{N}$ define polynomials $\psi_n(x, y)$ recursively as follows:

$$\begin{aligned}
\psi_0 &= 0,\ \psi_1 = 1, \psi_2 = 2y,\ \psi_3 = 3x^4 + 6ax^2 + 12bx - a^2, \\
\psi_4 &= 4y(x^6 + 5ax^4 + 20bx^3 - 5a^2x^2 - 4abx - 8b^2 - a^3), \\
\psi_{2m+1} &= \psi_{m+2}\psi_m^3 - \psi_{m-1}\psi_{m+1}^3, m \geq 2, \\
\psi_{2m} &= \frac{\psi_m(\psi_{m+2}\psi_{m-1}^2 - \psi_{m-2}\psi_{m+1}^2)}{2y}, m \geq 3\,.
\end{aligned}$$

For any point $P \in E(\overline{\mathbb{F}}_p)$ with $P \neq \mathcal{O}$, we then have that $n \cdot P = \mathcal{O}$ if and only if $\psi_n(x_P, y_P) = 0$. Furthermore, one can show by induction on n that ψ_n for n odd and $\psi_n/2y$ for n even, are polynomials in x only. We denote these polynomials by $\phi_n(x)$. It is easy to see that $n \cdot P = \mathcal{O}$ and $2 \cdot P \neq \mathcal{O}$ if and only if $\phi_n(x_P) = 0$.

3.2 Constructing Points with Low Order Neighbours

Given an elliptic curve E over $\mathbb{F}_p$ and an integer ℓ, we want to construct a point $P = (x_P, y_P)$ in $E(\mathbb{F}_p)$ with neighbour $P' = (x_P, y_P \oplus \epsilon)$ with $\epsilon = 2^k$ for some $k < \log_2(p)$ and $\ell \cdot P' = \mathcal{O}$. For $\ell > 2$, the points P and P' therefore have to satisfy the following non-linear system of equations:

$$\begin{cases} \boldsymbol{y_P}^2 - \boldsymbol{x_P}^3 - a \cdot \boldsymbol{x_P} - b = 0 & P \in E(\mathbb{F}_p) \\ (\boldsymbol{y_P} \oplus \epsilon)^2 - \boldsymbol{x_P}^3 - a \cdot \boldsymbol{x_P} - \boldsymbol{b'} = 0 & P' \in E'(\mathbb{F}_p) \\ \phi_\ell^{a,\boldsymbol{b'}}(\boldsymbol{x_P}) = 0 & \ell \cdot P' = \mathcal{O}, \end{cases}$$

where the unknown variables are printed in bold face. Since the $\oplus$-operation is not very algebraic, we will consider the following two cases that lead to equivalent results, namely we replace $y_p \oplus \epsilon$ by $y_p \pm \epsilon$ and then verify afterwards if an actual bit-flip occurred, i.e. that there was no carry.

Subtracting the first two equations expresses b' as a function of y_P, namely, $b' = \pm 2\epsilon y_P + \epsilon^2 + b$. Substituting this expression in the last equation leads to a bivariate polynomial in x_P and y_P, which we call $\Upsilon_\ell(x_P, y_P)$. The points P for the given ℓ and ϵ therefore are solutions of $E(x_P, y_P) = 0$ and $\Upsilon_\ell(x_P, y_P) = 0$. These solutions can be easily found by a Groebner basis [13] computation or by taking the resultant

$$R(x_p) = \text{Resultant}_{y_P}(E(x_P, y_P), \Upsilon_\ell(x_P, y_P)),$$

finding all possibilities for x_P as roots of R over $\mathbb{F}_p$ and the corresponding y_P from $E(x_P, y_P) = 0$. A final check is then necessary to only retain those (x_P, y_P) where the $\pm$ operation actually caused a bit-flip, in particular, in the $+\epsilon$-case (resp. $-\epsilon$-case) we only retain those results where the k-th bit is zero (resp. one).

To analyze the complexity of solving the above system, we simply need to figure out the degree of $\phi_\ell^{a,b'}$ in x_P and b'. The degree in x_P is easily seen to be $(\ell^2 - 1)/2$ since the full ℓ-torsion contains ℓ^2 points. The degree in b' can be seen to be upperbounded by $(\ell^2 - 1)/6$ since the same recursion holds and the degree in b' is three times smaller than for x_P in the initializations. This leads to a resultant of degree $\ell^2 - 1$ and an overall complexity of $\tilde{O}(\ell^4)$ to solve the non-linear system of equations.

The probability that a given curve admits a point with neighbour of order ℓ and bit-flip in position k (note that both ℓ and k are fixed) can be roughly approximated as follows. Denote by $P(n, p)$ the probability that a random polynomial of degree n has at least one root in $\mathbb{F}_p$. Assuming we can consider the resultant R as a random polynomial, then the probability is roughly the product of:

- the probability $P(\ell^2 - 1, p)$ that R has at least one root x_P in $\mathbb{F}_p$,
- the probability $1/2$ that the corresponding y_P is in $\mathbb{F}_p$ (and when it does, there are two roots y_P),
- the probability $3/4$ that of these two roots, at least one has a bit-flip in position k.

Note that we only analyze the case of $+\epsilon$ and not also $-\epsilon$, since the solvability of both systems of equations is not really independent. The overall probability

Table 2. Points with neighbours of low order on NIST P-192 curve

Order	P	bit-flip
2	xP = 0x6D9D789820A2C19237C96AD4B8D86B87FB49D4D6C728B84F yP = 0x1	0
3	xP = 0x8E1AEBDD6009F114490C7BC2C02509F8E432ED15F10C2D33 yP = 0x7A568946EFA602B3624A61E513E57869CAF2AE854E1A17B	2
4	xP = 0xB317D7BBD023E6293F1506221F5BC4A23D4BE2E05328C5F7 yP = 0xC70D48794F409831097620C0865B7D567329728C634CA6AE	0
5	xP = 0xCC9BCC0061F64371E3C3BDE165DAD5380A7DC1919765940 yP = 0xCC8B36B37928334B8AFD7A9FCCFB4B0773E94A4178093458	8
6	xP = 0xC3F76445E6A52138E283E485092F005BE0821C3F9E96B05E yP = 0x535DBCCB593D72E7885B66E57FD13A8FF9C57A8F8B91CE48	1
7	xP = 0x5C003567728CCBC9F4C06620B9973193837BAEC67A29E43A yP = 0x408D0C3135006B03EFF80961394D890F0E86D9FD1BA4EEC6	3
8	xP = 0x74FD6A1AD39479C75A85305FA786E1DBDC845E03754E723E yP = 0x6EF58ABFC0B71047BA4F425652B3EC1746EBE8FE16FEA1F5	1

therefore is roughly $3/8 \cdot P(\ell^2 - 1, p)$. A closed expression for $P(n, p)$ exists [32] and this can easily be shown to satisfy $P(n, p) > 1/2$, which leads to a lower bound of $3/16$. Note that this high probability stands in stark contrast with the probability that a given fixed point has a neighbour of order ℓ, which we expect to be in $O(1/p)$.

The above algorithm can be easily extended to any given fixed error pattern, such as multiple bit-flips, or setting certain bits to zero/one. Furthermore, errors in the x-coordinate can also be dealt with.

To illustrate the effectiveness of the above procedure, in Table 2 we provide several example points with low order neighbours for the NIST P-192 curve [35], i.e. the curve over $\mathbb{F}_p$ with $p = 2^{192} - 2^{64} - 1$, $a = -3$ and $b =$ 0x64210519E59C80E70FA7E9AB72243049FEB8DEECC146B9B1. For each small integer ℓ, the table gives a P with neighbour of order ℓ when a specific bit of the y-coordinate is flipped (bit 0 is the LSB). Each of these examples was generated in less than a second using Magma [11] on a standard laptop.

4 Combined Attack Using Low Order Neighbours

In this section we introduce a new combined attack using points with low order neighbours. The system under attack is the following: we have access to a target implementation that on input an elliptic curve point P computes $k \cdot P$ for some unknown secret k. The goal is to recover the secret k.

The basic version of our attack requires the following *two assumptions*. The realistic nature of these assumptions and the applicability of our attack will be analyzed in Section 5.

1. It is possible, e.g. using side channel information, to determine when an intermediate result in the computation becomes $\mathcal{O}$.

2. It is possible to inject a fault immediately after initial validity checks, resulting in a bit-flip in a predetermined position.

The attack then proceeds as follows: we input a point P with low order neighbour P' and, after the initial validity checks have passed, inject a fault that turns P into P'. The implementation then tries to compute $k \cdot P'$. Since P' has low order, it is highly likely that an intermediate computation will result in $\mathcal{O}$. This corresponds to the fact that the part of the secret scalar k that has been processed up to that point is divisible by the order of P'.

If and how the implementation continues to run depends solely on how the elliptic curve group operations are implemented, i.e. whether the implementation is full or partial domain correct.

4.1 Full Domain Correctness

The implementation will compute the scalar multiplication $k \cdot P'$ until the final validity checks, at which point it will abort since $k \cdot P'$ is not on the curve E. During the computation however, we will obtain a huge amount of information of the following form: assume the order of P' is ℓ, then every time an elliptic curve addition/doubling results in $\mathcal{O}$, we know that the part of the scalar processed up to that point is divisible by ℓ. Note that we also obtain extra information when $\mathcal{O}$ does not appear, since then the corresponding part of the scalar is not divisible by ℓ.

This attack is extremely powerful since in most cases one trace will suffice to recover (almost all of) k. In Section 5.3 we will show that the attack can recover ephemeral keys, blinded keys and randomly split keys.

Example. To illustrate the effectiveness of this attack in the full domain correctness case, we apply it to an implementation using Algorithm 1. If we choose to input a point P with neighbour P' of order 2, all occurring computations $(2 \cdot P', 2 \cdot \mathcal{O}, \mathcal{O} + P')$ are borderline cases, which may not be desirable. Therefore, we choose a point P with neighbour P' of order 4.

The computation of $\boldsymbol{R} \leftarrow 2 \cdot \boldsymbol{R}$ then either consists of $2 \cdot P'$, $2 \cdot (2P')$, $2 \cdot (3P')$ or $2 \cdot \mathcal{O}$. Note that the cases $2 \cdot (2P')$ and $2 \cdot \mathcal{O}$ are borderline cases, and thus distinguishable from the cases $2 \cdot P'$ and $2 \cdot (3P')$, which are ordinary doublings. The crucial point to note is that point addition always generates odd multiples of P' and thus will never result in $\mathcal{O}$. Furthermore, since P' has order 4, the point $\mathcal{O}$ will only occur after two consecutive doublings. Therefore, if $\mathcal{O}$ occurs during the processing of bit k_i, we know that bit k_{i+1} must have been zero. This uniquely identifies the zero key bits (except for possibly the LSB), which implies that the other key bits have to be one. As such, we easily obtain all of k with one trace only.

Table 3 shows the intermediate results for the computation of $k \cdot P'$ where $k = 5405$ and $\ell = 4$. Note that we assume that distinguishing point addition from point doubling is not possible. As such, the adversary sees a sequence of normal operations (additions or ordinary doublings), denoted by `Op`, and occurrences of

Table 3. Intermediate results in the computation of $5405 \cdot P'$ with $\ell = 4$ and view of the adversary when attacking the scalar multiplication

<table>
<tr><td>i</td><td colspan="2">11</td><td colspan="2">10</td><td>9</td><td colspan="2">8</td><td>7</td><td>6</td><td>5</td><td colspan="2">4</td><td colspan="2">3</td><td colspan="2">2</td><td>1</td><td colspan="2">0</td></tr>
<tr><td>k_i</td><td colspan="2">0</td><td colspan="2">1</td><td>0</td><td colspan="2">1</td><td>0</td><td>0</td><td>0</td><td colspan="2">1</td><td colspan="2">1</td><td colspan="2">1</td><td>0</td><td colspan="2">1</td></tr>
<tr><td>$\boldsymbol{R}$</td><td colspan="2">$2P'$</td><td colspan="2">$\mathcal{O},P'$</td><td>$2P'$</td><td colspan="2">$\mathcal{O},P'$</td><td>$2P'$</td><td>$\mathcal{O}$</td><td>$\mathcal{O}$</td><td colspan="2">$\mathcal{O},P'$</td><td colspan="2">$2P'$,$3P'$</td><td colspan="2">$2P'$,$3P'$</td><td>$2P'$</td><td colspan="2">$\mathcal{O},P'$</td></tr>
</table>

<table>
<tr><td>view</td><td>Op</td><td>$\mathcal{O}$</td><td>Op</td><td>Op</td><td>$\mathcal{O}$</td><td>Op</td><td>Op</td><td>$\mathcal{O}$</td><td>$\mathcal{O}$</td><td>$\mathcal{O}$</td><td>Op</td><td>Op</td><td>Op</td><td>Op</td><td>Op</td><td>Op</td><td>$\mathcal{O}$</td><td>Op</td></tr>
<tr><td>step 1</td><td>0</td><td></td><td></td><td>0</td><td></td><td></td><td>0</td><td>0</td><td>0</td><td></td><td></td><td></td><td></td><td></td><td></td><td>0</td><td></td><td></td></tr>
<tr><td>step 2</td><td>0</td><td colspan="2">1</td><td>0</td><td colspan="2">1</td><td>0</td><td>0</td><td>0</td><td colspan="2">1</td><td colspan="2">1</td><td colspan="2">1</td><td>0</td><td colspan="2">1</td></tr>
</table>

$\mathcal{O}$ as shown in the fourth row of Table 3. To recover the secret key, the adversary proceeds as follows: in step one, he puts a 0 in each cell to the left of an $\mathcal{O}$. Then in step two, he groups the empty cells in pairs of two, from left to right, merges them and writes a 1 in the resulting cell. If there is a single cell left in the end, he writes a zero in it.

4.2 Partial Domain Correctness

Partial domain correctness implies that we can only gather information up to the first occurrence of the point $\mathcal{O}$. Indeed, either the implementation simply crashes during the computation of $\mathcal{O}$ or it performs some nonsensical computations thereafter. The result is that for each point submitted, we can only obtain partial information about k. When k is fixed over several invocations, this is not a real problem since we can submit many points with neighbours of different order and then deduce all bits of k from this information. Note that due to the behavior of the implementation, i.e. no further information after occurrence of $\mathcal{O}$, the orders of the neighbouring points submitted do not have to be coprime.

The type of information gathered will be of the following form: let $k = \sum_{i=0}^{n-1} k_i 2^i$, then for each small integer ℓ we will obtain the index $I(\ell)$ such that the leftmost (or rightmost) $I(\ell)$ bits of k form an integer divisible by ℓ. By definition we set $I(\ell) = 0$ when no part of k is divisible by ℓ. As such we obtain a list of positive information `PosInfo`, consisting of pairs $[\ell, I(\ell)]$, and a list of negative information `NegInfo` containing those ℓ for which each $I(\ell) = 0$. The list `PosInfo` will be sorted according to $I(\ell)$.

A very simple incremental search algorithm is given in Algorithm 2. The algorithm keeps a list `PartialKeys` containing all possibilities for the `BitsScanned` leftmost bits of k. The procedure `ExpandPartialKeys` expands all partial keys in the list by appending (on the right) all possible bit sequences of length `PosInfo[j][2] - BitsScanned` and then only keeps those candidates divisible by `PosInfo[j][1]`. It furthermore updates `BitsScanned` to `PosInfo[j][2]`. The function `PrunePartialKeys` simply removes all elements from `PartialKeys` that violate one of the non-divisibility conditions for any of the integers in `NegInfo`.

We implemented this algorithm in Magma and ran several tests to evaluate its behaviour for the NIST P-192 curve. Given a fixed secret random k, we computed for each integer ℓ smaller than an upper bound B the value $I(\ell)$ and then tried to recover k from PosInfo and NegInfo. The tests show that even for $B \simeq 100$ we can typically recover a large part of the secret k (much more than 100 bits on average) and that for larger values of B like 192 or 384 we recover most of k bar a few least significant bits.

5 Analysis of the Attack

In this section we discuss the assumptions made in the previous section and analyze the attack for a wide range of implementation choices, such as coordinate systems and curves used, scalar multiplication algorithms and finally, common countermeasures against SSCA and DSCA attacks and validity checks against fault attacks.

5.1 Analysis of Assumptions

Chosen Input Point. The target implementation is assumed to compute $k \cdot P$ for any given input point P, where k is supposed to be secret. This setting arises for instance in ElGamal decryption [22], ECIES [1] and in static Diffie-Hellman key agreement [21]. In the latter case, one of the ephemeral keys is simply the long term public key. We note that the attack does not apply to ECDSA [41], where the ephemeral key is computed on a fixed base point P (unless P has neighbours of low order).

Recognizing $\mathcal{O}$ via Side Channels. In the case of partial domain correctness, the implementation either crashes during the computation of $\mathcal{O}$ or it ends up in $\mathcal{O}$ and remains there. We assume that either event can be detected through side channels. Indeed, if the implementation crashes it can for example stall or exit the scalar multiplication routine early, which should be clearly visible e.g. in power traces. If the implementation continues to run it will get stuck in $\mathcal{O}$, which should be visible as a repetitive pattern.

In the case of full domain correctness, the implementation does not crash because it correctly deals with all borderline cases. Most textbooks on ECC, e.g. Hankerson et al. [26], use checks and conditional branches in their code examples to ensure full domain correctness. It is well known that conditional branches can leak through side channels [31,17] and so it is clear that the occurrence of any borderline case can be easily detected. Even if we assume that these checks and

Algorithm 2. Recovering private key from PosInfo and NegInfo

```
PartialKeys ← ∅, BitsScanned ← 0
for j from 1 upto # PosInfo do
    ExpandPartialKeys (& PartialKeys, PosInfo[j], & BitsScanned)
    PrunePartialKeys (& PartialKeys, NegInfo)
end
return PartialKeys, BitsScanned.
```

branches are implemented with side channel resistance in mind (which is highly unlikely) the actual occurrence of $\mathcal{O}$ (a point with at least one coordinate equal to zero) in a point operation should be visible [2].

Fault Injection. The assumption that an adversary can flip a single chosen bit in an implementation is certainly strong. We can relax this assumption greatly using a trivial approach: by repeatedly faulting a specified byte (resulting in a random byte), after an average of 256 trials, the fault will be precisely the fault required. With overwhelming probability only the required fault will lead to a point of low order, thus the good case is easily distinguished from undesired faults. However, we still have to assume that an adversary can inject a fault with sufficiently precise timing, in this case after initial validity checks.

The construction of points with low order neighbours is also flexible enough to accommodate a more accurate fault model for the target implementation. Assume we have extra information on the most likely state of a byte after fault injection, then we can compute points specially crafted for this fault pattern.

Group Law Formulae. An implicit assumption, which is automatically satisfied when using the formulae given in Eqs. (1) and (2), is that the group law formulae do not depend on *all* coefficients of the curve equation. More formally, assume the elliptic curve equation $E(a_1, \ldots, a_k)$ depends on k coefficients, but that only the first $m < k$ appear in the group law formulae. Then the implementation can also be used to compute correctly on all elliptic curves with the same $a_1, \ldots, a_m$, but differing $a_{m+1}, \ldots, a_k$. Note that for *all* elliptic curve forms, it is always possible to write down group law formulae with $m < k$. However, for the most efficient formulae used in practice, our assumption seems only valid for Weierstrass forms, which are most widely used, and Hessian forms. In fact, using group law formulae involving all coefficients of the curve combined with initial and final validity checks, is a possible combination of countermeasures to our attack.

5.2 Scalar Multiplication

Many scalar multiplication algorithms have been proposed, either to speed up the computation or to aid resistance against simple side channel analysis. In this section we focus on scalar recoding [5], the Montgomery powering ladder [28], unified formulae [12,19] and side channel atomicity [14].

Scalar Representation. Apart from the usual binary representation of the scalar $k = \sum_{i=0}^{n-1} k_i 2^i$, several other representations are frequently used. The non-adjacent form (NAF) represents $k = \sum_{i=0}^{n-1} k_i 2^i$, where $k_i \in \{0, \pm 1\}$. More generally, a width w-NAF of an integer k is an expression $k = \sum_{i=0}^{n-1} k_i 2^i$ with each nonzero k_i odd, $|k_i| < 2^{w-1}$, $k_{n-1} \neq 0$ and at most one of any w consecutive digits nonzero. In all cases, we still obtain a similar type of information as in the basic attack: when $\mathcal{O}$ is encountered, we know that the part of the scalar processed up to that point is divisible by ℓ. However, since the number of intermediate points computed during the scalar multiplication is no longer n but n/w, the probability of hitting $\mathcal{O}$ is lower.

Montgomery Powering Ladder. The Montgomery powering ladder given in Algorithm 3 is a popular choice because it provides speed and a highly regular structure.

Algorithm 3. Montgomery powering ladder

Input: $\boldsymbol{P}$, $k = (k_{n-1}, k_{n-2}, \ldots, k_0)_2$
Output: $\boldsymbol{Q} = k \cdot \boldsymbol{P}$

$\boldsymbol{R}_0 \leftarrow P$, $\boldsymbol{R}_1 \leftarrow 2 \cdot P$;
for $i \leftarrow n-2$ **down to** 0 **do**
  $\boldsymbol{R}_{\neg k_i} \leftarrow \boldsymbol{R}_{k_i} + \boldsymbol{R}_{\neg k_i}$, $\boldsymbol{R}_{k_i} \leftarrow 2 \cdot \boldsymbol{R}_{k_i}$;
end
return $\boldsymbol{R}_0$

Attacking the Montgomery ladder is a bit more tricky because the sequence of operations is fixed and independent of the key. Nevertheless, the attack applies since it does not exploit the sequence of operations but the evolution of the intermediate values. Assume we input a point P with neighbour P' of order 4 and inject a fault after initial validity checks. The implementation will then try to compute $k \cdot P'$. Note that if two consecutive bits of k are equal, then the same point (either $\boldsymbol{R}_0$ or $\boldsymbol{R}_1$) will be doubled twice by the operation $\boldsymbol{R}_{k_i} \leftarrow 2 \cdot \boldsymbol{R}_{k_i}$ resulting in $\mathcal{O}$. On the other hand, if two consecutive bits differ, an ordinary doubling $2 \cdot P'$ or $2 \cdot (3P')$ will be computed. Finally, note that $\mathcal{O}$ can never be the result of the addition operation $\boldsymbol{R}_{k_i} + \boldsymbol{R}_{\neg k_i}$, since this is always an odd multiple of P'. As such, we obtain (almost all of) k with one trace only in the full domain correctness case.

Unified Formulae and Side Channel Atomicity. These countermeasures render point additions and doublings indistinguishable to prevent SSCA, and they can be implemented together with a possibly faster, irregular scalar multiplication algorithm like double-and-add (at the cost of leaking the Hamming weight of the exponent). It is clear that our attack is not affected by countermeasures of this kind because it does not require point operations to be distinguishable.

5.3 Common DSCA and FA Countermeasures

Random Scalar Splitting [15]. With this countermeasure, the scalar k is randomly split into two parts: $k = k_1 + k_2$. As such, $Q = k \cdot P$ can be computed as $k_1 \cdot P + k_2 \cdot P$ by two consecutive scalar multiplications and addition of their results. In the case of full domain correctness, (almost all of) k_1 and k_2 can be revealed (assuming that only the final output point is checked for validity), which immediately results in (a small number of candidates for) k. Otherwise the situation is similar to that of partial domain correctness. In the partial domain correctness case, the attack will no longer work, since we will only be able to recover a part of k_1 or k_2, but not both at the same time. Indeed, typically the implementation will stop working the first time it hits $\mathcal{O}$.

Scalar Randomization [20]. In this case, the scalar is blinded using a multiple of the curve order, i.e. k is replaced by $k' = k + r \cdot \#E$. It is easy to see that this countermeasure is useless in the full domain correctness case, where only a single trace is needed. For partial domain correctness, we do get partial information on k', but currently have no method to exploit this. Note that the same conclusion applies for ephemeral keys.

Coordinate Randomization [20]. This countermeasure assumes that some form of projective coordinates are being used and that the coordinates of the input point P are randomized before the scalar multiplication is started. For instance, when using projective coordinates, $P = (rX_P, rY_P, rZ_P)$ with r randomly chosen is used. It is easy to see that our attack remains valid if initial checks are performed *before* point randomization.

Random Elliptic Curve Isomorphisms [27]. This method first applies a random isomorphism of the form $\psi : (x, y) \mapsto (r^2 x, r^3 y)$ and then proceeds by computing $Q = k \cdot \psi(P)$ and outputting $\psi^{-1}(Q)$. Since an isomorphism does not change the order of a point, it is clear that the attack still applies if initial checks are performed *before* ψ is applied.

Isogeny Defence [40]. To prevent Goubin's attack using special points of large order, Smart proposed to use an isogeny $\mathcal{I}$ to map the input points to an isogenous curve without special points. Furthermore, for each curve in the main standards Smart provides a fixed isogeny that works for that curve. It is clear that our attack still applies if we look for points P with low order neighbours on the isogenous curve instead of on the original curve. The input to the target device will then be the points $\mathcal{I}^{-1}(P)$ and the fault will be injected after initial checks and isogeny have been applied.

Point Blinding [20]. With this countermeasure, the implementation contains a random point R and the corresponding multiple $k \cdot R$. The scalar multiplication $k \cdot P$ is computed by first computing $k \cdot (P + R)$ and then subtracting $k \cdot R$ from the result. Since we have no control over the point R, we cannot compute an appropriate point P such that we can fault the point $P + R$ into a point of low order. As such, this countermeasure does thwart our attack, both in the full and partial domain correctness case. Point blinding can be seen as an instance of infective computation [42].

Cofactor Multiplication [40]. To prevent small subgroup attacks, most protocols can be reformulated using cofactor multiplication. For instance, the

Diffie-Hellman protocol can be adapted as follows: a user first computes $Q \leftarrow h \cdot P$ and then $R \leftarrow k \cdot Q$ if $Q \neq \mathcal{O}$. It is easy to see that our attack still applies when we input a point with neighbour of order different from h.

Validity Checks [9,15]. To prevent fault attacks, Biehl et al. [9] and Ciet and Joye [15] recommend input/output point validity checks and curve parameter checks. These recommendations were part of the original motivation for our work and do not prevent the attack.

5.4 Curves over Finite Fields of Characteristic Two

Although the attack has mainly been described for elliptic curves in Weierstrass form over fields of large characteristic, we briefly touch on the characteristic two case. The short Weierstrass form is given by $E : y^2 + xy = x^3 + ax^2 + b$. The applicability of our attack then depends on the coordinate system being used. For affine and standard projective coordinates, the attack applies since only the a-coefficient is used in the group law formulae. For Jacobian coordinates the attack does not apply since both a and b are used in the group law formulae. For Lopez-Dahab formulae, only b is used in the group law formulae, but changing a only results in an isomorphic curve or its quadratic twist. As such it is impossible to find a point of given low order, since both the curve and its twist should not have many small subgroups.

6 Conclusions

We have described a novel attack that combines three ideas: fault injection, simple side channel analysis, and specially crafted, but valid input points that after a single fault injection have very low order. Our attack breaks ECC implementations that are protected by many of the known countermeasures such as initial and final point validity checks, curve parameter checks, cofactor multiplication check, SSCA countermeasures and bypasses many DSCA countermeasures. A secondary yet irritating result of our analysis is that proper, i.e. full domain correct implementations are more vulnerable to the attack and can be broken using one successful fault injection.

The attack does not apply to protocols that use a fixed point P (with no near neighbours of low order). For other applications, the attack can be prevented by physical fault injection sensors, concurrent point validity checks, using group law formulae that involve all curve coefficients, using randomized coordinates or randomized curve isomorphisms with randomization *before* the initial point validity check and by point blinding.

References

1. Abdalla, M., Bellare, M., Rogaway, P.: DHAES: An encryption scheme based on the Diffie-Hellman problem. Submission to P1363a: Standard specifications for Public-Key-Cryptography: Additional techniques (2000)

2. Akishita, T., Takagi, T.: Zero-value point attacks on elliptic curve cryptosystem. In: Boyd, C., Mao, W. (eds.) ISC 2003. LNCS, vol. 2851, pp. 218–233. Springer, Heidelberg (2003)
3. Amiel, F., Villegas, K., Feix, B., Marcel, L.: Passive and active combined attacks: Combining fault attacks and side channel analysis. In: FDTC 2007, pp. 92–102. IEEE Computer Society, Los Alamitos (2007)
4. Anderson, R., Kuhn, M.: Tamper resistance - a cautionary note. In: The Second USENIX Workshop on Electronic Commerce Proceedings, pp. 1–11. USENIX Association (1996)
5. Avanzi, R.M., Cohen, H., Doche, C., Frey, G., Lange, T., Nguyen, K., Vercauteren, F.: Handbook of elliptic and hyperelliptic curve cryptography. In: Discrete Mathematics and Its Applications. Chapman & Hall/CRC (2006)
6. Bar-El, H., Choukri, H., Naccache, D., Tunstall, M., Whelan, C.: The sorcerer's apprentice guide to fault attacks. Proceedings of the IEEE 94(2), 370–382 (2006)
7. Bernstein, D.J., Lange, T.: Faster Addition and Doubling on Elliptic Curves. In: Kurosawa, K. (ed.) ASIACRYPT 2007. LNCS, vol. 4833, pp. 29–50. Springer, Heidelberg (2007)
8. Bernstein, D.J., Lange, T., Farashahi, R.R.: Binary edwards curves. In: Oswald, E., Rohatgi, P. (eds.) CHES 2008. LNCS, vol. 5154, pp. 244–265. Springer, Heidelberg (2008)
9. Biehl, I., Meyer, B., Müller, V.: Differential fault attacks on elliptic curve cryptosystems. In: Bellare, M. (ed.) CRYPTO 2000. LNCS, vol. 1880, pp. 131–146. Springer, Heidelberg (2000)
10. Boneh, D., DeMillo, R.A., Lipton, R.J.: On the importance of checking cryptographic protocols for faults. In: Fumy, W. (ed.) EUROCRYPT 1997. LNCS, vol. 1233, pp. 37–51. Springer, Heidelberg (1997)
11. Bosma, W., Cannon, J., Playoust, C.: The Magma algebra system. I. The user language. J. Symbolic Comput. 24(3-4), 235–265 (1997)
12. Brier, E., Joye, M.: Weierstraß elliptic curves and side-channel attacks. In: Naccache, D., Paillier, P. (eds.) PKC 2002. LNCS, vol. 2274, pp. 335–345. Springer, Heidelberg (2002)
13. Buchberger, B.: Ein Algorithmus zum Auffinden der Basiselemente des Restklassenringes nach einem nulldimensionalen Polynomideal. PhD thesis, Universität Innsbruck (1965)
14. Chevallier-Mames, B., Ciet, M., Joye, M.: Low-cost solutions for preventing simple side-channel analysis: Side-channel atomicity. IEEE Trans. Computers 6(53), 760–768 (2004)
15. Ciet, M., Joye, M.: (Virtually) free randomization techniques for elliptic curve cryptography. In: Qing, S., Gollmann, D., Zhou, J. (eds.) ICICS 2003. LNCS, vol. 2836, pp. 348–359. Springer, Heidelberg (2003)
16. Ciet, M., Joye, M.: Elliptic curve cryptosystems in the presence of permanent and transient faults. Designs, Codes and Cryptography 36(1), 33–43 (2005)
17. Clavier, C., Coron, J.-S.: On the implementation of a fast prime generation algorithm. In: Paillier, P., Verbauwhede, I. (eds.) CHES 2007. LNCS, vol. 4727, pp. 443–449. Springer, Heidelberg (2007)
18. Clavier, C., Feix, B., Gagnerot, G., Roussellet, M., Verneuil, V.: Horizontal correlation analysis on exponentiation. In: Soriano, M., Qing, S., López, J. (eds.) ICICS 2010. LNCS, vol. 6476, pp. 46–61. Springer, Heidelberg (2010)

19. Clavier, C., Joye, M.: Universal exponentiation algorithm. In: Koç, Ç.K., Naccache, D., Paar, C. (eds.) CHES 2001. LNCS, vol. 2162, pp. 300–308. Springer, Heidelberg (2001)
20. Coron, J.-S.: Resistance against differential power analysis for elliptic curve cryptosystems. In: Koç, Ç.K., Paar, C. (eds.) CHES 1999. LNCS, vol. 1717, pp. 292–302. Springer, Heidelberg (1999)
21. Diffie, W., Hellman, M.E.: New directions in cryptography. IEEE Trans. Inform. Theory 22(6), 644–654 (1976)
22. El Gamal, T.: A public key cryptosystem and a signature scheme based on discrete logarithms. In: Blakely, G.R., Chaum, D. (eds.) CRYPTO 1984. LNCS, vol. 196, pp. 10–18. Springer, Heidelberg (1985)
23. Fan, J., Guo, X., De Mulder, E., Schaumont, P., Preneel, B., Verbauwhede, I.: State-of-the-art of secure ECC implementations: A survey on known side-channel attacks and countermeasures. In: HOST 2010, pp. 76–87 (2010)
24. Gandolfi, K., Mourtel, C., Olivier, F.: Electromagnetic analysis: Concrete results. In: Koç, Ç.K., Naccache, D., Paar, C. (eds.) CHES 2001. LNCS, vol. 2162, pp. 251–261. Springer, Heidelberg (2001)
25. Goubin, L.: A refined power-analysis attack on elliptic curve cryptosystems. In: Desmedt, Y.G. (ed.) PKC 2003. LNCS, vol. 2567, pp. 199–210. Springer, Heidelberg (2002)
26. Hankerson, D., Menezes, A.J., Vanstone, S.: Guide to Elliptic Curve Cryptography. Springer, Heidelberg (2004)
27. Joye, M., Tymen, C.: Protections against differential analysis for elliptic curve cryptography. In: Koç, Ç.K., Naccache, D., Paar, C. (eds.) CHES 2001. LNCS, vol. 2162, pp. 377–390. Springer, Heidelberg (2001)
28. Joye, M., Yen, S.-M.: The Montgomery Powering Ladder. In: Kaliski Jr., B.S., Koç, Ç.K., Paar, C. (eds.) CHES 2002. LNCS, vol. 2523, pp. 291–302. Springer, Heidelberg (2003)
29. Koblitz, N.: Elliptic curve cryptosystem. Math. Comp. 48, 203–209 (1987)
30. Kocher, P.C.: Timing Attacks on Implementations of Diffie-Hellman, RSA, DSS, and Other Systems. In: Koblitz, N. (ed.) CRYPTO 1996. LNCS, vol. 1109, pp. 104–113. Springer, Heidelberg (1996)
31. Kocher, P.C., Jaffe, J., Jun, B.: Differential power analysis. In: Wiener, M. (ed.) CRYPTO 1999. LNCS, vol. 1666, pp. 388–397. Springer, Heidelberg (1999)
32. Leontév, V.K.: Roots of Random Polynomials over a Finite Field. Mat. Zametki 80(2), 313–316 (2006)
33. Miller, V.S.: Use of elliptic curves in cryptography. In: Williams, H.C. (ed.) CRYPTO 1985. LNCS, vol. 218, pp. 417–426. Springer, Heidelberg (1986)
34. Montgomery, P.L.: Speeding up the Pollard and elliptic curve methods for factorizations. Mathematics of Computation 48, 243–264 (1987)
35. National Institute of Standards and Technology (NIST). Digital signature standard (DSS), FIPS PUB 186-3 (2009)
36. Quisquater, J.-J., Samyde, D.: ElectroMagnetic analysis (EMA): Measures and counter-measures for smart cards. In: Attali, S., Jensen, T. (eds.) E-smart 2001. LNCS, vol. 2140, pp. 200–210. Springer, Heidelberg (2001)
37. Rivest, R.L., Shamir, A., Adleman, L.: A method for obtaining digital signatures and public-key cryptosystems. Communications of the ACM 21, 120–126 (1978)
38. Skorobogatov, S.P., Anderson, R.J.: Optical fault induction attacks. In: Kaliski Jr., B.S., Koç, Ç.K., Paar, C. (eds.) CHES 2002. LNCS, vol. 2523, pp. 2–12. Springer, Heidelberg (2003)

39. Smart, N.P.: The Hessian Form of an Elliptic Curve. In: Koç, Ç.K., Naccache, D., Paar, C. (eds.) CHES 2001. LNCS, vol. 2162, pp. 118–125. Springer, Heidelberg (2001)
40. Smart, N.P.: An analysis of goubin's refined power analysis attack. In: Walter, C.D., Koç, Ç.K., Paar, C. (eds.) CHES 2003. LNCS, vol. 2779, pp. 281–290. Springer, Heidelberg (2003)
41. Vanstone, S.: Responses to NIST's proposal. Communications of the ACM 35, 50–52 (1992)
42. Yen, S.-M., Kim, S., Lim, S., Moon, S.-J.: RSA speedup with residue number system immune against hardware fault cryptanalysis. In: Kim, K.-c. (ed.) ICISC 2001. LNCS, vol. 2288, pp. 397–413. Springer, Heidelberg (2002)

Random Sampling for Short Lattice Vectors on Graphics Cards

Michael Schneider and Norman Göttert

Technische Universität Darmstadt, Germany
mischnei@cdc.informatik.tu-darmstadt.de

Abstract. We present a GPU implementation of the Simple Sampling Reduction (SSR) algorithm that searches for short vectors in lattices. SSR makes use of the famous BKZ algorithm. It complements an exhaustive search in a suitable search region to insert random, short vectors to the lattice basis. The sampling of short vectors can be executed in parallel.

Our GPU implementation increases the number of sampled vectors per second from 5200 to more than 120,000. With this we are the first to present a parallel implementation of SSR and we make use of the computing capability of modern graphics cards to enhance the search for short vectors even more.

Keywords: Lattice reduction, random sampling, SSR, BKZ.

1 Introduction

Lattices are discrete additive groups in the Euclidean vector space. They are known for hundreds of years in mathematics, but their use in cryptography and other fields of computer science started in the last decades of the twentieth century. Roughly speaking, lattice reduction is the search for short vectors with special geometric structure, i.e., vectors that are nearly orthogonal to each other. In 1982, the famous LLL algorithm was presented by Lenstra, Lenstra, and Lovász [16]. It set a starting point for developments and improvements of lattice reduction algorithms until today. In 1991, the BKZ algorithm (for Block-Korkine-Zolotarev reduction) which is a generalization of LLL was presented [27]. Today, BKZ is still the strongest and mostly used algorithm for lattice basis reduction. In 2003, the Random Sampling Reduction (RSR) algorithm was presented [26]. It is an adaption of BKZ, and applies BKZ together with the insertion of some randomly sampled vectors. In 2006, Simple Sampling Reduction (SSR) improved RSR by removing its heuristic assumptions [7].

In cryptology, lattice reduction has applications in cryptography as well as in cryptanalysis. The security of lattice based cryptosystems can be sustained by hard problems in lattices. The fact that makes lattice based cryptography special is the ability to base the security of cryptosystems on *worst case problems* in lattices, whereas usually security is only based on average case problems. This

B. Preneel and T. Takagi (Eds.): CHES 2011, LNCS 6917, pp. 160–175, 2011.

so-called *worst case to average case reduction* is unique for lattices and is not known in other fields.

For estimating the practical security of lattice based cryptosystems, it is necessary to know the strength of lattice reduction algorithms such as LLL, BKZ, and their revisions. Since there is a well-known gap between practical and theoretical strength of these algorithms, it is important to assess their practical borders. Since today, even desktop computers and laptops are equipped with multicore CPUs or graphics cards that support the CPU, this kind of special hardware must be taken into account when talking about security of cryptosystems. Due to the fact that supercomputers and new paradigms such as cloud computing gain more and more importance, the computing capabilities of attackers of cryptosystems rises as well. Therefore it is necessary to examine the strength of lattice reduction algorithms concerning parallelization potential.

The BKZ algorithm is the lattice reduction algorithm most commonly used in practice. It consists of two building blocks. One part is the LLL algorithm, the other part is an enumeration subroutine that performs exhaustive search for shortest vectors. No parallel version of BKZ is known to date. There are approaches of parallelizing LLL in the SIMD model, e.g. [30,2] and also for enumeration [9,15,10]. The combination of both however has not yet been tried.

It is apparent that SSR allows for distributed computing, since sampling short vectors can be performed independently in parallel. The authors of [7] state that most time of SSR is spent on sampling, which would allow for good parallelization.

1.1 Previous Results

Schnorr presented the first sampling algorithm called Random Sampling Reduction (RSR) in [26]. Ludwig and Buchmann refine the algorithm and promise to make sampling practical with their Simple Sampling Reduction (SSR) in [7]. They get rid of two RSR assumptions, namely the *Randomness Assumption* (RA) and the *Geometric Series Assumption* (GSA), which they claim both do not hold in practice. They replace the independent random sampling of vectors in the search space by a deterministic exhaustive search. This makes it impossible to sample the same vector multiple times, which was the case for RSR. Ludwig gives a more detailed view on SSR in [17]. The implementation of Ludwig is available upon request. Comparisons of his SSR implementation with BKZ on cryptographic lattices can be found, e.g., in [6,5].

1.2 Our Contribution

In this paper we present CUDA-SSR, a parallel variant of simple sampling reduction running on graphics cards using NVIDIAs CUDA framework. Our experiments are twofold. First we compare CUDA-SSR to BKZ, and second we compare it to our CPU-SSR implementation to show the strength of the GPU.

Although it is already mentioned in [7] that SSR is a good candidate for parallelizing, we are the first to present a distributed version of SSR. The authors of [7] mention a sampling rate of up to 5200 samples per second (on a 2.4GHz

Intel Pentium 4). On an NVIDIA GTX295 GPU (which was released in 2009) we get rates of more than 120,000 samples per second.

1.3 Organization of the Paper

The remainder of this paper is organized as follows. In Section 2, we present the required background knowledge concerning lattices, random sampling, and GPU computations. In Section 3, we develop a parallel version of SSR and explain how we implemented it on graphics cards. This is the main contribution of our work. Section 4 presents experimental results that show the strength of the GPU version of random sampling.

2 Preliminaries

Let $\|\mathbf{v}\|$ denote the Euclidean norm of the vector $\mathbf{v}$. Other norms are subscripted like $\|\mathbf{v}\|_\infty$.

Let $n, d \in \mathbb{N}$, $n \le d$, and let $\mathbf{b}_1, \ldots, \mathbf{b}_n \in \mathbb{R}^d$ be linearly independent. Then the set $\mathcal{L}(\mathbf{B}) = \{\sum_{i=1}^{n} x_i \mathbf{b}_i : x_i \in \mathbb{Z}\}$ is the lattice spanned by the basis column matrix $\mathbf{B} = [\mathbf{b}_1, \ldots, \mathbf{b}_n] \in \mathbb{Z}^{d \times n}$. The lattice $\mathcal{L}(\mathbf{B})$ is called n-dimensional. Its basis $\mathbf{B}$ is not unique, unimodular transformations lead to a different basis of the same lattice. The first successive minimum $\lambda_1(\mathcal{L}(\mathbf{B}))$ is the length of a shortest vector of $\mathcal{L}(\mathbf{B})$. The lattice determinant $\det(\mathcal{L}(\mathbf{B}))$ is defined as $\sqrt{\det(\mathbf{B}^t\mathbf{B})}$. It is invariant under basis changes. For full-dimensional lattices, where $n = d$, there is $\det(\mathcal{L}(\mathbf{B})) = |\det(\mathbf{B})|$ for every basis $\mathbf{B}$. In the remainder of this paper we will only be concerned with full-dimensional lattices.

Denote the Gram-Schmidt-orthogonalization (GSO) with $\mathbf{b}_i^* = \pi_i(\mathbf{b}_i)$ where $\pi_i(\mathbf{b}) \rightarrow \langle \mathbf{b}_1 \ldots \mathbf{b}_{i-1} \rangle^\perp$ is the orthogonal projection. The GSO is calculated via $\mathbf{b}_i^* = \mathbf{b}_i - \sum_{j=1}^{i-1} \mu_{i,j} \mathbf{b}_j^*$ for all $1 \le i \le n$, where $\mu_{i,j} = \mathbf{b}_i^T \mathbf{b}_j^* / \left\|\mathbf{b}_j^*\right\|^2$ for all $1 \le j \le i \le n$. The values $\mu_{i,j}$ are called *Gram-Schmidt (GS) coefficients.*

Roughly speaking, lattice reduction is the process of transforming a basis of a lattice into a second one consisting of short vectors which are nearly orthogonal. The LLL [16] and BKZ [27] algorithms are the most common algorithms for lattice reduction. BKZ is controlled by a blocksize parameter β, which allows for a trade-off between runtime and reduction quality. Higher values of β lead to better reduced bases at the expense of an exponentially (in β) increasing runtime. LLL is the special case of BKZ with $\beta = 2$. Both LLL and BKZ sort the basis vectors in increasing order, so that $\mathbf{b}_1$ is the shortest among the basis vectors after reduction. Applied to a basis $\mathbf{B}$, LLL provably finds a vector $\mathbf{b}_1$ with $\|\mathbf{b}_1\| \le 2^{(n-1)/2} \lambda_1(\mathcal{L}(\mathbf{B}))$. When LLL or BKZ is applied to a generator system of a lattice $\mathcal{L}$ it will output a basis of $\mathcal{L}$, so it will remove linear dependent vectors. The first basis vector found by BKZ with $\beta > 2$ is shorter than with LLL, i.e., it holds that $\|\mathbf{b}_1\| \le (\gamma_\beta)^{(n-1)/(\beta-1)} \lambda_1(\mathcal{L}(\mathbf{B}))$ [25], where γ_β is the β-dimensional Hermite constant. A practical comparison of LLL and BKZ can be found in [12]. Both LLL and BKZ are equipped with a parameter δ, which only slightly controls the reduction quality and is usually set to 0.99. For further information concerning lattices and lattice reduction we refer to [19,20,22].

2.1 Random Sampling

The idea of random sampling was presented by Schnorr in 2003 [26]. It was adopted and improved in [17,7]. The idea of random sampling is the following. Iteratively, it switches between reduction of the basis (using BKZ) and sampling a random short vector of norm $< 0.99 \left\|\mathbf{b}_1\right\|^2$, which is then prepended to the reduced basis (cf. Algorithm 1).

Every basis vector $\mathbf{v} = [v_1, \ldots, v_n]$ can be written in its orthogonalized form $\mathbf{v} = \sum_{i=1}^{n} \nu_i \mathbf{b}_i^*$. We can write its squared norm as

$$\left\|\mathbf{v}\right\|^2 = \sum_{i=1}^{n} \nu_i^2 \left\|\mathbf{b}_i^*\right\|^2 . \tag{1}$$

Therefore, shortening a vector $\mathbf{v}$ is done either by decreasing ν_i or by decreasing the $\|\mathbf{b}_i^*\|$.

For a reduced basis $\mathbf{B}$ (either LLL or BKZ reduced), it is known that the norm of the orthogonalized vectors $\|\mathbf{b}_i\|$ decreases for increasing index i. This implies that for higher indices, the influence of the coefficient ν_i in Equation (1) is less noticeable than for smaller indices. This fact helps interpreting the following definition of a search space. For a basis $\mathbf{B} \in \mathbb{Z}^{n \times n}$ and an integer u with $1 \leq u \leq n$ we define the set $\mathcal{S}_{u,\mathbf{B}}$ as the set of all lattice vectors $\mathbf{v} = \sum_{i=1}^{n} \nu_i \mathbf{b}_i^*$ with

$$|\nu_i| \leq \begin{cases} 0.5 & \text{for } 1 \leq i < n-u \\ 1 & \text{for } n-u \leq i < n \end{cases} , \quad \nu_n = 1 \tag{2}$$

and call it the search space. It is $\mathcal{S}_{u,\mathbf{B}} \subseteq \mathcal{L}(\mathbf{B})$, and this search space is supposed to contain short lattice vectors. The algorithm SAMPLE (Algorithm 2, original in [17]) uses as input a lattice basis $\mathbf{B}$ and an integer value x, and as output it computes a vector $\mathbf{v} \in \mathcal{S}_{u,\mathbf{B}}$ in the search space. The bit representation of the integer x controls the sampling deterministically. If the search space $\mathcal{S}_{u,\mathbf{B}}$ consists of 2^u many points, running SAMPLE with all values $x \in \{1, \ldots, 2^u\}$ guarantees that the complete search space is sampled.

Algorithm 1. SSR

Input: Lattice basis $\mathbf{B} \in \mathbb{Z}^{n \times n}$, GS-coefficients $\mathbf{R} \in \mathbb{Q}^{n \times n}$, bound $u_{max} \in \mathbb{N}$, blocksize β, norm bound A
Output: reduced basis $\mathbf{B}$ s.t. $\|\mathbf{b}_1\| < A$

$\mathbf{B} \leftarrow BKZ([\mathbf{b}_1, \ldots, \mathbf{b}_n], \beta)$
while $\|\mathbf{b}_1\| > A$ **do**
    for $x = 1$ *to* $2^{u_{max}}$ **do**
        $\mathbf{v} \leftarrow$ SAMPLE$(\mathbf{B}, \mathbf{R}, x)$
        if $\left\|\mathbf{v}\right\|^2 \leq 0.99 \left\|\mathbf{b}_1\right\|^2$ **then** break
    end
    if $x = 2^{u_{max}}$ **then** terminate("No short vector found")
    $\mathbf{B} \leftarrow BKZ([\mathbf{v}, \mathbf{b}_1, \ldots, \mathbf{b}_n], \beta)$
end

Algorithm 2. SAMPLE

Input: Lattice basis $\mathbf{B} \in \mathbb{Z}^{n\times n}$, GS-coefficients $\mathbf{R} \in \mathbb{Q}^{n\times n}$, $x \in \mathbb{Z}$
Output: vector $\mathbf{v} \in \mathcal{S}_{u,\mathbf{B}}$

$\mathbf{v} \leftarrow \mathbf{b}_n$, $\nu \leftarrow \mathbf{r}_n$
for $j = n-1$ *to* 1 **do**
    $y \leftarrow \lceil \nu_j - 0.5 \rceil$
    if $x = 1 \mod 2$ **then**
        if $\nu_j - y \leq 0$ **then** $y \leftarrow y - 1$
        else $y \leftarrow y + 1$
    end
    $x \leftarrow \lfloor x/2 \rfloor$, $\mathbf{v} \leftarrow \mathbf{v} - y\mathbf{b}_j$, $\nu \leftarrow \nu - y\mathbf{r}_j$
end
return v

Algorithm 1 shows a pseudo-code listing of SSR, Algorithm 2 shows a listing of SAMPLE. For more details on random sampling we refer to the works of [26,17,7].

2.2 GPU Computation

Graphical Processing Units (GPUs) were developed to perform huge numbers of graphical operations in parallel. The introduction of computing platforms such as CUDA by NVIDIA [23] and CTM by ATI [1] opened graphics cards equipped with GPUs for running custom user programs. The development of these computing frameworks where the starting point of the breakthrough these processing units had over the last years. The existence of standard libraries like BLAS [24] for linear algebra made GPUs interesting for cryptographic applications as well.

In the field of cryptography, there are (among others) implementations of AES [8,18,14] and RSA [21,29,11] available as well as implementations of the SHA3 hash competition finalists [4]. In cryptanalysis, Bernstein et al. use parallelization techniques on graphics cards to solve integer factorization using elliptic curves [3]. Concerning lattices and lattice reduction, there is an implementation of the ENUM algorithm on graphics cards [15]. We are not aware of other work in the field of lattice reduction.

Programming Model. We will be using the CUDA framework from NVIDIA on an NVIDIA GTX 295 card. The description might be slightly different for newer cards of the Fermi architecture. A CUDA-capable GPU is equipped with several multiprocessors, which contain small numbers of scalar processors each. The programmer can stick to the *single instruction - multiple thread* (SIMT) programming model. The programmer writes code for single threads, which is uploaded to the device and executed in parallel by multiple threads.

The threads altogether are organized in *blocks*, which again are organized in *grids*. A kernel is a program running on a graphics device. When a kernel (a grid) is executed, 32 threads are scheduled in a so-called *warp*. These 32 threads should perform the same computation, since otherwise the threads are handled in serial, not in parallel.

Memory Model. One big issue on NVIDIAs GPUs is the different types of memory available. There are registers, shared memory, global memory, texture, and constant memory. Registers and shared memory are on chip and close to the multiprocessor and can be accessed with low latency. The number of registers and shared memory is limited, since the number available for one multiprocessor must be shared among all threads in a single block. Global memory is slow, since it is off-chip and there is no cache for it. Constant and texture memory are parts of the global memory, but they are cached and can be used for specific types of data or special access patterns.

3 GPU Algorithm CUDA-SSR

The CUDA-SSR approach in Algorithm 3 is a slightly changed variant of the original SSR algorithm. In each outer while loop, up to $2^{u_{max}}$ vectors are sampled in parallel, and the m shortest samples are added to the basis. The main difference to the original SSR is the sampling of new vectors $\mathbf{v}$, which is done on the GPU and returns not only a single vector but multiple ones within a bound of m. The calculated vectors $[\mathbf{v}_1, \mathbf{v}_2, \ldots, \mathbf{v}_m]$ are added to the front of the lattice $\mathbf{B}$ in a sorted order, before the extended lattice is reduced by the BKZ algorithm. With the adding of multiple vectors we get a benefit of a more stabilized reduction, as we will see in the experiments section.

The algorithm terminates if a given norm of the first vector of $\mathbf{B}$ is undercut by a new vector $\mathbf{v}$ or if no smaller vector is found in the given search space.

The subroutine PAR-SAMPLE (which is now executed on GPU) is a slightly changed variant of SAMPLE (Algorithm 2). The original sample algorithm was

Algorithm 3. CUDA-SSR

Input: Lattice basis $\mathbf{B} \in \mathbb{Z}^{n \times n}$, GS-coefficients $\mathbf{R} \in \mathbb{Q}^{n \times n}$, bound $u_{max} \in \mathbb{N}$, blocksize β, norm bound A, add vector bound m
Output: BKZ-β reduced basis $\mathbf{B}$ s.t. $\|\mathbf{b}_1\| \leq A$

$\mathbf{B} \leftarrow BKZ([\mathbf{b}_1, \ldots, \mathbf{b}_n], \beta)$
foundSmaller $=$ *true*
xOffset $= 0$
while $\|\mathbf{b}_1\| > A$ *and foundSmaller* $=$ *true* **do**
  while *xOffset* $< 2^{u_{max}}$ **do**
    parallel $[i =$ *xOffset* $\ldots$ *xOffset* $+$ *maxSamplesPerCall*$]$ **do**
      $[\mathbf{v_1}, \mathbf{v_2}, \ldots, \mathbf{v_m}]$, foundSmaller $\leftarrow$ PAR-SAMPLE$(\mathbf{B}, \mathbf{R}, x_i, m)$
    end
    if *foundSmaller* $=$ *true* **then** break inner while loop
    xOffset $+=$ maxSamplesPerCall
    if *xOffset* $\geq 2^{u_{max}}$ **then** terminate
  end
  $\mathbf{B} \leftarrow BKZ([\mathbf{v}_1, \mathbf{v}_2, \ldots, \mathbf{v}_m, \mathbf{b}_1, \ldots, \mathbf{b}_n], \beta)$
end

parallelized, so that it computes a huge number of vectors per call. The possibility of parallelization is based on the independence of the samples. The only difference among two samples is the input value x, which can be interpreted as an unique identifier or seed.

One sample is stored in the shared memory of a CUDA block. The amount of shared memory, which is used for producing one sample, consists of memory for the vector $\mathbf{v}$ ($4Byte \cdot dimension$), for the vector ν ($4Byte \cdot dimension$), for y ($4Byte$), and for a valid-Byte ($1Byte$). For one CUDA block a number of

$$samplesPerBlock = \left\lfloor \frac{available\ shared\ memory}{(4 + 4Byte) \cdot dimension + 4Byte + 1Byte} \right\rfloor$$

vectors are produced. If we use all available CUDA blocks, the overall number of samples is $65535 \cdot samplesPerBlock$ per call. For example, at a dimension of 80 one call calculates $65535 \cdot \lfloor \frac{16344}{8 \cdot 80+5} \rfloor = 1,638,375$ samples.[1]

3.1 Parallel Implementation of Subroutine Sample

Here we describe how we implemented the sampling of $samplesPerBlock$ many vectors in $\mathcal{S}_{u,\mathbf{B}}$ on GPU. This is the main contribution of the paper.

The first step for determining $samplesPerBlock$ samples in one CUDA block is to copy the entries of the last vector of the matrices $\mathbf{B}$ and $\mathbf{R}$ to $\mathbf{v}$ and ν in parallel. The matrices $\mathbf{B}$ and $\mathbf{R}$ resist in the texture memory, because they are read multiple times and this memory is cached.

The second step is to compute the factor y for every sample and build new vectors inside a for-loop. A single y is processed by one CUDA thread, therefore all $y's$ of one CUDA block can be calculated in parallel. Afterwards the temporary new vectors $\mathbf{v}$ and ν are built, whereby all entries of a vector are assigned in one parallel step. If an integer overflow is noticed in this step, the sample will be indicated as invalid.

When the loop is finished, the square norms of the new samples are calculated with the common *vector reduction* approach, after squaring all entries of $\mathbf{v}$ in parallel. Figure 1 illustrates this procedure. Once a square norm of 2^x (with $x = \max\{y \in \mathbb{Z} : 2^y \leq dimension\}$) is determined, the result will be added to the first entry of the next interval. This procedure continues, until there is no more than one entry left.

Because the square norms of all vectors are calculated step by step, we can register the smallest square norm of a CUDA block. Therefore a CUDA block writes only the smallest vector back to the global memory, assumed that the square norm is less than 99% of $\|\mathbf{b}_1\|^2$ and the sample is valid. With this we save a lot of global memory. Instead of writing $65535 \cdot samplesPerBlock$ many vectors to global memory we use shared memory for $samplesPerBlock$ many vectors of each block and only write 65535 many vectors to the device.

[1] The shared memory of $16384Byte$ is decreased by the parameters of the kernel call, which are also stored in shared memory ($16Byte$ for dimBlock and dimGrid, $24Byte$ for 3 pointers). These values might change for other CUDA compute capabilities.

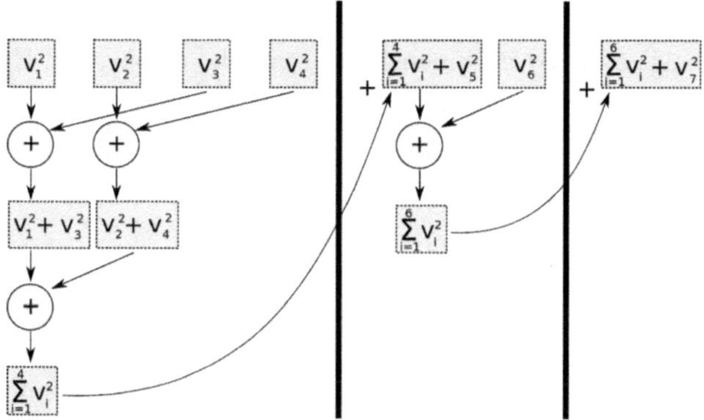

Fig. 1. Computation of the norm of a single vector $\mathbf{v}$ in parallel

For achieving higher performance we introduce a counter, which increases if a vector with a square norm less than 99% of $\|\mathbf{b}_1\|^2$ is found. When m vectors below this bound have been found, we break the parallel sampling. The counter is increased with so called *atomic operations*, which provides an exclusive *read-modify-write* operation for one CUDA thread. The parallel processing of the CUDA framework is only "semi-parallel", because only a part of all CUDA blocks are processed parallel for real (we have 65535 blocks but only 30 multiprocessors available). Therefore we can abort further calculations, if the counter m reached a defined value. A flow chart of our GPU algorithm of SAMPLE is shown in Appendix A. In order to remove serialization we also tested replacing the condition in Line 4 of SAMPLE by arithmetic computations, but recognized no speedups. Since there is no *else*-block, the fact that (on average) half of the threads are idle does not influence the total runtime.

For establishing the gain of parallel sampling we also implement a CPU version of the SSR algorithm (called CPU-SSR), which produces new vectors step by step. Our CPU as well as the GPU implementation are available online.[2]

4 Experimental Results

We are using an NVIDIA GTX 295 GPU for our experiments. The CPU that we use is an Intel Core2 Duo E8400 CPU running at 3GHz. The lattices we use are the SVP challenge lattices [13] with seed 0, so we use only one lattice per dimension. For LLL and BKZ reduction we use the NTL library [28] in version 5.5.2. The parameter δ is always set to the standard value 0.99. We run LLL with precision RR followed by BKZ with precision QP.

First we compare our results of CUDA-SSR to BKZ, and second we present experiments comparing CUDA-SSR to CPU-SSR.

[2] `http://www.cdc.informatik.tu-darmstadt.de/mitarbeiter/mischnei.html`

4.1 Comparison of CUDA-SSR and BKZ

Let $\mathbf{B}$ be the basis of $\mathcal{L}(\mathbf{B})$ in dimension n and c be a constant. Using BKZ with blocksize β, Gama and Nguyen [12] predict the average norm of the first basis vector after BKZ reduction to be

$$gn = c^n \det(\mathcal{L}(\mathbf{B}))^{1/n}, \tag{3}$$

where the *Hermite factor constant* c relies on the blocksize used. For BKZ-20, e.g., they experimentally gain a value of $c = 1.0128$.

Our experiments are performed as follows. First, we reduce a lattice basis with BKZ with increasing blocksize, until we reach a vector of desired goal norm gn, cf., Equation (3). We use a value of $c = 1.0129$ to calculate our goal norm. The resulting run times and the reached norms are shown in Figures 2 and 3. Second, we use CUDA-SSR with half the blocksize (rounded up) that BKZ needed to reach the goal norm and run CUDA-SSR on the same lattice; i.e., random sampling has to close the gap between BKZ with half blocksize and BKZ with full blocksize. We stop the GPU sampling when $m = 0.25 \cdot n$ vectors below $0.99 \cdot \|\mathbf{b}_1\|^2$ were found by PAR-SAMPLE.

Figure 4 shows the blocksize that BKZ needed to find the resulting vector. The picture shows that the blocksize is around 20 in most of the cases, as predicted by [12]. Figure 5 shows the speedup factor of CUDA-SSR compared to BKZ.

We notice that with both approaches, BKZ as well as CUDA-SSR, we find vectors of comparable length (Figure 2). CUDA-SSR is always faster (up to 40%). For comparison reason, Figure 3 includes the runtime of BKZ with blocksize $\lceil \beta/2 \rceil$, the pre-processing step of SSR (Line 1 of Algorithm 3). The picture shows that it takes a huge part of the random sampling time (dashed line). This implies that the later part of SSR (sampling - BKZ - sampling - ...) takes a lot less time (the time difference between the dotted and dashed curve) than the initial BKZ. Therefore, the total SSR runtime cannot profit too much from the parallel sampling part in this setting.

The runtime speedup factor (Figure 5) seems to increase with the dimension, from 1.1 in dimension 80 to a maximum value of 1.6 in dimension 160. The peek in dimension 150 is also apparent in Figure 4 and seems to result from special structure in the lattice (SSR is working less in this lattice).

4.2 Comparison of GPU and CPU Variant of SSR

Our second block of experiments is supposed to show the strength of parallelization on GPU of the SSR algorithm. For this, we run our CPU implementation and our GPU implementation of SSR for the same lattices until they undercut the goal norm. For pre-reduction, we use LLL only. We note the reached norm (cf. Figure 6) and the runtime (cf. Figure 7). Figure 8 shows the speedup factor gained by the GPU version. We prepend $m = 0.1 \cdot n$ vectors to the basis in each GPU iteration. Figure 9 compares a typical behaviour of SSR on GPU and CPU over time, concerning the norm of the sampled vectors.

On CPU, the sampling rate was about 160 samples per second for a 180-dimensional lattice. The GTX 295 GPU reached about $120,000$ samples per

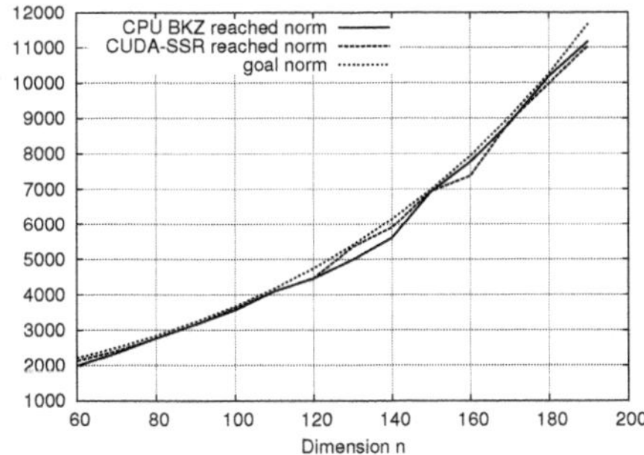

Fig. 2. Reached norm of BKZ and CUDA-SSR

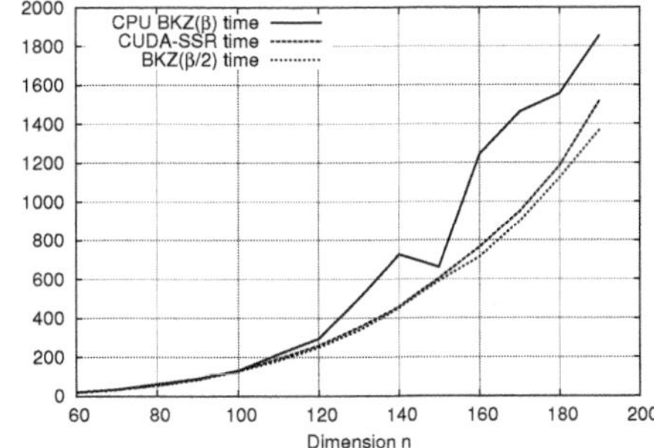

Fig. 3. Required time in seconds for BKZ with blocksize β and for CUDA-SSR to reach the same goal norm

Fig. 4. Required blocksize of BKZ to reach the desired goal norm

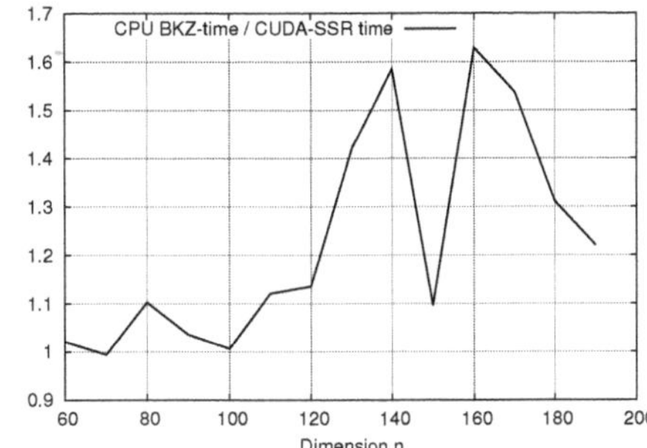

Fig. 5. Speedup factor of CUDA-SSR in comparison to BKZ

second for a 180 dimensional lattice. In smaller dimension, sampling rates of more than $250,000$ are possible, e.g. in dimension 60.

From Figure 7 we conclude that the runtime of SSR on GPU is very stable, whereas on CPU (solid curve), we see two different behavior patterns. In some dimensions, e.g. 90 or 110, SSR finds shorter vectors very early, and the runtime is comparable to the CUDA-SSR runtime. In other cases we see huge peeks in the runtime curve, e.g. in dimension 100 or 120, which suggest that on CPU it takes a long time until shorter vectors are found. We conclude that sampling multiple vectors in each iteration helps SSR to run much more stable.

The speedup factor shown in Figure 8 shows the potential of the CUDA version compared to the CPU version. In small dimension we gain speedup factors of up to 180. On GPU, in the first iteration a vector below the bound is already found, whereas on CPU multiple iterations have to be performed. In bigger dimensions, the speedup factor decreases, depending on the behaviour pattern.

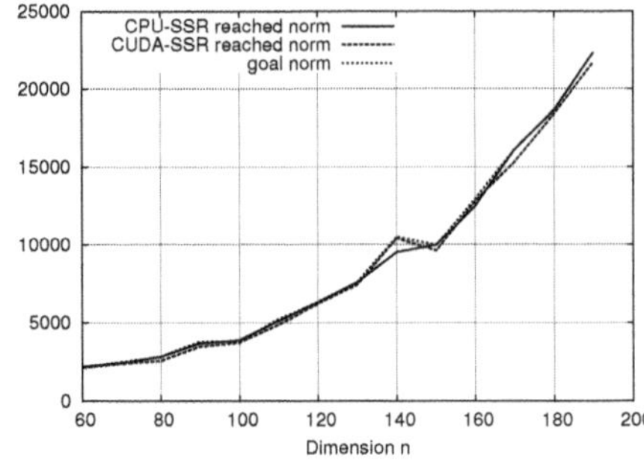

Fig. 6. Reached norm of CPU-SSR and CUDA-SSR

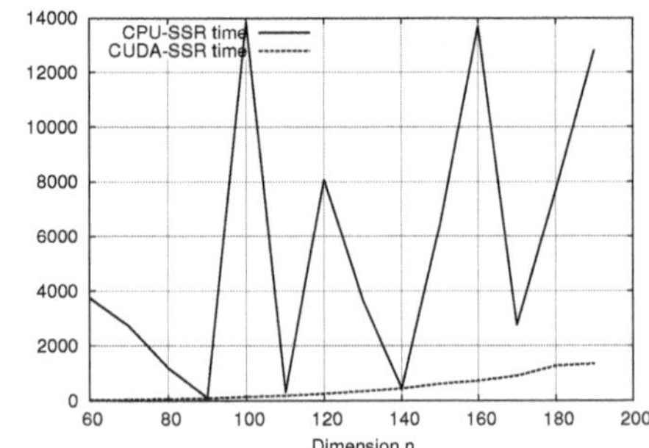

Fig. 7. Required time in seconds of CPU-SSR and CUDA-SSR to reach the same goal norm

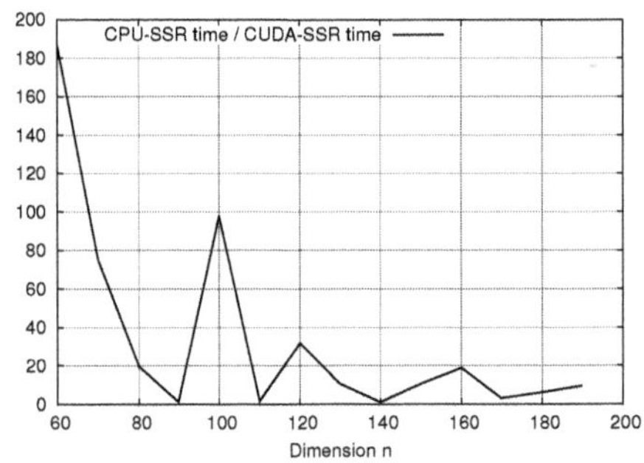

Fig. 8. Speedup factor of GPU compared to CPU variant of SSR

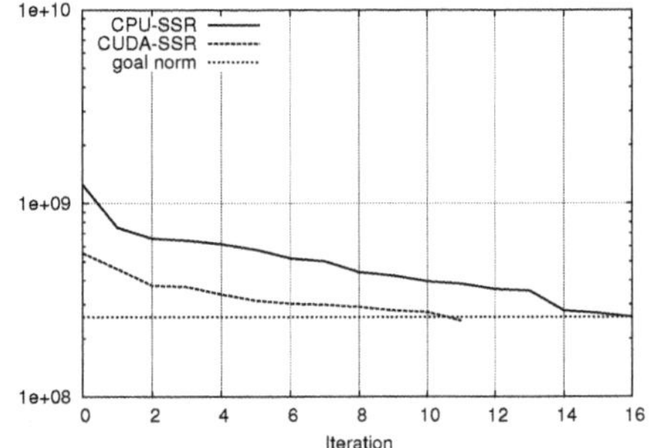

Fig. 9. Development of the squared norm of SSR over time, in a 190 dimensional lattice. The ordinate shows the squared norm of the vectors found by sampling.

Figure 9 shows a typical behaviour of SSR on CPU and GPU. CUDA-SSR starts with lower norm, which implies that the first iterations of SSR decreases the norm much more than on CPU. We noticed that in the first iterations, there exists a huge number of vectors below the $0.99 \|\mathbf{b}_1\|^2$ bound. Therefore, on GPU we have good chance to find a much shorter vector. On CPU only the first vector below the bound is picked, whereas on GPU multiple vectors are prepended to the basis, and all these vectors are potentially smaller than the CPU one.

To show the strength of our GPU version, Figure 10 shows the time needed by CUDA-SSR and CPU-SSR to sample the same amount of vectors, namely 2^{21} many. It is evident that on GPU, the sampling is much faster, with a maximum factor 14.5 in dimension 190.

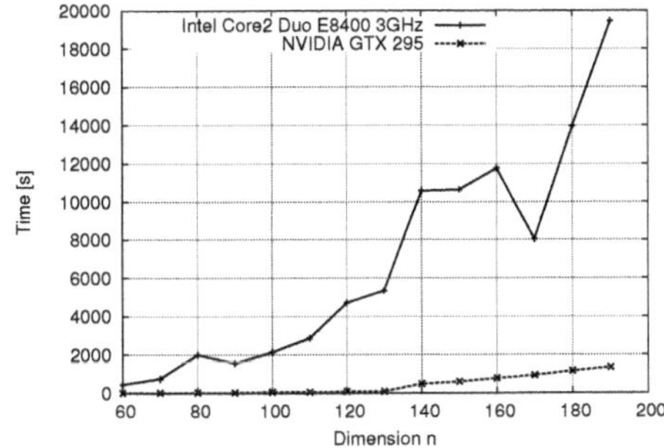

Fig. 10. Time to sample 2^{21} many vectors using CUDA-SSR and CPU-SSR

5 Conclusion and Further Work

We have presented a parallel version of random sampling and an implementation on GPU. Our results show the strength of parallelism for this type of algorithm. Our proposal CUDA-SSR allows for more than 120,000 sampled vectors per second, which is the maximum stated in literature. Unfortunately, the speedups compared to BKZ are not too impressive, due to the big fraction of the runtime that BKZ takes. The percentage of BKZ of the total runtime was up to 97%. This is not optimal, since BKZ does not apply the hardware acceleration of the graphics card. LLL took 67% of the total runtime in dimension 100. [7] mention that sampling takes most of the time, but we were not able to reproduce that.

The speedup in sampling rates is much higher than the speedups in runtime. So the potential of parallelization is visible, but SSR does not take full advantage of it.

The SVP challenge comes with a generator for lattices, to allow participants not only to download one lattice in each dimension but to generate multiple instances. To present smoother graphs it is necessary to run our experiments on multiple instances in each dimension.

In order to allow for good parallelization, we did not include new search spaces as proposed in [7]. Ludwig and Buchmann present *check search space size* (CSSS) functions in order to sample from smaller sets of vectors. It would be interesting to compare how this influences the rate of parallelism and if usage of CSSS could speed up CUDA-SSR even more.

Acknowledgements. Michael Schneider is supported by project BU 630/23-1 of the German Research Foundation (DFG). This work was supported by CASED (`www.cased.de`). We thank the anonymous reviewers for their comments.

References

1. Advanced Micro Devices. ATI CTM Guide. Technical report, ATI (2006)
2. Backes, W., Wetzel, S.: Parallel lattice basis reduction using a multi-threaded schnorr-euchner LLL algorithm. In: Sips, H., Epema, D., Lin, H.-X. (eds.) Euro-Par 2009. LNCS, vol. 5704, pp. 960–973. Springer, Heidelberg (2009)
3. Bernstein, D.J., Chen, T.-R., Cheng, C.-M., Lange, T., Yang, B.-Y.: ECM on graphics cards. In: Joux, A. (ed.) EUROCRYPT 2009. LNCS, vol. 5479, pp. 483–501. Springer, Heidelberg (2009)
4. Bos, J.W., Stefan, D.: Performance analysis of the SHA-3 candidates on exotic multi-core architectures. In: Mangard, S., Standaert, F.-X. (eds.) CHES 2010. LNCS, vol. 6225, pp. 279–293. Springer, Heidelberg (2010)
5. Buchmann, J., Lindner, R.: Secure parameters for SWIFFT. In: Roy, B., Sendrier, N. (eds.) INDOCRYPT 2009. LNCS, vol. 5922, pp. 1–17. Springer, Heidelberg (2009)
6. Buchmann, J., Lindner, R., Rückert, M.: Explicit hard instances of the shortest vector problem. In: Buchmann, J., Ding, J. (eds.) PQCrypto 2008. LNCS, vol. 5299, pp. 79–94. Springer, Heidelberg (2008)
7. Buchmann, J., Ludwig, C.: Practical lattice basis sampling reduction. In: Hess, F., Pauli, S., Pohst, M. (eds.) ANTS 2006. LNCS, vol. 4076, pp. 222–237. Springer, Heidelberg (2006)
8. Cook, D.L., Ioannidis, J., Keromytis, A.D., Luck, J.: CryptoGraphics: Secret key cryptography using graphics cards. In: Menezes, A. (ed.) CT-RSA 2005. LNCS, vol. 3376, pp. 334–350. Springer, Heidelberg (2005)
9. Dagdelen, Ö., Schneider, M.: Parallel enumeration of shortest lattice vectors. In: D'Ambra, P., Guarracino, M., Talia, D. (eds.) Euro-Par 2010. LNCS, vol. 6272, pp. 211–222. Springer, Heidelberg (2010)
10. Detrey, J., Hanrot, G., Pujol, X., Stehlé, D.: Accelerating lattice reduction with FPGAs. In: Abdalla, M., Barreto, P.S.L.M. (eds.) LATINCRYPT 2010. LNCS, vol. 6212, pp. 124–143. Springer, Heidelberg (2010)
11. Fleissner, S.: GPU-accelerated montgomery exponentiation. In: Shi, Y., van Albada, G.D., Dongarra, J., Sloot, P.M.A. (eds.) ICCS 2007. LNCS, vol. 4487, pp. 213–220. Springer, Heidelberg (2007)
12. Gama, N., Nguyen, P.Q.: Predicting lattice reduction. In: Smart, N.P. (ed.) EUROCRYPT 2008. LNCS, vol. 4965, pp. 31–51. Springer, Heidelberg (2008)
13. Gama, N., Schneider, M.: SVP Challenge (2010), `http://www.latticechallenge.org/svp-challenge`
14. Harrison, O., Waldron, J.: AES encryption implementation and analysis on commodity graphics processing units. In: Paillier, P., Verbauwhede, I. (eds.) CHES 2007. LNCS, vol. 4727, pp. 209–226. Springer, Heidelberg (2007)
15. Hermans, J., Schneider, M., Buchmann, J., Vercauteren, F., Preneel, B.: Parallel shortest lattice vector enumeration on graphics cards. In: Bernstein, D.J., Lange, T. (eds.) AFRICACRYPT 2010. LNCS, vol. 6055, pp. 52–68. Springer, Heidelberg (2010)
16. Lenstra, A., Lenstra, H., Lovász, L.: Factoring polynomials with rational coefficients. Mathematische Annalen 4, 515–534 (1982)
17. Ludwig, C.: Practical Lattice Basis Sampling Reduction. PhD thesis, Technische Universität Darmstadt (2005), `http://elib.tu-darmstadt.de/diss/000640/`
18. Manavski, S.A.: CUDA compatible GPU as an efficient hardware accelerator for AES cryptography. In: ICSPC, pp. 65–68. IEEE Computer Society Press, Los Alamitos (2007)

19. Micciancio, D., Goldwasser, S.: Complexity of Lattice Problems: a cryptographic perspective. Kluwer Academic Publishers, Dordrecht (2002)
20. Micciancio, D., Regev, O.: Lattice-based cryptography. In: Bernstein, D.J., Buchmann, J.A., Dahmen, E. (eds.) PQCrypto 2008. LNCS, vol. 5299, pp. 147–191. Springer, Heidelberg (2008)
21. Moss, A., Page, D., Smart, N.P.: Toward acceleration of RSA using 3D graphics hardware. In: Galbraith, S.D. (ed.) Cryptography and Coding 2007. LNCS, vol. 4887, pp. 364–383. Springer, Heidelberg (2007)
22. Nguyen, P.Q., Vallée, B.: The LLL Algorithm - Survey and Applications. Springer, Heidelberg (2010)
23. NVIDIA. Compute Unified Device Architecture Programming Guide. Technical report, NVIDIA (2007)
24. NVIDIA. CUBLAS Library (2007)
25. Schnorr, C.-P.: Block reduced lattice bases and successive minima. Combinatorics, Probability & Computing 3, 507–522 (1994)
26. Schnorr, C.-P.: Lattice reduction by random sampling and birthday methods. In: Alt, H., Habib, M. (eds.) STACS 2003. LNCS, vol. 2607, pp. 146–156. Springer, Heidelberg (2003)
27. Schnorr, C.-P., Euchner, M.: Lattice basis reduction: Improved practical algorithms and solving subset sum problems. Mathematical Programming 66, 181–199 (1994)
28. Shoup, V.: Number theory library (NTL) for C++, `http://www.shoup.net/ntl/`
29. Szerwinski, R., Güneysu, T.: Exploiting the power of gPUs for asymmetric cryptography. In: Oswald, E., Rohatgi, P. (eds.) CHES 2008. LNCS, vol. 5154, pp. 79–99. Springer, Heidelberg (2008)
30. Villard, G.: Parallel lattice basis reduction. In: ISSAC, pp. 269–277. ACM, New York (1992)

A Flow Chart of Parallel Sampling in CUDA-SSR

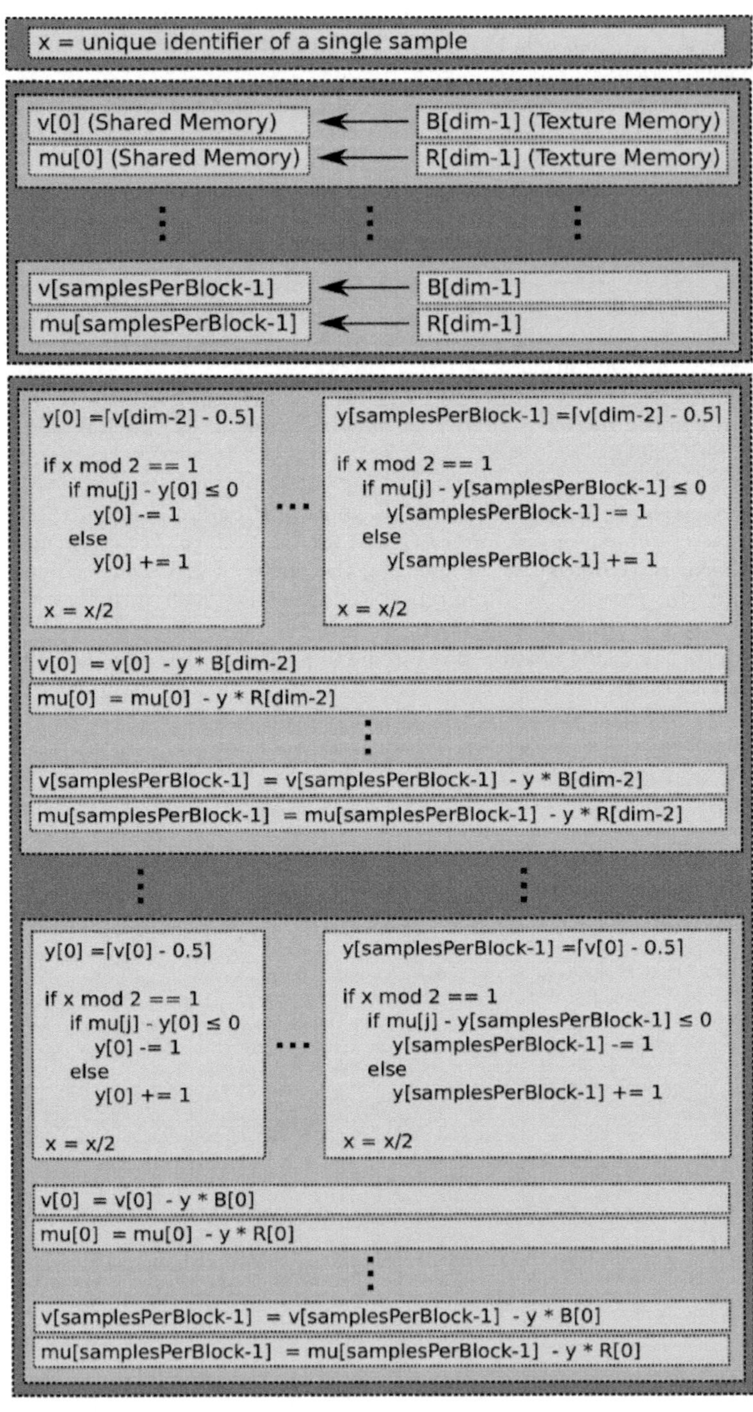

```
minNorm = ∞

calculate square norm of v[0] in parallel

if square norm of v[0] < minNorm
   minNorm = square norm of v[0]
   minIndex = 0

⋮

calculate square norm of v[samplesPerBlock-1] in parallel

if square norm of v[samplesPerBlock-1] < minNorm
   minNorm = square norm of v[samplesPerBlock-1]
   minIndex = samplesPerBlock-1

Result of the block is v[minIndex]
```

Extreme Enumeration on GPU and in Clouds

- How Many Dollars You Need to Break SVP Challenges -

Po-Chun Kuo[1], Michael Schneider[2], Özgür Dagdelen[3], Jan Reichelt[3], Johannes Buchmann[2,3], Chen-Mou Cheng[1], and Bo-Yin Yang[4]

[1] National Taiwan University, Taipei, Taiwan
[2] Technische Universität Darmstadt, Germany
[3] Center for Advanced Security Research Darmstadt (CASED), Germany
[4] Academia Sinica, Taipei, Taiwan

Abstract. The complexity of the Shortest Vector Problem (SVP) in lattices is directly related to the security of NTRU and the provable level of security of many recently proposed lattice-based cryptosystems. We integrate several recent algorithmic improvements for solving SVP and take first place at dimension 120 in the SVP Challenge Hall of Fame. Our implementation allows us to find a short vector at dimension 114 using 8 NVIDIA video cards in less than two days.

Specifically, our improvements to the recent Extreme Pruning in enumeration approach, proposed by Gama *et al.* in Eurocrypt 2010, include: (1) a more flexible bounding function in polynomial form; (2) code to take advantage of Clouds of commodity PCs (via the MapReduce framework); and (3) the use of NVIDIA's Graphics Processing Units (GPUs). We may now reasonably estimate the cost of a wide range of SVP instances in U.S. dollars, as rent paid to cloud-computing service providers, which is arguably a simpler and more practical measure of complexity.

Keywords: Shortest Vector Problem, GPU, Cloud Computing, Enumeration, Extreme Pruning.

1 Introduction

Lattice-based cryptography is a hot topic, with numerous submissions and publications at prestigious conferences in the last two years. The reasons that it might have become so popular include:

- lattice-based PKCs, unlike ECC, do not immediately succumb to large quantum computers (i.e., they are "post-quantum");
- lattice-based PKCs enjoy the (so far) unique property of being protected by a worst-case hardness assumption (i.e., they are unbreakable if *any* of a large class of lattice-based problem *at a lower dimension* is intractable);
- lattices can be used to create fully homomorphic encryptions.

One of the main problems in lattice-based cryptography is the *shortest vector problem* (SVP). As the name implies, it is a search for a non-zero vector with

B. Preneel and T. Takagi (Eds.): CHES 2011, LNCS 6917, pp. 176–191, 2011.

the smallest Euclidean norm in a lattice. The SVP is NP-hard under randomized reductions. The *approximate shortest vector problem* (ASVP) is the search for a short non-zero vector whose length is at most some given multiple of the minimum. It is easy in some cases, as shown by the LLL algorithm [LLL82]. Although LLL has polynomial running time, the approximation factor of LLL is exponential in the lattice dimension. The complexity of SVP (and ASVP) has been studied for decades, but practical implementations that take advantage of special hardware are not investigated seriously until recently [HSB+10, DS10, DHPS10].

In contrast, enumeration is another way to solve SVP and ASVP, which can be viewed as a depth-first search in a tree structure, going over all vectors in a specified search region deterministically. Typically, a basis transformation such as BKZ [SE94] is performed first to improve the basis to one likely to yield a short vector via enumeration.

At Eurocrypt 2010, Gama *et al.* proposed the *Extreme Pruning* approach to solving SVP and ASVP [GNR10] and showed that it is possible to speed up the enumeration exponentially by randomizing the algorithm. The idea is that, instead of spending a lot of time searching *one* tree, one generates *many* trees and only spends a small amount of time on each of them by aggressively pruning the subtrees unlikely to yield short vectors using a bounding function. That is, one focuses on the parts of the trees that are more "fruitful" in terms of the likelihood of producing short vectors per unit time spent.

In other words, one should try to maximize the success probability of finding a short vector *per unit of computing time spent* by choosing an appropriate bounding function in pruning. Therefore, which bounding function works better depends on the particular *implementation.*

In this paper, we make a practical contribution on several fronts.

1. We integrate the Extreme Pruning idea of Gama *et al.* [GNR10] into the GPU implementation of [HSB+10].
2. We extend the implementation by using multiple GPUs and run it on Amazon's EC2 in order to harness the immense computational power of such cloud services.
3. We extrapolate our average-case run times to estimate the run time of our implementation for solving ASVP instances of the SVP Challenge in higher dimensions.
4. Consequently, we set new records for the SVP challenge in dimensions 114, 116, and 120. The previous record was for dimension 112.

As a result, the average "cost" of solving ASVP (and breaking lattice-based cryptosystems) with our implementation can henceforth be measured directly in U.S. dollars, taking Lenstra's *dollarday* metric [Len05] to a next level[1]. That is, the cost will be shown literally as an amount on your invoice, e.g., the effort in our solving a 120-dimensional instance of the SVP Challenge translates to a 2300

[1] Before the final version went to press, it is brought to our attention that, unbeknownst to us, Kleinjung, Lenstra, Page, and Smart had also started to adopt a similar metric in an ePrint report [KLPS11] dated May 2011.

USD bill from Amazon. Moreover, this new metric is more practical in that the parallelizability of the algorithm or the parallelization of the implementation is *explicitly* taken into account, as opposed to being assumed or unspecified in the dollarday metric. Needless to say, such a cost should be understood as an upper bound obtained based on our implementation, which can certainly be improved, e.g., by using a better bounding function or better programming.

2 Preliminaries

2.1 Lattices, Algorithms, and SVP

Let $m, n \in \mathbb{Z}$ with $n \leq m$, and let $\mathbf{b}_i \in \mathbb{Z}^m$ for $1 \leq i \leq n$ be a set of linearly independent vectors. The set of all integer linear combinations of the vectors $\mathbf{b}_i$ is called a lattice Λ: $\Lambda = \{\sum_{i=1}^{n} x_i \mathbf{b}_i \mid x_i \in \mathbb{Z}\}$. The matrix $\mathbf{B} \in \mathbb{Z}^{m \times n}$ consisting of the column vectors $\mathbf{b}_i$ is called a basis of Λ, we write $\Lambda = \Lambda(\mathbf{B})$. Λ is an additive group in $\mathbb{Z}^m$. If $n = m$ the lattice is called full-dimensional. The basis of a lattice is not unique. The product of a basis with an unimodular matrix $\mathbf{M}$ ($|\det(\mathbf{M})| = 1$) does not change the lattice. The value $\lambda_1(\Lambda(\mathbf{B}))$ denotes the norm of a shortest non-zero vector in the lattice. It is called the first minimum. The determinant of a lattice is the value $\det(\Lambda(\mathbf{B})) = \sqrt{\det(\mathbf{B}^t\mathbf{B})}$. If $\Lambda(\mathbf{B})$ is full-dimensional, then the lattice determinant is equal to the absolute value of the determinant of the basis matrix ($\det(\Lambda(\mathbf{B})) = |\det(\mathbf{B})|$). In the remainder of this paper, we will only be concerned with full-dimensional lattices. The determinant of a lattice is independent of the basis; if the basis changes, the determinant remains the same.

The shortest vector problem in lattices is stated as follows. Given a lattice basis $\mathbf{B}$, output a vector $\mathbf{v} \in \Lambda(\mathbf{B}) \setminus \{0\}$ subject to $\|\mathbf{v}\| = \lambda_1(\Lambda(\mathbf{B}))$. The problem that we address in the remainder of this paper is the following: given a lattice basis $\mathbf{B}$ and a norm bound A, find a non-zero vector $\mathbf{v} \in \Lambda(\mathbf{B})$ subject to $\|\mathbf{v}\| \leq A$.

The Gaussian heuristic assumes that the number of lattice points inside a set S is approximately $\mathrm{vol}(S)/\det(\Lambda)$. Using this heuristic and the volume of a unit sphere in dimension n, we can compute an approximation of the first minimum of the lattice Λ: $FM(\Lambda) = \frac{\Gamma(n/2+1)^{1/n}}{\sqrt{\pi}} \cdot \det(\Lambda)^{1/n}$. Here $\Gamma(x)$ is the gamma-function. This estimate is used, among others, to predict the length of shortest vectors in the SVP challenge [GS10]. In our experiments as well as in the SVP challenge the heuristic shows to be a good estimate of a shortest vector length for the lattices used. Throughout the rest of this paper, our goal will always be to find a vector below $1.05 \cdot FM(\Lambda)$, the same as in the SVP challenge.

The Gram-Schmidt orthogonalization (GSO) of a matrix $\mathbf{B} \in \mathbb{Z}^{n \times n}$ is $\mathbf{B}^* = [\mathbf{b}_1^*, \ldots, \mathbf{b}_n^*] \in \mathbb{R}^{n \times n}$. It is computed via $\mathbf{b}_i^* = \mathbf{b}_i - \sum_{j=1}^{i-1} \mu_{i,j} \mathbf{b}_j^*$ for $i = 1, \ldots, n$, where $\mu_{i,j} = \mathbf{b}_i^T \mathbf{b}_j^* / \left\|\mathbf{b}_j^*\right\|^2$ for all $1 \leq j \leq i \leq n$. We have $\mathbf{B} = \mathbf{B}^* \mu^T$, where $\mathbf{B}^*$ is orthogonal and μ^T is an upper triangular matrix. Note that $\mathbf{B}^*$ is not necessarily a lattice basis. The values μ are called the Gram-Schmidt coefficients.

The LLL [LLL82] and the BKZ [SE94] algorithms can be used for pre-reduction of lattices, before running an SVP algorithm. Pre-reduction speeds up the enumeration, since the size of the enumeration tree is depending on the quality of the input basis. BKZ is controlled by a blocksize parameter β, and LLL is the special case of BKZ with parameter $\beta = 2$. Higher blocksize guarantees a better reduction quality, in the sense that vectors in the basis are shorter and the angles between basis vectors are closer to orthogonal. The gain in reduction quality comes at the cost of increasing runtime. The runtime of BKZ increases exponentially with the blocksize β. In the lattice dimension, the runtime of BKZ behaves polynomial in practice, whereas no proof of this runtime is known. The overall runtime of our SVP solver will include the BKZ pre-reduction run times as well as enumeration run times. It is an important issue to find suitable blocksize parameters for pre-reduction.

Algorithms for SVP. There are mainly three different approaches how to solve the shortest vector problem. First, there are probabilistic sieving algorithms [AKS01, NV08, MV10b]. They output a solution to SVP with high probability only, but allow for single exponential runtime. The most promising sieving candidate in practice at this time is the GaussSieve algorithm [MV10b]. Further, there exists an algorithm based on Voronoi cell computation [MV10a]. This is the first deterministic SVP algorithm running in single exponential time, but experimental results lack so far. Third, there is the group of enumeration algorithms that perform an exhaustive search over all lattice points in a suitable search region. Based on the algorithms by Kannan [Kan83] and Fincke/Pohst [FP83], Schnorr and Euchner presented the ENUM algorithm [SE94]. It was analyzed in more details in [PS08]. The latest improvement called extreme pruning providing for huge exponential speedups, was shown by Gama, Nguyen, and Regev [GNR10]. In the remainder of this paper, we will only be concerned with extreme pruned enumeration, since this variant of enumeration is the strongest SVP solver at this time.

Ideas for parallel enumeration for shortest vectors were presented in [HSB+10] for GPUs, in [DS10] for multicore CPUs, and in [DHPS10] for FPGAs. Concerning extreme pruning, there is no parallel version known to us to date, even no serial implementation is publicly available.

The lattices that we use for our tests throughout this paper are those of the SVP challenge [GS10]. They follow the ideas of the lattices from [GM03], and are used for testing SVP and lattice reduction algorithms, e.g., in [GN08, HSB+10].

2.2 Enumeration and Extreme Pruning

Here we will present the basic idea of enumeration for shortest vectors. n denotes the dimension of the full-dimensional lattices. To find a shortest non-zero vector of a lattice $\Lambda(\mathbf{B})$ with $\mathbf{B} = [\mathbf{b}_1, \ldots, \mathbf{b}_n]$, ENUM takes as input the Gram-Schmidt coefficients $(\mu_{i,j})_{1 \leq j \leq i \leq n}$, the quadratic norm of the Gram-Schmidt orthogonalization $\|\mathbf{b}_1^*\|^2, \ldots, \|\mathbf{b}_n^*\|^2$ of $\mathbf{B}$, and an initial search bound A.

The search space is the set of all coefficient vectors $\mathbf{u} \in \mathbb{Z}^n$ that satisfy $\|\sum_{t=1}^{n} u_t \mathbf{b}_t\| \leq A$. Starting with an LLL-reduced basis, it is common to set

$A = \|\mathbf{b}_1^*\|^2$ in the beginning. If the norm of the shortest vector is known beforehand, it is possible to start with a lower A, which limits the search space and reduces the runtime of the algorithm. In the equation

$$\left\|\sum_{t=1}^{n} u_t \mathbf{b}_t\right\| = \min_{x\in\mathbb{Z}^n} \left\|\sum_{t=1}^{n} x_t \mathbf{b}_t\right\|$$

we replace all $\mathbf{b}_t$ by their orthogonalization, i.e., $\mathbf{b}_t = \mathbf{b}_t^* + \sum_{j=1}^{t-1} \mu_{t,j}\mathbf{b}_j^*$ and get

$$\left\|\sum_{t=1}^{n} u_t \mathbf{b}_t\right\|^2 = \left\|\sum_{t=1}^{n} \left(u_t(\mathbf{b}_t^* + \sum_{j=1}^{t-1} \mu_{t,j}\mathbf{b}_j^*)\right)\right\|^2 = \sum_{t=1}^{n} (u_t + \sum_{i=t+1}^{n} \mu_{i,t} u_i)^2 \left\|\mathbf{b}_t^*\right\|^2 .$$

For index k, enumeration is supposed to check all coefficient vectors $\mathbf{u}$ with

$$\sum_{t=n+1-k}^{n} (u_t + \sum_{i=t+1}^{n} \mu_{i,t} u_i)^2 \cdot \|\mathbf{b}_t^*\|^2 < A \quad , \quad 1 \le k \le n . \tag{1}$$

For index t, the summand is independent of values with lower index $< t$. This means that changing the coefficient u for lower indices $< t$ does not affect the upper part of the sum with index $\ge t$. Therefore, the indices are arranged in a tree structure, where the root node contains values for coefficient u_n, intermediate nodes contain partly filled coefficient vectors $(\times, u_t, \ldots, u_n)$, and leaf nodes contain full linear combinations $(u_1 \ldots u_n)$. Here the symbol $\times$ denotes that the first values of the coefficient vector are not set. Since the $\|b_i^*\|$ are orthogonal, the sum can only increase when we step a layer down in the tree, the sum will never decrease. Therefore, when an inner node of the tree has extended the search norm A, we can cut off the whole subtree rooted at this node and skip enumerating the subtree.

Schnorr and Hörner already presented an idea to prune some of the subtrees that are unlikely to contain a shorter vector [SH95]. Their pruned enumeration runs deterministically with a certain probability to miss a shortest vector. The [SH95] pruning idea was analyzed and improved in [GNR10][2]. Instead of using the same norm bound A on every layer of the enumeration tree (Equation (1)), Gama et al. introduce a bounding vector $(R_1, \ldots, R_n) \in [0,1]^n$, with $R_1 \le \ldots \le R_n$. A on the right side of the testing condition (1) is replaced by $R_k \cdot A$. It can be shown that, assuming various heuristics [GNR10], the lattice vectors cut off by this approach only contain a shortest vector with low probability.

With this pruning technique, an exponential speedup compared to deterministic enumeration can be gained. In the original paper, various bounding function vectors were presented in theory. For the experiments, the authors use a numerically optimized function.

[2] The authors of [GNR10] also showed some flaws in the analysis of [SH95].

2.3 Cloud Computing, Amazon EC2, and GPU

Cloud computing is an emerging computing paradigm that allows data centers to provide large-scale computational and data-processing power to the users on a "pay-as-you-go" basis. Amazon Web Services (AWS) is one of the earliest and major cloud-computing providers, who provides, as the name suggests, web services platforms in the cloud. The Elastic Compute Cloud (EC2) provides compute capacity in the cloud as a foundation for the other products that AWS provides. With EC2, the users can rent large-scale computational power on demand in the form of "instances" of virtual machines of various sizes, which is charged on an hourly basis. The users can also use popular parallel computing paradigms such as the MapReduce framework [DG04], which is readily available as the AWS product "Elastic MapReduce." Furthermore, such a centralized approach also frees the users from the burden of provisioning, acquiring, deploying, and maintaining their own physical compute facilities.

Naturally, such a paradigm is economically very attractive for most users, who only need large-scale compute capacity occasionally. For large-scale computations, it may be advisable to buy machines instead of renting them because Amazon presumably expects to make a profit on renting out equipment, so our extrapolation might over-estimate the cost for long-term computations. However, we believe that these cloud-computing service providers will become more efficient in the years to come if cloud computing indeed becomes the mainstream paradigm of computing. Moreover, trade rumors has it that Amazon's profit margins are around 0% (break-even) as of mid-2011, and nowhere close to 100%, so we can say confidently that Amazon rent cannot be more than 2$\times$ what a large-scale user would have spent if he bought and maintained his own computers and networking. Thus, Amazon prices can still be considered a realistic measure of computing cost and a good yardstick for determining the strength of cryptographic keys.

In estimating complexity such that of solving (A)SVP or problems of the same or similar nature, Amazon EC2 can be used to provide a common measure of cost as a metric in comparing alternative or competing cryptanalysis algorithms and their implementations. Moreover, when using the Amazon EC2 metric, the parallelizability of the algorithm or the parallelization of the implementation is *explicitly* taken into account, as opposed to being assumed or unspecified. In addition to its simplicity, we argue that the EC2 metric is more practical than the *dollardays* metric of [Len05], and a recent report by Kleinjung, Lenstra, Page, and Smart [KLPS11] also agrees with us in taking a similar approach and measure with Amazon's EC2 cloud.

Graphics processing units (GPUs) represent another class of many-core architectures that are cost-effective for achieving high arithmetic throughput. The success of GPU has mainly been driven by the economy of scale in the video game industry. Currently, the most widely used GPU development toolchain is NVIDIA's CUDA (Compute Unified Device Architecture) [KH10]. At the core of CUDA are three key abstractions, namely, a hierarchy of thread groups, shared memories, and barrier synchronization, that are exposed to the programmers as a set of extensions to the C programming language. At the system level, the GPU is used as a coprocessor to the host processor for massively data-parallel computations, each of which is executed by a grid of GPU threads that must run the same program (the kernel). This is the SPMD (single program, multiple data) programming model, similar to SIMD but with more flexibility such as in changing of data size on a per-kernel-launch basis, as well as deviation from SIMD to MIMD at a performance penalty.

AWS offers several different compute instances for their customers to choose based on their computational needs. The one that interests us the most is the largest instance called "Cluster Compute Quadruple Extra Large" (`cc1.4xlarge`) which is designed for high-performance computing. Each such instance consists of 23 GB memory provide 33.5 "EC2 Compute Units" where each unit roughly "provides the equivalent CPU capacity of a 1.0–1.2 GHz 2007 Opteron or 2007 Xeon processor," according to Amazon.

Starting from late 2009, AWS also adds to its inventory a set of instances equipped with GPUs, which is called "Cluster GPU Quadruple Extra Large" (`cg1.4xlarge`), which is basically a `cc1.4xlarge` plus two NVIDIA Tesla "Fermi" M2050 GPUs. As of the time of writing, the prices for renting the above compute resources are shown in Table 1. The computation time is always rounded up to the next full hour for pricing purposes.

Table 1. Pricing information from `http://aws.amazon.com/ec2/pricing/`

	Elastic Compute Cloud	1 Year Reserved Pricing	Elastic MapReduce
`cc1.4xlarge`	1.60 USD/hour	4290 USD + 0.56 USD/hour	0.33 USD/hour
`cg1.4xlarge`	2.10 USD/hour	5630 USD + 0.74 USD/hour	0.42 USD/hour

For computations lasting less than 172 days it is cheaper to use on-demand pricing. For longer runs, there is an option to "reserve" an instance for 1 year (or even 3), which means that the user pays an up-front cost (see table above) to cut the on-demand cost of these instances.

3 Implementation

For each randomized basis, we use LLL-XD followed by BKZ-FP of the NTL Library [Sho] with $\delta = 0.99$, different blocksizes β, and pruning parameter $p = 15$. As already mentioned above, the problem we address is finding a vector below a search bound $1.05 \cdot FM$ that heuristically guesses the length of a shortest vector of the input lattice. Adapting our implementations to other goal values is straight forward. It will only change the success probability and the runtime, therefore, we have to fix the bound for this work.

3.1 Bounding Function

As mentioned above, selecting a suitable bounding function is an important part of extreme enumeration. It influences the runtime as well as the success probability of each enumeration tree. The bounding function we use is a polynomial

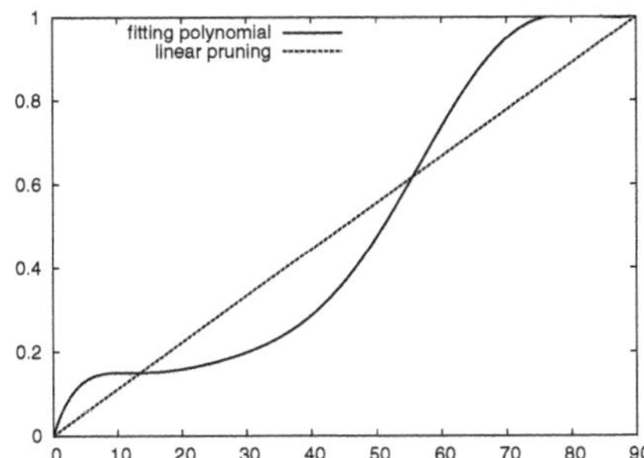

Fig. 1. Polynomial bounding function $p(x)$, scaled to lattice dimension 90. The dashed line shows a linear bounding function.

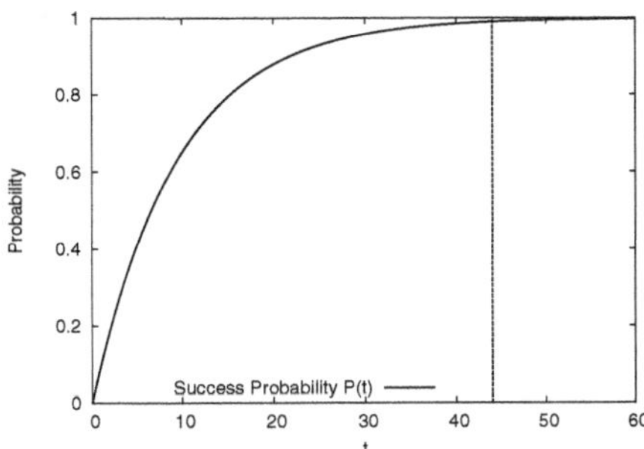

Fig. 2. Success probability of Extreme Enum assuming a success probability $p_{succ} = 10\%$ for one single tree. On average, we have to start 44 trees to finish with success probability $> 99\%$

$p(x)$ of degree eight that aims to fit the best bounding function of [GNR10] in dimension 110. We use

$$p(x) = \sum_{i=0}^{8} v_i x^i$$

where $\mathbf{v} = (9.1 \cdot 10^{-4}, 4 \cdot 10^{-2}, -4 \cdot 10^{-3}, 2.3 \cdot 10^{-4}, -6.9 \cdot 10^{-6}, 1.21 \cdot 10^{-7}, -1.2 \cdot 10^{-9}, 6.2 \cdot 10^{-12}, -1.29 \cdot 10^{-14})$ to fit the 110-dimensional bounding function. For dimension n we use $p(x \cdot 110/n)$. Figure 1 shows our polynomial bounding function $p(x)$, scaled to dimension 90.

Using an MPI-implementation for CPU we gained a success probability of finding a vector below $1.05 \cdot FM(\Lambda)$ of $p_{succ} > 10\%$. We use 10 lattice bases in each dimension and run BKZ and enumeration on up to 1000 randomized instances for each basis. We stop each lattice after 5 hours of computation, so that the total time is still manageable. In dimensions 96 we increase the maximum time from 5 to 20 hours. In total, we have up to 1000 trees in each dimension to compute the success probability of our bounding function. For a comparable bounding function, the authors of [GNR10] get a much smaller success probability. This is due to the fact that in our case we expect about 1.05^n many vectors below the larger search bound, whereas the analysis of Gama et al. assumes that only a single vector exists below their bound.

Figure 9 in Appendix A shows the expectation values of the success of BKZ and ENUM. More exactly, it shows the expectation value $E(X)$ of $P(X \leq t)$, which gives a success probability of $p = 1/E(X)$. For higher dimensions $m > 90$ the success probability of BKZ tends to zero in every tested case. $P(t) = 1 - (1 - p_{succ})^t$ is the success probability to find a shortest vector below $1.05 \cdot FM(\Lambda)$ when starting t enumeration trees in parallel. Figure 2 shows the success probability P for $p_{succ} = 10\%$. This implies that on average we have to start 44

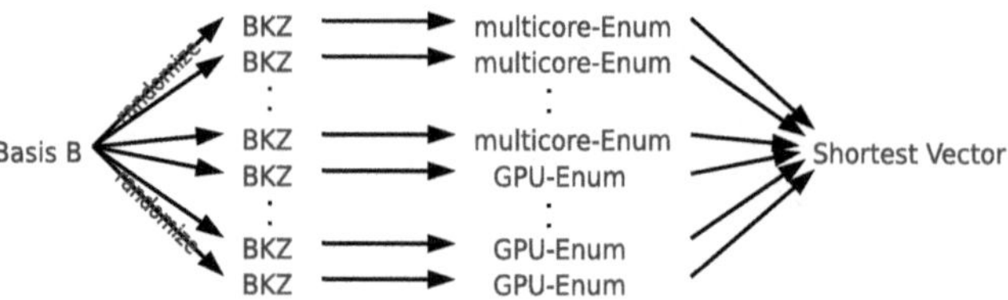

Fig. 3. The model of our parallel SVP solver. The basis **B** is randomized, and each instance is solved either on CPU or on GPU. In the end, the shortest of all found vectors is chosen as output. Since we use pruned enumeration, not all instances will find a vector below the given bound.

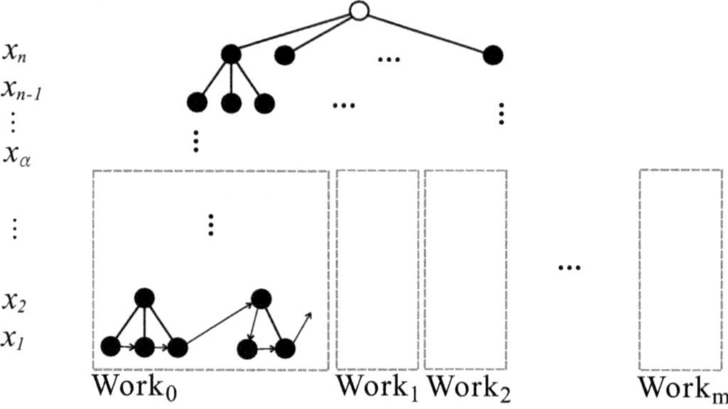

Fig. 4. Illustration of the parallel enumeration process. The top tree $x_n, x_{n-1}, ..., x_\alpha$ is enumerated on a single core, and the lower trees $x_{\alpha-1}, ..., x_2, x_1$ are explored in parallel on many mappers.

trees to find a vector below the given bound with probability $P(t) > 99\%$ (and not $1/p_{succ}$ many trees, as one could imagine).

3.2 Parallelization of Extreme Pruning Using GPU and Clouds

Our overall parallelization strategy follows the model shown in Figure 3. For success, it is sufficient if one randomized instance of ENUM finishes. The number of instances we start depends on the success probability of each instance, which itself is depending on the bounding function used. The high-level algorithm run by each multicore-Enum or GPU-Enum instance is illustrated in Figure 4.

For the calculation of the cost, it makes no difference if we use 8 cores for a multicore-tree or only one core. In practice, however, we can stop the whole computation if one of the trees has found a vector below the bound. Therefore, using multiple cores for enumeration may have some influence on the running time.

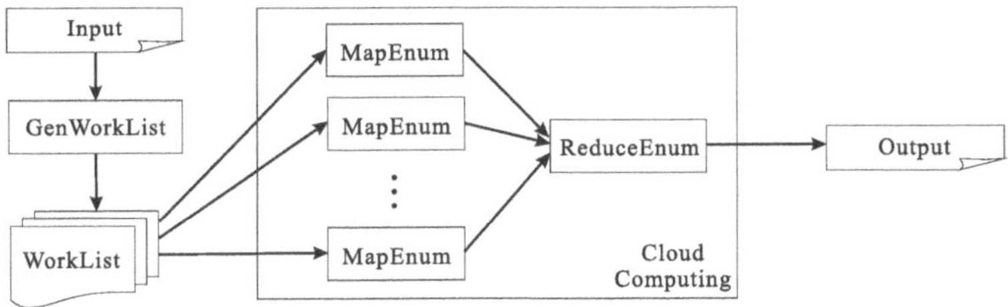

Fig. 5. Illustration of our MapReduce implementation of the enumeration algorithm

GPU Implementation. We used the implementation of [HSB$^+$10] and included pruning according to [GNR10]. The GPU enumeration uses enumeration on top of the tree, which is performed on CPU, to collect a huge number of *starting points*, as shown in Figure 4. These starting points are vectors $(\times, \ldots, \times, x_{n-\alpha+1}, \ldots, x_n)$, where only the last α coefficients are set. A starting point can be seen as the root of a subtree in the enumeration tree. All starting points are copied to the GPU and enumerated in parallel. Due to load balancing reasons, this approach is done iteratively, until no more start points exist on top of the tree (see [HSB$^+$10] for more details).

Since the code of extreme pruning only changes a few lines compared to usual enumeration, including pruning to the GPU implementation is straight forward. The improvement mentioned in [GNR10] concerning storage of intermediate sums was in parts already contained in the [HSB$^+$10] implementation, so only slight changes were integrated into the GPU ENUM.

The GPU implementation allows the usage of different bounding functions, but for simplicity reasons we stick to the polynomial function specified above. Our implementation is available online[3].

MapReduce Implementation. Our MapReduce implementation is also based on [HSB$^+$10]. The overall search process is illustrated in Figure 5. Specifically, we divide the search tree to top and lower trees. A top tree, which consists of levels x_n through x_α, is enumerated by a single thread in a DFS fashion, outputting all possible starting points $(x_\alpha, \ldots, x_n)$ to a WorkList. When a mapper receives a starting point $(x_\alpha, \ldots, x_n)$ from the WorkList, it first populates the unspecified coordinates $x_1, \ldots, x_{\alpha-1}$ and obtains the full starting point

$$(x_1 = \lceil -\sum_{k=2}^{n} \mu_{k,1}, x_k \rfloor, \ldots, x_{\alpha-1} = \lceil -\sum_{k=\alpha}^{n} \mu_{k,\alpha-1}, x_k \rfloor, x_\alpha, ..., x_n).$$

It then starts enumerating the lower tree from level 1 through $\alpha - 1$.

Because we scan the coefficients in a zigzag path, the lengths of the starting points usually show an increasing trend from the first to the last starting point. This can result uneven work distribution among the mappers. Therefore, we

[3] `http://homes.esat.kuleuven.be/~jhermans/gpuenum/index.html`

subdivide and randomly shuffle the WorkList so that each mapper gets many random starting points and hence have roughly equal amount of work among themselves. The effect is evident from the fact that the load-balancing factor, i.e., average running time divided by that of the slowest mapper, increases from 24% to 90%.

4 Experimental Results

In this section, we present the experimental results for our algorithmic improvements and parallel implementations on GPU and with MapReduce.

4.1 GPU Implementation

The GPU enumeration using extreme pruning solved the 114-dimensional SVP-challenge in about 40 hours using one single workstation with eight NVIDIA GeForce GTX 480 cards in parallel. Each GTX 480 has one GPU with 480 cores running at 1.4 GHz. The performance decreases from 200 Msteps/s to ≈ 100 Msteps/s using polynomial bounding function compared with an instance without pruning. With linear pruning, the decrease is less noticeable, but still apparent. This decrease is caused by the fact that subtrees are much thinner when pruning the tree. The number of starting points per second increases a lot, which coincides with the fact that subtrees, even though their dimension is much bigger now, are processed faster than without pruning.

We use 10 different lattices of the SVP challenge in each dimension 80–104 on the workstation equipped with eight GTX 480 cards to generate the timings of Figures 6 and 7.

Workload Distribution between BKZ and ENUM. We note that in general, if we spend more time in BKZ to produce a better basis, we would have a higher probability of finding a short vector in the subsequent ENUM phase. A natural question is, what is the optimal breakdown of workload between BKZ and ENUM?

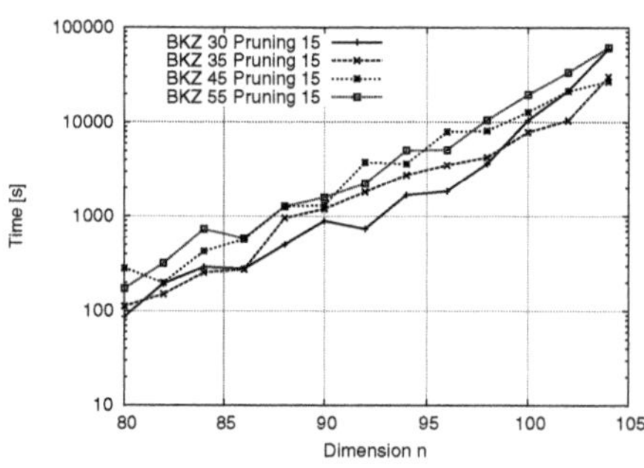

Fig. 6. Total running time for solving SVP instances from dimension 80 to 104

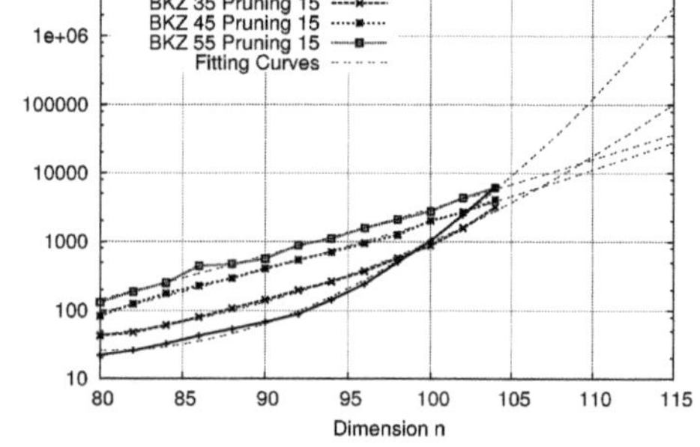

Fig. 7. Running time for one round of pruned ENUM, including fitting curves $t_{30}(n)$ to $t_{55}(n)$

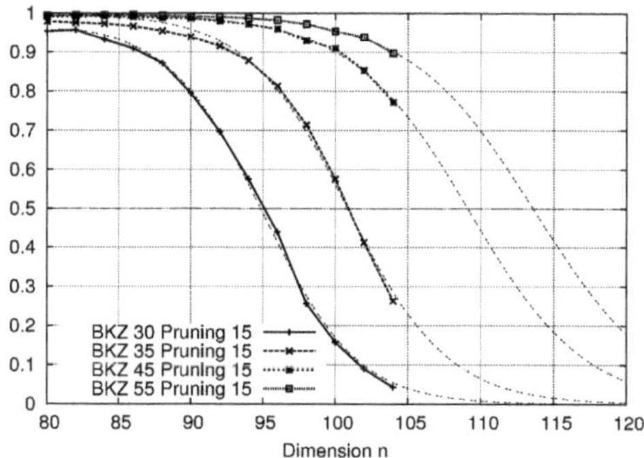

Fig. 8. Ratio of BKZ runtime to total runtime for a single enumeration tree

We conjecture that the distribution should be roughly equal, as is supported by empirical evidence that we obtained from our experiments (cf. Figure 8). In our experiments, BKZ 40 performs the best in 104-dimensional instances, whereas in Figure 8, it has a ratio that is the closest to 0.5. Similar trends can be observed for dimensions 86–97, for which the best BKZ block size is 30.

We use the data shown in Figure 8 to assess which of the curves from Figure 7 is the fastest one, and we use the extrapolation of this curve gained from data in dimension 80–104. This results in the cost function shown in Conjecture 1.

Conjecture 1 (GPU timing function). *Running BKZ and our implementation of pruned enumeration once on an NVIDIA GTX-480 GPU takes time*

$$time_{GPU}(n) = \begin{cases} t_{30}(n) = 2^{0.0138n^2-2.2n+93.2} & \text{for } n \leq 97 \\ t_{35}(n) = 2^{0.0064n^2-0.92n+38.4} & \text{for } 98 \leq n \leq 104 \\ t_{45}(n) = 2^{0.001n^2+0.034n-2.8} & \text{for } 105 \leq n \leq 111 \\ t_{55}(n) = 2^{0.00059n^2+0.11n-5.8} & \text{for } 112 \leq n \end{cases} \quad sec.$$

A more theoretic way to extrapolate the runtime would be to compute BKZ reduced bases, note the slope of the orthogonalized basis vectors, and use the runtime function of [GNR10] to compute the runtime. This approach ignores the runtime of BKZ (which is up to 50%) of the total runtime and relies on the Gaussian heuristic, while we are interested in practical runtime.

From the regression results shown in Figures 6 and 7, we can see that the run times for BKZ and ENUM are indeed polynomial and super-exponential, respectively. However, we notice that a larger BKZ block size does have a positive effect on the per-round running time of subsequent ENUM.

One difference is that Amazon uses M2050 GPU, not GTX480 (like in our experiments). The M2050 has better double precision performance. Since many operations in enumeration are performed using double precision operations, we expected a huge speed-up for enumeration. However, tests on M2050 GPUs did not show large speed-ups. One possible explanation is as follows. On the GPU, many additional operations have to be performed in integer-precision in order to split the work and reach a good load balancing. Therefore, double-precision operations are less than a fourth of the total number of operations, which makes the speed-ups on M2050 GPUs minor.

4.2 MapReduce Implementation

Our MapReduce implementation is compiled by `g++` version 4.4.4 x86_64 with the options `-O9 -ffast-math -funroll-loops -ftree-vectorize`. Using the MapReduce implementation, we are able to solve the 112-dimensional SVP-challenge in a few days. More exactly, we were using 10 nodes, 84 physical cores (totaling 140 virtual cores as some of the cores are hyperthreaded), which gives a total number of 334 GHz.

We note that the bounding function used in this computation is different from the polynomial bounding function described earlier. We were lucky in that only after 101 hours, or 1/9 of the estimated time, a shorter vector was found. We also noticed that the runtime scales linearly with the number of CPU cores used in total, meaning if we increase the number of CPU cores by a factor of 10, the runtime will decrease by factor 10.

Overall, from the test data of solving SVPs at dimension 100, 102, and 104 using the same set of seeds, we found that a GTX480 is roughly two to three times faster than a four-core, 2.4 GHz Intel Core i7 processor for running our SVP solvers. We conjecture that the running time for our MapReduce implementation is also similar to that of our GPU implementation, as shown in Conjecture 1.

4.3 Final Pricing

We use Conjecture 1 to derive the final cost function for solving SVP challenges in higher dimensions $n \geq 112$. Recall that Amazon instances have to be paid for complete hours, therefore we round the runtime in hours to the next highest integer value. Using 44 enumeration trees leads to a success probability of at least 99%.

Conjecture 2 (Final Pricing). *Solving an SVP challenge with our implementation in dimension $n \geq 112$ with a success probability of $\geq 99\%$ on Amazon EC2 (using on demand pricing) costs*

$$cost_{GPU}(n) = \lceil time_{GPU}(n)/3600 \rceil \cdot 44 \cdot 2.52 \ USD.$$

Following Conjecture 2 solving the 120-dimensional instance of the challenge costs 1,885 USD, which is a bit less than the amount we paid for practically solving it (due to conservative reservation of compute resources on EC2). We actually fired up 50 cg1.4xlarge instances for a total of 946 instance-hours, and incurred a bill of 2,300 USD. For instance, solving the 140-dimensional challenge would cost roughly 72,405 USD.

5 Concluding Remarks and Further Work

Cryptographic Key Sizes. The ability of solving SVP does not directly affect cryptographic schemes based on lattice problems. The hardness of lattice-based signature schemes is mostly based on the SIS problem, whereas the hardness of encryption schemes is mostly based on the LWE problem. Both the SIS and the LWE problem can be proven to be as hard as the SVP in lattices of a smaller dimension (so-called *worst-case to average-case reduction*). That means that a successful attacker of a cryptographic system is able to solve SVP in all lattices of a smaller dimension. This implies that our cost estimates for SVP can be used to assess the hardness of the basic problem of cryptosystems only.

Real attacks on cryptosystems mostly apply approximation algorithms, like BKZ. Since enumeration can be used as a subroutine there, speeding up enumeration also affects direct attacks on lattice based cryptosystems.

Further Work. For GPUs the need of finding new bounding functions seems apparent. Since trees are very thin when our polynomial bounding is applied the performance of the GPU decreases. Finding a new bounding function that allows for the same success probability but guarantees better performance will show the strength of the GPU even more. Besides that, it is an open problem which bounding function gives the best performance in practice, be it on CPU or GPU.

Acknowledgements. Schneider is supported by project BU 630/23-1 of the German Research Foundation (DFG). Dagdelen was supported by CASED (`www.cased.de`). Cheng, Kuo and Yang are supported by National Science Council, National Taiwan University and Intel Corporation under Grants NSC99-2911-I-002-201, 99R70600, and 10R70500, and Yang also by Academia Sinica Career Award. We thank Paul Baecher and Pierre-Louis Cayrel for their comments on an earlier version of this paper, Phong Nguyen and Oded Regev for their hints regarding the success probability of the bounding function, Pierre-Alain Fouque for his advice, and the anonymous reviewers for their helpful comments.

References

[AKS01] Ajtai, M., Kumar, R., Sivakumar, D.: A sieve algorithm for the shortest lattice vector problem. In: STOC 2001, pp. 601–610. ACM, New York (2001)

[DG04] Dean, J., Ghemawat, S.: MapReduce: Simplified data processing on large clusters. In: OSDI 2004: Sixth Symposium on Operating System Design and Implementation, San Francisco, CA, USA (December 2004)

[DHPS10] Detrey, J., Hanrot, G., Pujol, X., Stehlé, D.: Accelerating Lattice Reduction with FPGAs. In: Abdalla, M., Barreto, P.S.L.M. (eds.) LATINCRYPT 2010. LNCS, vol. 6212, pp. 124–143. Springer, Heidelberg (2010)

[DS10] Dagdelen, Ö., Schneider, M.: Parallel Enumeration of Shortest Lattice Vectors. In: D'Ambra, P., Guarracino, M., Talia, D. (eds.) Euro-Par 2010. LNCS, vol. 6272, pp. 211–222. Springer, Heidelberg (2010)

[FP83] Fincke, U., Pohst, M.: Michael Pohst. A procedure for determining algebraic integers of given norm. In: van Hulzen, J.A. (ed.) ISSAC 1983 and EUROCAL 1983. LNCS, vol. 162, pp. 194–202. Springer, Heidelberg (1983)

[GM03] Goldstein, D., Mayer, A.: On the equidistribution of Hecke points. Forum Mathematicum 15(2), 165–189 (2003)

[GN08] Gama, N., Nguyen, P.Q.: Predicting lattice reduction. In: Smart, N.P. (ed.) EUROCRYPT 2008. LNCS, vol. 4965, pp. 31–51. Springer, Heidelberg (2008)

[GNR10] Gama, N., Nguyen, P.Q., Regev, O.: Lattice Enumeration Using Extreme Pruning. In: Gilbert, H. (ed.) EUROCRYPT 2010. LNCS, vol. 6110, pp. 257–278. Springer, Heidelberg (2010)

[GS10] Gama, N., Schneider, M.: SVP Challenge (2010), `http://www.latticechallenge.org/svp-challenge`

[HSB+10] Hermans, J., Schneider, M., Buchmann, J., Vercauteren, F., Preneel, B.: Parallel Shortest Lattice Vector Enumeration on Graphics Cards. In: Bernstein, D.J., Lange, T. (eds.) AFRICACRYPT 2010. LNCS, vol. 6055, pp. 52–68. Springer, Heidelberg (2010)

[Kan83] Kannan, R.: Improved algorithms for integer programming and related lattice problems. In: STOC 1983, pp. 193–206. ACM, New York (1983)

[KH10] Kirk, D.B., Hwu, W.-m.: Programming Massively Parallel Processors: A Hands-on Approach, 1st edn. Morgan Kaufmann, San Francisco (2010)

[KLPS11] Kleinjung, T., Lenstra, A.K., Page, D., Smart, N.P.: Using the cloud to determine key strengths. Cryptology ePrint Archive, Report 2011/254 (2011), `http://eprint.iacr.org/`

[Len05] Lenstra, A.: Key lengths. In: Bidgoli, H. (ed.) Handbook of Information Security. Wiley, Chichester (2005)

[LLL82] Lenstra, A., Lenstra, H., Lovász, L.: Factoring polynomials with rational coefficients. Mathematische Annalen 4, 515–534 (1982)

[MV10a] Micciancio, D., Voulgaris, P.: A deterministic single exponential time algorithm for most lattice problems based on voronoi cell computations. In: STOC, pp. 351–358. ACM, New York (2010)

[MV10b] Micciancio, D., Voulgaris, P.: Faster exponential time algorithms for the shortest vector problem. In: SODA 2010, pp. 1468–1480. ACM/SIAM (2010)

[NV08] Nguyen, P.Q., Vidick, T.: Sieve algorithms for the shortest vector problem are practical. J. of Mathematical Cryptology 2(2) (2008)

[PS08] Pujol, X., Stehlé, D.: Rigorous and Efficient Short Lattice Vectors Enumeration. In: Pieprzyk, J. (ed.) ASIACRYPT 2008. LNCS, vol. 5350, pp. 390–405. Springer, Heidelberg (2008)

[SE94] Schnorr, C.-P., Euchner, M.: Lattice basis reduction: Improved practical algorithms and solving subset sum problems. Mathematical Programming 66, 181–199 (1994)

[SH95] Schnorr, C.-P., Hörner, H.H.: Attacking the chor-rivest cryptosystem by improved lattice reduction. In: Guillou, L.C., Quisquater, J.-J. (eds.) EUROCRYPT 1995. LNCS, vol. 921, pp. 1–12. Springer, Heidelberg (1995)

[Sho] Shoup, V.: Number theory library (NTL) for C++, version 5.5.2, `http://www.shoup.net/ntl/`

A Success Probability Using $p(x)$

We ran experiments using the SVP challenge lattices, in order to assess the practical success probability (the probability of a single ENUM run to find a short vector) of extreme pruning using the polynomial bounding function $p(x)$. Using a multicore CPU implementation we started extreme pruning on up to 10,000 lattices in each dimension (we stopped each experiment after 20 hours of computation). Figure 9 shows the average success rate of BKZ (with pruning parameter 15) and ENUM in dimensions 80 to 96 for different BKZ blocksizes. The values shown are the number of successfully reduced lattices divided by the number of started lattices in each dimension.

With BKZ blocksize 20, the pre-reduction was not strong enough, so neither BKZ nor ENUM could find a vector below the search bound in dimensions ≥ 96 within 20 hours. In dimension 100, the number of finished enumeration trees was already too small to derive a meaningful success rate.

The success rate of BKZ vanishes in higher dimensions. For each BKZ blocksize, the success rate of ENUM stabilizes at a value $> 10\%$. Since the success rate is higher than this value in almost every case, we assume a value of $p_{succ} = 10\%$ for our polynomial bounding function $p(x)$.

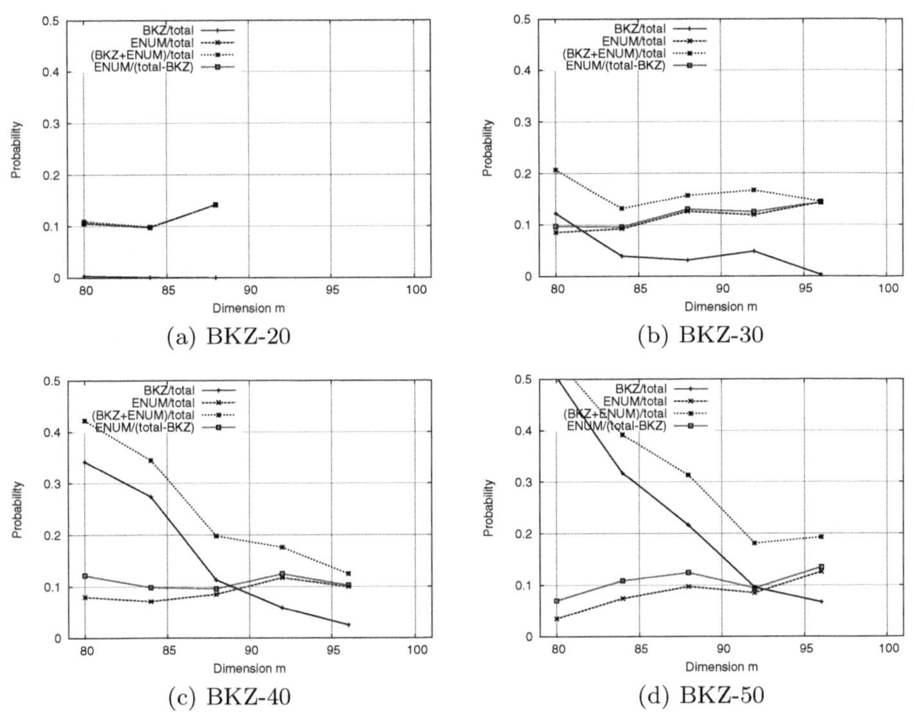

(a) BKZ-20

(b) BKZ-30

(c) BKZ-40

(d) BKZ-50

Fig. 9. Average values of success of the polynomial bounding function. total = number of samples; BKZ = number of samples solved by BKZ; ENUM = number of samples solved by pruned enumeration.

Modulus Fault Attacks against RSA-CRT Signatures

Éric Brier[1], David Naccache[2], Phong Q. Nguyen[2], and Mehdi Tibouchi[2]

[1] Ingenico
1, rue Claude Chappe, BP 346, F-07503 Guilherand-Granges, France
eric.brier@ingenico.com
[2] École normale supérieure
Département d'informatique, Groupe de Cryptographie
45, rue d'Ulm, F-75230 Paris CEDEX 05, France
{david.naccache,phong.nguyen,mehdi.tibouchi}@ens.fr

Abstract. RSA-CRT fault attacks have been an active research area since their discovery by Boneh, DeMillo and Lipton in 1997. We present alternative key-recovery attacks on RSA-CRT signatures: instead of targeting one of the sub-exponentiations in RSA-CRT, we inject faults into the *public modulus* before CRT interpolation, which makes a number of countermeasures against Boneh *et al.*'s attack ineffective.

Our attacks are based on orthogonal lattice techniques and are very efficient in practice: depending on the fault model, between 5 to 45 faults suffice to recover the RSA factorization within a few seconds. Our simplest attack requires that the adversary knows the faulty moduli, but more sophisticated variants work even if the moduli are unknown, under reasonable fault models. All our attacks have been fully validated experimentally with fault-injection laser techniques.

Keywords: Fault Attacks, Digital Signatures, RSA, CRT, Lattices.

1 Introduction

1.1 RSA-CRT Signatures

RSA [23] is the most widely used signature scheme. To sign a message m, the signer first applies an encoding function μ to m, and then computes the signature $\sigma = \mu(m)^d \bmod N$. To verify the signature σ, the receiver checks that $\sigma^e = \mu(m) \bmod N$. The Chinese Remainder Theorem (CRT) is often used to speed up signature generation by a factor of about 4. This is done by computing:

$$\sigma_p = \mu(m)^{d \bmod p-1} \mod p \quad \text{and} \quad \sigma_q = \mu(m)^{d \bmod q-1} \mod q$$

and deriving σ from (σ_p, σ_q) using the CRT.

B. Preneel and T. Takagi (Eds.): CHES 2011, LNCS 6917, pp. 192–206, 2011.

1.2 Fault Attacks on RSA-CRT Signatures

Back in 1997, Boneh, DeMillo and Lipton [6] showed that RSA-CRT implementations are vulnerable to fault attacks. Assuming that the attacker can induce a fault when σ_q is computed while keeping the computation of σ_p correct, one gets:

$$\sigma_p = \mu(m)^{d \bmod p-1} \mod p \quad \text{and} \quad \sigma_q \neq \mu(m)^{d \bmod q-1} \mod q$$

hence:

$$\sigma^e = \mu(m) \mod p \quad \text{and} \quad \sigma^e \neq \mu(m) \mod q$$

which allows the attacker to factor N by computing $\gcd(\sigma^e - \mu(m) \bmod N, N) = p$. This attack applies to any deterministic padding function μ, such as RSA PKCS#1 v1.5 or Full-Domain Hash [2], or probabilistic signatures where the randomizer used to generate the signature is sent along with the signature, such as PFDH [13]. Only probabilistic signature schemes such that the randomness remains unknown to the attacker may be safe, though some particular cases have been attacked as well [12].

In 2005, Seifert [24] introduced a new type of RSA fault attacks, by inducing faults on the RSA public modulus. The initial attack [24] only allowed to bypass RSA verification, but key-recovery attacks were later discovered by Brier *et al.* [8], and improved or extended in [17,5,3,4]. These key-recovery attacks only apply to RSA without CRT, and they require significantly more faults than Boneh *et al.*'s attack, at least on the order of 1000 faulty signatures.

1.3 Our Contribution

We present new alternative key-recovery attacks on RSA-CRT signatures: instead of targeting one of the RSA-CRT sub-exponentiations, we inject faults into the *public modulus* like in Seifert's attack. This makes typical countermeasures against Boneh *et al.*'s attack ineffective against the new attacks.

Our attacks are based on the orthogonal lattice techniques introduced by Nguyen and Stern [19] in 1997. They are very effective in practice: they disclose the RSA factorization within a few seconds using only between 5 to 45 faulty signatures. The exact running time and number of faulty signatures depend on the fault model.

For instance, in our simplest attack, the running time is a fraction of a second using only 5 faulty signatures, but the attacker is assumed to know the faulted moduli for the 5 different messages. However, our attack can be extended to the case where the attacker no longer knows the faulted moduli, using at most 45 faulty signatures, under the following two fault models: either the faulted moduli only differ from the public modulus on a single byte of unknown position and unknown value, or the faulted moduli may differ from the public modulus by many bytes, but the differences are restricted to the least significant bits, up to half of the modulus size.

All our attacks have been fully validated with physical experiments with laser shots on a RISC microcontroller.

1.4 Related Work

Many countermeasures have been proposed to protect against Boneh *et al.*'s attack and its numerous generalizations, but they often focus on the exponentiation process. The previously mentioned fault attacks [8,17,5,3,4] on RSA using faulty moduli only apply to standard RSA without CRT, and they use non-lattice techniques. Our attack seems to be the first attack on RSA-CRT with faulted moduli.

It should be pointed out, however, that a number of protected RSA-CRT implementations also protect the CRT recombination. This is for example the case of [1,10,14,7,26,22].

More generally, as we observe in §5, using the technique known as Garner's formula for CRT recombination does thwart the attack introduced in this paper. Since this formula is often used in practice, typical implementations conforming to RSA standards like PKCS#1 and IEEE P1363 should in principle be immune to this attack.

1.5 Roadmap

In §2, we describe the basic attack where the faulty moduli are assumed to be known to the attacker. In §3, we extend the attack to realistic fault models in which the faulty moduli are no longer known to the attacker. In §4, we describe physical experiments with laser shots on a RISC microcontroller to validate the attack. Finally, in §5, we suggest possible countermeasures against this attack.

2 The New Attack

2.1 Overview

Consider again the generation of RSA-CRT signatures. To obtain the signature σ of a message m padded as $\mu(m)$, the signer computes the mod-p and mod-q parts:

$$\sigma_p = \mu(m)^d \mod p \quad \text{and} \quad \sigma_q = \mu(m)^d \mod q$$

and returns the signature:

$$\sigma = \sigma_p \cdot \alpha + \sigma_q \cdot \beta \mod N \tag{1}$$

where α, β are the pre-computed Chinese Remainder coefficients $\alpha = q \cdot (q^{-1} \bmod p)$ and $\beta = p \cdot (p^{-1} \bmod q)$.

Assume that an adversary can obtain the correct signature σ, and also a signature σ' of the same padded message $\mu(m)$ after corrupting the modulus N before the CRT step (1). In other words, the attacker gets σ as before but also σ' defined as:

$$\sigma' = \sigma_p \cdot \alpha + \sigma_q \cdot \beta \mod N' \quad \text{for some } N' \neq N$$

Suppose further, for the moment, that the adversary is able to recover the faulty modulus N': we will see in §3 how this not-so-realistic hypothesis can be lifted in a more practical setting. Then, by applying the Chinese Remainder Theorem to σ and σ', the adversary can compute

$$v = \sigma_p \cdot \alpha + \sigma_q \cdot \beta \mod N \cdot N'.$$

But if we denote the bit length of N by n, then $N \cdot N'$ is a $2n$-bit integer, whereas α, β are of length n and σ_p, σ_q of length $n/2$, so v is really a linear combination of α and β in $\mathbb{Z}$:

$$v = \sigma_p \cdot \alpha + \sigma_q \cdot \beta.$$

That alone does not suffice to factor N, but several such pairs (σ, σ') provide multiple linear combinations of the (unknown) integers α, β with relatively small coefficients. Then lattice reduction techniques allow us to recover the coefficients σ_p and σ_q, and hence obtain the factorization of N by GCDs. The following sections describe this process in detail.

2.2 Applying Orthogonal Lattice Techniques

We assume that the reader is familiar with cryptanalysis based on lattices (see [18,21] for more information), particularly the orthogonal lattices introduced by Nguyen and Stern [19]: if L is a lattice in $\mathbb{Z}^n$, we let $L^\perp$ be the lattice formed by all vectors in $\mathbb{Z}^n$ which are orthogonal to all vectors of L. If an attacker obtains ℓ pairs (σ, σ'), he can compute as before a vector $\boldsymbol{v} = (v_1, \dots, v_\ell)$ of $3n/2$-bit integers satisfying an equation of the form:

$$\boldsymbol{v} = \alpha \boldsymbol{x} + \beta \boldsymbol{y} \tag{2}$$

where $\boldsymbol{x}, \boldsymbol{y}$ are unknown vectors with $n/2$-bit components and α, β are the (unknown) CRT coefficients relative to p and q. Lattice reduction can exploit such a hidden linear relationship as follows.

Using standard techniques [19,20], it is possible to compute a reduced basis $\{\boldsymbol{b}_1, \dots, \boldsymbol{b}_{\ell-1}\}$ of the lattice $\boldsymbol{v}^\perp \subset \mathbb{Z}^\ell$ of vectors orthogonal to $\boldsymbol{v}$ in $\mathbb{Z}^\ell$. In particular we get:

$$\alpha\langle \boldsymbol{b}_j, \boldsymbol{x}\rangle + \beta\langle \boldsymbol{b}_j, \boldsymbol{y}\rangle = 0 \quad \text{for } j = 1, 2, \dots, \ell - 1.$$

Now, observe that the smallest nonzero solution $(u, v) \in \mathbb{Z}^2$ of the equation $\alpha \cdot u + \beta \cdot v = 0$ is $\pm(\beta, -\alpha)/g$, where $g = \gcd(\alpha, \beta)$ is heuristically expected to be very small, which implies that $|u|, |v| \geq \Omega(N)$ where the $\Omega()$ constant is very small. For each $j = 1, 2, \dots, \ell - 1$, there are thus two possibilities:

Case 1: $\langle \boldsymbol{b}_j, \boldsymbol{x}\rangle = \langle \boldsymbol{b}_j, \boldsymbol{y}\rangle = 0$, in which case $\boldsymbol{b}_j$ belongs to the lattice $L = \{\boldsymbol{x}, \boldsymbol{y}\}^\perp$ of vectors in $\mathbb{Z}^\ell$ orthogonal to both $\boldsymbol{x}$ and $\boldsymbol{y}$;

Case 2: $\langle \boldsymbol{b}_j, \boldsymbol{x}\rangle$ and $\langle \boldsymbol{b}_j, \boldsymbol{y}\rangle$ have absolute value $\geq \Omega(N)$ with a very small $\Omega()$ constant. Since $\boldsymbol{x}, \boldsymbol{y}$ both have norm at most $\sqrt{\ell N}$, this implies $\|\boldsymbol{b}_j\| \geq \Omega(\sqrt{N/\ell})$ by Cauchy-Schwarz.

Since the lattice $L = \{\boldsymbol{x}, \boldsymbol{y}\}^\perp$ is of rank $\ell - 2$, Case 1 cannot hold for all $\ell - 1$ linearly independent vectors $\boldsymbol{b}_j$, so that the longest one $\boldsymbol{b}_{\ell-1}$ should be in Case 2, and hence $\|\boldsymbol{b}_{\ell-1}\| \geq \Omega(\sqrt{N/\ell})$. On the other hand, the other vectors form a lattice of rank $\ell - 2$ and volume:

$$V = \mathrm{vol}(\mathbb{Z}\boldsymbol{b}_1 \oplus \cdots \oplus \mathbb{Z}\boldsymbol{b}_{\ell-2}) \approx \frac{\mathrm{vol}(\boldsymbol{v}^\perp)}{\|\boldsymbol{b}_{\ell-1}\|} = \frac{\|\boldsymbol{v}\|}{\|\boldsymbol{b}_{\ell-1}\|} \leq \frac{\sqrt{\ell} \cdot N^{3/2}}{\Omega(\sqrt{N/\ell})} = O(\ell N)$$

which can heuristically be expected to behave like a random lattice. In particular, we should have:

$$\|\boldsymbol{b}_j\| = O(\sqrt{\ell - 2} \cdot V^{1/(\ell-2)}) = O(\ell^{1/2+1/(\ell-2)} \cdot N^{1/(\ell-2)}) \quad \text{for } j = 1, 2, \ldots, \ell - 2.$$

This length is much smaller than $\sqrt{N/\ell}$ as soon as $\ell \geq 5$. Assuming that this is case, $\boldsymbol{b}_j$ should thus be in Case 1 for $j = 1, 2, \ldots, \ell - 2$. This means that those vectors generate a sublattice $L' = \mathbb{Z}\boldsymbol{b}_1 \oplus \cdots \oplus \mathbb{Z}\boldsymbol{b}_{\ell-2}$ of full rank in $L = \{\boldsymbol{x}, \boldsymbol{y}\}^\perp$.

Taking orthogonal lattices, we get $(L')^\perp \supset L^\perp = \mathbb{Z}\boldsymbol{x} \oplus \mathbb{Z}\boldsymbol{y}$. Therefore, $\boldsymbol{x}$ and $\boldsymbol{y}$ belong to the orthogonal lattice $(L')^\perp$ of L'. Let $\{\boldsymbol{x}', \boldsymbol{y}'\}$ be a reduced basis of that lattice. We can enumerate all the lattice vectors in $(L')^\perp$ of length at most $\sqrt{\ell N}$ as linear combinations of $\boldsymbol{x}'$ and $\boldsymbol{y}'$. The Gaussian heuristic suggests that there should be roughly:

$$\frac{\pi(\sqrt{\ell N})^2}{\mathrm{vol}((L')^\perp)} = \frac{\pi \ell N}{V} = O(1)$$

such vectors, so this is certainly feasible. For all those vectors $\boldsymbol{z}$, we can compute $\gcd(\boldsymbol{v} - \boldsymbol{z}, N)$. We will thus quickly find $\gcd(\boldsymbol{v} - \boldsymbol{x}, N)$ among them, since $\boldsymbol{x}$ is a vector of length $\leq \sqrt{\ell N}$ in $(L')^\perp$. But by definition of $\boldsymbol{v}$ we have:

$$\boldsymbol{v} = \boldsymbol{x} \mod p \quad \text{and} \quad \boldsymbol{v} = \boldsymbol{y} \mod q$$

so $\gcd(\boldsymbol{v} - \boldsymbol{x}, N) = p$, which reveals the factorization of N.

2.3 Attack Summary

Assume that, for $\ell \geq 5$ padded messages $\mu(m_i)$, we know a correct signature σ_i and a signature σ'_i computed with a faulty modulus N'_i. Then, we can heuristically recover the factorization of N as follows.

1. For each i, compute the integer $v_i = \mathrm{CRT}_{N,N'_i}(\sigma_i, \sigma'_i)$. They form a vector $\boldsymbol{v} = (v_1, \ldots, v_\ell) \in \mathbb{Z}^\ell$.
2. Compute an LLL-reduced [15] basis $\{\boldsymbol{b}_1, \ldots, \boldsymbol{b}_{\ell-1}\}$ of the lattice $\boldsymbol{v}^\perp \subset \mathbb{Z}^\ell$ of vectors in $\mathbb{Z}^\ell$ orthogonal to $\boldsymbol{v}$. This is done by applying LLL to the lattice in $\mathbb{Z}^{1+\ell}$ generated by the rows of the following matrix:

$$\begin{pmatrix} \kappa v_1 & 1 & & 0 \\ \vdots & & \ddots & \\ \kappa v_\ell & 0 & & 1 \end{pmatrix}$$

   where κ is a suitably large constant, and removing the first component of each resulting vector [19].

3. The first $\ell - 2$ vectors $\boldsymbol{b}_1, \ldots, \boldsymbol{b}_{\ell-2}$ generate a lattice $L' \subset \mathbb{Z}^\ell$ of rank $\ell - 2$. Compute an LLL-reduced basis $\{\boldsymbol{x}', \boldsymbol{y}'\}$ of the orthogonal lattice $(L')^\perp$ to that lattice. Again, this is done by applying LLL to the lattice in $\mathbb{Z}^{\ell+2+\ell}$ generated by the rows of

$$\begin{pmatrix} \kappa' b_{1,1} & \cdots & \kappa' b_{\ell-2,1} & 1 & & 0 \\ \vdots & & \vdots & & \ddots & \\ \kappa' b_{1,\ell} & \cdots & \kappa' b_{\ell-2,\ell} & 0 & & 1 \end{pmatrix}$$

and keeping the last ℓ components of each resulting vector.
4. Enumerate the vectors $\boldsymbol{z} = a\boldsymbol{x}' + b\boldsymbol{y}' \in (L')^\perp$ of length at most $\sqrt{\ell N}$, and for each such vector $\boldsymbol{z}$, compute $\gcd(\boldsymbol{v} - \boldsymbol{z}, N)$ using all components, and return any nontrivial factor of N.

2.4 Simulation Results

Since the attack is heuristic, it is important to evaluate its experimental performances. To do so, we have implemented a simulation of the attack in SAGE [25]: for a given modulus N, we compute the vector $\boldsymbol{v}$ corresponding to a series of ℓ signatures on random messages and apply the lattice attack, attempting to recover a factor of N.

Table 1 shows the measured success probabilities for various values of ℓ and modulus sizes. It confirms the heuristic prediction that 5 faulty signatures should always suffice to factor N. It turns out that even 4 signatures are enough in almost half the cases.

Table 1. Attack success probability as a function of the number of faulty signatures and the size of N. Each parameter set was tested with random faults on 500 random moduli of the given size.

Number of faulty signatures ℓ	4	5	6
1024-bit moduli	48%	100%	100%
1536-bit moduli	45%	100%	100%
2048-bit moduli	46%	100%	100%

Table 2. Efficiency of the attack with $\ell = 5$ faulty signatures and various modulus sizes. Each parameter set was tested with random faults on 500 random moduli of the given size. Timings for a SAGE implementation, on a single 2.4 GHz Core2 CPU core.

Modulus size	1024	1536	2048
Average search space $\pi \ell N/V$	24	23	24
Average total CPU time	16 ms	26 ms	34 ms

Experimental running times are given in Table 2. The whole attack takes a few dozen milliseconds on a standard PC. The number of vectors to test as part of the final exhaustive search step is about 20 in practice, which is done very quickly.

3 Extending the Attack to Unknown Faulty Moduli

As mentioned in §2.1, in its basic form, the attack requires the recovery of the faulty moduli N_i' in addition to the corresponding faulty signatures σ_i'. This is not a very realistic assumption, since a typical implementation does not output the public modulus along with each signature.

To work around this limitation, we would like to reconstruct the vector $\boldsymbol{v}$ of integer values needed to run the attack from signatures alone, without the knowledge of the faulty moduli—possibly at the cost of requiring a few more faulty signatures.

This can actually be achieved in various ways depending on the precise form of the faults inflicted to the modulus. We propose solutions for the following two realistic fault models:

1. The faulty moduli N_i' differ from N on a single (unknown) byte. This is known to be possible using power glitches or laser shots.
2. The differences between the faulty moduli N_i' and N are located on the least significant half: the errors on the least significant bits can be up to half of the modulus size. It is easy to obtain such faults with a laser or a cold boot attack.

3.1 Single Byte Faults

In this model, the attacker is able to obtain a certain number $\ell' \geq 5$ of pairs (σ_i, σ_i') where $\sigma_i = \alpha x_i + \beta y_i \bmod N$ is a valid signature and $\sigma_i' = \alpha x_i + \beta y_i \bmod N_i'$ is the same signature computed with a faulty modulus. The faulty moduli N_i' are not known, but they only differ from N on a single byte whose position and value is unknown.

This type of fault can for example occur when attacking the transfer of the modulus to memory on a smart card with an 8-bit processor, or when using a laser attack with a sufficiently focused beam.

For a 1024-bit modulus N, for example, there are $128 \times 255 \approx 2^{15}$ possible faulty moduli. It can thus seem like a reasonable approach to try and run the attack with all possible faults. However, since this should be done with 5 signatures, this results in a search space of size $\approx (2^{15})^5 = 2^{75}$ which is prohibitive.

This kind of exhaustive search can be made practical, though, if we take into account the fact that the CRT value $v_i = \mathrm{CRT}_{N,N_i'}(\sigma_i, \sigma_i')$ satisfies:

$$v_i = \alpha x_i + \beta y_i \leq N \cdot (p+q) = N^{3/2}\left(\sqrt{\frac{p}{q}} + \sqrt{\frac{q}{p}}\right) < (2N)^{3/2}$$

Table 3. Exhaustive search space size for the vector $\boldsymbol{v}$ of CRT values, and expected attack running time, depending on the number of pairs (σ_i, σ'_i) available to the attacker. Measured for a family of random single byte faults on a 1024-bit modulus. Timings are given for the SAGE implementation as above.

Number of pairs ℓ'	5	7	10	15	20	25
Search space size (bits)	11.6	9.8	7.2	6.2	4.2	2.6
Total attack time (seconds)	49	14	2.4	1.2	0.29	0.10

since $p/q \in (1/2, 2)$. Now, for a given value of σ'_i, there are only very few possible target moduli N_i^* differing from N on a single byte such that $v_i^* = \mathrm{CRT}_{N,N_i^*}(\sigma_i, \sigma'_i) < (2N)^{3/2}$: often only one or two, and almost never more than 20. We only need to run the attack with those target v_i^*'s until we find a factor.

Experimentally, for a 1024-bit modulus, the average base 2 logarithm of the number of possible v_i^*'s is about 2.5, so if an attacker has 5 pairs (σ_i, σ'_i) in this model, they can expect to try all vectors $\boldsymbol{v}$ in a search space of around 12.5 bits, *i.e.* run the attack a few thousand times, for a total running time of under 2 minutes. This is already quite practical.

If more pairs are available, the attacker can keep the 5 pairs for which the number of possible v_i^*'s is the smallest. This reduces the search space accordingly. In Table 3, we show how the exhaustive search space size and the expected running time evolve with the number of signatures in a typical example.

3.2 Faults on Many Least Significant Bits

In this model, the attacker is able to obtain $\ell = 5$ signature families of the form $(\sigma_i, \sigma'_{i,1}, \ldots, \sigma'_{i,k})$, where the σ_i's are correct signatures:

$$\sigma_i = \alpha x_i + \beta y_i \bmod N$$

and the $\sigma'_{i,j}$'s are faulty signatures of the form:

$$\sigma'_{i,j} = \alpha x_i + \beta y_i \bmod N'_{i,j} \quad 1 \le i \le \ell,\ 1 \le j \le k.$$

In other words, for each one of the ℓ different messages, the attacker learns the reduction of the CRT value $v_i = \alpha x_i + \beta y_i$ modulo N, as well as modulo k different unknown faulty moduli $N'_{i,j}$. Additionally, it is assumed that all $N'_{i,j}$ differ from N only on the least significant bits, but the number of distinct bits can be as large as half of the modulus size: we assume that$|N - N'_{i,j}| < N^{\delta}$ for a certain constant $\delta < 1/2$.

This is a reasonable fault model for a laser attack: it suffices to target a laser beam on the least significant bits of N to produce this type of faults.

To run the attack successfully, the attacker needs to recover the CRT values v_i. This can be done with high probability when the number of available faults k for a given message is large enough. The simplest approach is based on a GCD computation.

Indeed, fix an index $i \in \{1, \ldots, \ell\}$, and write $N'_{i,j} = N + \varepsilon_j$, $v_i = u$, $\sigma_i = u_0$ and $\sigma'_{i,j} = u_j$. The attacker knows the u_j's and wants to recover u.

Now, observe that there are integers t_j such that u satisfies $u = u_0 + t_0 \cdot N$ and $u = u_j + t_j \cdot (N + \varepsilon_j)$. In particular, for $j = 1, \ldots, k$ we can write:

$$(t_j - t_0) \cdot N + (u_j - u_0) + t_j \cdot \varepsilon_j = 0. \tag{3}$$

This implies that $u_j - u_0 \equiv t_j \cdot \varepsilon_j \pmod{N}$. However, we have $t_j \cdot \varepsilon_j < N^{1/2+\delta} \ll N$, so that the congruence is really an equality in $\mathbb{Z}$. In view of (3), this implies that all t_j's are in fact equal, and hence:

$$t_0 \cdot \varepsilon_j = u_0 - u_j \quad 1 \leq j \leq k.$$

If the errors ε_j on the modulus are co-prime, which we expect to happen with probability $\approx 1/\zeta(k)$, we can then deduce t_0 as the GCD of all values $u_0 - u_j$, and this gives:

$$u = u_0 + t_0 \cdot N = u_0 + N \cdot \gcd(u_0 - u_1, \ldots, u_0 - u_k).$$

As seen in Table 4, the success probability is in practice very close to $1/\zeta(k)$ regardless of the size of errors.

It is probably possible to further improve the success probability by trying to remove small factors from the computed GCD $g = \gcd(u_0 - u_1, \ldots, u_0 - u_k)$ to find t_0 when $g > \sqrt{N}$, but we find that the number of required faults is already reasonable without this computational refinement.

Indeed, recall that $\ell = 5$ CRT values are required to run the attack. If k faults are obtained for each of the ℓ messages, the probability that these ℓ CRT values can be successfully recovered with this GCD approach is $\zeta(k)^{-\ell}$. This is greater than 95% for $k = 7$, and 99% for $k = 9$.

We can also mention an alternate, lattice-based approach to recovering the CRT value u. The relation between the different quantities above can be written in vector form as:

$$u_0 \mathbf{1} = \boldsymbol{u} + t_0 \boldsymbol{e}$$

Table 4. Success probabilities of the GCD method for CRT value recovery, depending on the number of available faults on a given message. Tested with random 1024-bit moduli. In the simulation, errors ε_j are modeled as uniformly random signed integers of the given size, and 10,000 of them were generated for each parameter set.

k (faults per message)	3	5	7	9
$1/\zeta(k)$	.832	.964	.992	.998
100-bit errors	83.2%	96.8%	99.0%	99.8%
200-bit errors	83.4%	96.2%	99.2%	99.8%
400-bit errors	82.7%	96.6%	99.1%	99.8%
Average CPU time	.73 ms	.75 ms	.79 ms	.85 ms

where $\mathbf{1} = (1, \dots, 1)$, $\boldsymbol{u} = (u_1, \dots, u_k)$ and $\boldsymbol{e} = (\varepsilon_1, \dots, \varepsilon_k)$.

Then, since $u_0 \approx N$ is much larger than $\|t_0\boldsymbol{e}\| \approx N^{1/2+\delta}$, short vectors orthogonal to $\boldsymbol{u}$ will be orthogonal to both $\mathbf{1}$ and $\boldsymbol{e}$. More precisely, we can heuristically expect that when k is large enough ($k \gtrsim 2/(1-2\delta)$), the first $k-2$ vectors of a reduced basis of $\boldsymbol{u}^\perp$ will be orthogonal to $\mathbf{1}$ and $\boldsymbol{e}$.

Taking orthogonal lattices again, we can thus obtain a reduced basis $\{\boldsymbol{x}, \boldsymbol{y}\}$ of a two-dimensional lattice containing $\mathbf{1}$ and $\boldsymbol{e}$ (and of course $\boldsymbol{u}$). Since $\mathbf{1}$ is really short, we always find that $\boldsymbol{x} = \mathbf{1}$ in practice. Then, it happens quite often that $\boldsymbol{y}$ can be written as $\lambda\mathbf{1} \pm \boldsymbol{e}$, in which case t_0 is readily recovered as the absolute value of the second coordinate of $\boldsymbol{u}$ in the basis $\{\boldsymbol{x}, \boldsymbol{y}\}$.

However, this fails when $\mathbb{Z}\mathbf{1} \oplus \mathbb{Z}\boldsymbol{e}$ is a proper sublattice of $\mathbb{Z}\boldsymbol{x} \oplus \mathbb{Z}\boldsymbol{y} = \mathbb{Z}^k \cap (\mathbb{Q}\mathbf{1} \oplus \mathbb{Q}\boldsymbol{e})$, namely, when there is some integer $d > 1$ such that all errors ε_j are congruent mod d. Thus, we expect the success probability of this alternate approach to be $1/\zeta(k-1)$, which is slightly less than with the GCD approach.

4 Practical Experiments

Practical experiments for validating the new attack were done on an 8-bit 0.35μm RISC microcontroller with no countermeasures. As the microprocessor had no arithmetic coprocessor the values σ_p and σ_q were pre-computed by an external program upon each fault-injection experience and fed into the attacked device. The target combined σ_p and σ_q using multiplications and additions (using Formula 1) as well as the final modular reduction.

The location and spread of the faults were controlled by careful beam-size and shot-instant tuning. The reader is referred to the full version of this paper [9] for a description of the physical setting (common to the experiments reported in [16]).

We conducted several practical experiments corresponding to three different scenarios, roughly corresponding to the fault models considered in §2.1, §3.1 and §3.2 respectively. Let us describe these experiments in order.

4.1 First Scenario: Known Modulus

In this case, we considered 5 messages for a random 1024-bit RSA modulus N. For each message m_i, we obtained a correct signature σ_i, as well as a faulty-modulus signature σ'_i where the faulty modulus N'_i was also read back from the microcontroller.

Therefore, we were exactly in the setting described in §2.1, and could apply the algorithm from §2.3 directly: apply the Chinese Remainder Theorem to construct the vector $\boldsymbol{v}$ of CRT values and run the lattice-based attack to recover a factor of N.

The implementation of the attack used the same SAGE code as the simulation from §2.4. In our experimental case, the ball of radius $\sqrt{N\ell}$ contained only about 10 vectors of the double orthogonal lattice, and the whole attack revealed a factor of N in less than 20 milliseconds.

4.2 Second Scenario: Unknown Single Byte Fault

In this case, we tried to replicate a setting similar to the one considered in §3.1. We considered 20 messages and a random 1024-bit RSA modulus N. For each message m_i, we obtained a correct signature σ_i, as well as faulty-modulus signatures σ'_i with undisclosed faulty modulus N'_i generated by targeting a single byte of N with the laser.

We had to eliminate some signatures, however, because in some cases, errors on the modulus turned out to exceed 8 bits.[1] After discarding those, we had 12 pairs (σ_i, σ'_i) left to carry out the approach described in §3.1.

The first step in this approach is to find, for each i, all values v_i^* of the form $\mathrm{CRT}_{N,N_i^*}(\sigma_i, \sigma'_i)$ (N_i^* differing from N only on one byte) that are small enough to be correct candidate CRT values. Unlike the setting of §3.1, we could not assume that bit-differences were aligned on byte boundaries: we had to test a whole 1016×255 candidate moduli[2] N_i^* for each i . Therefore, this search step was a bit costly, taking a total of 11 minutes and 13 seconds. Additionally, due to the higher number of candidate moduli, the number of candidate CRT values v_i^* was also somewhat larger than in §3.1, namely:

$$7, 17, 3, 9, 15, 5, 14, 44, 44, 17, 10, 55$$

for our 12 pairs respectively. Keeping only the 5 indices with the smallest number of candidates, we obtained $3 \times 5 \times 7 \times 9 \times 10 = 9450$ possible CRT value vectors $\boldsymbol{v}^*$.

We then ran the lattice-based attack on each of these vectors in order until a factor of N was found. The factor was found at iteration number 2120, after a total computation time of 43 seconds.

4.3 Third Scenario: Unknown Least Significant Bytes Faults

In this case, we considered 10 messages for a random 1024-bit N. For each message m_i, we obtained a correct signature σ_i, as well as 10 faulty-modulus signature $\sigma'_{i,j}$ with undisclosed faulty modulus N'_i. The laser beam targeted the lower order bytes of N but with a large aperture, generating multiple faults stretching over as much as 448 modulus bits.

In practice, we only used the data $(\sigma_i, \sigma'_{i,1}, \ldots, \sigma'_{i,10})$ for the first 5 messages, discarding the rest. Then, we reconstructed the CRT values v_i using the GCD technique of §3.2:

[1] Note that in a real-world attack, it might not be possible to detect such overly spread out faults: hence, this particular technique should be used preferably when faults are *known* to affect only single bytes (*e.g.* in a glitch attack), whereas the technique from the next section is better suited to laser attacks as aperture control is much less of an issue.

[2] There are duplicates among those, corresponding to perturbations of 7 consecutive bits or less, but we did not attempt to avoid testing them several times, as this can only improve the search by a small constant factor while introducing significant complexity in the code.

$$v_i = \sigma_i + N \cdot \gcd(\sigma_i - \sigma'_{i,1}, \ldots, \sigma_i - \sigma'_{i,10}) \quad 1 \leq i \leq 5$$

and applied the lattice-based attack on the resulting vector $\boldsymbol{v}$. This revealed a factor of N in 16 milliseconds.

We also tried the same attack using a fewer number of the $\sigma'_{i,j}$'s, and found that it still worked when taking only 4 of those values in the computation of v_i:

$$v_i = \sigma_i + N \cdot \gcd(\sigma_i - \sigma'_{i,1}, \ldots, \sigma_i - \sigma'_{i,4}) \quad 1 \leq i \leq 5$$

but failed if we took 3 instead. Considering that $1/\zeta(3)^5 \approx .40$ and $1/\zeta(4)^5 \approx .67$, this is quite in line with expectations.

5 Countermeasures and Further Research

Probabilistic and stateful signature schemes are usually secure against this attack, since they make it difficult to obtain two signatures on the same padded message. However, all deterministic schemes are typically vulnerable, including those in which the attacker doesn't have full access to the signed message, provided that the target device can be forced to compute the same signature twice.

A natural countermeasure is to use a CRT interpolation formula that does not require N, such as Garner's formula, computed as follows:

$t \leftarrow \sigma_p - \sigma_q$
if $t < 0$ then $t \leftarrow t + p$
$\sigma \leftarrow \sigma_q + (t \cdot \gamma \bmod p) \cdot q$
return(σ)

where we assume that $p > q$, and γ is the usual CRT coefficient $q^{-1} \bmod p$. Note that the evaluation of σ does not require a modular reduction because

$$\sigma = \sigma_q + (t \cdot \gamma \bmod p) \cdot q \leq q - 1 + (p - 1)q < N$$

Besides the obvious countermeasures consisting in checking signatures before release, it would be interesting to devise specific countermeasures for protecting Formula (1) (or Garner's formula) taking into account the possible corruption of all data involved.

Finally, in a number of special cases and particular settings (*e.g.* Appendix A) other fault attacks on the CRT recombination phase can be devised. A thorough analysis of such scenarios is also an interesting research direction.

Acknowledgments. We would like to thank the anonymous referees for helpful comments. The work described in this paper has been supported in part by the European Commission through the ICT program under contract ICT-2007-216676 ECRYPT II.

References

1. Aumüller, C., Bier, P., Fischer, W., Hofreiter, P., Seifert, J.-P.: Fault Attacks on RSA with CRT: Concrete Results and Practical Countermeasures. In: Kaliski Jr., B.S., Koç, Ç.K., Paar, C. (eds.) CHES 2002. LNCS, vol. 2523, pp. 260–275. Springer, Heidelberg (2003)
2. Bellare, M., Rogaway, P.: The exact security of digital signatures - how to sign with RSA and rabin. In: Maurer, U.M. (ed.) EUROCRYPT 1996. LNCS, vol. 1070, pp. 399–416. Springer, Heidelberg (1996)
3. Berzati, A., Canovas, C., Dumas, J.-G., Goubin, L.: Fault attacks on RSA public keys: Left-to-right implementations are also vulnerable. In: Fischlin, M. (ed.) CT-RSA 2009. LNCS, vol. 5473, pp. 414–428. Springer, Heidelberg (2009)
4. Berzati, A., Canovas, C., Goubin, L.: Public key perturbation of randomized RSA implementations. In: Mangard, S., Standaert, F.-X. (eds.) CHES 2010. LNCS, vol. 6225, pp. 306–319. Springer, Heidelberg (2010)
5. Berzati, A., Canovas, C., Goubin, L.: Perturbating RSA public keys: An improved attack. In: Oswald, E., Rohatgi, P. (eds.) CHES 2008. LNCS, vol. 5154, pp. 380–395. Springer, Heidelberg (2008)
6. Boneh, D., DeMillo, R.A., Lipton, R.J.: On the importance of eliminating errors in cryptographic computations. J. Cryptology 14(2), 101–119 (2001)
7. Boscher, A., Naciri, R., Prouff, E.: CRT RSA algorithm protected against fault attacks. In: Sauveron, D., Markantonakis, K., Bilas, A., Quisquater, J.-J. (eds.) WISTP 2007. LNCS, vol. 4462, pp. 229–243. Springer, Heidelberg (2007)
8. Brier, E., Chevallier-Mames, B., Ciet, M., Clavier, C.: Why one should also secure RSA public key elements. In: Goubin, L., Matsui, M. (eds.) CHES 2006. LNCS, vol. 4249, pp. 324–338. Springer, Heidelberg (2006)
9. Brier, E., Naccache, D., Nguyen, P.Q., Tibouchi, M.: Modulus Fault Attacks Against RSA-CRT Signatures. Full version of this paper. Cryptology ePrint Archive, `http://eprint.iacr.org/`
10. Ciet, M., Joye, M.: Practical fault countermeasures for Chinese remaindering based cryptosystems. In: Breveglieri, L., Koren, I. (eds.) FDTC, pp. 124–131 (2005)
11. Coppersmith, D.: Small solutions to polynomial equations, and low exponent RSA vulnerabilities. J. Cryptology 10(4), 233–260 (1997)
12. Coron, J.-S., Joux, A., Kizhvatov, I., Naccache, D., Paillier, P.: Fault attacks on RSA signatures with partially unknown messages. In: Clavier, C., Gaj, K. (eds.) CHES 2009. LNCS, vol. 5747, pp. 444–456. Springer, Heidelberg (2009)
13. Coron, J.-S.: Optimal security proofs for PSS and other signature schemes. In: Knudsen, L.R. (ed.) EUROCRYPT 2002. LNCS, vol. 2332, pp. 272–287. Springer, Heidelberg (2002)
14. Giraud, C.: An RSA implementation resistant to fault attacks and to simple power analysis. IEEE Trans. Computers 55(9), 1116–1120 (2006)
15. Lenstra, A.K., Lenstra Jr., H.W., Lovász, L.: Factoring polynomials with rational coefficients. Math. Ann. 261(4), 515–534 (1982)
16. Mirbaha, A.-P., Dutertre, J.M., Tria, A., Agoyan, M., Ribotta, A.-L., Naccache, D.: Study of single-bit fault injection techniques by laser on an AES cryptosystem. In: Gizopoulos, D., Chatterjee, A. (eds.) IOLTS (2010)
17. Muir, J.A.: Seifert's RSA fault attack: Simplified analysis and generalizations. In: Ning, P., Qing, S., Li, N. (eds.) ICICS 2006. LNCS, vol. 4307, pp. 420–434. Springer, Heidelberg (2006)

18. Nguyen, P.Q.: Public-key cryptanalysis. In: Luengo, I. (ed.) Recent Trends in Cryptography. Contemporary Mathematics, vol. 477. AMS–RSME (2009)
19. Nguyên, P.Q., Stern, J.: Merkle-Hellman Revisited: A Cryptanalysis of the Qu-Vanstone Cryptosystem Based on Group Factorizations. In: Kaliski Jr., B.S. (ed.) CRYPTO 1997. LNCS, vol. 1294, pp. 198–212. Springer, Heidelberg (1997)
20. Nguyên, P.Q., Stern, J.: Cryptanalysis of a fast public key cryptosystem presented at SAC 1997. In: Tavares, S., Meijer, H. (eds.) SAC 1998. LNCS, vol. 1556, pp. 213–218. Springer, Heidelberg (1999)
21. Nguyên, P.Q., Stern, J.: The two faces of lattices in cryptology. In: Silverman, J.H. (ed.) CaLC 2001. LNCS, vol. 2146, pp. 146–180. Springer, Heidelberg (2001)
22. Rivain, M.: Securing RSA against fault analysis by double addition chain exponentiation. In: Fischlin, M. (ed.) CT-RSA 2009. LNCS, vol. 5473, pp. 459–480. Springer, Heidelberg (2009)
23. Rivest, R.L., Shamir, A., Adleman, L.M.: A method for obtaining digital signatures and public-key cryptosystems. Commun. ACM 21(2), 120–126 (1978)
24. Seifert, J.-P.: On authenticated computing and rsa-based authentication. In: Atluri, V., Meadows, C., Juels, A. (eds.) ACM Conference on Computer and Communications Security, pp. 122–127. ACM, New York (2005)
25. Stein, W.A., et al.: Sage Mathematics Software (Version 4.4.2). The Sage Development Team (2010), `http://www.sagemath.org`
26. Vigilant, D.: RSA with CRT: A new cost-effective solution to thwart fault attacks. In: Oswald, E., Rohatgi, P. (eds.) CHES 2008. LNCS, vol. 5154, pp. 130–145. Springer, Heidelberg (2008)

A Using Dichotomy in the Absence of Padding

Consider again the setting of §2.1, in which an adversary is able to obtain both a correct signature σ on a message m, and a signature on the same message m computed with a faulty modulus, allowing him to deduce the non reduced value $v = \sigma_p \cdot \alpha + \sigma_q \cdot \beta \in \mathbb{Z}$. We can write:

$$v = (\sigma \bmod p) \cdot \alpha + (\sigma \bmod q) \cdot \beta = \left(\sigma - p \left\lfloor \frac{\sigma}{p} \right\rfloor\right) \cdot \alpha + \left(\sigma - q \left\lfloor \frac{\sigma}{q} \right\rfloor\right) \cdot \beta$$

Moreover, observe that $\alpha + \beta = N + 1$ (as is easily seen by reducing $\alpha + \beta$ modulo p and q). Therefore, we have:

$$v = \sigma \cdot (N+1) - p\alpha \left\lfloor \frac{\sigma}{p} \right\rfloor - q\beta \left\lfloor \frac{\sigma}{q} \right\rfloor$$

Hence, if we let $\omega = (\sigma \cdot (N+1) - v)/N$, we get:

$$\omega = \frac{\sigma \cdot (N+1) - v}{N} = \frac{\alpha}{q} \left\lfloor \frac{\sigma}{p} \right\rfloor + \frac{\beta}{p} \left\lfloor \frac{\sigma}{q} \right\rfloor \tag{4}$$

and this value ω is an integer since $v \equiv \sigma \pmod{N}$.

Now assume further that the adversary can ask signatures on messages m such that σ is small. This is the case, for example, when signatures are computed without padding and the physical device under consideration will answer arbitrary signature queries: then, the adversary can simply ask signatures on messages of the form σ^e for small values σ of his choice.

In such a setting, the adversary can pick a σ close to $N^{1/2}$, carry out the fault attack and compute the integer ω. By (4), he gets $\omega = 0$ if $\sigma < \min(p, q)$ and $\omega > 0$ otherwise. Trying this process again several times, the smallest prime factor of N can be recovered by dichotomy.

Breaking Mifare DESFire MF3ICD40: Power Analysis and Templates in the Real World*

David Oswald and Christof Paar

Horst Görtz Institute for IT Security
Ruhr-University Bochum, Germany
{david.oswald,christof.paar}@rub.de

Abstract. With the advent of side-channel analysis, implementations of mathematically secure ciphers face a new threat: by exploiting the physical characteristics of a device, adversaries are able to break algorithms such as AES or Triple-DES (3DES), for which no efficient analytical or brute-force attacks exist. In this paper, we demonstrate practical, non-invasive side-channel attacks on the Mifare DESFire MF3ICD40 contactless smartcard, a 3DES-based alternative to the cryptanalytically weak Mifare Classic [9,25]. We detail on how to recover the complete 112-bit secret key of the employed 3DES algorithm, using non-invasive power analysis and template attacks. Our methods can be put into practice at a low cost with standard equipment, thus posing a severe threat to many real-world applications that employ the DESFire MF3ICD40 smartcard.

Keywords: contactless smartcard, side-channel analyis, templates, DESFire.

1 Introduction

Radio Frequency Identification (RFID) technology has become the basis for numerous large-scale, security-relevant applications, including public transport, wireless payment, access control, or digital identification [39]. The information stored on RFID smartcards, e.g., personal data, or cash balance, is often highly sensitive — however, the access to the air interface and to the device itself is virtually impossible to control. Hence, most modern RFIDs feature cryptographic mechanisms, including encryption and authentication, in order to thwart attacks such as eavesdropping, manipulation, or cloning of a smartcard.

Mifare DESFire MF3ICD40 is a contactless smartcard featuring a cryptographic engine for authentication and encryption based on (Triple-)DES. The smartcard is employed in several large payment and public transport systems around the world, e.g., the Czech railway in-karta [7], the Australian myki card [36], or the Clippercard used in San Francisco [40]. In the course of our

* The work described in this paper has been supported in part by the European Commission through the ICT programme under contract ICT-2007-216676 ECRYPT II.

B. Preneel and T. Takagi (Eds.): CHES 2011, LNCS 6917, pp. 207–222, 2011.

research, we also noticed many smaller installations, e.g., for mobile payment or access control, that are based on the Mifare DESFire MF3ICD40. From a mathematical point of view, the employed 3DES cipher is secure, because no efficient cryptanalytical attacks are known. Thus, in this paper, we focus on *side-channel attacks*, i.e., methods that target the physical implementation of the cryptographic primitive in soft- or hardware. Using non-invasive and hence non-detectable measurement of the electro-magnetic (EM) emanations of the device, we are able to completely recover the secret 112-bit master key and thus to, for example, read out, manipulate, or duplicate the contents of a Mifare DESFire MF3ICD40 card.

1.1 Related Work

The idea of exploiting physical side-channels to attack hardware implementations of secure ciphers was first put forward in [20] in 1998. Since then, a lot of research has been conducted in this area, with important contributions including the analysis using the EM emanation of a device [1] or the application of the correlation coefficient in Correlation Power Analysis (CPA) to better model the physical behaviour of Integrated Circuits (ICs) [2]. At CHES 2002, the authors of [5] proposed the use of machine learning techniques such as pattern recognition for Side-Channel Analysis (SCA) and coined the notion of "template attacks". Several extensions and improvements for this approach have been suggested in the last few years, cf. [31,33,35].

The susceptibility of ciphers running on RFID devices towards SCA was initially shown in [12,30]: the authors present attacks on a white-box software implementation of the AES executed by a standard, unprotected microcontroller (μC) on a self-made prototype RFID, evaluating techniques to overcome problems such as misaligment of the measured signals.

With respect to the application of SCA to break commercial, *real-world* devices, few papers have been published, as most research in this field is carried out by evaluation labs behind closed doors. The potential impact of SCA in practice was demonstrated by the complete break of the proprietary KeeLoq system presented at CRYPTO 2008 [8]. Results for the black-box analysis of a contactless smartcard are given in [17], proposing a leakage model for RFIDs that forms the basis for our analyses and is outlined in Sect. 2. However, the authors are unable to recover the complete key and do not disclose to which device their attacks apply. In [18], the application of analog demodulation for SCA of RFIDs is presented for the first time. The measurement setup used in the present paper is an extension of the setup described in [18].

1.2 Contribution of this Paper

The work presented in this paper is of practical nature: we highlight the relevance of SCA in the real-world by demonstrating the first full key-recovery attack on the popular Mifare DESFire MF3ICD40 smartcard reported in the literature. Doing so, we point out problems and obstacles that occur when conducting SCA in practice which are often neglected in academic papers. In addition, we present the — to our knowledge — first application of template attacks to break cryptographic RFIDs, allowing for potentially very fast determination of the secret key. The remainder of this paper is structured as follows: in Sect. 2, we give the signal-theoretical background of our measurement setup for RFID devices, which is presented in Sect. 3. We then practically apply the developed techniques to analyze the smartcard in Sect. 4, detailing on the internal hardware structure of the device. In Sect. 5, we extend our findings and present a successful full key-recovery attack on the 3DES engine. After that, in Sect. 6, we demonstrate a different approach for obtaining the secret key based on template attacks to eavesdrop on the internal databus. Finally, we conclude in Sect. 7, discussing the implications of our findings for commercial applications and giving directions for further research.

2 Demodulation for SCA of Contactless Smartcards

For contactless smartcards, the energy for operation is supplied wirelessly using magnetic coupling. As proposed in [17,18], this gives rise to a different leakage mechanism compared to contact-based devices. In a similar manner as for regular data transmission, the 13.56 MHz field generated by the reader is load-modulated by the power consumption of an RFID[1].

Let the power consumption of the target device be given as $p(t) = P_{const} + p_{dyn}(t)$, where P_{const} is the constant part and $p_{dyn}(t)$ the fraction caused by internal operations, e.g., intermediate values being manipulated during a cryptographic operation. Usually, the dynamic portion of the power consumption is far weaker than the constant part, i.e., $|p_{dyn}(t)| << P_{const}$. The leakage exploitable for an SCA thus heavily depends on the quality of the isolation and amplification of $p_{dyn}(t)$. As mentioned, in an RFID setting, the amplitude of the reader signal is modulated by $p(t)$, i.e., $s(t) = p(t) \cdot \cos(\omega_r \cdot t) = (P_{const} + p_{dyn}(t)) \cdot \cos(\omega_r \cdot t)$.

where $\omega_r = 2\pi\ f_r$, $f_r = 13.56$ MHz is the standard carrier frequency. Clearly, the extraction of $p(t)$ (and especially of the weak dynamic portion) from $s(t)$ can be done using amplitude demodulation, cf. for instance [34]. In practice, "incoherent" techniques (i.e., for which a separate, unmodulated carrier signal is not necessary) based on rectification (often called envelope detection) are very common, and in this paper, we follow that approach as well. The principle due to which rectification can be used for demodulation is best understood in the frequency domain, following [27]. First note that, as stated above, $|p_{dyn}(t)| << P_{const}$ and hence, $|s(t)| = |P_{const} + p_{dyn}(t)| \cdot |\cos(\omega_r \cdot t)| = (P_{const} + p_{dyn}(t)) \cdot |\cos(\omega_r \cdot t)|$.

Let $P(j\omega) = \mathrm{DFT}\{p(t)\} = \mathrm{DFT}\{P_{const} + p_{dyn}(t)\}$ denote the frequency domain representation of the signal that is to be reconstructed. By expanding $|\cos(\omega_r \cdot t)|$ using its Fourier series, one obtains the spectrum of the rectified signal:

[1] However, for data transmission, the fluctuations of the EM field are intentional and far stronger in magnitude.

$$\begin{aligned}\mathrm{DFT}\left\{\left|s\left(t\right)\right|\right\} &= \mathrm{DFT}\left\{p\left(t\right)\cdot\left|\cos\left(\omega_r\cdot t\right)\right|\right\} = \mathrm{DFT}\left\{p\left(t\right)\cdot\frac{2}{\pi}\sum_{\nu=-\infty}^{\infty}\frac{(-1)^{\nu}}{1-4\nu^2}e^{j2\nu\omega_r t}\right\}\\ &= \frac{2}{\pi}\sum_{\nu=-\infty}^{\infty}\frac{(-1)^{\nu}}{1-4\nu^2}\mathrm{DFT}\left\{p\left(t\right)\cdot e^{j2\nu\omega_r t}\right\} = \frac{2}{\pi}\sum_{\nu=-\infty}^{\infty}\frac{(-1)^{\nu}}{1-4\nu^2}P\left(j\omega-j2\nu\omega_r\right)\end{aligned}$$

The rectified signal is essentially formed by the spectrum of $P_{const}+p_{dyn}\left(t\right)$, which, however, is (scaled and) repeated at all even multiples of the carrier frequency $\omega_r = 2\pi\ 13.56\,\mathrm{MHz}$. Thus, the first repetition occurs at 27.12 MHz.

Using a lowpass filter with a cutoff frequency less than 13.56 MHz isolates the desired signal[2] $p\left(t\right)$.

3 Measurement Setup

For the analysis of the DESFire MF3ICD40, we extended the measurement environment of [18]. Fig. 1a gives an overview over the components of our setup. A custom, freely programmable RFID reader [16] compliant to ISO 14443 [13,14] and ISO 15693 [15] supplies the contactless smartcard (from now on occasionally referred to as Device Under Test (DUT)) with power and handles the communication, for instance to trigger an encryption operation.

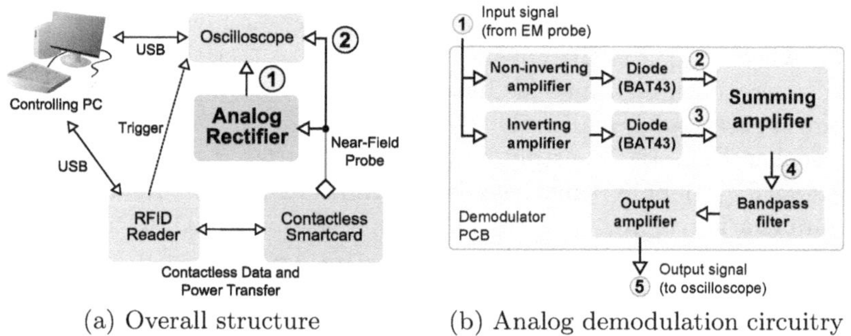

(a) Overall structure (b) Analog demodulation circuitry

Fig. 1. Measurement setup

A wide-band EM probe with a suitable pre-amplifier [21] captures the magnetic near-field in the proximity of the IC, resulting in a "raw" signal (denoted as ② in Fig. 1a) which is dominated by the 13.56 MHz carrier frequency of the reader. On the one hand, this signal is directly recorded and stored using a Picoscope 5204 Digital Storage Oscilloscope (DSO) [29] at a sample rate of 500 MHz, on the other hand, it is passed to an analog demodulator that performs the operations outlined in Sect. 2 to facilitate SCA, resulting in the signal ① in Fig. 1a. The central PC controls the measurement process, i.e., prepares and sends commands to the DUT via the RFID reader and acquires and stores the resulting side-channel signals ① and ② (from now on referred to as *traces*.

[2] The constant term P_{const} can be removed with a highpass filter that only blocks the DC and very low-frequency components.

As explained in Sect. 2, analog demodulation is required to separate the actual power consumption signal from the carrier signal and to thereby improve the quality of the (exploitable) side-channel leakage. Accordingly, we developed a custom Printed Circuit Board (PCB) comprising a full-wave rectifier and appropriate filter circuitry to perform the incoherent demodulation approach. Fig. 1b shows the basic structure of the demodulation circuitry. The full schematics are given in an appendix in the extended version of this paper [28]. The full-wave rectifier is formed by two isolated half-wave rectifiers, each employing an BAT43 Schottky diode [38]. To rectify the negative part of the input ①, the signal is first inverted and then rectified by the diode, yielding signal ③ in Fig. 1b. For the positive portion, the buffer amplifier only provides isolation of the input signal and driving of the corresponding diode, but does not perform inversion to produce signal ②. The two resulting parts ② and ③ are then added to form the full-wave rectified output ④.

A third-order LC bandpass filter extracts the baseband part, i.e., the portion of the spectrum centered around 0 Hz. In our case, the −3 dB frequency was specified to 12 Mhz. Additionally, the filter also suppresses frequency components below 10 kHz to remove the constant part of the modulating signal. Finally, the output amplifier adjusts the amplitude of the signal in order to optimally utilize the minimum input range of ±100 mV of the Picoscope and drives a 50 Ω load, i.e., a suitable coaxial cable.

In the case that a raw signal (i.e., ② in Fig. 1a) is used for SCA, it was shown in [17] that the demodulation has to be performed digitally in order to conduct a successful CPA, i.e., digital pre-processing is mandatory. For the output of the analog demodulator, digitally filtering the output signal ① is optional, however, might help to further reduce the 13.56 MHz frequency component still present due to certain characteristics of the analog circuits. For a more detailed description of the effects of the respective processing techniques, cf. [18].

4 Practical Results: Profiling of Mifare DESFire MF3ICD40

Mifare DESFire MF3ICD40 [26] is a contactless smartcard initially designed by the semicondutor division of Philips, which became the separate company NXP in 2006. The card is compliant to parts 1-4 of the ISO 14443A standard. A communication with the card can be performed in plain, with an appended Message Authentication Code (MAC), or with full data encryption using 3DES. The device offers 4 kByte of storage that can be assigned to up to 28 different applications, whereas each application may hold a maximum of 16 files. Depending on the configuration of the access rights, a mutual authentication protocol has to be carried out before accessing the card, ensuring that the symmetric 3DES keys of the card k_C and of the reader k_R are identical.

According to specifications found on the internet, the smartcard features several functions to thwart physical attacks such as SCA, fault injection, or reverse-engineering: the IC is built using asynchronous circuits and employs a

Reader		DESFire MF3ICD40	
	$\xrightarrow{\textit{begin}}$	Generate $n_c \in \{0, 1\}^{64}$ $B_0 = 3\text{DES}_{k_C}(n_c)$	Step 1
	$\xleftarrow{B_0}$		
Choose B_1, B_2	$\xrightarrow{B_1, B_2}$	$C_2 = 3\text{DES}_{k_C}(B_2)$ $C_1 = 3\text{DES}_{k_C}(B_1)$	Step 2

Fig. 2. Exerpt of the Mifare DESFire authentication protocol relevant for SCA

custom, asynchronous μC design based on the 8051 architectures. Besides, all digital units (i.e., control logic, cryptographic engine etc.) are "intermingled" so that no functional block are discernible, a technology called "glue logic" by the vendor. Note that all results in this paper do not directly apply to the newer AES-based variant DESFire EV1. The authentication protocol of the DESFire MF3ICD40 has been disclosed and can for instance be found in [19,4]. For the purpose of SCA, we refer to a simplified version in the following, given in Fig. 2. $k_C = (k_{C,1}, k_{C,2})$ is the 128-bit 3DES master key (including the parity bits) used by the DUT, whereas the two halfs are of size 64 bit each, i.e., $k_{C,1}, k_{C,2} \in \{0, 1\}^{64}$. $3\text{DES}_{k_C}(x) = \text{DES}_{k_{C,1}}\left(\text{DES}^{-1}_{k_{C,2}}\left(\text{DES}_{k_{C,1}}(x)\right)\right)$ denotes a 3DES encryption of a 64-bit value x in Encrypt-Decrypt-Encrypt (EDE) mode. The full command set[3] has been implemented for our custom reader mentioned in Sect. 3.

Initially, we are facing a *black-box* scenario, i.e., have (apart from the command set and the specifications in the datasheet) no further knowledge on the inner workings of the device. Hence, *profiling* to map different portions of a power trace to steps of the operation of the DUT (e.g., a data transfer or an encryption operation) is mandatory before attempting to perform real attacks on cryptographic operations. As a first step, we dismantled the IC, took magnified photographs of the silicon die, cf. Fig. 3a, and tried to distinguish the different parts of the circuit. The hypothetical structure depicted in Fig. 3b is a result of this optical inspection and the findings reported in the remainder of this section.

To prepare the actual SCA, we recorded side-channel traces for both steps of the authentication protocol, separately varying either the key of the card k_C or the values for B_1 and B_2 in step 2. To estimate the effect of our analog processing circuitry, we both store the "raw" signals before demodulation (② in Fig. 1a) and the result of the demodulation process (① in Fig. 1a).

We then perform several CPAs to locate the points in time in the power traces at which the known values for k_C, B_1 and B_2 (and the encryption results C_1, C_2[4]) are processed. Employing an 8-bit Hamming weight model, all mentioned

[3] Including the necessary commands for changing the key, performing a full authentication etc.

[4] As we know k_C during the profiling phase, we can predict these values that are never output by the DUT.

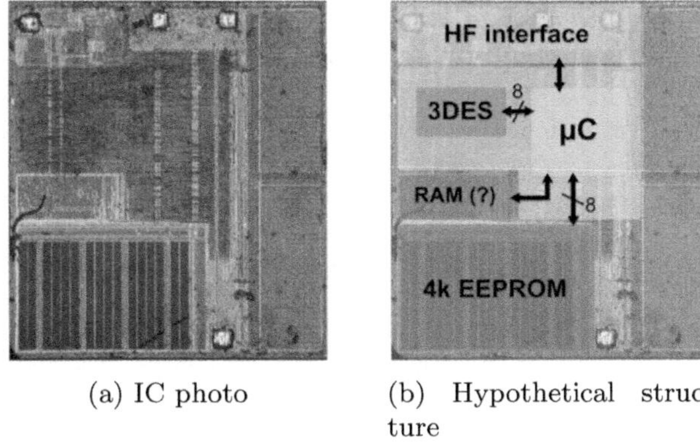

(a) IC photo (b) Hypothetical structure

Fig. 3. The DESFire MF3ICD40 IC

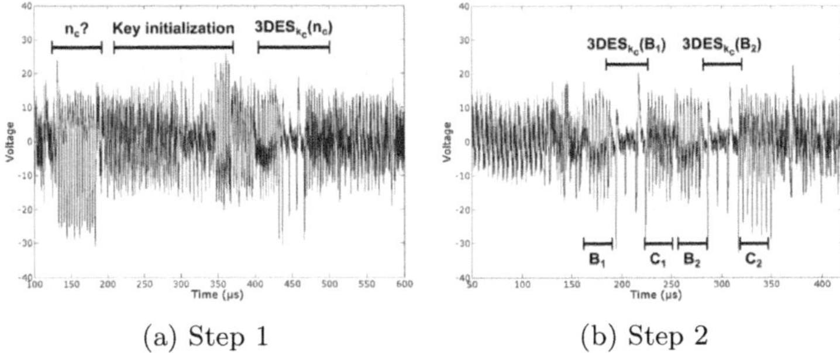

(a) Step 1 (b) Step 2

Fig. 4. Annotated traces during the authentication protocol (after analog processing)

values can be precisely pinpointed, cf. Fig. 4. We observed a stable value of ≈ 0.15 for the respective correlation coefficient after around 1,000 traces. This suggests that internally, an 8-bit data bus is used to connect the μC to the memory and the cryptographic engine, yielding the structure of Fig. 3b. For each byte transfered over this bus, a distinct peak appears in the power trace, whereas the distance between two such peaks indicates an internal bus frequency of $f_{bus} \approx 282.5\,\text{kHz} = {}^{13.56}/_{48}\,\text{MHz}$. Note that the peaks for data bus transfers later in a trace, e.g. for B_2 or C_2 in Fig. 4b, are often *misaligned*, i.e., their exact position slightly varies from execution to execution. The reason for this behaviour lies in the non-constant execution time of a 3DES operation, which is further detailed in Sect. 5. Hence, it is necessary to re-align the respective parts (for instance, using standard pattern matching approaches [23]) to obtain a significant correlation.

5 Practical Attack: CPA of the 3DES Engine

Having located the input and output values of the 3DES encryption, we now focus on this part to perform the recovery of the secret key. Comparing this part for several traces, we notice some interesting properties: first, the length of one DES operation varies from execution to execution, even if the input data and the key are kept constant. This hints at a countermeasure based on randomization in time being employed to thwart CPA. We further address this problem in Sect. 5.1. Second, the amplitude of the traces is significantly lower during the supposed encryption, which coincides with the statements in the available DESFire documentation that a dedicated low-power hardware engine performs the cryptographic operation.

To prepare the actual key-recovery, we first attempt to characterize the leakage of the 3DES engine and find a suitable power model by correlating with the full intermediate 64-bit states[5] using a known key. Conducting several experiments, we found the Hamming distance model to yield a significant correlation and were able to locate the first few rounds of the DES, as depicted in Fig. 5 for rounds 0→1, 5→6, 10→11, and 0→1 of the second DES iteration.

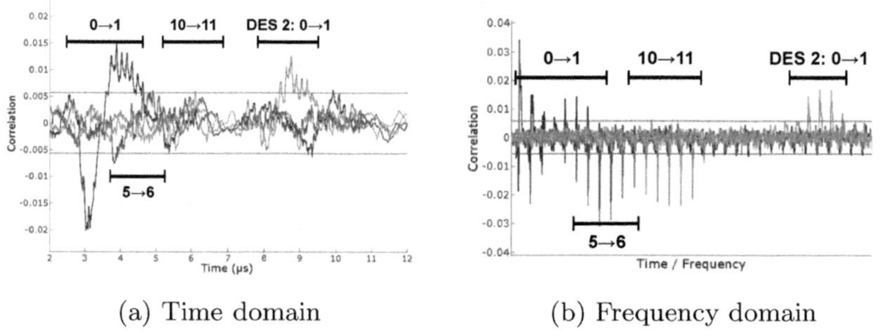

(a) Time domain (b) Frequency domain

Fig. 5. Correlation coefficients for the Hamming distances between rounds of the 3DES, 500,000 traces

However, as evident from Fig. 5a, this approach only is able to locate the first few rounds (with decreasing correlation), supposedly due to the randomization mentioned above. Statistically analyzing the length of the first DES iteration using 100,000 traces, we observe that one iteration takes 8.2 μs on average. This duration varies in discrete steps of 290 ns over a total range from 6.9 μs to 9.1 μs. This suggests that the cryptographic engine executes up to eight ($\lceil (9.1-6.9)/0.29 \rceil$) "dummy" rounds based on an internal Random Number Generator (RNG) to impede SCA.

[5] i.e., $\left(L_i^{(n)}, R_i^{(n)}\right)$, $0 \leq i \leq 16$, $n \in \{1, 2, 3\}$, where n denotes the Single-DES iteration within the complete 3DES, for details cf. [24].

To solve this problem, we tried out methods to overcome misalignment suggested in the literature, including comb filtering or windowing [6], Dynamic Time Warping (DTW) [37], and Differential Frequency Analysis (DFA) [10,30]. Our results show DFA to yield the best overall correlation, using the following steps: before correlating with the prediction of the power model, a trace is partitioned into (overlapping) segments, these segments are transformed to the frequency domain with the Discrete Fourier Transform (DFT), and the phase information is discarded by taking the absolute value of the DFT coefficients. The optimal value for the size of each segment was determined to be 1.5 μs, with an overlap of 75 % between adjacent segments. The strongest leakage occurs for low frequencies, hence, we limited the analysed spectral range to 0. . .16 MHz. Fig. 5b shows the according correlation coefficients for the respective rounds of the cipher — in contrast to the analysis in the time domain, all rounds are clearly distinguishable.

In order to quantify the improvement caused by the employed analog and digital processing methods, we compare the maximum correlation coefficient over the number of traces for the 32-bit Hamming distance $R_0 \rightarrow R_1$ (again, using a known key), with a detailed plot of the respective values given in an appendix in [28]. In all cases, the correlation converges rather quickly to a significant value far greater than $4/\sqrt{\text{No. of traces}}$, yet, a distinct gain due to both analog and digital processing is discernible: while the digitally demodulated traces without re-alignment by DFA result in a stable value of ≈ 0.015, the combination of analog demodulation with DFA yields ≈ 0.032, that is, an improvement by a factor of two. As a result, we utilize these pre-processing techniques for the full key-recovery presented in Sect. 5.1, taking the fact into account that in this case, we have to target each 4-bit S-Box output separately, so smaller overall correlations are to be expected.

5.1 Full Key-Recovery

Based on the findings of the profiling phase, a CPA can be mounted to obtain the full 3DES key by recovering the 6-bit part of the round key for each S-Box, starting with the first round of the first DES. To make use of all available information, a natural choice is to target the full 4-bit output of each S-Box in the Hamming distance $R_0 \rightarrow R_1$. However, for the case of the DESFire MF3ICD40, this turned out to be problematic: Fig. 6 shows the maximum correlation coefficients for the correct key candidate for a standard CPA in the time domain and DFA in the frequency domain, respectively. Although the complete key is discernible after $\approx$ 450,000 traces in Fig. 6b, the stable value for the correlation significantly differs depending on the S-Box, causing the attack to fail for five S-Boxes when performed without re-alignment by means of DFA, cf. Fig. 6a.Testing other prediction functions, a single-bit CPA (which is equivalent to the classic Differential Power Analysis (DPA)) proved to be the most successful approach. As depicted in Fig. 7, for each S-Box there is at least one bit providing sufficient leakage to allow our attack to succeed after approx. 250,000 traces and 350,000 traces with and without DFA, respectively.

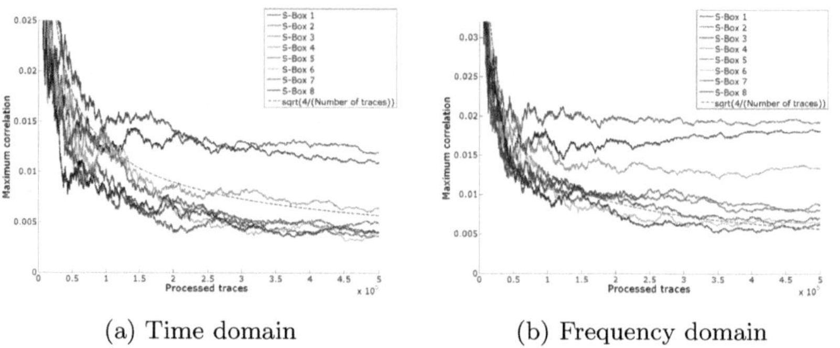

(a) Time domain (b) Frequency domain

Fig. 6. Maximum correlation coefficient for the correct key, 4-bit model, Hamming distance $R_0 \rightarrow R_1$ for all S-Boxes

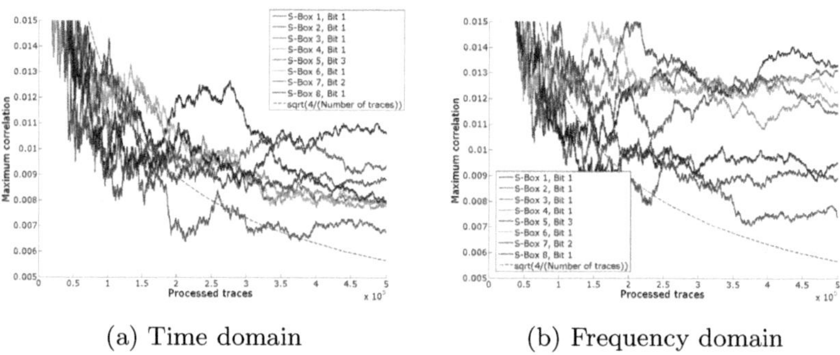

(a) Time domain (b) Frequency domain

Fig. 7. Maximum correlation coefficient for the correct key, 1-bit model, Hamming distance $R_0 \rightarrow R_1$ for all S-Boxes

For the sake of optical clarity, the maximum correlation for *wrong* key candidates has been omitted in the above figures. Yet, we performed the actual key-recovery computing these correlations as well and verified that in all cases, the correlation for the wrong candidates is below $4/\sqrt{\text{No. of traces}}$, i.e., there are no "ghost peaks" that might interfere with the retreival of the correct key. Besides, the results are not limited to the first round of the first DES: the analysis equivalently works for other rounds of the first DES (to recover the remaining eight bit of $k_{C,1}$) and for the second DES iteration[6] (to obtain $k_{C,2}$). In summary, as a result of this section, we conclude that the extraction of the complete secret 3DES key from a Mifare DESFire MF3ICD40 can be carried out with approx. 250,000 traces, which can be collected in approx. seven hours using our current measurement setup.

[6] In this case, alignment to the start pattern of this operation is necessary.

6 Practical Attack: Template Attack on the Key Transfer

As observed during the profiling phase described in Sect. 4, the internal databus of the DUT seems to be completely unprotected and exhibits a far stronger Hamming weight leakage than the cryptographic engine analyzed in Sect. 5. Thus, *template attacks* to obtain information on internal values transfered over this bus can be expected to work with a far lower number of traces compared to a CPA. Of special interest is the initialization of the cryptographic engine before the start of the actual 3DES operation: our analyses shows that the transfer of the *secret key* can be identified in the power trace of the DUT after the reader has sent the initial `begin` command in the authentication protocol (that is, during Step 1 in Fig. 2). Fig. 8a depicts a trace for the loading of the key and indicates

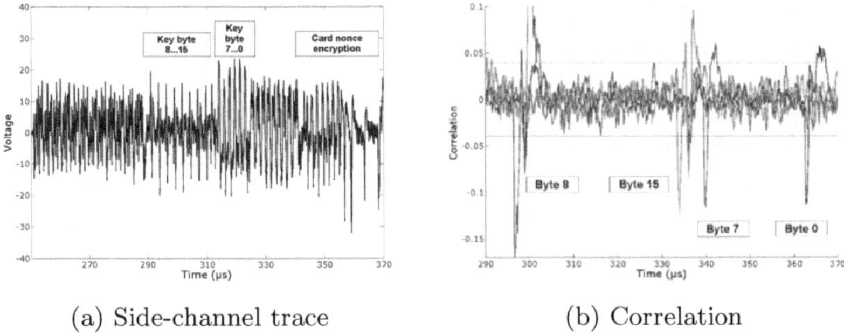

(a) Side-channel trace (b) Correlation

Fig. 8. Transfer of the 3DES key over the internal databus

the internal order of operation: by repeatedly changing the key and performing a CPA using the Hamming weight of each key byte, we found out that the 3DES key is initialized in two steps. First, the upper eight byte ($k_{C,2}$) are transfered, starting with the least significant byte. After that, the lower half $k_{C,1}$ (i.e., byte 0 ... 7) is transmitted, this time in reverse byte order. In both cases, the (redundant) parity bits are not removed prior to the key transfer, suggesting that they are discarded internally by the cryptographic engine. Fig. 8b exemplarily shows the corresponding correlation peaks for the key bytes 0 (blue), 7 (green), 8 (red) and 15 (cyan), allowing to exactly pinpoint the time instants at which information on a specific byte is leaking.

In contrast to CPA, template attacks require a profiling phase, i.e., a step during which the DUT is under full control of the adversary to estimate the statistical relation between the observable random variables — in our case the respective points in time of a trace — and the internal states that are to be distinguished (here, the value of a key byte). The resulting *training set* is then used to recover the desired values from a *test set*, i.e., traces for which the value of the key byte is considered unknown.

To systematically evaluate the success rate of template attacks for the case of the transfer of the key on the Mifare DESFire MF3ICD40, we obtain 8,000 traces for each possible value of a targeted key byte[7]. Here, we only address byte 0 and 15, however, our results hold for all other bytes as well. 4,000 traces are used for the training set, while the other 4,000 form the test set — in total, to cover all 256 possible values for a byte, we acquired $2 \cdot 256 \cdot 4{,}000 = 2{,}048{,}000$ traces. Again, we also compare the quality of analog demodulation compared to its digital equivalent and hence recorded traces both before and after the analog circuitry. Let $\mathcal{S}_b^{training} = \{\boldsymbol{t}_{b,0}, \ldots, \boldsymbol{t}_{b,3999}\}$ be the training set and $\mathcal{S}_b^{test} = \{\boldsymbol{t}_{b,4000}, \ldots, \boldsymbol{t}_{b,7999}\}$ the test set , where $\boldsymbol{t}_{b,n}$ denotes the n'th trace for a specific byte $0 \leq b < 256$, i.e., a $K \times 1$ vector of measured values. Given $\mathcal{S}^{test}$ for a fixed but unknown key — in our case, the test set for some key byte value b — the comparision to the training data is carried out as outlined in Alg. 1.

Algorithm 1. Template creation and matching procedure

> **for** $b = 0 \ldots 255$ **do**
> $(\boldsymbol{\mu}_b, \Sigma_b) \leftarrow \text{estimate}\left(\mathcal{S}_b^{training}\right)$
> **end for**
> $\overline{\Sigma} \leftarrow \frac{1}{256}\sum_{b=0}^{255} \Sigma_b$
> $(\boldsymbol{\mu}', \Sigma') \leftarrow \text{estimate}\left(\mathcal{S}^{test}\right)$
> **for** $b = 0 \ldots 255$ **do**
> $\delta_b \leftarrow \text{distance}\left(\boldsymbol{\mu}_b, \Sigma_b, \overline{\Sigma}, \boldsymbol{\mu}', \Sigma'\right)$
> **end for**
> **return** $\underset{b}{\operatorname{argmin}}\, \delta_b$

estimate $(\cdot)$ is an algorithm that estimates the (pointwise) sample mean and covariance matrix from the respective set of traces, e.g., using the standard empirical formulae [41]. distance $(\cdot)$ is a suitable distance measure based on the previously estimated statistical parameters. The value for the key byte b that minimizes the chosen distance measure is then returned as the most probable candidate for the given test traces. We exemplarily selected the following distance measures:

Difference of means. The simplest case only evaluates the norm of the pointwise difference of the class means, i.e., $\sum_{k=1}^{K} \left(\boldsymbol{\mu}_b(k) - \boldsymbol{\mu}'(k)\right)^2$, discarding any information on the (co-)variances

Euclidean. Assuming that the covariance matrix is diagonal, one obtains the Euclidean distance, $\sum_{k=1}^{K} {\left(\boldsymbol{\mu}_b(k) - \boldsymbol{\mu}'(k)\right)^2}/{\Sigma_b(k,k)}$, for which the differences are normalized using the pointwise variance

Mahalanobis. Taking all parameters of the distribution into account, the Mahalanobis distance [22] is given as $\left(\boldsymbol{\mu}_b(k) - \boldsymbol{\mu}'(k)\right)^T \overline{\Sigma^{-1}} \left(\boldsymbol{\mu}_b(k) - \boldsymbol{\mu}'(k)\right)$

Table 1 summarizes the results of our template analysis both with (Table 1a) and without analog preprocessing (Table 1b). The average bit error rates were

[7] The training and test sets were acquired in separate measurement campaigns to rule out effects due to slightly varying environmental conditions.

Table 1. Average bit error rates for the key recovery based on templates using 4,000 traces

(a) With analog processing

Keybyte	Distance	Bit error rate
0 ($k_{C,1}$)	DiffMeans	2.07
	Euclidean	2.14
	Mahalanobis	**1.77**
15 ($k_{C,2}$)	DiffMeans	0.55
	Euclidean	**0.51**
	Mahalanobis	0.64

(b) Without analog processing

Keybyte	Distance	Bit error rate
0 ($k_{C,1}$)	DiffMeans	2.89
	Euclidean	2.66
	Mahalanobis	**2.4**
15 ($k_{C,2}$)	DiffMeans	1.55
	Euclidean	**0.71**
	Mahalanobis	1.22

estimated by applying Alg. 1 for each byte, using the corresponding test set $\mathcal{S}_b^{test}$ and computing the Hamming distance between the detected and the actual value b. Evidently, the upper half $k_{C,2}$ can be recovered with significantly less error than $k_{C,1}$, which interestingly admits a rather different leakage characteristic. In either case, the remaining uncertainty can be accounted for using exhaustive search over the key candidates, starting with the ones having the smallest distance to the training set.

Limitations. Compared to the CPA presented in Sect. 5, the key recovery by means of templates might be carried out with far less traces and hence within a very short time[8], thus potentially posing a severe security threat in a scenario in which an adversary either has to extract many different keys (e.g., due to a key distribution mechanism) or faces a constant risk of being detected. However, due to the necessity for a profiling phase, implementing the approach in practice turns out to be highly problematic: for the results given in Table 1, we could employ the same DUT, whereas in a real-world attack, the profiling and the attack device are different. In our experiments with different cards, we observed significantly differing leakage characteristics, even if the measurement setup (i.e., the positions of the EM probe and the DUT on the antenna) was kept exactly fixed. At present, we are therefore not able to apply the profiling data to a different card, however, we are currently evaluating calibration approaches and improved classifiers (e.g., using Principal Component Analysis (PCA) [35]). We were already able to obtain correct matchings at least for a subset of all possible key values.

7 Conclusion

We show several SCA attacks to fully recover the 3DES key of the Mifare DESFire MF3ICD40, employing standard equipment in an academic measurement setup that can be built for approx. 3000 $. As we figured out the details of the

[8] In our current setup, recording 4,000 traces is a matter of minutes.

implementation of the DUT, the attacks can be realized within a few hours (e.g., to collect approx. 250,000 traces for a CPA), and hence pose a severe threat to the security of DESFire-based real-world systems.

System integrators should be aware of the new security risks that arise from the presented attacks and can no longer rely on the mathematical security of the used 3DES cipher. Hence, in order to avoid, e.g., manipulation or cloning of smartcards used in payment or access control solutions, proper actions have to be taken: on the one hand, multi-level countermeasures in the backend allow to minimize the threat even if the underlying RFID platform is insecure, cf. [32]. For long-term security and when developing new systems, we recommend to use certified smartcards, e.g., the AES-based Mifare DESFire EV1, which passed an EAL-4+ evaluation [3] and which comprises SCA countermeasures that thwart the attacks presented in this paper.

Having demonstrated the susceptibility of the DESFire MF3ICD40 towards SCA, there are several interesting directions for further research to consider: first, the SCA could be improved in order to work with a smaller number of traces, for instance, employing different alignment methods or model-independent distinguishers like Mutual Information Analyis (MIA) [11]. Apart from that, extensions of the proposed template attack may allow to reduce the error rate or to utilize the templates generated with a profiling device to recover the unknown key of another DESFire MF3ICD40 card. Also, a combination of CPA and templates could further reduce the required number of traces. Finally, the developed techniques can be applied in order to attempt attacks on different cryptographic RFIDs, possibly including (certified) high-security smartcards.

References

1. Agrawal, D., Archambeault, B., Rao, J.R., Rohatgi, P.: The EM side-channel(s). In: Kaliski Jr., B.S., Koç, Ç.K., Paar, C. (eds.) CHES 2002. LNCS, vol. 2523, pp. 29–45. Springer, Heidelberg (2003)
2. Brier, E., Clavier, C., Olivier, F.: Correlation Power Analysis with a Leakage Model. In: Joye, M., Quisquater, J.-J. (eds.) CHES 2004. LNCS, vol. 3156, pp. 16–29. Springer, Heidelberg (2004)
3. BSI – German Ministry of Security. Mifare DESFire8 MF3ICD81 Public Evaluation Documentation. Electronic resource (October 2008)
4. Carluccio, D.: Electromagnetic Side Channel Analysis for Embedded Crypto Devices. Master's thesis, Ruhr-University Bochum (2005)
5. Chari, S., Rao, J.R., Rohatgi, P.: Template attacks. In: Kaliski Jr., B.S., Koç, Ç.K., Paar, C. (eds.) CHES 2002. LNCS, vol. 2523, pp. 13–28. Springer, Heidelberg (2003)
6. Clavier, C., Coron, J.-S., Dabbous, N.: Differential Power Analysis in the Presence of Hardware Countermeasures. In: Paar, C., Koç, Ç.K. (eds.) CHES 2000. LNCS, vol. 1965, pp. 13–48. Springer, Heidelberg (2000)
7. Czech Railways. In-karta (March 2011), `http://www.inkarta.cz/`
8. Eisenbarth, T., Kasper, T., Moradi, A., Paar, C., Salmasizadeh, M., Shalmani, M.T.M.: On the Power of Power Analysis in the Real World: A Complete Break of the KeeLoq Code Hopping Scheme. In: Wagner, D. (ed.) CRYPTO 2008. LNCS, vol. 5157, pp. 203–220. Springer, Heidelberg (2008)

9. Garcia, F.D., de Koning Gans, G., Muijrers, R., van Rossum, P., Verdult, R., Schreur, R.W., Jacobs, B.: Dismantling MIFARE classic. In: Jajodia, S., Lopez, J. (eds.) ESORICS 2008. LNCS, vol. 5283, pp. 97–114. Springer, Heidelberg (2008)
10. Gebotys, C.H., Ho, S., Tiu, C.C.: EM Analysis of Rijndael and ECC on a Wireless Java-Based PDA. In: Rao, J.R., Sunar, B. (eds.) CHES 2005. LNCS, vol. 3659, pp. 250–264. Springer, Heidelberg (2005)
11. Gierlichs, B., Batina, L., Tuyls, P., Preneel, B.: Mutual Information Analysis – A Generic Side-Channel Distinguisher. In: Oswald, E., Rohatgi, P. (eds.) CHES 2008. LNCS, vol. 5154, pp. 426–442. Springer, Heidelberg (2008)
12. Hutter, M., Mangard, S., Feldhofer, M.: Power and EM Attacks on Passive 13.56 MHz RFID Devices. In: Paillier, P., Verbauwhede, I. (eds.) CHES 2007. LNCS, vol. 4727, pp. 320–333. Springer, Heidelberg (2007)
13. ISO. ISO/IEC 14443-3: Identification Cards – Contactless Integrated Circuit(s) Cards – Proximity Cards – Part 3: Initialization and Anticollision (February 2001)
14. ISO. ISO/IEC 14443-4: Identification cards – Contactless Integrated Circuit(s) Cards – Proximity Cards – Part 4: Transmission Protocol (February 2001)
15. ISO. ISO/IEC 15693-3: Identification Cards – Contactless Integrated Circuit Cards – Vicinity Cards – Part 3: Anticollision and Transmission Protocol (April 2009)
16. Kasper, T., Carluccio, D., Paar, C.: An Embedded System for Practical Security Analysis of Contactless Smartcards. In: Sauveron, D., Markantonakis, K., Bilas, A., Quisquater, J.-J. (eds.) WISTP 2007. LNCS, vol. 4462, pp. 150–160. Springer, Heidelberg (2007)
17. Kasper, T., Oswald, D., Paar, C.: EM Side-Channel Attacks on Commercial Contactless Smartcards Using Low-Cost Equipment. In: Youm, H.Y., Yung, M. (eds.) WISA 2009. LNCS, vol. 5932, pp. 79–93. Springer, Heidelberg (2009)
18. Kasper, T., Oswald, D., Paar, C.: Side-Channel Analysis of Cryptographic RFIDs with Analog Demodulation. Springer LNCS Proceedings of RFIDSec 2011, Northampton, USA (to appear)
19. Kasper, T., von Maurich, I., Oswald, D., Paar, C.: Chameleon: A versatile emulator for contactless smartcards. In: Rhee, K.-H. (ed.) ICISC 2010. LNCS, vol. 6829, pp. 189–206. Springer, Heidelberg (to appear)
20. Kocher, P.C., Jaffe, J., Jun, B.: Differential Power Analysis. In: Wiener, M. (ed.) CRYPTO 1999. LNCS, vol. 1666, pp. 388–397. Springer, Heidelberg (1999)
21. Langer EMV-Technik. Details of Near Field Probe Set RF 2. Website
22. Mahalanobis, P.C.: On the Generalised Distance in Statistics. In: Proceedings National Institute of Science, India, vol. 2, pp. 49–55 (April 1936)
23. Mangard, S., Oswald, E., Popp, T.: Power Analysis Attacks: Revealing the Secrets of Smart Cards. Springer, Heidelberg (2007)
24. NIST. FIPS 46-3 Data Encryption Standard (DES), `http://csrc.nist.gov/publications/fips/fips46-3/fips46-3.pdf`
25. Nohl, K., Evans, D., Plötz, H.: Reverse-Engineering a Cryptographic RFID Tag. In: USENIX Security Symposium, pp. 185–194. USENIX Association (2008)
26. NXP. Mifare DESFire Contactless Multi-Application IC with DES and 3DES Security MF3ICD40 (April 2004)
27. Ochs, K.: Transmission of Digital Signals. Lecture notes (2006)
28. Oswald, D., Paar, C.: Breaking Mifare DESFire MF3ICD40: Power Analysis and Templates in the Real World — Extended Version (2011), `http://www.emsec.rub.de/research/publications/`
29. Pico Technology. PicoScope 5200 USB PC Oscilloscopes (2008)

30. Plos, T., Hutter, M., Feldhofer, M.: Evaluation of Side-Channel Preprocessing Techniques on Cryptographic-Enabled HF and UHF RFID-Tag Prototypes. In: Dominikus, S. (ed.) Workshop on RFID Security 2008, pp. 114–127 (2008)
31. Rechberger, C., Oswald, E.: Practical Template Attacks. In: Lim, C.H., Yung, M. (eds.) WISA 2004. LNCS, vol. 3325, pp. 443–457. Springer, Heidelberg (2005)
32. Rohr, A., Nohl, K., Plötz, H.: Establishing Security Best Practices in Access Control (September 2010), `http://www.srlabs.de/pub/acs`
33. Schindler, W., Lemke, K., Paar, C.: A Stochastic Model for Differential Side Channel Cryptanalysis. In: Rao, J.R., Sunar, B. (eds.) CHES 2005. LNCS, vol. 3659, pp. 30–46. Springer, Heidelberg (2005)
34. Schwartz, M., Bennett, W.R., Stein, S.: Communication Systems and Techniques. Wiley, Chichester (1966)
35. Standaert, F.-X., Archambeau, C.: Using Subspace-Based Template Attacks to Compare and Combine Power and Electromagnetic Information Leakages. In: Oswald, E., Rohatgi, P. (eds.) CHES 2008. LNCS, vol. 5154, pp. 411–425. Springer, Heidelberg (2008)
36. State Government Victoria. myki (March 2011), `http://www.myki.com.au/`
37. van Woudenberg, J.G.J., Witteman, M.F., Bakker, B.: Improving Differential Power Analysis by Elastic Alignment. In: Kiayias, A. (ed.) CT-RSA 2011. LNCS, vol. 6558, pp. 104–119. Springer, Heidelberg (2011)
38. Vishay Semiconductors, Inc. BAT43 Schottky Diode Datasheet
39. Wikipedia. Contactless Smart Card — Wikipedia, The Free Encyclopedia (2011) (accessed March 5, 2011)
40. Wikipedia. MIFARE — Wikipedia, The Free Encyclopedia (2011) (accessed March 25, 2011)
41. Wikipedia. Sample Mean and Sample Covariance — Wikipedia, The Free Encyclopedia (2011) (accessed April 1, 2011)

Information Theoretic and Security Analysis of a 65-Nanometer DDSLL AES S-Box

Mathieu Renauld*, Dina Kamel,
François-Xavier Standaert**, and Denis Flandre

UCL Crypto Group, Université catholique de Louvain
Place du Levant 3, B-1348, Louvain-la-Neuve, Belgium

Abstract. In a recent work from Eurocrypt 2011, Renauld et al. discussed the impact of the increased variability in nanoscale CMOS devices on their evaluation against side-channel attacks. In this paper, we complement this work by analyzing an implementation of the AES S-box, in the DDSLL dual-rail logic style, using the same 65-nanometer technology. For this purpose, we first compare the performance results of the static CMOS and dual-rail S-boxes. We show that full custom design allows to nicely mitigate the performance drawbacks that are usually reported for dual-rail circuits. Next, we evaluate the side-channel leakages of these S-boxes, using both simulations and actual measurements. We take advantage of state-of-the-art evaluation tools, and discuss the quantity and nature (e.g. linearity) of the physical information they provide. Our results show that the security improvement of the DDSLL S-box is typically in the range of one order of magnitude (in terms of "number of traces to recover the key"). They also confirm the importance of a profiled information theoretic analysis for the worst-case security evaluation of leaking devices. They finally raise the important question whether dual-rail logic styles remain a promising approach for reducing the side-channel information leakages in front of technology scaling, as hardware constraints such as balanced routing may become increasingly challenging to fulfill, as circuit sizes tend towards the nanometer scale.

1 Introduction

Side-channel attacks are an important concern for the security of cryptographic devices. Since their apparition in the late 1990s, a significant attention has been paid to the development of various solutions to prevent them, at different abstraction levels (e.g. hardware, algorithmic, protocol). In this paper, we are concerned with technological countermeasures, usually denoted as dynamic and differential logic (DDL). DDL aims to solve the side-channel issue directly at the circuit level. For this purpose, such logic styles typically ensure that the switching activity of an implementation is independent of the data that it manipulates. However,

* PhD student supported by the Walloon region SCEPTIC project.
** Associate researcher of the Belgian Fund for Scientific Research (FNRS-F.R.S.).

B. Preneel and T. Takagi (Eds.): CHES 2011, LNCS 6917, pp. 223–239, 2011.

despite a constant switching activity, small data-dependent variations in the current traces can generally be observed, e.g. due to the unbalanced capacitances of the differential nodes and their interconnections. As a result, and similarly to other countermeasures against side-channel attacks, the design of DDL is mainly a tradeoff between performance and security. That is, logic styles such as SABL [26], WDDL [27], DyCML [1,13], MCML [4] or MDPL [19] (to name a few), have been introduced as different attempts to best reach the security objectives of DDL while limiting their performance overheads.

Performance evaluation of integrated circuits is a relatively well understood topic. Figures of merit such as the area, the delay, the area-delay product or the power consumption of an implementation can be used, depending on the applications. Additionally, one also cares about the design facilities. In this respect, circuits that can be taped out from standard cell libraries (such as WDDL, MCML, MDPL) offer a flexibility advantage compared to full custom logic styles (such as SABL, DyCML). This flexibility naturally comes with a cost, as using full custom logic offers more freedom for the designer to limit the information leakages.

By contrast, evaluating the resistance against side-channel attacks is more challenging, and many different tools have been introduced for this purpose in the literature. For example, early works on DDL used specific criteria such as the normalized energy deviation (NED) [13,26,27]. Some other logic styles have been analyzed using dedicated attacks, e.g. based on the correlation coefficient [19,21]. The main issue with such evaluation tools is that (at least in theory), they can possibly lead to a false sense of security, because they do not consider a worst-case scenario. The evaluation framework proposed at Eurocrypt 2009 was consequently introduced in order to relax this limitation [25]. It suggests to evaluate side-channel attacks in two steps. First, a (profiled) information theoretic analysis is performed, in order to quantify the physical leakages, independently of the adversary who exploits them. Second, a security analysis is performed, in order to measure the effectiveness of various (e.g. non-profiled) distinguishers. When designing countermeasures or logic styles, it is the information theoretic analysis that is most revealing, as it provides a (more) objective measure of their quality[1]. Unfortunately, while such an analysis has already been successfully applied in the case of different software implementations, no results have been published in the case of DDL, except evaluations based on simulated experiments, of which the practical relevance was left as an open problem [14].

Besides the development and evaluation of new countermeasures, the scaling of microelectronic technologies towards the nanometer scale also has a significant impact on the security of cryptographic devices. For example, it was recently observed that process variations imply changes in the typical models (based on the Hamming weight/distance) used in non-profiled side-channel attacks [12,22]. As a result, distinguishers performing "on-the-fly" estimations of the leakage models

[1] Naturally, the objectiveness of this evaluation still depends on the accurate estimation of the leakage probability density functions. The expectation in [25] is that it is significantly simplified by using profiled attacks rather than non-profiled ones. Yet, as will be discussed Section 4.3, our analysis still relies on certain assumptions.

have gained a particular interest. As discussed in [5], Schindler et al.'s stochastic approach [23] is a very convenient tool for this purpose, when the actual leakage function can be approximated by a linear function of the target bits in the attack. By contrast, as discussed in [28], none of the non-profiled distinguishers used so far, including the ones based on Mutual Information Analysis [6], can perform successful key recoveries when this function becomes highly non-linear. Hence, it is an interesting problem to determine whether such highly non-linear leakage functions can be observed in practice, e.g. for dual-rail logic styles.

The present paper brings two contributions related to this state-of-the-art:

First, we report and analyze the performances of a full custom designed AES S-box, implemented in a 65-nanometer Dynamic and Differential Swing Limited Logic (DDSLL) [8]. We compare it with a static CMOS S-box (full custom designed as well) in the same 65-nanometer technology. In both cases, our evaluations are based on simulations *and* measurements of a test chip (which allows us to discuss the relevance of simulations). These experiments put forward an interesting tradeoff, as the DDSLL S-box shows similar area cost as the CMOS one (contrary to most previous similar logic styles), and reduced power consumption. We explain its limited power consumption by a limited swing and its limited area by the exploitation of trees in the DDL. Interestingly, and despite this clear focus on performance, we also show that the resulting implementation has reduced information leakages compared to the CMOS one.

Second, and for the first time, we apply the information theoretic analysis of [25] to a DDL circuit. As the estimation of its leakage distributions turns out to be relatively simple (as the logic style is not masked), this allows us to provide a fair evaluation of its information leakage reduction. In addition to a fully profiled analysis using templates [3], we also pay attention to the linear nature of the leakages. We confirm that the stochastic approach from [23] is a very interesting tool for analyzing this linearity. We also suggest to use it as an informal criteria, reflecting the easiness of performing a successful non-profiled attack. These results allow us to put forward a variety of leakage samples where, depending on the cases: (1) a simple model based on the Hamming weight allows efficient key recoveries, (2) only the on-the-fly stochastic approach, or single-bit DPA attacks using well-chosen bits, are successful, (3) none of the non-profiled attacks attempted was successful (under our measurement constraints). They confirm the importance of profiled information theoretic evaluations if all the available information is to be exploited, in a worst-case security evaluation.

Finally, our results raise important open questions related to the impact of technology scaling on DDL. Informally speaking, the expectation for such logic styles is that they allow reducing the information leakage and to make basic assumptions (such as Hamming weight/distance models) invalid. In this respect, the prevailing intuition for advanced technologies may suggest that this clear advantage over CMOS vanishes as the technologies are shrinking, for two main concurrent reasons. On the one hand, CMOS circuits become harder to attack with side-channel analysis, as discussed in [22]. On the other hand, the constraints of DDL (e.g. the need of properly balanced capacitances) could become more difficult to fulfill in advanced technologies, because of variability. Nevertheless, our results highlight that well-designed DDL could remain an interesting alternative to CMOS for securing cryptographic hardware, in order to both reduce the information leakage and to increase its non-linearity.

Note that, because of place constraints, the variability issues are not discussed in this paper. However, we mention that the observations made for the static CMOS S-box in [22] essentially hold for the DDSLL one as well.

2 Previous Works

This section briefly surveys results related to DDL and side-channel attacks.

The first logic style purposed to prevent side-channel attacks was SABL. It is a full custom logic style, in which the problem of unbalanced intrinsic differential output capacitances is addressed by adding a transistor to each gate, in order to discharge all internal nodes independent of the data. Compared to CMOS, the area of a Kasumi S-box SABL implementation is increased by a factor of 1.8, and its energy per cycle by a factor of 2, in a 0.18μm 1.8V technology in [26].

WDDL was introduced shortly after SABL and aims to emulate the behavior of SABL gates using static CMOS standard cells. An implementation of a WDDL AES coprocessor, in a 0.18μm 1.8V CMOS technology, was proposed in [9]. It costs a 3 times increase in area, a 3.8 times decrease in throughput and a 3.7 times increase of the power consumption at 50 MHz, compared to its static CMOS counterpart. WDDL is also expected to provide less security margins, as it inherits from certain weaknesses of the CMOS library it is based on [14].

DyCML is a full custom, low-swing and self-timed current mode logic. It was introduced independently of power analysis concerns and constitutes an interesting alternative to SABL. SPICE simulation results using a 0.13μm 1.2V CMOS partially depleted SOI technology suggest that DyCML and SABL have similar NED, while the first one shows slightly better performance (e.g. a reduced power consumption) [13]. By combining complex functions into a differential pull-down network (DPDN), such logic styles can also implement cryptographic functionalities with limited circuit size. However, the design of these DPDN may contain unbalanced intrinsic capacitances, hence causing increased information leakage.

MCML is a CMOS current mode logic. It can be seen as a standard cell counterpart to DyCML and has been the focus of significant attention with respect to side-channel attacks [21]. As for WDDL, its main limitation is a significantly increased power consumption and an area increase by a factor around 2.

Finally, MDPL is masked and dual-rail logic style. It was introduced in order to get rid of the need of balanced capacitances in DDL. Experiments performed on a prototype chip showed that this logic style is affected by an "early propagation" effect [18]. It constitutes a good illustration of the difficulty to predict all types of leaking events that can occur in electronic circuits.

Summarizing, and as already mentioned, these previous works propose different tradeoffs between security and efficiency, and generally show large overheads when compared to CMOS. As a result, the next section first tackles the question whether it is possible to design a (full custom) DDL for which the performances better compare to CMOS. For this purpose, we investigate the implementation of a DDSLL AES S-box, that combines a reduced swing (hence, power consumption) and the combination of complex (here, 4-bit) functionalities into DPDN.

3 Performance Analysis

The DDSLL logic proposed in [8] is a low-power, dynamic & differential, self-timed, low-swing logic. Figure 1 shows the basic structure of a generic DDSLL gate. It mainly consists of a DPDN to realize the function of the gate, a dynamic current source made of transistors M_1, M_2, a feedback circuit made of transistors M_3, M_5, a precharge circuit featuring transistors M_6, M_7, M_{10} and M_{11}, a latch realized by transistors M_{12}, M_{13} and finally a self-timing buffer which is a simple inverter (transistors M_{14}, M_{15}). The DDSLL logic operates as a typical DDL, with two phases: precharge and evaluation. During precharge, the clock signal Clk_i is low charging the output nodes (out and $\overline{\mathsf{out}}$) to VDD via transistors M_{10} and M_{11}. Meanwhile, node S discharges to GND as transistor M_7 turns on, which subsequently switches off transistor M_3. At the same time, node ENO charges to VDD via transistor M_6, which in turn switches on transistor M_2. However, there is no DC current path from VDD to GND, as transistor M_1 is switched off. Next, during evaluation, the clock signal Clk_i is high, turning on transistor M_1, thus creating a current path from VDD to GND, through the DPDN. Simultaneously, the transistors of the precharge circuit are turned off (M_6, M_7, M_{10} and M_{11}), allowing the DPDN to evaluate. As a result, one of the output nodes will discharge, turning on one of the feedback transistors (M_4, M_5), which in turn charges node S to VDD. Hence, transistor M_3 turns on and starts discharging node ENO to GND. This will cause the dynamic current source to cut off the current supply of the DPDN, thereby limiting the voltage swing of the output node. Also, as node ENO discharges to GND, the output clock signal Clk_{i+1} charges to VDD, via transistor M_{14} of the self-timing buffer circuit, indicating the termination of the evaluation phase of the current block.

The S-box we considered in this work is taken from Mentens et al. [17]. The resulting DDSLL architecture is designed in such a way that complex functions, like the inversion in $GF(2^4)$, can be implemented with 4 DPDN, corresponding to the 4 output bits of this inversion. Such a design choice has clear advantages in terms of area cost, but potentially implies more side-channel leakage. In order to limit this leakage, we applied the methodology described in [15] for the implementation of the DPDN. It essentially exploits binary decision diagrams for choosing the representation of the DPDN that minimizes the power dependencies caused by variations of the number of internal capacitances that are charged/discharged in each cycle. The logic style additionally allows resources sharing (the dynamic current source, parts of the precharge circuit, the feedback circuit and the self-timing buffer of functions that evaluate at the same time). As illustrated in Figure 1, the internal clock of the DDSLL logic is driven from gate to gate, without any buffer added, and the block responsible of this derivation is

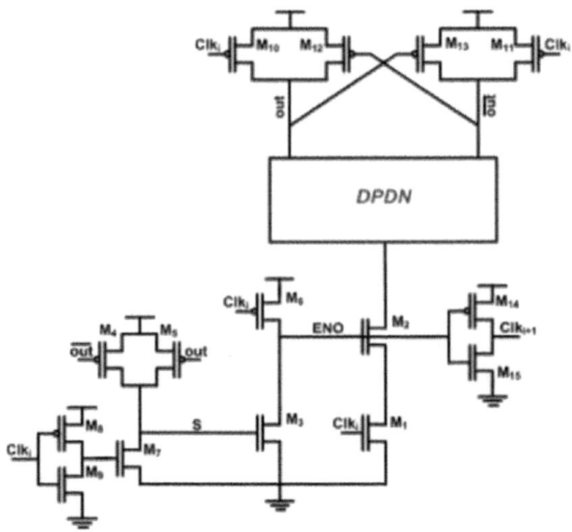

Fig. 1. Schematic of a generic DDSLL gate

considered as part of the logic in our different (simulated and measured) experiments. Hence, all the VDD nodes in the figure are included in these experiments. Note that the balanced routing, that is important for dual-rail logic styles to reduce side-channel leakages, has been hand-made as part of the full-custom S-box design. We checked the capacitors of the differential routes after extraction from the layout and, in the worst cases, found differences of 0.8 fF, corresponding to roughly 10% of imbalance. Eventually, the complete S-box used in the following accounts for 1275 transistors, with a logic depth of 13 gates.

Our test chip includes two versions of the AES S-box: the DDSLL one and a static CMOS one, used for reference and based on the design described in [22]. The chip was fabricated using a low-power 65-nanometer technology. All measurements were done at ambient temperature and using a nominal supply voltage of 1.2 V. None of the S-boxes uses flip flops and both are fed with buffered inputs. The DDSLL S-box additionally has a buffered clock. Each S-box is furnished by its own power supply, which is different than the input buffer power supply. Both S-boxes are full custom designed, with a target of minimizing the area. The measurement setup is standard and monitored the voltage variations over a resistor included in the S-box supply circuit, using a differential probe.

The performances of these S-boxes, obtained from actual measurements of the prototype chip, are summarized in Table 1. One can notice that the area of the DDSLL S-box is only 1.125 times that of the static CMOS S-box. This can be explained by two factors. First, the use of DPDN to implement complex (e.g. 4-bit) functions allows reduced sizes compared to the gate level approaches used in previous SABL or WDDL designs. Next, the logic style allows sharing the dynamic current sources and self-timing buffers of functions evaluating at the same time. In addition, the average power consumption of the DDSLL S-box at 100 kHz is 36 % less than that of the static CMOS. This results from the

Table 1. Comparison between the DDSLL S-box and the static CMOS S-box

S-box:	Static CMOS	DDSLL
Area	1000 μm^2	1125 μm^2
Avg. power @ 100kHz	128 nW	82 nW
Delay	3 ns	8 ns

low-swing logic (combined with the previously mentioned low area). These interesting figures come at the cost of a 2.6× increase in delay, which can be tolerated for low-cost applications (e.g. running at 100 KHz is reasonable for RFID).

4 Side-Channel Attacks

In this section, we aim to compare the DDSLL logic style with static CMOS implementations, from a side-channel attacks point of view. For this purpose, we will analyze both simulated traces and actual measurements, obtained from the test chip described in the previous section. Following [25], we will first consider a (worst-case) information theoretic analysis and next perform a security analysis, in order to evaluate how efficiently a non-profiled distinguisher can take advantage of the information leakages. We start by introducting our notations, metrics and tools, together with an informal investigation of our leakage traces.

4.1 Notations, Metrics and Tools

Notations. We use capital letters for random variables, lower cases for samples and sans serif fonts for functions. Let a power trace l be the output of a leakage function L. In our experiments, the leakage function output corresponds to the power consumption of the AES S-boxes we investigate and essentially depends on two input arguments: x and n. These sample values correspond to the discrete random variable X, representing the S-box input, and the continuous random variable N, representing the measurement noise. As a result, the continuous random variable corresponding to the leakage is noted $L(.,.)$, with the parameters written as capital letters if they are variable or as lower cases if they are fixed. For instance, $L(X,N)$ is the variable corresponding to a random input X with a random noise vector N, and $l(x,n)$ is a specific leakage for a fixed input x. Finally, the leakage variable at a specific time sample t is noted L_t. Our experiments considered three types of traces. First, simulated traces obtained from ELDO, to which we added a Gaussian noise, are denoted as $L^1(X,N) = L^{\text{simu}}(X) + N$. Second, real measurements are denoted as $L^2(X,N) = L^{\text{meas}}(X,N)$. Finally, we also considered a hybrid context, in which ELDO simulations are replaced by the average measurement traces $\overline{L^{\text{meas}}}(X) = \hat{\mathbf{E}}[L^{\text{meas}}(X,N)]$, with $\hat{\mathbf{E}}$ the sample mean operator. This final context is denoted as $L^3(X,N) = \overline{L^{\text{meas}}}(X) + N$.

Information Theoretic Metric. In order to evaluate the leakage of the CMOS and DDSLL S-boxes, we start by estimating the perceived information:

$$\mathrm{PI}(X;L) = \mathrm{H}[X] - \sum_{x\in\mathcal{X}} \Pr[x] \sum_{l\in\mathcal{L}} \hat{\Pr}_{\texttt{chip}}[l|x] \log_2 \hat{\Pr}_{\texttt{model}}[x|l]. \quad (1)$$

This metric essentially captures how accurately the leakage model used by an adversary (denoted as $\hat{\Pr}_{\texttt{model}}[x|l]$) can predict the actual leakage distribution of a target chip (denoted as $\hat{\Pr}_{\texttt{chip}}[l|x]$). The perceived information has been introduced in [22] in order to capture the fact that in certain contexts (e.g. when inter chip variability is significant), the adversary's model and actual chip's leakage distribution can strongly differ, possibly resulting in a negative perceived information. If the adversary's model is perfect (i.e. exactly corresponds to the chip's one), then the perceived information is equal to the mutual information metric from [25] and accurately captures the worst-case information leakage.

Tools. As a matter of fact, estimating the perceived information essentially requires to perform a good estimation of the leakage distributions. The better this estimation, the more accurate the evaluations. For this purpose, the following section will consider two types of estimation tools: the (Gaussian) template attacks introduced in [3] and the stochastic approach proposed in [23]. The template attacks are useful to estimate the worst-case scenario, with the most powerful adversary in the information theoretic sense. The stochastic approach with a linear model allows to evaluate the linearity of the leakage function.

Template attacks work as follows. First, during a profiling phase, the adversary builds 256 templates, corresponding to the 256 possible input values of the AES S-box. Each of these templates is a Gaussian distribution $\mathcal{N}(l|\hat{\mu}_{x,N}, \hat{\sigma}^2_{x,N})$, defined by two parameters: a sample mean $\hat{\mu}_{x,N}$ and a sample variance $\hat{\sigma}^2_{x,N}$. Profiling just means that the adversary (or evaluator) estimates these parameters for each S-box input[2]. Next, during the online phase, these templates are used to select the candidate input that has maximum likelihood:

$$\tilde{x} = \underset{x^*}{\operatorname{argmax}} \hat{\Pr}_{\texttt{model}}[x^*|l]. \quad (2)$$

The stochastic approach works in a slightly different fashion than template attacks. During profiling, the adversary chooses a basis $[g_0(x), g_1(x), ..., g_N(x)]$. This basis is usually made of monomials in the input and/or output bits of the target operation in the attack (e.g. the S-box in our case). Then, he performs a regression in order to find the model $\hat{\mathsf{L}}_t = \sum_i \beta_{i,t} \cdot g_i(x)$ that best matches the measured leakages. The output of this model can be used as a replacement of the sample means $\hat{\mu}_{x,N}$ in template attacks. Intuitively, the stochastic approach offers a tradeoff between the precision and robustness of the model. For example, a linear model obtained from a 9-element basis (corresponding to the S-box output bits and a constant term) can be estimated rapidly (i.e. its profiling is cheap),

[2] Template attacks can be directly generalized to a multivariate setting, by replacing the means and variances by mean vectors and covariance matrices.

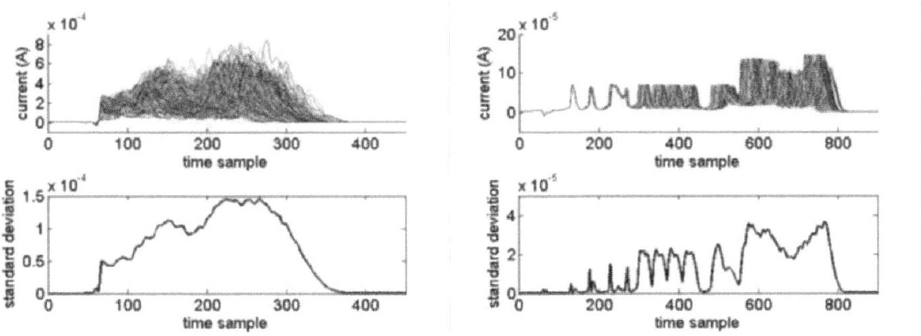

Fig. 2. Up: simulated power traces for the static CMOS (left) and DDSLL (right) S-boxes. Down: std. deviation over the inputs for different time samples.

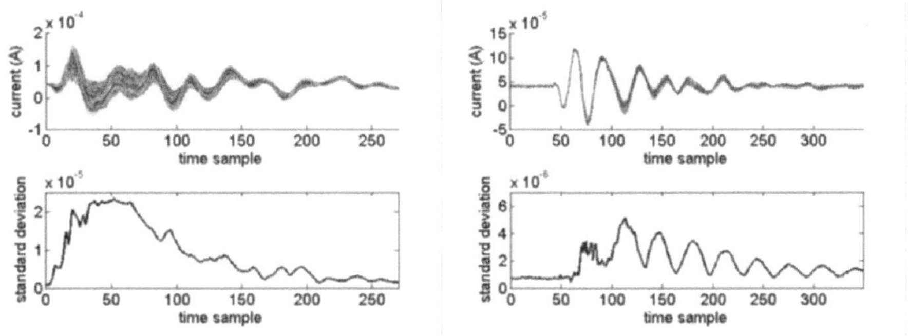

Fig. 3. Up: avg. measured power traces for the static CMOS (left) and DDSLL (right) S-boxes. Down: std. deviation over the inputs for different time samples.

but only provides a rough approximation of the leakage function if it is highly non-linear. Next, increasing the degree of the stochastic model allows refining the approximation at the cost of a more expensive profiling. Eventually, a stochastic model of degree 8 with 256 coefficients is equivalent to a template. A convenient feature of the stochastic approach is that it can be used to evaluate the non-linearity of a leakage function, by comparing the information leakage obtained from a linear basis model to the one obtained from exhaustive templates.

4.2 Leakage Traces

As a first (informal) step in our analysis, we observe the power traces in Figures 2 and 3. For the CMOS S-box, they represent the 256 transitions between 0 and an arbitrary S-box input. For the DDSLL logic style, they correspond to the evaluation phase of the S-box computation, for the same 256 possible inputs. The figures also display the standard deviations computed over the different inputs. Note that, for readability purposes, the scaling of the Y-axis is not the same in the different figures (i.e. we zoomed on the relevant parts). One can observe that the simulated and measured traces have significantly different shapes, in particular for the DDSLL case. For example, the simulated DDSLL traces are perfectly aligned at the beginning of the clock cycle and then begin to misalign. By contrast, this misalignment is much smaller in actual measurements, where the variance between the curves rather comes from an amplitude difference. The most likely reason to explain these differences is that our simulation environment does not include the specificities of the measurement setup and the filtering it implies. Modeling this setup and including it in our simulations is an interesting scope for further research. However, we also note that, despite these visual differences, the standard deviations for the static CMOS and DDSLL differ by one order of magnitude, both in measurements and simulations.

4.3 Information Theoretic Analysis

The second step in our analysis is to apply the information theoretic metric to the DDSLL and the CMOS logic styles, for simulations and real measurements, using the tools from the previous section (i.e. template attacks, stochastic approach). As mentioned in the introduction (Footnote 1), this information theoretic analysis still relies on a number of hypotheses that we now detail:

- H1. The measurement noise is assumed to be normally distributed.
- H2. Only 256 transitions are considered for the CMOS S-box (among the 256^2 possible ones). This restriction was mainly motivated by practical measurement constraints and is not expected to strongly affect the comparison between the logic styles[3]. In practice, when performing attacks in Section 4.4, we considered a pre-charge to zero before each S-box computation.
- H3. We focused our analysis on the leakage samples of the evaluation phase for the DDSLL S-box, because they exhibited the largest information leakages.
- H4. All our analyzes are univariate, i.e. they consider leakage samples one by one. This choice was motivated by the goal to compare the amount and nature of the leakage function observed for the different samples.

In order to estimate the perceived information, we use Equation 1 in which the three probability distributions are computed as follows. First, $\Pr[x]$ is the prior on the input value, i.e. the probability of each input hypothesis before taking into account the side-channel information. We considered a uniform prior $\Pr[x] = 1/256$. Next, $\hat{\Pr}_{\texttt{chip}}[l|x]$ is the estimated conditional probability of observing a leakage l given an input x. We considered two cases: either real measurements with simulated noise in which case we have $\hat{\Pr}_{\texttt{chip}}[l|x]$ computed from the leakage probability density functions, or real measurements with real noise, in which case this distribution is sampled from the actual chip, i.e. we use $\sum_x \Pr[x] \sum_l \hat{\Pr}_{\texttt{chip}}[l|x] = \sum_{(x,l)=1}^{q} \frac{1}{q}$, with q the number of traces measured.

[3] The standard deviation curves of Figures 2, 3, obtained from the transitions between a fixed value x and 256 byte values, looked essentially the same for different x.

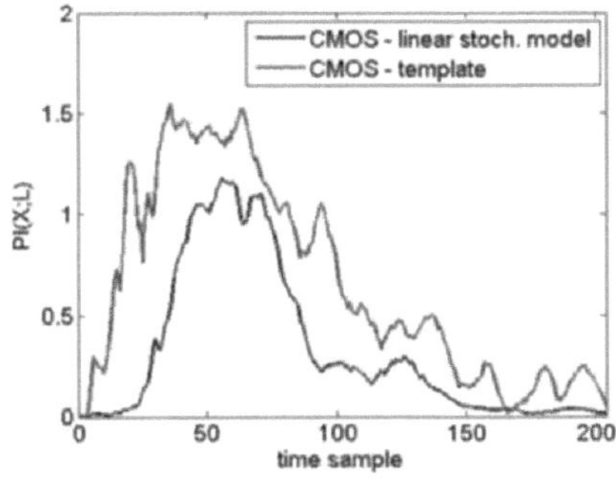

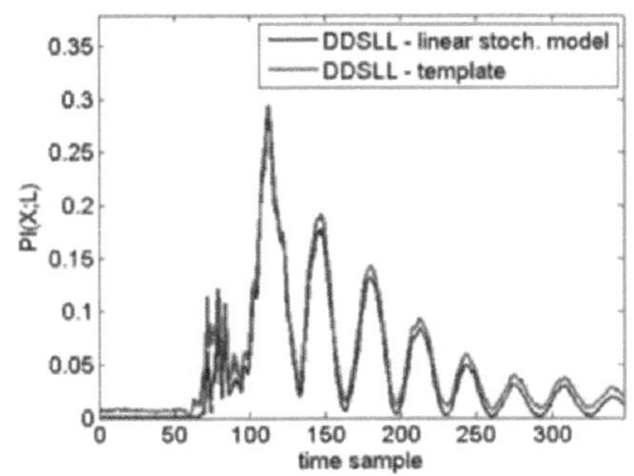

Fig. 4. Perceived information in function of the time samples in the actual measurement traces with real noise for the CMOS (left) and DDSLL (right) S-boxes

Finally, $\hat{\Pr}_{\texttt{model}}[x|l]$ is the conditional probability of the input given the leakages, derived using the adversary's model. This probability is computed from the template distributions estimated during the profiling. Both for the CMOS and the DDSLL S-box, our estimations are based on sets of 100 measurements for each of the input events x, both for the profiling and the attack phases.

We started by investigating the informativeness of the different time samples in the traces in Figure 4. Note again the different Y axes of the CMOS and DDSLL plots. As expected, one can see that the information extracted with templates is always higher than the one obtained from stochastic models using a linear basis (i.e. the S-box output bits). This is natural, as the templates capture all the information available, in an extensive manner, while the stochastic approach provides a "simpler to estimate" approximation [7,24]. Nevertheless, one can also observe that some of the samples are very accurately predicted by a linear leakage model. Interestingly, the template-based curve is also reasonably predicted by the standard deviation curves in Figure 3, confirming the analysis of standard DPA in [16]. As a result, these information curves can be directly used for selecting the points of interest for univariate attacks.

Next, we investigated the information theoretic metric for different noise levels. As an illustration, Figure 5 shows the perceived information for the two logic styles, computed from simulations (left) and actual measurements (right). In both cases, we selected the time sample that maximized the perceived information. For the real measurement curves, we used the hybrid scenario (i.e. $L^3(X,N) = \overline{L^{\text{meas}}}(X) + N$) as well as the real noise values (i.e. $L^2(X,N) = L^{\text{meas}}(X,N)$), represented by the dots on the figure). This experiment allows a number of useful observations that we now detail:

1. The Gaussian noise hypothesis is reasonably accurate, as the dots in Figure 5 are remarkably close to their corresponding (simulated) curves.
2. More importantly, the curves clearly illustrate the information leakage reduction obtained by implementing the AES S-box in the DDSLL logic style (rather than in CMOS), in front of a worst-case template adversary.

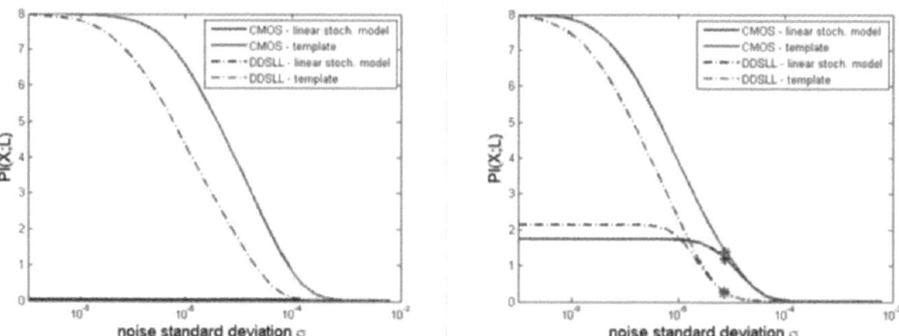

Fig. 5. Perceived information in function of the noise, for the simulations (left), and real measurements (right). The PI with real noise is marked with stars (*).

3. Both simulations and actual measurements show reduced information leakage. However, there is a noticeable reduction of the gap between logic styles when moving from simulations towards measurements. This confirms that, while ELDO (or Spice) simulations are a useful first step in order to analyze physical security issues, the quantified data that their analysis provides does not offer any formal guarantee for the security of the final test chip. This difference is essentially due to the previously mentioned difference between time shifts and amplitude shifts in the traces corresponding to different inputs.
4. In the context of simulated DDSLL traces, the linear stochastic models do not allow to extract any information (i.e. the leakage function corresponding to the selected time sample cannot be approximated accurately enough with a linear combination of the S-box output bits in this simulated case).
5. Finally, the perceived information can be higher for the DDSLL S-box than for the CMOS one, when a linear stochastic model is used in the estimations. It happens, e.g. in the right part of Figure 5, when the (simulated) noise standard deviation is extremely small. This (seemingly paradoxical) observation is explained by the impossibility to predict the actual leakages precisely using an (incomplete) linear basis. Note that this observation would vanish by extending the basis (as it is not observed with templates) and that it does not happen for the actual noise values observed in our measurements.

Combined with the performance analysis in Section 3, these results lead to contrasted conclusions. On the one hand, they confirm that DDL can be an efficient solution for improving security against side-channel attacks. Clearly, the focus of the investigated DDSLL is more on performance than on perfectly data-independent power consumption. But even in this case, experimental data shows that the security improvement over CMOS remains noticeable in a 65-nanometer technology. On the other hand, the information leakage reduction is also limited and not sufficient to provide security when used as a stand alone solution (i.e. without additional countermeasures). As a result, an interesting question is to determine whether the information leakage of the CMOS and DDSLL S-boxes

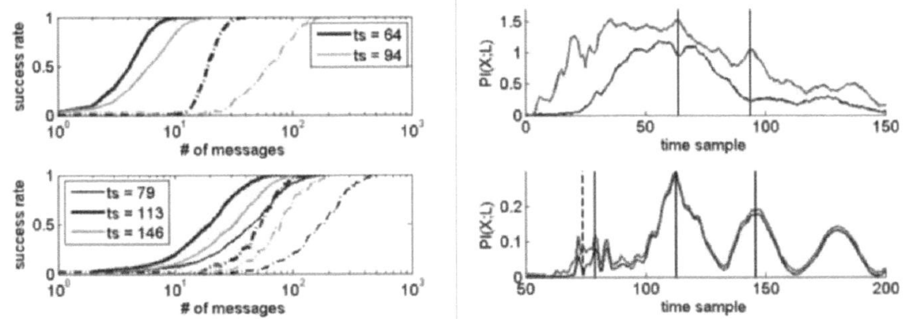

Fig. 6. Left: success rates for the template attacks (straight lines) and on-the-fly stochastic attacks (dotted lines) against the CMOS (top) and DDSLL (bottom) S-box. Right: perceived information for the selected samples (vertical lines).

are similarly easy to exploit with non-profiled side-channel attacks. We tackle this problem (i.e. the security analysis of our test chips) in the next section.

4.4 Security Analysis

The previous experiments considered the worst-case information theoretic analysis of two prototype S-boxes. In this section, we detail the second part of the evaluation framework in [25], namely the security analysis. For this purpose, we investigated the success rates of various profiled and non-profiled attacks.

As far as profiled attacks are concerned, we carried out the template attacks described in [3], as a worst-case (univariate) scenario. Next, and as a non-profiled complement, we started by applying a variant of the stochastic approach described in [11], in which the adversary builds a leakage model on-the-fly, using a linear basis. The underlying assumption of this variant is that a correct key hypothesis should give rise to the most accurate model, given that the base vectors used to build it reasonably match the actual leakages. In addition, we also performed a number of previously introduced attacks, namely the Correlation Power Analysis (CPA) using a Hamming weight leakage model introduced by Brier et al. in 2004 [2] and Kocher et al.'s single-bit Differential Power Analysis (DPA) [10]. However, we note that these attacks are redundant to some extent and that most intuitions can be extracted from the on-the-fly stochastic attacks[4]. We now detail a few meaningful observations from our experiments.

First, Figure 6 shows the success rates corresponding to the template and on-the-fly stochastic attacks, estimated using the CMOS and DDSLL

[4] Summarizing, when provided with a single-element basis that corresponds to the target bit of a single-bit DPA, or the model of a CPA, the on-the-fly stochastic approach is essentially similar to these CPA/DPA. Following a reasoning similar to to [16], one could show that minor differences in the success rates of these distinguishers are due to statistical artifacts. By contrast, when provided with a larger (e.g. 9-element linear) basis, it allows improved resistance against incorrect assumptions, at the cost of a more expensive estimation, reflected in a slightly higher data complexity.

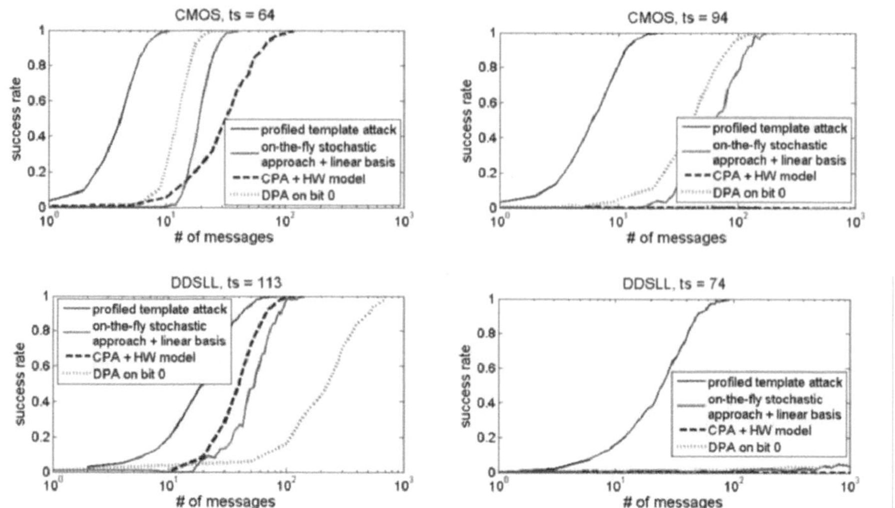

Fig. 7. Non-profiled attacks against the CMOS (up) and DDSLL (bottom) test chips, using the most informative samples (left) and less informative ones (right)

measurements, for different time samples, including the most informative ones. The template curves provide a worst-case estimate of the number of measurements needed to recover the key. They exhibit a security increase of approximately one order of magnitude, between the CMOS and DDSLL S-boxes. As expected, the number of texts required to perform a successful template attack is nicely correlated with the perceived information computed with profiled templates. Interestingly, the perceived information computed with profiled and linear stochastic models also provides an accurate prediction of the non-profiled and linear stochastic attacks for certain samples, although we have no formal guarantee in this case. These results underline that both for the CMOS and DDSLL S-boxes, there are samples in the traces that have sufficiently strong linear dependencies in the S-box output bits for being easily exploited with non-profiled attacks.

Next, the left parts of Figure 7 show the results of DPA and CPA attacks using the most informative samples in our traces. They confirm that these attacks, when based on a good assumption (e.g. single-bit DPA for the CMOS S-box, CPA with Hamming weight leakage model for the DDSLL one) can slightly outperform the on-the-fly stochastic approach (and naturally remain bounded by the worst-case template curves). Yet, as discussed in [5], the increase in data complexity of the on-the-fly stochastic approach with linear basis is very limited. Interestingly, we could verify that the S-box output bits having high weight in the stochastic models are the ones for which a single-bit DPA succeeds. In other words, all the intuition provided by a single-bit DPA could be extracted from a stochastic attack in this case. As illustrated in the upper right part of Figure 7, there also exist time samples for which a correlation attack using a Hamming weight leakage model does not succeed in recovering the key, while the non-profiled stochastic attack using a linear basis does. This confirms the previous observation in [22] that the latter distinguisher is a tool of choice for dealing with leakage in recent technologies (including possible variability issues). Finally, we could spot leakage samples, e.g. in the lower right part of Figure 7, for which only profiled side-channel attacks allow successful key recoveries (i.e. for which even the best single-bit DPA could not reach a high success rate).

Regarding the discussion about non-linear leakage functions in [28], these experiments again lead to contrasted conclusions. First, let us mention that with "non-linear" leakage samples, we simply denote the ones that could not be exploited with a linear leakage model. As a matter of fact, many leakage samples in the traces, including the most informative ones, were sufficiently linear. On the other hand, we could also spot a few pathologic samples (e.g. ts = 74 for the DDSLL S-box) for which these attacks do not succeed. As a result, and as far as the experiments in this work are concerned, one can conclude that the non-linearity issue is not a limitation for practical attacks, in which the existence of a few linear leakage samples is sufficient to perform a successful key recovery. But the existence of non-linear leakage samples also confirms that only a profiled information theoretic evaluation is theoretically able to evaluate all the information leaked by an implementation (i.e. in a worst-case scenario extended to the multivariate setting, by getting rid of H4 in Section 4.3).

5 Discussion and Open Questions

This paper brings two main contributions related to the implementation of the AES S-box in protected logic styles, using a 65-nanometer technology.

First, we show that optimizing such logic styles with performances in mind can strongly mitigate the overheads of DDL compared to CMOS. Arguably, our investigated DDSLL S-box is based on full custom design, which remains an important drawback. But it illustrates that for some primitives, it could be an acceptable solution, both in terms of area cost and power consumption.

Second, we put forward the information leakage reduction of such an S-box, when evaluated in front of various side-channel distinguishers. In the case of a worst-case template attack, it typically corresponds to one order of magnitude, in terms of number of measurements to recover the key. Hence, practically secure implementations would clearly require to combine the investigated logic style with other countermeasures, like masking. Yet, it remains that relying on DDL gives additional means for the designers to control the information leakage of an implementation. As a consequence, it is an interesting open problem to determine whether the gap between CMOS and such modified designs further reduces in smaller technologies. That is, do the traditional advantages of DDL vanish as the circuit sizes are shrinking, because of hardware constraints that become harder to fulfill? In this respect, a very interesting project would be to systematically compare technology nodes (e.g. 130nm, 90nm, 65nm, 45nm and 22nm) in terms of their respective resistance against side-channel attacks.

Eventually, one disappointing point of our DDSLL test chip is the strongly linear nature of its leakages (that goes against the expectations of ELDO/Spice simulations). Although stochastic models are very useful to discuss such questions, the precise understanding of this linearity remains difficult. Hence, an important question for further research is to determine if the use of DPDN in our designs (allowing improved efficiency) is not the main cause of this limitation. In other words, do other DDL such as SABL, WDDL or MCML provide more non-linear power consumption traces? More generally, is it possible to develop a logic style with non-linearity as a design guideline? In addition to the reduced information leakages, this would then provide a clear gain over CMOS devices, in view of the difficulty to exploit such leakages with non-profiled attacks [28].

References

1. Allam, M., Elmasry, M.: Dynamic current mode logic: a new low-power high-performance logic style. Journal of Solid State Circuits 36, 550–558 (2001)
2. Brier, E., Clavier, C., Olivier, F.: Correlation power analysis with a leakage model. In: Joye, M., Quisquater, J.-J. (eds.) CHES 2004. LNCS, vol. 3156, pp. 16–29. Springer, Heidelberg (2004)
3. Chari, S., Rao, J.R., Rohatgi, P.: Template attacks. In: Kaliski Jr., B.S., Koç, Ç.K., Paar, C. (eds.) CHES 2002. LNCS, vol. 2523, pp. 13–28. Springer, Heidelberg (2003)
4. Deniz, Z.T., Leblebici, Y.: Low-power current mode logic for improved dpa-resistance in embedded systems. In: ISCAS (2), pp. 1059–1062. IEEE, Los Alamitos (2005)
5. Doget, J., Prouff, E., Rivain, M., Standaert, F.-X.: Univariate side channel attacks and leakage modeling. Journal of Cryptographic Engineering (to appear)
6. Gierlichs, B., Batina, L., Tuyls, P., Preneel, B.: Mutual information analysis. In: Oswald, E., Rohatgi, P. (eds.) CHES 2008. LNCS, vol. 5154, pp. 426–442. Springer, Heidelberg (2008)
7. Gierlichs, B., Lemke-Rust, K., Paar, C.: Templates vs. Stochastic methods. In: Goubin, L., Matsui, M. (eds.) CHES 2006. LNCS, vol. 4249, pp. 15–29. Springer, Heidelberg (2006)
8. Hassoune, I., Macé, F., Flandre, D., Legat, J.-D.: Dynamic differential self-timed logic for robust and low-power security ics. Integration 40(3), 355–364 (2007)
9. Hwang, D.D., Tiri, K., Hodjat, A., Lai, B.-C., Yang, S., Schaumont, P., Verbauwhede, I.: Aes-based security coprocessor ic in 0.18um cmos with resistance to differential power analysis side-channel attacks. IEEE Journal of Solid-State Circuits 41(4), 781–792 (2006)
10. Kocher, P.C., Jaffe, J., Jun, B.: Differential power analysis. In: Wiener, M.J. (ed.) CRYPTO 1999. LNCS, vol. 1666, pp. 388–397. Springer, Heidelberg (1999)
11. Lemke-Rust, K.: Models and algorithms for physical cryptanalysis. PhD dissertation, University of Bochum (January 2007)
12. Lin, L., Burleson, W.P.: Analysis and mitigation of process variation impacts on power-attack tolerance. In: DAC, pp. 238–243. ACM, New York (2009)
13. Macé, F., Standaert, F.-X., Hassoune, I., Legat, J.-D.: A dynamic current mode logic to counteract power analysis attacks. In: DCIS, pp. 186–191 (2004)
14. Macé, F., Standaert, F.-X., Quisquater, J.-J.: Information theoretic evaluation of side-channel resistant logic styles. In: Paillier, P., Verbauwhede, I. (eds.) CHES 2007. LNCS, vol. 4727, pp. 427–442. Springer, Heidelberg (2007)

15. Macé, F., Standaert, F.-X., Quisquater, J.-J., Legat, J.-D.: A design methodology for secured iCs using dynamic current mode logic. In: Paliouras, V., Vounckx, J., Verkest, D. (eds.) PATMOS 2005. LNCS, vol. 3728, pp. 550–560. Springer, Heidelberg (2005)
16. Mangard, S., Oswald, E., Standaert, F.-X.: One for all - all for one: Unifying standard dpa attacks. IEEE Information Security 5(2), 100–110 (2011)
17. Mentens, N., Batina, L., Preneel, B., Verbauwhede, I.: A systematic evaluation of compact hardware implementations for the rijndael S-box. In: Menezes, A. (ed.) CT-RSA 2005. LNCS, vol. 3376, pp. 323–333. Springer, Heidelberg (2005)
18. Popp, T., Kirschbaum, M., Zefferer, T., Mangard, S.: Evaluation of the masked logic style mdpl on a prototype chip. In: CHES 2007 [19], pp. 81–94
19. Popp, T., Mangard, S.: Masked dual-rail pre-charge logic: Dpa-resistance without routing constraints. In: Rao and Sunar [20], pp. 172–186
20. Rao, J.R., Sunar, B. (eds.): CHES 2005. LNCS, vol. 3659. Springer, Heidelberg (2005)
21. Regazzoni, F., Eisenbarth, T., Poschmann, A., Großschädl, J., Gürkaynak, F.K., Macchetti, M., Deniz, Z.T., Pozzi, L., Paar, C., Leblebici, Y., Ienne, P.: Evaluating resistance of mcml technology to power analysis attacks using a simulation-based methodology. Transactions on Computational Science 4, 230–243 (2009)
22. Renauld, M., Standaert, F.-X., Veyrat-Charvillon, N., Kamel, D., Flandre, D.: A formal study of power variability issues and side-channel attacks for nanoscale devices. In: Paterson, K.G. (ed.) EUROCRYPT 2011. LNCS, vol. 6632, pp. 109–128. Springer, Heidelberg (2011)
23. Schindler, W., Lemke, K., Paar, C.: A stochastic model for differential side channel cryptanalysis. In: Rao and Sunar [20], pp. 30–46
24. Standaert, F.-X., Koeune, F., Schindler, W.: How to Compare Profiled Side-Channel Attacks? In: Abdalla, M., Pointcheval, D., Fouque, P.-A., Vergnaud, D. (eds.) ACNS 2009. LNCS, vol. 5536, pp. 485–498. Springer, Heidelberg (2009)
25. Standaert, F.-X., Malkin, T., Yung, M.: A unified framework for the analysis of side-channel key recovery attacks. In: Joux, A. (ed.) EUROCRYPT 2009. LNCS, vol. 5479, pp. 443–461. Springer, Heidelberg (2009)
26. Tiri, K., Verbauwhede, I.: A dynamic and differential cmos logic with signal indipendent power consumption to withstand differential power on smart cards. In: Proceedings of the 28th European Solid-State Circuits Conference (ESSCIRC 2002), Florence, Italy, pp. 403–406 (September 2002)
27. Tiri, K., Verbauwhede, I.: A logic level design methodology for a secure dpa resistant asic or fpga implementation. In: DATE, pp. 246–251 (2004)
28. Veyrat-Charvillon, N., Standaert, F.-X.: Generic side-channel distinguishers: Improvements and limitations. In: Rogaway, P. (ed.) CRYPTO 2011. LNCS, vol. 6841, pp. 354–372. Springer, Heidelberg (2011)

Thwarting Higher-Order Side Channel Analysis with Additive and Multiplicative Maskings

Laurie Genelle[1], Emmanuel Prouff[1], and Michaël Quisquater[2]

[1] Oberthur Technologies
{l.genelle,e.prouff}@oberthur.com
[2] University of Versailles
michael.quisquater@prism.uvsq.fr

Abstract. Higher-order side channel attacks is a class of powerful techniques against cryptographic implementations. Their complexity grows exponentially with the order, but for small orders (*e.g.* 2 and 3) recent studies have demonstrated that they pose a serious threat in practice. In this context, it is today of great importance to design software countermeasures enabling to counteract higher-order side channel attacks for any arbitrary chosen order. At CHES 2010, Rivain and Prouff have introduced such a countermeasure for the AES. It works for any arbitrary chosen order and benefits from a formal resistance proof. Until now, it was the single one with such assets. By generalizing at any order a countermeasure introduced at ACNS 2010 by Genelle *et al.* , we propose in this paper an alternative to Rivain and Prouff's solution. The new scheme can also be proven secure at any order and has the advantage of being at least 2 times more efficient than the existing solutions for orders 2 and 3, while maintaining the RAM consumption lower than 200 bytes.

1 Introduction

In the late nineties, attacks called *Side Channel Analysis* (SCA for short) have been exhibited against cryptosystems implemented in embedded devices. Since then, they have been refined and, in particular, their initial principle has been generalized in order to exploit several leakage points simultaneously. This led to the introduction of the *higher-order SCA* concept, which exploit leakage observations resulting from the handling of several intermediate variables during the cryptosystem processing. One way to make them ineffective at *order* d is to randomize the algorithm such that the probability distribution of any vector of d observations is independent of the key. To perform this randomization, a standard technique is based on *secret sharing* [23] and is often called *masking* in the context of side channel analysis. We talk about *additive* (resp. *multiplicative*) masking when any value is expressed as a sum (resp. product) of *shares*.

As side channel attacks, masking can be characterized by the number of random shares per variable. This number is called the *masking order*. A dth-order masking can always be theoretically defeated by a $(d+1)^{\text{th}}$-order SCA, but noise effects imply that the difficulty of carrying out such an attack in practice

B. Preneel and T. Takagi (Eds.): CHES 2011, LNCS 6917, pp. 240–255, 2011.

increases exponentially with its order [3,22]. For this reason, the masking order is today a well accepted security criterion and many works have studied how to apply d^{th}-order masking to protect any kind of cryptosystem at any order d. In particular, block cipher software implementations have been a privileged target either to demonstrate the efficiency of an attack [15] or to argue on the effectiveness of a countermeasure [4,6,16,19,21]. It is actually a matter of fact that any improvement of an attack against, or a countermeasure for, a standard block cipher such as AES has an important and direct impact on the (public or military) embedded security industry.

1.1 Related Works

Protecting a block cipher software implementation by masking at any order d reveals some issues which are very close to those tackled out in the *Multi-Party Computation* or *Private Circuits* area [2,5]. The main difficulty lies in performing all the algorithm steps by manipulating the shares separately, while being able to re-build the expected result. As we will see, non-linear layers – crucial for the block cipher security – are particularly difficult to protect. Only a few proposals have been made regarding this issue in the context of embedded security. For $d = 2$, there only exist three methods that perfectly thwart 2O-SCA [19,21,22]. For $d > 2$, several methods have been proposed [21,22], but except [21] all those attempts have been shown to be flawed, which has raised the need for solutions with formal resistance proof. Solution in [21], which is dedicated to the AES, benefits from such a proof and, when applied for $d = 2$, it is much more efficient than [22] and [19]. However, the time efficiency is still low (around 200 K-cycles in a classical smart card 8-bit architecture) and, even, becomes prohibitive when $d = 3$ (greater than 300 K-cycles). Alternative solutions are therefore missing, which would have equivalent security but would be more efficient. It is all the more important that second and third order SCA have been substantially improved during the last years and have even been successfully put into practice [12,14,15,18,25].

1.2 Our Results

In this paper, we are interested in masking to the order d, block ciphers whose design involves affine operations and power functions defined on a finite field. The classic strategy is to mask the message additively and to calculate the masks propagation through the various transformations. While calculating the propagation of additive masks through affine operations is easy, this is no longer the case for the power functions. The approach proposed in [21] is to express a power function in terms of squares and multiplications. The computation of the propagation of the additive masks through a multiplication requires little memory and can be managed regardless of the order d. However, this step is very time consuming (quadratic in the order d). A natural idea to achieve better performance is to mask affine functions additively and power functions multiplicatively. In this case, the calculation of the masks spreading is fast and requires little memory. When applied at order d, the only potentially costly part lies in the conversion of additives masks into multiplicative ones (and *vice versa*) since this conversion must be done without manipulating d-tuple of shares dependent on sensitive data. This strategy has already been followed to define implementations with assumed security at order $d = 1$ [1, 10, 24]. Unfortunately, none of them was perfectly thwarting first-order SCA and, even, [1] and [24] were shown to be flawed. Finally, Genelle *et al.* have proposed in [7] a satisfactory solution with formal security proof and good performances. This is an encouraging step but the extension of [7] to any order poses several problems. Firstly, it requires to calculate a Dirac function in a secure manner w.r.t. higher-order SCA. Secondly, it implies to generalize the conversion functions that map additive maskings into multiplicative ones and conversely. In a recent work, the authors of [8] have solved the first issue efficiently. In this paper, we solve the second one and we prove the security of our proposal. Having solved the two issues related to the generalization of Genelle *et al.* 's work at any order, we are now able to design a masking scheme for any block cipher combining affine transformations and power functions. When applied to secure the software implementation of the AES at order $d = 2$ (resp. $d = 3$), we achieve a time efficiency around 70K cycles (resp. 180K cycles) at the cost of a `RAM` memory consumption lower than 200 bytes in both cases. Since this amount of `RAM` is almost always acceptable in the nowadays embedded systems, this secure AES implementation is, to the best of our knowledge, the first one that makes 2nd and 3rd order security achievable, even in very constraint contexts.

1.3 Road Map

In Sect. 2, we first introduce a few basics and notations related to the additive and multiplicative maskings in finite fields. Then, in Sect. 3 we present the core principle of our approach, we recall how the computation of a Dirac function can be secured at any order d and we present two new conversion algorithms enabling to securely convert an additive masking into a multiplicative one and conversely. Eventually, in Sect. 4 we apply our masking scheme to the AES and compare its efficiency with that of the state of the art solutions.

2 Basics and Notations

2.1 Notations

The bit-length of the elements involved in the algorithmic description of the cryptosystem will be denoted by n. By default, any variables in this paper are assumed to be in a vector space of some dimension m over GF(2^n). The field addition in GF(2^n) is denoted by $\oplus$ and the field multiplication by $\otimes$. To operate on elements of GF$(2^n)^m$, the two previous laws are extended: the addition continues to be a bitwise addition and the multiplication between two vectors in GF$(2^n)^m$ corresponds to the *componentwise product*. For two vectors $(x_1, \ldots, x_m)$ and $(y_1, \ldots, y_m)$ in GF$(2^n)^m$, the result of the latter product is a vector $(z_1, \ldots, z_m)$ in GF$(2^n)^m$ whose coordinates satisfy $z_i = x_i \otimes y_i$. The inverse of an element

$(x_1, \ldots, x_m) \in (\mathrm{GF}(2^n)^\star)^m$ for the componentwise product will be simply defined as the vector $(x_1^{-1}, \ldots, x_m^{-1})$, where for every i, x_i^{-1} is the inverse of x_i for the multiplicative law $\otimes$ of $\mathrm{GF}(2^n)^\star$. For convenience, we will keep the notations $\oplus$ and $^{-1}$ for the extensions of the field operations $\oplus$ and $^{-1}$. On the other hand and to avoid any ambiguity, we will denote by $\dot{\otimes}$ the extension of the field operations $\otimes$ into a componentwise multiplication. To differentiate vectors in $\mathrm{GF}(2^n)^m$ from elements of $\mathrm{GF}(2^n)$, we shall write the vector in bold. Namely, by convention $\mathbf{x}$ shall denote a vector in $\mathrm{GF}(2^n)^m$, whereas x shall denote an element of $\mathrm{GF}(2^n)$.

2.2 Basics on Masking

When higher-order masking is involved to secure the physical implementation of a cryptographic algorithm, every sensitive variable $\mathbf{x}$ occurring during the computation is randomly split into $d+1$ shares $\mathbf{x}_0$, ..., $\mathbf{x}_d$ in such a way that the following relation is satisfied for a group operation $\perp$:

$$\mathbf{x}_0 \perp \mathbf{x}_1^{-1} \perp \cdots \perp \mathbf{x}_d^{-1} = \mathbf{x} \;, \tag{1}$$

where, $\mathbf{x}_i{}^{-1}$ denotes the inverse of $\mathbf{x}_i$ w.r.t. to $\perp$.

Usually, the d shares $\mathbf{x}_1, \ldots, \mathbf{x}_d$ (called *the masks*) are randomly picked up and the last one $\mathbf{x}_0$ is processed such that (1) is satisfied. When d random masks are involved per sensitive variable, the masking is said to be *of order d*. The so-called *additive masking* assumes that $\perp$ is the addition $\oplus$ in $\mathrm{GF}(2^n)^m$. In this case, we have $\mathbf{x}_i^{-1} = \mathbf{x}_i$ for every i. The $(d+1)$-tuple $(\mathbf{x}_0, \ldots, \mathbf{x}_d)$ is called a $(d+1)$-*additive sharing* of $\mathbf{x}$ and the transformation $(\mathbf{x}, (\mathbf{x}_i)_{i>1}) \mapsto \mathbf{x}_0 = \mathbf{x} \oplus \mathbf{x}_1 \oplus \cdots \oplus \mathbf{x}_d$ is called d^{th}-*order additive masking*. In *multiplicative masking*, the operation $\perp$ is the componentwise product $\dot{\otimes}$ in the group $(\mathrm{GF}(2^n)^\star)^m$. The $(d+1)$-tuple $(\mathbf{x}_0, \ldots, \mathbf{x}_d)$ is called $(d+1)$-*multiplicative sharing* of $\mathbf{x}$ and the transformation $(\mathbf{x}, (\mathbf{x}_i)_{i>1}) \mapsto \mathbf{x}_0 = \mathbf{x} \dot{\otimes} \mathbf{x}_1 \dot{\otimes} \cdots \dot{\otimes} \mathbf{x}_d$ is the dth-*order multiplicative masking* of $\mathbf{x}$. Note that the multiplicative masking is only defined for vectors $\mathbf{x}$ with only non-zero coordinates. In what follows, we shall simply say masking if there is no ambiguity on the nature of the operation or if the text is applicable for the two kinds of maskings.

When d^{th}-order masking is involved to secure an implementation composed of elementary transformations in the form $\mathbf{y} \leftarrow Op(\mathbf{x})$, a so-called d^{th}-*order masking scheme* must be designed to replace them by new transformations taking at input a sharing $(\mathbf{x}_0, \ldots, \mathbf{x}_d)$ of $\mathbf{x}$ and returning a sharing $(\mathbf{y}_0, \ldots, \mathbf{y}_d)$ of $\mathbf{y}$. The dth-order security of such a design holds if and only if it can be proved that every d-tuple of manipulated intermediate results during the computation is independent of any sensitive variable of the implementation (including $\mathbf{x}$ and $\mathbf{y}$).

3 Higher-Order Masking

We formally define a block cipher as a cryptographic algorithm that transforms a plaintext block into a ciphertext block from a secret key. The transformation is done by operating several elementary operations on a so-called *internal state*, viewed as a vector in $\mathrm{GF}(2^n)^m$ and initially filled with the plaintext. In this section, we show how to secure at any order d a block cipher composed of transformations Op that are either affine or are bijective power functions defined w.r.t. to the same field operation laws $\oplus$ and $\otimes$ over $\mathrm{GF}(2^n)$. Affine transformations will be assumed to be defined over the vector space $\mathrm{GF}(2^n)^m$. Power functions will be assumed to operate on a vector in $\mathrm{GF}(2^n)^m$ coordinate by coordinate.

3.1 Core Idea

As usually done when applying masking, each calculation $\mathbf{y} \leftarrow Op(\mathbf{x})$ is replaced by a sequence of elementary calculations that securely construct a $(d+1)$-sharing $(\mathbf{y}_0, \ldots, \mathbf{y}_d)$ of $\mathbf{y}$ from the $(d+1)$-sharing $(\mathbf{x}_0, \ldots, \mathbf{x}_d)$ of $\mathbf{x}$. To define those sequences of elementary operations we use the fact that linear transformations are automorphisms of $(\mathrm{GF}(2^n)^m, \oplus)$, while bijective power functions are automorphisms of $(\mathrm{GF}(2^n)^m, \dot{\otimes})$. Hence, depending of its (affine or multiplicative) nature, we involve either an additive or a multiplicative sharing of the internal state to secure the operation Op.

Affine Transformations processing. If Op is a linear transformation defined over $\mathrm{GF}(2^n)^m$, then the sensitive variable $\mathbf{x}$ is assumed to be represented by a $(d+1)$-additive sharing $(\mathbf{x}_0, \ldots, \mathbf{x}_d)$. In this case, securing the calculation $\mathbf{y} \leftarrow Op(\mathbf{x})$ simply consists in replacing it by $d+1$ applications of Op, one for each share $\mathbf{x}_i$. After denoting by $\mathbf{y}_i$ the value $Op(\mathbf{x}_i)$, we have $\bigoplus_{j=0}^{d} \mathbf{y}_j = \mathbf{y}$. We conclude that $(\mathbf{y}_0, \ldots, \mathbf{y}_d)$ is a $(d+1)$-sharing of $\mathbf{y}$. Moreover, it is obvious that no d-tuple of intermediate data is sensitive during this processing. For affine transformations the processing is done similarly, except for d even where only the linear part of Op is applied to the last share $\mathbf{x}_d$.

Power Functions processing. If Op is a power function over $\mathrm{GF}(2^n)^m$, then the sensitive variable $\mathbf{x}$ is assumed to be non-zero and represented by a $(d+1)$-multiplicative sharing $(\mathbf{z}_0, \ldots, \mathbf{z}_d)$. In this case, $\mathbf{y} \leftarrow Op(\mathbf{x})$ is simply replaced by $d+1$ elementary calculations of Op, one on each multiplicative share $\mathbf{z}_i$. This results in $d+1$ shares $\mathbf{y}_i = Op(\mathbf{z}_i)$ that satisfy $\mathbf{y}_0 \dot{\otimes} \dot{\bigotimes}_{j=1}^{d} \mathbf{y}_j^{-1} = \mathbf{y}$ and are thus a $(d+1)$-multiplicative sharing of $\mathbf{y}$. It can be easily checked that every d-tuple of intermediate variables involved in the processing is independent of any sensitive variable, since all the $\mathbf{z}_i$ (and $\mathbf{y}_i$) are manipulated independently.

The application of the most appropriate masking for each elementary operation enables to efficiently secure each (affine or non-linear) layer of the block cipher. Nevertheless, the mix of additive and multiplicative maskings arises the two following issues:

Issue 1: the proposed power functions processing involves multiplicative sharings and the latter ones can only be defined for an element $\mathbf{x}$ in $(\mathrm{GF}(2^n)^\star)^m$, whereas the block cipher internal state is defined in $\mathrm{GF}(2^n)^m$. A dth-order secure scheme for the mapping of an element of $\mathrm{GF}(2^n)^m$ into an element of $(\mathrm{GF}(2^n)^\star)^m$ (and *vice versa*) must therefore be defined. Moreover, the mapping must be reversible at any time during the block cipher processing.

Issue 2: since affine functions and power functions are processed alternatively, special transformations must be defined to convert additive sharings into multiplicative ones and conversely. Moreover those transformations must themselves be d^{th}-order secure to not decrease the overall security of the block cipher implementation.

The first issue has been solved in [8]. We give in Sect. 3.2 the outlines of the solution that essentially relies on the secure processing of a *Dirac function*. The second issue is tackled out in Sect. 3.3, where we propose two algorithms that transform an additive masking (AM for short) into a multiplicative masking (MM for short) and conversely. All those transformations are eventually combined in Sect. 3.4 to secure a block cipher round transformation according to the following diagram.

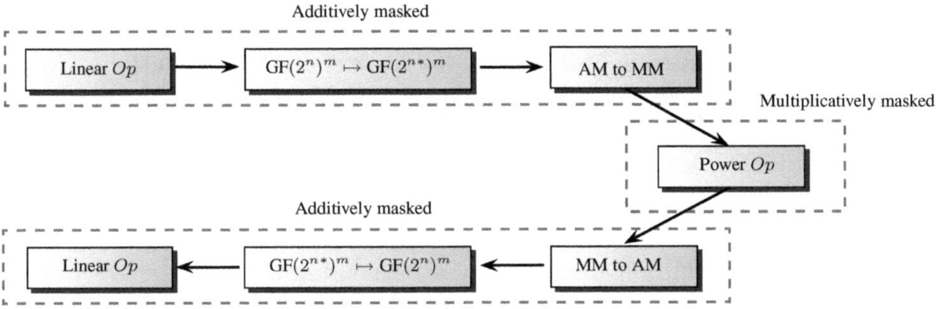

3.2 Issue 1: Mapping Elements of GF$(2^n)^m$ Into (GF$(2^n)^\star)^m$

The solution of Issue 1 proposed in [8] consists in transforming any zero value into a non-zero one, keeping track of this modification if applied. Let us denote by δ_0 the Dirac function defined in GF(2^n) by $\delta_0(x) = 1$ if $x = 0$ and $\delta_0(x) = 0$ otherwise. To map any $x \in \text{GF}(2^n)$ into GF$(2^n)^\star$, the element is simply added with its dirac value $\delta_0(x)$. After extending the Dirac function to GF$(2^n)^m$ by setting $\delta(\mathbf{x}) = (\delta(x_0), \ldots, \delta(x_{m-1}))$, we get a function $\mathbf{x} \mapsto \mathbf{x} \oplus \delta(\mathbf{x})$ mapping any element of GF$(2^n)^m$ into an element of (GF$(2^n)^\star)^m$. To secure the processing of the latter transformation against d^{th}-order SCA, the vector $\mathbf{x}$ is represented by a $(d+1)$-additive sharing $(\mathbf{x}_0, \ldots, \mathbf{x}_d)$ and a secure processing is applied to output an additive sharing $(\Delta_0, \ldots, \Delta_d)$ of $\delta(\mathbf{x})$. The details of the processing, as long a proof of its security against d^{th}-order SCA are given in [8]. We call this processing `SecDirac` in the following.

3.3 Issue 2: Conversion Functions

In this section, we show how to build d^{th}-order secure transformations passing from the $(d+1)$-additive sharing $(\mathbf{x}_0, \ldots, \mathbf{x}_d)$ of $\mathbf{x} \in (\text{GF}(2^n)^\star)^m$ to its $(d+1)$-multiplicative sharing $(\mathbf{z}_0, \ldots, \mathbf{z}_d)$ and conversely. These transformations are respectively called `AMtoMM`$(\cdot)$ and `MMtoAM`$(\cdot)$ and act as follows:

– $\texttt{AMtoMM}(\mathbf{x} \oplus \bigoplus_{i=1}^{d} \mathbf{x}_i, \mathbf{x}_1, \ldots, \mathbf{x}_d) \to (\mathbf{x} \otimes \dot{\bigotimes}_{i=1}^{d} \mathbf{z}_i, \mathbf{z}_1, \ldots, \mathbf{z}_d)$,
– $\texttt{MMtoAM}(\mathbf{x} \otimes \dot{\bigotimes}_{i=1}^{d} \mathbf{z}_i, \mathbf{z}_1, \ldots, \mathbf{z}_d) \to (\mathbf{x} \oplus \bigoplus_{i=1}^{d} \mathbf{x}_i, \mathbf{x}_1, \ldots, \mathbf{x}_d)$.

To process the AMtoMM transformation, the general strategy developed hereafter consists in converting sequentially each additive mask of $\mathbf{x}$ into a multiplicative one. To preserve the security order of the scheme at each step, an additive mask is added to the multiplicatively masked representation of $\mathbf{x}$ prior to remove one of the remaining multiplicative masks. The strategy followed for the MMtoAM is exactly the same, except that the roles of the additive and multiplicative masks are reversed.

In the hereafter detailed descriptions of the transformations we will use three ordered sets $\mathcal{S}_{MV}$, $\mathcal{S}_{AM}$ and $\mathcal{S}_{MM}$ that will be respectively dedicated to the storage of the masked value, the additive shares and the multiplicatives shares.

At the beginning of the AMtoMM processing, let us associate the $(d+1)$-additive sharing $(\mathbf{x}_0, \ldots, \mathbf{x}_d)$ of $\mathbf{x}$ with the sets $\mathcal{S}_{MV} = \{\mathbf{x}_0\}$, $\mathcal{S}_{AM} = \{\mathbf{x}_1, \ldots, \mathbf{x}_d\}$ and $\mathcal{S}_{MM} = \emptyset$. The conversion of the $(d+1)$-additive sharing to a multiplicative one $(\mathbf{z}_0, \ldots, \mathbf{z}_d)$ may be viewed as a sequence of updatings of those three sets such that, at the end, $\mathcal{S}_{MV} = \{\mathbf{z}_0\}$, $\mathcal{S}_{AM} = \emptyset$ and $\mathcal{S}_{MM} = \{\mathbf{z}_1, \ldots, \mathbf{z}_d\}$. To perform such a conversion, the following treatment is repeated for every $i \in [1; d]$:

1. Masking multiplicatively the element in $\mathcal{S}_{MV}$ and all the shares in $\mathcal{S}_{AM}$ by $\mathbf{z}_i$.
2. Inserting the multiplicative mask $\mathbf{z}_i$ at the end of $\mathcal{S}_{MM}$.
3. Removing the first element of $\mathcal{S}_{AM}$ and adding this value to the masked value in $\mathcal{S}_{MV}$.

For the MMtoAM method, the $(d+1)$-multiplicative sharing $(\mathbf{z}_0, \ldots, \mathbf{z}_d)$ of $\mathbf{x}$ is associated with the sets $\mathcal{S}_{MV} = \{\mathbf{z}_0\}$, $\mathcal{S}_{MM} = \{\mathbf{z}_1, \ldots, \mathbf{z}_d\}$ and $\mathcal{S}_{AM} = \emptyset$ and the conversion consists in repeating the following treatment for every $i \in [1; d]$:

1. Masking additively the element of $\mathcal{S}_{MV}$ with $\mathbf{x}_i$.
2. Inserting the mask $\mathbf{x}_i$ at the end of $\mathcal{S}_{AM}$.
3. Removing the first component, i.e. $\mathbf{z}_i$, of $\mathcal{S}_{MM}$ and multiplying by $\mathbf{z}_i^{-1}$ the element of $\mathcal{S}_{MV}$ and all the additive shares in $\mathcal{S}_{AM}$.

This straightforward strategy is d^{th}-order secure when $d = 1$ or $d = 2$ but not when d is higher. Indeed, it can be observed that the process of AMtoMM (resp. MMtoAM) leads to the computation of the value $\mathbf{x}_d \otimes \dot{\bigotimes}_{i=1}^{d} \mathbf{z}_i$ (resp. $\mathbf{x}_1 \otimes \dot{\bigotimes}_{i=1}^{d} \mathbf{z}_i^{-1}$). Hence, if $\mathbf{x}_d \neq 0$ (resp. $\mathbf{x}_1 \neq 0$), then the secret value $\mathbf{x}$ may be recovered from $\mathbf{x}_d$, $\mathbf{x}_d \otimes \dot{\bigotimes}_{i=1}^{d} \mathbf{z}_i$ and $\mathbf{x} \otimes \dot{\bigotimes}_{i=1}^{d} \mathbf{z}_i$ (resp. $\mathbf{x}_1$, $\mathbf{x}_1 \otimes \dot{\bigotimes}_{i=1}^{d} \mathbf{z}_i^{-1}$ and $\mathbf{x} \otimes \dot{\bigotimes}_{i=1}^{d} \mathbf{z}_i$). In both cases, this means that 3 shares are sufficient to recover $\mathbf{x}$ which implies that the straightforward schemes are never 3rd-order secure.

In order to solve this issue, we slightly modify our approach. In place of the third step in the sequence related to AMtoMM, we mask at order 1 all the shares in $\mathcal{S}_{AM}$ with new fresh random values, except for the last share which stays unchanged. We remove all those elements from $\mathcal{S}_{AM}$ and we add them to the element in $\mathcal{S}_{MV}$. Finally, we insert all the new fresh random masks into $\mathcal{S}_{AM}$. For the MMtoAM transformation, we do not replace the third step of the sequence and instead, we add a fourth step during which all the shares in $\mathcal{S}_{AM}$ are masked at order 1 with new fresh random values. We remove then all those values from $\mathcal{S}_{AM}$ and we add them to the element in $\mathcal{S}_{MV}$. Finally, we insert all the new fresh random masks into $\mathcal{S}_{AM}$.

We present in Alg. 1 the sequence of the different steps required for the conversion of an additive masking into a multiplicative one.

Algorithm 1. Secure AMtoMM($\cdot$)

INPUT(S): A $(d+1)$-additive sharing $(\mathbf{x}_0, \ldots, \mathbf{x}_d)$ of $\mathbf{x}$
OUTPUT(S): A $(d+1)$-multiplicative sharing $(\mathbf{z}_0, \ldots, \mathbf{z}_d)$ of $\mathbf{x}$

1. $\mathbf{z}_0 \leftarrow \mathbf{x}_0$
2. **for** $i = 1$ **to** d **do**
   $\mathbf{z}_i \leftarrow \textbf{rand}((\mathrm{GF}(2^n)^\star)^m)$
   $\mathbf{z}_0 \leftarrow \mathbf{z}_0 \dot{\otimes} \mathbf{z}_i$
3. **for** $j = 1$ **to** $d - i$ **do**
   $U \leftarrow \textbf{rand}(\mathrm{GF}(2^n)^m)$
   $\mathbf{x}_j \leftarrow \mathbf{z}_i \dot{\otimes} \mathbf{x}_j$
   ** *Refreshing of the additive share*
   $\mathbf{x}_j \leftarrow \mathbf{x}_j \oplus U$
   $\mathbf{z}_0 \leftarrow \mathbf{z}_0 \oplus \mathbf{x}_j$
   $\mathbf{x}_j \leftarrow U$
   $\mathbf{x}_{d-i+1} \leftarrow \mathbf{z}_i \dot{\otimes} \mathbf{x}_{d-i+1}$
   $\mathbf{z}_0 \leftarrow \mathbf{z}_0 \oplus \mathbf{x}_{d-i+1}$
4. **return** $(\mathbf{z}_0, \mathbf{z}_1, \ldots, \mathbf{z}_d)$

Alg. 2 describes the different steps to convert a multiplicative sharing into an additive one.

Remark 1. The security of AMtoMM and MMtoAM algorithms is not affected if additive masks are not refreshed during the two first steps. This optimization was not presented for the sake of clarity. Also, in our application (see Sect. 4), we will only handle the inverse of the multiplicative shares $(\mathbf{z}_1, \ldots, \mathbf{z}_d)$. Therefore, AMtoMM and MMtoAM can be input with $(\mathbf{z}_0, \mathbf{z}_1^{-1}, \ldots, \mathbf{z}_d^{-1})$ instead of $(\mathbf{z}_0, \ldots, \mathbf{z}_d)$, so that the inverse of the $\mathbf{z}_i$ does not need to be computed inside the algorithms.

The following propositions state the completeness and the security of AMtoMM and MMtoAM. There proofs are given in the extended version [9].

Proposition 1 (Completeness). *If* $(\mathbf{x}_0, \ldots, \mathbf{x}_d)$ *is a* $(d+1)$*-additive sharing of* $\mathbf{x}$*, then algorithm* AMtoMM$(\mathbf{x}_0, \ldots, \mathbf{x}_d)$ *is a* $(d+1)$*-multiplicative sharing of* $\mathbf{x}$*. If* $(\mathbf{z}_0, \ldots, \mathbf{z}_d)$ *is a* $(d+1)$*-multiplicative sharing of* $\mathbf{x}$*, then* MMtoAM$(\mathbf{z}_0, \ldots, \mathbf{z}_d)$ *is a* $(d+1)$*-additive sharing of* $\mathbf{x}$*.*

Algorithm 2. Secure MMtoAM(·)

INPUT(S): A $(d+1)$-multiplicative sharing $(\mathbf{z}_0, \ldots, \mathbf{z}_d)$ of $\mathbf{x}$
OUTPUT(S): A $(d+1)$-additive sharing $(\mathbf{x}_0, \ldots, \mathbf{x}_d)$ of $\mathbf{x}$

1. $\mathbf{x}_0 \leftarrow \mathbf{z}_0$
2. **for** $i = 1$ **to** d **do**
   $\mathbf{x}_i \leftarrow \mathbf{rand}(\mathrm{GF}(2^n)^m)$
   $\mathbf{x}_0 \leftarrow \mathbf{x}_0 \oplus \mathbf{x}_i$
   $\mathbf{x}_0 \leftarrow \mathbf{x}_0 \dot{\otimes} \mathbf{z}_i^{-1}$
3. **for** $j = 1$ **to** i **do**
   $\mathbf{x}_j \leftarrow \mathbf{x}_j \dot{\otimes} \mathbf{z}_i^{-1}$
   $U \leftarrow \mathbf{rand}(\mathrm{GF}(2^n)^m)$
   ** *Refreshing of the additive share*
   $\mathbf{x}_j \leftarrow \mathbf{x}_j \oplus U$
   $\mathbf{x}_0 \leftarrow \mathbf{x}_0 \oplus \mathbf{x}_j$
   $\mathbf{x}_j \leftarrow U$
4. **return** $(\mathbf{x}_0, \mathbf{x}_1, \ldots, \mathbf{x}_d)$

Proposition 2 (Security). AMtoMM(·) *and* MMtoAM(·) *are* d^{th}*-secure.*

3.4 Full Scheme

In this section we apply the principle presented in Sect. 3.1 and the functions introduced in Sect(s) 3.2 and 3.3 to secure the processing of a block cipher round. We assume that this round is parameterized by a secret round key $\mathbf{k} \in \mathrm{GF}(2^n)^m$ and operates a transformation of the form $\lambda' \circ \gamma \circ \lambda$ on an internal state $\mathbf{x} \in \mathrm{GF}(2^n)^m$. Functions λ and λ' are assumed to be automorphisms of $(\mathrm{GF}(2^n)^m, \oplus)$ and function γ is assumed to be an automorphism of $(\mathrm{GF}(2^n)^m, \dot{\otimes})$ (*e.g.* a transformation processing bijective power functions – not necessarily the same – to the n-bit coordinates of the input vector). In the following algorithm, we assume that the round key $\mathbf{k} \in \mathrm{GF}(2^n)^m$ and the internal state $\mathbf{x} \in \mathrm{GF}(2^n)^m$ have been previously additively shared into $(\mathbf{k}_0, \ldots, \mathbf{k}_d)$ and $(\mathbf{x}_0, \ldots, \mathbf{x}_d)$ respectively. In the right-hand column of the following algorithm description, we added an expression of the form $\cdot \leftarrow \cdot$ to explicit to which variable (on the left) relies the sharing (on the right).

The completeness of Alg. 3 is discussed in [9].

Security. The security of Alg. 3 w.r.t. d^{th}-order SCA can be deduced from the local resistance of its main steps. Steps 1, 2, 5-6, 8 and 9 operate a transformation or an operation on each share of the $(d+1)$-sharing of the internal state independently. They are therefore secure against d^{th}-order SCA. The security of SecDirac has been proved in [8] and is a direct consequence of the security proof in [11]. Eventually, transformations AMtoMM and MMtoAM have been proved to be secure against d^{th}-order SCA in [9]. We deduce that Alg. 3 thwarts d^{th}-order SCA for any d.

Algorithm 3. d^{th}-order secure processing of $\lambda' \circ \gamma \circ \lambda(\mathbf{x} \oplus \mathbf{k})$

INPUT(S): A $(d+1)$-additive sharing $(\mathbf{k}_0, \ldots, \mathbf{k}_d)$ of $\mathbf{k}$ and a $(d+1)$-additive sharing $(\mathbf{x}_0, \ldots, \mathbf{x}_d)$ of $\mathbf{x}$
OUTPUT(S): A $(d+1)$-additive sharing $(\mathbf{x}_0, \ldots, \mathbf{x}_d)$ of $\mathbf{x} = \lambda \circ \gamma \circ \lambda'(\mathbf{x} \oplus \mathbf{k})$

** *Secure processing of the round-key addition*
1. **for** $i = 0$ **to** d **do**
 $\mathbf{x}_i \leftarrow \mathbf{x}_i \oplus \mathbf{k}_i$
 $[\mathbf{x} \oplus \mathbf{k} \leftarrow (\mathbf{x}_0, \ldots, \mathbf{x}_d)]$

** *Secure processing of* λ
2. **for** $i = 0$ **to** d **do**
 $\mathbf{x}_i \leftarrow \lambda(\mathbf{x}_i)$
 $[\lambda(\mathbf{x} \oplus \mathbf{k}) \leftarrow (\mathbf{x}_0, \ldots, \mathbf{x}_d)]$

** *Secure mapping from* $\mathrm{GF}(2^n)^m$ *into* $(\mathrm{GF}(2^n)^\star)^m$
** *The* $(d+1)$*-additive sharing* $(\Delta_0, \ldots, \Delta_d)$ *of* $\delta(\mathbf{x} \oplus \mathbf{k})$ *is saved in memory*
3. $(\mathbf{x}_0, \ldots, \mathbf{x}_d) \leftarrow \texttt{SecDirac}(\mathbf{x}_0, \ldots, \mathbf{x}_d)$
 $[\lambda(\mathbf{x} \oplus \mathbf{k}) \oplus \delta(\lambda(\mathbf{x} \oplus \mathbf{k})) \leftarrow (\mathbf{x}_0, \ldots, \mathbf{x}_d)]$

** *Secure conversion of the additive masking into a multiplicative one*
4. $(\mathbf{z}_0, \ldots, \mathbf{z}_d) \leftarrow \texttt{AMtoMM}(\mathbf{x}_0, \ldots, \mathbf{x}_d)$
 $[\lambda(\mathbf{x} \oplus \mathbf{k}) \oplus \delta(\lambda(\mathbf{x} \oplus \mathbf{k})) \leftarrow (\mathbf{z}_0, \ldots, \mathbf{z}_d)]$

** *Secure processing of* γ
5. $\mathbf{z}_0 \leftarrow \gamma(\mathbf{z}_0)$
6. **for** $i = 1$ **to** d **do**
 $\mathbf{z}_i \leftarrow \gamma(\mathbf{z}_i)$
 $[\gamma \circ \lambda(\mathbf{x} \oplus \mathbf{k}) \oplus \delta(\lambda(\mathbf{x} \oplus \mathbf{k})) \leftarrow (\mathbf{z}_0, \mathbf{z}_1, \ldots, \mathbf{z}_d)]$

** *Secure conversion of the multiplicative masking into an additive one*
7. $(\mathbf{x}_0, \ldots, \mathbf{x}_d) \leftarrow \texttt{MMtoAM}(\mathbf{z}_0, \mathbf{z}_1 \ldots, \mathbf{z}_d)$
 $[\gamma \circ \lambda(\mathbf{x} \oplus \mathbf{k}) \oplus \delta(\lambda(\mathbf{x} \oplus \mathbf{k})) \leftarrow (\mathbf{x}_0, \mathbf{x}_1, \ldots, \mathbf{x}_d)]$

** *Secure mapping from* $(\mathrm{GF}(2^n)^\star)^m$ *into* $\mathrm{GF}(2^n)^m$
8. **for** $i = 0$ **to** d **do**
 $\mathbf{x}_i \leftarrow \mathbf{x}_i \oplus \Delta_i$
 $[\gamma \circ \lambda(\mathbf{x} \oplus \mathbf{k}) \leftarrow (\mathbf{x}_0, \mathbf{x}_1, \ldots, \mathbf{x}_d)]$

** *Secure processing of* λ'
9. **for** $i = 0$ **to** d **do**
 $\mathbf{x}_i \leftarrow \lambda'(\mathbf{x}_i)$
 $[\lambda' \circ \gamma \circ \lambda(\mathbf{x} \oplus \mathbf{k}) \leftarrow (\mathbf{x}_0, \mathbf{x}_1, \ldots, \mathbf{x}_d)]$

10. **return** $(\mathbf{x}_0, \ldots, \mathbf{x}_d)$

Complexity. We list in Tables 1 and 2 the complexity of Alg. 3 in terms of the following elementary operations: (n, n)-matrix transpositions $\mathbf{M}^\intercal$ required for the secure Dirac computation (see [8]), n-bit operations AND, XOR and $\otimes$ and transformations λ, γ and λ' over $\mathrm{GF}(2^n)^m$. To have interpretable results we consider separately the operations related to: (1) the secure mapping from $\mathrm{GF}(2^n)^m$ to

Table 1. Complexity of Algorithm SecDirac

Order	SecDirac		
	$M^{\intercal}$	XOR	AND
1	$2m/n$	$28m/n + 2m$	$28m/n$
2	$3m/n$	$84m/n + 3m$	$63m/n$
3	$4m/n$	$168m/n + 4m$	$112m/n$
d	$(d+1)m/n$	$(14d+n)(d+1)m/n$	$7(d+1)^2m/n$

Table 2. Complexity of the masking conversions and Algorithm 3

Order	AMtoMM		MMtoAM		Algorithm 3			
	XOR	$\otimes$	XOR	$\otimes$	λ	γ	λ'	XOR
1	m	$2m$	$3m$	$2m$	2	2	2	$4m$
2	$4m$	$5m$	$8m$	$5m$	3	3	3	$6m$
3	$9m$	$9m$	$15m$	$9m$	4	4	4	$8m$
d	md^2	$\frac{md}{2}(3+d)$	$\frac{md}{2}(2+d)$	$\frac{md}{2}(3+d)$	$d+1$	$d+1$	$d+1$	$2m(d+1)$

$(\mathrm{GF}(2^n)^\star)^m$ (*i.e.* SecDirac), (2) the conversion functions (Algorithms 1 and 2) and (3) the remaining transformations in Alg. 3 (right-hand column of Table 2).

Remark 2. For our implementations reported in Sect. 4 (in this case we had $m = 16$ and $n = 8$), we experimented that the cost of the n-bit operations XOR and AND was equal to 1 clock cycle. The cost of $\otimes$ was equal to 22 and that of $\mathbf{M}^{\intercal}$ was equal to 148. Moreover, we implemented the functions λ, λ' and γ thanks to ROM lookup tables and hence, each computation costed around m clock cycles (considering that one table access costs one clock cycle).

As it can be checked in Table 2, the complexity of the secure processing of the non-linear function γ (Steps 3 to 8 in Alg. 3) essentially corresponds to the sum of the complexities of SecDirac and Algorithms AMtoMM and MMtoAM. Neglecting the cost of the matrix transposition and assuming $m = 16$ and $n = 8$ (as it is the case for the AES), our secure processing of γ requires $74d^2+104d+30$ operations[1] XOR or AND and $16d^2 + 48d$ operations $\otimes$. For comparison, the cost of Rivain and Prouff's solution [21] to secure the AES non-linear layer (which corresponds to our transformation γ) is $128d^2 + 192d$ operations XOR and $64d^2 + 128d + 64$ operations $\otimes$, neglecting the cost of the look-up table accesses. In view of the two costs above, our solution is clearly more efficient than that in [21]. In particular, the number of operations $\otimes$ is divided by around 4, which is an important improvement considering that the latter operation is costly (around 20 times more costly than a XOR or a AND).

[1] The two operations are considered globally since they have the same cost.

4 Application to the AES

The AES-128 is a cryptosystem that iterates 10 times a same round transformation on a 16-bytes internal state initially filled with the plaintext (*i.e.* parameters n and m in Sect. 3.4 equal here 8 and 16 respectively). The round is composed of a key addition `AddRoundKey`, a nonlinear layer `SubBytes` which applies the same *substitution-box* (s-box) to every byte of the internal state and linear transformations `ShiftRows` and `MixColumns`. The s-box is defined as the left-composition of a linear transformation λ_A over GF(256) with the power function $f : x \in \mathrm{GF}(256) \mapsto x^{254} \in \mathrm{GF}(256)$, followed by the addition of a constant term. The `SubBytes` transformation can thus also be represented as the left composition of the two transformations $A = (\lambda_A, \ldots, \lambda_A)$ and `Inv` $= (f, \cdots, f)$, both defining a componentwise transformation of the internal state, followed by the bitwise addition of a constant term $\mathbf{c} \in \mathrm{GF}(2^n)^m$. While A, `ShiftRows` and `MixColumns` are automorphisms of $(\mathrm{GF}(2^n)^m, \oplus)$, the transformation `Inv` defines an automorphism of $(\mathrm{GF}(2^n)^\star)^m, \dot{\otimes})$. In view of this description, it is clear that the AES round can be rewritten as a composition of transformations satisfying the assumptions done in the introduction of Sect. 3: λ is defined as the identity function over $\mathrm{GF}(2^n)^m$, γ is the function `Inv` and λ' is the function `MixColumns` $\circ$ `ShiftRows` $\circ$ A. The masking scheme presented in Sect. 3.4 can thus be applied to protect the AES rounds.

In this section, we compare the efficiency of our proposal with that of the state of the art solutions when applied to secure the AES. All the implementations presented below involve the same code to process the linear transformations λ' and `AddRoundKey`. Namely, we use a $(d+1)$-additive masking such as presented previously in this article. We chose to protect all the rounds of the AES processing. To secure the γ transformation, we chose to select few methods from the literature for $d = \{1, 2, 3\}$. In what follows, we give details on the methods we chose in each category.

For $d = 1$, we selected four methods. First we chose the *table re-computation method* [13] since it achieves the best timing performance. The second chosen method is the *tower field method* [16], which offers the best memory performance. Then, since the work of this paper is the generalization of the 1st-order multiplicative masking scheme proposed in [7], we implemented it as well. Eventually, we chose to implement the dth-order SCA secure scheme proposed in [21]. Though it is less efficient than the others for $d = 1$, choosing it enables to compare our proposal with another method which can be applied generally for any order d.

For $d = 2$, only few methods exist that are perfectly SCA secure. Actually, only the works [22], [20] and [21] propose such kind of schemes. We chose to implement all of them.

For $d = 3$, only [21] proposes a solution in this category.

Table 3. Comparison of secure AES implementations

	Method	Reference	cycles (10^3)	RAM (bytes)
	Unprotected Implementation			
1.	No Masking	Na.	2	32
	First-Order Masking			
2.	Re-computation	[13]	10	256
3.	Tower Field in GF(2^2)	[16,17]	77	42
4.	Multiplicative Masking	[7]	22	256
5.	Secure exponentiation for $d = 1$	[21]	73	24
6.	Our scheme for $d = 1$	This paper	25	50
	Second-Order Masking			
7.	Double Re-computations	[22]	594	$512 + 28$
8.	Single Re-computation	[20]	672	$256 + 22$
9.	Secure exponentiation for $d = 2$	[21]	189	48
10.	Our scheme for $d = 2$	This paper	69	86
	Third-Order Masking			
11.	Secure exponentiation for $d = 3$	[21]	326	72
12.	Our scheme for $d = 3$	This paper	180	128

Table 3 lists the timing/memory performances of the different implementations. We wrote the codes in assembly language for an `8051` based 8-bit architecture with bit-addressable memory. `RAM` consumption related to implementation choices (*e.g.* use of some local variables, use of pre-computed values to speed-up some computations, etc.) are not taken into account in the performances reporting. Also, `ROM` consumptions (*i.e.* code sizes) are not listed since they are not prohibitive for almost all current embedded devices. Eventually, cycles numbers are multiple of 10^3.

Remark 3. For $d = 1$ (Implementations 2 to 5), improvements have been added to the original proposals. They essentially amount to preprocess a part of the masking material, which is possible since the latter one does not need to be changed during the algorithm processing when only first-order SCA are considered.

We observe that only two methods achieve better timing performances than our proposal and that this occurs only in the case $d = 1$. As expected, the re-computation remains the most efficient method when 256 bytes of `RAM` are available. We can also note that the original countermeasure involving multiplicative masking [7] stays better than our countermeasure (which merely generalizes it at any order). The difference is due to the tabulation of the Dirac function used in [7] which implies a faster processing than the algebraic implementation of this function but at the cost of memory. Except those two particular cases, it turns out that our proposal is the most efficient one: it is at least 2.9 times faster for $d = 1, 2$ and 1.8 times faster for $d = 3$. Even if our scheme requires more RAM than [21], the consumption stays lower than 200 bytes and is therefore acceptable for almost all embedded systems (even the low cost ones).

Memory and timing performances of the solution [21] and those of our proposal progress similarly as soon as the order increases. This is explained by the fact that both methods use the same approach to thwart SCA, that is to replace each transformation calculation by a sequence of elementary calculations. To secure them, the solution [21] involves additive maskings while our solution mixes additive and multiplicative maskings. Memory allocation differences between the two methods are merely due to the fact that additional vectors are required in our scheme since it involves more shares (multiplicative shares, dirac shares, etc.). The differences of timing performances come from the fact that solution [21] involve much more field multiplications than in our proposal (see Tables 1 and 2).

5 Conclusion

In this paper, we have introduced a new higher-order masking scheme dedicated to block ciphers mixing affine transformations with power functions. It is provably secure at any chosen order and can be implemented in software at the cost of a reasonable overhead. In particular, it is an efficient alternative to [21] in order secure the AES implementation at any order. For our construction, we have introduced conversion functions that can securely transform an additive masking into a multiplicative one. We think that those transformations could be interesting as secure primitives in other contexts where security against higher-order side channel attacks must be achieved and power functions are involved.

References

1. Akkar, M.-L., Giraud, C.: An implementation of DES and AES, secure against some attacks. In: Koç, Ç.K., Naccache, D., Paar, C. (eds.) CHES 2001. LNCS, vol. 2162, pp. 309–318. Springer, Heidelberg (2001)
2. Ben-Or, M., Goldwasser, S., Wigderson, A.: Completeness theorems for non-cryptographic fault-tolerant distributed computation. In: STOC 1988: Proceedings of the Twentieth Annual ACM Symposium on Theory of Computing, pp. 1–10. ACM, New York (1988)
3. Chari, S., Jutla, C.S., Rao, J.R., Rohatgi, P.: Towards Sound Approaches to Counteract Power-Analysis Attacks. In: Wiener, M.J. (ed.) CRYPTO 1999. LNCS, vol. 1666, pp. 398–412. Springer, Heidelberg (1999)
4. Coron, J.-S.: A New DPA Countermeasure Based on Permutation Tables. In: Ostrovsky, R., De Prisco, R., Visconti, I. (eds.) SCN 2008. LNCS, vol. 5229, pp. 278–292. Springer, Heidelberg (2008)
5. Faust, S., Rabin, T., Reyzin, L., Tromer, E., Vaikuntanathan, V.: Protecting Circuits from Leakage: the Computationally-Bounded and Noisy Cases. In: Gilbert, H. (ed.) EUROCRYPT 2010. LNCS, vol. 6110, pp. 135–156. Springer, Heidelberg (2010)

6. Fumaroli, G., Martinelli, A., Prouff, E., Rivain, M.: Affine Masking against Higher-Order Side Channel Analysis. In: Biryukov, A., Gong, G., Stinson, D.R. (eds.) SAC 2010. LNCS, vol. 6544, pp. 262–280. Springer, Heidelberg (2011)
7. Genelle, L., Prouff, E., Quisquater, M.: Secure Multiplicative Masking of Power Functions. In: Zhou, J., Yung, M. (eds.) ACNS 2010. LNCS, vol. 6123, pp. 200–217. Springer, Heidelberg (2010)
8. Genelle, L., Prouff, E., Quisquater, M.: Montgomery's trick and fast implementation of masked AES. In: Nitaj, A., Pointcheval, D. (eds.) AFRICACRYPT 2011. LNCS, vol. 6737, pp. 153–169. Springer, Heidelberg (2011)
9. Genelle, L., Prouff, E., Quisquater, M.: Thwarting Higher-Order Side Channel Analysis with Additive and Multplicative Masking. Cryptology ePrint Archive (to appear, 2011)
10. Golić, J., Tymen, C.: Multiplicative masking and power analysis of AES. In: Kaliski Jr., B.S., Koç, Ç.K., Paar, C. (eds.) CHES 2002. LNCS, vol. 2523, pp. 198–212. Springer, Heidelberg (2003)
11. Ishai, Y., Sahai, A., Wagner, D.: Private Circuits: Securing Hardware against Probing Attacks. In: Boneh, D. (ed.) CRYPTO 2003. LNCS, vol. 2729, pp. 463–481. Springer, Heidelberg (2003)
12. Joye, M., Paillier, P., Schoenmakers, B.: On second-order differential power analysis. In: Rao, J.R., Sunar, B. (eds.) CHES 2005. LNCS, vol. 3659, pp. 293–308. Springer, Heidelberg (2005)
13. Messerges, T.S.: Securing the AES Finalists Against Power Analysis Attacks. In: Schneier, B. (ed.) FSE 2000. LNCS, vol. 1978, pp. 150–164. Springer, Heidelberg (2001)
14. Oswald, E., Mangard, S.: Template Attacks on Masking—Resistance Is Futile. In: Abe, M. (ed.) CT-RSA 2007. LNCS, vol. 4377, pp. 243–256. Springer, Heidelberg (2006)
15. Oswald, E., Mangard, S., Herbst, C., Tillich, S.: Practical second-order DPA attacks for masked smart card implementations of block ciphers. In: Pointcheval, D. (ed.) CT-RSA 2006. LNCS, vol. 3860, pp. 192–207. Springer, Heidelberg (2006)
16. Oswald, E., Mangard, S., Pramstaller, N.: Secure and Efficient Masking of AES – A Mission Impossible? Cryptology ePrint Archive, Report 2004/134 (2004)
17. Oswald, E., Mangard, S., Pramstaller, N., Rijmen, V.: A Side-Channel Analysis Resistant Description of the AES S-Box. In: Gilbert, H., Handschuh, H. (eds.) FSE 2005. LNCS, vol. 3557, pp. 413–423. Springer, Heidelberg (2005)
18. Peeters, E., Standaert, F.-X., Donckers, N., Quisquater, J.-J.: Improved higher-order side-channel attacks with FPGA experiments. In: Rao, J.R., Sunar, B. (eds.) CHES 2005. LNCS, vol. 3659, pp. 309–323. Springer, Heidelberg (2005)
19. Rivain, M., Dottax, E., Prouff, E.: Block Ciphers Implementations Provably Secure Against Second Order Side Channel Analysis. Cryptology ePrint Archive, Report 2008/021 (2008), `http://eprint.iacr.org/`
20. Rivain, M., Dottax, E., Prouff, E.: Block Ciphers Implementations Provably Secure Against Second Order Side Channel Analysis. In: Baignères, T., Vaudenay, S. (eds.) FSE 2008. LNCS, vol. 5086, pp. 127–143. Springer, Heidelberg (2008)
21. Rivain, M., Prouff, E.: Provably Secure Higher-Order Masking of AES. In: Mangard, S., Standaert, F.-X. (eds.) CHES 2010. LNCS, vol. 6225, pp. 413–427. Springer, Heidelberg (2010)

22. Schramm, K., Paar, C.: Higher order masking of the AES. In: Pointcheval, D. (ed.) CT-RSA 2006. LNCS, vol. 3860, pp. 208–225. Springer, Heidelberg (2006)
23. Shamir, A.: How to Share a Secret. Communications of the ACM 22(11), 612–613 (1979)
24. Trichina, E., DeSeta, D., Germani, L.: Simplified adaptive multiplicative masking for AES. In: Kaliski Jr., B.S., Koç, Ç.K., Paar, C. (eds.) CHES 2002. LNCS, vol. 2523, pp. 187–197. Springer, Heidelberg (2003)
25. Waddle, J., Wagner, D.: Towards Efficient Second-Order Power Analysis. In: Joye, M., Quisquater, J.-J. (eds.) CHES 2004. LNCS, vol. 3156, pp. 1–15. Springer, Heidelberg (2004)

Extractors against Side-Channel Attacks: Weak or Strong?

Marcel Medwed* and François-Xavier Standaert**

UCL Crypto Group, Université catholique de Louvain
Place du Levant 3, B-1348, Louvain-la-Neuve, Belgium

Abstract. Randomness extractors are important tools in cryptography. Their goal is to compress a high-entropy source into a more uniform output. Beyond their theoretical interest, they have recently gained attention because of their use in the design and proof of leakage-resilient primitives, such as stream ciphers and pseudorandom functions. However, for these proofs of leakage resilience to be meaningful in practice, it is important to instantiate and implement the components they are based on. In this context, while numerous works have investigated the implementation properties of block ciphers such as the AES Rijndael, very little is known about the application of side-channel attacks against extractor implementations. In order to close this gap, this paper instantiates a low-cost hardware extractor and analyzes it both from a performance and from a side-channel security point of view. Our investigations lead to contrasted conclusions. On the one hand, extractors can be efficiently implemented and protected with masking. On the other hand, they provide adversaries with many more exploitable leakage samples than, e.g. block ciphers. As a result, they can ensure high security margins against standard (non-profiled) side-channel attacks and turn out to be much weaker against profiled attacks. From a methodological point of view, our analysis consequently raises the question of which attack strategies should be considered in security evaluations.

1 Introduction

Randomness extractors have recently been used as components of leakage-resilient cryptographic primitives such as stream ciphers [3,19], pseudorandom functions [2,16] and signatures [4]. They are also important in the design of public-key cryptosystems resistant to key leakage [12]. In this setting, the proofs of leakage-resilience usually rely on the fact that the amount of information leakage that is provided by one iteration of the extractor (i.e. when executed on one input) is bounded in some sense. As a result, an important requirement for these proofs to be meaningful in practice is that such a bounded leakage can actually be guaranteed by hardware designers. For this purpose, a first implementation and side-channel analysis of such a primitive was described in [14]. This work

* Postdoctoral researcher funded by the 7th framework European project TAMPRES.
** Associate researcher of the Belgian Fund for Scientific Research (FNRS-F.R.S.).

B. Preneel and T. Takagi (Eds.): CHES 2011, LNCS 6917, pp. 256–272, 2011.

analyzed an unprotected software implementation of an extractor. It was shown that, if no attention is paid, the extractor can actually lead to larger information leakages than an unprotected implementation of the AES Rijndael. Mainly, this happens because the extractor allows exploiting multiple leakage samples per plaintext. Thus, this previous work emphasized the importance of including the instantiation of cryptographic primitives in models of leakage resilience.

In this paper, we extend these preliminary investigations in two directions. First, we analyze a low-complexity extractor implemented in hardware (rather than software), and investigate the tradeoffs that such a design allows. Appealing design goals for the hardware implementation are a higher throughput and a leakage reduction due to parallelization. Second, we evaluate the impact of the masking countermeasure on the security of this extractor implementation. In particular, we exhibit an interesting homomorphic property that can be exploited to mask our design efficiently. The results of our hardware design-space evaluation show that the extractor can be masked up to unusually high orders while showing similar performance as a first-order masked block cipher implementation. As for the side-channel analysis results, they confirm part of the previous evaluations, showing that multi-sample per input attacks allow very efficient profiled side-channel attacks. Hence, depending on the adversarial strategies considered in the security evaluations, the implementation of a masked extractor may appear as weaker or stronger than the one of a block cipher. Positively, we show that hardware implementations of randomness extractors can guarantee a bounded leakage for bounded number of measurements. This validates their use as possible components of leakage-resilient constructions. Eventually, this work questions the methodologies for the evaluation of leaking devices in general, and underlines the large difference between profiled and non-profiled attacks that occurs for the extractor case.

The remainder of the paper is structured as follows. Sections 2 and 3 describe the analyzed low-complexity Hadamard extractor and its different hardware implementations. The side-channel attack scenario is detailed in Section 4. This is followed by an information theoretic analysis and security analysis in Sections 5 and 6. Finally, we draw conclusions in Section 7.

2 Low Complexity Extractor

In this section, we specify the instance of the low complexity Hadamard extractor, denoted as $\boxplus$, the implementation of which will be investigated in the remaining of the paper. It relies on the LFSR-based hashing technique from [8]. In order to compute $\boldsymbol{k} \boxplus \boldsymbol{x}$, one first expands $\boldsymbol{x} = s_0 s_1 \cdots s_{n-1}$ with the recurrence:

$$s_{i+n} \stackrel{def}{=} a_0 s_i + a_1 s_{i+1} + \cdots + a_{n-1} s_{i+n-1} \pmod 2, \tag{1}$$

where the public constants $a_{n-1} \cdots a_0$ are the coefficients of a primitive polynomial of degree n. Then, we simply compute m inner products (mod 2) between $\boldsymbol{k}$ and the rows of a matrix filled with the expanded $\boldsymbol{x}$:

$$\boxplus : \{0,1\}^n \times \{0,1\}^n \to \{0,1\}^m \ (m \leq n) :$$

$$\boldsymbol{k} \boxplus (\boldsymbol{x} \stackrel{def}{=} s_0 s_1 \cdots s_{n-1}) = \begin{bmatrix} s_{n-1} & \cdots & s_{k-1} & \cdots & s_1 & s_0 \\ s_n & \ddots & \cdots & \ddots & s_2 & s_1 \\ \vdots & \ddots & \ddots & \cdots & \ddots & \ddots \\ s_{n+m-2} & \cdots & s_n & s_{n-1} & \cdots & s_{m-1} \end{bmatrix} \cdot \boldsymbol{k}.$$

Hence, the function "$\boxplus$" is equivalent to:

$$\boxplus : \boldsymbol{k} \times \boldsymbol{x} \mapsto [\langle \boldsymbol{x} \cdot A^0, \boldsymbol{k} \rangle, \langle \boldsymbol{x} \cdot A^1, \boldsymbol{k} \rangle, \cdots, \langle \boldsymbol{x} \cdot A^{m-1}, \boldsymbol{k} \rangle], \tag{2}$$

where the matrix A is defined as follows:

$$A = \begin{bmatrix} 0 & 0 & \cdots & 0 & a_0 \\ 1 & 0 & \cdots & 0 & a_1 \\ 0 & 1 & \ddots & 0 & \vdots \\ \vdots & \ddots & \ddots & 0 & a_{n-2} \\ 0 & 0 & \cdots & 1 & a_{n-1} \end{bmatrix}. \tag{3}$$

This function is a 2-source extractor since the Toeplitz matrix (as in the definition of $\boxplus$) has full rank for any non-zero vector $\boldsymbol{x}$, which in turn follows from the properties of maximal length LFSR. Note, that this extractor directly inherits the homomorphic property of Krawczyk's hash function. Namely, we have that $\langle \boldsymbol{x} \cdot A^i, \boldsymbol{k} + \boldsymbol{m} \rangle + \langle \boldsymbol{x} \cdot A^i, \boldsymbol{m} \rangle = \langle \boldsymbol{x} \cdot A^i, \boldsymbol{k} \rangle$.

3 Hardware Implementation

Following the specification from the previous section, we now present the hardware architecture and the tradeoffs that we considered when implementing the Hadamard extractor. We also use this description of the hardware to list the different parameters that will be analyzed in our following side-channel evaluations. As indicated by the notations in Section 2, we will generally apply the extractor to an n-bit public value $\boldsymbol{x}$ and an n-bit secret key $\boldsymbol{k}$, in order to produce an l-bit random string y (which is the typical scenario in leakage resilient cryptography). Practical values that we consider in this work are $n = 192$ and $l = 128$. For simplicity, we start by describing a fully serial implementation (with $n = 8$), illustrated in Figure 1.

In this basic form, the extractor circuit mainly consists of two registers. One is used to store the current LFSR values (denoted as $r[0]$ to $r[7]$ in the figure), and consequently evolves as the implementation is running. The other one is used to store the key and remains static during the extraction process. Note that the decision to store the secret key in the static register is motivated by the minimization of the computations (hence, leakage) involving secret data. In a fully serial implementation, the r register is shifted by one position at each clock

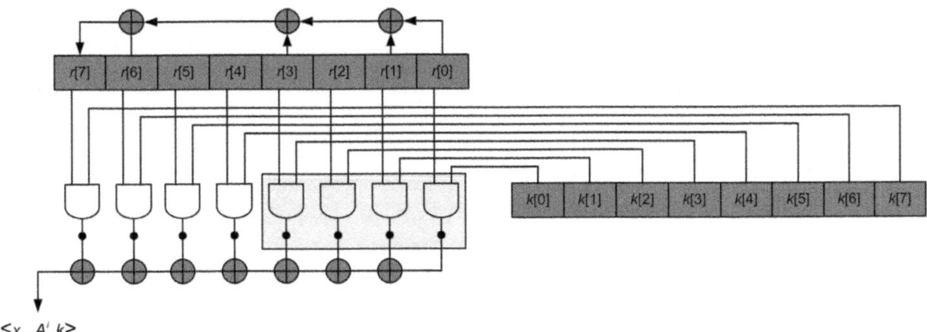

Fig. 1. Fully serial hardware implementation of the extractor

cycle. Next, n AND gates and $n-1$ XOR gates are added to the design, in such a way that one can extract one bit (i.e. compute one inner product $\langle \boldsymbol{x} \cdot A^i, \boldsymbol{k} \rangle$) per clock cycle. Thus, in order to extract 128 bits, we have to clock the circuit 128 times (while the registers are typically 192-bit long). Such a basic design can essentially be extended in two main directions that we now detail.

Parallelizing the Implementation. In general, hardware implementations are most efficient if they can take advantage of some inherent parallelism in algorithms. Fortunately, this is typically the case when considering our extractor. That is, as illustrated in Figure 2, one can easily double the throughput of the previous design, by extending the LFSR by one cell $r[8]$ and duplicating the combinatorial parts of the design (i.e. the XOR gates used in the LFSR recurrence and the inner product computation). This allows one to compute two inner products per clock cycle. Interestingly, the registers cells $r[1]$ to $r[7]$ and the key register can be shared by these two inner product combinations, which makes the parallelization quite efficient. Such a process can be further extended. In general, by multiplying the number of inner product combinations p times, we decrease the number of cycles to extract 128 bits by the same factor.

Masking the Implementation. Next, as detailed in Section 2, the proposed extractor inherently benefits from an additive homomorphic property. This implies that it can be easily masked, following the proposals of Goubin and Patarin [7] and Chari et al. [1]. In our setting, it is most natural to mask the key, as masking the plaintext would lead to a weakness similar to the "zero problem" when applying multiplicative masking to the AES S-box [6]. That is, the bitwise AND between a masked plaintext and a key would still allow distinguishing the zero key bits. From an implementation point of view, a masked computation $\langle \boldsymbol{x} \cdot A^i, \boldsymbol{k} + \boldsymbol{m} \rangle + \langle \boldsymbol{x} \cdot A^i, \boldsymbol{m} \rangle = \langle \boldsymbol{x} \cdot A^i, \boldsymbol{k} \rangle$ can be performed using essentially the same design as in the unmasked case. And it straightforwardly extends to higher-order masking, where the order o of the masking scheme refers to the number of n-bit masks consumed per extraction, as we now detail.

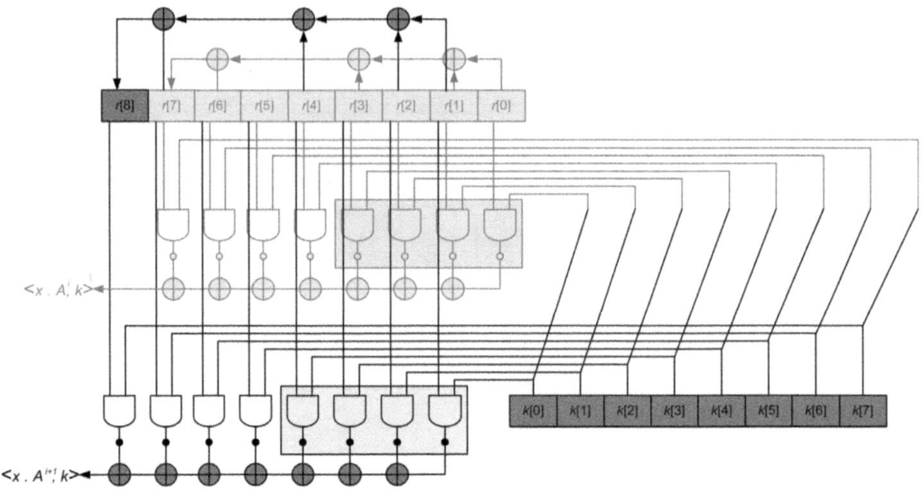

Fig. 2. Hardware implementation of the extractor with parallelism ($p = 2$)

First observe that, in the unprotected case, the result of an extraction is available after $128/p$ clock cycles. In the masked case, this performance decreases only linearly, since it is possible to operate on all shares independently. In other words, each mask can be discarded immediately after it has been processed. For this purpose, we first need 192 clock cycles to load the mask register and (at the same time) add the mask to the key. Next, 128 clock cycles are needed to extract from the mask. Finally, the (bidirectional) plaintext register is rewound during 128 clock cycles. Overall, every mask needs 448 clock cycles to be processed. And the final result is obtained by extracting from the masked key, which requires another 128 clock cycles. Summarizing, we have implemented the circuit such that the cycle count c increases linearly, following the formula $c = (128+o{\cdot}448)/p$, with o the masking order and p the degree of parallelization. Table 1 summarizes the performances of various extractor implementations. These numbers were obtained from post-synthesis results, using Cadence RTL compiler 2009 and the UMC F180GII standard-cell library. Note that, in a fully serial and unprotected implementation, the area cost is already dominated by the registers r and k. They alone account for 3.6 kGE. Roughly speaking, the hardware overhead of a masked implementation mainly corresponds to one additional register for storing the mask, and a 192-bit multiplexer in order to switch the AND gates' input between the key and the mask.

4 Adversarial Capabilities and Leakage Assumptions

The goal of this paper is to investigate the side-channel resistance of different versions of the implemented extractor, with and without parallelism and masking. For this purpose, we will apply the two parts of the framework in [15]. That is, we start with an information theoretic analysis (in the next section), in order

Table 1. Area in kilo gate equivalents (kGE) and cycle count (c) for extractor implementations with different datapath widths and different masking orders (the polynomial $x^{192} + x^{149} + x^{97} + x^{46} + 1$ has been used in the recurrence)

Parallelization	1	4	8
w/o masking	4.3 kGE 128 c	7.0 kGE 32 c	10.3 kGE 16 c
1st-order	7.3 kGE 576 c	10.1 kGE 144 c	13.6 kGE 72 c
2nd-order	7.3 kGE 1024 c	10.1 kGE 256 c	13.6 kGE 128 c
3rd-order	7.3 kGE 1472 c	10.1 kGE 368 c	13.6 kGE 184 c

to capture a worst case scenario. Next, we perform a security analysis, that considers the success rates of different adversaries. In general, such an evaluation requires to define the adversary's capabilities and leakage assumptions.

In our present context, the first question to answer is to determine the target operations for the side-channel adversary. For this purpose, one generally selects the operations where the known input $\boldsymbol{x}$ and secret key $\boldsymbol{k}$ are mingled. For the extractor implementations in Figures 1 and 2, this corresponds to the bitwise AND gates (the side-channel attacks against a software implementation of extractor in [14] were based on exactly the same assumption). Next, it is typically needed to determine the size of the key guess (i.e. the number of key candidates that will be enumerated in the attack). In the following, we will consider a 4-bit key guess, that is a convenient choice for limiting the time complexity of the evaluations. This choice is motivated by the fact that we aim to investigate many sets of parameters. In the Figures 1 and 2, it means that an adversary will typically try to predict the output of the AND gates that are included in the gray rectangles.

In addition, and more importantly, a central feature of the Hadamard extractor implementations is that the key register is used numerous times in order to extract l bits. For example, in the serial implementation of Figure 1, in which one extracts $l = 128$ bits, it implies that 128 leaking operations can potentially be exploited by the adversary. In the case of the parallel implementation of Figure 2, where $p = 2$, this amount of exploitable leakage points is decreased to 64. This is in strong contrast with traditional implementations of block ciphers, where one typically predicts a few leakage points, corresponding to the intermediate computations of the block cipher that can be easily enumerated. For example, in the block cipher PRESENT, a 4-bit guess allows predicting the first (or last) key addition and S-box computations of an encryption process. But following operations become hard to predict, because of the diffusion in the cipher. As implementations of extractors are not affected by such a strong diffusion property, a very important parameter is the number of leakage points exploited by the adversary. Our experiments will consider both single-sample attacks, that are similar to standard DPA attacks against block ciphers, and t-sample attacks, for which we aim at discussing their relevance in a side-channel evaluation context. Summarizing, our evaluations will investigate three main parameters:

1. the degree of parallelism in the implementation p,
2. the order of the masking scheme o,
3. the number of leakage samples per plaintext exploited in the attacks t.

Our experiments are based on simulated traces, which reflect the ideal power consumption of the previously described hardware architecture. In order to allow a systematic comparison between the level of security of the extractor implementation and the one of a masked S-box, we added Gaussian noise. More precisely, we used simulated traces to generate the mean value of the target leakages, with the Signal-to-Noise Ratio as a parameter ($\text{SNR} = 10 \cdot \log_{10}(\frac{\sigma_s^2}{\sigma_n^2})$). In other words, we extended the simulation environment of [17] to the context of an extractor. Note, that most of our following conclusions relate to the comparative impact of the parameters p, o and m. Hence, the possible deviations that one would observe between simulated traces and actual measurements would not affect these conclusions (i.e. they would essentially only cause some slight shifts of the information theoretic and security analysis curves in the following sections).

5 Information Theoretic Analysis

In this section, we aim to evaluate the security of the previously described extractor implementation, in function of the amount of parallelism, masking and leakage samples available to the adversary. For this purpose, we start with the information theoretic analysis advocated in [15], the goal of which is to analyze a worst case scenario, where the adversary has perfect knowledge of the leakage distribution (i.e. is able to perform a perfect profiling). For our simulated setting, this means that the adversary is provided with the leakage samples l_j^i, where the subscript j relates to the number of shares in a masking scheme and the superscript i relates to the number of samples used per input x in the attack. More specifically, in an unmasked implementation, the adversary is given the following leakage samples:

$$l_1^i = \mathrm{W_H}\left((\boldsymbol{x} \cdot A^i) \wedge \boldsymbol{k}\right) + n, \tag{4}$$

where $\mathrm{W_H}$ denotes the Hamming weight function and n is a Gaussian noise. From this definition, one can straightforwardly compute the following mutual information metric, for the fully serial (i.e. $p = 1$) single sample (i.e. $t = 1$) case:

$$\mathrm{I}(K; X, L_1^1) = \mathrm{H}[K] - \sum_{k \in \mathcal{K}} \Pr[k] \sum_{x \in \mathcal{X}} \Pr[x] \sum_{l_1^1 \in \mathcal{L}} \Pr[l_1^1 | x, k] \cdot \log_2 \Pr[k | x, l_1^1]. \tag{5}$$

Next, considering a masked implementation would change the previous analysis as follows. First, the adversary now has to exploit the leakage of several shares. For example, in the first-order case (i.e. $o = 1$), we have:

$$l_1^i = \mathrm{W_H}\left((\boldsymbol{x} \cdot A^i) \wedge \boldsymbol{m}\right) + n, \tag{6}$$

$$l_2^i = \mathrm{W_H}\left((\boldsymbol{x} \cdot A^i) \wedge (\boldsymbol{k} \oplus \boldsymbol{m})\right) + n. \tag{7}$$

Second, the computation of the mutual information metric is turned into:

$$\mathrm{I}(K; X, L_1^1, L_2^1) = \mathrm{H}[K] - \sum_{k \in \mathcal{K}} \Pr[k] \sum_{x \in \mathcal{X}} \Pr[x] \sum_{m \in \mathcal{M}} \Pr[m] \cdot \sum_{l_1^1, l_2^1 \in \mathcal{L}^2} \Pr[l_1^1, l_2^1 | x, m, k] \cdot \log_2 \Pr[k | x, l_1^1, l_2^1], \tag{8}$$

where $\Pr[k|x, l_1^1, l_2^1] = \sum_{m'} \Pr[m'|x, l_1^1] \Pr[k|x, m', l_2^1]$. That is, the mask is not given to the adversary, but its leakage allows building a bivariate conditional distribution that is key-dependent. This naturally extends towards larger o's.

These previous equations were considering single-sample-per-input attacks that are typically similar to the DPA against the AES S-box in [10] and masked AES S-box in [17]. When moving to the multi-sample context, the computation of the mutual information metric for the unmasked case is turned into:

$$\mathrm{I}\left(K; X, L_1^t\right) = \mathrm{H}[K] - \sum_{k \in \mathcal{K}} \Pr[k] \sum_{x \in \mathcal{X}} \Pr[x] \cdot \sum_{l_1^1, l_1^2, \ldots, l_1^t \in \mathcal{L}^t} \Pr[l_1^1, l_1^2, \ldots, l_1^t | x, k] \cdot \log_2 \Pr[k | x, l_1^1, l_1^2, \ldots, l_1^t]. \tag{9}$$

And for the masked case, it becomes:

$$\mathrm{I}\left(K; X, L_1^t, L_2^t\right) = \mathrm{H}[K] - \sum_{k \in \mathcal{K}} \Pr[k] \sum_{x \in \mathcal{X}} \Pr[x] \sum_{m \in \mathcal{M}} \Pr[m] \cdot \sum_{l_1^1, l_1^2, \ldots, l_1^t, l_2^1, l_2^2, \ldots, l_2^t \in \mathcal{L}^{2t}} \Pr[l_1^1, l_1^2, \ldots, l_1^t, l_2^1, l_2^2, \ldots, l_2^t | x, m, k] \cdot \log_2 \Pr[k | x, l_1^1, l_1^2, \ldots, l_1^t, l_2^1, l_2^2, \ldots, l_2^t]. \tag{10}$$

Interestingly, this multi-sample case implies that many samples can be used to "bias" the mask in the computation of the mixture distribution:

$$\Pr[k|x, l_1^1, l_1^2, \ldots, l_1^t, l_2^1, l_2^2, \ldots, l_2^t] = \sum_m \Pr[m|x, l_1^1, l_1^2, \ldots, l_1^t] \Pr[k|x, m, l_2^1, l_2^2, \ldots, l_2^t].$$

As will be seen in the following, this strongly reduces the security improvements of masking in this setting. Finally, independent of the parameters o and t, an increased parallelism is modeled by changing the leakage function. For example, in the $p = 2$ case of Figure 2, the adversary would obtain samples of the form:

$$l_1^{i,i+1} = \mathrm{W_H}\left((\boldsymbol{x} \cdot A^i) \wedge \boldsymbol{k}\right) + \mathrm{W_H}\left((\boldsymbol{x} \cdot A^{i+1}) \wedge \boldsymbol{k}\right) + n. \tag{11}$$

In general, modifying the parallelism has no impact on the previous equations. However, increasing p implies that the maximum number of samples that one can exploit per plaintext is more limited (to 64 if $p = 2$, 32 if $p = 4$, ...). Note again that, due to the weak diffusion of the extractor implementation, a 4-bit guess would then allow to predict several 4-bit parts of the inner product computations (e.g. one part of both $\mathrm{W_H}$ functions in Equation (11) can be predicted).

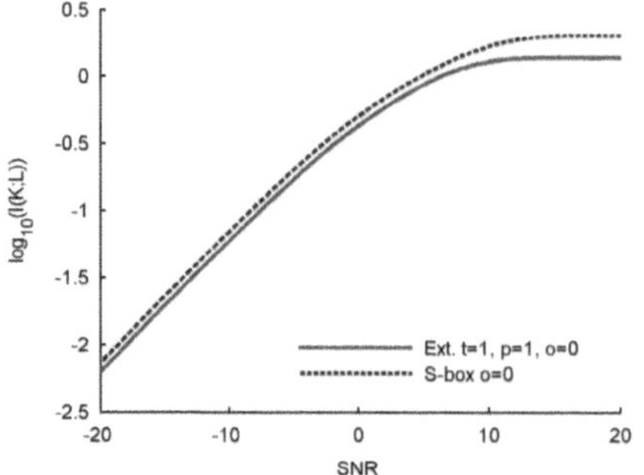

Fig. 3. Single sample attacks, serial implementation, without masking

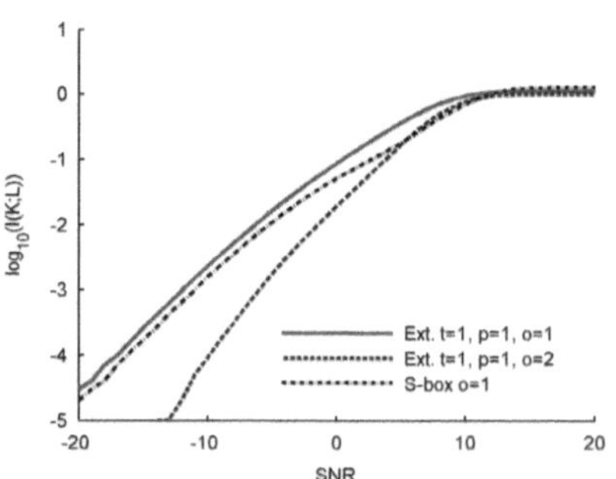

Fig. 4. Single sample attacks, serial implementation, with masking

5.1 Single Sample Attacks, Serial Implementation

For the first scenario, we assume an adversary who looks only at one leakage sample per side-channel trace. This can be seen as a naive attack, where the adversary applies exactly the same strategy as he would for an S-box. The implementation of the extractor is fully serialized and its masking order varies between zero and two. Figure 3 shows the unprotected case ($t = 1, p = 1, o = 0$) and Figure 4 shows the masked case ($t = 1, p = 1, o \in \{1, 2\}$). We compare those curves with the information curves for an unmasked and a masked PRESENT S-box. As higher-order masking schemes leading to efficient hardware implementations remain an open problem, we restrict the S-box evaluations to first-order masking[1]. For the unmasked case, the curves confirm what was already observed in [14]. Namely, a single extractor sample contains less information than a single S-box sample, on average. As for the masked case, the results follow the expectations in [1,17]. That is, for a sufficient amount of noise, increasing the order of the masking scheme implies an exponential security increase, reflected by the different slopes of the log scale curves in Figure 4. Note finally that, in this latter case and for similar orders, the information provided by a single sample of the extractor is now slightly higher than the one provided by an S-box sample (which can be explained by the shape of the masked leakage distributions).

5.2 Multi-sample Attacks, Serial Implementation

The results in the previous section suggest that the security of an extractor implementation can be strong when adversaries exploit a single sample per leakage trace. But as previously mentioned, extractors are different than standard block ciphers in the sense that a small key guess (here 4-bit) allows adversaries to predict multiple intermediate computation results (up to $t = 128$). In this section, we consider the worst case of a serial implementation where an adversary would exploit all this information. Applying this approach to the unmasked case implies

[1] The only straightforward solution is to use a large look-up table of size $2^{o \cdot n} \times n$ bits.

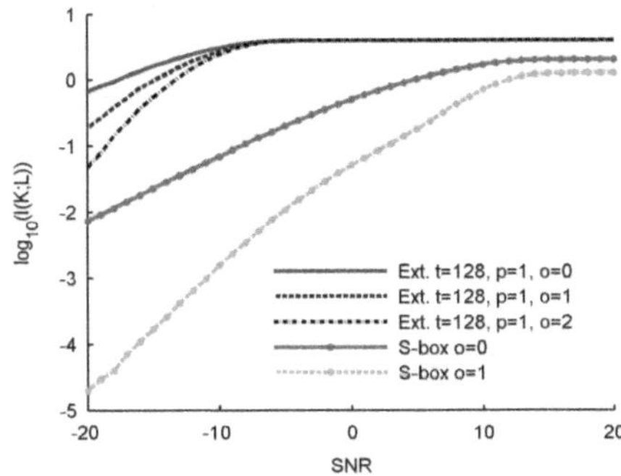

Fig. 5. Multi sample attacks, serial implementation

to compute the mutual information metric given in Equation (9). We mention that, since integrating over 128 dimensions is too complex, the following estimations are obtained by statistical sampling. Similarly, evaluating a multi-sample attack against a masked implementation requires to compute Equation (10). As previously mentioned, this equation suggests that the mask can be strongly biased because, in the multi-sample attack setting, an adversary can exploit 128 leakage points generated from the manipulation of the same mask value.

The results of the information theoretic analysis corresponding to this strongest possible adversary are depicted in Figure 5. It can be seen that the situation changes dramatically. Up to an SNR of -5, the remaining entropy of the key variable after seeing a single side-channel trace is zero (i.e. unbound leakage). For SNRs below -5 the recovered information eventually decreases and also masking starts to bring additional security. However, due to the mask biasing process, it also requires smaller SNRs until the impact of masking is fully released. Furthermore, even for an SNR as small as -20, the second-order masked extractor implementation reveals more information than the unprotected S-box one. Summarizing, while an extractor implementation provides strong security against standard univariate DPA attacks, the exploitation of multiple samples leads to an opposite conclusion. Roughly speaking, the exploitation of t samples per trace in this setting corresponds to the exploitation of t single-sample traces (all using the same mask) in the previous section. We now discuss strategies to relax this limitation.

5.3 Decreasing the Leakage by Reducing t

Following the previous section, one important objective for improving the security of an extractor implementation is to limit the number of leakage points exploitable per trace. In the full version of this paper [11], we investigate different strategies to achieve this goal, namely re-keying, re-masking and parallelism and compare them from an implementation and side-channel point of view. Here, we focus on the general effect of reducing the parameter t and in particular discuss parallelization as it is the most appealing approach from a performance point of view.

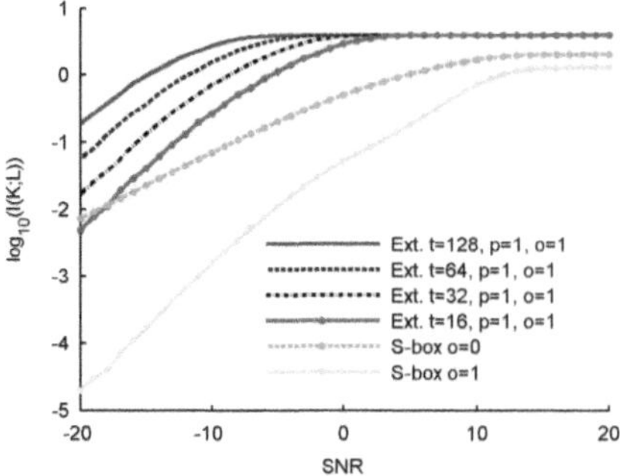

Fig. 6. Multi-sample attacks, reducing t for masking of order 1

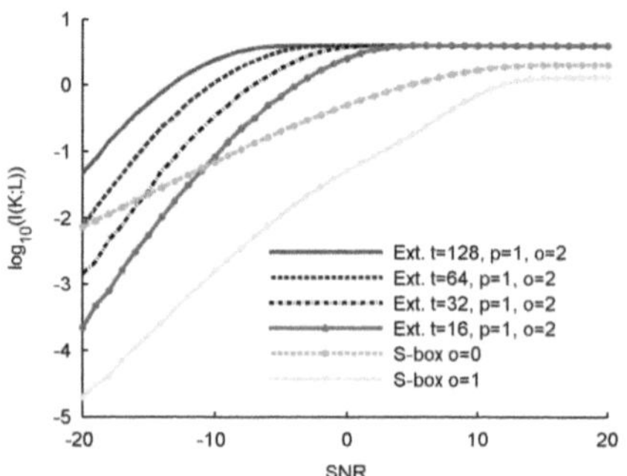

Fig. 7. Multi-sample attacks, reducing t for masking of order 2

Figures 6 and 7 show the impact of limiting t in such a way, for masking of order one and two. It can be seen that the information decreases exponentially with t. Furthermore, this exponential decrease is larger the higher the masking order is, essentially because we do not only limit the available samples for the key but also for the masks.

In general, parallelization has the same impact as just reducing t. However, as now more computations are performed per clock cycle, also the amount of information leakage contained per sample might be higher. As a simple example, let us denote two leakage samples generated by an unprotected serial implementation as l_1 and l_2. By parallelizing the operations corresponding to these leakage samples, one provides the adversary with a new sample $l' = l_1 + l_2$. If these two leakage samples are related to the same key guess, the information provided by one of them is generally less than the information provided by their sum, which is again less than their joint information. This is illustrated in Figures 8 and 9. However, as will be seen in the security analysis, not every distinguisher is capable of exploiting this extra information.

6 Security Analysis

The results of the previous IT analysis define upper bounds for the information which can be extracted by an adversary. In this section we discuss how well these upper bounds represent the capabilities of an adversary. In particular, we raise two questions: (1) Are t-sample attacks relevant in practice? (2) How large is the practical security gain of masking an extractor? The first question can be answered positively under some reasonable assumptions. As for the second question, it turns out that masked extractors can actually provide good security when looking at standard higher-order DPA attacks.

6.1 Identifying Multiple Samples

In general, the critical parameters for evaluating side-channel attacks are the data complexity (in the first place) and the time complexity (when the order of

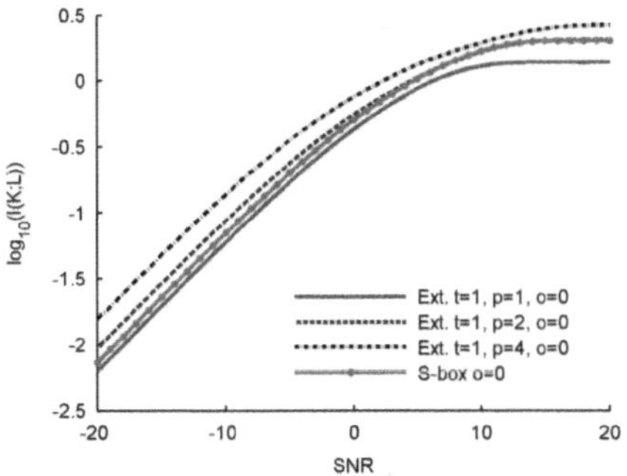

Fig. 8. Single sample attacks with parallelism, no masking

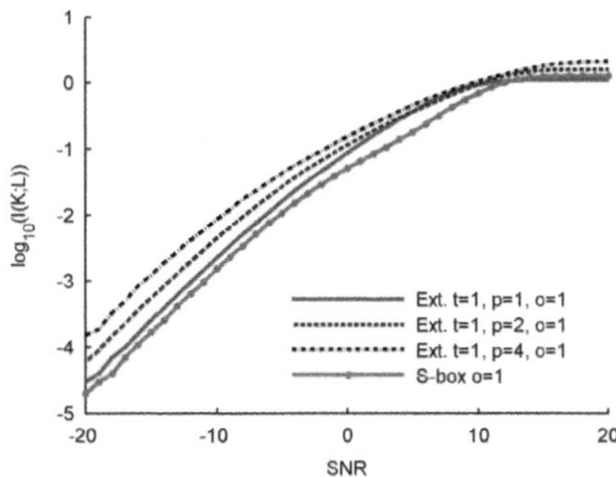

Fig. 9. Single sample attacks with parallelism, masking order 1

the attacks increases). The data complexity typically depends on the amount of information contained in each leakage trace. The time complexity typically depends on the number of samples of interest to identify in the traces. As a result, it is interesting to determine how efficiently the previous multi-sample attacks trade time and data. For this purpose, say an adversary has to identify t samples in an N-sample leakage trace. Without additional assumptions, the complexity of finding them is in $O(N^t)$. However, in the case of the hardware extractor, the adversary can assume that time samples are equidistantly distributed along the power trace. For instance, our hardware architecture produces p bits every clock cycle and the extraction process takes t clock cycles. Thus the distance between the interesting samples is one clock cycle. Given this information, one can sweep over the power trace like in a standard single-sample DPA, and directly launch a multi-sample attack exploiting the t equidistant samples with a complexity in $O(N)$. In other words, an adversary does not have to detect these samples of interest separately and in advance. This observation implies the interesting consequence that the order of a side-channel attack (usually defined as the number of samples exploited per trace) is not a generally good indicator of security. It is sometimes easy to launch high-order attacks. Note also that the assumptions on the underlying hardware to launch such low complexity attacks are generally easy to guess for adversaries. Hence, ruling out multi-sample attacks in this case implies to rely on "security by obscurity" in an excessive manner.

6.2 Attacking the Masking

An interesting feature of our extractor implementation is that its homomorphic property allows efficient masking of unusually large orders (compared, e.g. to the AES Rijndael). However, as previously discussed, the impact of masks can be strongly reduced in multi-sample attacks, as an adversary can theoretically bias the mask using all the available samples. In this section, we discuss the feasibility of such a biasing in a non-profiled attack setting.

Biasing the Mask. As detailed in Section 5.2, biasing the masks is straightforward in profiled side-channel attacks. The adversary just has to launch a template attack on the masks, prior to the attack, and can use the resulting distribution of the masks when launching the same template attack on the secret key. By contrast, when moving to a non-profiled attack setting, such a mask biasing becomes more difficult to exploit. The natural approach would again be to launch a DPA (e.g. correlation-based) on the mask extraction $\langle \boldsymbol{x} \cdot A^i, \boldsymbol{m} \rangle$. If the measurements are quite informative, then this DPA can lead to a complete recovery of the masks (i.e. the masked implementation becomes as easy to break as the unprotected one). But when measurements become noisy, the adversary ends up with a vector of key candidates ranked according to the output of the DPA distinguisher, e.g. a correlation coefficient. Exploiting this information in a non-profiled attack has to be based on one of the following approaches:

1. If only a couple of mask candidates remain likely after the biasing process, one solution is to test them exhaustively. But this strategy becomes intensive when combining multiple plaintexts, as the number of such tests scales exponentially in the number of plaintexts in the attack.
2. If the mask distribution obtained after the observation of a trace is not enough biased, or if the number of plaintexts required to perform a successful key recovery is too high, one then has to rely on heuristics. For example, one solution would be to normalize the correlation coefficient obtained after the DPA against the masks, and to interpret them as probabilities. As detailed in [18], this heuristic already implies significant efficiency losses compared to the application of a template attack. In addition, one then needs to exploit this vector of mask "probabilities" in the high-order attack against the extractor implementation, which is again not trivial. For example, one could try to exploit this information to improve the combination function used in a second-order DPA, as suggested in [13]. But it usually results in quite involved techniques that may not be easy to apply in practical settings.

As in the previous subsection, it is interesting to observe that the evaluation of the extractor implementation crucially relies on the strategy adopted by the adversary. But while the multi-sample approach is quite realistic in a hardware implementation context, even for non-profiled attacks, the biasing of the masks becomes much less realistic when profiling the chip and taking advantage of Bayesian key recovery is not possible. In fact, a more practical adversary will probably perform a higher-order DPA directly on the different time samples provided by the traces (for instance by performing a Pearson-correlation based attack after applying the normalized product combining function [9] or by directly using a multivariate distinguisher like MIA [5]). Interestingly, one can easily evaluate the information loss caused by this strategy, by replacing the last factor in Equation (10) by:

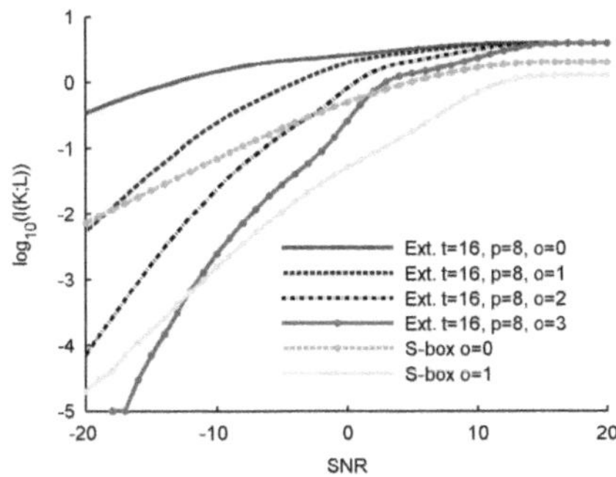

Fig. 10. Information leakage without mask biasing (p=8, t=16)

$$\Pr[k|x, l_1^1, l_1^2, \ldots, l_1^t, l_2^1, l_2^2, \ldots, l_2^t] = \frac{\prod_{i=1}^{t} \sum_m \Pr[m|x, l_1^i] \Pr[k|x, m, l_2^i]}{\sum_{k'} \prod_{i=1}^{t} \sum_m \Pr[m|x, l_1^i] \Pr[k'|x, m, l_2^i]}.$$

The result of such an evaluation for $t = 16$ and a parallelization of 8 is plotted in Figure 10. Intuitively, it represents the upper bound for the information which can be retrieved by a non-biased DPA attack. In a typical scenario where an adversary attacks 8 out of 128 bits at once, the SNR can be computed as $10 \log_{10}(8/120) = -11.761$. It can be seen from the figure that this is approximately the point where a profiled adversary does not gain more information from a multi-sample attack on the 3rd-order masked extractor than from a single-sample attack on a 1st-order masked S-box.

Higher-Order DPA Attacks. Eventually, it is interesting to evaluate how efficiently a standard higher-order DPA against the masked extractor implementation can take advantage of the information leakage in Figure 10. For this purpose, and based on the fact that our implementation exhibits a perfect Hamming weight leakage model, it is natural to apply a DPA based on Pearson's correlation coefficient, and using a normalized-product combining function. In Figure 11, we compare a 1st-order masked S-box and a 3rd-order masked extractor for SNRs of 0 and 10. One can observe that the results of this security evaluation do not directly reflect the outcome of the IT analysis. Namely, even in the low noise scenario, the multi-sample attack on the extractor is almost a magnitude less efficient than the single-sample attack on the S-box, due to the information loss caused by the normalized-product combining. Interestingly, even in an unprotected context, the sub-optimality of DPA attacks can be highlighted, e.g. in Figure 12, where parallelism always improves security[2].

We finally note that, more than the results of specific adversarial strategies used in this section (which may be suboptimal), it is the very observation that these strategies play a central role in the security evaluation of cryptographic

[2] In contrast with the result given in Section 5.3, Figure 8.

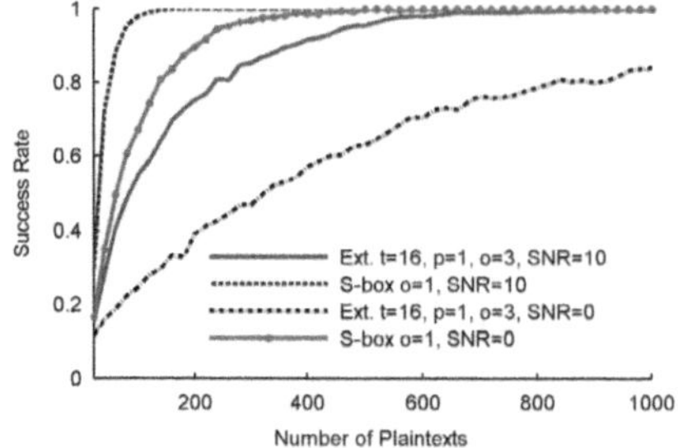

Fig. 11. Multi-sample correlation attacks against masked implementations, normalized product combining

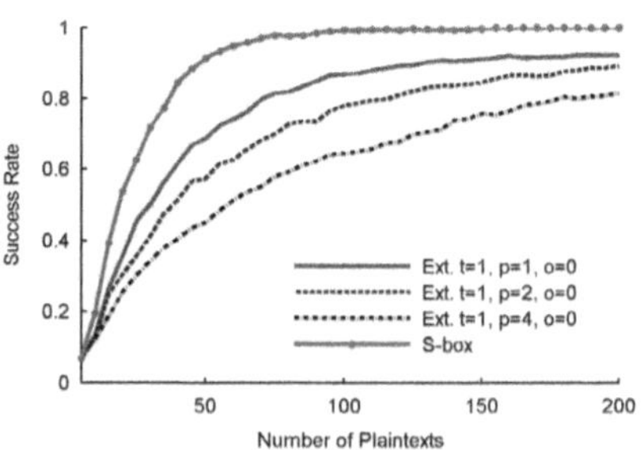

Fig. 12. Single-sample correlation attacks against unprotected implementations, with an SNR=-0.3

devices that is interesting. Especially, if the leakage of the analyzed primitive has flavors unknown from block ciphers, such as multiple samples per plaintext and inherent mask re-use, special care has to be taken during the evaluation and the interpretation of the results.

7 Conclusions

This paper first shows the interest of implementing randomness extractors in hardware. By taking advantage of different design tradeoffs (namely, parallelism, re-keying and re-masking), such implementations allow improving the security against side-channel attacks, in particular when compared to their software counterpart. Next, our results put forward the strong impact of adversarial strategies in the evaluation of extractor implementations (and leaking devices in general). Depending on the capabilities of an adversary, the information leakage provided by the extractor implementations considered in this paper range from large (in Figure 5) to much more limited (in Figure 10) and is sometimes difficult to exploit with standard DPA attacks (in Figure 11). From a methodological point of view, this observation suggests to always consider different capabilities in an evaluation, in order to avoid overestimating (or underestimating) the security of an implementation. Our results also show that the order of an attack, defined as the number of samples per trace, may not be a good indication of security if these samples do not corresponds to the different shares of a masking scheme. As a scope for further research, we notice that the main weakness of the extractor investigated in this paper derives from the ability to predict many intermediate computations from a small key guess, which reduces the interest in its nice homomorphic property. As a result, it would be interesting to design an extractor limiting this weakness, e.g. by introducing a type of "key-scheduling" in the algorithm, in order to add some diffusion during the extraction process.

References

1. Chari, S., Jutla, C.S., Rao, J.R., Rohatgi, P.: Towards Sound Approaches to Counteract Power-Analysis Attacks. In: Wiener, M.J. (ed.) CRYPTO 1999. LNCS, vol. 1666, pp. 398–412. Springer, Heidelberg (1999)
2. Dodis, Y., Pietrzak, K.: Leakage-Resilient Pseudorandom Functions and Side-Channel Attacks on Feistel Networks. In: Rabin, T. (ed.) CRYPTO 2010. LNCS, vol. 6223, pp. 21–40. Springer, Heidelberg (2010)
3. Dziembowski, S., Pietrzak, K.: Leakage-Resilient Cryptography. In: FOCS, pp. 293–302. IEEE Computer Society, Los Alamitos (2008)
4. Faust, S., Kiltz, E., Pietrzak, K., Rothblum, G.N.: Leakage-resilient signatures. In: Micciancio, D. (ed.) TCC 2010. LNCS, vol. 5978, pp. 343–360. Springer, Heidelberg (2010)
5. Gierlichs, B., Batina, L., Tuyls, P., Preneel, B.: Mutual Information Analysis. In: Oswald, E., Rohatgi, P. (eds.) CHES 2008. LNCS, vol. 5154, pp. 426–442. Springer, Heidelberg (2008)
6. Golic, J.D., Tymen, C.: Multiplicative Masking and Power Analysis of AES. In: Kaliski Jr., B.S., Koç, Ç.K., Paar, C. (eds.) CHES 2002. LNCS, vol. 2523, pp. 198–212. Springer, Heidelberg (2003)
7. Goubin, L., Patarin, J.: DES and Differential Power Analysis (The "Duplication" Method). In: Koç, Ç.K., Paar, C. (eds.) CHES 1999. LNCS, vol. 1717, pp. 158–172. Springer, Heidelberg (1999)
8. Krawczyk, H.: LFSR-Based Hashing and Authentication. In: Desmedt, Y. (ed.) CRYPTO 1994. LNCS, vol. 839, pp. 129–139. Springer, Heidelberg (1994)
9. Mangard, S., Oswald, E., Popp, T.: Power Analysis Attacks: Revealing the Secrets of Smart Cards. Springer, Heidelberg (2007)
10. Mangard, S., Oswald, E., Standaert, F.-X.: One for All - All for One: Unifying Standard DPA Attacks. Cryptology ePrint Archive, Report 2009/449 (2009), `http://eprint.iacr.org/` to appear in IET Information Security
11. Medwed, M., Standaert, F.-X.: Extractors Against Side-Channel Attacks: Weak or Strong? Cryptology ePrint Archive, Report 2011/348 (2011), `http://eprint.iacr.org/`
12. Naor, M., Segev, G.: Public-Key Cryptosystems Resilient to Key Leakage. In: Halevi, S. (ed.) CRYPTO 2009. LNCS, vol. 5677, pp. 18–35. Springer, Heidelberg (2009)
13. Prouff, E., Rivain, M., Bevan, R.: Statistical Analysis of Second Order Differential Power Analysis. IEEE Trans. Computers 58(6), 799–811 (2009)
14. Standaert, F.-X.: How Leaky Is an Extractor? In: Abdalla, M., Barreto, P.S.L.M. (eds.) LATINCRYPT 2010. LNCS, vol. 6212, pp. 294–304. Springer, Heidelberg (2010)
15. Standaert, F.-X., Malkin, T., Yung, M.: A Unified Framework for the Analysis of Side-Channel Key Recovery Attacks. In: Joux, A. (ed.) EUROCRYPT 2009. LNCS, vol. 5479, pp. 443–461. Springer, Heidelberg (2009)
16. Standaert, F.-X., Pereira, O., Yu, Y., Quisquater, J.-J., Yung, M., Oswald, E.: Leakage Resilient Cryptography in Practice. In: Basin, D., Maurer, U., Sadeghi, A.-R., Naccache, D. (eds.) Towards Hardware-Intrinsic Security, Information Security and Cryptography, pp. 99–134. Springer, Heidelberg (2010)

17. Standaert, F.-X., Veyrat-Charvillon, N., Oswald, E., Gierlichs, B., Medwed, M., Kasper, M., Mangard, S.: The World Is Not Enough: Another Look on Second-Order DPA. In: Abe, M. (ed.) ASIACRYPT 2010. LNCS, vol. 6477, pp. 112–129. Springer, Heidelberg (2010)
18. Veyrat-Charvillon, N., Standaert, F.-X.: Adaptive Chosen-Message Side-Channel Attacks. In: Zhou, J., Yung, M. (eds.) ACNS 2010. LNCS, vol. 6123, pp. 186–199. Springer, Heidelberg (2010)
19. Yu, Y., Standaert, F.-X., Pereira, O., Yung, M.: Practical leakage-resilient pseudo-random generators. In: Al-Shaer, E., Keromytis, A.D., Shmatikov, V. (eds.) ACM Conference on Computer and Communications Security, pp. 141–151. ACM, New York (2010)

Standardization Works for Security Regarding the Electromagnetic Environment

Tetsuya Tominaga

NTT Energy and Environment Systems Laboratories
3-9-11, Midori-cho, Musashino-shi, Tokyo, 180-8585

Abstract. Telecommunication functions of electronic devices have been and will continue to increase. The so called smart community, a society in which more advanced communications technology is used, will enable life to be increasingly convenient. Thus, telecommunications will become more and more important. However, when such functions become unavailable for some reason, it will negatively impact society. Therefore, device robustness and information leakage are serious issues that need to be addressed. Security regarding electromagnetic waves has been extensively studied in terms of electromagnetic compatibility. In particular, high power electromagnetic phenomena and information leakage due to electromagnetic waves have been discussed in IEEE EMC TC5, ITU-T SG5 and IEC SC77C. In this presentation, an overview of the results, trends, and future works are discussed. Recently developed recommendation ITU-T K.84 (test methods and guide against information leaks through unintentional EM emissions), a leakage mechanism, and protection methods are also discussed.

B. Preneel and T. Takagi (Eds.): CHES 2011, LNCS 6917, p. 273, 2011.

Meet-in-the-Middle and Impossible Differential Fault Analysis on AES

Patrick Derbez[1], Pierre-Alain Fouque[1], and Delphine Leresteux[2]

[1] École Normale Supérieure, 45 rue d'Ulm, F-75230 Paris CEDEX 05
[2] DGA Information Superiority, BP7, 35998 Rennes Armées
{patrick.derbez,pierre-alain.fouque}@ens.fr,
delphine.leresteux@dga.defense.gouv.fr

Abstract. Since the early work of Piret and Quisquater on fault attacks against AES at CHES 2003, many works have been devoted to reduce the number of faults and to improve the time complexity of this attack. This attack is very efficient as a single fault is injected on the third round before the end, and then it allows to recover the whole secret key in 2^{32} in time and memory. However, since this attack, it is an open problem to know if provoking a fault at a former round of the cipher allows to recover the key. Indeed, since two rounds of AES achieve a full diffusion and adding protections against fault attack decreases the performance, some countermeasures propose to protect only the three first and last rounds. In this paper, we give an answer to this problem by showing two practical cryptographic attacks on one round earlier of AES-128 and for all keysize variants. The first attack requires 10 faults and its complexity is around 2^{40} in time and memory, an improvement allows only 5 faults and its complexity in memory is reduced to 2^{24} while the second one requires either 1000 or 45 faults depending on fault model and recovers the secret key in around 2^{40} in time and memory.

Keywords: AES, Differential Fault Analysis, Fault Attack, Impossible Differential Attack, Meet-in-the-Middle Attack.

1 Introduction

Fault Analysis was introduced in 1996 by Boneh *et al.* [8] against RSA-CRT implementations and soon after Biham and Shamir described differential fault attack on the DES block cipher [4]. Several techniques are known today to provoke faults during computations such as provoking a spike on the power supply, a glitch on the clock, or using external methods based on laser, Focused Ion Beam, or electromagnetic radiations [18]. These techniques usually target hardware or software components of smartcards, such as memory, register, data or address bus, assembly commands and so on [1]. After a query phase where the adversary collects pairs of correct and faulty ciphertexts, a cryptographic analysis of these data allows to reveal the secret key. The knowledge of a small difference at an inner computational step allows to *reduce* the analysis to a small number

B. Preneel and T. Takagi (Eds.): CHES 2011, LNCS 6917, pp. 274–291, 2011.

of rounds of a block cipher for instance. On the AES block cipher, many such attacks have been proposed [6,14,17,23,27] and the first non trivial and the most efficient attack has been described by Piret and Quisquater in [27].

Related Works. The embedded software and hardware AES implementations are particularly vulnerable to side channel analysis [5,7,30]. Considering fault analysis, it exists actually three different categories of attacks. The first category is non cryptographic and allows to reduce the number of rounds by provoking a fault on the round counter [1,11]. In the second category, cryptographic attacks perform fault in the state during a round [6,14,17,23,27] and in the third category, the faults are performed during the key schedule [10,17,31].

Several fault models have been considered to attack AES implementations. The first one and the less common is the random bit fault [6], where a fault allows to switch a specific bit. The more realistic and widespread fault model is the random byte fault model used in the Piret-Quisquater attack [27], where a byte somewhere in the state is modified. These different fault models depend on the technique used to provoke the faults.

Piret and Quisquater described a general Differential Fault Analysis (DFA), against Substitution Permutation Network schemes in [27]. Their attack uses a single random byte fault model injected between the two last MixColumns of AES-128. They exploited only 2 pairs of correct and faulty ciphertexts. Since this article was published in 2003, many works have proposed to reduce the number of faults needed in [24,32], or to apply this attack to AES-192 and to AES-256 [20].

There exist two kinds of countermeasures to protect AES implementations against fault attacks. The first category detects fault injection with hardware sensors for instance. However, they are specifically designed for one precise fault injection mean and do not protect against all different fault injection techniques. The second one protects hardware implementation against fault effects. This kind of countermeasures increases the hardware surface requirement as well as the number of operations. As a consequence, there is a tradeoff between the protection and the efficiency and countermeasures essentially only protect from existing fault attacks by taking into account the known state-of-the-art fault analysis. Therefore, the first three and the last three rounds used to be protected [12]. The same kind of countermeasures has been performed on DES implementation and a rich literature has been devoted to increase the number of attacked rounds as it is done in [28]. Securing AES implementation consists in duplicating rounds, verifying operation with inverse operation for non-linear operations and with complementary property for linear ones, for example. Moreover, another approach computes and associates to each vulnerable intermediate value a cyclic redundancy checksum or, an error detection or correction code, for instance fault detection for AES S-Boxes [19] as it has been proposed at CHES 2008. Our attacks could target any operation between MixColumns at the 6th round and MixColumns at the 7th round. Another countermeasure consists in preventing from fault attack inside round [29]. However, it is possible to perform fault injection between rounds.

Our Results. We show that it is possible to mount realistic attacks between MixColumns at the 6th round and MixColumns at the 7th round on AES-128. In particular, we present one new attack and improve a second one at the 7th round on AES-128. We mount our attacks in two different fault models. The first attack corresponds of a strong adversary who could choose or know the attacked byte at the chosen round. The cryptographic analysis relies on a meet-in-the-middle and its complexity is around 2^{42} in time and memory. It only requires 10 pairs of correct and faulty ciphertexts. Recently, in [9], authors developed automatic tool that allows us to automatically recover an improved attack with only 5 pairs and 2^{24} in memory. The second attack describes an adversary that targets any byte among 16 bytes of the inner state at the targeted round. It uses ideas similar to impossible differential attack and allows to recover the secret key using around 2^{40} time. However, this attack requires 1000 pairs. If the position is fixed, the number of faults is reduced to 45. We have verified this attack experimentally using glitch fault on the clock on an embedded microprocessor board which contains an AES software and simulated these two last attacks. Finally, we extend all the attacks to AES-192 and AES-256.

Table 1. Summary of Differential Fault Analysis presented in this paper

Attack	Section	Fault model	# of faults	AES-128 cost	AES-192 & AES-256 cost
Meet-in-the-Middle	3.2	known byte	10	$\simeq 2^{40}$	$\simeq 2^{40}$
Meet-in-the-Middle	3.3	unknown byte	10	$\simeq 2^{60}$	$\simeq 2^{60}$
Meet-in-the-Middle	3.4	fixed unknown byte	5	$\simeq 2^{40}$	$\simeq 2^{40}$
Impossible	4.2	random unknown byte	1000	$\simeq 2^{40}$	$\simeq 2^{40}$
Impossible	4.3	fixed unknown byte	45	$\simeq 2^{40}$	$\simeq 2^{40}$

Organization of the Paper. In Section 2, we recall the backgrounds on AES and on the Piret-Quisquater attack. Then, we describe our meet-in-the-middle and our impossible differential attack on the 7th round in Sections 3 and 4 for AES-128. Finally, in Section 5, we extend these results to the other versions of AES.

2 Backgrounds and Previous Attacks

In this section, we recall the AES operations and we briefly explain how the Piret-Quisquater attack works.

2.1 Description of the AES

AES [15] has a 128-bit input block and can be used with three different key-sizes 128, 192 or 256-bit. It iterates 10 rounds (resp. 12 and 14) for the 128-bit version (resp. for the 192-bit version and for the 256-bit version). According to

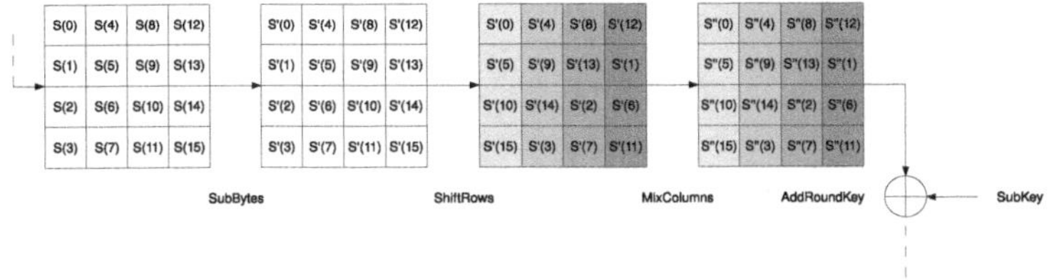

Fig. 1. SubBytes, ShiftRows and MixColumns operations [15]

the bitlength version, we define n as the number of rounds. In Figure 1, we describe one round of the AES which is a composition of the SubBytes, ShiftRows, MixColumns and AddRoundKey operations.

SubBytes (SB). This operation substitutes a value to another one according to the permutation table S-Box, which associates 256 input toward 256 output values. Its goal is to mix non-linearly the bits into one byte.

ShiftRows (SR). This operation changes byte order in the state depending on the row. Each row has its own permutation. The first row changes nothing, the second row is rotated by one position to the left, the third row is rotated by two positions to the left, the fourth row is rotated by three positions to the left.

MixColumns (MC). This operation linearly mixes state bytes by columns and consists in the multiplication of each columns of the state by an MDS matrix (Maximum Distance Separable) in the finite field $GF(2^8)$. We will use the property that, when the input column has one non-null difference in one byte, all the bytes after this operation have a non-null difference.

AddRoundKey Operation (ARK). This operation is only a XOR between intermediate state and the subkey generated by the key schedule.

KeySchedule. The key schedule, which derives the symmetric key K, is composed of two operations, RotWord and SubWord. RotWord is a circular permutation of four elements of one column. SubWord operation corresponds to SubBytes. It is well-known that one subkey of AES-128 allows to retrieve master key K and two consecutive subkeys of AES-192 and AES-256 allow to recover the whole key K. We denote by K_{10} the last subkey of AES-128 and by $K_{10}(0)$ the first byte of the last subkey of AES-128.

2.2 Previous Differential Fault Analysis

In [27], Piret and Quisquater assume a fault injection on one byte during the state computation between the 2 last MixColumns on AES-128 as it is represented in the Figure 2. This attack allows to recover the last subkey in 2^{40} in time and 2^{32} in memory. The idea of the attack consists of expressing 4 differential equation systems at the beginning of the last round state S_{12}. One system is described for each column like equation system (1), where X denotes a non-null

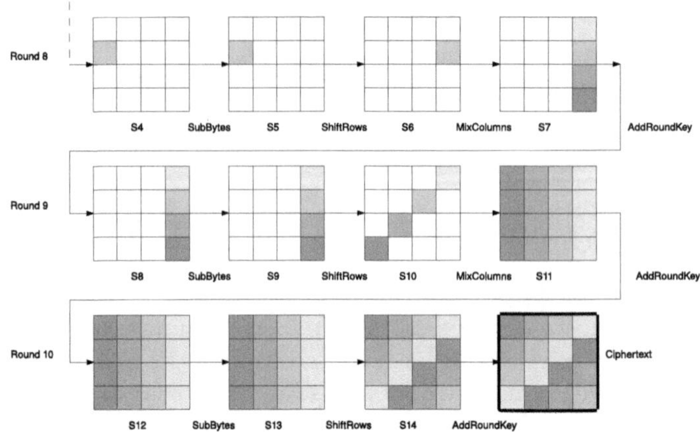

Fig. 2. State-of-the-art differential fault analysis on AES-128

byte difference in state S_{10}. After collecting two couples of correct and faulty ciphertexts, they entirely retrieve the subkey K_{10}.

$$\begin{cases} SB^{-1}(C(0) \oplus K_{10}(0)) \oplus SB^{-1}(\tilde{C}(0) \oplus K_{10}(0)) = X \\ SB^{-1}(C(13) \oplus K_{10}(13)) \oplus SB^{-1}(\tilde{C}(13) \oplus K_{10}(13)) = X \\ SB^{-1}(C(10) \oplus K_{10}(10)) \oplus SB^{-1}(\tilde{C}(10) \oplus K_{10}(10)) = 3X \\ SB^{-1}(C(7) \oplus K_{10}(7)) \oplus SB^{-1}(\tilde{C}(7) \oplus K_{10}(7)) = 2X \end{cases} \tag{1}$$

The right-hand side of the equation system is described one round earlier in annexe B. With only one couple of right and wrong associated results, these equations (1) allow to reduce the possible subkeys from $(2^8)^4 = 2^{32}$ to 2^8 for each equation system. Indeed, according to system (1), there are $(2^8)^4 = 2^{32}$ possible quadruplets of the whole K_{10}:$\{K_{10}(0), K_{10}(13), K_{10}(10), K_{10}(7)\}$. Moreover, there are 2^{40} candidates for $\{X, K_{10}(0), K_{10}(13), K_{10}(10), K_{10}(7)\}$, and the 4 equations give a 32-bit constraint, and consequently, the number of solution is $\frac{2^{40}}{2^{32}} = 2^8$. Then, instead of using another pair of faulty and correct ciphertext as it is done in [27], an exhaustive search can be performed at the end. In the following sections, we will present our differential fault analysis.

3 Meet-in-the-Middle Fault Analysis on AES-128

In our attack, we realize a fault injection on one byte between MixColumns at the 6th round and MixColumns at the 7th round on AES-128. The fault is totally diffused at the whole 10th round as the Figure 3 shows it. This fault analysis requires 10 pairs of correct and faulty ciphertexts. If the attacker knows exactly which byte is faulted, the complexity of the attack is around 2^{40} in time and memory. The overall attack consists in expressing the fault path from

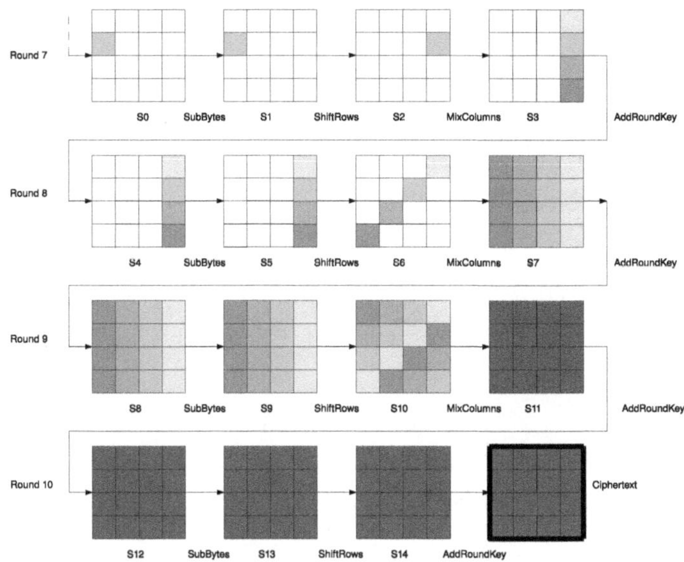

Fig. 3. Overall meet-in-the-middle fault attack on AES-128

the ciphertext to the beginning of the 9th round in the backward direction, and in the forward direction from the fault injection to the beginning of the 9th round. Figure 3 illustrates the error propagation. A classical cryptographic attack against AES, such as the square attack [13], allows to add two rounds after the distinguisher by guessing 5 key bytes. However, this allows to recover one byte of the state. Here, we need to know two bytes of the state, which depend each on 5 different key bytes. By using a clever meet-in-the-middle attack as in the attack of Gilbert and Minier in [16], we are able to recover the key using only 2^{40} space and time. In the following of this section, we explain our differential fault system, our method to retrieve all bytes of the last subkey of AES-128 and its complexity.

3.1 From Fault Path to Differential Fault Equations

The left-hand side of the equation (2) describes the fault path from the ciphertext C at the 10th round toward the state S_8 at the beginning of the 9th round. We obtain:

$$S_8 = SB^{-1}\left(SR^{-1}\left(MC^{-1}\left(ARK^{-1}\left(SB^{-1}\left(SR^{-1}\left(ARK^{-1}(C)\right)\right)\right)\right)\right) \quad (2)$$

We consider each equation byte by byte. The notation $S_8(x)$ denotes the value of the byte x at the state 8. We get the following relations (3) and (4) with $S_8(0)$ and the similar one with $\tilde{S}_8(0)$ as a function of faulty ciphertext $\tilde{C}$, where $MC|_0$ denotes the projection onto the state into the first byte 0.

$$S_8(0) = SB^{-1}\left(MC^{-1}|_0\left(SB^{-1}\left(C(0,7,10,13) \oplus K_{10}(0,7,10,13)\right) \oplus K_9(0,1,2,3)\right)\right) \tag{3}$$

If we define $U_9(0)$ for $\left(MC^{-1}|_0\left(K_9(0,1,2,3)\right)\right)$. Consequently, the byte $S_8(0)$ has the simple expression that depends on 5 unknown bytes, which come key bytes:

$$S_8(0) = SB^{-1}\left(MC^{-1}|_0\left(SB^{-1}\left(C(0,7,10,13) \oplus K_{10}(0,7,10,13)\right)\right) \oplus U_9(0)\right) \tag{4}$$

We obtain a differential equation from the difference between the correct and the faulty state at the end of MixColumns of the 8th round, for example, the first differential equation system (5).

$$\begin{cases} S_8(0) \oplus \tilde{S}_8(0) = X \\ S_8(1) \oplus \tilde{S}_8(1) = X \\ S_8(2) \oplus \tilde{S}_8(2) = 3X \\ S_8(3) \oplus \tilde{S}_8(3) = 2X \end{cases} \tag{5}$$

where X denotes for example the unknown difference of the first column in state S_6 (see Appendix B). We notice that the difference at the end of MixColumns of the 8th round is equal to the difference at the end of AddRoundKey of the 8th round for AES-128. As we mentioned before each equation depends on 5 unknown bytes. We can eliminate the unknown X by considering the following equation:

$$S_8(0) \oplus \tilde{S}_8(0) = S_8(1) \oplus \tilde{S}_8(1). \tag{6}$$

In the next subsection, we will explain how we solve this equation that depend on 80 key bits in time and memory 2^{40} using 10 pairs of faulty and correct ciphertexts.

3.2 Recovery K_{10}

We are interesting in solving 4 difference equations like (5). To simplify the exposition, we will assume that the fault is injected at a known position. Furthermore, the adversary has 10 pairs of correct and faulty ciphertexts and all faults are introduced between the MixColumns at the 6th round and the MixColumns at the 7th round. The constant $U_9(0)$ is invariant for $S_8(x)$ or $\tilde{S}_8(x)$ where $x \in \{0,1,2,3\}$ whatever the plaintext value is.

One idea to solve the system is the following. We consider equation (6) for the ten pairs. Then, we can compute the left hand side for the 2^{40} possible key bytes and store the ten bytes $S_8(0) \oplus \tilde{S}_8(0)$ and the key bytes in a first list. Then, we do the same with the right hand side and store the ten bytes $S_8(1) \oplus \tilde{S}_8(1)$ and the key bytes in a second list. We can merge the two lists, sort them and find collision for the ten bytes. If there is a collision between the two lists, the values of the key bytes gives a solution for the 80 key bits. This simple technique allows to recover the key bytes in time 2^{45} and memory 2^{41}. We can reduce the space complexity by storing and sorting the list and for each value computed for the second list, we look at if it is also in the first list.

In the following, we present a technique that avoids to increase the time complexity too much by using a hash table.

1. The differential state $S_8(0) \oplus \tilde{S}_8(0)$ is calculated for the 5 pairs of faulty and correct ciphertexts and the results are stored in one hash table according to the values $S_8^i(0) \oplus \tilde{S}_8^i(0)_{1 \le i \le 5}$in the one hand, and for the 5 others in the other hand for all possible values of $\{K_{10}(0), K_{10}(7), K_{10}(10), K_{10}(13), U_9(0)\}$. These two hash tables have for input index 5 values of $S_8(0) \oplus \tilde{S}_8(0)$ and for output $\{K_{10}(0), K_{10}(7), K_{10}(10), K_{10}(13),$ $U_9(0)\}$.
2. Then we calculate $\alpha(S_8(1) \oplus \tilde{S}_8(1))$ for all 10 couples of correct and faulty ciphertexts, for all possible hypotheses of $K_{10}(3)$, $K_{10}(6)$, $K_{10}(9)$, $K_{10}(12)$ and $U_9(13)$. Where $U_9(13) = MC^{-1}|_{13}(K_9(12, 13, 14, 15))$ and α is known because fault position is known, i.e $\alpha = 1$. Therefore, we have a relation (7) between $S_8(1)$ and $S_8(0)$ such as:
$$S_8(0) \oplus \tilde{S}_8(0) = \alpha(S_8(1) \oplus \tilde{S}_8(1)) \tag{7}$$
For each guess for $\{K_{10}(3), K_{10}(6), K_{10}(9), K_{10}(12), U_9(13)\}$, due to the 5 first $S_8(0) \oplus \tilde{S}_8(0)$ indexes and the 5 first $\alpha(S_8(1) \oplus \tilde{S}_8(1))$ calculations, we retrieve a very few potential number of solutions $\{K_{10}(0), K_{10}(7), K_{10}(10),$ $K_{10}(13), U_9(0)\}$ closed to 1 for the first hash table. For the second table, we obtain similar results too. For each table, we make and arrange a linked list for the results of $\{K_{10}(0), K_{10}(7), K_{10}(10), K_{10}(13), U_9(0)\}$. Due to these two arrangements and only for the right values of $\{K_{10}(0), K_{10}(7), K_{10}(10),$ $K_{10}(13), U_9(0)\}$, we have only one intersection between the two linked lists; that is why we only retrieve 8 bytes of K_{10} and the value of α is confirmed for each couple of correct and faulty ciphertexts.
3. We similarly compute $\beta(S_8(2) \oplus \tilde{S}_8(2))$ for the 10 couples of correct and faulty ciphertexts, and for all potential subkey bytes of $K_{10}(2)$, $K_{10}(5)$, $K_{10}(8)$, $K_{10}(15)$ and $U_9(10)$, where $U_9(10) = MC^{-1}|_{10}(K_9(8, 9, 10, 11))$. As step 3, β is known for known fault position, i.e $\beta = \frac{1}{3}$. We obtain the equation (8):
$$S_8(0) \oplus \tilde{S}_8(0) = \beta(S_8(2) \oplus \tilde{S}_8(2)) \tag{8}$$
Due to previous step, we have knowledge of the value $S_8(0) \oplus \tilde{S}_8(0)$ for the 10 pairs of cipher results. We reuse the previous method of two arranged linked lists. We retrieve $K_{10}(2)$, $K_{10}(5)$, $K_{10}(8)$, $K_{10}(15)$ and $U_9(10)$.
4. As $S_8(2)$, we compute $S_8(3) \oplus \tilde{S}_8(3)$ for the 10 correct and faulty ciphertexts for all possible subkey bytes of $\{K_{10}(1), K_{10}(4), K_{10}(11), K_{10}(14), U_9(7)\}$. Where $U_9(7) = MC^{-1}|_7(K_9(4, 5, 6, 7))$ and $\gamma = \frac{1}{2}$. We have the equation (9):
$$S_8(0) \oplus \tilde{S}_8(0) = \gamma(S_8(3) \oplus \tilde{S}_8(3)) \tag{9}$$
We also retrieve $K_{10}(1)$, $K_{10}(4)$, $K_{10}(11)$, $K_{10}(14)$, $U_9(7)$ as step 3.

3.3 Cost and Complexity

By the birthday paradox, we have two hash tables with 2^{40} values inside. The complexity of all the system is also 2^{80}. However each equation gives 8-bit constraints, so with ten equations we obtain 80-bit constraints. Consequently, with ten ciphertexts, there is only one solution in our system. Our meet-in-the-middle fault attack requires around 2^{40} in complexity for AES-128: 2^{40} in memory and 3×2^{40} in instructions.

Random Byte Fault Model. In equation (7), α takes on the values $\{\frac{1}{3}, 1, \frac{3}{2}, 2\}$ in case of unknown fault position. Several cases could be studied. In the first one, we know exactly for each faulty ciphertext byte faulty position, we have knowledge of α for each equation. In the second one, we use the same method to inject fault at the same time, we suppose that the same byte is faulted. For consequences, it multiplies by four the computations. In the third case, the worst, we make no assumptions on the location of the fault for each pair of correct and incorrect ciphertexts. In fact for each couple of correct and incorrect results, we need to compute 4 intermediate results. This operation costs 4^{10} values more, it costs too much, i.e 2^{60}.

3.4 Reduction of Memory Requirement

We suppose that an adversary has a *sixtuplet* of the correct message and five faulty ciphertexts, with all five faults on the same byte. In this case, the tool from [9] allows us to find a similar attack but it requires much less memory, 2^{24} instead of 2^{40}.
The previous attack can be schematized as follows :

- Build the four lists, the index 0 corresponds to the correct ciphertext :
  - $L_0 = \{(K_{10}(0), K_{10}(7), K_{10}(10), K_{10}(13), S_9^0(0))\}$
  - $L_1 = \{(K_{10}(3), K_{10}(6), K_{10}(9), K_{10}(12), S_9^0(1))\}$
  - $L_2 = \{(K_{10}(2), K_{10}(5), K_{10}(8), K_{10}(15), S_9^0(2))\}$
  - $L_3 = \{(K_{10}(1), K_{10}(4), K_{10}(11), K_{10}(14), S_9^0(3))\}$
- Each element of L_i allows to deduce unique values for $\Delta S_8^j(i)$, $j = 1, \ldots, 5$ is the index of jth faulty ciphertext.
- Look for collisions since the vector $\left(\Delta S_8^j(0), \ldots, \Delta S_8^j(3)\right)$ must be collinear with a column vector of the matrix of the MixColumn operation.

To reduce memory, we note that we can build each list in beginning with guessing $\Delta S_8^1(0)$ and $\Delta S_8^2(0)$. This operation allows us to partially build the lists and thus save memory.
Building, for example, the list L_0 by assuming that these values are known :

- Build the list $L'_0 = \{(K_{10}(0), S_8^0(0))\}$
- Each element of L'_0 allows to deduce unique values for :
  - $\Delta S_{10}^j(0)$, $j = 1, 2$
  - $\Delta S_{11}^j(0)$, $j = 1, \ldots, 5$

- Guess $K_{10}(7), K_{10}(10), K_{10}(13)$
  - Deduce $\Delta S_{11}^{j}(1,2,3)$, $j = 1, \ldots, 5$
  - Look in L'_0 corresponding values for $K_{10}(0)$ and $S_8^0(0)$ using $\Delta S_{11}^{j} = MC\left(\Delta S_{10}^{j}\right)$
  - Deduce $\Delta S_8^{j}(0)$, $j = 3, 4, 5$

L_1, L_2 and L_3 can be built in the same way.
This improvement makes the attack much more feasible. The implementation providing by the tool takes a little bit more than 13 days on a Core 2 Duo E8500 and 900MB of ram to test all possibilities but it can be improved by parallelizing the C code.

4 Impossible Differential Fault Attack on AES-128

In this section, we present a more efficient attack since we do not assume where the fault is provoked and the time complexity is reduced to 2^{41}. However, this fault attack needs more faulty ciphertexts, less than 1000 or 45 depending on the fault model. Our attack is based on the fact that it is impossible to have a zero-difference in state S_{10} in the 9th round just before MixColumns operation; as Phan and Yen mentioned this fact in [26] and developed with an example of the fault injected on the subkey K_7 in the key schedule. This fact is illustrated by the Figure 4. In this section, two principles are associated, the first one impossible differential, which is first published in [21,22], and the second one fault analysis, like [2,26]. Our impossible differential fault analysis corresponds to 5-round impossible differential cryptanalysis attack, which is described in [3]. We firstly present the differential inequation systems, then the retrieval algorithm and in the last part the comparison between the experimental, simulation and theoretical results.

4.1 From Impossible Differential to Inequation System

Due to a well-known property of the differential through the MixColumn operation, all differences between bytes are *not null* at the internal state S_{10} in (10).

$$S_{10}(C) \oplus S_{10}(\tilde{C}) \neq 0 \tag{10}$$

Moreover, we have the following equation (11):

$$S_{10}(C) = MC^{-1}\left(SB^{-1}\left(SR^{-1}(C \oplus K_{10})\right) \oplus K_9\right) \tag{11}$$

We obtain similar equation for $S_{10}(\tilde{C})$. Like the attack below, we have the same simplification with the subkey K_9. The differential equations have the following form (12):

$$S_{10}(C) \oplus S_{10}(\tilde{C}) = MC^{-1}\left(SB^{-1}\left(SR^{-1}(C \oplus K_{10})\right)\right) \oplus MC^{-1}\left(SB^{-1}\left(SR^{-1}(\tilde{C} \oplus K_{10})\right)\right) \tag{12}$$

We execute the same kind of computations as in the previous attack. We analyze column per column. We guess 4 key bytes of K_{10}. Due to the 4 inequalities, we

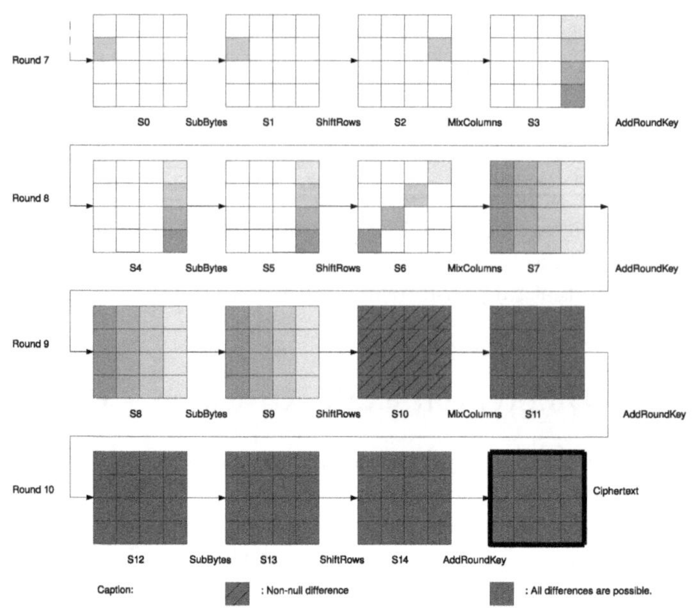

Fig. 4. Overall impossible differential fault attack on AES-128

can filter bad key byte candidates in a list of possible keys. Using many pairs of correct and faulty ciphertexts, we can reduce the possible key space. We reuse four times the no difference computation algorithm for each column of S_{10}. In this attack, the attacker does not use fault position to retrieve the last subkey bytes. The algorithm allows to recover all bytes of the subkey K_{10}. In the case of AES-128, it is enough to retrieve the secret key K.

4.2 Recovery Steps

1. For each pair of correct and incorrect results, we take four guesses for $\{K_{10}(0), K_{10}(13), K_{10}(10), K_{10}(7)\}$. Then we eliminate at each level the key quadruplets which do not satisfy the system (13). We test at each loop all not dismissed quadruplets among 2^{32} possible quadruplets at the beginning.

$$\begin{cases} MC^{-1}|_0(SB^{-1}(C(0) \oplus K_{10}(0))) \oplus MC^{-1}|_0(SB^{-1}(\tilde{C}(0) \oplus K_{10}(0))) \neq 0 \\ MC^{-1}|_1(SB^{-1}(C(13) \oplus K_{10}(13))) \oplus MC^{-1}|_1(SB^{-1}(\tilde{C}(13) \oplus K_{10}(13))) \neq 0 \\ MC^{-1}|_2(SB^{-1}(C(10) \oplus K_{10}(10))) \oplus MC^{-1}|_2(SB^{-1}(\tilde{C}(10) \oplus K_{10}(10))) \neq 0 \\ MC^{-1}|_3(SB^{-1}(C(7) \oplus K_{10}(7))) \oplus MC^{-1}|_3(SB^{-1}(\tilde{C}(7) \oplus K_{10}(7))) \neq 0 \end{cases} \quad (13)$$

2. We repeat previous steps and we retrieve the right quadruplets of K_{10} for each following column.

3. This research could be complemented by an exhaustive search, if only less than 2^{10} possible quadruplets for each column are left. Hence, a global complexity is 2^{40} research operations.

4.3 Property of Recombination

An interesting property of reusing incorrect ciphertexts is described here. The same plaintext is encrypted while fault injection targeted on the same byte. Only MixColumns operation generates collision in one byte, whereas the others do not. Furthermore, if two different inputs of MixColumns only vary on one byte, the two outputs of MixColumns do not collide. For instance, if two different random byte faults ϵ_1 and ϵ_2 are injected on state S_0.

$$\exists ! y \in [0,15], \forall \epsilon_1 \neq \epsilon_2, S_0(y) = x \oplus \epsilon_1 \quad S_0(y) = x \oplus \epsilon_2 \tag{14}$$

$\tilde{C}^{(1)}$ is the faulty ciphertext obtained where fault ϵ_1 is injected, similarly, $\tilde{C}^{(2)}$ the faulty ciphertext links to fault ϵ_2. We have the two following facts :

$$S_{10}(C) \oplus S_{10}(\tilde{C}^{(1)}) \neq 0 \tag{15}$$

$$S_{10}(C) \oplus S_{10}(\tilde{C}^{(2)}) \neq 0 \tag{16}$$

Due to equation (14) and the properties of the MixColumns described below, we obtain the following inequation:

$$S_{10}(\tilde{C}^{(1)}) \oplus S_{10}(\tilde{C}^{(2)}) \neq 0 \tag{17}$$

On our test platform, we collect with one correct ciphertext, 5 or 6 different faulty ciphertexts whose faulty bytes are the same.

4.4 Theoretical and Simulation Results

Theoretical Cost and Complexity. The impossible differential algorithm requires 2^{32} guesses as there are 4 unknown key bytes on each column. The probability that all 4 inequations are satisfied equals $(\frac{255}{256})^4$. With one pair of correct and faulty ciphertexts, we eliminate around 2^{26} subkeys of K_{10} amongst 2^{32} possible values of K_{10} for each column: $E = 2^{32} \times (1 - (\frac{255}{256})^4) \simeq 2^{26}$. Each couple could bring the same information about the key than another couple. The recombination of faulty results introduces collision too. Same quadruplets of key bytes are eliminated several times. Two couples of correct and incorrect ciphertexts create an overlap of $\frac{E^2}{2^{32}} \simeq \frac{(2^{26})^2}{2^{32}} = \frac{2^{52}}{2^{32}} = 2^{20}$. We define U_n as the number of rejected quadruplets with n pairs of correct and faulty ciphertexts with the following recursive formula, where $U_0 = 0$:

$$U_{n+1} = 2^{26} + U_n(1 - 2^{-6}). \tag{18}$$

In solving recurrence in previous equation 18, we obtain the following equation:

$$U_n = 2^{32} - 2^{32}(1 - 2^{-6})^n. \tag{19}$$

The recovery algorithm of the impossible difference stops where $U_n \geq 2^{32} - 2^{10}$. That is why, due to equation 19,

$$n \geq -22\log(2)/\log(1 - 2^{-6}) \Leftrightarrow n \geq 968. \tag{20}$$

Simulation Results. We obtain around 2^{26} eliminated quadruplets of bytes for each pair. We also retrieve the calculated overlap of 2^{20} between two pairs. Considering the random byte fault model, we need on average around 1000 couples of correct and faulty ciphertexts with performing an exhaustive search on 2^{40} possible subkeys at the end. In the case of recombination based on the fixed byte fault model, due to collision results, our attack only requires about 45 faulty ciphertexts with the same plaintext among the 256 possible ciphertexts: $\binom{45+1}{2} = \frac{45\times46}{2} = 1035 > 1000$. It is also possible to combine classical resolution with several recombinations.

5 Extension to AES-192 and AES-256

Introducing fault between the MixColumns of the 6th round and the MixColumns of the 7th round on AES-128 amounts to injecting fault between the MixColumns of the 8th round and MixColumns of the 9th round on AES-192, and between the 10th round and the 11th round on AES-256. Because faults are injected one round before all previous papers, we have access at the same time at subkeys K_n and K_{n-1} with the same differential path.

5.1 Meet-in-the-Middle Fault Analysis on AES-192 and AES-256

We extend the previous concepts for AES-192 and AES-256 without more faulty ciphertexts than AES-128. We use the meet-in-the-middle algorithm in order to recover: $\{K_n(4), K_n(1), K_n(14), K_n(11), U_{n-1}(7)\}$, $\{K_n(8), K_n(5), K_n(2), K_n(15), U_{n-1}(10)\}$, $\{K_n(0), K_n(7), K_n(10), K_n(13), U_{n-1}(0)\}$ and $\{K_n(3), K_n(6), K_n(9), K_n(12), U_{n-1}(13)\}$. We obtain 2 tables which contain $S_8(0) \oplus \tilde{S}_8(0)$ for 5 couples of correct and incorrect results. We compute $S_8(1) \oplus \tilde{S}_8(1)$, $S_8(2) \oplus \tilde{S}_8(2)$ and $S_8(3) \oplus \tilde{S}_8(3)$. By hypothesis, we know fault position for each faulty ciphertext, it means that α, β and γ are known for all equations. Due to these computations, we retrieve all bytes of the subkey K_n. We write the differential equations $S_8(5)$, $S_8(10)$ and $S_8(15)$ as a function of the same 4 bytes of K_{n-1}. Then we also write system of $S_8(6)$, $S_8(11)$ and $S_8(12)$ as a function of the same 4 bytes of K_{n-1}, $S_8(7)$, $S_8(8)$ and $S_8(13)$ as a function of 4 bytes of K_{n-1} and $S_8(4)$, $S_8(9)$ and $S_8(14)$ as a function of 4 bytes of K_{n-1}. We inject the 16 computed bytes of K_n in the previous equations like (5). We recognize the form of

Piret and Quisquater equations in our ones (21), that is why we apply Piret and Quisquater resolution in our recovery method.

$$\begin{cases} SB^{-1}(A \oplus U_{n-1}(4)) \oplus SB^{-1}(\tilde{A} \oplus U_{n-1}(4)) = Y \\ SB^{-1}(B \oplus U_{n-1}(1)) \oplus SB^{-1}(\tilde{B} \oplus U_{n-1}(1)) = 3Y \\ SB^{-1}(C \oplus U_{n-1}(14)) \oplus SB^{-1}(\tilde{C} \oplus U_{n-1}(14)) = 2Y \\ SB^{-1}(D \oplus U_{n-1}(11)) \oplus SB^{-1}(\tilde{D} \oplus U_{n-1}(11)) = Y \end{cases} \quad (21)$$

The values $\{A, B, C, D\}$ are known values at this stage and only depend on the correct ciphertext and K_n. The values $\{\tilde{A}, \tilde{B}, \tilde{C}, \tilde{D}\}$ are known values too and only depend on the faulty ciphertext and K_n. Using 3 generalizations of Piret and Quisquater equation systems allow to recover the subkey U_{n-1}, because we have already retrieved $\{U_{n-1}(0), U_{n-1}(13), U_{n-1}(10), U_{n-1}(7)\}$. Then we resolve 4 systems of 4 equations in using the Gauss' method. Each equation describes MixColumns inverse operation with unknown outputs, in order to recover all bytes of K_{n-1}. This scenario costs around 2^{40} in complexity for AES-192 or AES-256 divided in 2^{40} for memory and 3×2^{40} for operation code like AES-128, plus 2^{40} for Piret and Quisquater resolution.

5.2 Impossible Differential Fault Analysis on AES-192 and AES-256

In the cases of AES-192 and AES-256, we do not need more fault than AES-128 if no exhaustive search is realized. However, we have to collect couples until all bytes of the subkey K_n are retrieved. We reuse the equation systems (5) of the first attack, because both attacks consider fault injection between the same MixColumns. Now, we obtain as the previous subsection the systems (21), thanks to which we know all bytes of K_n. In order to retrieve all bytes of the subkey K_{n-1}, we use 4 Piret and Quisquater generalization. This fault attack is achieved with a complexity around 2^{42}, because Piret and Quisquater generalization has the same cost as Piret and Quisquater attack [27] described in the second part of this paper.

6 Conclusion

We have presented two different attacks on the $n-3^{th}$ round of AES as it is shown in Table 1. The first attack implies random fault byte on known or fixed position for AES-128, AES-192 or AES-256. The second attack involves random fault byte too with less complexity for AES-128. The first one costs around 2^{42} and requires 10 pairs of correct and faulty ciphertexts, its improvement 5 pairs and costs 2^{40} whereas the second one around 2^{40} deals with 1000 couples. Moreover, we can associate the first analysis to solve the second subpart of the second analysis. In this case, a differential fault analysis could be performed on AES-128, AES-192 and AES-256 with a random fault injected between the $n-4^{th}$ and the $n-3^{th}$ MixColumns. Current state-of-the-art countermeasure consists on protecting the three first rounds and the three last rounds of AES. All operations inside round need to be protected and state between rounds too. In order to defeat our fault analysis, all AES-128 rounds need to be protected against fault attacks. Considering AES-192 and AES-256, at least the last 5 rounds and the first 5 rounds need to be protected against fault analysis.

Acknowledgments. We would like to thank Nicolas Guillermin and the anonymous reviewers for their helpful and valuable comments and discussions.

References

1. Anderson, R.J., Kuhn, M.G.: Low Cost Attacks on Tamper Resistant Devices. In: Christianson, B., Lomas, M. (eds.) Security Protocols 1997. LNCS, vol. 1361, pp. 125–136. Springer, Heidelberg (1998)
2. Biham, E., Granboulan, L., Nguyen, P.Q.: Impossible fault analysis of RC4 and differential fault analysis of RC4. In: Gilbert, H., Handschuh, H. (eds.) FSE 2005. LNCS, vol. 3557, pp. 359–367. Springer, Heidelberg (2005)
3. Biham, E., Keller, N.: Cryptanalysis of Reduced Variants of Rijndael. In: 3rd AES Conference, New York, USA (2000)
4. Biham, E., Shamir, A.: Differential Fault Analysis of Secret Key Cryptosystems. In: Kaliski Jr., B.S. (ed.) CRYPTO 1997. LNCS, vol. 1294, pp. 513–525. Springer, Heidelberg (1997)
5. Biryukov, A., Khovratovich, D.: Two New Techniques of Side-Channel Cryptanalysis. In: Paillier, P., Verbauwhede, I. (eds.) CHES 2007. LNCS, vol. 4727, pp. 195–208. Springer, Heidelberg (2007)
6. Bloemer, J., Seifert, J.-P.: Fault Based Cryptanalysis of the Advanced Encryption Standard (AES). In: Wright, R.N. (ed.) FC 2003. LNCS, vol. 2742, pp. 162–181. Springer, Heidelberg (2003)
7. Bogdanov, A.: Improved Side-Channel Collision Attacks on AES. In: Adams, C., Miri, A., Wiener, M. (eds.) SAC 2007. LNCS, vol. 4876, pp. 84–95. Springer, Heidelberg (2007)
8. Boneh, D., DeMillo, R.A., Lipton, R.J.: On the Importance of Checking Cryptographic Protocols for Faults (Extended Abstract). In: Fumy, W. (ed.) EUROCRYPT 1997. LNCS, vol. 1233, pp. 37–51. Springer, Heidelberg (1997)
9. Bouillaguet, C., Derbez, P., Fouque, P.-A.: Automatic Search of Attacks on Round-Reduced AES and Applications. In: Rogaway, P. (ed.) CRYPTO 2011. LNCS, vol. 6841, pp. 169–187. Springer, Heidelberg (2011)
10. Chen, C.-N., Yen, S.-M.: Differential Fault Analysis on AES Key Schedule and Some Countermeasures. In: Safavi-Naini, R., Seberry, J. (eds.) ACISP 2003. LNCS, vol. 2727, pp. 118–129. Springer, Heidelberg (2003)
11. Choukri, H., Tunstall, M.: Round Reduction Using Faults. In: Proceedings of the Workshop on Fault Diagnosis and Tolerance in Cryptography, FDTC 2005, pp. 13–24 (2005)
12. Clavier, C., Feix, B., Gagnerot, G., Roussellet, M.: Passive and Active Combined Attacks on AES Combining Fault Attacks and Side Channel Analysis. In: FDTC, pp. 10–19 (2010)

13. Daemen, J., Rijmen, V.: The Design of Rijndael: AES - The Advanced Encryption Standard. Springer, Heidelberg (2002)
14. Dusart, P., Letourneux, G., Vivolo, O.: Differential Fault Analysis on A.E.S. In: Zhou, J., Yung, M., Han, Y. (eds.) ACNS 2003. LNCS, vol. 2846, pp. 293–306. Springer, Heidelberg (2003)
15. FIPS. Advanced Encryption Standard (AES). pub-NIST (November 2001)
16. Gilbert, H., Minier, M.: A Collision Attack on 7 Rounds of Rijndael. In: AES Candidate Conference. LNCS, pp. 230–241. Springer, Heidelberg (2000)
17. Giraud, C.: DFA on AES. In: Dobbertin, H., Rijmen, V., Sowa, A. (eds.) AES 2005. LNCS, vol. 3373, pp. 27–41. Springer, Heidelberg (2005)
18. Hamid, H.B.-E., Choukri, H., Tunstall, D.N.M., Whelan, C.: The Sorcerer's Apprentice Guide to Fault Attacks (2004), `http://eprint.iacr.org/2004/100.pdf`
19. Kermani, M.M., Reyhani-Masoleh, A.: A Lightweight Concurrent Fault Detection Scheme for the AES S-Boxes Using Normal Basis. In: Oswald and Rohatgi [25], pp. 113–129
20. Kim, C.H.: Differential Fault Analysis against AES-192 and AES-256 with Minimal Faults. In: Workshop on Fault Diagnosis and Tolerance in Cryptography, pp. 3–9 (2010)
21. Knudsen, L.R.: DEAL - a 128 bit block cipher. In: Technical report 151, Departement of Informatics, University of Bergen, Norway (1998)
22. Knudsen, L.R.: DEAL - a 128 bit block cipher. In: AES Round 1 Technical Evaluation, NIST (1998)
23. Moradi, A., Shalmani, M.T.M., Salmasizadeh, M.: A Generalized Method of Differential Fault Attack Against AES Cryptosystem. In: Goubin, L., Matsui, M. (eds.) CHES 2006. LNCS, vol. 4249, pp. 91–100. Springer, Heidelberg (2006)
24. Mukhopadhyay, D.: An Improved Fault Based Attack of the Advanced Encryption Standard. In: Preneel, B. (ed.) AFRICACRYPT 2009. LNCS, vol. 5580, pp. 421–434. Springer, Heidelberg (2009)
25. Oswald, E., Rohatgi, P. (eds.): CHES 2008. LNCS, vol. 5154. Springer, Heidelberg (2008)
26. Phan, R.C.-W., Yen, S.-M.: Amplifying Side-Channel Attacks with Techniques from Block Cipher Cryptanalysis. In: Domingo-Ferrer, J., Posegga, J., Schreckling, D. (eds.) CARDIS 2006. LNCS, vol. 3928, pp. 135–150. Springer, Heidelberg (2006)
27. Piret, G., Quisquater, J.-J.: A Differential Fault Attack Technique against SPN Structures, with Application to the AES and KHAZAD. In: Walter, C.D., Koç, Ç.K., Paar, C. (eds.) CHES 2003. LNCS, vol. 2779, pp. 77–88. Springer, Heidelberg (2003)
28. Rivain, M.: Differential Fault Analysis on DES Middle Rounds. In: Clavier, C., Gaj, K. (eds.) CHES 2009. LNCS, vol. 5747, pp. 457–469. Springer, Heidelberg (2009)
29. Satoh, A., Sugawara, T., Homma, N., Aoki, T.: High-Performance Concurrent Error Detection Scheme for AES Hardware. In: Oswald and Rohatgi [25], pp. 100–112
30. Schramm, K., Leander, G., Felke, P., Paar, C.: A Collision-Attack on AES: Combining Side Channel- and Differential-Attack. In: Joye, M., Quisquater, J.-J. (eds.) CHES 2004. LNCS, vol. 3156, pp. 163–175. Springer, Heidelberg (2004)
31. Takahashi, J., Fukunaga, T., Yamakoshi, K.: DFA Mechanism on the AES Key Schedule. In: FDTC 2007: Proceedings of the Workshop on Fault Diagnosis and Tolerance in Cryptography, pp. 62–74. IEEE Computer Society, Los Alamitos (2007)
32. Tunstall, M., Mukhopadhyay, D.: Differential Fault Analysis of the Advanced Encryption Standard using a Single Fault. Cryptology ePrint Archive, Report 2009/575 (2009), `http://eprint.iacr.org/`

A Difference Path from the 10th to the 9th Round for AES-128

Due to fault path from the ciphertext to the beginning of the 9th round, we give the following relations between bytes at different steps for the Meet-in-the-Middle attack. We obtain the following system of 4 equations, where $U_9(a, b, c, d) = MC^{-1}(K_9(a, b, c, d))$, for AES-128 from the 10th to the 9th round, for AES-192 from the 12th to the 11th and for AES-256 from the 14th to the 13th:

$$S_8(0,5,10,15) = SB^{-1}\left(MC^{-1}\left(SB^{-1}\left(C(0,7,10,13) \oplus K_{10}(0,7,10,13)\right)\right) \oplus U_9(0,1,2,3)\right) \quad (22)$$

$$S_8(1,6,11,12) = SB^{-1}\left(MC^{-1}\left(SB^{-1}\left(C(3,6,9,12) \oplus K_{10}(3,6,9,12)\right)\right) \oplus U_9(12,13,14,15)\right) \quad (23)$$

$$S_8(2,7,8,13) = SB^{-1}\left(MC^{-1}\left(SB^{-1}\left(C(2,5,8,15) \oplus K_{10}(2,5,8,15)\right)\right) \oplus U_9(8,9,10,11)\right) \quad (24)$$

$$S_8(3,4,9,14) = SB^{-1}\left(MC^{-1}\left(SB^{-1}\left(C(1,4,11,14) \oplus K_{10}(1,4,11,14)\right)\right) \oplus U_9(4,5,6,7)\right) \quad (25)$$

B Difference Path from the 7th towards the 8th Round on AES-128

Fault on one byte among bytes $\{0, 5, 10, 15\}$ at the 7th round on AES-128 produces case 1, fault on one byte among $\{3, 4, 9, 14\}$ produces case 2, fault on one byte among $\{2, 7, 8, 13\}$ produces case 3 and fault on one byte among $\{1, 6, 11, 12\}$ produces case 4. All different cases are presented in Figure 5. We obtain same behavior with fault injected at the 9th round of AES-192 and at the 11th round of AES-256.

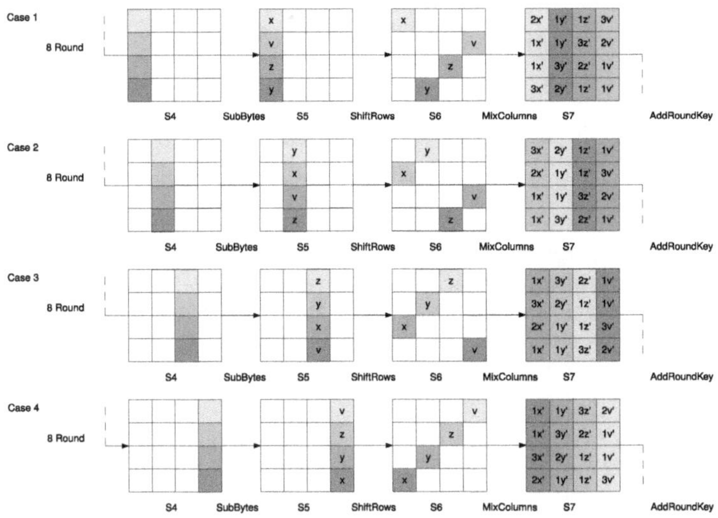

Fig. 5. Difference path during the 8th round for the four different AES-128 cases

On the Power of Fault Sensitivity Analysis and Collision Side-Channel Attacks in a Combined Setting

Amir Moradi[1], Oliver Mischke[1], Christof Paar[1], Yang Li[2], Kazuo Ohta[2], and Kazuo Sakiyama[2]

[1] Horst Görtz Institute for IT Security, Ruhr University Bochum, Germany
{moradi,mischke,cpaar}@crypto.rub.de

[2] Department of Informatics, The University of Electro-Communications
1-5-1 Chofugaoka, Chofu, Tokyo 182-8585, Japan
liyang@ice.uec.ac.jp, {ota,saki}@inf.uec.ac.jp

Abstract. At CHES 2010 two powerful new attacks were presented, namely the Fault Sensitivity Analysis and the Correlation Collision Attack. This paper shows how these ideas can be combined to create even stronger attacks. Two solutions are presented; both extract leakage information by the fault sensitivity analysis method while each one applies a slightly different collision attack to deduce the secret information without the need of any hypothetical leakage model. Having a similar fault injection method, one attack utilizes the non-uniform distribution of faulty ciphertext bytes while the other one exploits the data-dependent timing characteristics of the target combination circuit. The results when attacking several AES ASIC cores of the SASEBO LSI chips in different process technologies are presented. Successfully breaking the cores protected against DPA attacks using either gate-level countermeasures or logic styles indicates the strength of the attacks.

1 Introduction

Since the last decade designers of cryptographic devices have to deal with the problem that embedded secret information, e.g., the used encryption key in a symmetric cipher, is leaking through physical side channels. The side-channel leakage can be the timing [11], the power consumption [12], the electro-magnetic radiation [8,21] and so on. Also fault analysis attacks such as differential fault analysis (DFA) was demonstrated to be effective against implementations of block ciphers such as DES [5] and AES [20].

At CHES 2010 a new fault attack called Fault Sensitivity Analysis (FSA) [14] was proposed. The authors used fault injections by means of clock glitches to measure the calculation time of the S-boxes as side-channel leakage. This information, whose data dependency is caused by the underlying gates, is then used to recover the secret information. Using this attack the secret key of an AES ASIC implementation could be completely extracted, but it was stated that masking might be a countermeasure against such a kind of attack since the

B. Preneel and T. Takagi (Eds.): CHES 2011, LNCS 6917, pp. 292–311, 2011.

randomization of the S-box input makes it difficult to repeat fault injections and to measure the timings of specific calculations.

Another contribution to CHES 2010 was a collision attack enhanced by correlation [15]. Compared to classical power analysis attacks, its main feature is that it does not rely on the knowledge of an underlying (hypothetical) power model. Instead, it directly correlates power traces to each other and – by finding colliding S-box computations – is able to recover the relation between key parts. Using such an attack a complete break of a masked FPGA implementation of AES has been demonstrated.

The combination and improvement of these two ideas is the main contribution of this article. From [14] we will use the fault injection method and utilize the fact that the timing characteristics of each S-box can be independently observed, while the correlation-collision approach from [15] is used to find the relations between key bytes without the need to have any knowledge how the observed characteristics relate to the inputs. Two options to combine these two schemes are presented:

- First we present an attack which exploits the finding that given a fixed clock glitch period and a fixed unmasked S-box input byte the distribution of the resulting faulty ciphertext byte will be data dependent. This is achieved by setting the period of the clock glitch so that about 50% of the executions lead to faulty values. Repeating this measurements for all possible differences between two targeted ciphertext bytes, one can identify the correct key difference if the distributions of two faulty ciphertext bytes collide.
- In the second attack we take a closer look at the fault rate of each S-box instance over a large range of clock glitch periods, thereby getting very detailed information about the timing characteristics of the targets. Again using the concept of the correlation collision attack, several collisions between these timing sets can be detected at the same time. This allows us to fully recover all relations between the key bytes, i.e., shrinking the key space to 2^8.

Similarly to [14], we have chosen the SASEBO-R [2] board as the evaluation platform. The board can hold different ASICs, and we have analyzed the SASEBO LSI2 [3], both in 130nm and 90nm technology, as well as the SASEBO LSI3 [4] in 65nm technology. Each of the LSIs contain the same 14 different implementations of AES, therefore the only difference is the process technology, which has a big influence on the timing characteristics. The implementations themselves differ in the style of the S-box realization and in side-channel countermeasures. Using the attacks presented in this paper we will provide detailed results showing the successful key recovery for a large number of cores including the ones applying gate-level DPA countermeasures and DPA-resistant logic styles.

In the later parts of this article the prerequisites, including a review of fault sensitivity analysis and the correlation collision attack, are given in Section 2. Our first proposed attack using collisions of faulty ciphertext distributions and its results on two SASEBO LSI2 cores are presented in Section 3. The second proposed attack, namely the collision timing attack, is expressed in Section 4, which also contains the practical evaluation results on several of the 130nm, 90nm, and 65nm SASEBO LSI2 and LSI3 cores. Finally, Section 5 concludes the paper.

2 Preliminaries

This section summarizes the underlying attacks, namely the fault sensitivity analysis [14] and correlation collision attack [15], which are the basis to the attacks presented here. We also address how these two methods can be combined to develop more sophisticated attacks improving their efficiency and relaxing the requirements. The experimental setup used in all the practical results shown in this work are also introduced in this section.

2.1 Fault Sensitivity Analysis

A new type of fault attack called Fault Sensitivity Analysis (FSA) was proposed in [14]. Unlike some of the previous fault attacks, e.g., DFA [5], the FSA attack does not require the value of the faulty outputs in the key recovery process. Instead, the attack works by increasing the fault intensity until a distinguishable characteristic can be observed, e.g., the first appearance of a faulty output. The concept of the FSA attack was verified by attacking the unprotected AES_PPRM1 core of the SASEBO LSI2 [3] using only 50 different plaintexts[1]. By successfully revealing three key bytes of the AES_WDDL implementation of the same ASIC using 1200 plaintexts it was also shown that the FSA attack can bypass some known countermeasures against DFA attacks.

The presented method of increasing the fault sensitivity in [14] is the shortening of the clock glitch, whereby the glitch period can be gradually decreased until a faulty output occurs or the fault becomes stable. Since the critical path of some gates, e.g., AND and OR gates, is data dependent, knowing the underlying model for this data dependency helps revealing the secret. For example, the simulation results ascertained that the timing delay of a PPRM S-box correlates to the Hamming weight (HW) of its input. For AES_WDDL, which in theory should be immune against set-up time violation attacks, by profiling with a known key it was shown that at least some bits correlate to the timing delay which lead to the aforementioned recovery of three key bytes. Moreover, this issue has been carefully studied later in [13].

In addition to the fact that FSA does not require faulty ciphertexts, another difference to DFA attacks is that the fault does not need to be restricted to a small sub-space. In contrary, by for example attacking the last round of the AES_PPRM1 implementation, each faulty output byte can be independently observed and therefore the same complete faulty output can be used to attack all key bytes simultaneously. On the other hand, as stated in [14], while countermeasures like masking are only of limited use against DFA attacks, they may

[1] The fault sensitivity leakage for each plaintext has been recovered using around 200 faulty executions.

have a great impact on FSA attacks since the critical path is affected by the random mask bits. Indeed, this is an issue which we demonstrate in this paper to be incorrect by providing the result of successful attacks on the masked implementations.

2.2 Correlation Collision Attack

The correlation collision attack was introduced in [15]. Its major advantage compared to classical power analysis attacks is that it neither relies on a hypothetical power model nor requires a profiling phase. Enhancing linear collision attacks [6] by the methods of correlation-based DPAs, it is able to overcome side-channel countermeasures as long as a minimal first order leakage remains.

A linear collision occurs if two instances of combinational circuits or one instance at two different points in time process(es) the same value, i.e., for AES two 8-bit outputs and thereby the inputs of the S-boxes must be the same. It is therefore possible to recover the key relation between the attacked bytes since rearranging of $\mathrm{Sbox}(i_1 \oplus k_1) = \mathrm{Sbox}(i_2 \oplus k_2)$ leads to $\Delta = i_1 \oplus i_2 = k_1 \oplus k_2$, where both i_1 and i_2 are known.

The correlation collision attack on AES works similarly, but starts by computing sets of mean traces for each possible input byte in case the attack is performed on the first round. To do this for two input bytes, namely i_1 and i_2, all traces are sorted based on the corresponding input byte value, and traces with the same value are averaged, thereby creating 256 different mean traces $M_1(i_1)$ and $M_2(i_2)$ for each of the two input bytes. Computing the variances for each set of mean traces will reveal the point in time where the corresponding bytes are processed by the S-box, which is necessary to align the mean traces for the attack.

If the power consumption of two S-box computations are highly similar, comparing pairs of mean sets also shows a high similarity between certain mean traces. Therefore, when attacking the input bytes i_1 and i_2 and $\Delta = k_1 \oplus k_2$, then $M_1(i_1) \approx M_2(i_2 = i_1 \oplus \Delta)$. The correct Δ can be found by computing the correlation between the two sets of mean traces for each of the 256 candidates of Δ. This yields a very high correlation coefficient since no hypothetical model is applied but instead the averaged real power consumptions are used and only a low number of points contributes to the estimation of the correlation coefficient.

2.3 Combinations

In order to avoid the need of a hypothetical model matching the fault sensitivity leakages, the two above attacks can be combined in several different ways. Two options for such a combination are expressed in this article. It should be noted that each of these two options has been independently developed by each group of the authors, and both have been submitted in parallel to CHES 2011. These two works have been merged as requested by the program committee. The first option, which is developed by the team of the University of

Electro-Communications (Japan), is expressed in Section 3. It extracts the distribution of the faulty ciphertext bytes and tries to find the collision within the distributions to recover the linear difference between the corresponding key bytes. The feasibility of this attack is practically confirmed by breaking two masked AES cores. The second option is developed by the team of the Ruhr University of Bochum (Germany) and is illustrated in Section 4. It extracts the precise timing characteristics of combinational circuits, e.g., S-boxes, and applies the correlation collision attack on timings to detect the colliding cases which reveal, similar to the first option, the linear difference between the corresponding key bytes. The shown practical results of this attack on several different AES cores in different process technologies highlight the strength of this attack.

2.4 Experimental Setup

All the practical results shown in this work are based on the AES implementations of three ASIC chips built for the SASEBO-R board, namely the SASEBO LSI2 (130nm), LSI2 (90nm), and LSI3 (65nm). Each chip contains the same 14 different AES cores including unprotected, DPA protected, and fault attack protected ones. The similar approach for fault injection as in [14] is used to inject the faults or extract the timing characteristics of the target circuit. An additional external clock, generated by an programmable digital function generator, is fed into the SASEBO-R control FPGA where it is multiplied using a Digital Clock Management (DCM) unit. This fast clock signal is then used together with some logic to shape the glitchy clock signal. An internal circuit controls the clock signal of the LSI to infer the glitchy clock at the preferred instance of time synchronized to the AES computation of the target core.

We have first tried to generate the glitchy clock inside the control FPGA without using an external function generator, but the width of the glitchy clock could only be adjusted in large steps (e.g., of around 170ps [7]), which were not small enough to reach the desired results. Therefore, we had to use a function generator to externally provide the precise clock frequencies. As it is represented in the following, we change the width of the glitchy clock in steps of 25ps to 5ps. Also, the multiplication of the clock frequency is necessary because of the limitation (maximum frequency of 15 MHz) of the function generator we have used, while the frequencies necessary to inject a fault in the combinational circuit are up to the range of 300MHz. Also, the DCMs inside the Virtex-II control FPGA of the SASEBO-R can, when fed with a low frequency input signal, only generate output frequency up to 210 MHz. Since some of the cores, especially of the 65nm LSI3, require a higher frequency for fault injection, for these cores it was necessary to daisy chain two DCMs, one for generating a high frequency signal out of the function generator output and another one to reach the maximum supported output frequency which can only be generated by the DCM using a high frequency input [26].

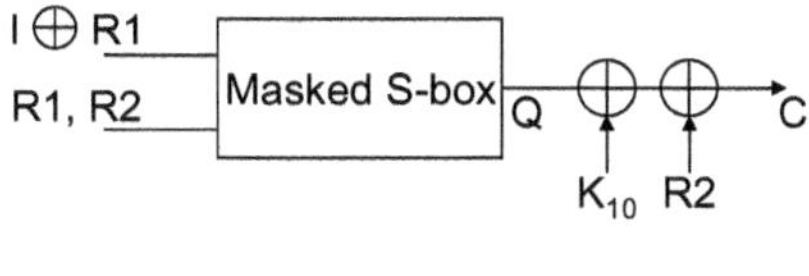

Fig. 1. A combinational circuit in the final round of a masked AES

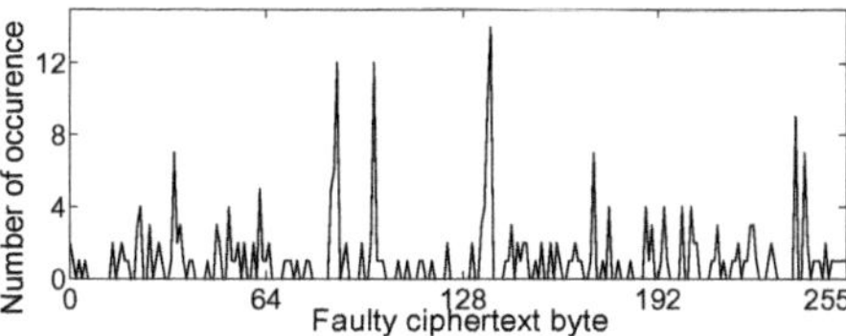

Fig. 2. A faulty ciphertext byte distribution

3 Option 1: Colliding Faulty Ciphertext Distributions

For the implementation with masking countermeasures, attackers cannot keep the device repeating the same calculation due to the randomization of the internal calculations in each trial. Thus, the fault sensitivity is difficult to be measured for a specific calculation, e.g., an S-box calculation with a fixed unmasked input and masks.

As shown in [15], the statistical observation of the side-channel leakage of a masked implementation may recover some sensitive information, e.g., the unmasked inputs, when there is still a first order leakage. We note that when faults are injected, the faulty ciphertexts can be used as an information source in the context of attacking the masked implementation. This section shows that the faulty ciphertext distribution is data dependent and can be used to detect the collision between unmasked intermediate values. This attack has been verified by successfully attacking two AES implementations masked using the Masked-AND gates and a form of threshold implementation.

3.1 Model and Attack Concept

As shown in Fig. 1, we make a simple model of a combinational circuit in the final AES encryption round of a masked implementation. A masked intermediate result $I \oplus R1$ goes through the substitution in a masked S-box calculation, the addition with the final round key K_{10} and the unmasking procedure (XOR with $R2$) to become a ciphertext byte C. In Fig. 1, Q denotes the output of the masked S-box. Hereafter, we use Q' and C' to denote the faulty masked S-box output and the faulty ciphertext byte, respectively. The attack procedure can be divided in two steps consisting of *i*) classifying the random numbers, and *ii*) detecting the colliding unmasked S-box inputs.

Classifying the Random Numbers. A general security requirement for a masking countermeasure is the uniform distribution of the used random numbers, while we use a variation of fault sensitivity to classify the used random numbers.

Given a plaintext, the value of I in Fig. 1 is fixed. For all the possible random numbers, the calculations in the S-box circuit are different, and more importantly the critical delay timings are different. Attackers can focus on a specific S-box calculation and trigger the setup-time violation in the final AES round. The fault injection intensity can be adjusted by modifying the period of the clock glitch.

In our attack, the clock glitch is set to a level where about 50% of the executions generate a faulty output. In this case, the executions are divided into two groups according to whether or not the output is faulty. Furthermore, these two groups of executions are corresponding to two groups of random numbers whose corresponding intermediate values, as we see later, are not uniformly distributed.

Detecting the Colliding Unmasked S-box Inputs. After using the fault sensitivity as a leakage to classify the random intermediate values which are non-uniformly distributed, the next step is to find another information source to effectively identify the sensitive intermediate value. According to Fig. 1, since both K_{10} and $R2$ are the inputs of the XOR gates, these part of the circuit can be seen as a set of fixed (per clock cycle) Buffers and Inverters. According to the architecture of our experimental setup, the round key K_{10} and $R2$ get available at the start of the corresponding clock cycle (last encryption round). Therefore, the computation of the masked S-box (Q), which is definitely longer than two following XORs, is interrupted by the clock glitch. So, we can assume that the faulty ciphertext C' is calculated as

$$C' = Q' \oplus K_{10} \oplus R2. \tag{1}$$

If we suppose that the faulty S-box output Q' has a fixed non-uniform distribution when I is fixed and the fault injection intensity is fixed at 50% success rate, those values of $R2$ which are corresponding to the 50% faulty executions should follow a fixed non-uniform distribution. On the other hand, the values of Q' and $R2$ are not independent of each other. As a result, we expect that the value of $Q' \oplus R2$ follows a fixed non-uniform distribution corresponding to the value of the fixed I. At last, the distribution of C' is permuted based on the value of K_{10}. An example of the distribution of C' is shown in Fig. 2.

The main idea of the attack is to check the similarity between two faulty ciphertext distributions, e.g., of two ciphertext bytes, each of which corresponds to a masked S-box followed by a fixed key addition, i.e., in the last round of the AES encryption. According to the linear collision in AES [6], the difference between key bytes equals to the difference between the corresponding fault-free ciphertext bytes when such a collision occurs.

3.2 Attack Scheme

The attack target is the linear difference ΔK between two bytes of K_{10}, i.e., $\Delta K = K_{10}^i \oplus K_{10}^j$, where $i, j = 1, 2, \cdots, 16$. After guessing the value of ΔK as ΔK_g, the attacker provides one plaintext so that the corresponding ciphertext satisfies $C^i \oplus C^j = \Delta K_g$. Therefore, if the current key difference guess is correct, the unmasked input I of the corresponding masked S-boxes collide. Such a case can be detected by examining the similarity of the distributions of the corresponding faulty ciphertext bytes. The distributions of the faulty ciphertext byte for the targeted S-boxes can be collected using Algorithm 1.

Algorithm 1. Collecting the distribution of the faulty ciphertext byte

Inputs: A plaintext P, number of executions N, position of the target: j
Outputs: The count of all the faulty ciphertext byte: $Cnt(i), i = 0, 1, 2 \ldots 255$.
Set the plaintext as P, $Cnt(i) \leftarrow 0$ for $i = 0, 1, 2 \ldots 255$
Obtain the fault-free ciphertext byte C^j by running the fault-free encryption on P
for $i = 1$ to N **do**
  Obtain the j-th byte of the output C'^j by running the faulty encryption on P
  if $C'^j \neq C^j$ **then**
    $Cnt(C'^j) \leftarrow Cnt(C'^j) + 1$
  end if
end for

Algorithm 2. Attack algorithm (colliding faulty ciphertext distributions)

Inputs: Position of the target key bytes: i and j
Output: Most probable key difference $\Delta K = K^i \oplus K^j$
for $\Delta K = 0$ to 255 **do**
  Select randomly plaintext P so that ciphertext bytes $C^i \oplus C^j = \Delta K$
  Obtain faulty ciphertext distributions Cnt^i and Cnt^j using Algorithm 1
  Set Cnt'^j as the rearranged form of Cnt^j based on ΔK
  $Cor(\Delta K) = \rho(Cnt^i, Cnt'^j)$
end for
return $\arg\max_{\Delta K} Cor(\Delta K)$

Given two distributions of the faulty ciphertext bytes Cnt^i and Cnt^j, one can use the current $\Delta K_g = C^i \oplus C^j$ to rearrange Cnt^j as

$$Cnt'^j(i \oplus \Delta K_g) = Cnt^j(i), \;\; i = 0, 1, 2 \ldots 255,$$

and check the similarity using the correlation coefficient as $\rho(Cnt^i, Cnt'^j)$, where ρ denotes the calculation of the Pearson product-moment correlation coefficient. Repeating this procedure for all possible key differences, the ΔK_g corresponding to the largest correlation coefficient is expected to be the correct key difference. For clarification we have provided a pseudo code of the attack shown by Algorithm 2.

3.3 Practical Results

Results on AES_MAO. The first attack target is the AES core protected against power analysis attacks using the masked-AND gates [24] in 130nm technology. In our experiments, the total number of executions to obtain the faulty ciphertext byte distribution is set to $N = 400$.[2] In order to cover all possible key byte differences, we collected the distributions for 256 plaintexts which

[2] We should mention that N is the total number of executions including faulty and fault-free ones.

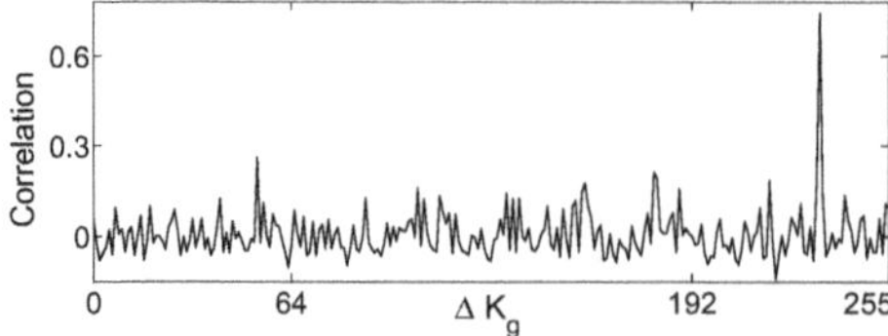

Fig. 3. Correlation vs. Key byte difference for AES_MAO (130nm)

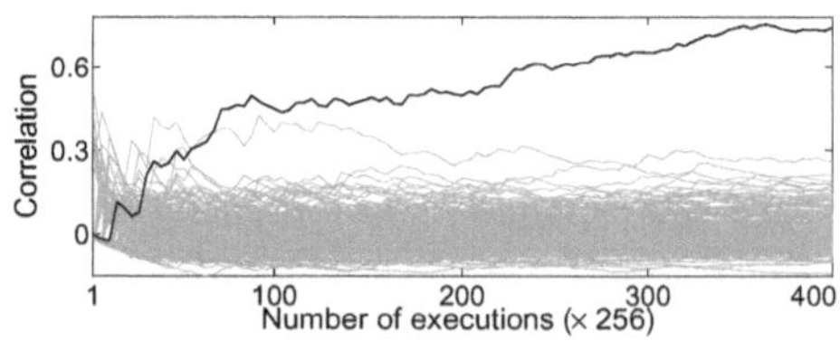

Fig. 4. Correlation evolution vs. Number of executions for AES_MAO (130nm)

correspond to 256 differences between the first two ciphertext bytes. Running the attack algorithm, which is given by Algorithm 2, led to the results shown in Fig. 3 where the correct key byte difference can be clearly identified. Furthermore, Fig. 4 shows that 150 × 256 executions of AES_MAO is sufficient to identify the correct key byte difference. The successful attack experiments have been also confirmed recovering the difference between other key bytes.

Results on AES_TI. The next target is the AES core in 130nm technology realized using the threshold implementation scheme which is a high-order masking countermeasure based on secret sharing [18]. Even in the presence of signal glitches, its resistance against power analysis attacks has been theoretically proven [19]. We should emphasize that the threshold implementation is an algorithmic-level countermeasure which needs to fulfill certain properties including correctness, non-completeness, and uniformity. In contrary to [17], our targeted AES_TI core has been made without considering the later two properties. This core has been realized by modifying a plain AES core at the gate level. The non-linear gates are provided by only 2-input AND gates, every signal is represented by four shares, and finally the AND gates are replaced with the 4-shared threshold implementation of a 2-input AND gate which is available in [18]. This can be verified by examining the source code of this core available at [1].

The attack procedure is the same as the one applied to the AES_MAO core even with the same number of executions, i.e., 400, to obtain the distribution of the faulty ciphertext bytes. The attack result on the key byte difference between the first two key bytes is shown in Fig. 5. The peak corresponding to the correct key byte difference can be clearly identified. Figure 6 also shows that at around 200 × 256 executions are required to identify the correct key byte difference.

3.4 Observations

Relaxing Fault Requirements. One important observation from experiments is that the setting of the fault injection success rate is not as strict as we expected. In our experiments, we could collect the distributions of faulty ciphertext bytes simultaneously for parallel S-boxes. Due to the difference between the inherent delay of parallel S-boxes, the fault injection success rates were different from 40% to 60%. Surprisingly, the key difference can still be clearly recovered. In

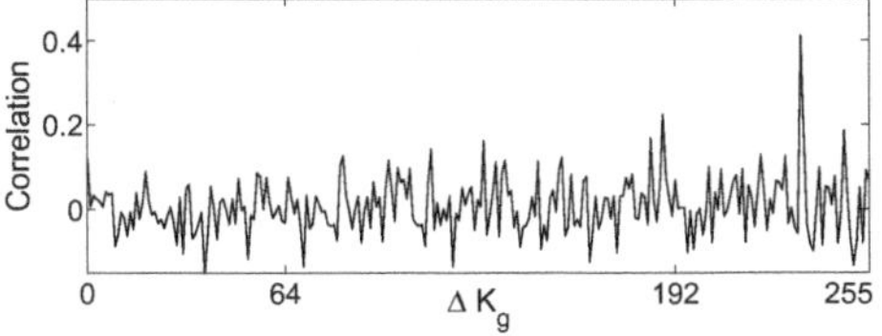

Fig. 5. Correlation vs. Key byte difference for AES_TI (130nm)

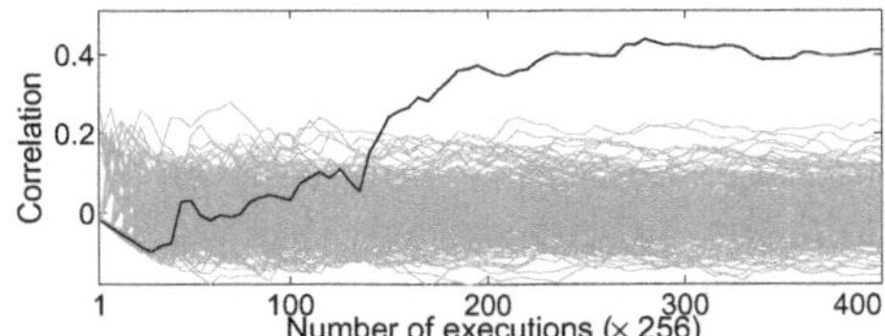

Fig. 6. Correlation evolution vs. Number of executions for AES_TI (130nm)

other words, we found that the distribution of the faulty ciphertext byte is not very sensitive to the fault injection intensity.

As a result, compared to the one-byte fault injection in the DFA attacks or the accurate intensity management in the FSA attack, our proposed attack has the fewest requirements about the fault injection.

Reducing the Number of Executions. In our experiments, we have collected the distributions for 256 ciphertexts to identify a linear difference between two key bytes. The attack efficiency regarding to the number of executions of AES can be easily improved by specifically selecting ciphertexts (plaintexts). For example, one can collect 16 distributions for the first fault-free ciphertext byte as 0x00, 0x01 ... 0x0F and 16 distributions for the second fault-free ciphertext byte as 0x00, 0x10 ... 0xF0, respectively. The combinations of these two groups of 16 distributions already cover all the possible linear key differences. Therefore, with a delicate selection of the ciphertexts (plaintexts), collecting the distributions for the 16 selected ciphertexts are enough to identify a key byte difference.

4 Option 2: Colliding Timing Characteristics

This section expresses the second combination of fault sensitivity analysis and correlation collision attack where the timing characteristics of combinational circuits like S-boxes are analyzed. In the following the fundamental concepts which are essential for the attack are explained, and later practical results of the attack breaking a couple of ASIC AES cores are presented.

4.1 How to Measure the Timing

As explained in [14], when the input of a combinational circuit changes, its output stops toggling after a certain time (so-called Δt). The maximum value of Δt for different inputs is known as the longest critical path of the circuit, and defines the maximum frequency of the clock signal which triggers the flip-flops providing the input and storing the output of the considered combinational circuit. Timing characteristics of a circuit are therefore defined as a set of Δt ($\{\Delta t^1, \Delta t^2, \ldots, \Delta t^n\}$), where Δt^i is the minimum Δt for the given input i.

Let us suppose that the target combinational circuit is a part of a bigger circuit, e.g., a co-processor, which provides some I/O signals for communication.

If the output of the target combinational circuit is stored into registers which are triggered by a clock signal that can be controlled from the outside, as shown in [14], one can steadily shorten the time interval between the input transition and the output storage (known as *setup time*) till an incorrect value is stored into the registers while input i is given to the combinational circuit. The minimum time interval when the considered register stores the correct value can be concluded to Δt^i. Note that this procedure is similar to the scheme explained in Section 3. However, measuring Δt in this case does not deal with the faulty outputs; once a faulty output is detected, Δt^i can be concluded.

It should be noted that, because of the environmental noise, it might be required to repeat the same procedure and shorten the clock glitch period until the probability of detecting faulty output gets higher than a threshold. Also, if the target combinational circuit is not a single-bit function and it is possible to detect which output bit is faulty, one can measure Δt^i for each output bit independently.

Therefore, we define the adversary model and define his capabilities in order to be able to measure Δt^i of the target combinational circuit for the given input i:

- The adversary has access to and can control the clock signal which triggers the registers providing the input and saving the output of the target combinational circuit.
- He knows in which clock cycle the target combinational circuit processes the desired data, e.g., known or guessed input or output.
- He can control the target device in a way that the same input value i is repeatedly processed by the target combinational circuit during shortening the time interval of the clock glitch.
- He is equipped with appropriate instruments to shorten the duration of the clock glitch with suitable accuracy.

4.2 Definitions

Bitwise Capture: $\text{BitCap}^i_{b,\Delta t}$ is the result of a *Bernoulli trial* whether the output of the target combinational circuit at bit b is faulty while processing the input i and when Δt is the time interval of the clock glitch. Correspondingly, $p^i_{b,\Delta t}$ is defined as the probability of "success" in independently repeated $\text{BitCap}^i_{b,\Delta t}$ trials.

Capture: $\text{Cap}^i_{\Delta t} = \bigvee_b \text{BitCap}^i_{b,\Delta t}$. In other words, $\text{Cap}^i_{\Delta t}$ is the same as the above defined trial regardless of a certain output bit, and is meaningful when differentiating between different faulty output bits is not possible, e.g., if a circuit is equipped with a fault detection scheme and prevents the propagation of faulty results. $p^i_{\Delta t}$ is also the probability of "success" in independently repeated $\text{Cap}^i_{\Delta t}$ trials.

Time: To represent the timing characteristics of the target combinational circuit, we define $T^i_b = \Delta t;\ p^i_{b,\Delta t} \approx p_{TH}$ as the time required to compute the corresponding output bit b when input i is given, where p_{TH} is a threshold for the probability and is defined based on physical characteristics of the target circuit and is also based on the maximum probability achieved by shortening Δt. Accordingly, the time required to complete the computation of all bits when processing input i is defined as $T^i = \Delta t;\ p^i_{\Delta t} \approx p_{TH}$.

Remark: Depending on the target device, its architecture, and the role of the target combinational circuit inside the target device, it might not be possible to know the input i processed. However, if the output of the target combinational circuit is accessible, one can make all the above defined terms based on the fault-free output o, i.e., $\mathrm{BitCap}^o_{b,\Delta t}$, $p^o_{b,\Delta t}$, $\mathrm{Cap}^o_{\Delta t}$, $p^o_{\Delta t}$, T^o_b, and T^o.

4.3 Attack Scheme

For simplicity let us suppose that the target combinational circuit is an S-box of the first round of an AES encryption, i.e., $\mathrm{Sbox(i \oplus k)}$, where i is the corresponding input plaintext byte and k the target key byte.

If (bitwise) timing characteristics of an S-box, i.e., T^i (T^i_b), show a diversity of Δt depending on input i, one can perform an attack and recover the secret knowing how the secret k contributes in T^i (T^i_b). In other words, if the timing characteristics of an S-box itself regardless of k and prior key addition ($\oplus$) are known as an extra information or are obtained by profiling using a circuit similar to the target, one can make a hypothetical leakage function and examine its similarity to T^i (T^i_b) for each key guess. A similar approach has been presented in [14], where the timing characteristics of an AES S-box implementation were profiled and an attack similar to a correlation power analysis using a HW model was successfully performed. In fact, a set of $\mathrm{Cap}^o_{\Delta t}$ for a specific Δt is used in [14] to mount the attack at the last round of the AES encryption.

One may also try using information theoretic tools, e.g., mutual information analysis [9], to relax the leakage model. However, it is necessary to use a suitable leakage model that cannot be selected without extra knowledge about the (timing) characteristics of the target combinational function [25], or several different models must be examined to find a suitable one. It is noteworthy that the leakages ($\mathrm{Cap}^i_{\Delta t}$) consist of only two values ("fail" and "success"). This causes probability distributions (used in e.g., mutual information analysis) to be represented by only two bins in a histogram, and using other schemes to estimate the probability distributions, e.g., kernel density estimation, in this case leads to increasing the noise. Here, when using histograms, mutual information will also be identical to the variance of means.

In contrast to a correlation attack or mutual information analysis using a leakage model, we apply a correlation collision attack [15] to avoid the necessity of considering any such model. Here the correlation collision attack compares the timing characteristics T^i (T^i_b) of two S-box instances running on two input sets, each of which is previously XORed by a secret key byte. Suppose that $T1^i$ and $T2^i$ (or their corresponding bitwise versions) are the timing characteristics of the S-box when processing $\mathrm{Sbox}(i \oplus k1)$ and $\mathrm{Sbox}(i \oplus k2)$ respectively. As stated

Algorithm 3. Correlation Timing Attack (the last round of the AES encryption)

Input: $T1^o : \left(\Delta t^{o=0}, \Delta t^{o=1}, \ldots, \Delta t^{o=255}\right); \, o = \text{Sbox}(i) \oplus k1$
Input: $T2^o : \left(\Delta t^{o=0}, \Delta t^{o=1}, \ldots, \Delta t^{o=255}\right); \, o = \text{Sbox}(i) \oplus k2$
1: **for** $0 \leq \Delta \leq 255$ **do**
2: $\quad Cor(\Delta) = \text{Correlation}(T1^o, T2^{o \oplus \Delta})$
3: **end for**
4: **return** $\arg\max_{\Delta} Cor(\Delta)$

in [15], the aim of a correlation collision attack is to find the linear difference between $k1$ and $k2$, i.e., $\Delta = k1 \oplus k2$.

This can be extended when attacking the last round of the AES encryption, thanks to the absence of the MixColumns in the last round. For example, suppose that $T1^o$ and $T2^o$ are the timing characteristics of the S-box followed by the key addition when calculating $o = \text{Sbox}(i) \oplus k1$ and $o = \text{Sbox}(i) \oplus k2$. Then, the correlation collision attack can – exactly as in the previous case – recover $\Delta = k1 \oplus k2$ comparing $T1^o$ and $T2^o$ for all possible guesses of Δ. For clarification of the attack scheme see Algorithm 3.

4.4 Practical Results

In all cores of the LSIs 16 instances of the S-box are implemented to perform the complete SubBytes operation in each clock cycle. According to [3,4], all cores – except the one supporting a counter mode and the fault-protected one – realize a round based architecture, i.e., S-boxes and MixColumns are performed consecutively in each clock cycle except for the last round where MixColumns is absent. Therefore, extracting the timing characteristics of the S-boxes in the first 9 rounds is not easily possible. So, one needs to inject and play with the width of the clock glitches in the last round, when the target cores only compute the SubBytes operation followed by the final key addition and the result is stored in registers (similar scheme as used in [14]). In addition, one can see from the design architecture of the cores (see Fig. 5.1 of [3] and Fig. 13.1 of [4]) that the round key of the last round is already computed in the previous round and is stored into a register. The glitchy clock at the last round, hence, does not affect the key scheduling computations.

In the following the results of the attacks on the different cores and different LSIs are presented. Because of the high number of broken cores, only a subset of the performed attacks are presented in detail, giving additional information about the differences to the not mentioned cores as required.

Attacking the Unprotected Cores. We start by showing the results of the attack on the first AES core of the 130nm chip, namely AES_Comp, whose S-boxes have been made using a composite field approach. As stated before, 16 separate S-box instances have been implemented which are active at the same time. Therefore, it is not possible to compare the timing characteristics of one S-box instance when processing e.g., two values with different key bytes,

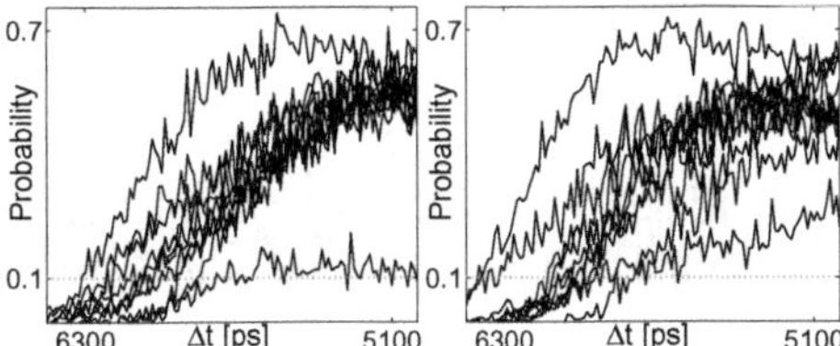

Fig. 7. First 10 $p^o_{b=0,\Delta t}$ curves for S-box instances no. (left) 0 and (right) 4 of AES_Comp (130nm)

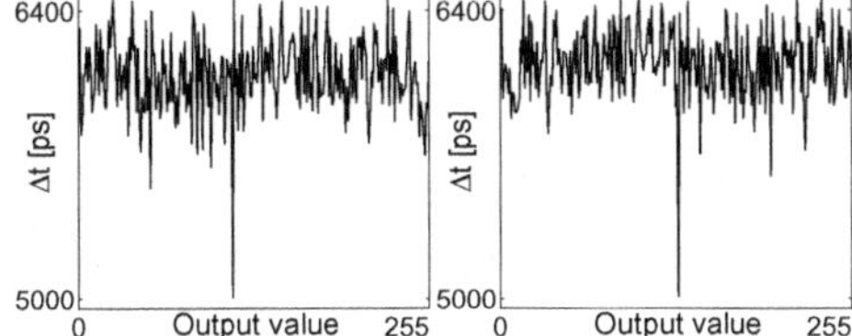

Fig. 8. Bitwise timing characteristics $T^o_{b=0}$ of S-box instances no. (left) 0 and (right) 4 of AES_Comp (130nm).

that would be an ideal case for a collision timing attack. In contrast, the timing characteristics of different S-box instances must be compared, which may slightly vary because of different placement and routing even when being based on the same netlist.

Since changing the glitchy clock width in our setup requires reseting the DCM(s), we have collected $\text{BitCap}^o_{b,\Delta t}$ for a specific Δt while random plaintexts are given to the core. This was repeated shortening Δt by steps of 10ps and finally exploiting the bitwise timing characteristics T^o_b. Figure 7 shows $p^o_{b,\Delta t}$ of the LSB (i.e., $b = 0$) for some output byte values of two S-box instances[3] extracted from their corresponding bitwise captures (10 000 captures for each Δt). Also, Fig. 8 presents the bitwise timing characteristics $T^o_{b=0}$ of these two S-box instances obtained by defining $p_{TH} = 0.1$ (as can be seen in Fig. 7). The diversity of Δt for these two S-boxes shows the dependency between the timing characteristics and the output values. Performing the attack Algorithm 3 on $T^o_{b=0}$ of S-box instance number 0 and all other instances led to recovering all 15 independent relations between the 16 bytes of the last round key; part of the result is shown in Fig. 9. The attack works the same considering other output bits to derive T^o_b as well as on other LSI chips.

Carefully study of the timing characteristics shown in Fig. 8 revealed that Δt is much smaller than the other cases when the S-box input is zero, that is a known issue since the zero-input power model has been defined [10] to mount CPA attacks on AES S-box leakages. In fact, it is not needed to mount the collision timing attack in this case, and the key bytes can be recovered observing $T^o_{b=0}$ of each S-box instance separately. However, as it is shown later this property does not hold for the other cores realized by different S-boxes, and mounting our proposed attack is essential to reveal the secrets.

In order to perform the attack on the cores AES_PPRM1, AES_PPRM3, AES_Comp_ENC_top, and AES_PKG the same procedures as explained above have been repeated. As a reference for the timing characteristics and the number of captures collected to mount the attack on different cores in different LSIs, we have provided a list shown in Table 1. Attacking the AES_TBL core, where

[3] S-box instance numbers start from 0 and are corresponding to ciphertext byte indexes.

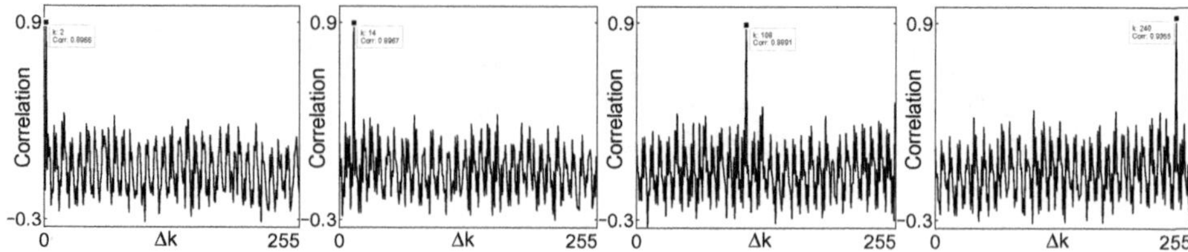

Fig. 9. Result of the attack on the last round of AES_Comp (130nm) recovering Δk between key bytes (from left to right) (0,1), (0,2), (0,3), and (0,4)

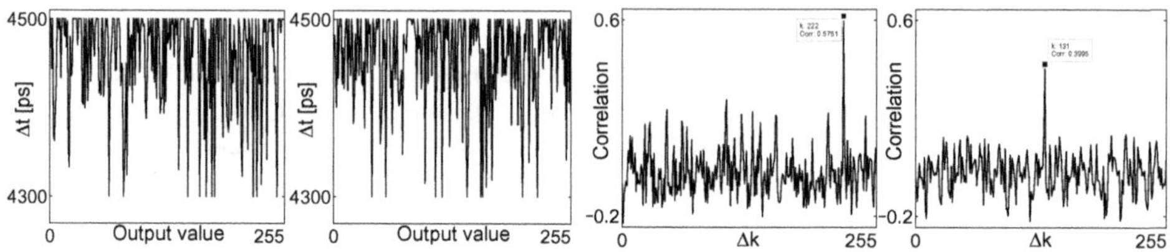

Fig. 10. Bitwise timing characteristics $T^{o}_{b=0}$ of S-box instances no. (left) 5 and (right) 6 of AES_MAO (65nm)

Fig. 11. Result of the attack on the last round of AES_MAO (65nm) recovering Δk between key bytes (left) (5,6), (right) (5,7)

S-boxes have been realized by look-up tables (case statements), is different to the aforementioned cores. We illustrate this case when explaining how to mount the attack on the WDDL and MDPL cores.

Attacking the DPA-Protected Cores. Most of the DPA-protected cores can be attacked in the same way as the unprotected ones. In Fig. 10 one can see that even when using the masked AND-gates of the AES_MAO (65nm) core, the timing characteristics for different outputs still differ. Consequently, it is possible to extract the relation between the key bytes, which is depicted in Fig. 11. Interestingly the randomness provided by the masked gates does not have much impact on the timing characteristics, as shown in Fig. 11, where the results after obtaining 10 000 captures (the same technique as used to attack the unprotected cores) while shortening Δt with steps of 25ps are presented.

Attacking the other DPA-protected cores is the same except on those realizing WDDL and MDPL logic styles. The result of the attack on the AES_WO core, which is implemented using an Pseudo-RSL [22] logic style, is shown in Fig. 12 and Fig. 13. Although we have used 10 000 captures for each Δt in steps of 10ps to attack the AES_WO core, attacking the AES_PR core (which is another realization of Pseudo-RSL) and the AES_TI (which has been discussed in Section 3.3) required considerably more captures. As stated in Table 1 we have used 1 000 000 captures for each of these cores to successfully mount the attack. To the best of our knowledge it is due to the amount of randomness provided

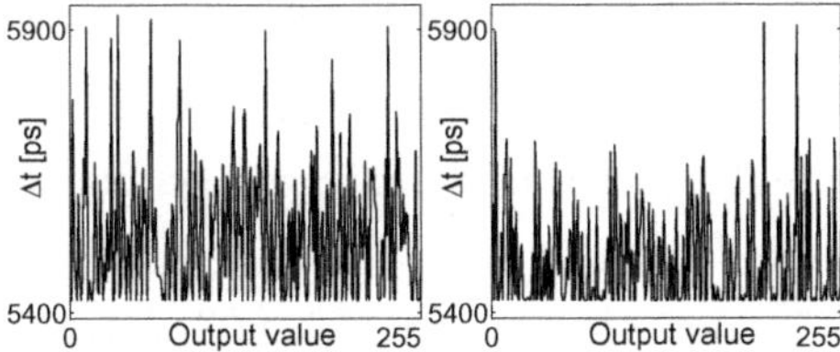

Fig. 12. Bitwise timing characteristics $T^o_{b=1}$ of S-box instances no. (left) 7 and (right) 8 of AES_WO (90nm)

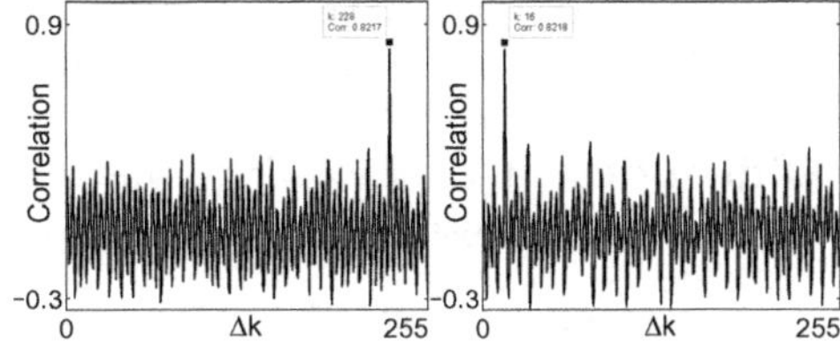

Fig. 13. Result of the attack on the last round of AES_WO (90nm) recovering Δk between key bytes (left) (7,8), (right) (7,9)

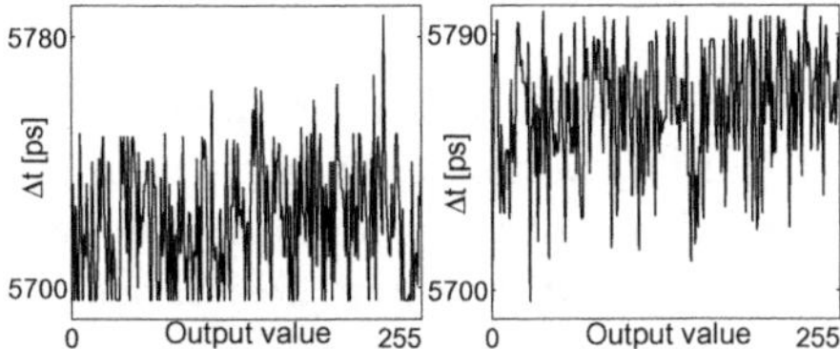

Fig. 14. Timing characteristics T^o of S-box instances no. (left) 2 and (right) 3 of AES_MDPL (65nm)

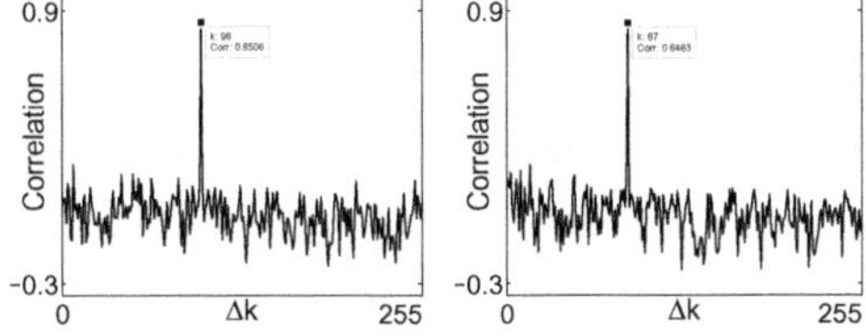

Fig. 15. Result of the attack on the last round of AES_MDPL (65nm) recovering Δk between key bytes (left) (2,3), (right) (2,6)

by the DPA countermeasures. For instance, third-order masking is used in the AES_TI core compared to first-order masking in AES_MAO.

The AES_WDDL and AES_MDPL cores require an slightly adjusted approach, since they need two clock cycles per round because of the used master-slave flip-flops, i.e., four flip-flops to store a single bit value. Also, an injected fault by a clock glitch at the evaluation phase can only lead to a bit flip from 1 to 0, not vice versa, because of the pre-discharge phase of both WDDL and MDPL styles. This issue has been also addressed in [13] where a successful attack is performed on an AES_WDDL core. Interestingly, we have seen – for reasons unknown to us – the same behavior when attacking the AES_TBL core. Therefore, bitwise timing characteristics T^o_b does not provide any information for those output values o in which bit b is zero. Our solution is to avoid using bitwise characteristics, and apply the attack on timing characteristics T^o, e.g., those shown in Fig. 14, which are of the AES_MDPL (65nm) core. Two attack results are also shown in Fig. 15. It should be noted that, to attack the AES_MDPL and AES_WDDL cores, we only used 10 000 captures for each Δt with steps of 5ps. On the other hand, a successful attack on the AES_TBL required around 1 000 000 captures, which might be because of marginal differences between the critical paths of the circuit realizing the look-up table.

We should emphasize that this attack has been successfully performed on all AES cores including one equipped by a fault attack countermeasure [23]. Because of the page restriction the details of the attack on the other cores are left for the extended version which can be found in [16].

Difficulties. In the following some of the difficulties that we experienced during our practical investigations are explained in detail.

- As stated in [3] and [4], each core has its own clock tree which had a strong impact on glitchy clocks. The capacitive and resistive features of the clock line of the LSIs, which is supplied by the control FPGA, changed the glitchy clock shape and modified the situation whether the registers are triggered two times, or if one of the positive edges is filtered by the clock tree elements. Therefore, we had to put different capacitors and resistors in the SASEBO-R board to change the rising and falling slopes of the clock signal to reach the desired situation.
- Since the temperature has an effect on the critical path and the speed of the ASICs, some values are given in Table 4.7 of [3], we not only kept the room temperature constant during capturing, but also kept the board and the LSIs in different temperatures while playing with capacitances and resistances to solve the problem mentioned above.
- Since DCM outputs have an increased jitter when they are used to multiply the clock inputs, the glitchy clock width also had significant jitter that made the capturing process noisy. The situation got worse when we had to cascade two DCMs for some cores, especially in 65nm technology, to reach the desired Δt. In this case, the second DCM often could not get locked because of the high jitter of the first DCM. So, we had to provide another circuit controlled by the PC to automatically reset the control FPGA (in fact the DCMs) until the DCMs get locked and provide the requested high frequencies.
- Since we have required high amounts of captures, e.g., 1 000 000, for different Δt values to successfully mount the attacks, we have developed a special design for the control FPGA to speed up the capturing process. Our control FPGA communicates with the target LSI, makes the glitchy clock on the desired clock cycle, and finally after performing a couple of capturing process sends the result back to the PC. In this way we could efficiently increase the speed of the capturing up to couple of thousands per second.
- According to [3] and [4], the clock signal of the interface circuit of the LSIs is separated from the core clocks. So, the glitchy clock does not appear on the interface circuit which makes the attacks easier. It might be a challenging case when the interface circuit sees the glitches, and the control flow of the target core gets infected.

5 Conclusions

We have presented two collision attacks which utilize certain kinds of side-channel leakage which is made possible by the fault injection method of [14]. One is a major improvement of the attack idea of [14] since by applying techniques of correlation-based DPAs to find collisions it does not require any knowledge about the characteristics of the target combinational circuit. The other one exploits a newly observed leakage which is the fact that given a fixed fault intensity the distribution of the resulting faulty ciphertext bytes is not completely random but data dependent.

It is indicated in [14] that while masking does not prevent DFA attacks, it may actually provide security against FSA-based attacks because of the randomized inputs of the combinational functions. However, by breaking all DPA-protected cores of the mentioned ASICs we have shown that randomizing countermeasures itself cannot prevent data-dependent timing of the combinational circuit, and they therefore remain vulnerable to the attacks introduced here.

Using the attack exploiting the faulty ciphertext byte distributions two DPA protected cores could be broken. Furthermore, using the attack focusing on the timing of the combinational circuits all SASEBO LSI2 and LSI3 cores could be broken, including the one applying an algorithmic fault detection scheme. In short, the results shown in this work imply the need for a special unit in the – especially side-channel protected – designs in order to detect the clock glitches to thwart such kind of attacks.

Acknowledgment. The authors would like to thank Akashi Satoh and the Research Center for Information Security (RCIS) of Japan for the prompt and kind help in obtaining SASEBOs and cryptographic LSIs. The authors of the Ruhr University of Bochum (Germany) have been supported in part by the European Commission through the ICT programme under contract ICT-2007-216676 ECRYPT II. The authors of the University of Electro-Communications (Japan) have been supported by the Strategic International Cooperative Program (Joint Research Type), Japan Science and Technology Agency.

References

1. Cryptographic Circuits with Logic Level Countermeasures against DPA. Information and Physical Security Research Group, YOKOHAMA National University, `http://ipsr.ynu.ac.jp/circuit/`
2. Side-channel Attack Standard Evaluation Board (SASEBO-R). Further information are available via, `http://staff.aist.go.jp/akashi.satoh/SASEBO/en/board/sasebo-r.html`
3. ISO/IEC 18033-3 Standard Cryptographic LSI – with Side Channel Attack Countermeasures – Specification, ver 1.0 (2009), `http://staff.aist.go.jp/akashi.satoh/SASEBO/resources/crypto_lsi/CryptoLSI2_Spec_Ver1.0_English.pdf`
4. Standard Cryptographic LSI Specification – Countermeasures against Side Channel Attacks (65nm) – Specification, ver 0.9 (2010), `http://staff.aist.go.jp/akashi.satoh/SASEBO/resources/crypto_lsi/CryptoLSI3_Spec_Ver0.9_English.pdf`
5. Biham, E., Shamir, A.: Differential Fault Analysis of Secret Key Cryptosystems. In: Kaliski Jr., B.S. (ed.) CRYPTO 1997. LNCS, vol. 1294, pp. 513–525. Springer, Heidelberg (1997)
6. Bogdanov, A.: Multiple-Differential Side-Channel Collision Attacks on AES. In: Oswald, E., Rohatgi, P. (eds.) CHES 2008. LNCS, vol. 5154, pp. 30–44. Springer, Heidelberg (2008)

7. Endo, S., Sugawara, T., Homma, N., Aoki, T., Satoh, A.: An on-chip glitchy-clock generator and its application to safe-error attack. In: COSADE 2011, pp. 175–182 (2011)
8. Gandolfi, K., Mourtel, C., Olivier, F.: Electromagnetic Analysis: Concrete Results. In: Koç, Ç.K., Naccache, D., Paar, C. (eds.) CHES 2001. LNCS, vol. 2162, pp. 251–261. Springer, Heidelberg (2001)
9. Gierlichs, B., Batina, L., Tuyls, P., Preneel, B.: Mutual Information Analysis. In: Oswald, E., Rohatgi, P. (eds.) CHES 2008. LNCS, vol. 5154, pp. 426–442. Springer, Heidelberg (2008)
10. Golic, J.D., Tymen, C.: Multiplicative Masking and Power Analysis of AES. In: Kaliski Jr., B.S., Koç, Ç.K., Paar, C. (eds.) CHES 2002. LNCS, vol. 2523, pp. 198–212. Springer, Heidelberg (2003)
11. Kocher, P.C.: Timing Attacks on Implementations of Diffie-Hellman, RSA, DSS, and Other Systems. In: Koblitz, N. (ed.) CRYPTO 1996. LNCS, vol. 1109, pp. 104–113. Springer, Heidelberg (1996)
12. Kocher, P.C., Jaffe, J., Jun, B.: Differential Power Analysis. In: Wiener, M. (ed.) CRYPTO 1999. LNCS, vol. 1666, pp. 388–397. Springer, Heidelberg (1999)
13. Li, Y., Ohta, K., Sakiyama, K.: Revisit Fault Sensitivity Analysis on WDDL-AES. In: HOST 2010, pp. 148–153. IEEE Computer Society, Los Alamitos (2010)
14. Li, Y., Sakiyama, K., Gomisawa, S., Fukunaga, T., Takahashi, J., Ohta, K.: Fault Sensitivity Analysis. In: Mangard, S., Standaert, F.-X. (eds.) CHES 2010. LNCS, vol. 6225, pp. 320–334. Springer, Heidelberg (2010)
15. Moradi, A., Mischke, O., Eisenbarth, T.: Correlation-Enhanced Power Analysis Collision Attack. In: Mangard, S., Standaert, F.-X. (eds.) CHES 2010. LNCS, vol. 6225, pp. 125–139. Springer, Heidelberg (2010); The extended version is available on ePrint Archive, Report 2010/297, `http://eprint.iacr.org/`
16. Moradi, A., Mischke, O., Paar, C.: Collision Timing Attack when Breaking 42 AES ASIC Cores. Cryptology ePrint Archive, Report 2011/162 (2011), `http://eprint.iacr.org/`
17. Moradi, A., Poschmann, A., Ling, S., Paar, C., Wang, H.: Pushing the Limits: A Very Compact and a Threshold Implementation of AES. In: Paterson, K.G. (ed.) EUROCRYPT 2011. LNCS, vol. 6632, pp. 69–88. Springer, Heidelberg (2011)
18. Nikova, S., Rechberger, C., Rijmen, V.: Threshold Implementations Against Side-Channel Attacks and Glitches. In: Ning, P., Qing, S., Li, N. (eds.) ICICS 2006. LNCS, vol. 4307, pp. 529–545. Springer, Heidelberg (2006)
19. Nikova, S., Rijmen, V., Schläffer, M.: Secure Hardware Implementation of Nonlinear Functions in the Presence of Glitches. In: Lee, P.J., Cheon, J.H. (eds.) ICISC 2008. LNCS, vol. 5461, pp. 218–234. Springer, Heidelberg (2009)
20. Piret, G., Quisquater, J.-J.: A Differential Fault Attack Technique against SPN Structures, with Application to the AES and KHAZAD. In: Walter, C.D., Koç, Ç.K., Paar, C. (eds.) CHES 2003. LNCS, vol. 2779, pp. 77–88. Springer, Heidelberg (2003)
21. Quisquater, J.-J., Samyde, D.: ElectroMagnetic Analysis (EMA): Measures and Counter-Measures for Smart Cards. In: Attali, S., Jensen, T. (eds.) E-smart 2001. LNCS, vol. 2140, pp. 200–210. Springer, Heidelberg (2001)
22. Saeki, M., Suzuki, D., Shimizu, K., Satoh, A.: A Design Methodology for a DPA-Resistant Cryptographic LSI with RSL Techniques. In: Clavier, C., Gaj, K. (eds.) CHES 2009. LNCS, vol. 5747, pp. 189–204. Springer, Heidelberg (2009)
23. Satoh, A., Sugawara, T., Homma, N., Aoki, T.: High-Performance Concurrent Error Detection Scheme for AES Hardware. In: Oswald, E., Rohatgi, P. (eds.) CHES 2008. LNCS, vol. 5154, pp. 100–112. Springer, Heidelberg (2008)

24. Trichina, E.: Combinational Logic Design for AES SubByte Transformation on Masked Data. Cryptology ePrint Archive, Report 2003/236 (2003), http://eprint.iacr.org/
25. Veyrat-Charvillon, N., Standaert, F.-X.: Generic Side-Channel Distinguishers: Improvements and Limitations. In: Rogaway, P. (ed.) CRYPTO 2011. LNCS, vol. 6841, pp. 354–372. Springer, Heidelberg (2011); The extended version is available on ePrint Archive, Report 2011/149, http://eprint.iacr.org/
26. XILINX. Virtex-II Pro and Virtex-II Pro X FPGA User Guide. Technical report version 4. 2 (2007), http://www.xilinx.com/support/documentation/user_guides/ug012.pdf

Appendix

Table 1. Specification of AES cores of the three targeted LSIs including the Δt ranges and the number of captures used to mount the attacks

IP core	Description	LSI2 130nm		LSI2 90nm		LSI3 65nm	
		Δt range [ps]	No. of Captures	Δt range [ps]	No. of Captures	Δt range [ps]	No. of Captures
AES_Comp	composite field S-box	6450 Δ : 10 5000	10 000	5320 Δ : 10 5130	10 000	3650 Δ : 10 3370	10 000
AES_TBL	table look-up S-box by case statement	5475 Δ : 25 4900	1 000 000	3960 Δ : 20 3550	1 000 000	3570 Δ : 10 3420	1 000 000
AES_PPRM1	S-box by 1-stage AND-XOR	11350 Δ : 25 7775	10 000	6135 Δ : 20 5555	10 000	5325 Δ : 25 5000	10 000
AES_PPRM3	S-box by 3-stage AND-XOR	6425 Δ : 25 5150	10 000	5230 Δ : 10 5130	10 000	3650 Δ : 10 3420	10 000
AES_Comp ENC_top	composite field S-box, only encryption	6325 Δ : 25 5100	10 000	5200 Δ : 10 5130	10 000	3700 Δ : 10 3410	10 000
AES_PKG	composite field S-box, precomp. roundkeys	6325 Δ : 25 5100	10 000	5360 Δ : 10 5130	10 000	3850 Δ : 10 3370	10 000
AES_MAO	DPA count. by Masked And Operation	8475 Δ : 25 6250	10 000	5900 Δ : 5 5850	10 000	4500 Δ : 25 4300	10 000
AES_MDPL	DPA count. by MDPL logic style	12825 Δ : 25 10850	10 000	9350 Δ : 25 8050	10 000	5800 Δ : 5 5260	10 000
AES_TI	DPA count. by Threshold Implementation	10860 Δ : 20 9800	1 000 000	5900 Δ : 5 5850	1 000 000	6340 Δ : 20 5940	1 000 000
AES_WDDL	DPA count. by WDDL logic style	6750 Δ : 10 5730	10 000	5250 Δ : 5 5150	50 000	3835 Δ : 5 3675	10 000
AES_PR	DPA count. by pseudo RSL logic style	31685 Δ : 10 31055	1 000 000	14400 Δ : 20 13840	1 000 000	6650 Δ : 25 6150	1 000 000
AES_WO	DPA count. by pseudo RSL (evaluation)	7575 Δ : 25 6475	10 000	5910 Δ : 10 5430	10 000	3900 Δ : 25 3600	10 000

SPONGENT: A Lightweight Hash Function*

Andrey Bogdanov[1], Miroslav Knežević[1,2], Gregor Leander[3], Deniz Toz[1], Kerem Varıcı[1], and Ingrid Verbauwhede[1]

[1] Katholieke Universiteit Leuven, ESAT/COSIC and IBBT, Belgium
{andrey.bogdanov,deniz.toz,kerem.varici, ingrid.verbauwhede}@esat.kuleuven.be
[2] NXP Semiconductors, Leuven, Belgium
miroslav.knezevic@nxp.com
[3] DTU Mathematics, Technical University of Denmark
g.leander@mat.dtu.dk

Abstract. This paper proposes SPONGENT – a family of lightweight hash functions with hash sizes of 88 (for preimage resistance only), 128, 160, 224, and 256 bits based on a sponge construction instantiated with a PRESENT-type permutation, following the hermetic sponge strategy. Its smallest implementations in ASIC require 738, 1060, 1329, 1728, and 1950 GE, respectively. To our best knowledge, at all security levels attained, it is the hash function with the smallest footprint in hardware published so far, the parameter being highly technology dependent. SPONGENT offers a lot of flexibility in terms of serialization degree and speed. We explore some of its numerous implementation trade-offs.

We furthermore present a security analysis of SPONGENT. Basing the design on a PRESENT-type primitive provides confidence in its security with respect to the most important attacks. Several dedicated attack approaches are also investigated.

Keywords: Hash function, lightweight cryptography, low-cost cryptography, low-power design, sponge construction, PRESENT, SPONGENT, RFID.

1 Introduction

1.1 Motivation

As crucial applications go pervasive, the need for security in RFID and sensor networks is dramatically increasing, which requires secure yet efficiently implementable cryptographic primitives including secret-key ciphers and hash functions. In such constrained environments, the area and power consumption of

* Andrey Bogdanov is a postdoctoral fellow of the Fund for Scientific Research - Flanders (FWO). This work is supported in part by the IAP Programme P6/26 BCRYPT of the Belgian State, by FWO project G.0300.07, by the European Commission under contract number ICT-2007-216676 ECRYPT NoE phase II, by K.U.Leuven-BOF (OT/08/027 and OT/06/40), and by the Research Council K.U.Leuven: GOA TENSE.

B. Preneel and T. Takagi (Eds.): CHES 2011, LNCS 6917, pp. 312–325, 2011.

a primitive usually comes to the fore and standard algorithms are often prohibitively expensive to implement.

Once this research problem was identified, the cryptographic community designed a number of tailored lightweight cryptographic algorithms to specifically address this challenge: stream ciphers like Trivium [12,10], Grain [13,14], and Mickey [2] as well as block ciphers like SEA [26], DESL, DESXL [21], HIGHT [16], mCrypton [22], KATAN/KTANTAN [11], and PRESENT [5] — to mention only a small selection of the lightweight designs.

Rather recently, some significant work on lightweight hash functions has been also performed: [6] describes ways of using the PRESENT block cipher in hashing modes of operation and [1] takes the approach of designing a dedicated lightweight hash function QUARK based on a sponge construction [9,3]. However, while for the stream and block ciphers, the designs have already closely approached the minimum ASIC hardware footprint theoretically attainable, it does not seem the case for lightweight hash functions so far. This paper illustrates this point by proposing the lightweight hash function SPONGENT with a considerably smaller footprint than SHA-2, SHA-3 finalists, PRESENT in hashing modes, and QUARK. Similarly to QUARK, a part of this advantage comes from a reduced level of preimage and second preimage security, while maintaining the standard level of collision resistance.

1.2 Design Considerations for a Lightweight Hash Function

The standard security requirements for a hash function with an n-bit output size are collision resistance of $2^{n/2}$ as well as preimage and second-preimage resistance of 2^n.

The footprint of a hash function is mainly determined by

1. the number of state bits (incl. the key schedule for block cipher based designs) as well as
2. the size of functional and control logic used in a round function.

For highly serialized implementations (usually used to attain low area and power), the logic size is normally rather small and the state size dominates the total area requirements of the design.

As shown in [6], using a lightweight block cipher in a hashing mode (single block length such as Davies-Meyer or double block length such as Hirose) is not necessarily an optimal choice for reducing the footprint, the major restriction being the doubling of the datapath storage requirement due to the feed-forward operation. At the same time, no feed-forward is necessary for the sponge construction.

In a permutation-based sponge construction, let r be the *rate* (the number of bits input or output per one permutation call) and c be the *capacity* (internal state bits not used for input or output). The design of [1] as well as the works [3,4,9] convincingly demonstrate that a permutation-based sponge construction can allow to almost halve the state size for $n \geq c$ and reasonably small r. In this case, if the underlying permutation does not have any structural distinguishers (thus, the sponge construction being *hermetic*), the preimage and second-preimage resistances are reduced to 2^{n-r} and $2^{c/2}$, correspondingly, while the collision resistance remains at the level of $2^{c/2}$. As in most embedded scenarios, where a lightweight hash function is likely to be used, the full second-preimage security is not a necessary requirement, we will take this approach in the design of SPONGENT. For relatively small rate r, the loss of preimage security is limited.

However, while using this novel idea of reducing the state size to minimize (1), the QUARK hash function does not appear to provide an optimal logic size, which is mainly due to the Boolean functions with many inputs used in its round transform. SPONGENT keeps the round function very simple which reduces the logic size close to the smallest theoretically possible, thus, minimizing (2) and resulting in a significantly more compact design.

As to the output hash size n, we opt for 5 variants of SPONGENT covering most security applications in the field. SPONGENT-88 is designed for extremely restricted scenarios and low preimage security requirements. It can be used e.g. in some RFID protocols and for PRNGs. SPONGENT-128 and SPONGENT-160 might be used in highly constrained applications with low and middle requirements for collision security. The latter also provides compatibility to the SHA-1 interfaces. The parameters of SPONGENT-224 and SPONGENT-256 correspond to those of a subset of SHA-2 and SHA-3 to make SPONGENT compatible to the standard interfaces in usual lightweight embedded scenarios.

1.3 Organization of the Paper

The remainder of the paper is organized as follows. Section 2 describes the design of SPONGENT and gives a design rationale. Section 3 presents some results of security analysis, including proven lower bounds on the number of differentially active S-boxes, best differential characteristics found, rebound attacks, and linear attacks. In Section 4, the implementation results are given for a range of trade-offs. We conclude in Section 5.

2 The Design of SPONGENT

SPONGENT is a sponge construction based on a wide PRESENT-type permutation. Given a finite number of input bits, it produces an n-bit hash value. A design goal for SPONGENT is to follow the hermetic sponge strategy (no structural distinguishers for the underlying permutation are allowed).

2.1 Permutation-Based Sponge Construction

SPONGENT relies on a sponge construction – a simple iterated design that takes a variable-length input and can produce an output of an arbitrary length based on a permutation π_b operating on a state of a fixed number b of bits. The size of the internal state $b = r + c \geq n$ is called *width*, where r is the *rate* and c the *capacity*.

The sponge construction proceeds in three phases (see also Figure 1):

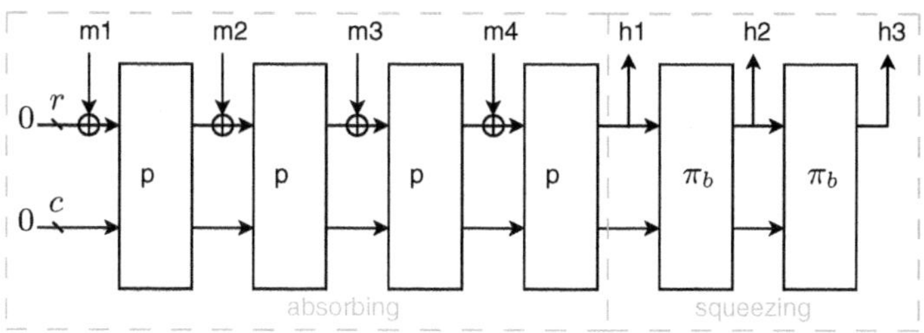

Fig. 1. Sponge construction based on a b-bit permutation π_b with capacity c bits and rate r bits. m_i are r-bit message blocks. h_i are parts of the hash value.

- **Initialization phase:** the message is padded by a single bit 1 followed by a necessary number of 0 bits up to a multiple of r bits (e.g., if $r = 8$, then the 1-bit message '0' is transformed to '01000000'). Then it is cut into blocks of r bits.
- **Absorbing phase:** the r-bit input message blocks are xored into the first r bits of the state, interleaved with applications of the permutation π_b.
- **Squeezing phase:** the first r bits of the state are returned as output, interleaved with applications of the permutation π_b, until n bits are returned.

In SPONGENT, the b-bit 0 is taken as the initial value before the absorbing phase. In all SPONGENT variants, except SPONGENT-88, the hash size n equals capacity c. The message chunks are xored into the r rightmost bit positions of the state. The same r bit positions form parts of the hash output.

Let a permutation-based sponge construction have $n \geq c$ and $c/2 > r$ which is fulfilled for the parameter choices of all SPONGENT variants. Then the works [3,4,9] imply the preimage security of 2^{n-r} as well as the second preimage and collision securities of $2^{c/2}$ if this construction is hermetic (that is, if the underlying permutation does not have any structural distinguishers). The best preimage attack we are aware of in this case has a computational complexity of $2^{n-r}+2^{c/2}$.

2.2 Parameters

We propose five variants of SPONGENT with five different security levels:

	n (bit)	b (bit)	c (bit)	r (bit)	R number of rounds	security(bit) preimage	2nd preimage	collision
SPONGENT-88	88	88	80	8	45	80	40	40
SPONGENT-128	128	136	128	8	70	120	64	64
SPONGENT-160	160	176	160	16	90	144	80	80
SPONGENT-224	224	240	224	16	120	208	112	112
SPONGENT-256	256	272	256	16	140	240	128	128

2.3 PRESENT-type Permutation

The permutation $\pi_b : \mathbb{F}_2^b \to \mathbb{F}_2^b$ is an R-round transform of the input STATE of b bits that can be outlined at a top-level as:

for $i = 1$ to R **do**
  STATE $\leftarrow$ ɹǝʇunoƆl$_b(i) \oplus$ STATE $\oplus$ lCounter$_b(i)$
  STATE $\leftarrow$ sBoxLayer$_b$(STATE)
  STATE $\leftarrow$ pLayer$_b$(STATE)
end for

where sBoxLayer$_b$ and pLayer$_b$ describe how the STATE evolves. For ease of design, only widths b with $4|b$ are allowed. The number R of rounds depends on block size b and can be found in Subsection 2.2. lCounter$_b(i)$ is the state of an LFSR dependent on b at time i which yields the round constant in round i and is added to the rightmost bits of STATE. ɹǝʇunoƆl$_b(i)$ is the value of lCounter$_b(i)$ with its bits in reversed order and is added to the leftmost bits of STATE.

The following building blocks are generalizations of the PRESENT structure to larger b-bit widths:

1. sBoxLayer$_b$: This denotes the use of a 4-bit to 4-bit S-box $S : \mathbb{F}_2^4 \to \mathbb{F}_2^4$ which is applied $b/4$ times in parallel. The S-box fulfills the PRESENT S-box criteria [5]. The action of the S-box in hexadecimal notation is given by the following table:

x	0	1	2	3	4	5	6	7	8	9	A	B	C	D	E	F
$S[x]$	E	D	B	0	2	1	4	F	7	A	8	5	9	C	3	6

2. pLayer$_b$: This is an extension of the (inverse) PRESENT bit-permutation and moves bit j of STATE to bit position $P_b(j)$, where

$$P_b(j) = \begin{cases} j \cdot b/4 \mod b-1, & \text{if } j \in \{0, \ldots, b-2\} \\ b-1, & \text{if } j = b-1. \end{cases}$$

3. lCounter$_b$: This is one of the three $\lceil \log_2 R \rceil$-bit LFSRs. The LFSR is clocked once every time its state has been used and its final value is all ones. If ζ is the root of unity in the corresponding binary finite field, the 6-bit LFSR used in SPONGENT-88 is defined by the primitive trinomial $\zeta^6 + \zeta^5 + 1$ (initialized with '000101'). The 7-bit LFSR with a primitive trinomial of $\zeta^7 + \zeta^6 + 1$ is used in SPONGENT-128, SPONGENT-160, and SPONGENT-224 and respectively initialized with '1111010', '1000101', and '0000001'. SPONGENT-256 uses an 8-bit LFSR based on the pentanomial $\zeta^8 + \zeta^4 + \zeta^3 + \zeta^2 + 1$ and it is initialized with '10011110'.

2.4 Design Rationale

Permutation. The 4-bit S-box is the major block of functional logic in a serial low-area implementation of SPONGENT, the bit permutation requiring some additional space in silicon. Its simplicity and small size minimize the area and power consumption on the logic side. The structures of the bit permutation and the S-box in SPONGENT make it possible to prove

Theorem 1. *Any 5-round differential characteristic of the underlying permutation in* SPONGENT-{*88, 128, 160, 224, 256*} *has a minimum of* 10 *active S-boxes.*

Proof. The statements for SPONGENT-{88, 128, 160, 224, 256} can directly be proven by applying the same technique used in [5, Appendix III].

The concept of counting active S-boxes is central to the differential cryptanalysis. The minimum number of active S-boxes relates to the maximum differential characteristic probability of the construction. Since in the hash setting there are no random and independent key values added between the rounds, this relation is not exact (in fact that it is even not exact for most practical keyed block ciphers). However, differentially active S-boxes are still the major technique used to evaluate the security of SPN-based hash functions.

An important property of the SPONGENT S-box is that its maximum differential probability is 2^{-2}. This fact and the assumption of the independency of difference propagation in different rounds yield an upper bound on the differential characteristic probability of 2^{-20} over 5 rounds for SPONGENT-$\{88, 128, 160, 224, 256\}$, which follows from the claims of Theorem 1.

Theorem 1 is used to determine the number R of rounds in permutation π_b: R is chosen in a way that π_b provides at least b active S-boxes. Other types of analysis are performed in the next section.

3 Security Analysis

In this section, we discuss the security of SPONGENT against the currently known cryptanalytic attacks by applying the most important state-of-the-art methods of cryptanalysis and investigating their complexity.

3.1 Resistance against Differential Cryptanalysis

Here we analyze the resistance of SPONGENT against differential attacks where Theorem 1 plays a key role providing a lower bound on the number of active S-boxes in a differential characteristic. The similarities of the SPONGENT permutations and the basic PRESENT cipher allow to reuse some of the results obtained for PRESENT in [5]. More precisely, the results on the number of differentially active S-boxes over 5 rounds will hold for all SPONGENT variants which is reflected in Theorem 1.

For all SPONGENT variants, we found that those 5-round bounds are actually tight. We present the characteristics attaining them in Table 1 as well as in Appendix A.

3.2 Collision Attacks

A natural approach to obtain a collision for a sponge construction is to inject a difference in a message block and then cancel the propagated difference by a difference in the next message block, i.e., $(0 \ldots 0||\Delta m_i) \xrightarrow{\pi} (0 \ldots 0||\Delta m_{i+1})$.

Table 1. Differential characteristics with lowest numbers of differentially active S-boxes (ASN). The probabilities are calculated assuming the independency of round computations.

# of	SPONGENT-88		SPONGENT-128		SPONGENT-160		SPONGENT-224		SPONGENT-256	
rounds	ASN	Prob	ASN	Prob	ASN	Prob	ASN	Prob	ASN	Prob
5	10	2^{-21}	10	2^{-22}	10	2^{-21}	10	2^{-21}	10	2^{-20}
10	20	2^{-47}	29	2^{-68}	20	2^{-50}	20	2^{-43}	–	–
15	30	2^{-74}	-	-	30	2^{-79}	30	2^{-66}	–	–

For this purpose, we follow a narrow trail strategy using truncated differential characteristics. We start from a given input difference (some difference restricted to S-boxes that the message block is xored into) and look for all paths that go to a fixed output difference (also located in the bitrate part of the state). Based on our experiments, even by using truncated differential characteristics, the probability of such a path is quite low and it is not possible to attack the full number of rounds (see Appendix A).

The rebound attack [24], a recent technique for cryptanalysis of hash functions, is applicable to both block cipher based and permutation based hash constructions. It consists of two main steps: the inbound phase where the freedom is used to connect the middle rounds by using the match-in-the-middle technique and the outbound phase where the connected truncated differentials are calculated in both forward and backward directions.

Dedicated Rebound Attack on 6 Rounds of SPONGENT-88. Here we describe a dedicated rebound attack on SPONGENT-88. For other hash sizes, a similar method is applicable. The path for this attack is shown in Figure 2.

- **Inbound phase:** In the forward direction, we start from the input of the second round. We generate 2^{28} structures as follows: We restrict the input of the active S-boxes to the eight values such that the difference $\Delta S : [\{\texttt{0x2}, \texttt{0x3}\} \rightarrow \texttt{0xf}]$ and to the four values such that the difference $\Delta S : [\texttt{0x4} \rightarrow \texttt{0xf}]$ and we generate the passive bits at random. For each fixed value of the passive part, we can obtain $(2^3)^2 \cdot (2^2)^2 = 2^{10}$ pairs, so we repeat the procedure 2^{18} times. At the input of round 3, we have 16 active S-boxes and 6 passive S-boxes, and it is guaranteed to have all S-boxes active at the input of round 4. In the backward direction, we start from the input of the S-boxes in round 6. Similarly, we generate 2^{28} structures by restricting the values of the active S-boxes to four values such that the difference $\Delta S^{-1} : [\texttt{0x1} \rightarrow \{\texttt{0xb}, \texttt{0xd}\}]$. Again, we can generate $(2^2)^{11} = 2^{22}$ pairs for each value of the passive part, hence we choose random 2^6 values. Then at the input of the fifth round all S-boxes are active and with high probability they will still be active after the permutation layer.
- **Merging phase:** We look for a matching input/output difference of the S-box layer in round 4 using the precomputed difference distribution table. We

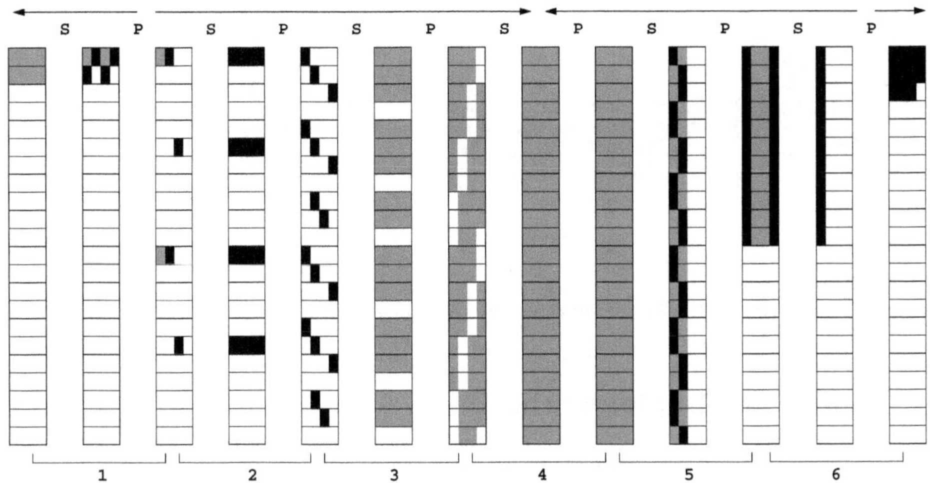

Fig. 2. Differential path for the rebound attack (black: starting S-boxes with conditions, grey: active S-boxes, S: sBoxLayer$_{88}$, P: pLayer$_{88}$)

can find a match with probability of $2^{-2.4}$ for each word that has one fixed bit with zero difference and a match with probability $2^{3.6}$ for each word that has two fixed bit zero. Therefore, we can find an entire differential path with probability $(2^{-2.4})^{20} \cdot (2^{-3.6})^2 = 2^{-55.2}$. Hence we expect to find at least one solution.

- **Outbound phase:** We further extend the differential path backwards (from round 2 to round 1) and forwards (to the end of round 6) with probability 1.

3.3 Linear Attacks

The most successful attacks, the attacks that can break the highest number of rounds, for the block cipher PRESENT are attacks based on linear approximations. In particular the multi-dimensional linear attack [7] and the statistical saturation attack [8] claim to break up to 26 rounds. It was shown in [20] that both attacks are closely related. Moreover, the main reason why these attacks are the most successful attacks on PRESENT so far, is the existence of many linear trails with only one active S-box in each round. It is not immediately clear how linear distinguishers on the SPONGENT permutation π_b could be transferred into collision or (second) pre-image attacks on the hash function. However, as we claim that SPONGENT is a hermetic sponge construction, the existence of such distinguishers has to be excluded. So the SPONGENT S-box was chosen in a way that allows for at most one trail with this property given a linear approximation.

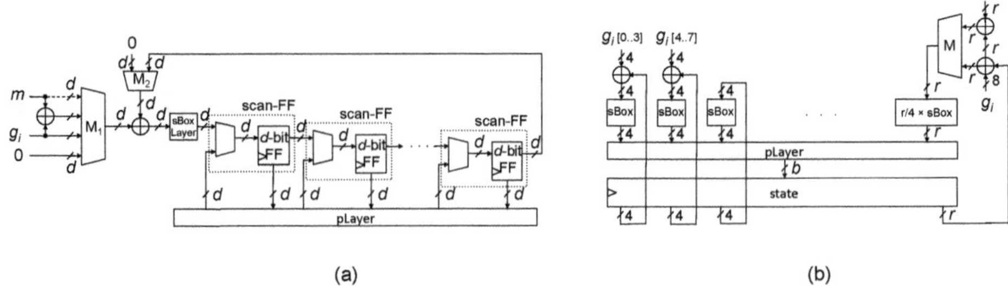

Fig. 3. Hardware architecture representing (a) serial datapath (b) parallel datapath of the SPONGENT variants

4 Hardware Implementations

Using the most serialized implementation, the hash functions SPONGENT-{88, 128, 160, 224, 256} can be implemented with 738 GE, 1060 GE, 1329 GE, 1728 GE, and 1950 GE, respectively, which is smaller than the most compact QUARK designs [1] of respective sizes. Furthermore, even the SPONGENT-256 hash function is more compact than S-QUARK having a hash output of 224 bits. Though some of this advantage is at the expense of a performance reduction, also less serialized (and, thus, faster) implementations result in area requirements significantly lower than those of the corresponding QUARK variants.

In order to provide very compact implementations, we first focus on serialized designs. We explore different datapath sizes (d) for each of the SPONGENT variants: for SPONGENT-{88, 128, 160} we implement $d \in \{4, 8, 16, 32\}$, $d \in \{4, 8, 16, 68\}$, $d \in \{4, 8, 16, 44, 88\}$, respectively, while for SPONGENT-{224, 256} we implement $d \in \{4, 8, 16, 32, 64\}$. An architecture representing our serialized datapath is depicted in Fig. 3(a). The control logic consists of a single counter for the cycle count and some extra combinational logic to drive the select signals of the multiplexers. In order to further reduce the area we use so-called scan registers (6.25 GE in our library), which act as a combination of two input multiplexer and an ordinary register[1]. Instead of providing a reset signal to each register separately, we use two zero inputs at the multiplexers M_1 and M_2 to correctly initialize all the registers. This additionally reduces hardware resources, as the scan registers with a reset input approximately require additional GE per bit of storage. With g_i we denote the value of $\text{lCounter}_b(i)$ in round i. $\text{lCounter}_b(i)$ is implemented as an LFSR as explained in Subsection 2.3. The input of the message block m, denoted with dashed line, is omitted in some cases, i.e. $d \geq r$. The pLayer module requires no additional logic except some extra wiring.

Additionally, we implement all the SPONGENT variants as depicted in Fig. 3(b). Every round now requires a single clock cycle, therefore resulting in faster, yet rather compact designs.

[1] Scan registers are typically used to provide scan-chain based testability of the circuit. Due to the security issues of scan-chain based testing [28], other methods such as Built-In-Self-Test (BIST) are recommended for testing the cryptographic hardware.

Table 2. Hardware performance of the SPONGENT family and comparison with state-of-the-art lightweight hash designs. The nominal frequency of 100 kHz is assumed in all cases and the power consumption is therefore adjusted accordingly.

Hash function	Security (bit) Pre.	Coll.	2nd Pre.	Hash (bit)	Cycles	Datapath (bit)	Process (μm)	**Area** (GE)	**Throughput** (kbps)	**Power** (μW)
SPONGENT-88	80	40	40	88	990	4	0.13	738	0.81	1.57
					45	88	0.13	1127	17.78	2.31
SPONGENT-128	120	64	64	128	2380	4	0.13	1060	0.34	2.20
					70	136	0.13	1687	11.43	3.58
SPONGENT-160	144	80	80	160	3960	4	0.13	1329	0.40	2.85
					90	176	0.13	2190	17.78	4.47
SPONGENT-224	208	112	112	224	7200	4	0.13	1728	0.22	3.73
					120	240	0.13	2903	13.33	5.97
SPONGENT-256	240	128	128	256	9520	4	0.13	1950	0.17	4.21
					140	272	0.13	3281	11.43	6.62
U-QUARK [1]	120	64	64	128	544	1	0.18	1379	1.47	2.44
					68	8	0.18	2392	11.76	4.07
D-QUARK [1]	144	80	80	160	704	1	0.18	1702	2.27	3.10
					88	8	0.18	2819	18.18	4.76
S-QUARK [1]	192	112	112	224	1024	1	0.18	2296	3.13	4.35
					64	16	0.18	4640	50.00	8.39
DM-PRESENT-80 [6]	64	32	64	64	547	4	0.18	1600	14.63	1.83
					33	64	0.18	2213	242.42	6.28
DM-PRESENT-128 [6]	64	32	64	64	559	4	0.18	1886	22.90	2.94
					33	128	0.18	2530	387.88	7.49
H-PRESENT-128 [6]	128	64	64	128	559	8	0.18	2330	11.45	6.44
					32	128	0.18	4256	200.00	8.09
C-PRESENT-192 [6]	192	96	192	192	3338	12	0.18	4600	1.90	-
					108	192	0.18	8048	59.26	9.31
KECCAK-f[400] [17]	160	80	160	160	1000	16	0.13	5090	14.40	11.50
					20	16	0.13	10560	720.00	78.10
KECCAK-f[200] [17]	128	64	128	128	900	8	0.13	2520	8.00	5.60
					18	8	0.13	4900	400.00	27.60
SHA-1 [18]	160	80	160	160	450	32	0.25	6812	113.78	11.00
SHA-256 [19]	256	128	256	256	490	32	0.25	8588	104.48	11.20
BLAKE [15]	256	128	256	256	816	32	0.18	13575	62.79	11.16
Grøstl [27]	256	128	256	256	196	64	0.18	14622	261.14	221.00

Next, we present our hardware figures of all the SPONGENT variants. For the purpose of extensive hardware evaluation we use Synopsys Design Compiler version D-2010.03-SP4 and target the High-Speed UMC 0.13 μm CMOS process provided by Faraday Technology Corporation (fsc0h_d_tc). We provide synthesis results only. Our reasoning follows a simple design decision: we provide a large design space, focusing on multitude of design choices and discuss in detail our implementation strategy. Therefore, we rather spend our efforts by exploring a large set of hardware designs than by performing a time-consuming place and route process for each of the design separately. Moreover, the physical size of the designs (in terms of gate equivalences) is expected to remain the same even after the place and route is performed. We expect slightly worse results only with respect to the overall power consumption.

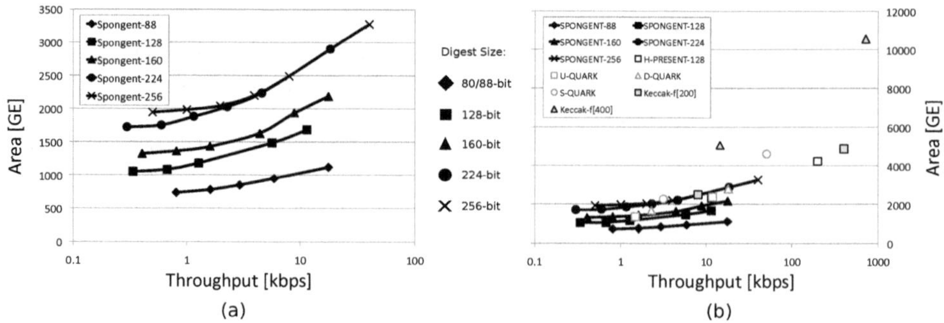

Fig. 4. (a) Area versus throughput trade-off of the SPONGENT hash family. (b) Comparison with state-of-the-art lightweight hash functions.

The power is estimated by observing the internal switching activity of the complete design. Using Mentor Graphics ModelSim version 10.0 SE, we simulate the circuits' behavior for very long messages and generate the VCD (Value Change Dump) files. The VCD files are then converted to the backward SAIF (Switching Activity Interchange Format) files and used within Synopsys Design Compiler for the accurate estimation of the mean power consumption. A typical frequency of 100 kHz is used for all measurements.

Table 2 reports hardware figures obtained using the aforementioned methodology. Besides having a very small footprint, another remarkable result is that the most serialized versions of SPONGENT-{88, 128, 160, 224, 256} are built of 89.3%, 92.5%, 93.8%, 95%, and 96% sequential logic, respectively. For the sake of comparison, we include figures for several state-of-the-art lightweight hash functions. We also include two out of five SHA-3 finalists for which the data of compact hardware implementations is publicly available. We do not compare our design with software-like solutions that benefit from using an external memory for storing the intermediate data. Figure 4(a) illustrates the wide spectrum of our explored design space, where a typical trade-off between speed and area is scrutinized. Using the same metrics, we compare our design with state-of-the-art lightweight hash functions (Fig. 4(b)). For the same level of security, the SPONGENT family tends to require much smaller area than its counterparts.

5 Conclusion

In this paper, we have proposed the family of lightweight hash functions SPONGENT with hash sizes 88, 128, 160, 224, and 256 bits. Its serialized implementations in ASIC hardware require 738, 1060, 1329, 1728, and 1950 GE, respectively. Thus, SPONGENT has the smallest footprint among all hash functions published so far at all security levels it attains, though area requirements are highly dependent on technology used.

References

1. Aumasson, J.P., Henzen, L., Meier, W., Naya-Plasencia, M.: Quark: A lightweight hash. In: Mangard and Standaert [23], pp. 1–15
2. Babbage, S., Dodd, M.: The MICKEY Stream Ciphers. In: Robshaw and Billet [25], pp. 191–209
3. Bertoni, G., Daemen, J., Peeters, M., Assche, G.V.: On the Indifferentiability of the Sponge Construction. In: Smart, N.P. (ed.) EUROCRYPT 2008. LNCS, vol. 4965, pp. 181–197. Springer, Heidelberg (2008)
4. Bertoni, G., Daemen, J., Peeters, M., Assche, G.V.: Sponge-Based Pseudo-Random Number Generators. In: Mangard and Standaert [23], pp. 33–47
5. Bogdanov, A., Knudsen, L.R., Leander, G., Paar, C., Poschmann, A., Robshaw, M.J.B., Seurin, Y., Vikkelsoe, C.: PRESENT: An Ultra-Lightweight Block Cipher. In: Paillier, P., Verbauwhede, I. (eds.) CHES 2007. LNCS, vol. 4727, pp. 450–466. Springer, Heidelberg (2007)
6. Bogdanov, A., Leander, G., Paar, C., Poschmann, A., Robshaw, M.J.B., Seurin, Y.: Hash Functions and RFID Tags: Mind the Gap. In: Oswald, E., Rohatgi, P. (eds.) CHES 2008. LNCS, vol. 5154, pp. 283–299. Springer, Heidelberg (2008)
7. Cho, J.Y.: Linear Cryptanalysis of Reduced-Round PRESENT. In: Pieprzyk, J. (ed.) CT-RSA 2010. LNCS, vol. 5985, pp. 302–317. Springer, Heidelberg (2010)
8. Collard, B., Standaert, F.-X.: A Statistical Saturation Attack against the Block Cipher PRESENT. In: Fischlin, M. (ed.) CT-RSA 2009. LNCS, vol. 5473, pp. 195–210. Springer, Heidelberg (2009)
9. Daemen, J., Peeters, M., Assche, G.V.: Sponge Functions. In: Ecrypt Hash Workshop 2007 (2007), `http://www.csrc.nist.gov/pki/HashWorkshop/PublicComments/2007May.html`
10. De Cannière, C.: TRIVIUM: A Stream Cipher Construction Inspired by Block Cipher Design Principles. In: Katsikas, S.K., López, J., Backes, M., Gritzalis, S., Preneel, B. (eds.) ISC 2006. LNCS, vol. 4176, pp. 171–186. Springer, Heidelberg (2006)
11. De Cannière, C., Dunkelman, O., Knežević, M.: KATAN and KTANTAN — A Family of Small and Efficient Hardware-Oriented Block Ciphers. In: Clavier, C., Gaj, K. (eds.) CHES 2009. LNCS, vol. 5747, pp. 272–288. Springer, Heidelberg (2009)
12. De Cannière, C., Preneel, B.: Trivium. In: Robshaw and Billet [25], pp. 244–266
13. Hell, M., Johansson, T., Maximov, A., Meier, W.: The Grain Family of Stream Ciphers. In: Robshaw and Billet [25], pp. 179–190
14. Hell, M., Johansson, T., Meier, W.: Grain: a stream cipher for constrained environments. IJWMC 2(1), 86–93 (2007)
15. Henzen, L., Aumasson, J.P., Meier, W., Phan, R.C.W.: LSI Characterization of the Cryptographic Hash Function BLAKE (2010), `http://131002.net/data/papers/HAMP10.pdf`
16. Hong, D., Sung, J., Hong, S., Lim, J., Lee, S., Koo, B., Lee, C., Chang, D., Lee, J., Jeong, K., Kim, H., Kim, J., Chee, S.: HIGHT: A New Block Cipher Suitable for Low-Resource Device. In: Goubin, L., Matsui, M. (eds.) CHES 2006. LNCS, vol. 4249, pp. 46–59. Springer, Heidelberg (2006)
17. Kavun, E., Yalcin, T.: A Lightweight Implementation of Keccak Hash Function for Radio-Frequency Identification Applications. In: Ors Yalcin, S.B. (ed.) RFIDSec 2010. LNCS, vol. 6370, pp. 258–269. Springer, Heidelberg (2010)

18. Kim, M., Ryou, J.: Power Efficient Hardware Architecture of SHA-1 Algorithm for Trusted Mobile Computing. In: Qing, S., Imai, H., Wang, G. (eds.) ICICS 2007. LNCS, vol. 4861, pp. 375–385. Springer, Heidelberg (2007)
19. Kim, M., Ryou, J., Jun, S.: Efficient Hardware Architecture of SHA-256 Algorithm for Trusted Mobile Computing. In: Yung, M., Liu, P., Lin, D. (eds.) Inscrypt 2008. LNCS, vol. 5487, pp. 240–252. Springer, Heidelberg (2009)
20. Leander, G.: On Linear Hulls, Statistical Saturation Attacks, PRESENT and a Cryptanalysis of PUFFIN (to appear, 2011)
21. Leander, G., Paar, C., Poschmann, A., Schramm, K.: New Lightweight DES Variants. In: Biryukov, A. (ed.) FSE 2007. LNCS, vol. 4593, pp. 196–210. Springer, Heidelberg (2007)
22. Lim, C.H., Korkishko, T.: mCrypton – A Lightweight Block Cipher for Security of Low-Cost RFID Tags and Sensors. In: Song, J.-S., Kwon, T., Yung, M. (eds.) WISA 2005. LNCS, vol. 3786, pp. 243–258. Springer, Heidelberg (2006)
23. Mangard, S., Standaert, F.-X. (eds.): CHES 2010. LNCS, vol. 6225. Springer, Heidelberg (2010)
24. Mendel, F., Rechberger, C., Schläffer, M., Thomsen, S.S.: The Rebound Attack: Cryptanalysis of Reduced Whirlpool and `Grøstl`. In: Dunkelman, O. (ed.) FSE 2009. LNCS, vol. 5665, pp. 260–276. Springer, Heidelberg (2009)
25. Robshaw, M.J.B., Billet, O. (eds.): New Stream Cipher Designs. LNCS, vol. 4986. Springer, Heidelberg (2008)
26. Standaert, F.X., Piret, G., Gershenfeld, N., Quisquater, J.J.: SEA: A Scalable Encryption Algorithm for Small Embedded Applications. Presented at the Workshop on RFID and Light-Weight Crypto in Graz, Austria (2005)
27. Tillich, S., Feldhofer, M., Issovits, W., Kern, T., Kureck, H., Mühlberghuber, M., Neubauer, G., Reiter, A., Köfler, A., Mayrhofer, M.: Compact Hardware Implementations of the SHA-3 Candidates ARIRANG, BLAKE, Grøstl, and Skein. Cryptology ePrint Archive, Report 2009/349 (2009), `http://eprint.iacr.org/2009/349`
28. Yang, B., Wu, K., Karri, R.: Scan Based Side Channel Attack on Dedicated Hardware Implementations of Data Encryption Standard. In: International Test Conference, pp. 339–344 (2004)

A Some Differential Paths

Table 3. Sample differential paths for SPONGENT–{88, 128}

	SPONGENT-88		SPONGENT-128	
Round	Difference	Prob	Difference	Prob
0	0500000000000009000000	2^{-21}	0000000000000000900090000000000000	2^{-22}
5	0000000400800000000000		0000000000008000400000000000000000	
0	9000900000000000000000	2^{-47}	0000000550000000000000330000000000	2^{-68}
10	20000000000080000A0000		0000040040000000100000000000001010	
0	9000900000000000000000	2^{-74}		
15	0000000080100000000000			

Table 4. Sample differential paths for SPONGENT–{160, 224, 256}

SPONGENT-160		
Round	Difference	Prob
0	06000000060000000000000000000000000000000000	2^{-21}
5	00000000000000000400800000000000000000000000	
0	90000000000090000000000000000000000000000000	2^{-50}
10	00800000000000000000000020000000000A00000000	
0	00000000000000000000000000003000300000000000	2^{-79}
15	00000000000000000080100000000000000000000000	
SPONGENT-224		
Round	Difference	Prob
0	090000000000090000000000000000000000000000000000000000000000	2^{-21}
5	000000020004000000000000000000000000000000000000000000000000	
0	000000000000000000000000000000000009000000000009000000000000	2^{-43}
10	008000000000000000000000000000001000000000000009000000000000	
0	000000001001000000000000000000000000000000000000000000000000	2^{-66}
15	008000000000000000000000000000001000000000000009000000000000	
SPONGENT-256		
Round	Difference	Prob
0	00000000000000000000000000000000000000000000000000000660000000000660	2^{-20}
5	00000000020001000000000000000000000000000000000000000000000020001000	

The LED Block Cipher

Jian Guo[1], Thomas Peyrin[2,⋆], Axel Poschmann[2,⋆], and Matt Robshaw[3,⋆⋆]

[1] Institute for Infocomm Research, Singapore
[2] Nanyang Technological University, Singapore
[3] Applied Cryptography Group, Orange Labs, France
{ntu.guo,thomas.peyrin}@gmail.com,
aposchmann@ntu.edu.sg,
matt.robshaw@orange-ftgroup.com

Abstract. We present a new block cipher LED. While dedicated to compact hardware implementation, and offering the smallest silicon footprint among comparable block ciphers, the cipher has been designed to simultaneously tackle three additional goals. First, we explore the role of an ultra-light (in fact non-existent) key schedule. Second, we consider the resistance of ciphers, and LED in particular, to related-key attacks: we are able to derive simple yet interesting AES-like security proofs for LED regarding related- or single-key attacks. And third, while we provide a block cipher that is very compact in hardware, we aim to maintain a reasonable performance profile for software implementation.

Keywords: Lightweight, block cipher, RFID tag, AES.

1 Introduction

Over past years many new cryptographic primitives have been proposed for use in RFID tag deployments, sensor networks, and other applications characterised by highly-constrained devices. The pervasive deployment of tiny computational devices brings with it many interesting, and potentially difficult, security issues.

Chief among recent developments has been the evolution of lightweight block ciphers where an accumulation of advances in algorithm design, together with an increased awareness of the likely application, has helped provide important developments. To some commentators the need for yet another lightweight block cipher proposal will be open to question. However, in addition to the fact that many proposals present some weaknesses [2,10,45], we feel there is still more to be said on the subject and we observe that it is in the "second generation" of work that designers might learn from the progress, and omissions, of "first generation" proposals. And while new proposals might only slightly improve on

⋆ The authors were supported in part by the Singapore National Research Foundation under Research Grant NRF-CRP2-2007-03.

⋆⋆ The author gratefully acknowledges the support of NTU during his visit to Singapore. This work is also supported in part by the European Commission through the ICT program under contract ICT-2007-216676 ECRYPT II.

B. Preneel and T. Takagi (Eds.): CHES 2011, LNCS 6917, pp. 326–341, 2011.

successful initial proposals in terms of a single metric, *e.g.* area, they might, at the same time, overcome other important security and performance limitations. In this paper, therefore, we return to the design of lightweight block ciphers and we describe L*ight* E*ncryption* D*evice*, `LED`.

During our design, several key observations were uppermost in our mind. Practically all modern block cipher proposals have reasonable security arguments; but few offer much beyond (potentially thorough) *ad hoc* analysis. Here we hope to provide a more complete security treatment than is usual. In particular, related-key attacks are often dismissed from consideration for the application areas that typically use such constrained devices, *e.g.* RFID tags. In practice this is often perfectly reasonable. However, researchers will continue to derive cryptanalytic results in the related-key model [18,2] and there has been some research on how to modify or strengthen key schedules [35,15,39]. So having provable levels of resistance to such attacks would be a bonus and might help confusion developing in the cryptographic literature.

In addition, our attention is naturally focused on the performance of the algorithm on the tag. However, there can be constraints when an algorithm is also going to be implemented in software. This is something that has already been discussed with the design of `KLEIN` [22] and in the design of `LED` we have aimed at very compact hardware implementation while maintaining some software-friendly features.

Our new block cipher is based on `AES`-like design principles and this allows us to derive very simple bounds on the number of active Sboxes during a block cipher encryption. Since the key schedule is very simple, this analysis can be done in a related-key model as well; *i.e.* our bounds apply even when an attacker tries to mount a related-key attack. And while `AES`-based approaches are well-suited to software, they don't always provide the lightest implementation in hardware. But using techniques presented in [23] we aim to resolve this conflict.

While block ciphers are an important primitive, and arguably the most useful in a constrained environment, there has also been much progress in the design of stream ciphers [14,25] and even, very recently, in lightweight hash functions [23,4]. In fact it is this latter area of work that has provided inspiration for the block cipher we will present here.

2 Design Approach and Specifications

Like so much in today's symmetric cryptography, an `AES`-like design appears to be the ideal starting point for a clean and secure design. The design of `LED` will inevitably have many parallels with this established approach, and features such as `Sboxes`, `ShiftRows`, and (a variant of) `MixColumns` will all feature and take their familiar roles.

For the key schedule we chose to do-away with the "schedule", *i.e.* the user-provided key is used repeatedly *as is*. As well as giving obvious advantages in hardware implementation, it allows for simple proofs to be made for the security of the scheme even in the most challenging attack model of related keys. At first sight the re-use of the encryption key without variation appears dangerous, certainly to those familiar with slide attacks and some of their advanced variants [7,8]. But we note that such a simple key schedule is not without precedent [42] though the treatment here is more complete than previously.

The LED cipher is described in Section 2.1. It is a 64-bit block cipher with two primary instances taking 64- and 128-bit keys. The cipher state is conceptually arranged in a (4×4) grid where each nibble represents an element from $GF(2^4)$ with the underlying polynomial for field multiplication given by $X^4 + X + 1$.

Sboxes. LED cipher re-uses the PRESENT Sbox which has been adopted in many lightweight cryptographic algorithms. The action of this box in hexadecimal notation is given by the following table.

x	0	1	2	3	4	5	6	7	8	9	A	B	C	D	E	F
$S[x]$	C	5	6	B	9	0	A	D	3	E	F	8	4	7	1	2

MixColumnsSerial. We re-use the tactic adopted in [23] to define an MDS matrix for linear diffusion that is suitable for compact serial implementation. The MixColumnsSerial layer can be viewed as four applications of a hardware-friendly matrix A with the net result being equivalent to using the MDS matrix M where

$$(A)^4 = \begin{pmatrix} 0 & 1 & 0 & 0 \\ 0 & 0 & 1 & 0 \\ 0 & 0 & 0 & 1 \\ 4 & 1 & 2 & 2 \end{pmatrix}^4 = \begin{pmatrix} 4 & 2 & 1 & 1 \\ 8 & 6 & 5 & 6 \\ B & E & A & 9 \\ 2 & 2 & F & B \end{pmatrix} = M.$$

The basic component of LED will be a sequence of four identical rounds used without the addition of any key material. This basic unit, that we later call "step", makes it easy to establish security bounds for the construction.

2.1 Specification of LED

For a 64-bit plaintext m the 16 four-bit nibbles $m_0 \| m_1 \| \cdots \| m_{14} \| m_{15}$ are arranged (conceptually) in a square array:

$$\begin{bmatrix} m_0 & m_1 & m_2 & m_3 \\ m_4 & m_5 & m_6 & m_7 \\ m_8 & m_9 & m_{10} & m_{11} \\ m_{12} & m_{13} & m_{14} & m_{15} \end{bmatrix}$$

This is the initial value of the cipher STATE and note that the state (and the key) are loaded row-wise rather than in the column-wise fashion we have come to expect from the AES; this is a more hardware-friendly choice, as pointed out in [38].

The key is viewed nibble-wise and loaded nibble-by-nibble into one or two arrays, K_1 and K_2, depending on the key length. Our primary definition is for

64- or 128-bit keys, but other key lengths, *e.g.* the popular choice of 80 bits, can be padded to give a 128-bit key thereby giving a 128-bit key array. By virtue of the order of loading the tables, any key that is padded (with zeros) to give a 64- or 128-bit key array will effectively set unused nibbles of the key array to 0.

$$\begin{bmatrix} k_0 & k_1 & k_2 & k_3 \\ k_4 & k_5 & k_6 & k_7 \\ k_8 & k_9 & k_{10} & k_{11} \\ k_{12} & k_{13} & k_{14} & k_{15} \end{bmatrix} \quad \text{for 64-bit keys giving } K_1$$

$$\begin{bmatrix} k_0 & k_1 & k_2 & k_3 \\ k_4 & k_5 & k_6 & k_7 \\ k_8 & k_9 & k_{10} & k_{11} \\ k_{12} & k_{13} & k_{14} & k_{15} \end{bmatrix} \begin{bmatrix} k_{16} & k_{17} & k_{18} & k_{19} \\ k_{20} & k_{21} & k_{22} & k_{23} \\ k_{24} & k_{25} & k_{26} & k_{27} \\ k_{28} & k_{29} & k_{30} & k_{31} \end{bmatrix} \quad \text{for 128-bit keys giving } K_1 \| K_2$$

The operation `addRoundKey(`STATE`,`K_i`)` combines nibbles of subkey K_i with the state, respecting array positioning, using bitwise exclusive-or. There is no key schedule, or rather this is the sum total of the key schedule, and the arrays K_1 and, where appropriate, K_2 are repeatedly used without modification. Encryption is described using the previously mentioned `addRoundKey(`STATE`,`K_i`)` and a second operation, `step(`STATE`)`. This is illustrated in Figure 1.

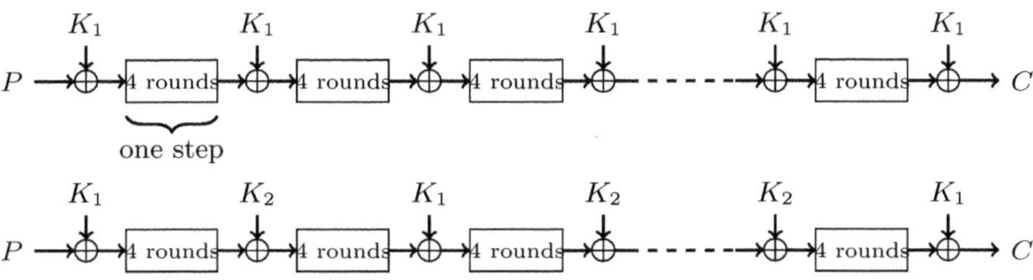

Fig. 1. The use of key arrays K_1 and K_2 in LED showing both a 64-bit key array (top) and a 128-bit key array (bottom)

The number of steps during encryption depends on whether there are one or two key arrays.

```
for i = 1 to 8 do {
    addRoundKey(STATE, K1)
    step(STATE)
}
addRoundKey(STATE, K1)
```

for 64-bit key arrays

```
for i = 1 to 6 do {
    addRoundKey(STATE, K1)
    step(STATE)
    addRoundKey(STATE, K2)
    step(STATE)
}
addRoundKey(STATE, K1)
```

for 128-bit key arrays

The operation `step(STATE)` consists of four rounds of encryption of the cipher state. Each of these four rounds uses, in sequence, the operations **`AddConstants`**, **`SubCells`**, **`ShiftRows`**, and `MixColumnsSerial` as illustrated in Figure 2.

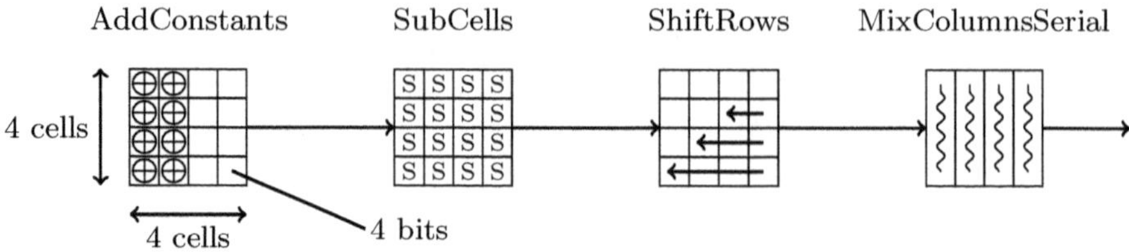

Fig. 2. An overview of a single round of LED

`AddConstants`. A round constant is defined as follows. At each round, the six bits (rc_5, rc_4, rc_3, rc_2, rc_1, rc_0) are shifted one position to the left with the new value to rc_0 being computed as $rc_5 \oplus rc_4 \oplus 1$. The six bits are initialised to zero, and updated *before* use in a given round. The constant, when used in a given round, is arranged into an array as follows:

$$\begin{bmatrix} 0 & (rc_5\|rc_4\|rc_3) & 0 & 0 \\ 1 & (rc_2\|rc_1\|rc_0) & 0 & 0 \\ 2 & (rc_5\|rc_4\|rc_3) & 0 & 0 \\ 3 & (rc_2\|rc_1\|rc_0) & 0 & 0 \end{bmatrix}$$

The round constants are combined with the state, respecting array positioning, using bitwise exclusive-or.

`SubCells`. Each nibble in the array STATE is replaced by the nibble generated after using the PRESENT Sbox.

`ShiftRow`. Row i of the array STATE is rotated i cell positions to the left, for $i = 0, 1, 2, 3$.

`MixColumnsSerial`. Each column of the array STATE is viewed as a column vector and replaced by the column vector that results after post-multiplying the vector by the matrix M (see earlier description in this section).

The final value of the STATE provides the ciphertext with nibbles of the "array" being unpacked in the obvious way. Test vectors for LED are provided at `https://sites.google.com/site/ledblockcipher/`.

3 Security Analysis

The LED block cipher is simple to analyze and this allows us to precisely evaluate the necessary number of rounds to ensure proper security.

Our scheme is meant to be resistant to classical attacks, but also to the type of related-key attacks that have been effective against AES-256 [9] and other ciphers [2]. We will even study the security of LED in a hash function setting, *i.e.* when it is used in a Davies-Meyer or similar construction with a compression function based on a block cipher. In other words, we will consider attackers that have full access to the key(s) and try to distinguish the fixed permutations from randomly chosen ones. While this analysis provides additional confidence in the security of LED, it is not our intent to propose a hash function construction.

We chose a conservative number of rounds for LED. For example, when using a 64-bit key array we use 32 AES-like rounds that are grouped as eight "big" add-key/apply-permutation steps that are each composed of four AES-like rounds. Further, our security margins are even more conservative if one definitively disregards related-key attacks; as will be seen with the following proofs.

3.1 The Key Schedule

The LED key schedule has been chosen for its simplicity and security. Because it is very simple to analyze, it allows us to directly derive a bound on the minimal number of active Sboxes, even in the scenario of related-key attacks. The idea is to first compute a bound on the number of active big steps (each composed of 4 AES-like rounds). Then, using the well known 4-round proofs for the AES, one can show that one active big step will contain at least 25 active Sboxes. Note that this bound is tight as we know 4-round differential paths containing exactly this number of active Sboxes.

When not considering related-key attacks, we directly obtain that any differential path for LED will contain at least $\lfloor r/4 \rfloor \cdot 25$ active Sboxes. For related-key attacks, we have to distinguish between the different key-size versions.

64-Bit Key Version. If we assume that differences are inserted in the key input, then every subkey K_1 in the 64-bit key variant of LED will be active. Therefore, one can easily see that it is impossible to force two consecutive non-active big steps and we are ensured that for every two big steps at least one is active. Overall, this shows that any related-key differential path contains at least $\lfloor r/8 \rfloor \cdot 25$ active Sboxes.

128-Bit Key Version. If we assume that differences are inserted in the key input, then we have to separate two cases. If the two independent parts K_1 and K_2 composing the key both contain a difference, then we end up with exactly the same reasoning as for the 64-bit key variant: at least $\lfloor r/8 \rfloor \cdot 25$ active Sboxes will be active. If only one of the two independent parts composing the key contains a difference, then subkeys with and without differences are alternatively incorporated after each big step. The non-active subkeys impact on the differential paths is completely void and thus in this case one can view LED as being composed of even bigger steps of 8 AES-like rounds instead. The very same reasoning then applies again: it is impossible to force two consecutive of these new bigger steps to be inactive and therefore we have at least $\lfloor r/16 \rfloor \cdot 50$ active Sboxes ensured

Table 1. Minimal number of active Sboxes and upper bounds on the best differential path and linear approximation probability for the 64-bit key array and 128-bit key array versions of LED (in both the single-key (SK) and related-key (RK) settings)

	LED-64 SK	LED-64 RK	LED-128 SK	LED-128 RK
minimal no. of active Sboxes	200	100	300	150
differential path probability	2^{-400}	2^{-200}	2^{-600}	2^{-300}
linear approx. probability	2^{-400}	2^{-200}	2^{-600}	2^{-300}

for any differential path (since the best differential path for 8 rounds trivially contains 50 active Sboxes).

We summarize in Table 1 the results obtained for the two main versions of LED, both for single-key attacks and related-key attacks. Note that the bounds on the number of active Sboxes are tight as we know differential paths meeting them (for example the truncated differential path for each active big step can simply be any of the 4-round path for AES-128 with 25 active Sboxes).

For LED-128, since we are using two independent key parts one can peel off the first and last key addition (which is always the first key part K_1). Thus, an attacker can remove one big step on each side of the cipher, for a total of 8 rounds, with a complexity of 2^{64} tries on K_1. This partially explains why the versions of LED using two independent key parts have 16 more rounds than for LED-64.

3.2 Differential/Linear Cryptanalysis

Since LED is an AES-like cipher, one can directly reuse extensive work that has been done on the AES. We will compute a bound on the best differential path probability (where all differences on the input and output of all rounds are specified) or even the best differential probability (where only the input and output differences are specified), in both single- and related-key settings.

As the best differential transition probability of the PRESENT Sbox is 2^{-2}, using the previously proven minimal number of active Sboxes we deduce that the best differential path probability on 4 active rounds of LED is upper bounded by $2^{-2 \cdot 25} = 2^{-50}$. By adapting the work from [40], the maximum differential probability for 4 active rounds of LED is upper bounded by

$$\max\left\{\max_{1\leq u\leq 15}\sum_{j=1}^{15}\{DP^S(u,j)\}^5, \max_{1\leq u\leq 15}\sum_{j=1}^{15}\{DP^S(j,u)\}^5\right\}^4 = 2^{-32}$$

where $DP^S(i,j)$ stands for the differential probability of the Sbox to map the difference i to j. The duality between linear and differential attacks allows us to similarly apply the same approaches to compute a bound on the best linear approximation. Over four rounds the best linear approximation probability is upper bounded by 2^{-50} and the best linear hull probability is upper bounded by 2^{-32}.

Since we previously proved that all rounds will be active in the single-key scenario and half of them will be active in the related-key scenario, we can easily compute the upper bounds on the best differential path probability and the best linear approximation probability for each version of LED (see Table 1). Note that this requires that random subkeys be used at each round to make the Sbox inputs independant. In the case of LED the subkeys are simulated by the addition of round constants and the derived bounds give a very good indication of the quality of the LED internal permutation with regards to linear and differential cryptanalysis.

3.3 Cube Testers and Algebraic Attacks

We applied the most recent developed cube testers [3] and its zero-sum distinguishers to the LED fixed-key permutation, the best we could find within practical time complexity is at most three rounds (with the potential to be doubled under a meet-in-the-middle scenario). Note, in case of AES, "zero-sum" property is also referred as "balanced", found by the AES designers [16], in which 3-round balanced property is shown. To the best of our knowledge, there is no balanced property found for more than 3 AES rounds.

The PRESENT Sbox used in LED has algebraic degree 3 and one can check that $3 \cdot \lfloor r/4 \rfloor \cdot 25 \ggg 64$ for all LED variants. Moreover, the PRESENT Sbox is described by $e = 21$ quadratic equations in the $v = 8$ input/output-bit variables over $GF(2)$. The entire system for a fixed-key LED permutation therefore consists of $(16 \cdot r \cdot e)$ quadratic equations in $(16 \cdot r \cdot v)$ variables. For example, in the case of the 64-bit key version, we end up with 10752 equations in 4096 variables. In comparison, the entire system for a fixed-key AES permutation consists of 6400 equations in 2560 variables. While the applicability of algebraic attacks on AES remains unclear, those numbers tends to indicate that LED offers a higher level of protection.

3.4 Other Cryptanalysis

The slide attack is a block cipher cryptanalysis technique [7] that exploits the degree of self-similarity of a permutation. In the case of LED, all rounds are made different thanks to the round-dependent constants addition, which makes the slide attack impossible to perform.

Integral cryptanalysis is a technique first applied on SQUARE [17] that is particularly efficient against block ciphers based on substitution-permutation networks, like AES or LED. The idea is to study the propagation of sums of values; something which is quite powerful on ciphers that only use bijective components. As for AES, the best integral property can be found on three rounds, or four rounds with the last mixing layer removed. Thus, two big LED steps avoid any such observation. Considering the large number of rounds of LED, we believe integrals attacks are very unlikely to be a threat.

Rotational cryptanalysis [28] studies the evolution of a rotated variant of some input words through the round process. It was proven to be quite successful against some Addition-Rotation-XOR (ARX) block ciphers and hash functions.

LED is an Sbox-oriented block cipher and any rotation property in a cell will be directly removed by the application of the Sbox layer. Even if one looks for a rotation property of cell positions, this is unlikely to lead to an attack since the constants used in a LED round are all distinct and any position rotation property between columns or lines is removed after the application of two rounds.

Methods to find better bounds on the algebraic degree were recently published in [12]. With the first two rounds combined as Super-Sboxes, the best algebraic degree we can find for fixed-key LED permutation and its inverse are $3, 11, 33, 53, 60, 62$, for r rounds with $r = 1, \ldots, 6$. Using this technique, one can distinguish up to 12 rounds with complexity bounded by 2^{63}, in the known key model.

3.5 LED in a Hash Function Setting

Studying a block cipher in a hash function setting is a good security test since it is very advantageous for the attacker. In this scenario he will have full control on all inputs. In the so-called known-key [29] or chosen-key models, the attacker can have access or even choose the key(s) used, and its goal is then to find some input/output pairs having a certain property with a complexity lower than what is expected for randomly chosen permutation(s). Typically, the property is that the input and output differences or values are fixed to a certain subset of the whole domain.

While we conduct an analysis of the security of LED in a hash function setting, we would like to emphasize that our goal is not to build a secure hash function. However, we believe that this section adds further confidence in the quality of our block cipher proposal.

Rebound and Super-Sbox Attacks. The recent rebound attack [37] and its improved variants (start-from-the-middle attack [36] and Super-Sbox cryptanalysis [21,31]) have much improved the best known attacks on many hash functions, especially for AES-based schemes. The attacker will first prepare a differential path and then use the available freedom degrees to the most costly part of the trail (often in the middle) so as to reduce the overall complexity. The costly part is called the controlled rounds, while the rest of the trail are the uncontrolled rounds and they are verified probabilistically. The rebound attack and its variants allows the attacker to nicely use the freedom degrees so that the controlled part is as big as possible. At the present time, the most powerful technique in the known-key setting allows the attacker to control three rounds and no method is known to control more rounds, even if the key is chosen by the attacker.

In order to ease the analysis, we assume pessimistically that the attacker can control four rounds, that is one full active big step, with a negligible computation/memory cost (even if one finds a method to control four AES-like rounds in the chosen-key model, it will not apply here since no key is inserted during four consecutive rounds). In the case of 64-bit key LED, the attacker can control two independent active big steps and later merge them by freely fixing the key value. However, even in this advantageous scenario for the attacker we are ensured that at least two big steps will be active and uncontrolled, and this seems sufficient to resist distinguishing attacks. Indeed, for two active big steps of LED, the upper bound for the best differential path probability and the best linear approximation probability (respectively the best differential probability and the best linear hull probability) is 2^{-100} (respectively 2^{-64}).

For the 128-bit key version, we can again imagine that the attacker to control and merge two active big steps with a negligible computation/memory cost. Even if so, with the same reasoning we are ensured that at least four big steps will be active and uncontrolled, and again this seems sufficient since for four active big steps of LED, the upper bound for the best differential path probability and the best linear approximation probability (respectively the best differential probability and the best linear hull probability) is 2^{-200} (respectively 2^{-128}).

Integral Attacks. One can directly adapt the known-key variant of integral attacks from [29] to the LED internal permutation. However, this attack can only reach seven rounds with complexity 2^{28}, which is worse than what can be obtained with previous rebound-style attacks.

4 Performance and Comparison

4.1 Hardware Implementation

We used *Mentor Graphics ModelSimXE 6.4b* and *Synopsys DesignCompiler A-2007.12-SP1* for functional simulation and synthesis of the designs to the *Virtual Silicon* (VST) standard cell library *UMCL18G212T3*, which is based on the *UMC L180 0.18μm 1P6M* logic process with a typical voltage of 1.8 V. For synthesis and for power estimation (using *Synopsys Power Compiler* version *A-2007.12-SP1*) we advised the compiler to keep the hierarchy and use a clock frequency of 100 KHz, which is a widely cited operating frequency for RFID applications. Note that the wire-load model used, though it is the smallest available for this library, still simulates the typical wire-load of a circuit with a size of around 10, 000 GE.

To substantiate our claims on the hardware efficiency of our LED family, we have implemented LED-64 and LED-128 in VHDL and simulated their post-synthesis performance. As can be seen in Figure 3, our serialized design consists of seven modules: MCS, State, AK, AC, SC, Controller, and Key State.

State comprises a $4 \cdot 4$ array of flip-flop cells storing 4 bits each. Every row constitutes a shift-register using the output of the last stage, *i.e.* column 0, as the input to the first stage (column 3) of the same row and the next row. Using this feedback functionality *ShiftRows* can be performed in 3 clock cycles with no additional hardware costs. Further, since *MixColumnsSerial* is performed on column 0, also a vertical shifting direction is required for this column. Consequently, columns 0 and 3 consist of flip-flop cells with two inputs (6 GE), while columns 1 and 2 can be realized with flip-flop cells with only one input (4.67 GE).

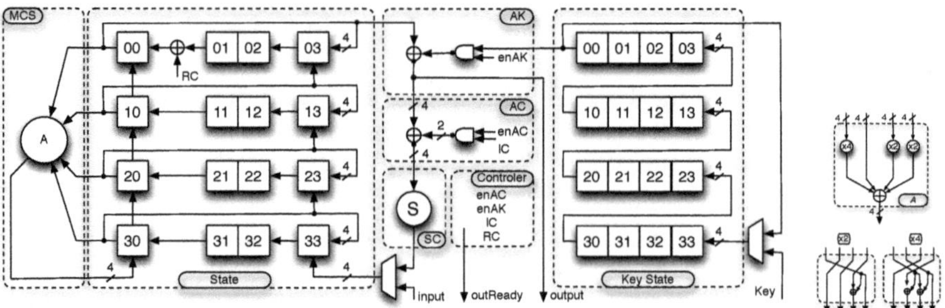

Fig. 3. Serial hardware architecture of LED (left) and A with its sub-components (right)

The key is stored in `Key State`, which comprises of a 4-bit wide simple shift register of the appropriate length, i.e. 64 or 128. Please note that the absence of a key-schedule of LED has two advantages: it allows 1) to use the most basic, and thus cheapest, flip-flops (4.67 GE per bit); and 2) to hardwire the key in case no key update is required. In the latter case additional combinational logic is required to select the appropriate key chunk, which reduces the savings to 278 GE and 577 GE for LED-64 and LED-128, respectively. For arbitrary key lengths the area requirements grow by 4.67 GE per bit. An LED-80 with the same parameters as PRESENT-80 would thus require approximately 1,040 GE with a flexible key and around 690 GE with fixed key.

MCS calculates the last row of A in one clock cycle. The result is stored in the `State` module, that is in the last row of column 0, which has been shifted upwards at the same time. Consequently, after 4 clock cycles the *MixColumnsSerial* operation is applied to an entire column. Then the whole state array is rotated by one position to the left and the next column is processed. As an example of the hardware efficiency of *MCS* we depict A in the upper and its sub-components in the lower right part of Figure 3. In total only 40 GE and 20 clock cycles are required to perform *MCS*, which is 4 clock cycles slower but 85% smaller than a serialized implementation of the AES *MixColumns* [24]. If we take into account that AES operates on 8 bits and not like LED on 4 bits, the area savings are still more than 40%.

AK performs the *AddRoundKey* operation by XORing the roundkey every fourth round. For this reason the input to the XNOR gate is gated with a NAND gate.

AC performs one part of the *AddConstant* operation by XORing the first column of the round constant matrix (a simple arithmetic 2-bit counter) to the first column of the state matrix. For this reason, the input to the XNOR gate is gated with a NAND gate. In order to use a single control signal for the addition of the round constants, which span over the first two columns, the addition of the second column of the round constant matrix to the second column of the state array is performed in the `State` module.

SC performs the *SubCells* operation and consists of a single instantiation of the corresponding Sbox. We used an optimized Boolean representation of the

PRESENT Sbox,[1] which only requires 22.33 GE. It takes 16 clock cycles to perform *AddConstant* and *SubCells* on the whole state.

Controller uses a Finite State Machine (FSM) to generate all control signals required. The FSM consists of one idle state, one init state to load the initial values, one state for the combined execution of *AC* and *SC*, 3 states for *ShR* and two states for *MCS* (one for processing one column and another one to rotate the whole state to the left). Several LFSR-based counters are required: 6-bit for the generation of the second column of the round constants matrix, 4-bit for the key addition scheduling and 2-bit for the transition conditions of the FSM. Besides, a 2-bit arithmetic counter is required for the generation of the first column of the round constants matrix. Its LSB is also used to select either the 3 MSB $rc_5||rc_4||rc_3$ or the 3 LSB $rc_2||rc_1||rc_0$ of the 6-bit LFSR-based counter. In total the control logic sums up to 199 GE.

It requires 39 clock cycles to perform one round of LED, resulting in a total latency of 1248 clock cycles for LED-64 and 1872 clock cycles for LED-128. The estimated power consumption at a frequency of 100 KHz and a supply voltage ov 1.8V is 1.67μW for LED-64 (1.11μW with a hard-wired key) and 2.2μW for LED-128 (1.11μW). It is a well-known fact that at low frequencies, as typical for low-cost applications, the power consumption is dominated by its static part, which is proportional to the amount of transistors involved. Furthermore, the power consumption strongly depends on the used technology and greatly varies with the simulation method. To address these issues and to reflect the time-area-power trade-off inherent in any hardware implementation a new figure of merit (FOM) was proposed by [5]. In order to have a fair comparison, we omit the power values in Table 2 and only compare cycles per block, throughput at 100 KHz (in kilo bits per second), the area requirements (in GE), and FOM (in nano bits per clock cycle per GE squared).

Table 2 compares our results to previous work, sorted according to key flexibility and increasing security levels. Note that we have not been able to include all recent proposals and we have restricted ourselves to block ciphers for our comparison. Other techniques such as HUMMINGBIRD [19] and ARMADILLO [5] are of some interest in the literature, though attacks on early versions have lead to some redesign [45,1,20]. As can be seen from Table 2, the block cipher LED is the smallest when compared to other block ciphers with similar key and block size.

4.2 Software Implementation

We have made two implementations of LED; one for reference and clarity with the second being optimized for performance (by using table lookups). The measurements were taken on an Intel(R) Core(TM) i7 CPU Q 720 clocked at 1.60GHz.

In the optimised implementation, we represent the LED state as a single 64-bit word and we build eight lookup tables each with 256 64-bit entries. This is similar to many AES implementations, except we treat two consecutive nibbles

[1] Due to Dag Arne Osvik.

Table 2. Hardware implementation results of some block ciphers. [44] also synthesized the same architecture of PRESENT and yielded a lower gate count of 1,000 GE. However, the number quoted below is from the same library used here and hence is a fairer choice for comparison. * denotes estimated values.

	Algorithm	Ref.	key size	block size	cycles/ block	T'put (@100 KHz)	Tech. [μm]	**Area** [GE]	FOM $[\frac{bits \times 10^9}{clk \cdot GE^2}]$
Flexible Keys	DESL	[32]	56	64	144	44.4	0.18	**1,848**	130
	LED-64		64	64	1,248	5.1	0.18	**966**	55
	KLEIN-64	[22]	64	64	207	N/A	0.18	**1,220**	N/A
	LED-80*		80	64	1,872	3.4	0.18	**1,040**	32
	PRESENT-80	[44]	80	64	547	11.7	0.18	**1,075**	101
	PRESENT-80	[11]	80	64	32	200.0	0.18	**1,570**	811
	KATAN64	[13]	80	64	255	25.1	0.13	**1,054**	226
	KLEIN-80	[22]	80	64	271	N/A	0.18	**1,478**	N/A
	LED-96*		96	64	1,872	3.4	0.18	**1,116**	27
	KLEIN-96	[22]	96	64	335	N/A	0.18	**1,528**	N/A
	mCrypton	[33]	96	64	13	492.3	0.13	**2,681**	685
	SEA	[34]	96	96	93	103.0	0.13	**3,758**	73
	LED-128		128	64	1,872	3.4	0.18	**1,265**	21
	PRESENT-128	[41]	128	64	559	11.4	0.18	**1,391**	59
	PRESENT-128	[11]	128	64	32	200.0	0.18	**1,886**	562
	HIGHT	[26]	128	64	34	188.0	0.25	**3,048**	203
	AES	[38]	128	128	226	56.6	0.13	**2,400**	98
	DESXL	[32]	184	64	144	44.4	0.18	**2,168**	95
Hard-wired Keys	LED-64		64	64	1,280	5.13	0.18	**688**	108
	PRINTcipher-48	[30]	80	48	768	6.2	0.18	**402**	387
	KTANTAN64	[13]	80	64	255	25.1	0.13	**688**	530
	LED-80*		80	64	1,872	3.4	0.18	**690**	72
	LED-96*		96	64	1,872	3,42	0.18	**695**	71
	LED-128		128	64	1,872	3.42	0.18	**700**	70
	PRINTcipher-96	[30]	160	96	3072	3.13	0.18	**726**	59

(2×4 bits) as a unit for the lookup table. Hence SubCells, ShiftRows and MixColumnsSerial can all be achieved using eight table lookups and XORs.

Overall, we need to access $8 \times 32 \times 2 = 512$ 32-bit words of memory (or $8 \times 32 = 256$ 64-bit words of memory). In contrast, an AES implementation with four tables of 256 entries would require $(16+4) \times 10 = 200$ accesses. This suggests that LED-64 should be about 2.5 times slower than AES on 32-bit platforms with table-based implementations, and similarly LED-128 will be 3.8 slower than AES, while the optimized table-based implementation runs 57 and 86 cycles per byte for LED-64 and LED-128, respectively.

5 Conclusion

In this paper we have presented the block cipher LED. Clearly, given its novelty, the cipher should not be used in applications until there has been sufficient independent analysis. Nevertheless, we hope that our design is of some interest and we have focused our attention on what seem to be the neglected areas of key schedule design and protection against related-key attacks. Furthermore, we have done so while working in one of the more challenging design spaces—that of constrained hardware implementation—and we have proposed one of the smallest block ciphers in the literature (for comparable choices of parameters) while striving to maintain a competitive performance in software. Additional information on `LED` will be made available via `https://sites.google.com/site/ledblockcipher/` and we welcome all comments and analysis.

References

1. Abdelraheem, M., Blondeau, C., Naya-Plasencia, M., Videau, M., Zenner, E.: Cryptanalysis of Armadillo-2, `http://eprint.iacr.org/2011/160.pdf`
2. Ågren, M.: Some Instant- and Practical-Time Related-Key Attacks on KTANTAN32/48/64, `http://eprint.iacr.org/2011/140`
3. Aumasson, J.-P., Dinur, I., Meier, W., Shamir, A.: Cube Testers and Key Recovery Attacks on Reduced-Round MD6 and Trivium. In: Dunkelman, O. (ed.) FSE 2009. LNCS, vol. 5665, pp. 1–22. Springer, Heidelberg (2009)
4. Aumasson, J.-P., Henzen, L., Meier, W., Naya-Plasencia, M.: Quark: A Lightweight Hash. In: Mangard, S., Standaert, F.-X. (eds.) CHES 2010. LNCS, vol. 6225, pp. 1–15. Springer, Heidelberg (2010)
5. Badel, S., Dagtekin, N., Nakahara, J., Ouafi, K., Reffé, N., Sepehrdad, P., Susil, P., Vaudenay, S.: ARMADILLO: A Multi-purpose Cryptographic Primitive Dedicated to Hardware. In: Mangard, S., Standaert, F.-X. (eds.) CHES 2010. LNCS, vol. 6225, pp. 398–412. Springer, Heidelberg (2010)
6. Barreto, P., Rijmen, V.: The Whirlpool Hashing Function. Submitted to NESSIE (September 2000), `http://www.larc.usp.br/~pbarreto/WhirlpoolPage.html` (revised May 2003)
7. Biryukov, A., Wagner, D.: Slide Attacks. In: Knudsen, L.R. (ed.) FSE 1999. LNCS, vol. 1636, pp. 245–259. Springer, Heidelberg (1999)
8. Biryukov, A., Wagner, D.: Advanced Slide Attacks. In: Preneel, B. (ed.) EUROCRYPT 2000. LNCS, vol. 1807, pp. 589–606. Springer, Heidelberg (2000)
9. Biryukov, A., Khovratovich, D.: Related-Key Cryptanalysis of the Full AES-192 and AES-256. In: Matsui, M. (ed.) ASIACRYPT 2009. LNCS, vol. 5912, pp. 1–18. Springer, Heidelberg (2009)
10. Blondeau, C., Naya-Plasencia, M., Videau, M., Zenner, E.: Cryptanalysis of ARMADILLO2, `http://eprint.iacr.org/2011/160`
11. Bogdanov, A., Knudsen, L.R., Leander, G., Paar, C., Poschmann, A., Robshaw, M.J.B., Seurin, Y., Vikkelsoe, C.: PRESENT: An Ultra-Lightweight Block Cipher. In: Paillier, P., Verbauwhede, I. (eds.) CHES 2007. LNCS, vol. 4727, pp. 450–466. Springer, Heidelberg (2007)
12. Boura, C., Canteaut, A., De Cannière, C.: Higher-Order Differential Properties of Keccak and *Luffa*. In: Joux, A. (ed.) FSE 2011. LNCS, vol. 6733, pp. 252–269. Springer, Heidelberg (2011)
13. De Cannière, C., Dunkelman, O., Knežević, M.: KATAN and KTANTAN — A Family of Small and Efficient Hardware-Oriented Block Ciphers. In: Clavier, C., Gaj, K. (eds.) CHES 2009. LNCS, vol. 5747, pp. 272–288. Springer, Heidelberg (2009)

14. De Cannière, C., Preneel, B.: Trivium. In: Robshaw and Billet [43], pp. 244–266
15. Choy, J., Zhang, A., Khoo, K., Henricksen, M., Poschmann, A.: AES variants secure against related-key differential and boomerang attacks. In: Ardagna, C.A., Zhou, J. (eds.) WISTP 2011. LNCS, vol. 6633, pp. 191–207. Springer, Heidelberg (2011), http://eprint.iacr.org/2011/072
16. Daemen, J., Rijmen, V.: AES Proposal: Rijndael. NIST AES proposal (1998)
17. Daemen, J., Knudsen, L.R., Rijmen, V.: The Block Cipher SQUARE. In: Biham, E. (ed.) FSE 1997. LNCS, vol. 1267, pp. 149–165. Springer, Heidelberg (1997)
18. Dunkelman, O., Keller, N., Shamir, A.: A Practical-Time Related-Key Attack on the KASUMI Cryptosystem Used in GSM and 3G Telephony. In: Rabin, T. (ed.) CRYPTO 2010. LNCS, vol. 6223, pp. 393–410. Springer, Heidelberg (2010)
19. Engels, D., Fan, X., Gong, G., Hu, H., Smith, E.M.: Ultra-Lightweight Cryptography for Low-Cost RFID Tags: Hummingbird Algorithm and Protocol, http://www.cacr.math.uwaterloo.ca/techreports/2009/cacr2009-29.pdf
20. Engels, D., Saarinen, M.-J.O., Smith, E.M.: The Hummingbird-2 Lightweight Authenticated Encryption Algorithm, http://eprint.iacr.org/2011/126.pdf
21. Gilbert, H., Peyrin, T.: Super-Sbox Cryptanalysis: Improved Attacks for AES-Like Permutations. In: Hong and Iwata [27], pp. 365–383
22. Gong, Z., Nikova, S., Law, Y.-W.: A New Family of Lightweight Block Ciphers. In: Juels, A., Paar, C. (eds.) RFIDSec 2011. Springer, Heidelberg (to appear, 2011), http://www.rfid-cusp.org/rfidsec/files/RFIDSec2011DraftPapers.zip
23. Guo, J., Peyrin, T., Poschmann, A.: The PHOTON Family of Lightweight Hash Functions. In: Rogaway, P. (ed.) CRYPTO 2011. LNCS, vol. 6841, pp. 222–239. Springer, Heidelberg (2011)
24. Hämäläinen, P., Alho, T., Hännikäinen, M., Hämäläinen, T.D.: Design and Implementation of Low-Area and Low-Power AES Encryption Hardware Core. In: DSD, pp. 577–583 (2006)
25. Hell, M., Johansson, T., Maximov, A., Meier, W.: The Grain Family of Stream Ciphers. In: Robshaw and Billet [43], pp. 179–190
26. Hong, D., Sung, J., Hong, S., Lim, J., Lee, S., Koo, B.S., Lee, C., Chang, D., Lee, J., Jeong, K., Kim, H., Kim, J., Chee, S.: HIGHT: A New Block Cipher Suitable for Low-Resource Device. In: Goubin, L., Matsui, M. (eds.) CHES 2006. LNCS, vol. 4249, pp. 46–59. Springer, Heidelberg (2006)
27. Hong, S., Iwata, T. (eds.): FSE 2010. LNCS, vol. 6147. Springer, Heidelberg (2010)
28. Khovratovich, D., Nikolic, I.: Rotational Cryptanalysis of ARX. In: Hong and Iwata [27], pp. 333–346
29. Knudsen, L.R., Rijmen, V.: Known-Key Distinguishers for Some Block Ciphers. In: Kurosawa, K. (ed.) ASIACRYPT 2007. LNCS, vol. 4833, pp. 315–324. Springer, Heidelberg (2007)
30. Knudsen, L.R., Leander, G., Robshaw, M.J.B.: PRINTCIPHER: A Block Cipher for IC-Printing. In: Mangard, S., Standaert, F.-X. (eds.) CHES 2010. LNCS, vol. 6225, pp. 16–32. Springer, Heidelberg (2010)
31. Lamberger, M., Mendel, F., Rechberger, C., Rijmen, V., Schläffer, M.: Rebound Distinguishers: Results on the Full Whirlpool Compression Function. In: Matsui, M. (ed.) ASIACRYPT 2009. LNCS, vol. 5912, pp. 126–143. Springer, Heidelberg (2009)
32. Leander, G., Paar, C., Poschmann, A., Schramm, K.: New Lightweight DES Variants. In: Biryukov, A. (ed.) FSE 2007. LNCS, vol. 4593, pp. 196–210. Springer, Heidelberg (2007)

33. Lim, C., Korkishko, T.: mCrypton – A Lightweight Block Cipher for Security of Low-Cost RFID Tags and Sensors. In: Kwon, T., Song, J., Yung, M. (eds.) WISA 2005. LNCS, vol. 3786, pp. 243–258. Springer, Heidelberg (2006)
34. Mace, F., Standaert, F.-X., Quisquater, J.-J.: ASIC Implementations of the Block Cipher SEA for Constrained Applications. In: RFID Security - RFIDsec 2007, Workshop Record, Malaga, Spain, pp. 103–114 (2007)
35. May, L., Henricksen, M., Millan, W.L., Carter, G., Dawson, E.: Strengthening the Key Schedule of the AES. In: Batten, L., Seberry, J. (eds.) ACISP 2002. LNCS, vol. 2384, pp. 226–240. Springer, Heidelberg (2002)
36. Mendel, F., Peyrin, T., Rechberger, C., Schläffer, M.: Improved Cryptanalysis of the Reduced Grøstl Compression Function, ECHO Permutation and AES Block Cipher. In: Jacobson Jr., M.J., Rijmen, V., Safavi-Naini, R. (eds.) SAC 2009. LNCS, vol. 5867, pp. 16–35. Springer, Heidelberg (2009)
37. Mendel, F., Rechberger, C., Schläffer, M., Thomsen, S.S.: The Rebound Attack: Cryptanalysis of Reduced Whirlpool and Grøstl. In: Dunkelman, O. (ed.) FSE 2009. LNCS, vol. 5665, pp. 260–276. Springer, Heidelberg (2009)
38. Moradi, A., Poschmann, A., Ling, S., Paar, C., Wang, H.: Pushing the Limits: A Very Compact and a Threshold Implementation of AES. In: Paterson, K. (ed.) EUROCRYPT 2011. LNCS, vol. 6632, pp. 69–88. Springer, Heidelberg (2011)
39. Nikolić, I.: Tweaking AES. In: Biryukov, A., Gong, G., Stinson, D.R. (eds.) SAC 2010. LNCS, vol. 6544, pp. 198–210. Springer, Heidelberg (2011)
40. Park, S., Sung, S.H., Lee, S., Lim, J.: Improving the Upper Bound on the Maximum Differential and the Maximum Linear Hull Probability for SPN Structures and AES. In: Johansson, T. (ed.) FSE 2003. LNCS, vol. 2887, pp. 247–260. Springer, Heidelberg (2003)
41. Poschmann, A.: Lightweight Cryptography - Cryptographic Engineering for a Pervasive World. Number 8 in IT Security. Europäischer Universitätsverlag, Published: Ph.D. Thesis, Ruhr University Bochum (2009)
42. Robshaw, M.J.B.: Searching for Compact Algorithms: CGEN. In: Nguyen, P. (ed.) VIETCRYPT 2006. LNCS, vol. 4341, pp. 37–49. Springer, Heidelberg (2006)
43. Robshaw, M.J.B., Billet, O. (eds.): New Stream Cipher Designs. LNCS, vol. 4986. Springer, Heidelberg (2008)
44. Rolfes, C., Poschmann, A., Leander, G., Paar, C.: Ultra-Lightweight Implementations for Smart Devices – Security for 1000 Gate Equivalents. In: Grimaud, G., Standaert, F.-X. (eds.) CARDIS 2008. LNCS, vol. 5189, pp. 89–103. Springer, Heidelberg (2008)
45. Saarinen, M.-J.O.: Cryptanalysis of Hummingbird-1. In: Joux, A. (ed.) FSE 2011. LNCS, vol. 6733, pp. 328–341. Springer, Heidelberg (2011)

Piccolo: An Ultra-Lightweight Blockcipher

Kyoji Shibutani, Takanori Isobe, Harunaga Hiwatari, Atsushi Mitsuda, Toru Akishita, and Taizo Shirai

Sony Corporation
1-7-1 Konan, Minato-ku, Tokyo 108-0075, Japan
{Kyoji.Shibutani,Takanori.Isobe,Harunaga.Hiwatari,Atsushi.Mitsuda, Toru.Akishita,Taizo.Shirai}@jp.sony.com

Abstract. We propose a new 64-bit blockcipher *Piccolo* supporting 80 and 128-bit keys. Adopting several novel design and implementation techniques, *Piccolo* achieves both high security and notably compact implementation in hardware. We show that *Piccolo* offers a sufficient security level against known analyses including recent related-key differential attacks and meet-in-the-middle attacks. In our smallest implementation, the hardware requirements for the 80 and the 128-bit key mode are only 683 and 758 gate equivalents, respectively. Moreover, *Piccolo* requires only 60 additional gate equivalents to support the decryption function due to its involution structure. Furthermore, its efficiency on the energy consumption which is evaluated by energy per bit is also remarkable. Thus, *Piccolo* is one of the competitive ultra-lightweight blockciphers which are suitable for extremely constrained environments such as RFID tags and sensor nodes.

Keywords: blockcipher, generalized Feistel networks, related-key differential attacks, meet-in-the-middle attacks, ultra-lightweight.

1 Introduction

Background and Motivation. Blockciphers are essential primitives for cryptographic applications such as data integrity, confidentiality, and protection of privacy. At the same time, with the large deployment of low resource devices such as RFID tags and sensor nodes and increasing need to provide security among such devices, lightweight cryptography has become a hot topic. Hence, recently, research on designing and analyzing lightweight blockciphers has received a lot of attention. In fact, there have been several blockciphers designed for a lightweight hardware implementation such as mCrypton [28], HIGHT [20], DESL/DESXL [27], PRESENT [11], KATAN/KTANTAN [13] and PRINTcipher [25]. The structures of these ciphers are generally categorized into two structures: Substitution Permutation Networks (SPNs) and Feistel-type structures[1].

SPNs are known as the basic structure of the current U.S. encryption standard AES [16]. Also, several lightweight blockciphers based on an SPN have been

[1] KATAN/KTANTAN is exceptional, which is based on a stream cipher.

B. Preneel and T. Takagi (Eds.): CHES 2011, LNCS 6917, pp. 342–357, 2011.

published. PRESENT consisting of an SPN is supposed to be competitive ciphers among them, since its required gate is comparable with compact stream ciphers such as Grain and Trivium[2] [19,15]. Recently, PRINTcipher was designed for IC-printing, which is also an instantiation of an SPN. It achieves remarkably compact implementation, though it has uncommon block size, i.e., 48 or 96 bits. mCrypton, which is a miniature of Crypton [29], also adopts an SPN.

On the other hand, Feistel-type structures including Feistel networks and generalized Feistel networks (GFNs) are the other most widely used structure and known as the basic structure of the former U.S. encryption standard DES [17]. Though a lot of lightweight blockciphers instantiated by the Feistel-type structure have also been published, most of them have security problems in contrast to the SPN based designs. HIGHT was designed for low resource devices, which is a variant of GFN. While it is relatively light, it has been theoretically broken by a related-key differential attack [26]. GOST is known as the former Soviet encryption standard, and has Feistel network [32]. Since the compact implementation result on GOST requiring 651 GE has been published [35], it is considered as one of the ultra-lightweight blockciphers. However it has also been theoretically broken by an improved three-subset meet-in-the-middle (MITM) attack [21].

These attacks basically rely on the slow diffusion of the Feistel-type structures and high controllability of round keys caused by a simple key schedule. Thus, to avoid those attacks, the Feistel-type structures generally require a larger number of rounds than an SPN based construction. Since this reduces the efficiency on the energy consumption, the Feistel-type structure does not seem to be suitable for lightweight blockciphers. However, it has a lot of distinct features from those of SPNs. For instance, the Feistel-type structure has a smaller round function than SPNs, since only half of the data are updated per one round. Moreover the Feistel-type structure can support a decryption function without much implementation cost. As discussed in [11], by using the counter-mode, any encryption-only ciphers can support decryption function. Yet, if the cipher itself supports decryption function, it can be used for more applications, e.g., an application requiring CBC-mode. Also, a diversity of designs is considered to be important. Thus, it is meaningful to think about design possibilities of a Feistel-type structure based lightweight blockcipher that is not only efficient but also secure against known attacks including the above explained powerful attacks.

Efficiency Metrics. While hardware efficiency can be measured in many different ways, both the *energy* consumption and the *power* consumption are important measure for lightweight applications. The energy consumption is considered as a metric for active devices which have an own power supply, and the power consumption for passive devices which do not have an own power supply. Though the power consumption heavily depends on the used technology and the EDA tool, it is well known that it is proportional to the area requirement at low frequencies, e.g., 100 kHz [25]. Thus, we adopt the area requirement, i.e., *gate equivalents* (GE) as the measure to evaluate the efficiency with respect to

[2] Note that the expected security of them against distinguish attacks is substantially higher than that of 64-bit lightweight blockciphers.

Table 1. Comparative results in hardware implementations

Algorithm	block size [bit]	key size [bit]	type	serialized arch. area [GE]	serialized arch. cycles/ block	round-based arch. area [GE]	round-based arch. cycles/ block	round-based arch. energy/*1 bit	round-based arch. FOM*2
DESXL [27]	64	184	Feistel	2,168	144	-	-	-	-
†HIGHT [20]⋆	64	128	GFN	-	-	3,048	34	1,620	202
mCrypton-96 [28]	64	96	SPN	-	-	2,681	13	545	684
mCrypton-128 [28]	64	128	SPN	-	-	2,949	13	600	566
PRESENT-80 [36,11]	64	80	SPN	1,000	547	1,570	32	785	811
KATAN64 [13]	64	80	stream	1,054	254	-	-	-	-
‡KTANTAN64 [13]	64	80	stream	688	254	-	-	-	-
‡GOST-PS [35]	64	256	Feistel	651	264	1,017	32	509	1,933
‡GOST-FB [35]	64	256	Feistel	800	264	1,000	32	500	2,000
Piccolo-80	64	80	GFN	**683**	432	**1,136**	27	**480**	**1,836**
Piccolo-128	64	128	GFN	**758**	528	**1,197**	33	**618**	**1,353**
Piccolo-80⋆	64	80	GFN	**743**	432	**1,274**	27	**538**	**1,460**
Piccolo-128⋆	64	128	GFN	**818**	528	**1,362**	33	**703**	**1,045**
AES-128 [31],[38]⋆	128	128	SPN	2,400	226	12,454*3	11	1,071	75
CLEFIA-128 [1],[40]⋆	128	128	GFN	2,488	328	5,979	18	841	202
PRINTcipher-48 [25]	48	80	SPN	402	768	503	48	503	3,952
PRINTcipher-96 [25]	96	160	SPN	726	3,072	967	96	967	1,069

†: Theoretically broken under related-key setting [26].
‡: Theoretically broken under single-key setting [12,21].
⋆: Including decryption function. The others support encryption-mode only.
∗1: energy / bit = (area [GE] × required cycles for one block process [cycle]) / block size [bit].
∗2: FOM = (nanobit per cycles) / area squared [GE^2].
∗3: This implementation is not intended to be high efficiency but high throughput.

the power consumption in this work. The energy consumption is the power consumption over a certain time period, and for one block process, it is evaluated by multiplying the area requirements with the required cycles for one block. Then, by dividing the power estimation for one block process by the block size, we obtain *energy per bit* as the fair measure for the energy consumption. *FOM* (in nano bits per clock cycle per GE squared) proposed by [4] is known as another metric for energy consumption. In this work, we mainly adopt the above mentioned measures *area requirement*, *energy per bit* and *FOM* for the efficiency comparison.

Contributions and Outline. In this paper, we propose a new lightweight blockcipher *Piccolo* which is optimized for extremely constrained devices. *Piccolo* supports 64-bit block with 80 or 128-bit keys, and has an iterative structure which is a variant of a generalized Feistel network. We demonstrate that *Piccolo* offers a sufficient security level against known analyses including recent related-key differential and MITM attacks. Moreover, we present that *Piccolo* achieves remarkably compact implementation in hardware. In our smallest implementation, the area requirements for the 80 and the 128-bit key mode are only 683 and 758 GE with 432 and 528 cycles per block, respectively. The efficiency on the energy consumption evaluated by energy per bit is 480 for the 80-bit key mode, which is the smallest class among current lightweight blockciphers in literature. Furthermore, *Piccolo* requires only 60 additional GE to support decryption function. Therefore, *Piccolo* supporting both encryption and decryption functions is still comparable to other encryption-only lightweight blockciphers. These

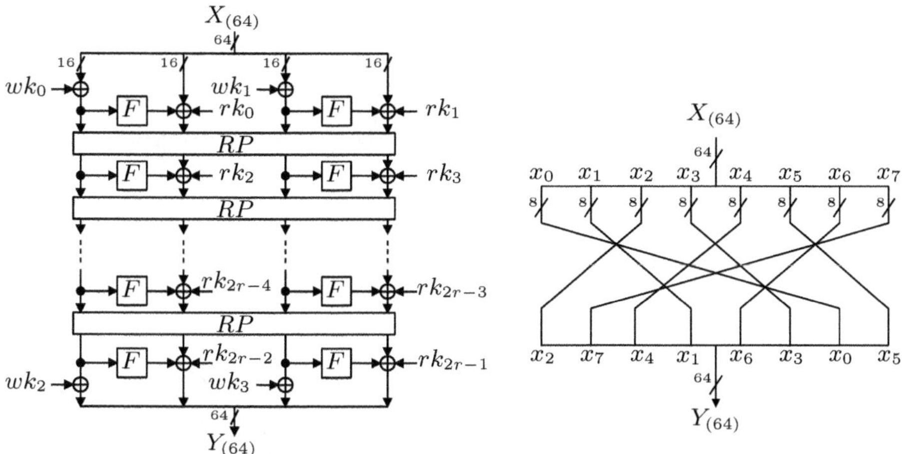

Fig. 1. Encryption function G_r

Fig. 2. Round permutation RP

comparative results regarding the hardware efficiency for lightweight blockciphers whose key size is more than 80 bits are summarized in Table 1. Note that, in our implementations, a key input is assumed to hold its value during the block process. Thus, *Piccolo* achieves both high security and extremely compact implementation unlike the other Feistel-type structure based lightweight blockciphers.

This paper is organized as follows. The specification of *Piccolo* is given in Section 2. Section 3 describes the design rationale. Sections 4 and 5 provide results on security and hardware implementation, respectively. Finally, we conclude in Section 6.

2 Specification

This section provides the specification of *Piccolo*. *Piccolo* is a 64-bit blockcipher supporting 80 and 128-bit keys. The 80 and the 128-bit key mode are referred as *Piccolo*-80 and *Piccolo*-128, respectively. Both ciphers consist of a data processing part and a key scheduling part. The differences between two key modes lie in the number of rounds for the data processing part and the key scheduling part. We first give notations used throughout this paper, then define each part.

2.1 Notations

$a_{(b)}$: b denotes the bit length of a.
$a|b$ or $(a|b)$: Concatenation.
$a \leftarrow b$: Updating a value of a by a value of b.
${}^t\boldsymbol{a}$: Transposition of a vector or a matrix $\boldsymbol{a}$.
$\{a\}_b$: Representation in base b.

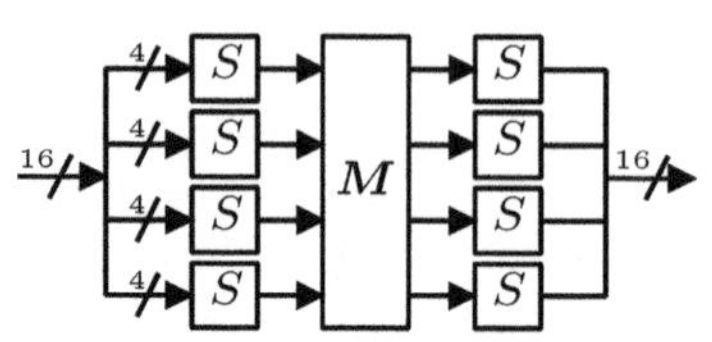

Fig. 3. F-function

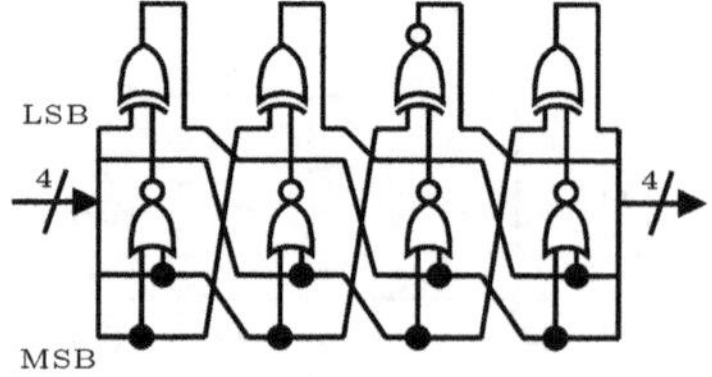

Fig. 4. S-box

2.2 Data Processing Part

The data processing part of *Piccolo* consisting of r rounds, G_r, takes a 64-bit data $X \in \{0,1\}^{64}$, four 16-bit whitening keys $wk_i \in \{0,1\}^{16} (0 \leq i < 4)$ and $2r$ 16-bit round keys $rk_i \in \{0,1\}^{16} (0 \leq i < 2r)$ as the inputs, and outputs a 64-bit data $Y \in \{0,1\}^{64}$. G_r is defined as follows:

$$G_r : \begin{cases} \{0,1\}^{64} \times \{\{0,1\}^{16}\}^4 \times \{\{0,1\}^{16}\}^{2r} \rightarrow \{0,1\}^{64} \\ (X_{(64)}, wk_{0(16)}, ..., wk_{3(16)}, rk_{0(16)}, ..., rk_{2r-1(16)}) \mapsto Y_{(64)} \end{cases}$$

Algorithm $G_r(X_{(64)}, wk_0, ..., wk_3, rk_0, ..., rk_{2r-1})$ **:**
$X_{0(16)}|X_{1(16)}|X_{2(16)}|X_{3(16)} \leftarrow X_{(64)}$
$X_0 \leftarrow X_0 \oplus wk_0,\ X_2 \leftarrow X_2 \oplus wk_1$
for $i \leftarrow 0$ to $r-2$ do
  $X_1 \leftarrow X_1 \oplus F(X_0) \oplus rk_{2i},\ X_3 \leftarrow X_3 \oplus F(X_2) \oplus rk_{2i+1}$
  $X_0|X_1|X_2|X_3 \leftarrow RP(X_0|X_1|X_2|X_3)$
$X_1 \leftarrow X_1 \oplus F(X_0) \oplus rk_{2r-2},\ X_3 \leftarrow X_3 \oplus F(X_2) \oplus rk_{2r-1}$
$X_0 \leftarrow X_0 \oplus wk_2,\ X_2 \leftarrow X_2 \oplus wk_3$
$Y_{(64)} \leftarrow X_0|X_1|X_2|X_3$

where F is a 16-bit F-function and RP is a 64-bit permutation defined in the following sections. The decryption function G_r^{-1} is obtained from G_r by simply changing the order of whitening and round keys as follows:

$$G_r^{-1} : \begin{cases} \{0,1\}^{64} \times \{\{0,1\}^{16}\}^4 \times \{\{0,1\}^{16}\}^{2r} \rightarrow \{0,1\}^{64} \\ (Y_{(64)}, wk_{0(16)}, ..., wk_{3(16)}, rk_{0(16)}, ..., rk_{2r-1(16)}) \mapsto X_{(64)} \end{cases}$$

Algorithm $G_r^{-1}(Y_{(64)}, wk_0, ..., wk_3, rk_0, ..., rk_{2r-1})$ **:**
$wk'_0 \leftarrow wk_2, wk'_1 \leftarrow wk_3, wk'_2 \leftarrow wk_0, wk'_3 \leftarrow wk_1$
for $i \leftarrow 0$ to $r-1$ do
  $rk'_{2i}|rk'_{2i+1} \leftarrow \begin{cases} rk_{2r-2i-2}|rk_{2r-2i-1} \text{ (if } i \bmod 2 = 0) \\ rk_{2r-2i-1}|rk_{2r-2i-2} \text{ (if } i \bmod 2 = 1) \end{cases}$
$X_{(64)} \leftarrow G_r(Y, wk'_0, ..., wk'_3, rk'_0, ..., rk'_{2r-1})$

The number of rounds, r, is 25 and 31 for *Piccolo*-80 and -128, i.e., G_{25} and G_{31} for *Piccolo*-80 and -128, respectively (See Fig. 1).

Table 2. 4-bit bijective S-box S in hexadecimal form

x	0	1	2	3	4	5	6	7	8	9	a	b	c	d	e	f
$S[x]$	e	4	b	2	3	8	0	9	1	a	7	f	6	c	5	d

F-Function. F-function $F : \{0,1\}^{16} \rightarrow \{0,1\}^{16}$ consists of two S-box layers separated by a diffusion matrix (See Fig. 3). The S-box layer consists of four 4-bit bijective S-boxes S given by Table 2, and updates a 16-bit data $X_{(16)}$ as follows:

$$(x_{0(4)}, x_{1(4)}, x_{2(4)}, x_{3(4)}) \leftarrow (S(x_{0(4)}), S(x_{1(4)}), S(x_{2(4)}), S(x_{3(4)})),$$

where $X_{(16)} = x_{0(4)}|x_{1(4)}|x_{2(4)}|x_{3(4)}$. The diffusion matrix $\boldsymbol{M}$ is defined as

$$\boldsymbol{M} = \begin{pmatrix} 2 & 3 & 1 & 1 \\ 1 & 2 & 3 & 1 \\ 1 & 1 & 2 & 3 \\ 3 & 1 & 1 & 2 \end{pmatrix}.$$

Then the diffusion function updates a 16-bit data $X_{(16)}$ as follows:

$${}^t(x_{0(4)}, x_{1(4)}, x_{2(4)}, x_{3(4)}) \leftarrow \boldsymbol{M} \cdot {}^t(x_{0(4)}, x_{1(4)}, x_{2(4)}, x_{3(4)}),$$

where the multiplications between matrices and vectors are performed over $\mathrm{GF}(2^4)$ defined by an irreducible polynomial $x^4 + x + 1$.

Round Permutation. The round permutation $RP : \{0,1\}^{64} \rightarrow \{0,1\}^{64}$ divides a 64-bit input $X_{(64)}$ into eight 8-bit data as $X_{(64)} = x_{0(8)}|x_{1(8)}|...|x_{7(8)}$, then permutes them by the following manner:

$$RP : (x_{0(8)}, x_{1(8)}, ..., x_{7(8)}) \leftarrow (x_{2(8)}, x_{7(8)}, x_{4(8)}, x_{1(8)}, x_{6(8)}, x_{3(8)}, x_{0(8)}, x_{5(8)}).$$

Finally, the round permutation concatenates $(x_{0(8)}, x_{1(8)}, ..., x_{7(8)})$ into $X_{(64)}$ (See Fig. 2).

2.3 Key Scheduling Part

The key scheduling part of *Piccolo* supports 80 and 128-bit keys, and outputs 16-bit whitening keys $wk_{i(16)} (0 \leq i < 4)$ and round keys $rk_{j(16)} (0 \leq j < 2r)$ for the data processing part. The key scheduling functions for *Piccolo*-80 and -128 are referred as KS_r^{80} and KS_r^{128}, respectively. We first define 16-bit constants con_i^{80} and con_i^{128}, then describe each key schedule.

Constant Values. The constants con_i^{80} and con_i^{128} used in KS_r^{80} and KS_r^{128}, respectively, are generated as follows:

$$\begin{cases} (con_{2i}^{80}|con_{2i+1}^{80}) \leftarrow (c_{i+1}|c_0|c_{i+1}|\{00\}_2|c_{i+1}|c_0|c_{i+1}) \oplus \{\texttt{0f1e2d3c}\}_{16}, \\ (con_{2i}^{128}|con_{2i+1}^{128}) \leftarrow (c_{i+1}|c_0|c_{i+1}|\{00\}_2|c_{i+1}|c_0|c_{i+1}) \oplus \{\texttt{6547a98b}\}_{16}, \end{cases}$$

where c_i is a 5-bit representation of i, e.g., $c_{11} = \{\texttt{01011}\}_2$.

Key Schedule for 80-Bit Key Mode (KS_r^{80}). The key scheduling function for the 80-bit key mode, KS_r^{80}, divides an 80-bit key $K_{(80)}$ into five 16-bit sub-keys $k_{i(16)}$ $(0 \le i < 5)$ and provides $wk_{i(16)} (0 \le i < 4)$ and $rk_{j(16)} (0 \le j < 2r)$ as follows:

Algorithm $KS_r^{80}(K_{(80)})$:
$wk_0 \leftarrow k_0^L | k_1^R, wk_1 \leftarrow k_1^L | k_0^R, wk_2 \leftarrow k_4^L | k_3^R, wk_3 \leftarrow k_3^L | k_4^R$
for $i \leftarrow 0$ to $(r-1)$ do

$$(rk_{2i}, rk_{2i+1}) \leftarrow (con_{2i}^{80}, con_{2i+1}^{80}) \oplus \begin{cases} (k_2, k_3) \text{ (if } i \bmod 5 = 0 \text{ or } 2) \\ (k_0, k_1) \text{ (if } i \bmod 5 = 1 \text{ or } 4) \\ (k_4, k_4) \text{ (if } i \bmod 5 = 3), \end{cases}$$

where k_i^L and k_i^R are left and right half 8 bits of k_i, respectively, i.e., $k_{i(16)} = k_{i(8)}^L | k_{i(8)}^R$ and $k_{i(8)}^R$ contains the least significant bit of $k_{i(16)}$.

Key Schedule for 128-Bit Key Mode (KS_r^{128}). The key scheduling function for the 128-bit key mode, KS_r^{128}, divides a 128-bit key $K_{(128)}$ into eight 16-bit sub-keys $k_{i(16)}$ $(0 \le i < 8)$ and provides $wk_{i(16)} (0 \le i < 4)$ and $rk_{j(16)} (0 \le j < 2r)$ as follows:

Algorithm $KS_r^{128}(K_{(128)})$:
$wk_0 \leftarrow k_0^L | k_1^R,\ wk_1 \leftarrow k_1^L | k_0^R,\ wk_2 \leftarrow k_4^L | k_7^R,\ wk_3 \leftarrow k_7^L | k_4^R$
for $i \leftarrow 0$ to $(2r-1)$ do
  if $(i+2) \bmod 8 = 0$ then
    $(k_0, k_1, k_2, k_3, k_4, k_5, k_6, k_7) \leftarrow (k_2, k_1, k_6, k_7, k_0, k_3, k_4, k_5)$
  $rk_i \leftarrow k_{(i+2) \bmod 8} \oplus con_i^{128}$

3 Design Rationale

In this section, we briefly describe design rationale of *Piccolo*.

Structure. *Piccolo* supports 64-bit block to fit standard applications, and 80 and 128-bit keys to achieve moderate security levels. The underlying structure is a variant of GFN that can easily support decryption function without much implementation cost and has light round functions.

Key Schedule. We adopt a permutation based key schedule which can significantly reduce the required number of gates. For instance, the registers for storing keys are not required and it leads the almost same gate requirement for each key size, in contrast to a key schedule requiring key state. While the drawback is security concern, by carefully choosing the permutation, it has enough immunity against attacks exploiting weakness of the key schedule such as related-key differential and MITM attacks. Note that, in our evaluation, key inputs are not required to be hard-wired, but are assumed to hold its values during the block operation.

Round Permutation. In order to improve diffusion property, *Piccolo* utilizes an 8-bit word based permutation between rounds instead of a 16-bit word based cyclic shift used in the standard GFN. Moreover, it demolishes the 16-bit word structure and thus improves the security against cryptanalysis exploiting strong word-based structure such as saturation attacks. We choose the specific one among several possibilities not to destroy the involution property in which the encryption process is identical to the decryption process when whitening and round keys are not introduced.

F-Function. The F-function consists of two S-box layers separated by a diffusion matrix without key additions before the second S-box layer. The S-box in the F-function has a 4-round iterative structure like GFN, and is extremely light. As shown in Fig. 4, each S-box consists of only four NOR gates, three XOR gates and one XNOR gate. Both the maximum differential probability (MDP) and the maximum linear probability (MLP) of the S-box are 2^{-2} which are optimal, and it has no fixed point. Moreover, it is suitable for efficient threshold implementation as discussed in Section 5. Furthermore, by using a standard PC, we obtain $2^{-9.3}$ and $2^{-8.0}$ as MDP and MLP of the F-function, respectively. While those figures are not optimal for a 16-bit bijective function, it is sufficient for our design, since *Piccolo* has enough differentially and linearly active F-functions over a certain number of rounds.

4 Security Analysis

In this section, we provide results on security analysis for *Piccolo*.

Differential Attack / Linear Attack [7,30]. We first show the minimum numbers of differentially and linearly active F-functions of G_r up to 30 rounds in Table 3. The figures in the table are obtained by an exhaustive search based on the algorithm given by [39]. Note that the minimum numbers for differentially and linearly active F-functions are the same due to the duality of differential and linear attacks and the similarity of G_r and G_r^{-1}. As explained in Section 3, MDP and MLP of the F-function are $2^{-9.3}$ and $2^{-8.0}$, respectively. Combining those results, *Piccolo* consisting of at least 7 or 8 rounds provide at least 7 or 8 active F-functions, and have no differential or linear trails whose probabilities are more than 2^{-64}, respectively. Thus, we expect that the full-round of *Piccolo* (25 and 31 rounds for *Piccolo*-80 and -128) has enough immunity against differential and linear attacks, since it has large security margin.

Boomerang-Type Attacks [42,23,6]. The boomerang-type attacks (including the boomerang, amplified boomerang and rectangle attacks) first divide the cipher into two sub-ciphers, then find a boomerang quartet with high probability. The probability of constructing a boomerang quartet is denoted as $\hat{p}^2\hat{q}^2$, where $\hat{p} = \sqrt{\sum_\beta \Pr^2[\alpha \rightarrow \beta]}$, and α and β are input and output differences for the first sub-cipher, and $\hat{q}$ for the second sub-cipher. $\hat{p}^2$ is bounded by the maximum differential trail probability, i.e., $\hat{p}^2 \leq \max_\beta \Pr[\alpha \rightarrow \beta]$, and $\hat{q}^2$ as well. Let p, q

Table 3. Min. # differentially and linearly active F-functions (single-key setting)

rounds	1	2	3	4	5	6	7	8	9	10	11	12	13	14	15
min. # active F-functions	0	1	2	3	4	6	7	8	9	10	11	12	13	14	15
rounds	16	17	18	19	20	21	22	23	24	25	26	27	28	29	30
min. # active F-functions	16	17	18	19	20	21	22	23	24	25	26	27	28	29	30

be the maximum differential trail probability for the first and the second sub-ciphers. Then, p, q are bounded by multiplying the minimum number of active F-functions in each sub-cipher with MDP of the F-function. From Table 3, any combination of two sub-ciphers for *Piccolo* consisting of at least 9 rounds has at least 7 active F-functions in total. Hence, we conclude that the full-round of *Piccolo* is sufficiently secure against boomerang-type attacks.

Impossible Differential Attack [5]. An impossible differential attack is likely to be applied to a variant of GFN due to its slow diffusion. However, *Piccolo* utilizes the round permutation RP to achieve faster diffusion compared to a standard type-II GFN. Then, for both encryption and decryption sides, *Piccolo* requires only four rounds to be full diffusion, which is a property that all outputs are affected by all inputs. This implies that there exists at most 9-round impossible differential using a 16-bit truncated differential from the observation in [41]. We also search the longest impossible differential by modified $\mathcal{U}$-method [24] algorithm and found a 7-round impossible differential exploiting a 4-bit truncated differential. Therefore, we conclude that the full-round of *Piccolo* is expected to be secure against the impossible differential attack.

Related-Key Differential Attacks [9,8]. In the related-key setting, a distinguisher is allowed to use related-keys and usually uses key differentials to cancel out differentials in a data processing part. While the practical impact of related-key differential attacks is still controversial, we care about it from a pessimistic (designers') point of view. To evaluate the resistance to it, we follow an approach presented in [10]. In other words, we evaluate the immunity against related-key differential attacks by counting the minimum number of differentially active F-functions in the related-key setting. Table 4 shows the minimum numbers of differentially active F-functions for the 80 and the 128-bit key modes up to 20 rounds. Unlike the attacks under the single-key setting, the total number of active F-functions for the related-key differential attacks may vary according to the starting round. However, in our evaluations, those differences are at most 2 active F-functions, even if the starting round is changed. Consequently, we obtain that over 14 and 16 rounds for *Piccolo*-80 and -128 have at least 7 differentially active F-functions in the related-key setting, respectively.

Moreover, we consider related-key boomerang/rectangle attacks [8]. Similarly to non related-key boomerang-type attacks, we evaluate the security in the worst case that an attacker can use pq instead of $\hat{p}^2\hat{q}^2$ for the probability of a boomerang quartet. As a result, we confirmed that over 17 and 21 rounds of *Piccolo*-80 and -128 provide enough (seven) differentially active F-functions in this setting.

Table 4. Min. # differentially active F-functions (related-key setting)

starting round i \ rounds	1	2	3	4	5	6	7	8	9	10	11	12	13	14	15	16	17	18	19	20
for *Piccolo*-80 encryption																				
$i \bmod 5 = 0$	0	0	0	0	0	2	3	4	4	5	5	6	7	7	7	8	9	10	11	11
$i \bmod 5 = 1$	0	0	0	0	1	2	3	4	4	5	6	6	7	7	8	9	9	10	11	11
$i \bmod 5 = 2$	0	0	0	0	1	2	3	3	4	6	6	6	7	7	9	9	9	10	10	12
$i \bmod 5 = 3$	0	0	0	0	1	2	2	3	4	5	5	6	6	7	8	8	9	9	10	11
$i \bmod 5 = 4$	0	0	0	0	0	0	2	3	4	5	6	6	7	7	7	7	9	10	11	11
for *Piccolo*-80 decryption																				
$i \bmod 5 = 0$	0	0	0	0	1	2	2	3	4	5	5	6	6	7	8	8	9	9	10	11
$i \bmod 5 = 1$	0	0	0	0	1	2	3	3	4	6	6	6	7	7	9	9	9	10	10	12
$i \bmod 5 = 2$	0	0	0	0	1	2	3	4	4	5	6	6	7	7	8	9	9	10	11	11
$i \bmod 5 = 3$	0	0	0	0	0	2	3	4	4	5	5	6	7	7	7	8	9	10	11	11
$i \bmod 5 = 4$	0	0	0	0	0	0	2	3	4	5	6	6	7	7	7	7	9	10	11	11
for *Piccolo*-128 encryption																				
$i \bmod 4 = 0$	0	0	0	0	0	0	0	1	3	3	4	5	5	6	7	7	8	9	10	10
$i \bmod 4 = 1$	0	0	0	0	0	0	1	2	3	3	4	5	5	6	7	7	8	9	10	11
$i \bmod 4 = 2$	0	0	0	0	0	0	1	2	2	3	4	4	5	6	6	7	7	9	9	9
$i \bmod 4 = 3$	0	0	0	0	0	1	1	1	2	3	4	5	5	6	7	7	8	9	9	10
for *Piccolo*-128 decryption																				
$i \bmod 4 = 0$	0	0	0	0	0	1	1	2	3	3	4	5	5	6	6	7	8	9	9	11
$i \bmod 4 = 1$	0	0	0	0	0	0	1	2	3	3	4	4	5	6	7	7	8	9	9	9
$i \bmod 4 = 2$	0	0	0	0	0	0	0	1	2	3	4	5	5	6	7	7	7	9	10	10
$i \bmod 4 = 3$	0	0	0	0	0	0	1	1	2	3	4	5	5	6	7	7	8	9	10	10

Furthermore, we take related-key impossible differential attacks [22] into account. Consequently, by using modified $\mathcal{U}$-method, we found an 11 and a 17-round impossible differential distinguisher using an 8-bit truncated differential for *Piccolo*-80 and -128 in the related-key setting, respectively, and they are the longest in our evaluation. Therefore, we conclude that the full-round *Piccolo* is expected to be resistant to those attacks.

Meet-in-the-Middle Attack [12]. Three-subset meet-in-the-middle (MITM) cryptanalysis [12] is a recent attack on blockciphers. This attack works well for blockciphers having a simple key schedule and slow diffusion. Indeed, KTANTAN and GOST have been theoretically broken by this attack [12,21]. Since *Piccolo* consists of the permutation based key scheduling and a variant of GFN, evaluating the resistance against this attack is important.

Similarly to data difference, *Piccolo* requires 4 rounds to non-linearly diffuse any round-key difference to all output data in the data processing part, i.e., any round-key bits of the i-th round non-linearly affect all input of the $(i-3)$-th round and all output of the $(i+3)$-th round. Thus, we assume that an attacker might construct an 8-round *indirect-partial matching* [3] and a 4-round *initial structure* [37] in the worst case. Besides, we even allow the attacker to use *code book* and *splice and cut* techniques [2]. In this worst setting, *Piccolo*-80 and -128 without whitening keys have neutral words up to 19 and 23 consecutive rounds, respectively. We expect that the attacked rounds obtained by this observation are upper bounds on the security against the three-subset MITM attack, since the given assumptions are sufficiently strong. Moreover, we attempt to construct actual attacks to obtain the lower bounds on the security. As a result, the *Piccolo*-80 and -128 without whitening keys reduced to 14 and 21 rounds can be attacked by the three-subset MITM attacks, respectively. Since *Piccolo* actually has whitening keys, it is obviously stronger than the variants evaluated above. Thus, we conclude that *Piccolo* has enough immunity against the three-subset MITM attack.

Other Attacks. We also consider other attacks including a slide, a saturation, an interpolation, a higher order differential, a truncated differential, and an algebraic attack. Though the details of the evaluations for those attacks are omitted due to the page limitation, consequently, we expect that none of them work better than the previously explained attacks.

5 Implementation Aspects

This section provides results on compact hardware implementation of *Piccolo* with novel implementation techniques, showing two types of implementations: a round-based implementation and a serialized implementation. While one round function is processed within one clock cycle in a round-based implementation, only a fraction of one round is treated in a clock cycle in a serialized implementation to realize the low-power and low-area implementation.

5.1 Optimization in Key Scheduling Part

The key scheduling part of *Piccolo* can be implemented by using multiplexers without flip-flops which have high area requirement, in a way similar to the implementation of GOST and KTANTAN [35,13]. Actually, our round-based implementation of *Piccolo*-80 needs only 32-bit wide 3-to-1 MUX to select the appropriate round key. For a serialized implementation, we require a 4-bit wide 20-to-1 MUX to select the right chunk of the round key.

In our evaluation, key inputs are assumed to hold those values during the block process, but are not required to be hard-wired. Therefore, our results do not contain registers for storing keys. If such registers are needed, around 360 and 576 extra GE are required for Piccolo-80 and -128, respectively. Moreover, if we use hard-wired key, we can reduce around 85 and 114 GE from the round-based implementations, also about 67 and 104 GE from the serialized implementations for *Piccolo*-80 and -128, respectively.

5.2 Optimization in Data Processing Part

A round-based implementation of *Piccolo* can be done straightforwardly. Note that we use scan flip-flops for the data state, which take both an input and an output of a round function as inputs.

On the other hand, a serialized implementation has many variety. Our serialized implementation is based on 4-bit shift registers in the similar way as [18]. The 4-bit data path for *Piccolo*-80 is described in Fig. 5.

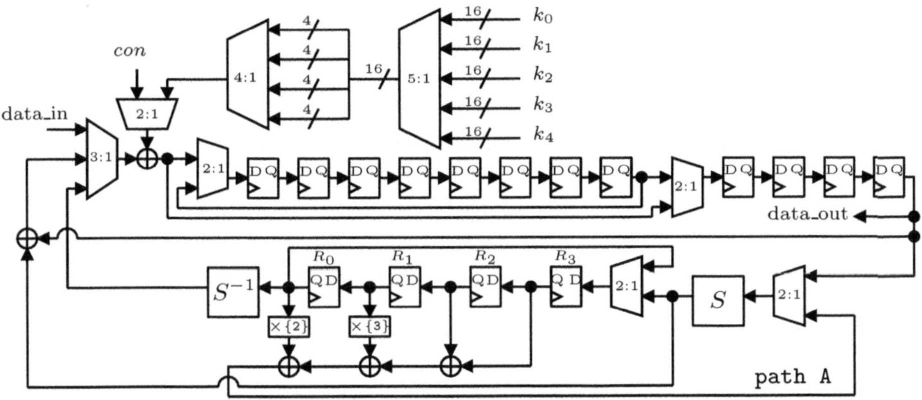

Fig. 5. Data path of our serialized implementation

In our serialized implementation, firstly outputs of the first S-box are set to the registers (R_0, R_1, R_2, R_3) described in Fig. 5. In the next four clock cycles, each row of the diffusion matrix is updated in order by rotating the registers (R_0, R_1, R_2, R_3). Simultaneously, the outputs of the matrix are input to S-box S through `path A`, then the outputs of the F-function are obtained. In the next four clock cycles, the inputs of the F-function are recovered in order through S^{-1} which is the inversion of S. At the same time, the outputs of the first S-box layer of the next F-function are set to the registers (R_0, R_1, R_2, R_3). Therefore, this implementation requires 8 clock cycles per F-function, and thus 16 clock cycles per round. We emphasize that our serialized implementation does not require additional registers for storing intermediate values of the F-functions by appending S^{-1} which costs only 12 GE.

5.3 Hardware Performance

Table 5 shows the detailed implementation figures of the round-based and the serialized implementations of *Piccolo*-80 and -128.

We designed hardware implementations of *Piccolo* in Verilog-HDL and synthesized the designs to a 0.13 μm standard cell library. We used *VCS version 2006.06* for simulation and *Design Compiler version 2007.03-SP3* for synthesis. One GE is equivalent to the area of a 2-way NAND.

In a recent trend, the implementation of lightweight blockciphers uses a scan flip-flop instead of a combination of a D flip-flop and a 2-to-1 MUX [13,35,36] to reduce the gate requirement. In our evaluation environment, a D flip-flop and a 2-to-1 MUX cost 4.5 and 2.0 GE, respectively, while a scan flip-flop costs 6.25 GE. Thus, we can save 0.25 GE per bit of storage by using this implementation technique. Moreover, the library we used has the 4-input AND-NOR and 4-input OR-NAND gates with two inputs inverted as described in Fig. 6. The outputs of these cells are corresponding to those of XOR or XNOR when the inputs X, Y are set as shown in Fig. 6. Thus, we can use these cells instead of XOR or

Table 5. Implementation figures for Piccolo

		Piccolo-80		*Piccolo*-128	
		serial	round	serial	round
cycles per block		432	27	528	33
throughput @ 100 kHz (kbps.)		14.81	237.04	12.12	193.94
Area [GE]	sum	683.00	1,135.25	757.75	1,196.50
	Key scheduling	95	72	135	120
	Data state	309	344	309	344
	S-box/S-box^{-1}	24	192	24	192
	Matrix	34	208	34	208
	Key XOR	8*	64	8*	64
	Constants XOR	-*	40	-*	40
	F-func. output XOR	8	64	8	64
	MUX	24	72	24	72
	Others/Control	181.00	79.25	215.75	92.50

∗: XOR for round keys and constants is shared

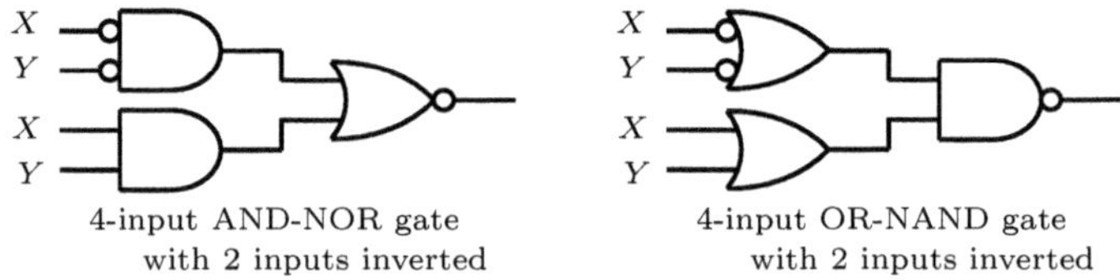

Fig. 6. 4-input AND-NOR and 4-input OR-NAND gates with 2 inputs inverted, which correspond to XOR and XNOR gate

XNOR cells. Since both cells cost 2 GE instead of 2.25 GE required for XOR or XNOR, we can save 0.25 GE per an XOR or XNOR gate. We employed the above mentioned implementation techniques in our evaluation.

5.4 Security against Side Channel Attacks

A provably secure countermeasure against first order side-channel attacks called threshold implementations [33,34] can be applied to *Piccolo*. In threshold implementations, at least three shares are necessary for any nonlinear function. The S-box of *Piccolo* defined in Section 2 is chosen to belong to the alternating group A_{16}, where a 4×4 bijection can be decomposed using quadratic bijections [14]. Therefore, for the S-box of *Piccolo*, the masking method can be applied using only three shares, which leads efficient threshold implementations of *Piccolo*.

6 Conclusion

In this paper, we have presented a lightweight blockcipher consisting of a variant of generalized Feistel network with a permutation based key schedule. Despite several desirable implementation properties for a combination of Feistel-type structure with a permutation based key schedule, the ciphers having such structures are likely to be vulnerable to attacks. The proposed cipher *Piccolo* employs several new design approaches including the half-word based round permutation and the effective permutation for key expanding to avoid known attacks without loosing efficiency on both power and energy consumptions. Consequently, *Piccolo* achieves not only notably compact implementation but also high security.

Acknowledgments. The authors would like to thank the anonymous reviewers for their helpful comments.

References

1. Akishita, T., Hiwatari, H.: Very compact hardware implementations of the blockcipher CLEFIA. Sony corporation (June 2011), `http://www.sony.co.jp/Products/cryptography/clefia/download/data/clefia-hw-compact-20110615.pdf`
2. Aoki, K., Sasaki, Y.: Preimage attacks on one-block MD4, 63-step MD5 and more. In: Avanzi, R.M., Keliher, L., Sica, F. (eds.) SAC 2008. LNCS, vol. 5381, pp. 103–119. Springer, Heidelberg (2009)
3. Aoki, K., Sasaki, Y.: Meet-in-the-middle preimage attacks against reduced SHA-0 and SHA-1. In: Halevi, S. (ed.) CRYPTO 2009. LNCS, vol. 5677, pp. 70–89. Springer, Heidelberg (2009)
4. Badel, S., Dagtekin, N., Nakahara, J., Ouafi, K., Reffé, N., Sepehrdad, P., Susil, P., Vaudenay, S.: ARMADILLO: A multi-purpose cryptographic primitive dedicated to hardware. In: Mangard, S., Standaert, F.-X. (eds.) CHES 2010. LNCS, vol. 6225, pp. 398–412. Springer, Heidelberg (2010)
5. Biham, E., Biryukov, A., Shamir, A.: Cryptanalysis of Skipjack Reduced to 31 Rounds Using Impossible Differentials. In: Stern, J. (ed.) EUROCRYPT 1999. LNCS, vol. 1592, pp. 12–23. Springer, Heidelberg (1999)
6. Biham, E., Dunkelman, O., Keller, N.: The rectangle attack - rectangling the Serpent. In: Pfitzmann, B. (ed.) EUROCRYPT 2001. LNCS, vol. 2045, pp. 340–357. Springer, Heidelberg (2001)
7. Biham, E., Shamir, A.: Differential Cryptanalysis of the Data Encryption Standard. Springer, Heidelberg (1993)
8. Biham, E., Dunkelman, O., Keller, N.: Related-key boomerang and rectangle attacks. In: Cramer, R. (ed.) EUROCRYPT 2005. LNCS, vol. 3494, pp. 507–525. Springer, Heidelberg (2005)
9. Biham, E., Dunkelman, O., Keller, N.: A unified approach to related-key attacks. In: Nyberg, K. (ed.) FSE 2008. LNCS, vol. 5086, pp. 73–96. Springer, Heidelberg (2008)
10. Biryukov, A., Nikolić, I.: Automatic Search for Related-Key Differential Characteristics in Byte-Oriented Block Ciphers: Application to AES, Camellia, Khazad and Others. In: Gilbert, H. (ed.) EUROCRYPT 2010. LNCS, vol. 6110, pp. 322–344. Springer, Heidelberg (2010)
11. Bogdanov, A., Knudsen, L., Leander, G., Paar, C., Poschmann, A., Robshaw, M.J.B., Seurin, Y., Vikkelsoe, C.: PRESENT: An ultra-lightweight block cipher. In: Paillier, P., Verbauwhede, I. (eds.) CHES 2007. LNCS, vol. 4727, pp. 450–466. Springer, Heidelberg (2007)
12. Bogdanov, A., Rechberger, C.: A 3-subset meet-in-the-middle attack: Cryptanalysis of the lightweight block cipher KTANTAN. In: Biryukov, A., Gong, G., Stinson, D.R. (eds.) SAC 2010. LNCS, vol. 6544, pp. 229–240. Springer, Heidelberg (2011)

13. De Cannière, C., Dunkelman, O., Knežević, M.: KATAN and KTANTAN — A family of small and efficient hardware-oriented block ciphers. In: Clavier, C., Gaj, K. (eds.) CHES 2009. LNCS, vol. 5747, pp. 272–288. Springer, Heidelberg (2009)
14. De Cannière, C., Nikov, V., Nikova, S., Rijmen, V.: S-box decompositions for SCA-resisting implementations. In: Poster Session of CHES 2010 (2010)
15. De Cannière, C., Preneel, B.: TRIVIUM. In: Robshaw, M.J.B., Billet, O. (eds.) New Stream Cipher Designs. LNCS, vol. 4986, pp. 244–266. Springer, Heidelberg (2008)
16. FIPS, Advanced Encryption Standard (AES). Federal Information Processing Standards Publication 197
17. FIPS, Data Encryption Standard. Federal Information Processing Standards Publication 46
18. Hämäläinen, P., Alho, T., Hännikäinen, M., Hämäläinen, T.D.: Design and implementation of low-area and low-power AES encryption hardware core. In: DSD, pp. 577–583. IEEE Computer Society, Los Alamitos (2006)
19. Hell, M., Johansson, T., Maximov, A., Meier, W.: The Grain Family of Stream Ciphers. In: Robshaw, M.J.B., Billet, O. (eds.) New Stream Cipher Designs. LNCS, vol. 4986, pp. 179–190. Springer, Heidelberg (2008)
20. Hong, D., Sung, J., Hong, S., Lim, J., Lee, S., Koo, B., Lee, C., Chang, D., Lee, J., Jeong, K., Kim, H., Kim, J., Chee, S.: HIGHT: A new block cipher suitable for low-resource device. In: Goubin, L., Matsui, M. (eds.) CHES 2006. LNCS, vol. 4249, pp. 46–59. Springer, Heidelberg (2006)
21. Isobe, T.: A single-key attack on the full GOST block cipher. In: Joux, A. (ed.) FSE 2011. LNCS, vol. 6733, pp. 290–305. Springer, Heidelberg (2011)
22. Jakimoski, G., Desmedt, Y.: Related-key differential cryptanalysis of 192-bit key AES variants. In: Matsui, M., Zuccherato, R.J. (eds.) SAC 2003. LNCS, vol. 3006, pp. 208–221. Springer, Heidelberg (2004)
23. Kelsey, J., Kohno, T., Schneier, B.: Amplified boomerang attacks against reduced-round MARS and Serpent. In: Schneier, B. (ed.) FSE 2000. LNCS, vol. 1978, pp. 75–93. Springer, Heidelberg (2001)
24. Kim, J., Hong, S., Sung, J., Lee, C., Lee, S.: Impossible differential cryptanalysis for block cipher structures. In: Johansson, T., Maitra, S. (eds.) INDOCRYPT 2003. LNCS, vol. 2904, pp. 82–96. Springer, Heidelberg (2003)
25. Knudsen, L., Leander, G., Poschmann, A., Robshaw, M.J.B.: PRINTcipher: A block cipher for IC-printing. In: Mangard, S., Standaert, F.-X. (eds.) CHES 2010. LNCS, vol. 6225, pp. 16–32. Springer, Heidelberg (2010)
26. Koo, B., Hong, D., Kwon, D.: Related-key attack on the full HIGHT. In: Pre-Proceedings of ICISC 2010. Springer, Heidelberg (2010)
27. Leander, G., Paar, C., Poschmann, A., Schramm, K.: New lightweight DES variants. In: Biryukov, A. (ed.) FSE 2007. LNCS, vol. 4593, pp. 196–210. Springer, Heidelberg (2007)
28. Lim, C.H., Korkishko, T.: mCrypton – A lightweight block cipher for security of low-cost RFID tags and sensors. In: Song, J.-S., Kwon, T., Yung, M. (eds.) WISA 2005. LNCS, vol. 3786, pp. 243–258. Springer, Heidelberg (2006)
29. Lim, C.H.: A Revised Version of CRYPTON - CRYPTON V1.0 -. In: Knudsen, L.R. (ed.) FSE 1999. LNCS, vol. 1636, pp. 31–45. Springer, Heidelberg (1999)
30. Matsui, M.: Linear cryptanalysis of Data Encryption Standard. In: Helleseth, T. (ed.) EUROCRYPT 1993. LNCS, vol. 765, pp. 386–397. Springer, Heidelberg (1994)
31. Moradi, A., Poschmann, A., Ling, S., Paar, C., Wang, H.: Pushing the limits: A very compact and a threshold implementation of AES. In: Paterson, K.G. (ed.) EUROCRYPT 2011. LNCS, vol. 6632, pp. 69–88. Springer, Heidelberg (2011)

32. National Soviet Bureau of Standards, Information Processing System - Cryptographic Protection - Cryptographic Algorithm GOST 28147-89
33. Nikova, S., Rechberger, C., Rijmen, V.: Threshold implementations against side-channel attacks and glitches. In: Ning, P., Qing, S., Li, N. (eds.) ICICS 2006. LNCS, vol. 4307, pp. 529–545. Springer, Heidelberg (2006)
34. Nikova, S., Rijmen, V., Schläffer, M.: Secure hardware implementation of non-linear functions in the presence of glitches. In: Lee, P.J., Cheon, J.H. (eds.) ICISC 2008. LNCS, vol. 5461, pp. 218–234. Springer, Heidelberg (2009)
35. Poschmann, A., Ling, S., Wang, H.: 256 bit standardized crypto for 650 GE – GOST revisited. In: Mangard, S., Standaert, F.-X. (eds.) CHES 2010. LNCS, vol. 6225, pp. 219–233. Springer, Heidelberg (2010)
36. Rolfes, C., Poschmann, A., Leander, G., Paar, C.: Ultra-lightweight implementations for smart devices – security for 1000 gate equivalents. In: Grimaud, G., Standaert, F.-X. (eds.) CARDIS 2008. LNCS, vol. 5189, pp. 89–103. Springer, Heidelberg (2008)
37. Sasaki, Y., Aoki, K.: Finding preimages in full MD5 faster than exhaustive search. In: Joux, A. (ed.) EUROCRYPT 2009. LNCS, vol. 5479, pp. 134–152. Springer, Heidelberg (2009)
38. Satoh, A., Morioka, S.: Hardware-Focused Performance Comparison for the Standard Block Ciphers AES, Camellia, and Triple-DES. In: Boyd, C., Mao, W. (eds.) ISC 2003. LNCS, vol. 2851, pp. 252–266. Springer, Heidelberg (2003)
39. Shirai, T., Araki, K.: On generalized Feistel structures using the diffusion switching mechanism. IEICE Trans. Fundamentals E91-A(8), 2120–2129 (2008)
40. Shirai, T., Shibutani, K., Akishita, T., Moriai, S., Iwata, T.: The 128-Bit Blockcipher CLEFIA (Extended Abstract). In: Biryukov, A. (ed.) FSE 2007. LNCS, vol. 4593, pp. 181–195. Springer, Heidelberg (2007)
41. Suzaki, T., Minematsu, K.: Improving the generalized Feistel. In: Hong, S., Iwata, T. (eds.) FSE 2010. LNCS, vol. 6147, pp. 19–39. Springer, Heidelberg (2010)
42. Wagner, D.: The boomerang attack. In: Knudsen, L.R. (ed.) FSE 1999. LNCS, vol. 1636, pp. 156–170. Springer, Heidelberg (1999)

A Test Vectors

We give test vectors of *Piccolo* for each key length. The data are represented in hexadecimal form.

80-bit key:

key	`00112233 44556677 8899`
plaintext	`01234567 89abcdef`
ciphertext	`8d2bff99 35f84056`

128-bit key:

key	`00112233 44556677 8899aabb ccddeeff`
plaintext	`01234567 89abcdef`
ciphertext	`5ec42cea 657b89ff`

Lightweight and Secure PUF Key Storage Using Limits of Machine Learning

Meng-Day (Mandel) Yu[1], David M'Raihi[1], Richard Sowell[1], and Srinivas Devadas[2]

[1] Verayo Inc., San Jose, CA, USA
{myu,david,rsowell}@verayo.com
[2] MIT, Cambridge, MA, USA
devadas@mit.edu

Abstract. A lightweight and secure key storage scheme using silicon Physical Unclonable Functions (PUFs) is described. To derive stable PUF bits from chip manufacturing variations, a lightweight error correction code (ECC) encoder / decoder is used. With a register count of 69, this codec core does not use any traditional error correction techniques and is 75% smaller than a previous provably secure implementation, and yet achieves robust environmental performance in 65*nm* FPGA and 0.13μ ASIC implementations. The security of the syndrome bits uses a new security argument that relies on what *cannot* be learned from a machine learning perspective. The number of *Leaked Bits* is determined for each Syndrome Word, reducible using *Syndrome Distribution Shaping*. The design is secure from a min-entropy standpoint against a machine-learning-equipped adversary that, given a ceiling of leaked bits, has a classification error bounded by ϵ. Numerical examples are given using latest machine learning results.

Keywords: Physical Unclonable Functions, Key Generation, Syndrome Distribution Shaping, Machine Learning, FPGA, ASIC.

1 Introduction

Gassend et al. introduced silicon-based Physical Unclonable Functions (PUFs) in [5], [6]; PUFs generate responses based on device manufacturing variations. Given a challenge as input, a PUF produces a response that is based on manufacturing variations on a particular instance of a silicon device. As such, PUF responses are noisy; although most of the response bits stay the same from run to run, some of the bits may flip. PUF noise increases with a change in voltage, temperature, and age between the *provisioning* condition, where a reference snapshot of the response is taken, and the *regeneration* condition. To derive stable PUF bits, some form of error correction code (ECC) or equivalent function is required. A set of *syndrome* bits is generated during provisioning, to help correct the regenerated PUF response back to the provisioned snapshot. There have been relatively few works that explicitly address the issue of information leaked via syndrome bits in the context of key storage using PUFs.

B. Preneel and T. Takagi (Eds.): CHES 2011, LNCS 6917, pp. 358–373, 2011.

Security arguments for the syndrome bits have taken several forms. The idea is to construct an argument that quantifies the amount of secrecy remaining in the provisioned PUF secret given that the syndrome is known to the adversary. One frequently cited work is by Dodis et al. [4], which contains an often-used result that Code-Offset Generic Syndrome with a code word length of n has an entropy loss of $n - \log A(n, 2t + 1)$, where $A(n, 2t + 1)$ represents the size of the largest code for a Hamming space with minimal distance of $2t + 1$. Assuming an optimal code, the value $n - \log A(n, 2t + 1)$ is also the size of the parity encoded as the syndrome. This result is useful to the extent that the number of parity bits in the error correction code does not exceed the min-entropy of the PUF bits that are used to derive the n-bit codeword. To safely account for the case where a large number of syndrome bits are used (a seemingly necessary tradeoff to reduce ECC complexity), additional security arguments may be useful. Yu and Devadas in [21] developed an alternative to Code-Offset Syndrome, using a technique called Index-Based Syndrome (IBS) coding, which was proven to be information-theoretically secure under the assumption that the PUF output bits are independent and identically distributed (i.i.d.); the security arguments apply even for a heavily biased (and thus min-entropy reduced) PUF.

Although the work of [21] achieved a quadratic reduction in *ECC complexity*, the work does not explicitly describe the *PUF complexity* required for producing the i.i.d. PUF output bits beyond a brief mention of using disjoint oscillator pairs. The PUF complexity (number of PUF elements, and in this case disjoint oscillator pairs per key bit) is $2520/128 = 19.7$, taking [21] at face value.[1] The current work, by contrast, describes a lightweight key storage mechanism that is lightweight both in terms of its ECC complexity *and* PUF complexity, by using the indexing scheme in [21] as a starting point and eliminating the BCH coding. It achieves a 75% reduction in ECC complexity compared to [21] and achieves a PUF complexity of 5 (using ten 64-sum PUFs, see Figure 1 for a description of k-sum PUFs) to as little as 1 (using two 64-sum PUFs); this is a 4x to 20x improvement. The PUF complexity reduction derives from a machine-learning-based security argument that each additional syndrome bit does not require a linear increase in the number of PUF elements (e.g., disjoint oscillator pair [18] [21] or a memory cell [1] [12] [7] [17] [8]) but instead relies on assumptions on what *cannot* be learned about a challengeable physical system.

Machine learning theory as pioneered in [20] is interested primarily in what *can* be learned. For example, the number of training samples needed to learn reasonably well a hypothesis class grows linearly with the Vapnik-Chervonenkis dimension of that class. In the context of learning a k-sum PUF, the theory suggests that the number of training samples required for learning grows linearly in the number of parameters in the learning model, which corresponds to the number of summation stages in the PUF. Empirical results in [13] [14] [15] show

[1] Using the example parameters described in [21], a 128-bit key would require 5 BCH(63,30,t=6) blocks, or 315 bits. Since each bit is coded using a 3-bit index, picking the best value out of 8 PUF output bits, a total of 2520 disjoint oscillator pairs are needed.

that the required number of training samples does in fact grow linearly with the number of delay sums. The current work turns this (apparent) weakness of a PUF into a strength by taking advantage of what cannot be learned; this results in reduction in the PUF complexity required. PUF complexity is the number of PUF elements per key bit; for a k-sum PUF (Figure 1), a PUF element is a pair of ring oscillators whose frequency difference effectively forms a stage in the PUF.

1.1 Contributions

The main contributions of the current work include the following:

- 75% reduction in *ECC complexity* because no traditional error correction codes are used. We note that [21] mentions this possibility; in this paper we provide extensive supporting experiments;
- 4x to 20x reduction in *PUF complexity*;
- ASIC implementation results. We believe this paper to be the first to give results on reliable key generation in an ASIC with integrated ECC;
- Accelerated aging result on stable PUF keys, well beyond published PUF aging results;
- A new metric, *Leaked Bits*, which entropically computes leakage per Syndrome Word;
- A new technique called *Syndrome Distribution Shaping*, to minimize *Leaked Bits*;
- A new security argument relating machine learning classification error ϵ to the average min-entropy remaining in the PUF-derived secret.

1.2 Related Works

To the best knowledge of the authors, the current work is the first to marry results from two fields: machine learning [13] [14] [15] and PUF-based key generation [1] [5] [11] [12] [18] [21]. [5] pioneered the use of error correction on PUF outputs using 2D Hamming codes. [4] provides a security framework for using Code-Offset Syndrome. [18] took a more robust approach to account for environmental noise using a single stage BCH(255) code. [1] used a two-stage coding approach, with the use of heavy first-stage repetition coding to reduce second-stage ECC complexity. [12] introduced the use of soft-decision decoding. The Code-Offset Syndrome in [4], however, yields at best very little (or in some cases negative) remaining min-entropy for some of the more efficient recent approaches where a large number of syndrome bits are produced (e.g., by a repetition coding stage) to reduce ECC complexity.[2] [21] introduced an alternative to Code-Offset Syndrome; under the assumption that PUF output bits

[2] Consider a PUF with a min-entropy = 0.8. If a (5,1,5) repetition code is used, 4 bits are leaked via syndrome (assume code is optimal). This is also the min-entropy of the 5 PUF output bits ($5 \times 0.8 = 4$) used to form the code word. No secrecy remains from a min-entropy standpoint.

are i.i.d., Index-Based Syndrome Coding (IBS) results in syndrome bits that are information-theoretically secure. The current work uses IBS as a starting point and eliminates the BCH coding of [21], and uses a security framework that eliminates the need for the i.i.d. PUF output assumption in [21] by deriving syndrome security based on what cannot be learned by a machine-learning-equipped adversary. This new security framework based on limits of machine learning has the effect of reducing PUF complexity.[3] The machine-learning-based syndrome security framework differs from and complements the syndrome security arguments derived in [4] and [21]; for example, operating in the regime where machine learning classification error $\epsilon = 0.5$ is essentially equivalent in security to using an i.i.d. PUF output assumption. The current work is also one of few published works which contains results of an integrated ASIC PUF + ECC implementation under environmental stresses, and complements FPGA results obtained in [21].

1.3 Organization

We describe the implementations of k-sum PUFs with associated error correction in Section 2. Section 3 establishes the empirical viability of the lightweight error correction scheme with respect to stability results (against voltage, temperature, and aging) and implementation complexity. Section 4 uses the empirically viable building blocks, consisting of the lightweight error correction coder plus one or more 64-sum PUFs, to derive Secure Constructions.

2 PUFs with Lightweight Error Correction

In this section, we describe the FPGA and ASIC implementations of PUFs and the error correction schemes that we are evaluating in Section 3.

A simplified high-level block diagram is shown in Figure 1. The basic 64-sum PUF looks at the difference between two delay terms, each produced by the sum of 64 ring oscillator delay values. The challenge bit C_i for each of the 64 stages determines which ring oscillator is used to compute the top delay term, and which is used to compute the bottom delay term. The sign bit of the difference between the two delay terms determines whether the PUF produces a '1' output bit or a '0' output bit for the 64-bit challenge $C_0 \cdots C_{63}$. The remaining bits of the difference determine the confidence level of the '1' or the '0' output bit. The k-sum PUF can be thought of as a k-stage Arbiter PUF [11] with a real-valued output (as opposed to a single bit output) that contains both the output bit as well as its confidence level. This information is used by the downstream lightweight error correction block, using the indexing scheme described in [21], coupled with a Syndrome Distribution Shaper (to be described in Section 4) to minimize syndrome leakage. The indexing scheme uses index sizes between 4 bits (choosing best out of 16 output bits) and 5 bits (best out of 32). If a '1' bit is

[3] Reducing PUF complexity is more difficult to achieve with an i.i.d. PUF output assumption, where an increase in syndrome length requires no less than a linear increase in the number of PUF elements.

to be encoded, the location of the maximum (out of 16 for a 4-bit index, and out of 32 for a 5-bit index) is chosen and written out as the Syndrome Word; alternatively, if a '0' bit is to be encoded, the location of the minimum is chosen and written out as the Syndrome Word.

The 0.13μ ASIC implementation contains multiple banks of 64-sum PUFs, a lightweight error correction engine (including Indexing algorithm and Syndrome Distribution Shaping algorithm), universal hashing [10], cryptographic functions, and various other logic. Various Xilinx FPGA versions were created as the design evolved; the final FPGA version included all the functionality mentioned above for the ASIC.

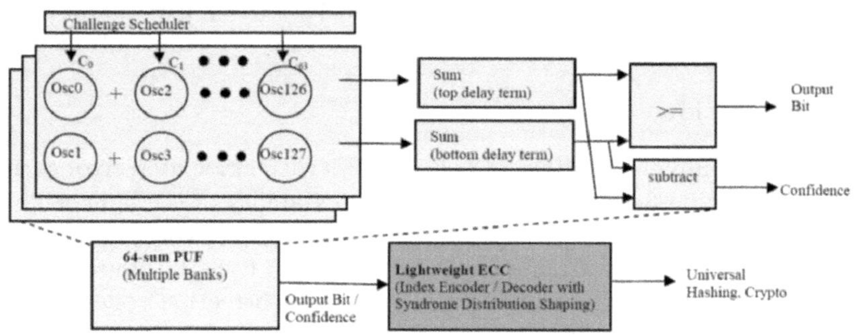

Fig. 1. Lightweight PUF Key Storage Block Diagram

3 Empirical Viability of Lightweight Error Correction

This section establishes the empirical viability of the lightweight error correction scheme, which is derived from the Index-only coding approach in [21], without the use of traditional ECC. The results in this section show a 75% reduction in error correction implementation complexity. Yet, when the indexing parameters are properly selected and applied in the context of the 64-sum PUF shown in Figure 1, stable PUF bits are derived under very extreme environmental conditions in FPGAs and ASICs.

3.1 Implementation Complexity

This section compares ECC complexity for three main classes of PUF error correction schemes using representatives from each:

1. Lightweight (Indexing only)
2. 2-stage ECC (Indexing + BCH63)
3. Large Block ECC (BCH255)

The analysis includes both encoder and decoder complexity and does not include I/O buffering, host interface logic, and other peripheral logic. Lightweight ECC has an implementation complexity that is estimated to be 75% smaller than the two-stage scheme published in [21] (secure based on i.i.d. PUF output assumption) and an estimated 98% smaller than the single-stage scheme published in [18] (secure based on Dodis' framework). The results are summarized in Table 1 below. The SLICE utilization is minimal (1.2% of a modestly-sized Xilinx Virtex-5 LX50 SLICE count), containing only 69 registers.

Table 1. Three Classes of PUF Error Correction and Relative Complexities

Lightweight (This work)	**2-stage ECC** (From [21])	**Large Block** (From [18])
69 registers	471 registers	6400 registers (est. 16x)
1.2% SLICE count*(99/7200)	5% SLICE count*(393/7200)	65% SLICE count*

*Utilization of a modestly-sized Xilinx Virtex-5 LX50 device as a benchmark.

3.2 Stability

This section describes the performance of the Lightweight ECC with a 64-sum PUF. The results show that a 4-bit index is capable of achieving parts-per-million (ppm) level performance when provisioning is performed under nominal temperature and voltage (25°C, V_{nom}), and regeneration is performed under a fast-fast temperature-voltage corner (-55°C, V_{nom} + 10%) and a slow-slow temperature-voltage corner (125°C, V_{nom} - 10%). Figure 2 shows representative results for each corner, where a total of 1M+ error correction blocks using 4-bit indexing ran without errors for each corner using empirical data collected from Xilinx Virtex-5 FPGAs, with the 4-bit indexing post-processed in software using empirical PUF data. The data illustrates that ppm level stability is feasible, and better performance is achievable with either a larger index size (choosing best out of > 16) or using retry mechanisms if a failure is observed [21]. The average number of noisy bits is about 6 bits out of 63 in both cases, with the maximum number of noisy bits (for 1M+ blocks) at 9 out of 63 bits, and every single noisy bit was error corrected for all the cases that were run.

The empirical results also showed that under higher stress, a larger index size was required. For example, in the context of accelerated aging (Figure 3) where provisioning was performed at 25°C, 1.0V and regeneration at 125°C, 1.1V, an increase in index size by 0.25 bit is necessary (choosing best out of 20 instead of best out of 16) to achieve error-free performance. The analysis was performed using empirical PUF data from a Xilinx Virtex-5 FPGA device aged under high temperature and high voltage stress, with empirical PUF data extracted *in-situ* and 4.25-bit indexing (best out of 20) emulated as a post-processing step (in practice, this can be implemented using a 5-bit index, and choosing best out of 20 instead of best out of 32). Test parameters for accelerated aging were derived from *MIL-STD-883G Method 1005.8 Steady State Life* as well as accelerated aging parameters obtained from Xilinx. Specifically, 0.70eV activation energy was assumed, at a confidence level of 60% (same assumptions as those used by Xilinx). Over 80M+ blocks of PUF data were corrected, representing an accelerated life of 260+ years at 25°C and 20+ years at 55°C, with every single block error corrected using a 4.25-bit index for that entire dataset; this has an

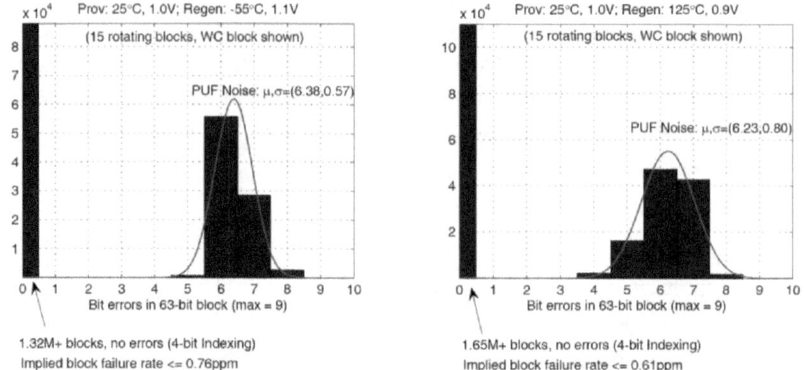

Fig. 2. Lightweight ECC performance, **WC Temperature / Voltage corners** (4-bit index). The right distribution in each plot is the PUF noise histogram before ECC, and the left distribution (at 0 errors) is the histogram after lightweight ECC.

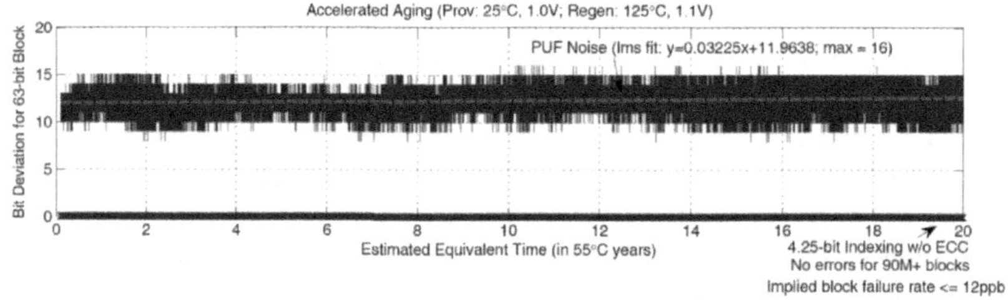

Fig. 3. Lightweight ECC performance, **Accelerated Aging** (4.25 bit index)

implied error rate of less than 12 parts per *billion*. As shown below, the average number of bits in error prior to indexing ranges from about 8 bits to 16 bits for a block size of 63 over 20 years at 55°C (or equivalently 260+ years at 25°C). The least mean square fit shows a slight upward slope of the PUF noise over this time. Yet with 4.25-bit indexing all the errors were corrected.

The FPGA results are consistent with results from a 0.13μ ASIC implementation (Figure 4), which has multiple 64-sum PUFs as well as the lightweight encoding / decoding algorithm integrated into a single device. The results in Figure 4 show that under extreme voltage conditions, 4-bit indexing (best out of 16) results in a 2.5ppm block failure rate, whereas 5-bit indexing (best out of 32) results in error-free performance. The integrated ASIC device (unlike the FPGA results above which emulated the indexing with empirical PUF data as a post-processing step to help algorithmic derivation) does not allow for fractional index sizes.

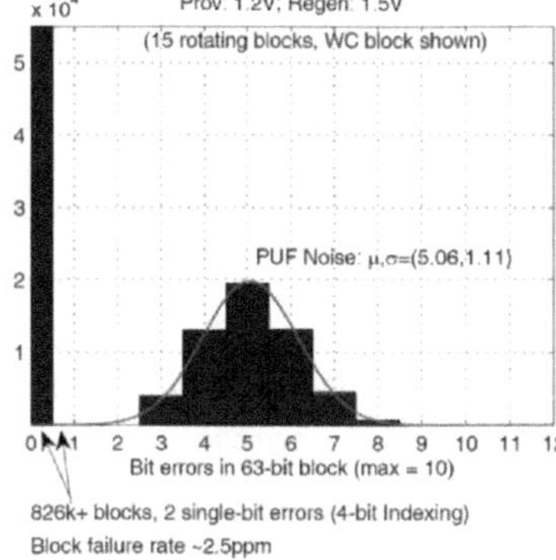

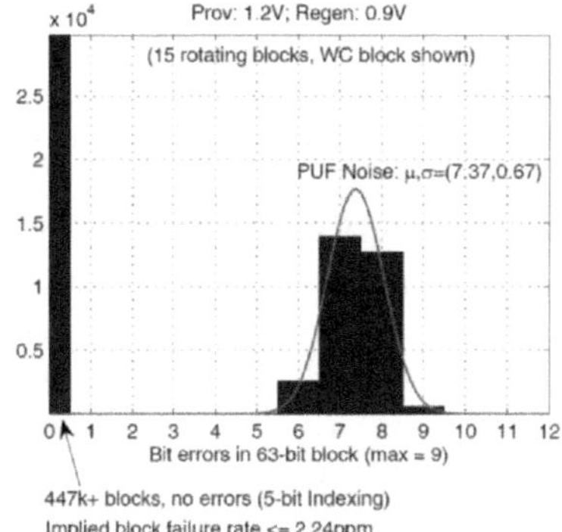

Fig. 4. 0.13μ ASIC with PUF + Lightweight ECC, **Extreme Voltage** performance. The right distribution in each plot is the PUF noise histogram before ECC, and the left distribution (near 0 errors) is the histogram after lightweight ECC.

4 Secure Constructions

The previous section demonstrated the empirical viability of a PUF + lightweight ECC combination. This section derives several Secure Constructions consisting of lightweight ECC and one or more 64-sum PUF blocks in the context of deriving a 128-bit key.[4] By adopting a machine-learning-based security argument instead of an i.i.d. PUF output argument, the number of ring oscillator pairs is reduced from 2520 to 640 (Secure Construction #1) or as little as 128 (Secure Construction #4). The PUF complexity reduction resulting from the machine-learning-based security argument is a result of the fact that each additional syndrome bit does not require a linear increase in the number of ring oscillators but instead relies on what cannot be learned about a challengeable physical system.

4.1 Unlearnable Bits

To determine what cannot be learned from a k-sum PUF, consider what is required to learn the delay differences of each pair of oscillators. A machine-learning-equipped adversary using a physical model of the PUF for learning starts with a model consisting of k parameters. The adversary also needs access to challenge/response pairs, for example, pairs consisting of k-bit challenges and 1-bit responses. Ruhrmair et al. in [14] derived an empirical equation relating the number of challenge/response (C/R) pairs N_{CRP}, number of parameters k, and the classification error rate ϵ as follows:

$$N_{CRP} \approx 0.5 \frac{k+1}{\epsilon}$$

[4] If additional key bits are required, then the entropy has to be increased, e.g., by doubling the number of 64-sum PUFs used for a 256-bit key.

The equation was derived using the best of results obtained from using Support Vector Machine (SVM), Logistical Regression (LR), and Evolution Strategy (ES) algorithms corresponding to an Arbiter delay PUF [11], including the case where $k = 64$. (While our PUF is not an Arbiter PUF per se, it has a very similar structure.) According to the equation, if k C/R pairs are known to the adversary for a k-parameter PUF, the adversary cannot do much better than guessing since the error rate $\epsilon = 0.5$. Intuitively, the results make sense; a k-parameter PUF would have at least k or more bits worth of parameter information (if each parameter is 1-bit, there would be k bits of information, and the parameter size likely needs to be a few bits for the machine learning to converge). As a result, if no more than k C/R pairs (each response is a single bit) are given out, no more than k bits of information are derived, and therefore the machine learning algorithm cannot infer much information.

4.2 Leaked Bits (LB)

We analyze several PUF Syndrome Coding algorithms, and describe their behavior with respect to *Leaked Bits*, the number of bits leaked per *Syndrome Word* for a particular Syndrome Coding algorithm, as defined below.

$$\mathbf{LB}(\underline{S}^{alg}) \equiv \mathbf{I}(\underline{S}^{alg}; \underline{M}^{\infty}) = \mathbf{H}(\underline{S}^{alg}) - \mathbf{H}(\underline{S}^{alg}|\underline{M}^{\infty})$$ [5]

where,

- $\underline{S}^{alg}$ is a random variable representing the Syndrome Word. Its variability comes from a particular syndrome coding algorithm used, denoted by the superscript *alg*.
- $\underline{M}^{\infty}$ is a random variable representing a PUF model that is perfect in predicting the PUF output bits (superscript ∞ denotes its perfect predicting ability). Its variability comes from PUF manufacturing variations.
- $\mathbf{H}$ is the *Shannon entropy* measure [2]

$$\mathbf{H}(\underline{X}) = -\sum_{x} p(x) \log_2 p(x)$$

  $\underline{X}$ is a random variable with a probability mass function $p(x)$, and the summation is taken over x over its entire alphabet
- $\mathbf{I}$ is the *mutual information* measure [2]

$$\mathbf{I}(\underline{Y}; \underline{X}) = \mathbf{H}(\underline{Y}) - \mathbf{H}(\underline{Y}|\underline{X})$$

  where $H(\underline{Y}|\underline{X}) = -\sum_{x} p(x) \sum_{y} p(y|x) \log_2 p(y|x)$. Note: $\mathbf{I}(\underline{Y}; \underline{X}) = \mathbf{I}(\underline{X}; \underline{Y})$.

Mutual information $\mathbf{I}$ computes the amount of information shared between two random variables. For example, in a cryptographic encryption system, the amount of information shared between the Ciphertext (denoted CT) and Key (i.e., information leaked by the Ciphertext about the Key from an information-theoretic standpoint) is

$$\mathbf{I}(\underline{CT}^{alg}; \underline{Key}) = \mathbf{I}(\underline{Key}; \underline{CT}^{alg}) = \mathbf{H}(\underline{Key}) - \mathbf{H}(\underline{Key}|\underline{CT}^{alg}).$$

To determine the amount of information leaked by a Syndrome Word about the PUF, we use the same concept, and call the result Leaked Bits, as defined above.

We now describe several syndrome coding algorithms that have been published in open literature, and analyze their information leakage with respect to Leaked Bits. We also analyze the use of a new technique called *Syndrome Distribution Shaping* to reduce Leaked Bits while preserving the average error correction power.

Code-Offset. In the Code-offset Method [4], the syndrome bits generated correspond to the XOR mask for a sequence of PUF output bits required to form a valid error correction code codeword. Consider a simple example of a binary 3x repetition code. Let the random variable $\underline{B}$ be a bit we want to store in a PUF.[6] Let the random variable $\underline{O}$ represent a sequence of PUF output bits corresponding to the error correction word size (3-bits in the 3x repetition coding example). Let the random variable $\underline{S}$ represent the corresponding Syndrome Word (3-bits in the 3x repetition coding example). The valid code words are (000) and (111). If we want to store a bit $\underline{B} = 0$, the valid code word (000) is used. Alternatively, if we want to store a bit $\underline{B} = 1$, the valid code word (111) is used. To generate the syndrome, three PUF output bits $O = o_0o_1o_2$ are required. The syndrome $\underline{S}$ using Code-Offset is the XOR mask required to make the PUF output bits $\underline{O} = o_0o_1o_2$ a valid code word, i.e., $\underline{S} = \underline{C} \wedge \underline{O}$, where $\wedge$ is the bitwise XOR operator.

Now, let's compute Leaked Bits using this example. In the 3x repetition example above, $H(\underline{S}^{3x}) = \log_2(\#(\underline{S}^{3x})) = 3$ bits, where # operator is the cardinality of the random variable, or the number of possibilities that the random variable can take. This is the amount of uncertainty of the 3-bit syndrome *not conditioned on any other knowledge*. Recall that $\underline{M}^{\infty}$ represents a perfect PUF model in that it has perfect knowledge in predicting PUF output bit $\underline{O}$. Given a perfect PUF model, the uncertainty remaining in $\underline{S}$ reduces to the uncertainty as to whether the valid code word is (000) or (111). That is, $\mathbf{H}(\underline{S}^{3x}|\underline{M}^{\infty}) = \log_2(\#(\underline{B})) = 1$ bit. Putting it together:

$$\begin{aligned}\mathbf{LB}(\underline{S}^{3x}) \equiv \mathbf{I}(\underline{S}^{3x}; \underline{M}^{\infty}) &= \mathbf{H}(\underline{S}^{3x}) - \mathbf{H}(\underline{S}^{3x}|\underline{M}^{\infty}) \\ &= \log_2(\#(\underline{S}^{3x})) - \log_2(\#(\underline{B})) = 3 - 1 = 2 \text{ bits.}\end{aligned}$$

Two bits of information are leaked for each Syndrome Word derived from a 3x repetition codeword.

[6] Here, we consider the case where the PUF is used as a generalized key-store: the keying bit $\underline{B}$ can come from a source external to the PUF chip (e.g., user chosen key), or it can be derived from a PUF on the same chip.

Index-Based Syndrome (IBS) Coding. In Index-Based Syndrome coding [21], the Syndrome Word is an index lookup into a sequence of PUF output bits. Consider a simple example of a syndrome index with a width of 3 (i.e., $W = 3$, or a 3-bit index). $\mathbf{H}(\underline{S}^{3i}) = \log_2(\#(\underline{S}^{3i})) = 3$ bits. The 3-bit index takes on the value of the best out of $\#(\underline{S}^{3i}) = 8$ choices, requiring 8 PUF output bits $\underline{O} = o_0o_1o_2o_3o_4o_5o_6o_7$. If we want to store $\underline{B} = 0, \underline{S}^{3i} = \arg(\min_{j\in 0,1,\cdots,J-1}(o_j^r))$, where $J = 2^{W=3}$. (Superscript r of o^r denotes the real-valued output of the PUF, not just the binary '1' or '0' portion of the PUF output.) If we want to store $\underline{B} = 1, \underline{S}^{3i} = \arg(\max_{j\in 0,1,\cdots,J-1}(o_j^r))$. Now consider $\underline{M}^{\infty}$, which a perfect PUF model that predicts $\underline{O} = o_0o_1o_2o_3o_4o_5o_6o_7$ with perfect accuracy. Given a perfect PUF model, the uncertainty remaining in $\underline{S}$ reduces to the uncertainty as to whether the maximum or the minimum value is picked. That is, $\mathbf{H}(\underline{S}^{3i}|M^{\infty}) = \log_2(\#(\underline{B})) = 1$ bit. The amount of information leaked by $\underline{S}^{3i}$ about $\underline{M}^{\infty}$ is the Leaked Bits for $\underline{S}^{3i}$:

$$\begin{aligned}\mathbf{LB}(\underline{S}^{3i}) \equiv \mathbf{I}(\underline{S}^{3i};\underline{M}^{\infty}) &= \mathbf{H}(\underline{S}^{3i}) - \mathbf{H}(\underline{S}^{3i}|\underline{M}^{\infty})\\ &= \log_2(\#(\underline{S}^{3i})) - \log_2(\#(\underline{B})) = 3 - 1 = 2 \text{ bits.}\end{aligned}$$

Two bits of information are leaked for each 3-bit index.[7]

Syndrome Distribution Shaping (SDS). Now, we present a new technique where we shape the syndrome distribution to minimize Leaked Bits while preserving, on average, error correction power. In its simplest form, the main idea is to enlarge the number of bits generated by $\underline{O}$ and to randomly select which of those bits would be used in forming a Syndrome Word and which would not be. As an example, visualize a case with a 3-bit Index-Based Syndrome. Let us order $o_0^ro_1^ro_2^ro_3^ro_4^ro_5^ro_6^ro_7^r$ from minimum to maximum, as $t_0^rt_1^rt_2^rt_3^rt_4^rt_5^rt_6^rt_7^r = \pi(o_0^ro_1^ro_2^ro_3^ro_4^ro_5^ro_6^ro_7^r)$, where π is a minimum-to-maximum sorting permutation. Here, t_0^r is the smallest $o^r_{j,j\epsilon 0\ldots J-1}$ value, t_1^r is the next smallest $o^r_{j,j\epsilon 0\ldots J-1}$ value, t_7^r is the largest $o^r_{j,j\in 0\ldots J-1}$ value. If there is equality in any of the comparisons, a random ordering among those is chosen.

Now, look at the unconditional probability of a particular 3-bit index value being selected. If we have no knowledge of the model or anything else, the probability of any index being selected is 1/8, i.e., $\mathbf{H}(\underline{S}^{3i}) = 3$ bits.

$$pr(o_0^r \text{ selected}) = 1/\#(\underline{S}^{3i}) = 1/8$$

$$\ldots$$

$$pr(o_7^r \text{ selected}) = 1/\#(\underline{S}^{3i}) = 1/8$$

Now, look at the probabilities conditioned upon $\underline{M}^{\infty}$, a perfect model, which allows us to sort the PUF output bits $t_0^rt_1^rt_2^rt_3^rt_4^rt_5^rt_6^rt_7^r = \pi(o_0^ro_1^ro_2^ro_3^ro_4^ro_5^ro_6^ro_7^r)$ and obtain:

$$pr(t_0^r \text{ selected}) = 1/\#(\underline{B}) = 1/2$$

[7] We are not making the assumption that PUF output bits are i.i.d., as in [21], under which the indices are provably secure.

$$pr(t_1^r \text{ selected}) = 0$$
$$\dots$$
$$pr(t_6^r \text{ selected}) = 0$$
$$pr(t_7^r \text{ selected}) = 1/\#(\underline{B}) = 1/2$$

Here, $\mathbf{H}(\underline{S}^{3i}|\underline{M}^{\infty}) = 1$ bit; either $o^r_{j,j\epsilon 0\dots J-1} = t_0^r$ or $o^r_{j,j\in 0\dots J-1} = t_7^r$ will be selected, depending on whether $\underline{B} = 0$ or $\underline{B} = 1$. Now, we want a randomization mapping that flattens the probability distribution while on the average preserving the error correction power. In the example above, a 3-bit index is used to choose the best PUF output value out of 8 choices. The distribution peaks on the ends: either the maximum or the minimum value will be selected. To flatten the distribution, one possibility is to use a 4-bit index, and randomly clobber or skip over half of the 16 choices such that the 4-bit index still chooses the best out of 8 values (same as the 3-bit index case).

This distribution will not peak at the ends (and be zero elsewhere), as is the case for a 3-bit index, but will be flatter (i.e., more uniformly distributed). More generally, consider an independent bit generator with output $\underline{R}$ where $pr(\underline{R} = 1) = p$ and $pr(\underline{R} = 0) = 1 - p = q$. Now, consider a 4-bit index ($W = 4$), where each of the 16 values ($J = 2^W = 16$) have a probability of p of being clobbered (i.e., skipped or not used). On the average, if $p = 0.5$, 8 values out of 16 will not be clobbered, and the maximum or the minimum out of 8 will still be selected, thus giving on the average a similar error correction power as the 3-bit index case (which also selects the maximum or minimum out of 8). More generally,

$$pr(t_{j\epsilon 0\dots J-1} \text{ selected}) = 1/2(p^j q + p^{J-1-j} q)$$
$$pr(\text{none selected}) = p^J$$

Assume in this example that if all values are clobbered we randomly choose a value. Practically, this distinction does not make a difference in this example, since the probability is very small:

$$pr(t_{j\epsilon 0\dots J-1} \text{ selected}) = 1/2(p^j q + p^{J-1-j} q) + p^J/J$$

Here, the amount of uncertainty remaining when applying Syndrome Distribution Shaping (SDS) given a perfect PUF model is:

$$\mathbf{H}(\underline{S}|\underline{M}^{\infty}) = \mathbf{H}(pr(t_{j\in 0\dots J-1} \text{ selected}))$$

Now, compute the Leaked Bits of 4-bit SDS index:

$$\begin{aligned}
\mathbf{LB}(\underline{S}^{W=4,p=.5}) &= \mathbf{I}(\underline{S}^{W=4,p=.5}; \underline{M}^{\infty}) \\
&= \mathbf{H}(\underline{S}^{W=4,p=.5}) - \mathbf{H}(\underline{S}^{W=4,p=.5}|\underline{M}^{\infty}) \\
&= \log_2(\#(\underline{S}^{W=4,p=.5})) - \mathbf{H}(pr(t_{j\in 0..J-1} \text{ selected})) \\
&= 4 - 2.98 = 1.02 \text{ bits}
\end{aligned}$$

Note that a $W = 4$-bit SDS index with a clobbering rate of 0.5 has a similar average error correction capability as a $W = 3$-bit index-based syndrome (i.e., both, on the average, select the strongest out of 8), and yet the Leaked Bits has been reduced by 50%, from 2 bits to 1.02 bits. By expanding the number of PUF output bits $\underline{O}$ and randomly eliminating them so that on the average we select the index from the same number of choices (i.e., preserving on average the same error correction power), we have lowered the number of bits leaked via each Syndrome Word from 2 bits to about 1 bit. Moving on to $W = 5$-bit syndrome, and $p = 0.75$, to preserve on the average the same error correction power, and beyond we have the following results:

$$\begin{aligned}
\mathbf{I}(\underline{S}^{3i}, M^{\infty}) &= 2 \text{ bits} \\
\mathbf{I}(\underline{S}^{W=4,p=1/2}, M^{\infty}) &= 1.02 \text{ bits} \\
\mathbf{I}(\underline{S}^{W=5,p=3/4}, M^{\infty}) &= 0.80 \text{ bits} \\
\mathbf{I}(\underline{S}^{W=6,p=7/8}, M^{\infty}) &= 0.71 \text{ bits} \\
\mathbf{I}(\underline{S}^{W=7,p=15/16}, M^{\infty}) &= 0.67 \text{ bits}
\end{aligned}$$

Note that between $W = 3$ and $W = 6$ there is almost a 3x improvement. Beyond $W = 6$ there are diminishing returns. Alternative SDS algorithms include ones that yield an even lower Leaked Bits, while others guarantee a certain minimum number of un-clobbered choices available by using only one side of the un-clobbered binomial distribution.

4.3 Secure Construction Examples

Now, armed with a metric (LB) to determine the number of bits leaked from each Syndrome Word, we describe different methodologies to derive secure constructions using the 64-sum PUF as a building block. The methodology requires an assumption describing the relationship between the classification error ϵ and Leaked Bits (and more precisely, a Leaked Bits sum ΣLB). In other words, we are establishing secure constructions assuming an $(\epsilon, \Sigma\text{LB})$ machine-learning-equipped adversary. This adversary cannot produce a classification error better than ϵ given that a total leaked bits of ΣLB. For the numerical examples below, we use the machine learning results in [14] as a proxy for the relationship between classification error and a sum of Leaked Bits;[8] this use has been preliminarily affirmed by the authors of this work using SVM and Simulated Annealing methods, and shall be further developed as future work.

[8] Formally, Leaked Bits for raw PUF responses can be computed using a null Syndrome notation: $\mathbf{LB}(\underline{S}^{\text{null}}) \equiv \mathbf{I}(\underline{S}^{\text{null}}; \underline{M}^{\infty}) = \mathbf{H}(\underline{S}^{\text{null}}) - \mathbf{H}(\underline{S}^{\text{null}}|\underline{M}^{\infty}) = 1 - 0 = 1$ bit, since the unconditional entropy of $\underline{S}^{\text{null}}$ (raw PUF response bit) is $\mathbf{H}(\underline{S}^{\text{null}}) = 1$ bit, and when conditioned with a perfect PUF model, $\underline{S}^{\text{null}}$ is completely known, i.e., $\mathbf{H}(\underline{S}^{\text{null}}|\underline{M}^{\infty}) = 0$ bit. As such, each bit leaked in the context of Leaked Bits (as defined) can also be interpreted as a leak of one equation for the PUF system with a 1-bit outcome.

Secure Construction #1: ΣLB *well within* $\epsilon = 0.5$. In this construction, we conservatively operate *well within* the regime of $\epsilon = 0.5$. Using the machine learning results in [14], we can choose to operate at a point where the Leaked Bits sum is no more than *half* the number of parameters in the PUF.

$$\Sigma LB \approx \frac{1}{2} 0.5 \left. \frac{k+1}{\epsilon} \right|_{\epsilon=0.5,k=64} = 32.5 \text{ bits per 64-sum PUF}$$

As an example, using a 6-bit index with a clobber rate of 5/8, the average error correction power is between 4 and 5-bit index (best out of 64 x 3/8 = 24 bits on the average). The LB is:

$$LB = \mathbf{I}(\underline{S}^{W=6,p=5/8}, \underline{M}^{\infty}) = 2.45 \text{ bits per Syndrome Word}$$

With $k = 64$ sum stages, we allow $\Sigma LB = k/2 = 32$ bits to be leaked per PUF; this translates to the use of 32 / 2.45 = 13 SDS indices. To generate a 128-bit key, we need ten (128/13) 64-sum PUFs, using 640 delay parameters to keep secret 128-bits worth of information (PUF complexity = 640/128 = 5). *It is likely a safe assumption that if less than $k/2$ equations are leaked from a k parameter PUF, a machine-learning-equipped adversary cannot learn much about the PUF since there remain $k/2$ degrees of freedom. This construction is formally equivalent to security obtained using an i.i.d. PUF output assumption, with a 2x margin on the certainty of* $\epsilon = 0.5$.

Secure Construction #2: ΣLB at $\epsilon = 0.5$ boundary. We remove the 2x margin on the certainty of $\epsilon = 0.5$, thus requiring half the number of PUFs. This construction is formally equivalent to security obtained using an i.i.d. PUF output assumption.

Secure Construction #3: *Challenge Modification on Block Boundary.* Here, we operate across multiple blocks in the range where $\epsilon \leq 0.5$, and compute the average min-entropy to account for cases where $\epsilon \neq 0.5$. Continuing the example from above, the first block of 26 SDS indices leaks 64 bits worth of information, but $\epsilon = 0.5$. Now, let's assume that the results of the first block (consisting of 26 data bits) are used to modify the challenge bits for the second block; that is, we use a *Challenge Modification Schedule* at the block boundary so that the 26 Syndrome indices for the second block cannot be used by the machine learning algorithm unless the first 26 bits are guessed or estimated correctly. The machine learning algorithm requires input / output sets, i.e., Challenge/Syndrome sets in our case, in order to train the delay parameters and the Challenges are known for the second block 0.5^{26} of the time. Now, let's compute average min-entropy of this chained scheme given that after the first block $\epsilon \neq 0.5$.

First, we recall the definition of min-entropy $\mathbf{H}_\infty(.) \equiv -\log_2(Pr_{max}(.))$ and average min-entropy $\tilde{\mathbf{H}}_\infty(\underline{X}|\underline{Y}) \equiv -\log_2(E_{y \leftarrow \underline{Y}}[2^{-\mathbf{H}_\infty(\underline{X}|\underline{Y}=y)}])$ using the definitions and notation from [4].

In our context: $\tilde{\mathbf{H}}_\infty(\underline{P}|\underline{CS}) \equiv -\log_2(E_{cs \leftarrow \underline{CS}}[2^{-\mathbf{H}_\infty(\underline{P}|\underline{CS}=cs)}])$ where $\underline{P}$ is a random variable predicting all $\#(P)$ PUF-derived bits, and $\underline{CS} = cs$ a *subset* of the available Challenge/Syndrome sets used for regenerating $\underline{P}$. The subset is based on the number of Challenge/Syndrome sets that is known to the adversary *at any one time* as the result of the Challenge Modification Schedule.

Now let's consider the case where two blocks are generated using a 64-sum PUF, with syndrome modification at the block boundary. There is $\gamma = 0.5^{\#(P)/2}$ probability that the syndrome for the second block is useful; this is for the case where the 26 bits of the first block are guessed or estimated correctly.

$$\begin{aligned}\tilde{\mathbf{H}}_\infty(\underline{P}|\underline{CS}) &\equiv -\log_2(E_{cs\leftarrow \underline{CS}}[2^{-\mathbf{H}_\infty(\underline{P}|\underline{CS}=cs)}]) \\ &= -\log_2\{[(1-\gamma)\max(\epsilon, 1-\epsilon)^{\#(P)}]|_{\epsilon=0.5,\#(P)=52,\gamma=0.5^{26}} \\ &+ [\gamma \max(\epsilon, 1-\epsilon)^{\#(P)}]|_{\epsilon=0.25,\#(P)=52,\gamma=0.5^{26}}\} = 47.51 \text{ bits}\end{aligned}$$

The number of PUFs required is reduced to three (128 / 47.51) for 128-bit secret (PUF complexity = 3 x 64 / 128 = 1.5), if we assume that the machine learning result in [14] is a good proxy in estimating ϵ.

Secure Construction #4: *Challenge Modification on Syndrome Word Boundary.* Here, the challenge schedule is deviated once per Syndrome Word (e.g., index) vs. once per block. Due to space constraints, the derivation is omitted. Results for different number of blocks extracted per 64-sum PUF are below:

- 1 Block: $\tilde{\mathbf{H}}_\infty(\underline{P}|\underline{CS})|_{\#(P)=26} = 26$ bits
- 2 Blocks: $\tilde{\mathbf{H}}_\infty(\underline{P}|\underline{CS})|_{\#(P)=52} = 52$ bits
- 3 Blocks: $\tilde{\mathbf{H}}_\infty(\underline{P}|\underline{CS})|_{\#(P)=78} = 70$ bits

Two PUFs (128/70) are required for a 128-bit secret (PUF complexity = 1).

5 Conclusions

A PUF-based key storage is built using a lightweight ECC, without the use of traditional error correction techniques, and one or more 64-sum PUFs. The ECC complexity is low, with a register count of 69 for the encoder / decoder core, yet producing robust environmental stability results on FPGAs and ASICs. To our knowledge, this is the first time an integrated key generator ASIC implementation has been evaluated. We presented a new security argument that relies on what *cannot* be learned from a machine learning perspective, allowing a large reduction in PUF complexity. Future work includes further validation and refinements of the machine learning results in [14], applying the machine learning security method to XOR'ed PUFs (that are more difficult to learn), and methods to de-rate the ϵ vs. leaked bits curve to account for side channel leaks.

References

1. Bösch, C., Guajardo, J., Sadeghi, A.-R., Shokrollahi, J., Tuyls, P.: Efficient Helper Data Key Extractor on FPGAs. In: Oswald, E., Rohatgi, P. (eds.) CHES 2008. LNCS, vol. 5154, pp. 181–197. Springer, Heidelberg (2008)
2. Cover, T., Thomas, J.: Elements of Information Theory, 2nd edn. (2006)

3. Devadas, S., Suh, E., Paral, S., Sowell, R., Ziola, T., Khandelwal, V.: Design and Implementation of PUF-Based 'Unclonable' RFID ICs for Anti-Counterfeiting and Security Applications. In: Proc. RFID 2008, pp. 58–64 (May 2008)
4. Dodis, Y., Ostrovsky, R., Reyzin, L., Smith, A.: Fuzzy Extractors: How to Generate Strong Keys from Biometrics and Other Noisy Data (2008)
5. Gassend, B.: Physical Random Functions, Master's Thesis, EECS, MIT (2003)
6. Gassend, B., Clarke, D., van Dijk, M., Devadas, S.: Silicon Physical Random Functions. In: Proc. ACM CCS, pp. 148–160. ACM Press, New York (2002)
7. Guajardo, J., Kumar, S., Schrijen, G., Tuyls, P.: FPGA intrinsic pUFs and their use for IP protection. In: Paillier, P., Verbauwhede, I. (eds.) CHES 2007. LNCS, vol. 4727, pp. 63–80. Springer, Heidelberg (2007)
8. Holcomb, D., Burleson, W., Fu, K.: Initial SRAM State as a Fingerprint and Source of True Random Numbers for RFID Tags. In: Conf. RFID Security (2007)
9. Kocher, P., Jaffe, J., Jun, B.: Differential power analysis. In: Wiener, M. (ed.) CRYPTO 1999. LNCS, vol. 1666, pp. 388–397. Springer, Heidelberg (1999)
10. Krawczyk, H.: LFSR-based hashing and authentication. In: Desmedt, Y.G. (ed.) CRYPTO 1994. LNCS, vol. 839, pp. 129–139. Springer, Heidelberg (1994)
11. Lim, D.: Extracting Secret Keys from Integrated Circuits, MS Thesis, MIT (2004)
12. Maes, R., Tuyls, P., Verbauwhede, I.: A Soft Decision Helper Data Algorithm for SRAM PUFs. In: IEEE ISIT 2009. IEEE Press, Los Alamitos (2009)
13. Ruhrmair, U.: On the Foundations of Physical Unclonable Functions (2009)
14. Ruhrmair, U., Sehnke, F., Solter, J., Dror, G., Devadas, S., Schmidhuber, J.: Modeling Attacks on Physical Unclonable Functions. In: Proc. ACM CCS (October 2010)
15. Sehnke, F., Osendorfer, C., Sölter, J., Schmidhuber, J., Rührmair, U.: Policy gradients for cryptanalysis. In: Diamantaras, K., Duch, W., Iliadis, L.S. (eds.) ICANN 2010. LNCS, vol. 6354, pp. 168–177. Springer, Heidelberg (2010)
16. Skorobogatov, S.P.: Semi-Invasive Attacks: A New Approach to Hardware Security Analysis. Univ. Cambridge, Computer Lab.: Tech. Report (April 2005)
17. Su, Y., Holleman, J., Otis, B.: A 1.6pJ/bit 96 (percent) Stable Chip ID Generating Circuit Using Process Variations. In: ISSCC 2007, pp. 200–201 (2007)
18. Suh, G.: AEGIS: A Single-Chip Secure Processor, PhD thesis, EECS, MIT (2005)
19. Suh, G., Devadas, S.: Physical Unclonable Functions for Device Authentication and Secret Key Generation. In: DAC 2007, pp. 9–14 (2007)
20. Vapnik, V., Chervonenkis, A.: On the uniform convergence of relative frequencies of events to their probabilities. Theory of Prob. and its App. (1971)
21. Yu, M., Devadas, S.: Secure and Robust Error Correction for Physical Unclonable Functions. IEEE D&T 27(1), 48–65 (2010)

Recyclable PUFs: Logically Reconfigurable PUFs

Stefan Katzenbeisser[1], Ünal Koçabas[1], Vincent van der Leest[2], Ahmad-Reza Sadeghi[3], Geert-Jan Schrijen[2], Heike Schröder[1], and Christian Wachsmann[1]

[1] Technische Universität Darmstadt (CASED), Germany
{katzenbeisser,busch}@seceng.informatik.tu-darmstadt.de,
{unal.kocabas,christian.wachsmann}@trust.cased.de
[2] Intrinsic-ID, Eindhoven, The Netherlands
{vincent.van.der.leest,geert.jan.schrijen}@intrinsic-id.com
[3] Technische Universität Darmstadt and Fraunhofer SIT Darmstadt, Germany
ahmad.sadeghi@trust.cased.de

Abstract. We introduce the concept of Logically Reconfigurable Physical Unclonable Functions (LR-PUFs). In contrast to classical Physically Unclonable Functions (PUFs) LR-PUFs can be dynamically 'reconfigured' after deployment such that their challenge/response behavior changes in a random manner. To this end, we amend a conventional PUF with a stateful control logic that transforms challenges and responses of the PUF. We present and evaluate two different constructions for LR-PUFs that are simple, efficient and can easily be implemented. Moreover, we introduce a formal security model for LR-PUFs and prove that both constructions are secure under reasonable assumptions. Finally, we demonstrate that LR-PUFs enable the construction of securely recyclable access tokens, such as electronic tickets: LR-PUFs enhance security against manipulation and forgery, while reconfiguration allows secure re-use of tokens for subsequent transactions.

1 Introduction

In the last decades we are witnessing a rapid development and enhancement as well as an evolution of information technologies: On the one hand, computing and communication devices tend to become increasingly smaller and physically highly integrated. On the other hand, the growing usage and interconnection of millions of devices processing sensitive information raises many new trust and security challenges. Hence enabling technologies that can uniquely identify an (embedded) device and use the corresponding identity as a trust anchor in higher level security architectures are highly desirable. Although modern cryptography provides many useful tools for authentication and secure channels, it cannot guarantee the device's integrity, in particular in presence of hardware attacks.

In this context, Physically Unclonable Functions (PUFs) seem to be promising primitives that aim to exploit (random) physical variations to extract unique features of the underlying hardware to uniquely identify a device. The assumed properties of PUFs such as unclonability, unpredictability and tamper-evidence

B. Preneel and T. Takagi (Eds.): CHES 2011, LNCS 6917, pp. 374–389, 2011.

make them very appealing for deployment in cryptographic applications. Since their introduction by Pappu [38,39], PUFs have been proposed for secure generation and storage of strong cryptographic keys (see, e.g., [52,26]), and for emerging hardware-entangled cryptography [3], where the security of the cryptographic scheme is based on the physical properties of PUFs instead of mathematical problems. Moreover, today, there are already PUF-based security products aimed for the market (e.g., RFID, IP-protection, anti-counterfeiting solutions) [51,16].

So far, most existing PUFs exhibit a static behavior while a variety of applications greatly benefits from the availability of PUFs whose characteristics can be changed dynamically, i.e., *reconfigured*, after deployment: For instance, PUF-based key storage [52,26] and PUF-based cryptographic primitives [3] may require that previous secrets derived from the PUF cannot be retrieved any more. Another example are solutions to prevent downgrading of software [20] by binding the software to a certain hardware configuration, e.g., a PUF, which require the PUF behavior to be irreversibly altered upon installation of a software update. Moreover, when PUF-based wireless access tokens[1] (e.g., [40,49,37,42,51]) are re-used/recycled, the new users of the token shall not be able to retrieve access rights and/or to obtain privacy-sensitive information of the previous users of the token (see, e.g., [53,17,4]).

Unfortunately, all known implementations of physically reconfigurable PUFs rely on optical mechanisms, reconfigurable hardware (i.e., FPGAs), or novel memory technologies [20], which all have serious drawbacks in practice. In particular, optical PUFs cannot easily be integrated into integrated circuits and require expensive and error-prone evaluation equipment while FPGA-based solutions cannot be realized with non-reconfigurable hardware (e.g., ASICs) that is commonly used in practice [29].

Our goal and contributions. In this paper, we propose *Logically Reconfigurable PUFs* (LR-PUFs), an alternative construction to physically reconfigurable PUFs. LR-PUFs augment a physical PUF with a stateful control logic that changes the challenge/response behavior of the LR-PUF according to its internal state.[2] In particular, our contributions are as follows:

- *New constructions:* We propose two different constructions for logically reconfigurable PUFs (LR-PUFs). Our performance measurements show that the implementation overhead of the logical reconfiguration on top of a physical PUF is rather small.
- *Security model:* We introduce a formal security model for LR-PUFs and prove that both of our constructions are secure. More precisely, we show that, when instantiated by an appropriate physical PUF under reasonable assumptions,

[1] PUFs provide a lightweight and cost-effective solution to the problem of detecting counterfeit or cloned access tokens (e.g., RFID-based electronic tickets) by cryptographically binding a user's access rights to the physical characteristics of the token.

[2] A similar concept has been independently proposed by Lao et al. [22]. However, they do not provide a (formal) security model and do not discuss the adversary model and assumptions underlying their constructions.

our LR-PUFs can achieve both *forward-* and *backward-unpredictability*: The former assures that responses measured before the reconfiguration event are invalid thereafter, while the latter assures that an adversary with access to a reconfigured PUF cannot estimate the PUF behavior before reconfiguration.
– *Applications:* We demonstrate how LR-PUFs could be deployed for re-usable (recyclable) access tokens, such as electronic transit tickets, and discuss other envisaged applications of LR-PUFs.

Note that, although the constructions of LR-PUFs as proposed in this paper seem to be similar to Controlled PUFs [11], LR-PUFs and Controlled PUFs have very different objectives: In contrast to Controlled PUFs, LR-PUFs do not aim to prevent modeling attacks on PUFs but provide a practical way to *enable reconfigurability* for existing, typically static PUF constructions. We will elaborate on this aspect in Section 3.

Outline. The rest of the paper is structured as follows: After providing background information on Physically Unclonable Functions (PUFs) in Section 2, we present the concept of Logically Reconfigurable PUFs (LR-PUFs) in Section 3. We show two concrete LR-PUF constructions in Section 4, describe their implementation and evaluate their performance in Section 5, and formally prove their security in Section 6. In Section 7, we show how LR-PUFs could be used to realize recyclable access tokens and discuss several other potential use cases of LR-PUFs. Finally, we conclude in Section 8.

2 Background: Physically Unclonable Functions (PUFs)

A Physically Unclonable Function (PUF) is a noisy function that is embedded into a physical object, e.g., an integrated circuit [39,2]. When queried with a *challenge* w, a PUF generates a *response* $y \leftarrow \texttt{PUF}(w)$ that depends on both w and the unique device-specific intrinsic physical properties of the object containing $\texttt{PUF}()$. Since PUFs are subject to noise (e.g., environmental variations), they return slightly different responses when queried with the same challenge multiple times.

In literature, PUFs are typically assumed to be *robust*, *physically unclonable*, *unpredictable* and *tamper-evident*, and several approaches to heuristically quantify and formally define their properties have been proposed (see [2] for a comprehensive overview). Robustness means that, when queried with the same challenge multiple times, the same PUF will always return the same response. Physical unclonability means that it is infeasible to produce two PUFs that cannot be distinguished based on their challenge/response behavior, which cannot be achieved by (cryptographic) algorithms. Unpredictability requires that it is infeasible to predict the PUF response to a given unknown challenge, even if the PUF can be adaptively queried for a certain number of times. Since this is the most interesting property for cryptographic applications of PUFs [2], we will formally define unpredictability later, when we prove the security of our LR-PUF constructions. Tamper-evidence means that any attempt to physically access the

PUF irreversibly changes its challenge/response behavior. This is an important issue for practical deployment since it allows the detection of invasive hardware attacks, to which embedded devices are typically exposed to in practice.

A broad variety of different PUF constructions exists (see [29] for an overview). The most appealing ones for integration into electronic circuits are electronic PUFs. The most prominent examples of electrical PUFs include *delay-based PUFs* that exploit race conditions (arbiter PUFs [23,37,27]) and frequency variations (ring oscillator PUFs [12,48,30]) that can be found in integrated circuits; *memory-based PUFs* that are based on the instability of volatile memory cells like SRAM [14,15], flip-flops [28,24] and latches [47,19]; and *coating PUFs* [50], which are based on the capacitance caused by a special dielectric coating applied to the chip that houses the PUF.

Note that the amount of unique responses of a memory-based PUF is limited by the number of its memory cells. Moreover, it has been shown that most delay-based PUFs are subject to model building attacks that allow simulating the PUF in software (see, e.g., [23,37,27,41]). To counter this problem, additional primitives must be used: Controlled PUFs [11] use cryptography in hardware to hide the actual response of the underlying PUF, which prevents model building attacks. However, this requires the link between the PUF and the crypto component as well as the crypto component itself to be protected against invasive and/or side channel attacks.

3 Logically Reconfigurable PUFs

A logically reconfigurable PUF (LR-PUF) is a PUF whose challenge/response behavior depends on both the physical properties of the PUF and the logical state maintained by a control logic, as shown in Figure 1(a). The challenge/response behavior of the LR-PUF can be dynamically changed after it has been deployed by updating its state.

3.1 System Model

An LR-PUF combines a conventional physically unclonable function $\texttt{PUF}()$ and a control logic circuit. As shown in Figure 1(b), the control logic maintains a S, which is stored in non-volatile memory, provides an algorithm $\texttt{query}_S()$ for querying, and $\texttt{rcnf}()$ for reconfiguring the LR-PUF. The algorithm $\texttt{query}_S()$ consists of an input transformation function $\texttt{mapin}_S()$ and an output transformation function $\texttt{mapout}_S()$: $\texttt{query}_S(x)$ computes $w \leftarrow \texttt{mapin}_S(c)$, evaluates $y \leftarrow \texttt{PUF}(w)$, and returns $r \leftarrow \texttt{mapout}_S(y)$. The algorithm implementing $\texttt{rcnf}()$ reconfigures the LR-PUF by changing the current state S to a new independent state $S' \leftarrow \texttt{rcnf}()$.

Note that the generic LR-PUF construction depicted in Figure 1(b) can be seen as a generalization of controlled PUFs [11]. Controlled PUFs aim to *hide* the challenge/response behavior of the underlying PUF to the adversary to prevent model building attacks [11,41] by applying an appropriate $\texttt{mapin}()$ and/or

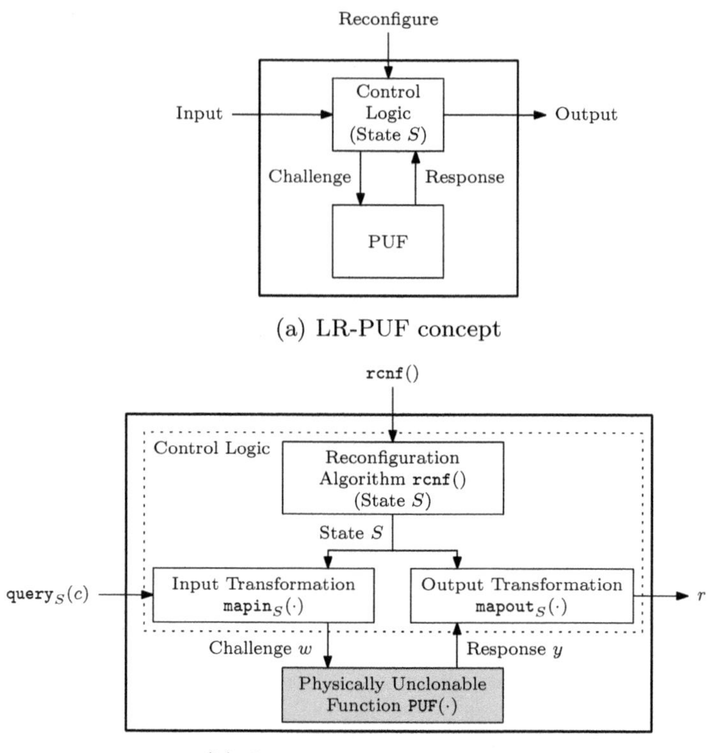

(a) LR-PUF concept

(b) Generic LR-PUF construction

Fig. 1. Logically Reconfigurable PUFs: Concept and generic construction

mapout() function. In contrast, LR-PUFs aim to *enable reconfigurability* for conventional non-reconfigurable PUFs after they have been deployed by entangling an updatable state with the challenges and/or responses of the underlying PUF.

3.2 Assumptions and Adversary Model

We assume that the underlying PUF is physically unclonable and unpredictable (see Section 2). The algorithms mapin(), mapout(), and rcnf() are publicly known. Moreover, the adversary $\mathcal{A}$ is assumed to *know* the current and all previous states S of the LR-PUF, e.g., by performing hardware attacks like side-channel or invasive attacks. However, we assume that $\mathcal{A}$ cannot force the control logic to set the LR-PUF state to a specific value, i.e., $\mathcal{A}$ *cannot change* the state S of the LR-PUF to a value of its choice (e.g., an old LR-PUF state).

For this, it must be assured that (1) rcnf() cannot be manipulated such that it generates predictable states, and that (2) the non-volatile memory cells storing the LR-PUF state cannot be set to specific values (e.g., by hardware attacks). The first requirement can be achieved by implementing the reconfiguration function using a fault injection aware design at a reasonable performance penalty [31,1]. Moreover, although fault injection attacks against non-volatile memory (e.g., EEPROM or Flash) have been shown [44], it seems to be difficult in practice to perform invasive attacks that change the content of specific non-volatile memory cells without affecting the content of the surrounding cells [45]. Hence, in practice it should be infeasible for an adversary to write a specific value (e.g., an old LR-PUF state) into the non-volatile memory of the LR-PUF. In particular, due to the increasing complexity of modern embedded systems and the fact that technology nodes are progressively getting smaller, the amount of precision and the quality of the equipment required to successfully perform such attacks renders them uneconomical in most practical applications (e.g., electronic ticketing).

3.3 Security Objectives

As pointed out in Section 2, physical unclonability and unpredictability are fundamental security requirements for PUF-based applications. Ideally, an LR-PUF should resemble a physically reconfigurable PUF. This implies that it should be infeasible for an adversary $\mathcal{A}$ to predict the response to a challenge of an LR-PUF for some state, even if $\mathcal{A}$ knows the responses to this challenge of the *same* LR-PUF but for *other* (e.g., old) states. Here, we must distinguish between the case where $\mathcal{A}$ aims to predict the responses of the LR-PUF for the current state (e.g., to forge a PUF response in an authentication protocol) or for a previous LR-PUF state (e.g., to recover an old key bound to the previous LR-PUF state). Moreover, in most applications of reconfigurable PUFs, it must be infeasible to set the state of the LR-PUF to a specific value, which would allow resetting the LR-PUF to a previous state and may help the adversary to predict LR-PUF responses. We first informally summarize the security requirements of LR-PUFs below and later give formal definitions and proofs for two different LR-PUF constructions in Section 6.

- *Backward unpredictability:* The adversary $\mathcal{A}$ cannot predict the response of the LR-PUF for a *previous* state S (i.e., before reconfiguration) to a challenge that has not been queried for the *previous* state, even if $\mathcal{A}$ knows an adaptively chosen set of challenge/response pairs of the LR-PUF for the previous state and can adaptively obtain challenge/response pairs of the LR-PUF for the current state.
- *Forward unpredictability:* The adversary $\mathcal{A}$ cannot predict the response of an LR-PUF for the *current* state S to a challenge that has not yet been queried for the *current* state, even if $\mathcal{A}$ knows an adaptively chosen set of challenge/response pairs of the LR-PUF for the previous state and can adaptively obtain challenge/response pairs of the LR-PUF for the current state (except for the challenge in question).
- *Non-resettability:* The adversary cannot set the state of the LR-PUF to a specific value.

Alg. 1 Speed-optimized LR-PUF

```
query_S(c)
  w ← Hash(S||c)          // mapin_S(c)
  y ← PUF(w)
  r ← y                   // mapout_S(y)
  Return r
rcnf()
  S ← Hash(S)
```

Alg. 2 Area-optimized LR-PUF

```
query_S(c)
  for j = 0 to n do                // mapin_S(c)
    w_j ← Hash(S||c||j)            // mapin_S(c)
    y_j ← PUF(w_j)
  endfor
  r ← (y_0|| ... ||y_n)            // mapout_S(y)
  Return r
rcnf()
  S ← Hash(S)
```

4 Constructions

In this section, we present two instantiations of our generic LR-PUF construction described in Section 3. The first construction is optimized for the fast generation of responses, while the second construction aims for the area constraints of low-cost devices (e.g., RFID chips) and provides a tradeoff between response generation time and the amount of area required.

4.1 Speed-Optimized LR-PUF Construction

Our first construction uses a PUF with a large challenge and a large response space and implements the control logic based on a single collision-resistant hash function. The challenge space must be large since otherwise it may be possible to create a complete challenge/response pair (CRP) database, which allows simulating the PUF. A large response space is a fundamental security requirement in many applications such as PUF-based identification/authentication [52,26] and hardware-entangled cryptography [3], where it is crucial that the PUF response to a formerly unknown challenge can be guessed with negligible probability only.

Our first construction is specified in Algorithm 1 and works as follows: When challenged with $\mathtt{query}_S(c)$, the control logic computes $w \leftarrow \mathsf{Hash}(S\|c)$ and returns $y \leftarrow \mathtt{PUF}(w)$, i.e., $\mathtt{mapin}_S(c) := \mathsf{Hash}(S\|c)$ and $\mathtt{mapout}_S(y) := y$. To reconfigure the LR-PUF, $\mathtt{rcnf}()$ sets the LR-PUF state to $S \leftarrow \mathsf{Hash}(S)$.

Since most PUF constructions that support a large challenge space (e.g., arbiter PUFs [23,37,27]) typically have only a small response space, several of these PUFs can be evaluated in parallel on the same challenge, which, however, significantly increases the amount of area required for their implementation. The collision-resistance property of the hash function assures the unpredictability property of the LR-PUF (see Section 2), as we will show later in the formal security analysis. Note that the LR-PUF state is just used to parameterize the hash function and thus needs not to be secret. Hence, to reconfigure the LR-PUF it is sufficient to hash the previous LR-PUF state to obtain a new and independent state (assuming the hash function implements a random oracle).

4.2 Area-Optimized LR-PUF Construction.

Our first LR-PUF construction described in Section 4.1 typically requires multiple parallel PUFs. Hence, we propose a second construction using just one single PUF that is evaluated sequentially n times to generate an n bit LR-PUF response, providing a tradeoff between area consumption and response generation speed. The intuition of this second construction is very similar as for the speed-optimized construction described in Section 4.1. Note that the underlying PUF must be queried with different challenges to generate a large response consisting of different (ideally) independent bits. This can be achieved by including a counter j as additional input to the hash function that now generates a sequence of PUF challenges w_j from the LR-PUF challenge c and the current LR-PUF state S. The corresponding PUF responses y_j are then concatenated to form the response r of the LR-PUF.

Our second construction is specified in Algorithm 2 and works as follows: On $\texttt{query}_S(c)$, the control logic computes $\texttt{mapin}_S(c)$ as $w_j \leftarrow \mathsf{Hash}(S\|w\|j)$ for $j \in \{0, \dots, n\}$, evaluates $y_j \leftarrow \texttt{PUF}(w_j)$, and $\texttt{mapout}_S()$ returns $r \leftarrow (y_0\| \dots \|y_n)$. To reconfigure the LR-PUF, $\texttt{rcnf}()$ sets the LR-PUF state to $S \leftarrow \mathsf{Hash}(S)$.

5 Implementation and Performance Evaluation

Both constructions presented in Section 4 are based on PUFs with a large challenge space. The only existing electronic PUFs that provide this feature seem to be arbiter PUFs [23,13]. The hash function of the control logic can be implemented efficiently by using a lightweight block cipher.

We implemented a prototype of both of our LR-PUF constructions on a Xilinx Spartan-6 FPGA board. We instantiated the underlying PUF based on arbiter PUFs that support 64 bit challenges and generate 1 bit responses, following the approach in [46]. The hash function of the control logic is based on the PRESENT block cipher [5] in Davies-Meyer mode [21]. Both resulting LR-PUF implementations use 80 bit challenges and generate 64 bit responses.

Table 1. Performance results of the LR-PUF constructions presented in Section 4

Optimization	Response time in clock cycles	Area consumption in slices (gate equivalents)		
		Control logic	Arbiter PUF	Total
Speed	1069	166 (1162 GE)	4288 (29056 GE)	4454 (30218 GE)
Area	64165	358 (2506 GE)	67 (454 GE)	425 (2960 GE)

We evaluated our implementation with regard to response generation speed and area consumption. Our results are summarized in Table 1. The second column shows the time in number of clock cycles required to compute an LR-PUF response r. The remaining columns show the number of slices and gate equivalents (GE) required to implement the control logic, the PUF, and the overall construction. The area estimation does not include the non-volatile memory for storing the LR-PUF state, which cannot be implemented on FPGA. Our results show that the area-optimized construction requires only about 10% of the area of the speed-optimized construction but is 60 times slower.

Note that our implementation is meant to demonstrate the feasibility of our approach and to obtain performance results. Due to the technical constraints of FPGAs, our implementation does not cover the non-volatile memory for storing the LR-PUF state, which is emulated by providing the state as an input to the FPGA. Moreover, our implementation is based on arbiter PUFs, which do not have the unpredictability property [41] that is required for the security of our constructions. To securely implement our LR-PUF constructions, the underlying PUF must be unpredictable (e.g., a Controlled PUF [11] can be used) and the non-volatile memory and control logic should be protected against fault-injection attacks, e.g., by applying the techniques described in [31,1].

6 Security Definitions and Evaluation

In this section we formally define the LR-PUF security properties of *forward-* and *backward-unpredictability* and show that both are fulfilled by the constructions proposed in Section 4. To this end, we first formalize the security property of unpredictability of a standard PUF.

Unpredictability of a PUF. Along the lines of [2] we define unpredictability of a PUF in terms of an *unpredictability game* between an adversary $\mathcal{A}$ and a challenger $\mathcal{C}$. $\mathcal{A}$ is first given a PUF and is allowed to query it at most q times. This step allows to model adversaries that are able to "learn" challenge/response pairs (CRPs) either by direct physical access to the interface of the PUF or by eavesdropping on messages containing PUF challenges and responses. At the end of the game, $\mathcal{A}$ is required to output a (non-trivial) valid pair of a PUF challenge and response.

Unpredictability Game of a PUF

Setup: The challenger $\mathcal{C}$ issues the $\texttt{PUF}$ to the adversary $\mathcal{A}$.

Queries: Proceeding adaptively, $\mathcal{A}$ queries the $\texttt{PUF}$ at most q times on challenges w_i (for $1 \leq i \leq q$). For each query, $y_i \leftarrow \texttt{PUF}(w_i)$ is given to $\mathcal{A}$.

Output: Eventually, $\mathcal{A}$ outputs a challenge/response pair (w^*, y^*).

Let Q denote the set of all challenges issued by $\mathcal{A}$. We say that $\mathcal{A}$ wins the game, if y^* is a valid PUF response to $\texttt{PUF}(w^*)$ and $w^* \notin Q$. Conversely, a PUF is unpredictable, if no efficient adversary $\mathcal{A}$ is able to win the game with significant success probability:

Definition 1. *A PUF is (q, ε)-unpredictable, if no probabilistic polynomial adversary $\mathcal{A}$ that makes at most q queries to the LR-PUF is able to win the unpredictability game with a probability greater than ε.*

Backward- and Forward-Unpredictability of an LR-PUF. We define backward- and forward-unpredictability in terms of a two-stage game between an adversary $\mathcal{A}$ and a challenger $\mathcal{C}$. In the first stage, $\mathcal{A}$ is given oracle access (i.e., access to the interface) of the LR-PUF, from which $\mathcal{A}$ can obtain challenge/response pairs (CRPs) at will. This stage models the ability of $\mathcal{A}$ to obtain challenges and responses (with respect to a fixed internal LR-PUF state) by passive eavesdropping. We also give $\mathcal{A}$ access to the internal LR-PUF state S in order to model hardware attacks against the LR-PUF implementation. Once $\mathcal{A}$ has learned enough CRPs, the challenger performs the reconfiguration operation and finally gives $\mathcal{A}$ oracle access to the reconfigured LR-PUF such that $\mathcal{A}$ can obtain CRPs of the reconfigured LR-PUF. At the end of the game, $\mathcal{A}$ outputs a prediction (c^*, r^*) of an LR-PUF challenge/response pair.

More formally, $\mathcal{A} = (\mathcal{A}_L, \mathcal{A}_C)$ consists of two probabilistic polynomial time algorithms, where $\mathcal{A}_L$ interacts with the LR-PUF before reconfiguration and $\mathcal{A}_C$ thereafter. $\mathcal{A}$ engages in the following experiment:

Backward- and Forward-Unpredictability Game of an LR-PUF

Setup: The challenger $\mathcal{C}$ sets up an LR-PUF by choosing a random state S, which is given to the adversary $\mathcal{A} = (\mathcal{A}_L, \mathcal{A}_C)$.

Phase I: $\mathcal{A}_L$ is allowed to call $\texttt{query}_S()$ of the LR-PUF up to q_L times. At the end of phase I, $\mathcal{A}_L$ stops and outputs a log file *st* that is used as input to $\mathcal{A}_C$. We denote with Q_L the set of challenges issued by $\mathcal{A}_L$ during phase I.

Reconfiguration: $\mathcal{C}$ reconfigures the LR-PUF by calling $\texttt{rcnf}()$, which updates the internal LR-PUF state to S'.

Phase II: $\mathcal{A}_C$ is initialized with log file *st* from $\mathcal{A}_L$ and the LR-PUF state S'. $\mathcal{A}_C$ is allowed to query the reconfigured LR-PUF $\texttt{query}_{S'}()$ up to q_C times on arbitrary challenges. We denote with Q_C the set of challenges issued by $\mathcal{A}_C$ during phase II.

Output: $\mathcal{A}_C$ outputs a challenge/response pair (c^*, r^*) of the LR-PUF.

Depending on whether we consider backward- or forward-unpredictability, we can state different conditions of an adversary being successful: $\mathcal{A}$ wins the *backward-unpredictability* game if r^* is a valid LR-PUF response to $\texttt{query}_{S'}(c^*)$ and $c^* \notin Q_C$. Thus, once the LR-PUF has been reconfigured, the adversary cannot output a (non-trivial) challenge/response pair for the *reconfigured* LR-PUF. Conversely, $\mathcal{A}$ wins the *forward-unpredictability* game if r^* is a valid LR-PUF response to $\texttt{query}_S(c^*)$ and $c^* \notin Q_L$. Thus, an adversary, who has access to a reconfigured LR-PUF cannot predict (non-trivial) responses of the LR-PUF *before* reconfiguration happened. We say that an LR-PUF is backward- (resp. forward-) unpredictable, if no efficient adversary $\mathcal{A}$ is able to win the game with significant success probability:

Definition 2 (Backward- and Forward-Unpredictability). *An LR-PUF is (q_L, q_C, ε)-backward unpredictable (resp. forward-unpredictable), if no probabilistic polynomial adversary $\mathcal{A}$ that makes at most q_L queries in phase I and at most q_C queries in phase II, is able to win the backward-unpredictability (resp. forward-unpredictability) game with a probability greater than ε.*

Both constructions of Section 4 achieve backward- and forward- unpredictability:

Proposition 1. *The speed-optimized LR-PUF construction shown in Section 4.1 is (q_L, q_C, ε)-backward unpredictable (resp. forward-unpredictable), if* Hash() *is collision-resistant and the underlying PUF is $(q_L + q_C, \varepsilon)$-unpredictable.*

Proposition 2. *The area-optimized LR-PUF construction shown in Section 4.2 is (q_L, q_C, ε)-backward unpredictable (resp. forward-unpredictable), if* Hash() *is collision-resistant and the underlying PUF is $(n(q_L + q_C), \varepsilon)$-unpredictable.*

The proofs of both propositions follow from the standard reductionist approach and can be found in the full version of this paper [18]. In particular, we show that any adversary $\mathcal{A}$ against the LR-PUF can be converted into an adversary $\mathcal{B}$ that either breaks the collision resistance of the hash function or the unpredictability of the underlying physical PUF. To this end, $\mathcal{B}$ simulates $\mathcal{A}$: Whenever $\mathcal{A}$ makes an LR-PUF query, $\mathcal{B}$ simulates this query by help of his PUF oracle, i.e., $\mathcal{B}$ transforms the challenges received from $\mathcal{A}$ by using the (known) internal LR-PUF state, queries the physical PUF on the transformed challenge and returns the obtained response to $\mathcal{A}$. Once the simulation stops, it can easily be seen that either a hash collision or a valid prediction of a challenge/response pair of the physical PUF can be extracted from $\mathcal{A}$'s output.

7 Applications

7.1 LR-PUF-Based Authentication Tokens

Electronic payment and ticketing systems have been gradually introduced in many countries over the past few years (see, e.g., [35,7,33]). Typically, these systems are using RFID-enabled tokens and provide different types of electronic transit tickets. Given the typically large number of tickets used in an electronic transit ticket system and the costs per token (typically between 1-3 Euro), from an economic perspective it may be worthwhile to consider recycling of RFID-based tickets. In fact, some systems (e.g., the Dutch transportation system [36]) allow recharging RFID-based tickets with money and to returning used tickets to the vendor with possible restitution of preloaded money left on the ticket. Moreover, many U.S. and European governments make manufacturers and importers of electronic products responsible for the disposal of their products when discarded by the consumer (see, e.g., [6,9]). In this context, recyclable tokens can help to save waste disposal costs and to reduce the amount of electronic waste. In this section, we discuss how LR-PUFs could be used to enhance the security of electronic ticketing and payment systems while at the same time enabling secure and privacy-preserving recycling of used RFID-tickets.

There are several proprietary solutions for electronic tickets in practice. Most of them are based on widely used RFID tokens, where the most prominent example is the MiFare family produced by NXP Semiconductors [34]. There are several hard- and software attacks against MiFare Classic tokens [32,43,10], which use a proprietary encryption algorithm that has been completely broken [8]. However, other MiFare products are claimed not to be affected. A recent attack on MiFare Classic 4K chipcards concerns the Dutch electronic payment and transit ticket system [36]: Using a MiFare compatible card reader and a software from the Internet, an average user can add debit to his RFID-based transit ticket without being detected [54,25].

In this context, PUFs could provide a cost-effective security mechanism: Authentication based on PUFs can prevent copying and manipulating the information (i.e., the debit of the RFID-based ticket and/or the user's rights) by cryptographically binding this data to the physical characteristics of the underlying RFID chip. Existing PUF-based authentication schemes (see, e.g., [40,15,37,42,51]) typically assume each device, i.e., each RFID-based token $\mathcal{T}$, to be equipped with a PUF, whereas the verifier $\mathcal{V}$ maintains a database $\mathcal{D}$, i.e., a set if challenge/response pairs (CRPs) of each ticket. In the authentication protocol, $\mathcal{V}$ chooses a random challenge from $\mathcal{D}$ and sends it to $\mathcal{T}$, which then returns some response. $\mathcal{V}$ accepts if the response of $\mathcal{T}$ matches the one in $\mathcal{D}$.

Using LR-PUFs instead of non-reconfigurable PUFs would allow for cost-effective, secure and privacy-preserving recycling of RFID-based tickets: By reconfiguring the LR-PUF all information and access rights bound to $\mathcal{T}$ are securely "erased", which cannot be achieved with non-reconfigurable PUFs. However, reconfiguring the LR-PUF invalidates the CRP database $\mathcal{D}$ of $\mathcal{V}$, which means that after each reconfiguration of $\mathcal{T}$ a new CRP database must be established. To counter this problem, $\mathcal{V}$ could know the LR-PUF state S of each token and maintain a CRP database $\mathcal{D}'$ of the PUF underlying the LR-PUF, which can be seen as the "authentication secrets" of the token. This is common in ticketing applications because usually the verifier is the ticket issuer who typically knows the authentication secrets of all tokens. Since the algorithms of the control unit, i.e., the input and output transition functions `mapin`() and `mapout`(), respectively, and the state update algorithm `rcnf`(), are publicly known, $\mathcal{V}$ could use $\mathcal{D}'$ to recompute the LR-PUF response for any state of $\mathcal{T}$ and compare it to the response sent by $\mathcal{T}$. $\mathcal{V}$ accepts if the response of $\mathcal{T}$ matches the one recomputed based on $\mathcal{D}'$ and S.

7.2 Other Applications Envisaged

Many airlines have started to move from paper-based tickets to electronic tickets. However, they still print luggage tags, which are increasingly equipped with disposable RFID chips. The purpose of these chips is to ease the tracking of individual luggage in the process of loading. However, RFID-enabled labels could be read out even without visual contact. This may allow several attacks ranging from copying luggage tags to smuggle in additional luggage in the name of another passenger. Moreover, RFID-enabled luggage tags may disclose personal information on their owner (e.g., name, number of luggage pieces, luggage weight), which could be used to track the user on the airport or provide useful information to luggage thieves. To solve these problems, travellers could purchase or rent a more powerful LR-PUF-enabled RFID token that is put into the luggage or that could even be embedded into new generations of suitcases. Each time the traveller checks in, his RFID-based tag is reconfigured by the airline attendant, which securely erases the previous information stored on it. This prevents tracking the traveler for more than one flight and impedes misrouting of luggage due to old travel information. Further, to avoid illegitimate tracking of travellers, the RFID-enabled luggage tag could be reconfigured or temporarily disabled once the passenger leaves the baggage claim area.

One can find many other applications that could take advantage of LR-PUFs. Examples include, secure deletion and/or update of cryptographic secrets in PUF-based key storage [52,26] and hardware-entangled cryptography [3], where the reconfiguration of the PUF ensures that old secrets cannot be retrieved any more. Another example are solutions to prevent downgrading of software [20] by binding the software to the PUF, where reconfiguring the PUF invalidates the old software version such that only the latest version can be used.

8 Conclusion

In this paper, we have proposed the concept of logically reconfigurable PUFs, which utilize a control logic to enable dynamic reconfigurability for existing, typically static PUFs. We introduced two different constructions to realize LR-PUFs: Our first construction is optimized for response generation speed, while our second construction aims for resource-constrained embedded devices (like RFID tags). Furthermore, we have shown that both constructions achieve the security properties of backward- and forward unpredictability, which are two desirable properties in the context of PUF-based cryptographic applications like key storage, device identification, and hardware-entangled cryptography. Finally, we showed how LR-PUFs could be applied in the context of recyclable (access) tokens to enhance the security properties of existing solutions while providing a means for secure recycling of PUF-based access tokens.

Acknowledgements. We thank our anonymous reviewers for their helpful comments, Patrick Koeberl and Jérôme Quevremont for several useful discussions on hardware attacks and use cases, and Timm Korte for providing us his implementation of PRESENT. This work has been supported in part by the European Commission under grant agreement ICT-2007-238811 UNIQUE.

References

1. Akdemir, K.D., Wang, Z., Karpovsky, M.G., Sunar, B.: Design of cryptographic devices resilient to fault injection attacks using nonlinear robust codes. In: Fault Analysis in Cryptography (2011)
2. Armknecht, F., Maes, R., Sadeghi, A.R., Standaert, F.X., Wachsmann, C.: A formal foundation for the security features of physical functions. In: IEEE Symposium on Security and Privacy, pp. 397–412. IEEE Computer Society, Los Alamitos (2011)

3. Armknecht, F., Maes, R., Sadeghi, A.R., Sunar, B., Tuyls, P.: Memory leakage-resilient encryption based on physically unclonable functions. In: Matsui, M. (ed.) ASIACRYPT 2009. LNCS, vol. 5912, pp. 685–702. Springer, Heidelberg (2009)
4. Armknecht, F., Sadeghi, A.R., Visconti, I., Wachsmann, C.: On RFID privacy with mutual authentication and tag corruption. In: Zhou, J., Yung, M. (eds.) ACNS 2010. LNCS, vol. 6123, pp. 493–510. Springer, Heidelberg (2010)
5. Bogdanov, A., Knudsen, L., Leander, G., Paar, C., Poschmann, A., Robshaw, M., Seurin, Y., Vikkelsoe, C.: PRESENT: An ultra-lightweight block cipher. In: Paillier, P., Verbauwhede, I. (eds.) CHES 2007. LNCS, vol. 4727, pp. 450–466. Springer, Heidelberg (2007)
6. Californians Against Waste: E-waste laws in other states (April 2011), `http://www.cawrecycles.org/issues/ca_e-waste/other_states`
7. Calypso Networks Association: Website (April 2011), `http://www.calypsonet-asso.org/`
8. Courtois, N.T., Nohl, K., O'Neil, S.: Algebraic attacks on the Crypto-1 stream cipher in MiFare Classic and Oyster Cards. Cryptology ePrint Archive, Report 2008/166 (2008)
9. European Commission: Waste electrical and electronic equipment website (April 2011), `http://ec.europa.eu/environment/waste/weee/index_en.htm`
10. Garcia, F.D., de Koning Gans, G., Muijrers, R., van Rossum, P., Verdult, R., Schreur, R.W., Jacobs, B.: Dismantling MIFARE classic. In: Jajodia, S., Lopez, J. (eds.) ESORICS 2008. LNCS, vol. 5283, pp. 97–114. Springer, Heidelberg (2008)
11. Gassend, B., Clarke, D., van Dijk, M., Devadas, S.: Controlled physical random functions. In: Computer Security Applications Conference, pp. 149–160. IEEE Computer Society, Los Alamitos (2002)
12. Gassend, B., Clarke, D., van Dijk, M., Devadas, S.: Silicon physical random functions. In: ACM Conference on Computer and Communications Security (ACM CCS), pp. 148–160 (2002)
13. Gassend, B., Lim, D., Clarke, D., van Dijk, M., Devadas, S.: Identification and authentication of integrated circuits. Concurrency and Computation: Practice and Experience 16(11), 1077–1098 (2004)
14. Guajardo, J., Kumar, S.S., Schrijen, G.J., Tuyls, P.: FPGA intrinsic PUFs and their use for IP protection. In: Paillier, P., Verbauwhede, I. (eds.) CHES 2007. LNCS, vol. 4727, pp. 63–80. Springer, Heidelberg (2007)
15. Holcomb, D.E., Burleson, W.P., Fu, K.: Initial SRAM state as a fingerprint and source of true random numbers for RFID tags. In: Conference on RFID Security (RFIDSec) (2007)
16. Intrinsic ID: Product webpage (April 2011), `http://www.intrinsic-id.com/products.htm`
17. Juels, A.: RFID security and privacy: A research survey. Journal of Selected Areas in Communication 24(2), 381–395 (2006)
18. Katzenbeisser, S., Ünal Kocabas, van der Leest, V., Sadeghi, A.R., Schrijen, G.J., Schröder, H., Wachsmann, C.: Recyclable PUFs: Logically reconfigurable PUFs (full version) (June 2011), `http://www.trust.cased.de/`
19. Kumar, S., Guajardo, J., Maes, R., Schrijen, G.J., Tuyls, P.: Extended abstract: The butterfly PUF protecting IP on every FPGA. In: IEEE Workshop on Hardware-Oriented Security and Trust (HOST), pp. 67–70 (2008)
20. Kursawe, K., Sadeghi, A.R., Schellekens, D., Tuyls, P., Scoric, B.: Reconfigurable physical unclonable functions — Enabling technology for tamper-resistant storage. In: IEEE International Workshop on Hardware-Oriented Security and Trust (HOST), pp. 22–29. IEEE Computer Society, San Francisco (2009)

21. Lai, X., Massey, J.: Hash functions based on block ciphers. In: Rueppel, R.A. (ed.) EUROCRYPT 1992. LNCS, vol. 658, pp. 55–70. Springer, Heidelberg (1993)
22. Lao, Y., Parhi, K.K.: Novel reconfigurable silicon unclonable functions. In: Workshop on Foundations of Dependable and Secure Cyber-Physical Systems (FDSCPS) (April 11, 2011)
23. Lee, J.W., Lim, D., Gassend, B., Suh, G.E., van Dijk, M., Devadas, S.: A technique to build a secret key in integrated circuits for identification and authentication application. In: Symposium on VLSI Circuits, pp. 176–179 (2004)
24. van der Leest, V., Schrijen, G.J., Handschuh, H., Tuyls, P.: Hardware intrinsic security from D flip-flops. In: ACM Workshop on Scalable Trusted Computing (ACM STC), pp. 53–62 (2010)
25. Letter from Dutch minister on OV-chipkaart, https://zoek.officielebekendmakingen.nl/dossier/32440/kst-23645-415.html
26. Lim, D., Lee, J.W., Gassend, B., Suh, G.E., van Dijk, M., Devadas, S.: Extracting secret keys from integrated circuits. IEEE Transactions on VLSI Systems 13(10), 1200–1205 (2005)
27. Lin, L., Holcomb, D., Krishnappa, D.K., Shabadi, P., Burleson, W.: Low-power sub-threshold design of secure physical unclonable functions. In: ACM/IEEE International Symposium on Low Power Electronics and Design (ISLPED), pp. 43–48 (2010)
28. Maes, R., Tuyls, P., Verbauwhede, I.: Intrinsic PUFs from flip-flops on reconfigurable devices. In: Workshop on Information and System Security (WISSec), p. 17 (2008)
29. Maes, R., Verbauwhede, I.: Physically unclonable functions: A study on the state of the art and future research directions. In: Sadeghi, A.R., Naccache, D. (eds.) Towards Hardware-Intrinsic Security. Information Security and Cryptography, pp. 3–37. Springer, Heidelberg (2010)
30. Maiti, A., Casarona, J., McHale, L., Schaumont, P.: A large scale characterization of RO-PUF. In: IEEE Symposium on Hardware-Oriented Security and Trust (HOST), pp. 94–99 (2010)
31. Monnet, Y., Renaudin, M., Leveugle, R.: Designing resistant circuits against malicious faults injection using asynchronous logic. IEEE Trans. Comput. 55, 1104–1115 (2006), http://dx.doi.org/10.1109/TC.2006.143
32. Nohl, K., Plötz, H.: MiFare — Little security despite obscurity (2007), http://events.ccc.de/congress/2007/Fahrplan/events/2378.en.html
33. NXP Semiconductors: MiFare applications (April 2008), http://www.mifare.net/applications/
34. NXP Semiconductors: MiFare smartcard ICs (February 2011), http://www.mifare.net/products/smartcardics/
35. Octopus Holdings: Website (April 2011), http://www.octopus.com.hk/en/
36. OV-Chipkaart: Website (April 2011), http://www.ov-chipkaart.nl/
37. Öztürk, E., Hammouri, G., Sunar, B.: Towards robust low cost authentication for pervasive devices. In: IEEE International Conference on Pervasive Computing and Communications (PERCOM 2008). IEEE Computer Society, Los Alamitos (2008)
38. Pappu, R.S.: Physical one-way functions. Ph.D. thesis, Massachusetts Institute of Technology (March 2001)
39. Pappu, R.S., Recht, B., Taylor, J., Gershenfeld, N.: Physical one-way functions. Science 297, 2026–2030 (2002)

40. Ranasinghe, D.C., Engels, D.W., Cole, P.H.: Security and privacy: Modest proposals for low-cost RFID systems. In: Auto-ID Labs Research Workshop (September 2004)
41. Rührmair, U., Sehnke, F., Sölter, J., Dror, G., Devadas, S., Schmidhuber, J.: Modeling attacks on physical unclonable functions. In: ACM conference on Computer and communications security (ACM CCS), pp. 237–249 (2010)
42. Sadeghi, A.R., Visconti, I., Wachsmann, C.: PUF-enhanced RFID security and privacy. In: Workshop on Secure Component and System Identification, SECSI (2010)
43. Schreur, R.W., van Rossum, P., Garcia, F., Teepe, W., Hoepman, J.H., Jacobs, B., de Koning Gans, G., Verdult, R., Muijrers, R., Kali, R., Kali, V.: Security flaw in MiFare Classic (March 2008), `http://www.sos.cs.ru.nl/applications/rfid/pressrelease.en.html`
44. Skorobogatov, S.: Semi-invasive attacks — A new approach to hardware security analysis. Technical Report UCAM-CL-TR-630, University of Cambridge, 15 JJ Thomson Avenue, Cambridge CB03 0FD, UK (April 2005)
45. Skorobogatov, S.: Local heating attacks on Flash memory devices. In: IEEE International Workshop on Hardware-Oriented Security and Trust (HOST 2009), pp. 1–6. IEEE, Los Alamitos (July 27, 2009)
46. Soybali, M., Ors, B., Saldamli, G.: Implementation of a PUF circuit on an FPGA. In: IFIP International Conference on New Technologies Mobility and Security (2011)
47. Su, Y., Holleman, J., Otis, B.: A 1.6pJ/bit 96% stable chip-ID generating circuit using process variations. In: IEEE International Solid-State Circuits Conference (ISSCC), pp. 406–611 (2007)
48. Suh, G.E., Devadas, S.: Physical unclonable functions for device authentication and secret key generation. In: Design Automation Conference, pp. 9–14 (2007)
49. Tuyls, P., Batina, L.: RFID-tags for anti-counterfeiting. In: Pointcheval, D. (ed.) CT-RSA 2006. LNCS, vol. 3860, pp. 115–131. Springer, Heidelberg (2006)
50. Tuyls, P., Schrijen, G.-J., Škorić, B., van Geloven, J., Verhaegh, N., Wolters, R.: Read-proof hardware from protective coatings. In: Goubin, L., Matsui, M. (eds.) CHES 2006. LNCS, vol. 4249, pp. 369–383. Springer, Heidelberg (2006)
51. Verayo, Inc.: Product webpage (April 2011), `http://www.verayo.com/product/products.html`
52. Škorić, B., Tuyls, P., Ophey, W.: Robust key extraction from physical uncloneable functions. In: Ioannidis, J., Keromytis, A.D., Yung, M. (eds.) ACNS 2005. LNCS, vol. 3531, pp. 407–422. Springer, Heidelberg (2005)
53. Weis, S.A., Sarma, S.E., Rivest, R.L., Engels, D.W.: Security and privacy aspects of low-cost radio frequency identification systems. In: Hutter, D., Müller, G., Stephan, W., Ullmann, M. (eds.) Security in Pervasive Computing. LNCS, vol. 2802, pp. 50–59. Springer, Heidelberg (2004)
54. Wikipedia: OV-Chipkaart, `http://en.wikipedia.org/wiki/OV-chipkaart`

Uniqueness Enhancement of PUF Responses Based on the Locations of Random Outputting RS Latches

Dai Yamamoto[1], Kazuo Sakiyama[2], Mitsugu Iwamoto[2], Kazuo Ohta[2], Takao Ochiai[1], Masahiko Takenaka[1], and Kouichi Itoh[1]

[1] FUJITSU LABORATORIES LTD
4-1-1, Kamikodanaka, Nakahara-ku, Kawasaki-shi, Kanagawa 211-8588, Japan
{ydai,tochiai,takenaka,kito}@labs.fujitsu.com
[2] The University of Electro-Communications
1-5-1, Chofugaoka, Chofu, Tokyo 182-8585, Japan
{saki,mitsugu,ota}@inf.uec.ac.jp

Abstract. Physical Unclonable Functions (PUFs) are expected to represent an important solution for secure ID generation and authentication etc. In general, PUFs are considered to be more secure the larger their output entropy. However, the entropy of conventional PUFs is lower than the output bit length, because some output bits are random numbers, which are regarded as unnecessary for ID generation and discarded. We propose a novel PUF structure based on a Butterfly PUF with multiple RS latches, which generates larger entropy by utilizing location information of the RS latches generating random numbers. More specifically, while conventional PUFs generate binary values (0/1), the proposed PUF generates ternary values (0/1/random) in order to increase entropy. We estimate the entropy of the proposed PUF. According to our experiment with 40 FPGAs, a Butterfly PUF with 128 RS latches can improve entropy from 116 bits to 192.7 bits, this being maximized when the frequency of each ternary value is equal. We also show the appropriate RS latch structure for satisfying this condition, and validate it through an FPGA experiment.

Keywords: PUF, Butterfly PUF, RS latch, Metastable, Random number, FPGA, ID Generation, Authentication.

1 Introduction

Secure identification/authentication technology using Integrated Circuit (IC) chips is very important for secure information infrastructure. It is used for anti-counterfeiting devices on medical supplies, prepaid-cards and public ID cards such as passports and driver's licenses. The IC card is a well-known solution for this kind of application. Counterfeiting is prevented by storing a secret key on the IC card and using a secure cryptographic protocol to make the key invisible from outside. In theory, however, the possibility of counterfeiting still remains

B. Preneel and T. Takagi (Eds.): CHES 2011, LNCS 6917, pp. 390–406, 2011.

if its design is revealed and reproduced by the counterfeiter. Naturally, this is difficult because current IC cards are equipped with several highly-developed tamper-proofing technologies. However, further anti-counterfeiting technologies are desirable to meet future developments in reverse-engineering techniques.

Recently, interest has been focused on Physical Unclonable Functions (PUFs) as a solution [1]. In a PUF, the output value (response) to the input value (challenge) is unique for each individual IC. This uniqueness is provided by the process variations of each individual IC [2] [3]. It is expected that PUFs will represent breakthrough in technology for anti-counterfeiting devices, through its use for ID generation, key generation and authentication protocol, which make cloning impossible even when the design is revealed.

The PUFs on ICs are classified into two categories [4]. One uses the characteristics of memory cells such as SRAM-PUFs [5] [6] and Butterfly PUFs (BPUFs) [7]. SRAM-PUFs are based on the unstable power-up values of SRAM cells on ICs such as ASIC and FPGA. However, a device power-up operation is required for the generation of every response. To counter this drawback, BPUFs are composed of cross-coupled latches which behave similarly to an SRAM cell. The output of the BPUF is triggered by a clock edge signal applied to the latches, without an actual device power-up. The other uses the characteristics of delay variations such as Arbiter PUFs [8], Glitch PUFs [9] and Ring Oscillator (RO) PUFs [10]. Arbiter PUFs have an *arbiter* circuit that generates a response determined by the difference in the signal delay between two paths, which is mixed by a challenge. However, a machine learning attack can predict challenge-response pairs by using a large number of past pairs [11]. The Glitch PUF [9] was proposed to solve this problem of ease of prediction. It generates a response by utilizing glitch waveforms and delay variations between logic gates. Since its response to challenges behaves like a non-linear function, machine learning attacks are prevented. RO PUFs derive entropy from the difference in oscillator frequencies.

Today, PUFs in the former category are some of the most feasible and secure because there have already been implementations of error correcting codes (ECCs) and universal hash functions [12] for randomness extraction optimized for the PUFs, which are needed for Fuzzy Extractors [13]. In addition, BPUFs implemented in ASIC seem to have many advantages over SRAM-PUFs, such as not requiring a power-up operation. This paper therefore focuses on BPUFs, which generate n-bit responses based on n outputs from n RS latches.

The PUFs in both categories need to eliminate the randomness of responses in order to generate stable responses. For example, the Glitch PUF can generate very stable responses because it selects available challenges to output stable responses by a masking process. However, as pointed out by the designers of the Glitch PUF, the masking process causes entropy loss. In conventional PUFs, the outputs of random latches are not used to generate stable responses; however, in this paper we make efficient use of random latches.

The responses from PUFs need to have extremely high *uniqueness*. This paper defines uniqueness as the independence among multiple PUFs of responses to the same challenge. In order to prevent clones of cryptographic hardware, it is important for manufacturers to make sure that multiple PUFs with the same challenge-response pairs do not exist. However, this is very difficult in terms of cost because there are a huge number of manufactured PUFs and challenge-response pairs. Therefore, one of the most practical solutions is to increase the number and range of responses as much as possible. We must note that a large number of responses are not necessarily equivalent to a high level of entropy in those responses. PUFs that output responses with high entropy are capable of generating completely unpredictable responses. Consequently, the probability of multiple PUFs that output unpredictable responses having the same challenge-response pairs is extremely small. Hence, it is also important for PUFs to increase the entropy of responses so as to have extremely high uniqueness.

In addition, the response needs to have high *reliability*. This paper defines reliability as the consistency of PUF challenge-response pairs for repeated measurements. That is, ideally, a PUF always generates the same response to a given challenge. The BPUF has some RS latches that generate random numbers (i.e. "random latches"). This randomness causes a problem in that the reliability of the response is reduced. This is because the values of the response corresponding to the random latches change every time a response is generated. In the conventional approach - in order to maintain the reliability of responses - the outputs of the random latches are discarded, similar to the masking process in the Glitch PUF, which is a widely known technique for the generation of responses. However, the number of responses becomes lower as the number of random latches increases, which reduces the entropy and uniqueness of responses.

Our Contributions. This paper proposes a novel PUF structure for generating high-entropy responses using randomness. Note that our proposed methods can be applied to any PUFs. As an example, our paper focuses on a BPUF with random latches. The use of random latches dramatically increases entropy and uniqueness. Also, the construction can maintain the reliability of responses even if random latches are used for the generation of responses. In specific terms, responses are generated based on the location information of the random latches. The proposed PUF generates approximately 3^n responses with ternary value (0/1/random), which is maximized when the frequency of each ternary value is equal. Here, 3^n is not accurate, but is intuitively easy-to-understand, and so a rigorous discussion is given below. We also propose a suitable RS latch structure to satisfy this equality condition to the maximum extent. We evaluate the performance of the proposed PUF with 40 FPGAs. A BPUF with 128 RS latches based on our RS latch construction increases the average number of random latches from 12 to 32, approaching around 43 (=128/3). The proposed PUF with ternary values improves the number of responses from 2^{116} to 2^{196}. From the actual responses generated by 40 PUFs, the entropy of responses is evaluated as 192.7 bits, which indicates that the proposed PUF has extremely high uniqueness.

Organization of the Paper. The rest of the paper is organized as follows. Section 2 gives an outline of the BPUF with RS latches, and the conventional methods for implementing RS latches on FPGAs. Section 3 proposes our original BPUF, which generates responses by using the location information of the random latches. In addition, new methods of implementing RS latches are proposed that maximize the performance of our PUF. Section 4 evaluates the performance of our PUF on an FPGA platform. Finally, in Section 5, we give a summary and comment on future directions.

2 Conventional Methods

2.1 Conv. Mtd (1): Generation of Responses from a BPUF

This paper focus on a BPUF using RS latches. First, we describe the circuit and behavior of an RS latch, shown in Fig. 1. An RS latch can be created from two NAND gates, and is in a stable state with output $(B, C) = (1, 1)$ when input $A = 0$. When input A changes from 0 to 1 (= rising edge), the RS latch temporarily enters a metastable state. It then enters a stable state with either output $(B, C) = (1, 0)$ or $(B, C) = (0, 1)$. Ideally, the probability of transition to either of these states is equal. In fact, however, many RS latches have a high probability of entering one specific state. This is because the drive capabilities of the two NAND gates and the wire length between them are not exactly the same. Hence, the output B from RS latches fall into three patterns: all 0's, all 1's, or a mixture of 0's and 1's (= random number) when a clock signal is applied to input A.

We now describe the BPUF, shown in Fig. 2. Challenges to the BPUF are equivalent to choosing $m(\leq n)$ RS latches from n implemented RS latches. The BPUF can generate m-bit responses corresponding to ${}_n\mathrm{C}_m$ challenges. Here, ${}_n\mathrm{C}_m$ is defined as the number of combinations of n elements taken m at a time. The BPUF in Fig. 2 generates an n-bit response $RES[n-1:0]$ because m is set equal to n. Note that, in order to simplify discussion in this paper, the more significant bits of the response correspond to the outputs of RS latches with bigger latch labels. BPUFs, which generate only a response, can be used for applications such as authentication. For example, a random number S is sent from an authentication server to a PUF as a new challenge, and a response R from the PUF is newly defined by equation $R = \mathrm{H}(S \parallel RES)$. Here, H() indicates a mixing function, such as various hash functions. The value of response R changes depending on the challenge S, so BPUFs provide security when used for this application. The PUF in Fig. 2 has some RS latches that generate random numbers such as LATCH_2 and LATCH_{n-2}. These random numbers cause a problem in that the reliability of the response RES is reduced since its value changes every time it is generated.

There are two widely known conventional approaches to response generation aimed at solving this problem. In the first approach ("conventional method (1-A)"), random latches are not used for the generation of responses. This approach maintains the reliability of responses, but reduces their uniqueness, and requires

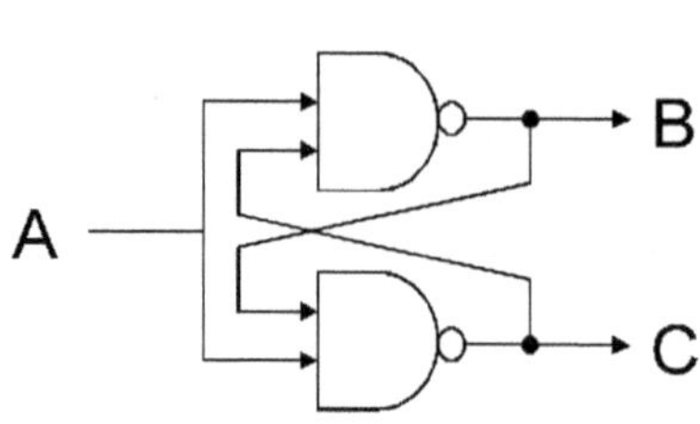

Fig. 1. NAND-based RS latch

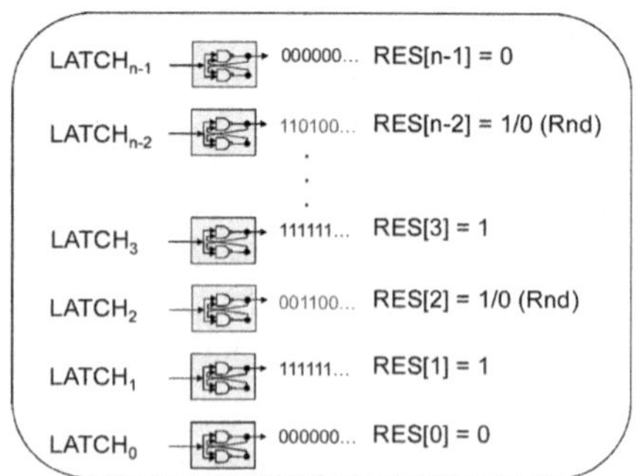

Fig. 2. Butterfly PUF

a mechanism to detect random latches. For example, the BPUF with 128 RS latches ($n = 128$) in Fig. 2 has 40 random latches. The bit-length of the responses is reduced from 128 bits to 88 bits, so their entropy and uniqueness are also reduced. Hence, it is necessary to implement extra RS latches in the PUF in accordance with the number of random latches. This PUF is, however, not suitable for embedded systems with limited hardware resources such as smart cards because, while also maintaining the uniqueness of responses, it is necessary for PUFs in embedded systems to have an RS latch area size and peripheral circuit that are as small as possible. In the second approach ("conventional method (1-B)"), ECCs are used to correct the variation in the responses resulting from the random latches. This approach requires larger redundant data for response correction as the number of random latches increases. In addition, it also suffers from the disadvantage of necessitating increased hardware resources and processing time for the ECCs. A BPUF with 128 RS latches generates no more than 2^{128} responses even if ECCs are used. From the above, it can be seen that the first approach, in which random latches are not used for responses, is not suitable. Furthermore, it is not sufficient to use only ECCs, as in the second approach. In Section 3, we propose a method for generating responses based on the locations of random latches. The proposed method maintains the reliability of responses, and dramatically improves their uniqueness.

2.2 Conv. Mtd (2): Implementation of RS Latches on FPGAs

A method for implementing RS latches as a true random number generator on Xilinx FPGAs ("conventional method (2-A)") is proposed in Ref. [14], [15]. Flip-Flops (FFs) are positioned in front of the two NAND gates, as shown in Fig. 3. This minimizes the difference in signal arrival time between the two gates, enabling the RS latch to enter the metastable state more readily and improving the probability of the RS latches outputting random numbers. A Xilinx FPGA consists of a matrix of configurable logic blocks (CLBs). Some kinds of Xilinx FPGA devices have four slices per CLB. A slice includes two pairs of LookUp Tables (LUTs) and FFs. The right and left slices of the CLB are different. The

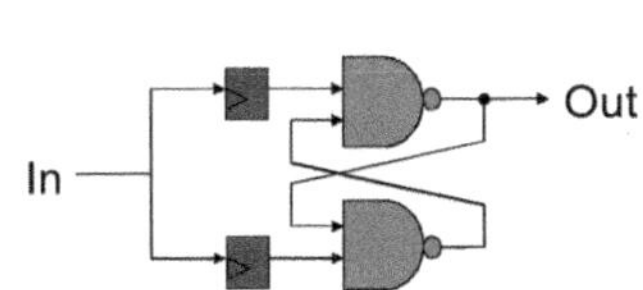

Fig. 3. Conv. mtd (2-A): RS latch circuit [14]

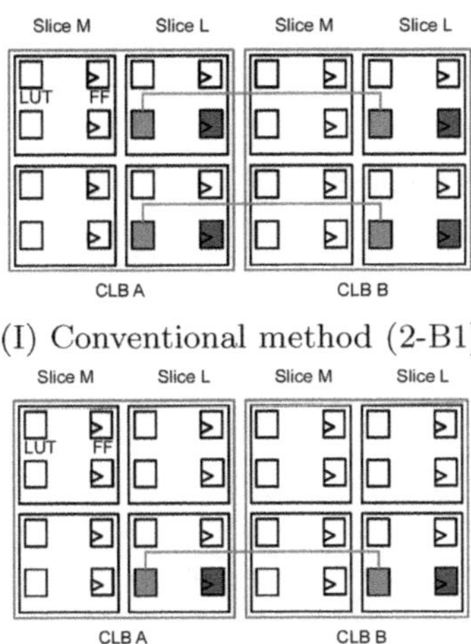

(I) Conventional method (2-B1)

(II) Conventional method (2-B2)

Fig. 4. Implementation of RS latches on Xilinx FPGAs [14]

right slice (SliceL) is available only for logic, while the left one (SliceM) is for both memory and logic. Two types of implementation for an RS latch are reported in Ref. [14]. In one type ("conventional method (2-B1)"), two RS latches are implemented on two CLBs, as shown in Fig. 4(I). In the other ("conventional method (2-B2)"), only one RS latch is implemented on two CLBs, as shown in Fig. 4(II). Both methods implement the NAND gates of an RS latch by using the same kind of slice (SliceL in Fig. 4) on different CLBs. The conventional method (2-B1) uses two CLBs per two RS latches, leading to reasonable circuit efficiency. However, Ref. [14] points out that multiple RS latches which have NAND gates implemented on the same CLB, as shown in Fig. 4(I), have a low probability of outputting random numbers. RS latches based on conventional method (2-B2) have some probability of generating random numbers, but result in low circuit efficiency because an RS latch requires two CLBs. The next section proposes an implementation method that gives the RS latches a high probability of outputting random numbers. In addition, the proposed method gives higher circuit efficiency than in the conventional methods.

3 Proposed Methods

3.1 Proposed Mtd (1): Use of the Locations of Random Latches

The conventional BPUF in Fig. 2 generates responses based only on RS latches outputting fixed numbers such as 0's or 1's (i.e. "fixed latches"). Our proposed BPUF uses the location information of random latch X, rather than the random numbers from the random lathes. If a BPUF with N RS latches has T random latches, then the number of locations of random latches equals to ${}_N\mathrm{C}_T$, which

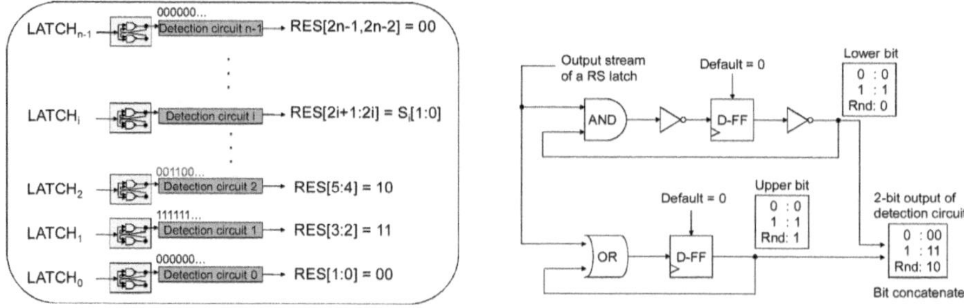

Fig. 5. Proposed method (1)

Fig. 6. Proposed detection circuit

generates the entropy due to random latch locations. Hence, the PUF based on our method utilizes the entropy for uniqueness of responses. However, this kind of BPUF requires complex controls to associate the location of RS latch X with the output number, which leads to a large circuit size. In this paper, we propose a simple and efficient method of solving this problem ("proposed method (1)"). Proposed method (1) regards the three types of output patterns from the RS latches (0's, 1's, and random numbers) as ternary values (00/11/10). Our method can generate responses with much larger patterns than conventional approaches. We describe the details of the proposed method with reference to Fig. 5. When a clock signal is applied to the inputs of the RS latches in our BPUF, they generate three types of outputs: 0's, 1's, and random numbers. The PUF based on our method has new detection circuits - shown in Fig. 6 - located after the RS latches which distinguish these three types. The detection circuit i outputs a 2-bit value (00/11/10) depending on the output of the RS latch i (0's/1's/random numbers). If the output stream of RS latch i includes a transition from 0(1) to 1(0), detection circuit i regards RS latch i as a random latch, and from that point onwards continues outputting the 2-bit value '10' regardless of RS latch i's subsequent output stream. Stated more rigorously, let $S_i[1:0]$ be the 2-bit output of detection circuit i located after RS latch i, and $RES[2n-1:0]$ be the $2n$-bit response of our BPUF. Then

$$RES[2n-1:0] = \sum_{i=0}^{n}\{S_i \cdot 2^{2i}\}. \tag{1}$$

The gate size of the detection circuit shown in Fig. 6 is estimated to be around 28 gates, which is definitely compact enough for embedded systems. Here, we use the equivalencies 1 FF = 12 NAND gate, 1 AND = 1.5 NAND gate, 1 OR = 1.5 NAND gate, and 1 INV = 0.5 NAND gate, introduced in [16]. Naturally, in order to distinguish three types of outputs, CPU-based software approach is able to be used instead of the detection circuit. The reason why we propose the detection circuit as hardware approach is that it is essential when our proposed PUF is implemented on ASIC.

Next, for the PUF based on our proposed method, we theoretically estimate the number of responses. Let N be the number of implemented RS latches, and T

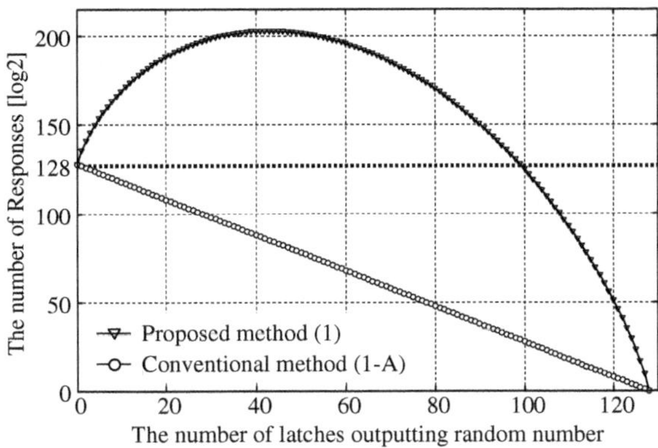

Fig. 7. The number of responses against the number of random latches (Estimate)

be the number of random latches. The number of responses arising from the fixed latches is 2^{N-T}, while the number of responses arising from the random latches is ${}_N C_T$. Therefore, the number of responses for a given value of T is estimated to be $2^{N-T} \cdot {}_N C_T$. The PUF based on the proposed method generates ternary values (00/11/10), so the total number of responses is 3^N. This total number is estimated in consideration of all the possible values of T ($0 \leq T \leq N$). However, the value of T is in fact determined by the kind of PUF device and the way in which the RS latches are implemented. Therefore, the PUF generates less than 3^N responses. To be specific, the number of responses for given T corresponds to the T-th term of the binomial expansion of $3^N = (2+1)^N$, which is $2^{N-T} \cdot {}_N C_T$, the same as the above estimate. Figure 7 shows a comparison between the number of responses for the conventional method (1-A) without random latches and the number of responses using our proposed method with various T values and given $N(= 128)$. The conventional method (1-A) generates 2^{N-T} responses, so the number of responses decreases as the number of random latches increases. Even conventional method (1-B), which uses ECCs, generates no more than 2^{128} responses. In contrast, the proposed method (1) dramatically increases the number of responses. The number of responses takes on its maximum value ($\approx 2^{203}$) when the numbers of the three types of RS latches are equal, that is, when T is around 43 ($\approx 128/3$). Hence, the proposed method dramatically improves the uniqueness of responses.

3.2 Proposed Mtd (2): Increasing the Number of Random Latches

This section proposes new methods ("proposed methods (2-A) and (2-B)") to give a higher probability of RS latches outputting random numbers than those obtained with the conventional methods in Sect. 2.2. The proposed methods increase the number of random latches to 1/3 of the total number of RS latches, which improves the effectiveness of the proposed method (1).

In proposed method (2-A), a shared FF is positioned in front of two NAND gates, as shown in Fig. 3. This FF sharing between two NAND gates eliminates

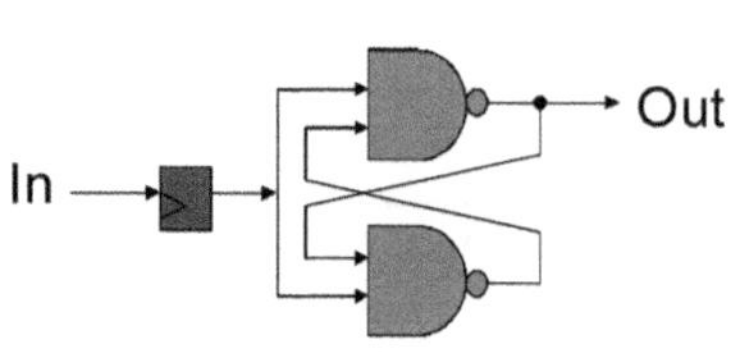

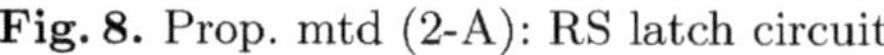

Fig. 8. Prop. mtd (2-A): RS latch circuit

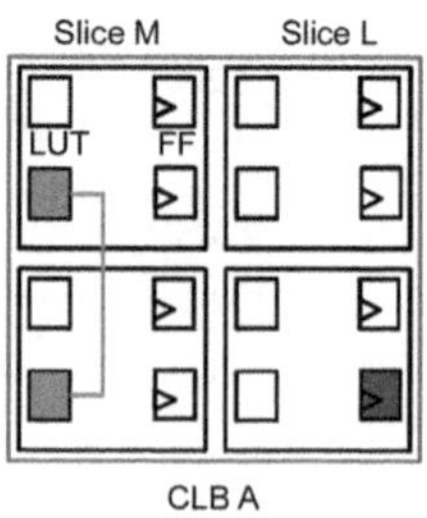

Fig. 9. Prop. mtd (2-B): Implementation of RS latches on Xilinx FPGAs

clock skew in FFs. Consequently, the signal arrival times for the two NAND gates are much closer, allowing the RS latches to become metastable more easily, and increasing the probability of the RS latches outputting random numbers. Proposed method (2-A) also reduces the FF gate size per RS latch by FF sharing.

In proposed method (2-B), one RS latch is implemented on a CLB in a Xilinx FPGA, as shown in Fig. 9. In Ref. [14], an RS latch is implemented on two different CLBs, as described in Sect. 2.2, because FPGA synthesis tools cannot implement two NAND gates of an RS latch on 'different' kinds of Slices (SliceM and SliceL) on the same CLB. To avoid this problem, proposed method (2-B) implements two NAND gates by using the 'same' kind of Slice on the same CLB. Proposed method (2-B) uses only one CLB (two slices) per RS latch, giving high circuit efficiency. In addition, it is anticipated that the probability of RS latches becoming metastable and outputting random numbers would increase since the signal arrival times for the two NAND gates are much closer due to shortening of the wire length between the gates. The concepts behind proposed methods (2-A) and (2-B) can be applied not only to FPGAs but also to ASICs.

4 Performance Evaluation

4.1 Experimental Environment

Figure 10 shows our experimental evaluation system, which uses a Spartan-3E starter kit board [17] with a Xilinx FPGA (XC3S500E). A 50-MHz clock signal generated by an on-board oscillator is applied to a Digital Clock Manager (DCM) primitive, which divides it into a 2.5-MHz clock signal that is applied to 128 RS latches. The output stream from each RS latch is switched by a multiplexer (MUX), and stored into a block RAM. Finally, the raw stream data from all the RS latches are transmitted to the PC through an RS232C port. In our evaluation, a software on the PC detects whether or not the streams contain random numbers rather than this being done with detection circuits. We regard that the detection technique does not influence PUF performance because the latter depends only on the output of the RS latches. We implement 128 RS latches on a 16×8 matrix of FPGA CLBs in accordance with proposed methods (2-A) and (2-B), this

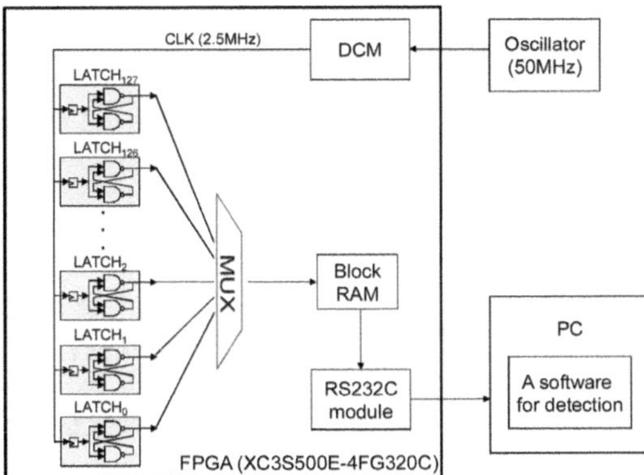

Fig. 10. Experimental evaluation system

being done manually with the FPGA synthesis tools in Xilinx ISE Design Suite 11.1. We regard one FPGA board as four virtual boards, since the RS latches are implemented at four completely different locations in the CLB matrixes for each FPGA. The evaluation uses 10 actual FPGA boards, but in the following discussion, we take the number of FPGA boards to be 40.

4.2 Experimental Results

Reliability and Uniqueness. Before we represent an evaluation of the effectiveness of proposed method (1), we show the basic performance of our BPUF, reliability and uniqueness. Our BPUF with 128 RS latches - based on proposed methods (2-A) and (2-B) - gives the results for reliability and uniqueness shown in Fig. 11 and Fig. 12, respectively. In our experiment, the PC is used to measure a 1000-bit output stream from each RS latch. The 2-bit partial response generated by each RS latch is '00(11)' if the 1000-bit bitstream is identically zero (one), or '10' if it includes a transition from 0(1) to 1(0). As a result, our BPUF with 128 RS latches can generate a 256-bit response. The reliability evaluation generates 40 responses using only a single specific FPGA selected at random. Figure 11 shows a histogram of normalized hamming distance between two arbitrary responses among the 40 responses (i.e. ${}_{40}C_2 = 780$ combinations). The average error rate is approximately 2.4% with a standard deviation of 0.75%, which is much less than the 15% assumed in Ref. [18] for stable responses based on a Fuzzy Extractor with a reasonable size of redundant data. Hence, our PUF gives responses that are of high reliability. Next, the uniqueness evaluation generates a total of 40 responses using all 40 FPGAs (one response per FPGA). Figure 12 shows a histogram of normalized hamming distance between two arbitrary responses among the 40 responses. This evaluation is a general way of showing the extent to which the responses of the chips are different. The difference in the responses of two arbitrary PUFs is approximately 46% with a standard deviation of 3.8%. The ideal difference is 50%, so our PUF gives responses with a high level of uniqueness.

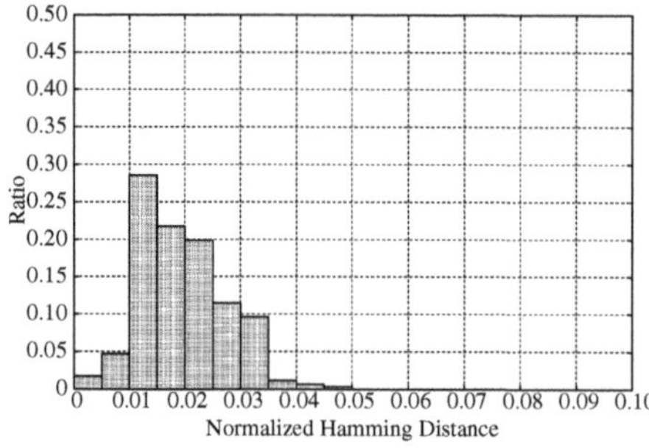

Fig. 11. Reliability

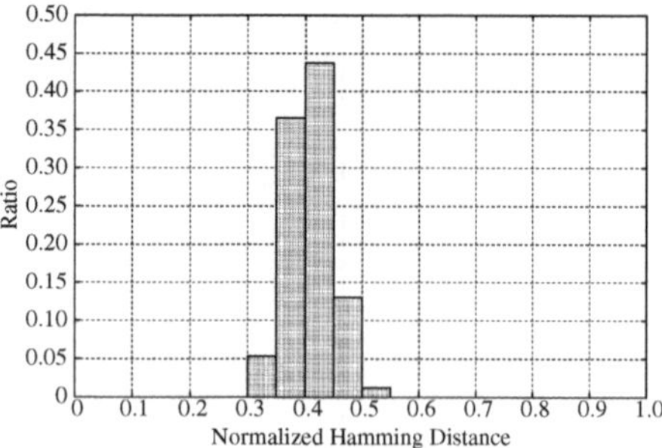

Fig. 12. Uniqueness

Table 1. Gate size and processing time of our PUF (not including detection circuits)

Gate size	SLICEs used	532/9312
	BRAM16s used	16/20
Processing time		0.4 ms (1000 cycles @ 2.5 MHz)

Cost. Table 10 indicates the gate size and processing time of our PUF evaluation system, shown in Fig. 10. In the FPGA evaluation system, a software on the PC is used instead of detection circuits. Our PUF (not including detection circuits) uses only 5% of the total slices in a FPGA, and the gate size is expected to be very small in ASICs. However, our PUF implemented in ASICs requires 128 detection circuits, and the gate size is estimated to be about 5.4K gates, using the gate equivalencies introduced in [16]. The gate size of our PUF is comparable to that of compact hardware for common key block ciphers such as AES. Hence, our PUF is sufficiently small to be implemented in embedded systems. The gate size can be reduced by a shared detection circuit switched by a multiplexer. The processing time is around 0.4ms, this being the total time taken to generate a response. One way of improving the processing time is to reduce the bitstream length for detection (1000 bits in our experiment). However, too short a length may result in misdetection. For example, RS latches outputting a large number of 0's and very few 1's might be detected not as random, but as fixed latches. This misdetection leads to loss of reliability, so our PUF makes a tradeoff between reliability and processing time. Our proposed PUF has advantages in terms of low noise because RS latches are allowed to become non-metastable through RS latch clock gating except when generating responses. In addition, our PUF can generate responses at anytime, unlike SRAM PUFs which can only generate them during power activation.

Effectiveness of Proposed Methods. Figure 13, a histogram showing the number of random latches per FPGA, represents an evaluation of the effectiveness of proposed methods (2-A) and (2-B). The results show that the proposed methods increase the number of random latches. This is because these methods allow the RS latches to become readily metastable, and increase their

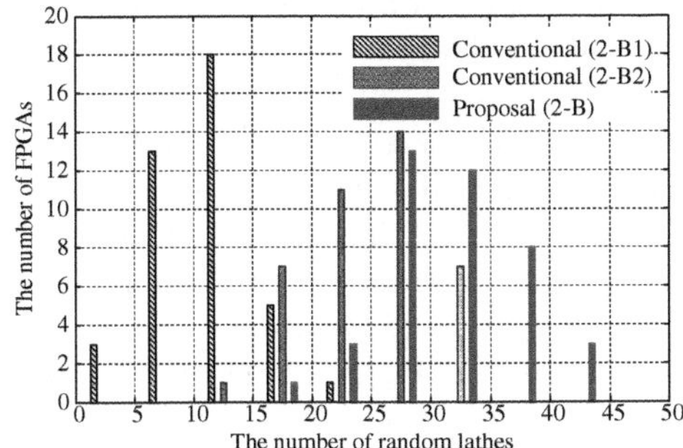

Fig. 13. Histogram for the number of random latches per FPGA

Table 2. The average number of random latches and number of responses

Implementation	Average # of random latches (experimental value)	# of responses (theoretical value)
Conv.(1-A)+Conv.(2-B1)	$\approx 12/128$	$\approx 2^{116}$
Prop.(1)+Conv.(2-B1)	$\approx 12/128$	$\approx 2^{170}$
Prop.(1)+Conv.(2-B2)	$\approx 26/128$	$\approx 2^{192}$
Prop.(1)+Prop.(2-B)	$\approx 32/128$	$\approx 2^{196}$

probability of outputting random numbers. In proposed method (1), the number of responses for 128 RS latches takes its maximum value when the number of random latches is around 43. Hence, the proposed methods improve the uniqueness of responses by increasing the number of random latches to as close to 43 as possible.

Table 2 shows the average number of random latches and number of responses for various implementation methods. Here, the number of responses is calculated theoretically based on the average number of random latches and Fig. 7. The number of responses is estimated to be $2^{116}(= 2^{128-12})$ when PUFs implemented by conventional method (2-B1) generate responses without 12 random latches. The PUFs based on proposed method (1) can generate 2^{170} responses using the location information entropy of 12 random latches. Moreover, PUFs based on both proposed methods (1) and (2-B) generate approximately 2^{196} responses with 32 random latches. Our proposed methods therefore dramatically increase the number of responses.

Entropy of Responses. Here, we perform a rigorous evaluation of the entropy of responses. The number of responses estimated in Table 2 is based on the assumption that RS latches output 0's, 1's, or random numbers with equal probability. If an RS latch (e.g. LATCH_0) outputs 0's independently of FPGA, then the 2-bit partial response corresponding to that latch must be '00', so the response cannot have a particular value. As a result, the actual number of responses is much smaller than the above estimated number. Table 2 therefore shows the theoretical upper bound on the number of responses. By following the 2 steps below, we rigorously calculate the entropy of responses from our PUFs using the experimental results with 40 FPGAs.

In the first step, we show the ratios of RS latches outputting 0's, 1's, and random numbers, shown in Fig. 14. We explain how to read the figure with the specific example in Fig. 15, as follows. First, the 40 RS latches at the same physical CLB location (e.g. LATCH_0) on the 40 FPGAs are called "a latch group". Hence, in our experiment, there are 128 latch groups corresponding to the range from LATCH_0 to LATCH_{127}. The 40 RS latches labeled as LATCH_0 include 15 latches outputting 0's, 20 outputting 1's, and 5 outputting random numbers. The ratios are therefore 0.375, 0.500 and 0.125, respectively. A plot of LATCH_0 is obtained by relating the ratios to the three sides of a triangle, and 128 plots are obtained, corresponding to the 128 latch groups in Fig. 14. A plot is located at the central point of the triangle if the ratios are equal, which is the ideal. Given the limited number of FPGAs (i.e. 40) in our experiment, it is desirable as a practical criterion that a large proportion of plots are located in the small central triangle illustrated by thick line. If the plot is in the small triangle, the three ratios fall within a range of 0.20 to 0.60. In conventional method (2-B1), it can be seen that all of the RS latches in each latch group have a low probability of outputting random numbers since many of the plots are located on the right side of the triangle. In addition, most RS latches in each latch group have a one-sided probability of outputting 0's or 1's since many of the plots are located throughout the whole of the right side. Conventional method (2-B2) improves the ratios, making them roughly equal, but requires a large number of CLBs to implement the RS latches shown in Fig. 4. In addition, there are not so many random latches (around 26), so the number of responses is not very large. In contrast, proposed method (2-B) improves the ratios such that they are almost equal since as many as 93 plots are located in the small central triangle. Furthermore, no latch groups have RS latches outputting ternary values at a high (> 0.9) or low (< 0.1) probability. The number of plots in the small triangle is significantly higher than with conventional methods, which implies that the proposed method makes many of the RS latches readily metastable, so that the ratios become almost equal as a favorable side effect. Hence, using the proposed methods, the number of responses is expected to be close to the upper bound shown in Table 2.

In the second step, we rigorously calculate the Shannon entropy of responses based on the ratios discussed in the first step. Let the ratios of the RS latches labeled as LATCH_n outputting 0's, 1's, or random numbers be $P_n(0)$, $P_n(1)$ and $P_n(R)$, respectively. The Shannon entropy E_n derived from LATCH_n is defined as

$$E_n = -P_n(0) \cdot \log_2 P_n(0) - P_n(1) \cdot \log_2 P_n(1) - P_n(R) \cdot \log_2 P_n(R). \quad (2)$$

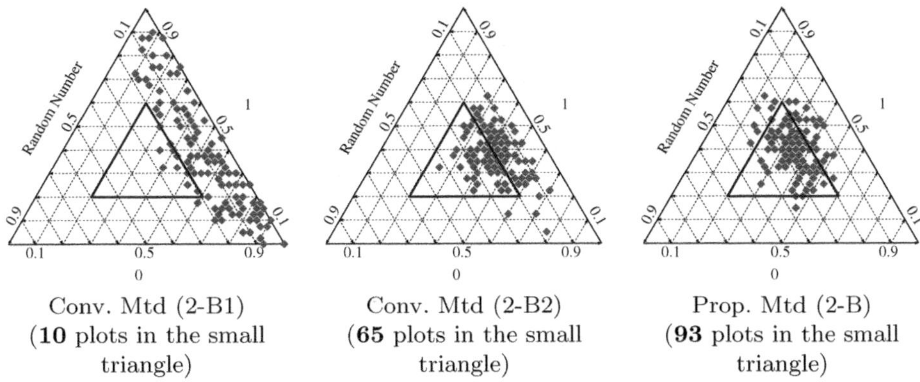

Fig. 14. Ratios of RS latches outputting 0's, 1's, or random nums in 128 latch groups

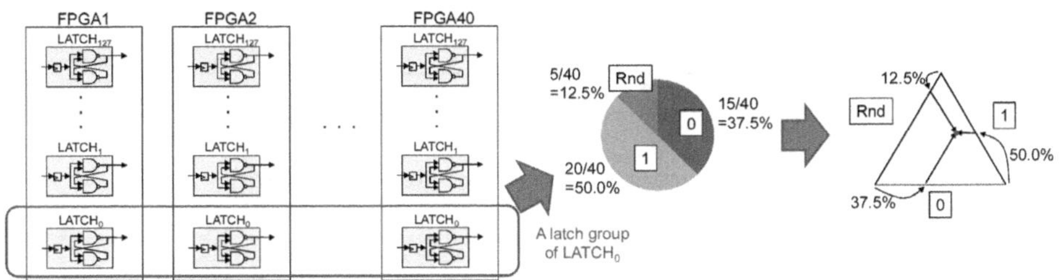

Fig. 15. How to read Fig. 14

Hence, the total Shannon entropy derived from LATCH_0 to LATCH_{127} is $\sum_{n=0}^{N-1} E_n$, where $N = 128$. Here, the total entropy can also be given as

$$\sum_{n=0}^{N-1} E_n = \log_2(2^{N-T} \cdot {}_N C_T) \tag{3}$$

by Stirling's approximation ($\log_2 x! \approx x \log_2 x - x \log_2 e$) under the ideal condition that $P_n(0) = P_n(1) = \frac{1}{2}(1 - T/N)$ and $P_n(R) = T/N$ $(0 \leq n \leq N, N = 128)$. Equation 3 shows that the total Shannon entropy corresponds to the number of responses estimated in Sect. 3.1. Therefore, in consideration of the ratios of RS latches outputting 0's, 1's, and random numbers in first step, the number of responses can be rigorously calculated on the basis of the Shannon entropy in Eq. 2. Table 3 shows the Shannon entropy for responses. A PUF with 128 RS latches based on conventional method (2-B1) generates $2^{126.6}$ responses even if proposed method (1) is applied. This is because the number of random latches is small, and the ratios are not equal. In contrast, the PUF based on proposed method (2-B) generates $2^{192.7}$ responses, which is almost same as the upper bound in Table 2, and is larger than for PUFs based on conventional methods. Hence, a PUF based on both proposed methods reduces circuit size and dramatically improves the entropy (i.e. uniqueness) of responses.

Table 3. Shannon entropy of responses

	Entropy of responses (bits)
Conv.(1-A)+Conv.(2-B1)	97.2
Prop.(1)+Conv.(2-B1)	126.6
Prop.(1)+Conv.(2-B2)	187.7
Prop.(1)+Prop.(2-B)	192.7

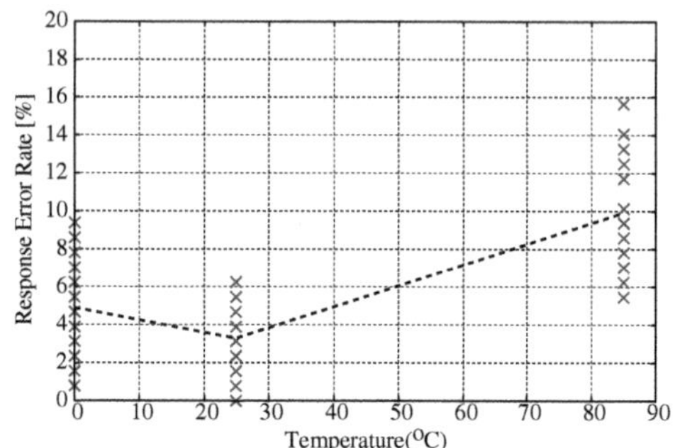

Fig. 16. Error rates against various temperatures

The entropy per unit area (gate size) of proposed method (1) is expected to be higher than that of conventional methods (1-A) and (1-B). Both proposed and conventional methods (1-A) requires a mechanism to detect random latches, so their area sizes are almost same, while the entropy of proposed method (1) is higher from Table 3. In contrast, conventional method (1-B) does not require the mechanism, so the area size is smaller than proposed method (1). Hence, by implementing more RS latches, the entropy of conventional method (1-B) seems to be higher. In fact, however, conventional method (1-B) needs to correct the variation resulting from all the random latches, which requires larger ECC redundant data for stable responses. In contrast, proposed method (1) regards random numbers as the third stable value, which leads to a reasonable size of redundant data. Therefore, in consideration of the area size for redundant data, the proposed method is expected to generate higher entropy per unit area.

Temperature Resistance. Figure 16 shows the change in error rate for various temperatures ranging from 0 °C to 85 °C, which is within the rated temperature of the FPGA (XC3S500E-4FG320C). Here, error rate is the ratio of the number of 2-bit partial responses that are different from those at 25 °C. Figure 16 plots the error rates of 40 FPGAs at 0 °C, 25 °C, and 85 °C. The bigger the temperature difference from 25 °C - as the standard temperature - the higher the error rate. The error rate is less than around 15% regardless of temperature, so stable responses are generated based on a Fuzzy Extractor with a reasonable size of redundant data [18].

5 Conclusion

This paper proposed a method for generating responses from a BPUF based on the location information of RS latches outputting random numbers. Our proposed detection circuit generates ternary values (00/11/10) in accordance with the three types of output bitstream from RS latches. This increases the number of responses from 2^{128} to around 3^{128} with 128 RS latches, thereby dramatically improving the uniqueness of the responses. In addition, with its small circuit size, the new implementation method increases the number of random latches, and equalizes the ratios of RS latches outputting 0's, 1's, and random numbers, thereby enhancing the effectiveness of the proposed method. According to our experiment with 40 FPGAs, a BPUF with 128 RS latches based on the proposed methods is able to generate responses with 193-bit entropy, which is larger than the 116-bit entropy achieved by conventional methods. The proposed methods can be applied to other PUFs, such as the Arbiter PUF. Unstable (random) outputs from the PUF can be used for generating highly-unique responses without the necessity of selecting available challenges. Future work will include discussion of voltage resistance, performance evaluation on ASIC, and the application of the proposed methods to other kinds of PUFs than BPUFs.

References

1. Pappu, R.S.: Physical one-way functions. PhD thesis, Massachusetts Institute of Technology (March 2001)
2. Gassend, B., Clarke, D., van Dijk, M., Devadas, S.: Silicon physical random functions. In: Proceedings of CCS 2002, pp. 148–160 (2002)
3. Gassend, B., Clarke, D., Lim, D., van Dijk, M., Devadas, S.: Identification and authentication of integrated circuits. In: Concurrency and Computation: Practice and Experiences, pp. 1077–1098 (2004)
4. Maes, R., Verbauwhede, I.: Physically unclonable functions: A study on the state of the art and future research directions. In: Towards Hardware Intrinsic Security: Foundation and Practice. Information Security and Cryptography, pp. 3–37. Springer, Heidelberg (2010)
5. Guajardo, J., Kumar, S.S., Schrijen, G.J., Tuyls, P.: FPGA intrinsic pUFs and their use for IP protection. In: Paillier, P., Verbauwhede, I. (eds.) CHES 2007. LNCS, vol. 4727, pp. 63–80. Springer, Heidelberg (2007)
6. Holcomb, D.E., Burleson, W.P., Fu, K.: Initial SRAM state as a fingerprint and source of true random numbers for RFID tags. In: Proceedings of the Conference on RFID Security (July 2007)
7. Kumar, S.S., Guajardo, J., Maes, R., Schrijen, G.J., Tuyls, P.: The butterfly puf: Protecting ip on every fpga. In: HOST, pp. 67–70 (2008)
8. Jae, W., Lee, D., Lim, B., Gassend, G.E., Suh, M., Van Dijk, M., Devadas, S.: A technique to build a secret key in integrated circuits with identification and authentication applications. In: Proceedings of the IEEE VLSI Circuits Symposium, pp. 176–179 (2004)
9. Suzuki, D., Shimizu, K.: The glitch PUF: A new delay-PUF architecture exploiting glitch shapes. In: Mangard, S., Standaert, F.-X. (eds.) CHES 2010. LNCS, vol. 6225, pp. 366–382. Springer, Heidelberg (2010)

10. Suh, G.E., Devadas, S.: Physical unclonable functions for device authentication and secret key generation. In: Proceedings of DAC 2007, pp. 9–14 (2007)
11. Rührmair, U., Sölter, J., Sehnke, F.: On the foundations of physical unclonable functions. Cryptology ePrint Archive, Report 2009/277 (2009)
12. Carter, J.L., Wegman, M.N.: Universal classes of hash functions. In: ACM Symposium on Theory of Computing (1977)
13. Dodis, Y., Ostrovsky, R., Reyzin, L., Smith, A.: Fuzzy extractors: How to generate strong keys from biometrics and other noisy data. SIAM J. Comput. 38, 97–139 (2008)
14. Hata, H., Ichikawa, S.: Fpga implementation of metastability-based true random number generator. In: IEICE Tech. Rep., RECONF2008-59 (January 2009)
15. Ichikawa, S., Hata, H.: True random number generator based on metastability of rs latch. In: SCIS 2009, pages 2F1–5 (2009)
16. Batina, L., Lano, J., Mentens, N., Ors, S.B., Preneel, B., Verbauwhede, I.: Energy, performance, area versus security trade-offs for stream ciphers. In: The State of the Art of Stream Ciphers, Workshop Record, ECRYPT 2004, pp. 302–310 (2004)
17. Spartan-3E starter kit board, `http://www.xilinx.com/products/devkits/HW-SPAR3E-SK-US-G.htm`
18. Maes, R., Tuyls, P., Verbauwhede, I.: Low-overhead implementation of a soft decision helper data algorithm for SRAM pUFs. In: Clavier, C., Gaj, K. (eds.) CHES 2009. LNCS, vol. 5747, pp. 332–347. Springer, Heidelberg (2009)

MECCA: A Robust Low-Overhead PUF Using Embedded Memory Array

Aswin Raghav Krishna, Seetharam Narasimhan, Xinmu Wang, and Swarup Bhunia

Case Western Reserve University, Cleveland OH-44106, USA
ark70@case.edu

Abstract. The generation of unique keys by Integrated Circuits (IC) has important applications in areas such as Intellectual Property (IP) counter-plagiarism and embedded security integration. To this end, Physical Unclonable Functions (PUF) have been proposed to build tamper-resistant hardware by exploiting random process variations. Existing PUFs suffer from increased overhead to the original design due to their specific functions for generating unique keys and/or routing constraints. In this paper, we propose a novel memory-cell based PUF (MECCA PUF), which performs authentication by exploiting the intrinsic process variations in read/write reliability of cells in static memories. The reliability of cells is characterized after manufacturing by inducing temporal failures, such as write and access failures in the cells using a programmable word line duty cycle controller. Since most modern designs already have considerable amount of embedded memory, the proposed approach incurs very little overhead (<1%) compared to existing PUF designs. Simulation results for 1000 chips with 10% inter-die variations show that the PUF provides large choice of challenge-response pairs with high uniqueness (49.9% average inter-die Hamming distance) and excellent reproducibility (0.85% average intra-die Hamming distance).

Keywords: Physical Unclonable Function (PUF), IC authentication, Memory failures, Negative Bias Temperature Instability (NBTI).

1 Introduction

In recent years, shorter product-cycle marketing requirements in the semiconductor industry have driven chip vendors to reuse their hardware designs and outsource Integrated Circuits (IC) production to external foundries shared by many companies. Apart from reuse, the intellectual property (IP) is often an additional source of income to a vendor through external licensing to other companies who can include the design in their products. However, production outsourcing and IP licensing have exposed the designs to theft and cloning and it is estimated that counterfeit electronics cost the industry upto US$100 billion every year [1]. Counterfeiting attacks on IP/IC can occur at the manufacturing site, e.g. an untrusted foundry makes several copies of the design, or during

B. Preneel and T. Takagi (Eds.): CHES 2011, LNCS 6917, pp. 407–420, 2011.

deployment in the field. These attacks can be broadly classified into two categories: (i) *invasive attacks*, e.g, by delayering the IC through reverse engineering and obtaining circuit function from physical layout; and (ii) *non-invasive attacks* which in turn can be classified as *passive* and *active* attacks. Passive attacks are mounted by observing side-channel information such as power consumption [2], delay or electromagnetic radiation, to obtain secret keys or sensitive information. Active attacks, on the other hand, are induced by the introduction of a fault followed by a passive attack [3]. IP designs can also be stolen from FPGAs during power-up by reading their bitstream information which is stored in an external memory. Building a tamper-proof hardware that is resistant to all forms of attacks is, thus, crucial for securing IP/IC against counterfeit attacks.

Authentication plays an important role in detecting counterfeit products. Simply put, the role of authentication is to check the identity of a product and to validate that it comes from a genuine source. The common practice is to embed a digital secret/ID in a non-volatile memory, e.g. in a RFID tag, and use digital key comparison and encryption for authentication and protection of secret information. However, since the secret information is stored in digital form, it is vulnerable to invasive attacks and providing high tamper resistance environment is very expensive. Furthermore, since each product contains only one unique identifier, it is possible for an attacker to obtain it by intercepting the communication of the key between an authorized reader and a tag and use it for cloning or mounting replay attacks.

Physical Unclonable Functions (PUFs) are rapidly becoming the preferred method for IC/IP identification, authentication and secure system design. They are secure, low-cost, and robust functions built into a design that implement a challenge-response protocol by exploiting the inherent random variations in the manufacturing process to generate unique signatures [4,5]. Inevitable variations in the device paramaters (e.g. threshold voltage) make it practically impossible to clone the original PUF even with the same mask set, foundry and manufacturing process. Typically, the challenge response pairs for each device are stored by the vendor after production in a database and is given to a trusted party who wishes to use the device. The trusted party applies a challenge and checks the corresponding response with the database to verify the authenticity of the device in any environment [6]. PUFs have several advantages that make them robust to cloning and replay attacks. Firstly, since a PUF response is based on random process variations, it is not possible for an attacker to 'predict' the response before or after production. Secondly, since PUFs have a very large set of challenge-response pairs, an attacker must obtain all the pairs to make an identical copy of the PUF - merely obtaining only a few pairs is not useful, since a different pair may be used for authentication. Thirdly, unlike conventional approaches in which the stored keys are preserved digitally in non-volatile memory even after power-down, PUFs generate signatures only when they are powered-up, thus forcing attackers to mount attacks to extract signatures only when the PUF is in operation [6].

In this paper, we propose a novel (ME)mory (C)ell-based (C)hip (A)uthentication PUF - MECCA PUF for authentication and key generation based on the concept of failure mechanisms in the memory array. It is observed that more than 50% of System-On-Chip (SOC) area is used for memory with estimates indicating that the number could increase to 90% in 2013 [7]. The proposed PUF leverages on the fact that most designs already contain embedded SRAM memory array for their operation and hence can also be used for generating signatures. The basic idea is to control the word line duty cycle of the SRAM cells to determine their vulnerability to failures during read/write access. Word line controllability allows us to generate multiple responses from the array and hence increase the number of challenge-response pairs. The random process variations of the cells' parameters across the chip determine the reliability (low or high) of the cells; the cells' reliability is translated to a digital response. We analyze the effectiveness of MECCA PUF in detail and show that it provides excellent unclonability, uniqueness and robustness of signatures. Since environmental effects such as temperature and device ageing effects (such as bias temperature instability or BTI [22]) affect the repeatability of signatures, we propose an ageing-tolerant scheme to make the cells generate highly stable responses. Simulation results show that the proposed PUF offers several advantages: 1) very less area overhead (<1%); 2) high uniqueness (49.9% average inter-die Hamming distance); and 3) high reproducibility. Additionally, the delay controller circuit can also be integrated with a Design-for-Test (DFT) technique [8] for detecting stability faults in memory, thus allowing it to serve dual purposes and to further reduce the cost per function.

The remainder of the paper is organized as follows: Section 2 describes prior works on PUF circuits. Section 3 describes the methodology of the proposed MECCA PUF along with theoretical analysis of PUF properties. Simulation results and analysis are presented in Section 4. Finally, we conclude in Section 5.

2 Related Work

Several silicon PUFs have been published in literature. Silicon PUFs can be classified as memory PUFs and delay PUFs [9]. Delay PUFs such as Ring Oscillator PUF [6, 10] and Arbiter PUF [11, 12] translate the process variations into random delay variations to produce a digital signature. These PUFs involve introduction of the circuits which are solely used for key generation and authentication and hence present substantial area overhead. Existing memory based PUFs rely on the random initializations of the cells due to process variations for generating signatures [13, 14, 15]. However, these PUFs only provide a single bit response per cell and have limited challenge-response pairs [9]. Furthermore, these PUFs are prone to cloning attacks in the foundry as the entire random initialization memory map can be copied to produce the signatures as the original PUF. Another type of PUF, known as Butterfly PUF [16], is based on exploiting interconnect variations in cross-coupled latches during startup. A major disadvantage with this PUF is that attaining the metastable point for

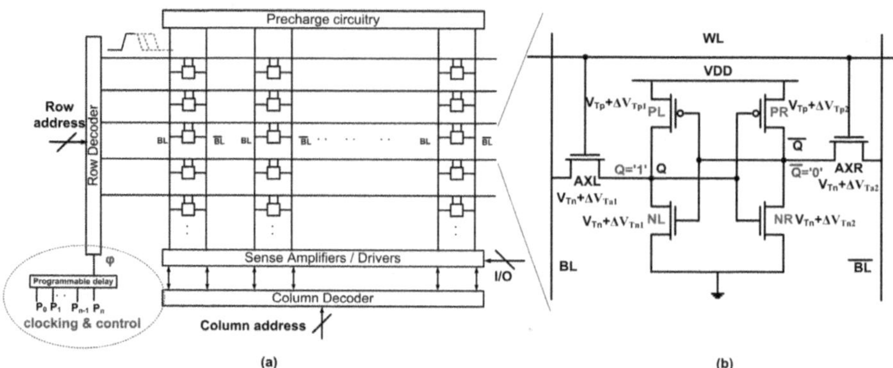

Fig. 1. MECCA PUF architecture: (a) Memory block with peripheral circuitry and programmable delay circuit, (b) Schematic of an SRAM cell

each cross-coupled latch prior to key generation is difficult due to the finite delays of the latches and interconnects which causes the outputs of the latches to oscillate. This oscillation imposes precise timing requirements of the control (excite) signal for reproducible keys. Finally, PE-PUFs [9] couple process variations with environmental effects, such as temperature, power supply noise and noise due to circuit activity, for generating signatures. However, PE-PUFs require long interconnects and placement over the entire chip which can result in significant area overhead/routing constraints in modern technologies.

3 MECCA PUF

The concept of the proposed MECCA PUF architecture can be explained with the help of Fig. 1. The PUF contains an SRAM array along with peripherals and a programmable delay generator. Most modern designs already contain one or more memory block(s) for their normal operation and the delay generator introduces only a minor area penalty. In the core array, inter-die and intra-die variations in the device parameters cause a mismatch in the strengths of transistors which can be exploited to cause failures in cells. However, some cells are more prone to failure than others because of the random effect of process variations.

The failure mechanisms [20, 21] observed in a memory cell are as below:

Write failure: Occurs when the internal node in an SRAM cell cannot be discharged through the access transistors during the word line's active duration.

Read failure: Flipping of the data in SRAM cell during a read operation.

Access failure: Occurs when the voltage difference between the bitlines is lower than the offset voltage of the sense amplifier when it is activated.

Hold failure: When the supply voltage is lowered during standby, leakage currents through the NMOS transistors can cause internal node voltage to reduce below the switching threshold of the inverter for a data flip.

The idea of using memory failures in PUFs has been investigated earlier in [17, 18] by inducing read/write collisions or using metastability in the cross-coupled

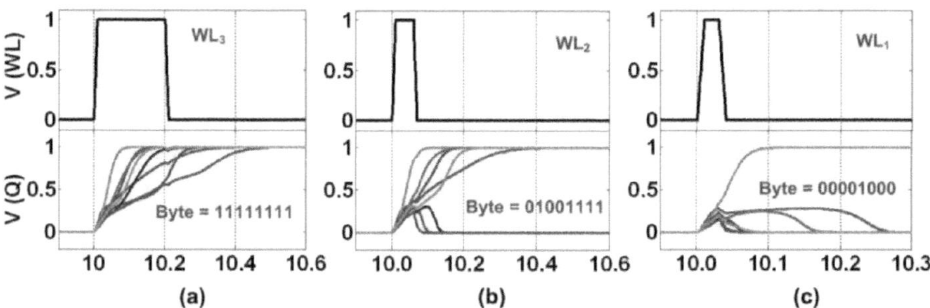

Fig. 2. Reliability of 8 cells for different word line duty cycles; $WL_3{>}WL_2{>}WL_1$ (WL_3 is used for normal memory operation)

loop to generate responses. In the MECCA PUF, we use a different approach of evaluating the reliability of a single SRAM cell by inducing a write failure by changing the word line duration. Assume that the cell stores a '0' and we wish to write a '1'. This is accomplished by setting BL to '1' and $\overline{BL}$ to '0' which causes the cross-coupled inverters to change states. As shown in Fig. 1(b), the different transistors have varying ΔV_T components imposed on nominal V_T due to process variations. Equating the dc currents through the transistors AXL and NL, we compute the internal node voltage, V_Q as shown in eqn. (1), where the required pull-up ratio of the cell, PUR, is decided such that V_Q is below the switching point of the inverter.

$$V_Q = V_{DD} - (V_{Tn} + \Delta V_{AXL}) - [((V_{DD} - (V_{Tn} + \Delta V_{AXL}))^2 - 2 \times PUR \times (\frac{\mu_p}{\mu_n})((V_{DD} - |(V_{Tp} + \Delta V_{PR})|) \times V_{DSATp}) - (\frac{V_{DSATp}^2}{2}))]^{0.5} \quad (1)$$

In a realistic scenario, the V_T and ΔV_T components of the other transistors also play an important role as the cell starts switching due to regenerative feedback. For normal memory operation, the word line (WL) duration is selected such that a write operation can be successfully performed under all process corners. When the memory is used as a PUF, we purposely reduce the WL duration such that a stable cell at a normal WL length may or may not be stable at a reduced length as determined by process variations. For example, by using a programmable delay word line, the WL duty-cycle is shortened which will randomly cause some of the cells to have write failure. The effect of reducing WL duration is shown in Fig. 2 where the values of 8 cells are compared for 3 durations. For each WL duration, by selecting a set of R cells, we obtain a signature consisting of both good cells (which are written correctly even with shorter WL) and defective cells (which have write failure). Access failure may also be exploited in the MECCA PUF to evaluate reliability of a cell by reducing the WL duration required for discharging one of the bit-lines for a read operation.

In contrast, read and hold failures are static failures which cannot be controlled by WL duration and hence are not useful for evaluating the reliability of

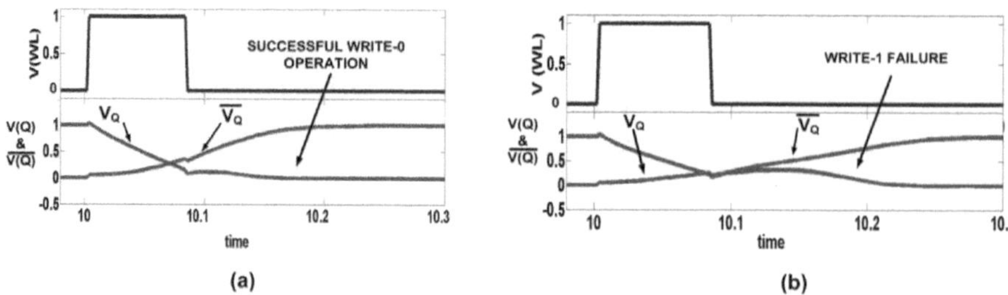

Fig. 3. Dependence of stability of a cell on write value (In both cases, the reduced word length ON time is the same) (a) Successful write-0 operation. (b) Write-0 failure.

a cell in our PUF. For example, for a read-1 failure, the node voltage ($V_{\bar{Q}}$) which is determined by voltage division across the resistances of transistors AXR and NR, must rise above the switching threshold (V_S) of the subsequent inverter for the cell to flip its value. The voltage division is independent of the WL duration and hence read failures cannot be induced by varying the WL duration. Hold failures occur due to leakage currents and hence cannot be induced by controlling the WL duration. The reliability of a cell also depends on the logic value being written into the cell at reduced word-line duration. As shown in Fig. 3, the cell has high reliability for a write-1 but a low reliability (write failure) for a write-0 operation for the same WL. This is due to the fact that different PMOS and access NMOS transistors (and hence different V_T variations) are involved at the initialization of write-0 and write-1 operations before the other transistors play a role in engaging positive feedback. This dependence on write value is very useful in increasing the number of unique signatures since a write-0 success (failure) does not imply a write-1 success (failure) at the same WL.

The major steps for generating unique signatures are as follows.

1. We choose the address of the R-cells as part of the input challenge.
2. A background write operation of 0 (or 1) in cell(s) is performed at the normal word line duration depending on whether we want to exploit the reliability for a write-1 or write-0. This is required to ensure that the cells are in a known initialized state, thus ensuring that a possible successful write is not due to random initialization. Next, the bitlines are precharged to their respective values depending on the values to be written.
3. We reduce the word line duration using the programmable delay circuit and perform the write operation at this reduced duration for all the chips.
4. Finally, we read out the values stored in the cells to give an R-bit response.

Word Line Duty Cycle Controller: Fig. 4 illustrates a programmable word line duty cycle controller that is used for inducing write failures in the SRAM array. The circuit consists of a chain of inverters with the outputs of k subgroups of odd number of inverters connected to the inputs of a k X 1 mux. The programmable select bits, which become part of the challenge, choose one out of k possible duty cycles to generate a shortened word line, e.g. by acting as the

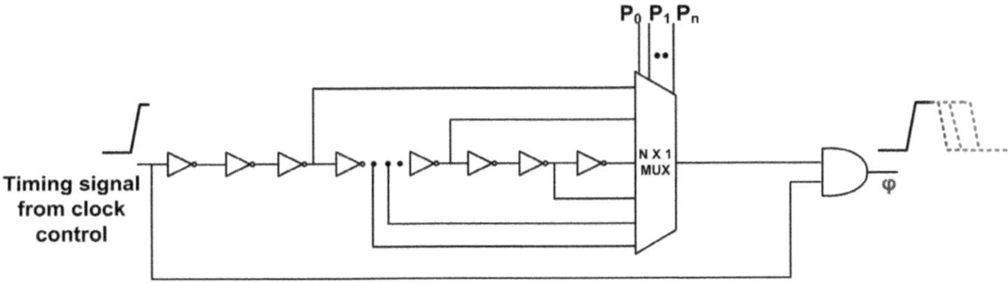

Fig. 4. Programmable word line duty cycle controller

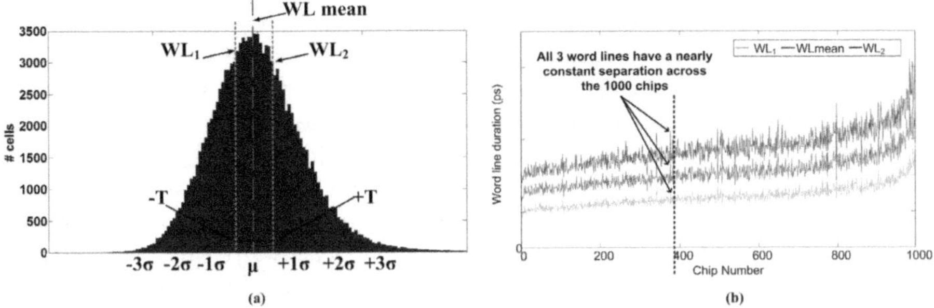

Fig. 5. (a) Write-time distribution of 1000 chips, (b) Programmable delay variation across 1000 chips

enable signal for the address row decoder. The k duty cycles consist of $n = k-1$ duty cycles used for PUF operation and one for normal memory operation.

The word line durations can be chosen from the distribution of write-time (Fig. 5(a)) of all the cells in the chips. For PUF operation, we choose the mean word line to be the nominal, μ, of the write-time distribution since the cells will have equal probability of write failure and success. The inverter chain is then tapped for outputs to obtain n duty cycles such that they fall within $\pm T$ from the nominal. The n levels must be separated such that they don't overlap with each other due to V_T variations in the duty cycle controller as shown in Fig. 5(b) where three WL durations are sorted in increasing order of inter-die V_T. Also, it is interesting to note that since the controller and the SRAM array are at the same inter-die V_T for a particular chip, all n levels will move in the same direction (albeit by different amounts) and only the intra-die variations need be considered for choosing T. For a C chosen based on the accepted distribution of '1's and '0's in the responses, i.e., $P(X < \mu - T) < C$, T can be computed in terms of σ. As an added layer of security, a challenge to address-duty cycle mapping or other well-known techniques (e.g. controlled PUFs [19]) can be used to make it harder for an attacker to model the PUF. We investigate the following properties of PUFs:

Unclonability: From a security perspective, a PUF itself must not be prone to cloning by an attacker by observing a few challenge response pairs. In the case of our PUF, an attacker must obtain the addresses of the cells, the word line duration and the values being written into each cell for each challenge. In this context, we are referring to an attacker in the foundry who has complete access to the chips. Choosing an arbitrary WL and observing the values of all the cells in the array to clone a copy by skewing will not be useful to the attacker since the values in all the cells in the original MECCA PUF will be different from the values in the skewed design for different word line durations; the attacker must skew the design such that each cell has the same response at each WL for a write-0 and write-1 - a significantly difficult challenge.

Entropy: Different signatures can be obtained by measuring the reliability of different sets of cells, i.e., by using a different set of addresses (challenges). Moreover, as the WL duration can be controlled to generate different sets of low and high reliability cells for a chosen set of addresses, the WL duration is also part of the challenge and can be used to produce many keys from a given set of addresses. As an example, for a set of chosen addresses (A_0, A_1, A_2,. . . A_n), by setting WL duration to, say WL_1, we obtain a signature as (D_0, D_1, D_2 . . . D_n) where D_i is 0 or 1 depending on whether the cell at A_i is a high reliability cell (with no write failure) or low reliability cell (with write failure) respectively. By changing the WL duration to WL_2, we can obtain another signature set (S_0, S_1, S_2,. . . S_n) for the same addresses (A_0, A_1, A_2,. . . A_n), with $D_k \neq S_k$ for some cells for $0 \leq k < n$. For example, the address set (900, 100, 500, 825,. . ., 1024) in a 1024 SRAM array for a write-1 can have a signature

- (1, 0, 0, 1,. . .,1) for word line length, WL_1, where cells at 900, 825 and 1024 are high reliability cells (1s) and cells at 100, 500 are low reliability cells (0s)
- (1, 0, 1, 1,. . ., 1) for word line length, WL_2, (such that $WL_2 > WL_1$) where cells at 900, 500, 825 and 1024 are high reliability cells (1s) and the cell at 100 is a low reliability cell (0)

As mentioned before, measurement of reliability is relative among cells and word line durations, i.e., a cell (such as A_2 in the above example) which has high reliability for a given WL duration can have low reliability for a shorter WL. However, since the delay circuit is designed to produce n-levels of durations close to the nominal required write-time of the cells, the values at many bit positions in an R-bit response will be different for a shorter WL duration. Additionally, the dependence of reliability on the value being written for a given WL duration also increases entropy by choosing different write values for the given set of addresses.

4 Simulation Results and Analysis

The proposed PUF has been implemented with an SRAM array designed for the 45nm *Predictive Technology Model* (PTM) [28]. Simulations were carried out using Synopsys *HSPICE* for 1000 chips for generating 128 bit responses. The effect of process variations for the chips was introduced by running Monte Carlo simulations for inter-die variations with $\sigma = 10\%$ and random intra-die

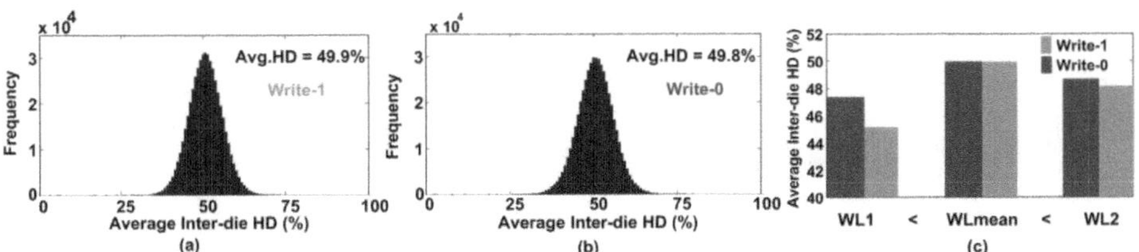

Fig. 6. HD distribution of inter-die responses of 1000 chips for (a) write-1 at mean WL (b) write-0 at mean WL (c) Avg. inter-die HD for 3 WL_S for write-0 and write-1

variations with $\sigma = 6\%$. The duty cycle controller was implemented as shown in Fig. 4 to produce n=3 duty cycles. The SRAM cells were brought to a known initialization state with a background write at the normal WL duration. The challenge consisted of the addresses of the cells along with the select inputs to the duty cycle controller. After a pattern was applied, the values of the 128 cells were extracted as a signature at a short WL duration for all the chips.

Uniqueness Analysis: To determine the uniqueness of the MECCA PUF, we plotted the distribution of the inter-die Hamming distance (HD) of 1000 MECCA PUFs for write-0 and write-1 operations for the 3 WL durations as shown in Fig. 6; the horizontal axis represents the number of bits differing between responses of any two chips for a given challenge while the vertical axis represents the number of comparisons among the 1000 chips corresponding to a HD. To quantify the uniqueness property, we computed the average inter-die Hamming Distance (HD_{Avg}) [29] of the signatures of m=1000 chips with percentage HD (out of r response bits) between any two chips m_1 and m_2 as follows:

$$HD_{Avg} = \frac{2}{m \times (m-1)} \sum_{i=1}^{m-1} \sum_{j=i+1}^{m} HD_{perc}(m_1, m_2). \quad (2)$$

The average inter-die HD was found to be close to the ideal 50% for write-0 and write-1 operations at the mean WL but reduced by a maximum of 2.5% (for a write-0 operation) for WL_1 and WL_2. The reduction in inter-die HD is due to bit-skewing at some bit positions in the responses from the 1000 chips. Ideally, each bit in the responses from all the chips should have 50% probability of being a 0 or a 1 to center the inter-die HD distribution about 50%. However, if some bit has higher probability towards 0 or 1, then the inter-die HD for that bit becomes close to zero. In our case, the skewing is due to the fact that a cell failure at a longer WL implies a failure at a shorter WL for a given write value (Fig. 7).

Robustness Analysis: The robustness of a PUF shows how reproducible are the signatures from the chips in changing operating conditions. The HD between responses from the same PUF in different operating conditions must be as low as possible for high reproducibility. We estimated the intra-die HD among the

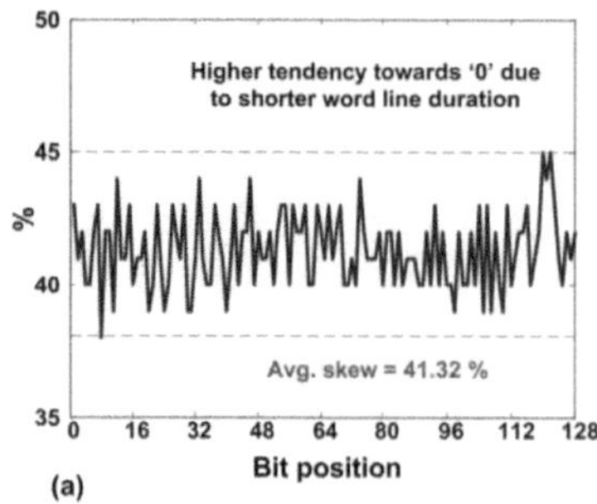

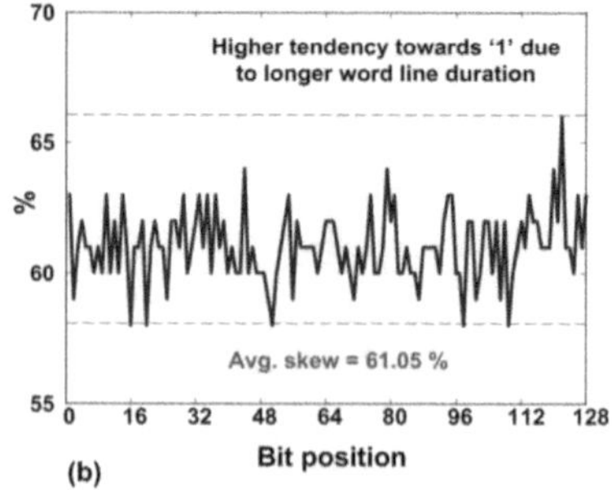

Fig. 7. Bit-skewing for write-1 at (a) $WL_1 <$ mean WL (b) $WL_2 >$ mean WL

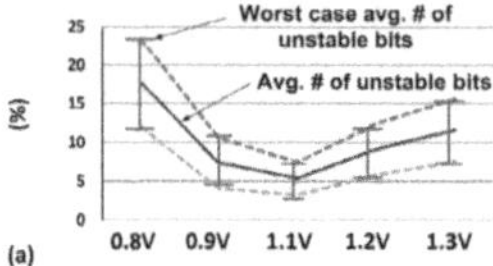

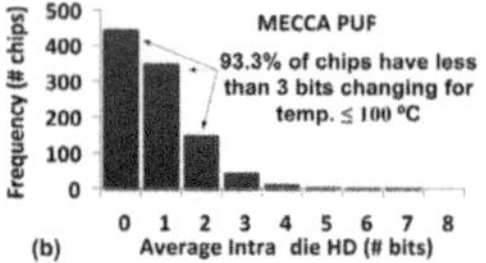

Fig. 8. Average intra-die HD from 1000 chips for MECCA PUF (a) for supply voltage variation (compared to nominal Vdd=1V), and (b) for temperature variation

128-bit responses for each chip for supply voltage variations and compared them with that obtained at nominal voltage (Fig. 8(a)). At the worst-case voltage of 0.8V, the intra-die HD is as high as 23 bits ($\approx$ 18%). Fig. 8(b) shows the intra-die HD for temperature variations for each chip at room temperature compared with that obtained at 5 temperatures (for the same challenge) till 100°C in steps of 15°C. Most of the responses ($\approx$ 93.3%) from the 1000 chips change by less than 3 bits with very few responses changing by 5 bits or more ($\approx$ 1%) for mean WL showing that the PUF is very stable even at high temperatures. From the two figures, the overall conclusion is that the PUF showed lower reliability due to voltage variations than due to temperature variations.

Ageing Effects: Ageing effects due to temporal variations in device parameters also affect the reliability of a device over its lifetime [23, 24, 25]. With continuous scaling in device dimensions, stronger electric fields have resulted in an increase in the number of interface traps in PMOS transistors over time at high temperatures. The increase in traps has resulted in an increase in the V_T of transistors causing reliability issues due to negative bias temperature instability (NBTI) [26, 27]. In the case of a PUF, V_T degradation of the transistors can affect the uniqueness and reproducibility of signatures. In the MECCA PUF, the transistor marked PL in the cell suffers from NBTI due to a strong electric field across gate-source ($|V_{gs}| = V_{dd}$), as shown in Fig. 9(a). Due to temporal variations in V_T, a cell A_1 (refer Fig. 9(b)) which fails at word length WL_2 can pass through that level and become successful at WL_2 (unfilled brown (dark) circle) and hence produce an incorrect bit output.

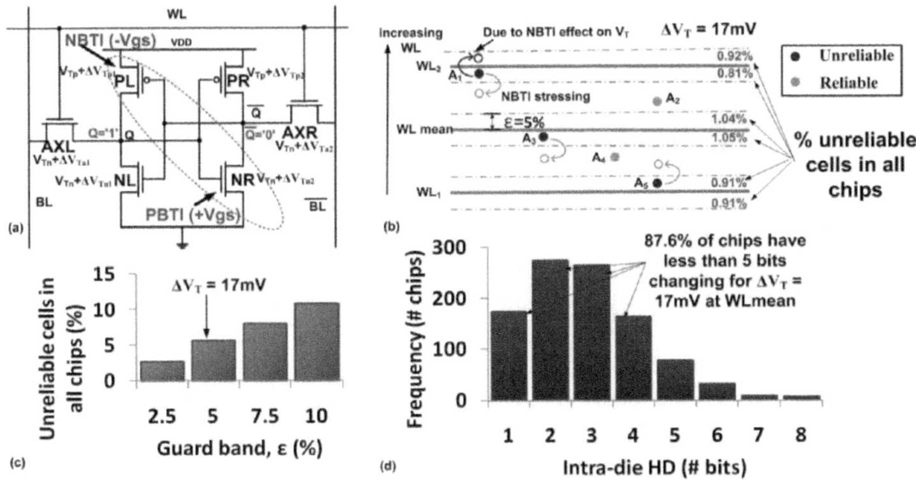

Fig. 9. Ageing effects in memory. (a) NBTI effect in SRAM cell. (b) Characterization of cells as reliable and unreliable cells for expected ΔV_T=17mV. (c) % unreliable cells in all the chips as a function of ε (and ΔV_T). (d) Number of unreliable bits per chip at mean WL.

To quantify the temporal reliability of cells, we design a guard band around the WLs based on an expected V_T shift of 17mV based on results in [22] over a 10-year lifetime of the product.Accordingly, we included the V_T shift (on top of process variations) in our PMOS model file and performed monte-carlo simulations to obtain the number of temporally unreliable cells and change in intra-die HD. The word line duty cycle controller can be used to produce additional WLs to characterize the cells post-production and identify unreliable cells (within a guard band). The numbers on the right in Fig. 9(b) show the distribution of the unreliable cells for guard band, ε=5% for ΔV_T=17mV. It can be seen that approximately only 6% of the cells (out of all cells in all chips) fall within the guard band and can potentially produce an incorrect output over the products' lifetime. The actual number of unreliable cells per chip at mean WL is shown in Fig. 9(d), from which it can be seen that 87.6% of the chips have less than 4 bits for a ΔV_T=17mV (corresponding to ≈2% unreliable cells around mean WL in Fig. 9(b)). We propose the following solutions to ageing: 1) discarding the temporally unreliable cells (by choosing more cells); 2) if the number of unreliable bits is tolerable (low avg. intra-die HD), they can be used in the signature generation; 3) intentional ageing can be done to those few unreliable cells (at high temperature and high supply voltage) to ensure that the cells are moved out of the guard band to make it a temporally reliable cell (dotted green circle in Fig. 9(b)). Additionally, since the V_T shift can be reduced by flipping contents of the cells [23, 22], it is possible that fewer cells can produce incorrect outputs due to V_T recovery during normal operation of the memory.

Table 1. Area Overhead Comparison

PUF	RO-PUF	MECCA PUF	
		including memory	w/o memory
Area $(\mu m)^2$	3122	520	21

Overhead: We computed the area overhead of the MECCA PUF and compared it with that of a RO-PUF (as an example of delay based PUF) to generate a 128-bit key. For the RO-PUF, assuming all orderings of the ROs are likely, 35 ROs are required. This would require two 35 X 1 MUXs, two 32-bit counters and one 32-bit comparator along with the 35 ROs (5 stages) to produce the key. For the MECCA PUF, we used a 128 cell SRAM array along with the peripherals (a 4 X 16 row address decoder, 8-bit I/O with buffers, sense amplifiers and precharge circuitry) and the programmable delay circuit. Both PUFs were synthesized using Synopsis *Design Compiler* and Table 1 shows the area comparison of the MECCA PUF and RO-PUF. Even if the chip has no memory, i.e. (the memory has to be implemented), the total overhead of the MECCA PUF is small compared to the RO-PUF ($\approx$ 16.6%). As mentioned earlier, since most modern designs already contain embedded memory which can be used as the PUF, the area overhead is only due to the programmable delay circuit and is extremely small compared to the RO-PUF ($\approx$ 0.6%).

5 Conclusion

In this paper, we have presented MECCA, a novel memory based PUF that exploits the intrinsic variations in static memory cells by inducing failures for cryptographic operations. We have shown that even moderate variations in the device parameters provides high-quality signatures (in terms of uniqueness, reproducibility and entropy) while incurring significantly less hardware overhead compared to other PUFs by using the embedded memory array already present in most designs. Furthermore, we analyze the effect of voltage/temperature variations as well as ageing effects on the robustness of the PUF outputs and propose solutions using temperature induced stressing to further improve the reliability. With increasing effect of parameter variations in nanoscale memory, effectiveness of the proposed method is expected to increase in future technology nodes. Extending the proposed approach to other forms of memory, e.g. flash, would be subject of future research.

References

1. ORS-LABS: Counterfeit Electronic Components - An Overview (2007), `http://www.ors-labs.com/pdf/MASH07CounterfeitDevice.pdf`
2. Kocher, P., Jaffe, J., Jun, B.: Differential power analysis. In: Wiener, M. (ed.) CRYPTO 1999. LNCS, vol. 1666, pp. 388–397. Springer, Heidelberg (1999)

3. Kulikowski, K.J., Karpovsky, M.G., Taubin, A.: DPA on faulty cryptographic hardware and countermeasures. In: Breveglieri, L., Koren, I., Naccache, D., Seifert, J.-P. (eds.) FDTC 2006. LNCS, vol. 4236, pp. 211–222. Springer, Heidelberg (2006)
4. Gassend, B., et al.: Controlled Physical Random Functions. In: Proceedings of 18th Annual Computer Security Applications Conference (2002)
5. Gassend, B., et al.: Silicon Physical Random Functions. In: Proceedings of the Computer and Communication Security Conference (2002)
6. Suh, G.E., Devadas, S.: Physical Unclonable Functions for Device Authentication and Secret Key Generation. In: Proc. DAC, pp. 9–14 (2007)
7. Semiconductor Industry Association (SIA), International Technology Roadmap for Semiconductors (ITRS) (2005)
8. Ney, A., et al.: A New Design-For-Test Technique for SRAM Core-Cell Stability Faults. In: Proc. DATE, pp. 1344–1348 (2009)
9. Wang, X., Tehranipoor, M.: Novel Physical Unclonable Function with Process and Environmental Variations. In: Proc. DATE, pp. 1065–1070 (2010)
10. Maiti, A., Schaumont, P.: Improved Ring Oscillator PUF: An FPGA-friendly Secure Primitive. Journal of Cryptology, 1–23 (2010)
11. Pappu, R.: Physical One-Way Functions, Phd thesis, Massachusetts Institute of Technology (2001)
12. Ozturk, E., Hammouri, G., Sunar, B.: Physical Unclonable Function with Tristate Buffers. In: Proc. ISCAS, pp. 3194–3197 (2008)
13. Guajardo, J., Kumar, S.S., Schrijen, G.J., Tuyls, P.: FPGA Intrinsic PUFs and Their Use for IP Protection. In: Paillier, P., Verbauwhede, I. (eds.) CHES 2007. LNCS, vol. 4727, pp. 63–80. Springer, Heidelberg (2007)
14. Holcomb, D., Burleson, W., Fu, K.: Initial SRAM State as a Fingerprint and Source of True Random Numbers for RFID Tags. In: Proc. the Conference on RFID Security (2007)
15. Su, Y., Holleman, J., Otis, B.: A Digital 1.6 pJ/bit Chip Identification Circuit Using Process Variations. In: Proc. ISSCC, pp. 15–17 (2007)
16. Kumar, S., Guajardo, J., Maes, R., Schrijen, G., Tuyls, P.: The Butterfly PUF: Protecting IP on Every FPGA. In: Proc. HOST (2008)
17. Guajardo, J., Kumar, S., Tuyls, P., Schrijen, G. : Identification Of Devices Using Physically Unclonable Functions. WIPO Patent Application WO/2009/024913 A2
18. Guyensu, T.: Using Data Contention in Dual-ported Memories for Security Applications. Journal of Signal Processing Systems (2010)
19. Gassend, B., Clarke, D., van Dijkm, M., Devadas, S.: Controlled Physical Random Functions. In: Proc. ACSAC (2002)
20. Mukhopadhyay, S., Mahmoodi, H., Roy, K.: Modeling of Failure Probability and Statistical Design of SRAM Array for Yield Enhancement in Nanoscaled CMOS. In: IEEE TCAD, pp. 1859–1880 (2005)
21. Mukhopadhyay, S., Mahmoodi, H., Roy, K.: Reduction of Parametric Failures in Sub-100-nm SRAM Array Using Body Bias. In: IEEE TCAD, pp. 174–183 (2008)
22. Luo, H., Wang, Y., He, K., Luo, R., Yang, H., Xie, Y.: Modeling of PMOS NBTI Effect Considering Temperature Variation. In: Proc. ISQED, pp. 139–144 (2007)
23. Kumar, S.V., Kim, K.H., Sapatnekar, S.S.: Impact of NBTI on SRAM Read Stability and Design for Reliability. In: Proc. ISQED (2006)
24. Paul, B.C., Kang, K., Kufluoglu, H., Alam, M.A., Roy, K.: Impact of NBTI on the Temporal Performance Degradation of Digital Circuits. IEEE Electron Devices, 560–562 (2005)

25. Kang, K., Gangwal, S., Park, S.P., Roy, K.: NBTI Induced Performance Degradation in Logic and Memory Circuits: How Effectively Can we Approach a Reliability Solution? In: Proc. ASP-DAC (2008)
26. Bansal, A., et al.: Impacts of NBTI and PBTI on SRAM Static/Dynamic Noise Margins and Cell Failure Probability. In: Microelectronics Reliability, pp. 642–649 (2009)
27. Yang, S., Yang, H., Chuang, C., Hwang, W.: Timing Control Degradation and NBTI/PBTI Tolerant Design for Write-Replica Circuit in Nanoscale CMOS SRAM. In: Proc. VLSI-DAT, pp. 162–165 (2009)
28. Predictive Technology Model, `http://www.eas.asu.edu/~ptm/`
29. Maiti, A., Casarona, J., McHale, L., Schaumont, P.: A Large Scale Characterization of RO-PUF. In: Proc. HOST, pp. 94–99 (2010)

FPGA Implementation of Pairings Using Residue Number System and Lazy Reduction*

Ray C.C. Cheung[1], Sylvain Duquesne[2], Junfeng Fan[4], Nicolas Guillermin[2,3], Ingrid Verbauwhede[4], and Gavin Xiaoxu Yao[1]

[1] Department of Electronic Engineering
City University of Hong Kong, Hong Kong SAR
r.cheung@cityu.edu.hk, gavin.yao@student.cityu.edu.hk
[2] IRMAR, UMR CNRS 6625, Université Rennes 1
Campus de Beaulieu, 35042 Rennes cedex, France
[3] DGA.IS, La Roche Marguerite - 35170 - Bruz, France
sylvain.duquesne@univ-rennes1.fr, nicolas.guillermin@m4x.org
[4] Katholieke Universiteit Leuven, COSIC & IBBT
Kasteelpark Arenberg 10, B-3001 Leuven-Heverlee, Belgium
{junfeng.fan,ingrid.verbauwhede}@esat.kuleuven.be

Abstract. Recently, a lot of progress has been made in the implementation of pairings in both hardware and software. In this paper, we present two FPGA-based high speed pairing designs using the Residue Number System and lazy reduction. We show that by combining RNS, which is naturally suitable for parallel architectures, and lazy reduction, which performs one reduction for multiple multiplications, the speed of pairing computation in hardware can be largely increased. The results show that both designs achieve higher speed than previous designs. The fastest version computes an optimal ate pairing at 126-bit security level in 0.573 ms, which is 2 times faster than all previous hardware implementations at the same security level.

Keywords: Optimal Pairing, Residue Number System, Lazy Reduction, FPGA.

1 Introduction and Motivation

Bilinear pairings on elliptic curves have been introduced in cryptography in the middle of 90's for cryptanalysis [18, 33]. In 2000, Joux introduced the first constructive use of pairings with a tripartite key exchange protocol [25]. In the last decade many pairing-based schemes such as identity-based encryption [10],

* This work was supported in part by the European Commission's ECRYPT II NoE (ICT-2007-216676), by the Belgian State's IAP program P6/26 BCRYPT, by the K.U. Leuven-BOF (OT/06/40), by the Research Council K.U. Leuven: GOA TENSE (GOA/11/007), by French ANR projects no. 07-BLAN-0248 "ALGOL" and 09-BLAN-0020-01 "CHIC", and by City University of Hong Kong Start-up Grant 7200179.

B. Preneel and T. Takagi (Eds.): CHES 2011, LNCS 6917, pp. 421–441, 2011.

identity-based signatures [12] and short signatures [11] have been proposed and studied. Compared with other popular public key cryptosystems, e.g. Elliptic Curve Cryptography (ECC) [30, 34] and RSA [41], pairing computation is much more complicated. For this reason, efficient implementation of cryptographic pairings has received increasing interests [3, 8, 16, 20, 23, 27, 36].

The computation of a pairing can be broken down into modular operations in the underlying fields. For example, one optimal ate pairing [44] defined on a 256-bit Barreto-Naehrig (BN) curve [7] requires around 10^4 modular multiplications [3]. Thus, having an efficient modular multiplier is the key step to a high performance pairing processor. In this work, we are interested in hardware implementation of pairings over large characteristic fields. In this case one possible optimization is to use lazy reduction. Lazy reduction in pairing computation was introduced by Scott [42] and then generalized by Aranha *et al.* in [3]. In short, it performs one reduction for expressions like $\sum A_i B_j$, where $A_i, B_j \in \mathbb{F}_p$. Aranha *et al.* have shown that lazy reduction can significantly speed up optimal ate pairings in software [3].

As suggested by Duquesne [14], lazy reduction can be combined with the Residue Number System (RNS) to further reduce the complexity. Besides, the RNS distributes computation over a group of small integers, and is naturally suitable for parallel implementations [22, 29]. In this paper, we propose two FPGA-based pairing processors that use RNS representation and lazy reduction. The first design, referred as Design I, is based on a general (but enhanced) RNS data-path. Design I is scalable in terms of security level and flexible in terms of target devices. On an Altera Stratix III FPGA, Design I computes a pairing at 126-bit security level in 1.07 ms. The second design, referred as Design II, has an optimized architecture for pairings over 254-bit curves. We use a set of parameters that leads to a reduced complexity and an optimized datapath that benefits from the reduction. Design II computes a pairing at 126-bit security level in 0.573 ms on a Xilinx Virtex-6 FPGA. To the best of our knowledge, these are the first hardware designs of pairing using RNS, and the designs outperform all previous hardware implementations at a similar security level [16, 19, 27].

The rest of the paper is organised as follows. Section 2 and Section 3 provide the background on optimal ate pairing and RNS, respectively. In Section 4 and Section 5 we describe the architecture of Design I and Design II. Section 6 describes the control of the data-path and the high level scheduling. We give a detailed analysis of the performance in Section 7. Finally, Section 8 concludes the paper.

2 Optimal Ate Pairings

2.1 Pairings on Barreto-Naehrig Curves

A bilinear pairing is a non-degenerate map from $\mathbb{G}_1 \times \mathbb{G}_2$ to $\mathbb{G}_T$ which is linear in both components. Popular pairings such as Tate pairing [6], ate pairing [24], R-ate pairing [32], optimal pairing [44] choose $\mathbb{G}_1$ and $\mathbb{G}_2$ to be specific cyclic subgroups of $E(\mathbb{F}_{p^k})$, and $\mathbb{G}_T$ to be a subgroup of $\mathbb{F}^*_{p^k}$.

Let $\mathbb{F}_p$ be a finite field and let E be an elliptic curve defined over $\mathbb{F}_p$. Let ℓ be a large prime dividing $\#E(\mathbb{F}_p)$ and k the embedding degree with respect to ℓ, namely, the smallest positive integer k such that $\ell | p^k - 1$. Small embedding degrees can be easily obtained using supersingular curves. However, it is too small ($k \leq 2$) if large characteristic base fields are used. Thus, we use ordinary curves with prescribed embedding degrees constructed via the complex multiplication method as surveyed in [17]. We focus on the most popular one to date, namely the Barreto-Naehrig curves [7]. The reason for their popularity is that they are well-suited for 128-bit security level and they have degree 6 twist.

Let $u \in \mathbb{Z}$ such that $p = 36u^4 + 36u^3 + 24u^2 + 6u + 1$ and $\ell = 36u^4 + 36u^3 + 18u^2 + 6u + 1$ are prime. A BN curve is an elliptic curve defined over $\mathbb{F}_p$ by

$$E : y^2 = x^3 + b,$$

where $b \neq 0$ such that $\#E = \ell$, and it has 12 as an embedding degree.

In this paper, we mainly focus on the discussion of optimal ate pairing [36] because it is the most efficient to date for BN curves. Let $r = 6u + 2$, an optimal ate pairing on BN curves is defined as follows [2, 36]:

$$\begin{aligned} a_{opt} : E(\mathbb{F}_{p^{12}}) \cap \mathrm{Ker}(\pi_p - p) \times E(\mathbb{F}_p)[\ell] &\rightarrow \mathbb{F}^*_{p^{12}} / \left(\mathbb{F}^*_{p^{12}}\right)^{\ell} \\ (Q, P) &\mapsto \left(f_{(r,Q)}(P) \cdot g_{(rQ,\pi_p(Q))}(P) \cdot g_{(rQ+\pi_p(Q),-\pi_p^2(Q))}(P)\right)^{\frac{p^{12}-1}{\ell}} \end{aligned}$$

where π_p is the Frobenius map on the curve ($\pi_p(x, y) = (x^p, y^p)$), and $g_{(Q_1,Q_2)}$ is the line through Q_1 and Q_2.

2.2 Pairing Computation and Parameter Selection

Pairing Computation. The computation consists of two main functions, $f_{(r,Q)}$ and $f^{\frac{p^{12}-1}{\ell}}$. The function $f_{(r,Q)}$ has the following divisor:

$$\mathrm{div}(f_{(r,Q)}) = r(Q) - (rQ) - (r - 1)(\mathcal{O}).$$

It is normally computed using a double-and-add method (also known as Miller's loop [35]). Concerning $f^{\frac{p^{12}-1}{\ell}}$, also known as the final exponentiation, Koblitz and Menezes show in [31] that it can be split in two steps due to the integer factorization

$$\frac{p^{12} - 1}{\ell} = \left(p^6 - 1\right)\left(p^2 + 1\right)\left(\frac{p^4 - p^2 + 1}{\ell}\right).$$

The first step is powering to $p^6 - 1$ and to $p^2 + 1$. This is easily obtained via cheap Frobenius computations and an inversion. The second step is powering to $\frac{p^4-p^2+1}{\ell}$ which is called the hard part of the final exponentiation.

Parameter Selection. The selection of parameters has an essential impact on the security and the performance of a pairing computation. It is explained in [39] how to generate BN curves with nice properties. For this work we choose two curves both with $b = 2$. The first one, BN_{126}, is defined by $u = -(2^{62} + 2^{55} + 1)$ and has already been used in [2, 39]. It ensures only 126 bits of security, but is well suited to registers on a general-purpose CPU when lazy reduction is used. However, FPGA architectures have no such constraints, since multipliers of larger size can always be constructed with DSP slices. Hence, we also consider a curve BN_{128} defined by $u = -(2^{63}+2^{22}+2^{18}+2^{7}+1)$ which ensures 128 bits of security. Finally, we propose BN_{192}, the BN curve defined by $u = -(2^{160} + 2^{74} + 2^{12} + 1)$. Optimal pairing defined on this curve provides 192 bits of security [38].

In the three cases, the extension fields are defined as follows:

- $\mathbb{F}_{p^2} = \mathbb{F}_p[\mathbf{i}]/(\mathbf{i}^2 + 1)$
- $\mathbb{F}_{p^6} = \mathbb{F}_{p^2}[\boldsymbol{\beta}]/(\boldsymbol{\beta}^3 - (1 + \mathbf{i}))$
- $\mathbb{F}_{p^{12}} = \mathbb{F}_{p^6}[\boldsymbol{\Gamma}]/(\boldsymbol{\Gamma}^2 - \boldsymbol{\beta}) = \mathbb{F}_{p^2}[\gamma]/(\gamma^6 - (1 + \mathbf{i}))$

This tower of extensions has many advantages, the most important being an efficient multiplication algorithm for the canonical polynomial base.

Note that BN curves always have degree 6 twists. This means E is isomorphic over $\mathbb{F}_{p^{12}}$ to a curve E' defined by $y^2 = x^3 + \frac{b}{\zeta}$, where ζ is neither a square nor a cube in $\mathbb{F}_{p^2}$. In our case we take $\zeta = 1 + \mathbf{i}$ so that it defines both the sextic extension of $\mathbb{F}_{p^2}$ and the twist. Then we can define twisted versions of pairings on $E'(\mathbb{F}_{p^2}) \times E(\mathbb{F}_p)[\ell]$. In other words, the coordinates of Q can be written as $(x_Q\zeta^{\frac{1}{3}}, y_Q\zeta^{\frac{1}{2}})$ where x_Q and y_Q are in $\mathbb{F}_{p^2}$. For u selected as a negative integer, the optimal ate pairing for BN curves is computed by Algorithm 1 [2] where `dbl`, `add` and `hard-part` are given in appendix.

3 Residue Number System

A Residue Number System (RNS) represents a large integer using a set of smaller integers. Let $\mathfrak{B} = \{b_1, b_2, \ldots, b_n\}$ be a set of pairwise co-prime integers, and $M_{\mathfrak{B}} = \prod_{i=1}^{n} b_i$. For any integer X, $0 \leq X < M_{\mathfrak{B}}$, there is a unique RNS representation on $\mathfrak{B}$: $\{X\}_{\mathfrak{B}} = \{x_1, x_2, \ldots, x_n\}$, where $x_i = |X|_{b_i}$, $1 \leq i \leq n$. Throughout the paper we use $|a|_b$ to denote $a \bmod b$. Given $\{X\}_{\mathfrak{B}}$, one can recover X using the Chinese Remainder Theorem (CRT):

$$X = \left| \sum_{i=1}^{n} \left| x_i \cdot B_i^{-1} \right|_{b_i} \cdot B_i \right|_{M_{\mathfrak{B}}} \quad \text{where}\, B_i = \frac{M_{\mathfrak{B}}}{b_i}. \tag{1}$$

The set $\mathfrak{B}$ is also known as a *base*, and each element b_i, $1 \leq i \leq n$, is called an RNS modulus or an RNS channel.

RNS representation admits efficient parallel computations. Consider two integers X, Y and their RNS representations $\{X\}_{\mathfrak{B}} = \{x_1, x_2, \ldots, x_n\}$ and $\{Y\}_{\mathfrak{B}} = \{y_1, y_2, \ldots, y_n\}$, then we have

$$\{|X \odot Y|_{M_{\mathfrak{B}}}\}_{\mathfrak{B}} = \{|x_1 \odot y_1|_{b_1}, \ldots, |x_n \odot y_n|_{b_n}\}, \odot \in \{+, -, \times, /\}. \tag{2}$$

Algorithm 1. Optimal ate pairing on BN curves for $u < 0$

Require: $P \in E(\mathbb{F}_p)[\ell], Q = (x_Q\gamma^2, y_Q\gamma^3) \in E(\mathbb{F}_{p^{12}}) \cap \mathrm{Ker}(\pi_p - p)$ with x_Q and $y_Q \in \mathbb{F}_{p^2}, r = |6u+2| = \sum_{i=0}^{s-1} r_i 2^i$, where $u < 0$.
Ensure: $a_{opt}(Q, P) \in \mathbb{F}_{p^{12}}$

$T = (X_T\gamma^2, Y_T\gamma^3, Z_T) \leftarrow (x_Q\gamma^2, y_Q\gamma^3, 1), f \leftarrow 1$
for $i = s-2$ downto 0 **do**
  $T, g \leftarrow$ `dbl`$(T, P), f \leftarrow f^2 \cdot g$
  if $r_i = 1$ **then**
    $T, g \leftarrow$ `add`$(T, Q, P), f \leftarrow f \cdot g$
  end if
end for
$T \leftarrow -T, f \leftarrow f^{p^6}$ (f^{p^6} is equivalent to f^{-1} as noticed in [3])
$Q_1 \leftarrow \pi_p(Q), Q_2 \leftarrow -\pi_p(Q_1)$
$T, g \leftarrow$ `add`$(T, Q_1, P), f \leftarrow f \cdot g$
$T, g \leftarrow$ `add`$(T, Q_2, P), f \leftarrow f \cdot g$
$f \leftarrow \left(f^{p^6-1}\right)^{p^2+1}$
$f \leftarrow$ `hard-part`$(f, |u|)$
return f

Note that the division is available only if Y is co-prime with $M_{\mathfrak{B}}$. For all these operations, computations between x_i and y_i have no dependency on other channels, which makes RNS naturally suitable for parallel implementations.

The computation of $|X|_{b_i}$ is called a channel reduction. To accelerate this operation, pseudo-Mersenne numbers of the form $b_i = 2^{\mathrm{w}} - \varepsilon_i$, where $\varepsilon_i < 2^{\mathrm{w}/2}$, are typically selected as RNS moduli. Hence, the computation of $|X|_{b_i}$ is performed using 2 times of $X \leftarrow \lfloor X/2^{\mathrm{w}} \rfloor \cdot \varepsilon_i + (X \bmod 2^{\mathrm{w}})$, and a correction step in the end to bring the result back to the range $[0, b_i)$.

3.1 RNS Montgomery Reduction

Using RNS representation ensures efficient computation in $\mathbb{Z}/M_{\mathfrak{B}}\mathbb{Z}$. Unfortunately, it can't be applied directly in $\mathbb{F}_p$ since $M_{\mathfrak{B}}$ is not prime. One way to utilize RNS for field multiplication is to combine RNS and Montgomery reduction [22, 29]. This is shown in Algorithm 2.

RNS Montgomery reduction requires two bases, $\mathfrak{B}$ and $\mathfrak{C}$, with $M_{\mathfrak{C}}$ co-prime to $M_{\mathfrak{B}}$. The reason of including $\mathfrak{C}$ is that division by $M_{\mathfrak{B}}$ is not possible in $\mathfrak{B}$. Note that the size of $M_{\mathfrak{B}}$ and $M_{\mathfrak{C}}$, compared to p, determine the upper bound of input X. Guillermin found that if $X < \alpha p^2$, $M_{\mathfrak{B}} > \alpha p$ and $M_{\mathfrak{C}} > 2p$, then Algorithm 2 has output $S < 2p$ [22, Proposition 1]. This is an important principle for base selection.

3.2 Base Extension

The operation to transform the representation in one RNS base to another base is called Base Extension (BE). To compute $\{X\}_{\mathfrak{C}} = \{x'_1, x'_2, \dots, x'_n\}$ from $\{X\}_{\mathfrak{B}} =$

Algorithm 2. RNS Montgomery reduction [5]

Require: RNS bases $\mathfrak{B}$ and $\mathfrak{C}$ with $M_{\mathfrak{B}} > \alpha p, M_{\mathfrak{C}} > 2p$, p coprime with $M_{\mathfrak{B}} M_{\mathfrak{C}}$, $\{X\}_{\mathfrak{B}}$ and $\{X\}_{\mathfrak{C}}$ being the RNS representations of $X < \alpha p^2$.
Precomputed: $\{|-p^{-1}|_{M_{\mathfrak{B}}}\}_{\mathfrak{B}}$, $\{|M_{\mathfrak{B}}^{-1}|_{M_{\mathfrak{C}}}\}_{\mathfrak{C}}$ and $\{p\}_{\mathfrak{C}}$.
Ensure: $\{S\}_{\mathfrak{B}}$, $\{S\}_{\mathfrak{C}}$ such that $|S|_p = |XM_{\mathfrak{B}}^{-1}|_p$ and $S < 2p$

```
{Q}_B ← {X}_B × {−p^{−1}}_B
{Q}_B --Base Extension--> {Q}_C
{S}_C ← ({X}_C + {Q}_C × {p}_C) × {M_B^{−1}}_C
{S}_B <--Base Extension-- {S}_C
```

$\{x_1, x_2, \ldots, x_n\}$, one can use the Posch-Posch method [40]. Given $\{X\}_{\mathfrak{B}}$, for (1), there must exist an integer $\lambda < n$ such that:

$$X = \left| \sum_{i=1}^{n} \left| x_i \cdot B_i^{-1} \right|_{b_i} \cdot B_i \right|_{M_{\mathfrak{B}}} = \left| \sum_{i=1}^{n} \xi_i \cdot B_i \right|_{M_{\mathfrak{B}}} = \sum_{i=1}^{n} \xi_i \cdot B_i - \lambda \cdot M_{\mathfrak{B}} \quad (3)$$

where $\xi_i = \left| x_i \cdot B_i^{-1} \right|_{b_i}$, $1 \le i \le n$. In the Posch-Posch method, λ can be calculated by the following equation:

$$\lambda = \left\lfloor \sum_{i=1}^{n} \frac{\xi_i \cdot B_i}{M_{\mathfrak{B}}} \right\rfloor = \left\lfloor \sum_{i=1}^{n} \frac{\xi_i}{b_i} \right\rfloor \quad (4)$$

In [29], ξ_i / b_i is further approximated by $\xi_i / 2^{\mathrm{w}}$ as b_i is chosen as a pseudo-Mersenne number near 2^{w}. Once λ is obtained, $\{X\}_{\mathfrak{C}} = \{x'_1, \ldots, x'_n\}$ can be computed as follows:

$$x'_j = \left| \sum_{i=1}^{n} \xi_i \cdot B_i - \lambda \cdot M_{\mathfrak{B}} \right|_{c_j} = \left| \sum_{i=1}^{n} \xi_i \cdot |B_i|_{c_j} - \lambda \cdot |M_{\mathfrak{B}}|_{c_j} \right|_{c_j} . \quad (5)$$

$|B_i|_{c_j}$ and $|M_{\mathfrak{B}}|_{c_j}$, $1 \le i, j \le n$, can be precomputed once $\mathfrak{B}$ and $\mathfrak{C}$ are fixed. The following algorithm describes the computation of $|X|_{c_k}$.

4 Design I: A Scalable Architecture

In this section we propose an enhancement of the Cox-Rower architecture first proposed in [29] such that it is suitable for pairing computation.

4.1 Cox-Rower Architecture

The Cox-Rower architecture was first proposed by Kawamura *et al.* in [29]. It was first implemented in a VLSI design of an RSA cryptosystem [37]. It was later enhanced by Guillermin [22] to support all arithmetic operations in $\mathbb{F}_p$ and fast Elliptic Curve Scalar Multiplications (ECSM).

Algorithm 3. Base extension algorithm for k-th element of X [29]

Require: $|X|_{b_i}$ for $i \in \{1, .., n\}$
Ensure: $|X|_{c_k}$
    Precomputed: $|B_i^{-1}|_{b_i}, |B_i|_{c_k}, 1 \leq i \leq n, |M_{\mathfrak{B}}|_{c_k}$
1: $\xi_i \leftarrow |x_i \cdot B_i^{-1}|_{b_i}$ for $i \in \{1, .., n\}$
2: $\psi \leftarrow 0, z \leftarrow 0$
3: **for** $i = 0$ to $(n-1)$ **do**
4:     $\psi \leftarrow \psi + \xi_i$
5:     $\rho \leftarrow \lfloor \psi / 2^{\mathbf{w}} \rfloor$ $// \rho \in \{0, 1\}$
6:     $\psi \leftarrow |\psi|_{2^{\mathbf{w}}}$
7:     $z \leftarrow z + (\xi_i \cdot |B_i|_{c_k}) - \rho \cdot |M_{\mathfrak{B}}|_{c_k}$
8: **end for**
9: **return** z

Fig. 1 shows our Cox-Rower implementation. The top-level structure is similar to the one proposed by [29]. It is consists of n similar Rower units performing in parallel operations in one RNS channel of $\mathfrak{B}$ or $\mathfrak{C}$. The Cox unit is only used during the base extension (Algorithm 3) to generate the ρ. The Rower unit normally consists of a multiplier and channel reduction logic, and serves as the workhorse of both multiplication (operation $\times$ in (2)) and reduction (step 1, 3 in Algorithm 2 and step 1, 7 in Algorithm 3). The Cox-Rower is driven by a microcoded sequencer, and the sequencer can be easily reprogrammed to provide support for different algorithms.

While the basic structure of our design resembles the Cox-Rower architecture in Guillermin's ECC processor [22], our architecture has an optimized pipeline structure and a more aggressive memory organization specifically designed for pairing support.

4.2 Cox-Rower Parametrization for Pairing

First we need to parametrize the Cox-Rower to support the pairings on the three curves defined in Section 2.2. Popular Altera and Xilinx FPGAs have embedded 18×18 multipliers or even 25×18 multipliers (Virtex-5 and higher). Moreover, a full 36×36 multiplier can be built by efficiently combining several such small multipliers. It is thus natural to select $\mathbf{w}$=18 or $\mathbf{w}$=36. For the implementation of BN_{126} or BN_{128}, $\mathbf{w}$=36 gives the best trade-off. Indeed the gain in frequency brought by smaller multipliers and adders does not compensate the necessary cycle surplus of reductions. The parameter n can then be set to 8 for BN_{126} and BN_{128}, and 19 for BN_{192}. Because of our chosen arithmetic (see Section 6), the value α defined in Section 3.1 reaches 198 for all the curves. The worst case is reached during the schoolbook multiplication in $\mathbb{F}_{p^{12}}$. One can verify that, in order to have $M_{\mathfrak{B}} > \alpha p$, $\mathbf{w}$=33 is enough to support BN_{126} and w=34 for both BN_{126} and BN_{128}. For BN_{192}, $\mathbf{w}$ is set to 35. In this architecture, we used the same method to select bases as in [22], with pseudo-Mersennes of the form $t_i = 2^{\mathbf{w}} - \varepsilon_i$, $t_i \in \mathfrak{B} \cup \mathfrak{C}$ and ε_i positive.

An adaptation of Guillermin's architecture is necessary to provide pairing support. As the number of local variables and precomputations is much larger for pairings than that of ECSM (in fact, the Miller's loop has a built-in ECSM), we use a single triple port RAM of 256 words instead of the ROM and a group of 16 registers to store precomputed values and temporary results. This is enough to support all curves listed in Section 2.2.

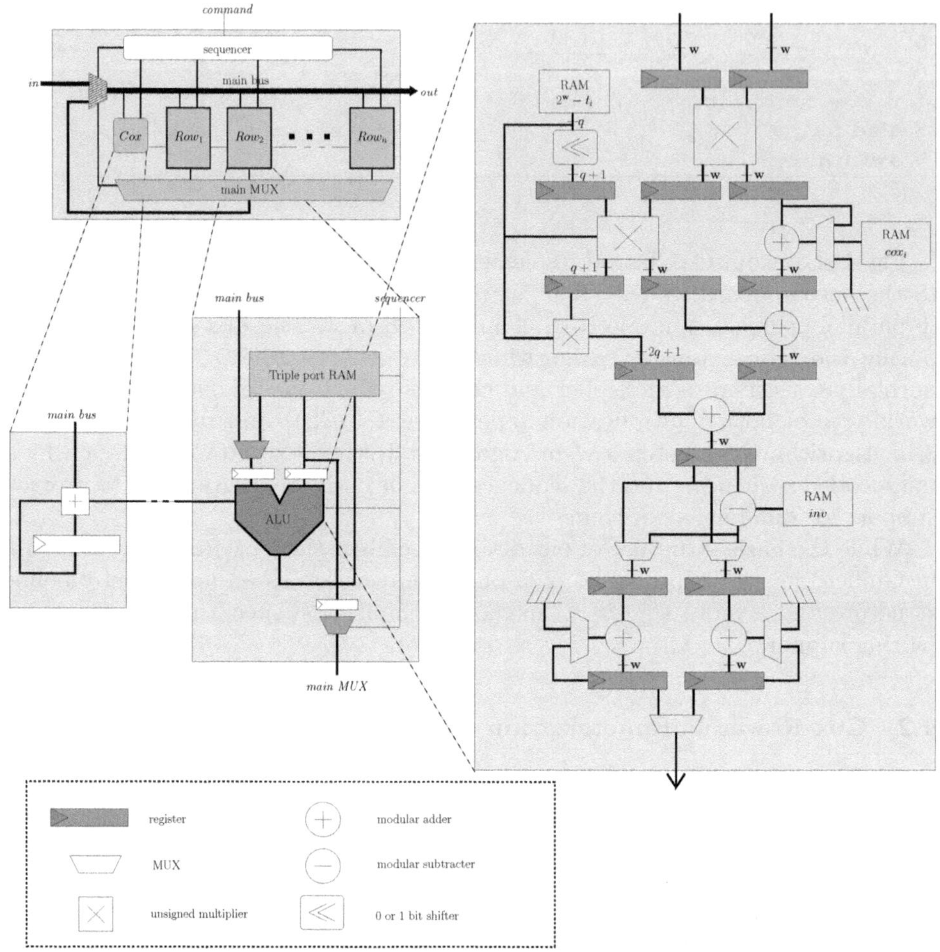

Fig. 1. Design I: architecture of the Cox-Rower and its pipeline structure

4.3 Pipeline Architecture

Our goal is to keep the maximal frequency already available in [22]. Additive hardware must be carefully introduced, to keep the critical path under control. On the other hand, we can easily raise the pipeline depth without a lot of cycle loss. Indeed, pairing computation has more parallelizable operations than classical elliptic curve scalar multiplication over $\mathbb{F}_p$. The architecture in [37] and [22] uses 2-stage and 5-stage pipelines, respectively. For the implementation of pairings, we found that a pipeline of up to 10 stages can still be efficiently filled during the whole pairing computation except for $\mathbb{F}_p$ inversion.

Based on Guillermin's architecture, two accumulators are included in the pipeline to efficiently support pairing computation. Fig. 1 shows the adapted pipeline architecture. The first 4 stages perform $|a \times b|_{t_i}$, where $t_i \in \mathfrak{B} \cup \mathfrak{C}$. We also introduce shift and MUX units in the first 4 stages to compute $|2 \times a \times b|_{t_i}$, which are utilized to accelerate squarings in the extension fields. Three multipliers ($\mathbf{w} \times \mathbf{w}$, $\mathbf{w}\times(\mathbf{q}+1)$ and $(\mathbf{q}+1)\times\mathbf{q}$) are used, where $\mathbf{q}=\lceil \log_2 \varepsilon_i \rceil$.

In the 6th stage we implemented two independent accumulators. They are preceded by a subtracter (which computes $|-a\times b|_{t_i}$) and a MUX in the 5th stage. The output of the first 4 stages can then be independently added, subtracted or ignored by both accumulators. These two accumulators, together with the use of tower extensions, save a lot of cycles during the pairing computation. See Section 6 for details.

On this architecture, an RNS multiplication costs 2 cycles and the results can be accumulated immediately. An RNS reduction costs $2n+3$ cycles as previously described in [22].

5 Design II: Hardware/Algorithm Co-optimization

In this section, we propose an optimized design that achieves an even higher throughput. The improvement comes mainly from two tricks: a set of *good* bases that admits a less-expensive base extension, and a fine-tuned pipeline structure that allows a higher frequency.

5.1 Base Selection Revisited

The core observation here is that the complexity of base extension can be reduced if the moduli in the two bases are close to each other. The base extension is the most computational expensive operation in the RNS Montgomery algorithm. It requires n^2 times of $\mathbf{w}\times\mathbf{w}$ multiplications. Indeed, (5) can be written as a matrix multiplication below.

$$\begin{pmatrix} x'_1 \\ \vdots \\ x'_n \end{pmatrix} \equiv \begin{pmatrix} |B_1|_{c_1} & \cdots & |B_n|_{c_1} \\ \vdots & \ddots & \vdots \\ |B_1|_{c_n} & \cdots & |B_n|_{c_n} \end{pmatrix} \begin{pmatrix} \xi_1 \\ \vdots \\ \xi_n \end{pmatrix} - \gamma \begin{pmatrix} |M_{\mathfrak{B}}|_{c_1} \\ \vdots \\ |M_{\mathfrak{B}}|_{c_n} \end{pmatrix} \tag{6}$$

Note that the elements in the matrix, $|B_i|_{c_j}$, $1 \le i, j \le n$, are constants and are generated as follows:

$$|B_i|_{c_j} = \left| \prod_{k=1,k\neq i}^{n} b_k \right|_{c_j} = \left| \prod_{k=1,k\neq i}^{n} (b_k - c_j) \right|_{c_j} \tag{7}$$

Define $\tilde{B}_{i,j} := \prod_{k=1,k\neq i}^{n}(b_k - c_j)$. When b_k and c_j are close to each other, the difference $b_k - c_j$ is small. In practice, $|\tilde{B}_{i,j}|$ could be much smaller than $|c_j|$ if n is relatively small. Note that using $|B_i|_{c_j}$ or $\tilde{B}_{i,j}$ makes no difference in the final results due to the channel reduction on the products.

Furthermore, $\tilde{B}_{i,j}$ for $1 \leq i, j \leq n$ will be predictably divisible by 2^{n-2} if c_j is odd. Consider the two bases $\mathfrak{B} = \{b_1, b_2, \cdots, b_n\}$ and $\mathfrak{C} = \{c_1, c_2, \cdots, c_n\}$. Since there is at most one even number in $\mathfrak{B} \cup \mathfrak{C}$, $(b_i - c_j)$ is divided by 2 unless b_i or c_j is even. In practice, $\tilde{B}_{i,j}$ can have more than $n - 2$ zero bits in the least significant bits (LSBs). To shrink the operand size of the matrix multiplications, we can use truncate the least significant zeros of $\tilde{B}_{i,j}$, and restore the correct results by a simple left-shift. We denote $\tilde{B}'_{i,j}$ the $\tilde{B}_{i,j}$ after truncation.

Considering the size of p and the Cox-Rower architecture, we again select $n = 8$ and thus b_i (and c_j) close to 2^{33}. The selection of moduli also takes into account the size of $\tilde{B}'_{i,j}$ for $1 \leq i, j \leq n$. The following 16 moduli ($\mathbf{w} = 33$) were selected as the bases.

$$\begin{aligned}
\mathfrak{B} &= \{2^{\mathbf{w}} - 1, 2^{\mathbf{w}} - 9, 2^{\mathbf{w}} + 3, 2^{\mathbf{w}} + 11, 2^{\mathbf{w}} + 5, \ \ 2^{\mathbf{w}} + 9, \ \ 2^{\mathbf{w}} - 31, 2^{\mathbf{w}} + 15\},\\
\mathfrak{C} &= \{2^{\mathbf{w}}, \qquad 2^{\mathbf{w}} + 1, 2^{\mathbf{w}} - 3, 2^{\mathbf{w}} + 17, 2^{\mathbf{w}} - 13, 2^{\mathbf{w}} - 21, 2^{\mathbf{w}} - 25, 2^{\mathbf{w}} - 33\}.
\end{aligned}$$

After applying the truncation of zero bits, we manage to reduce the bitlength of all $|B_{i,j}|$ (actually, $\tilde{B}'_{i,j}$) from standard 34 to 25. While this complexity reduction seems negligible, it admits notable savings in hardware design. A detailed analysis of the base selection is given in Appendix A.2.

5.2 A Fine-tuned Rower for Pairing Computation

Our refinement focuses solely on the design of Rowers. Fig. 2(c) shows the overview of a Rower. It consists of a dual mode multiplier, 3 accumulators, a channel reduction module, a channel adder, a triple port RAM for multiplier inputs and a RAM for adder inputs.

Dual Mode Multiplier. The multiplier in the Rower is built to support the bases selected in Section 5.1. There are two types of multiplication executed in an RNS multiplication: 34×34 multiplication (step 1, 3 of Algorithm 2) and 25×35 multiplication (step 1, 7 of Algorithm 3). The DSP slices in recent Xilinx FPGAs are made of a signed 25×18 bit multiplier, a 17-bit left shifter, and an accumulator. The dual mode multiplier is built with four DSP slices, which supports either 34×34 unsigned multiplication or two signed 35×25 multiplications in parallel. Fig. 2(a) illustrates the structure of the dual mode multiplier. For pairing computation, multiplication by small constants (2, 3, 6) is widely employed. Therefore, a constant multiplier is also included in the pipeline.

Because two multiplications are executed simultaneously in the base extension, the number of cycles to perform the Cox-Rower algorithm (Algorithm 3) is reduced from n to $n/2$. This is a significant speedup of the base extension.

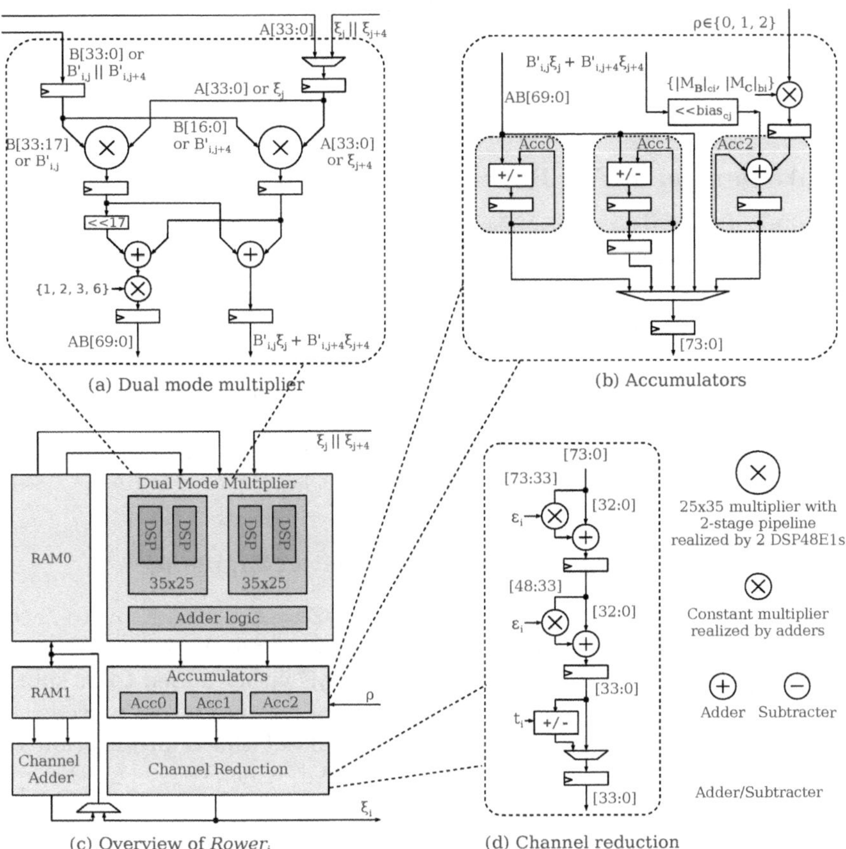

Fig. 2. The fine-tuned Rower architecture

Other Optimizations. The architecture of channel reduction is shown in Fig. 2(d). As the Hamming weights of all the ε_i are either equal to or less than 3 in non-adjacent form, multiplication by ε_i can be realized by 4 adders instead of multipliers. In order to maintain a high operating frequency, we use a three-stage pipeline to realize the channel reduction.

To achieve the maximum usage of the multiplier, a separate adder together with a dedicated RAM is included. Because of the lazy reduction inside the Rower, the write port of RAM0 is not always occupied by the multiplier. Therefore, the addition and the multiplication can run in parallel. While most of the additions in the pairing computation are performed by the accumulators, the separate adder is useful in point additions and doublings.

While this architecture requires less cycles for reduction than Design I, it is less scalable. Indeed, in order to support larger p, we need to choose larger n. The bitlength of $B_{i,j}$ increases quickly when n goes up, and two multiplications in parallel becomes impossible on the current Rower.

6 Scheduling the Pairing Algorithm

In this section we present the implementation choices for every step of the pairing computation.

6.1 Arithmetic in $\mathbb{F}_{p^2}$: Back to the Schoolbook Method

Karatsuba and derived interpolation methods have been used intensively in field operations in the literature to save the expensive multiplication [3, 8, 23, 42]. Karatsuba uses 3 instead of 4 multiplications at the cost of 3 extra additions. In a normal positional number system, this method saves computation power, as multiplications are much more expensive than additions. However, in RNS, the complexities of a multiplication and an addition are the same. Hence, the schoolbook method involves less operations (counting both additions and multiplications) and is preferred. On both architectures, a full $\mathbb{F}_{p^2}$ multiplication finishes in 8 cycles, and a squaring in only 6 cycles.

6.2 Arithmetic in $\mathbb{F}_{p^{12}}$: Interpolation with Parsimony

Let $X = \{x_1, \cdots, x_{12}\}$, $Y = \{y_1, \cdots, y_{12}\}$ and $Z = X \times Y \in \mathbb{F}_{p^{12}}$. To accelerate $\mathbb{F}_{p^{12}}$ arithmetic, we aim to execute only once $x_i \times y_j$, for all $\{i, j\} \in \{1, \cdots, 12\}^2$. As $\mathbf{i}^2 = -1$ and $\gamma^6 = (1 + \mathbf{i})$, the result of $x_i \times y_j$ is either added to or subtracted from at most two components of Z. This is the reason for the presence of two independent accumulators. Moreover, the result of the multiplication can be multiplied by 2 on the fly before it is accumulated, which speeds up the squaring in $\mathbb{F}_{p^{12}}$.

We estimate if the interpolation techniques at higher levels may save cycles or not. We exclude the reduction step, therefore the cycle count is equivalent on the flexible as well as the optimized design. We only consider Karatsuba in $\mathbb{F}_{p^{12}}$ over $\mathbb{F}_{p^6}$ which is the best case for these techniques (interpolation at other levels $\mathbb{F}_{p^{12}}/\mathbb{F}_{p^4}$, $\mathbb{F}_{p^6}/\mathbb{F}_{p^2}$ or $\mathbb{F}_{p^4}/\mathbb{F}_{p^2}$ would give worse results). Four different types of multiplications and squaring are needed during pairing, each one requiring a different algorithm:

- Squaring during the Miller loop: the operand of this squaring has no specific structure. We propose to use schoolbook squaring. Thanks to the multiplication by 2 included in the pipeline of both Design I and Design II, a squaring costs 156 cycles. The interpolation on $\mathbb{F}_{p^{12}}/\mathbb{F}_{p^6}$ leads to the following formula: $(X_H + \boldsymbol{\Gamma} X_L)^2 = \big((X_H + X_L)(X_H + \boldsymbol{\Gamma}^2 X_L) - (1 + \boldsymbol{\Gamma}^2) X_H X_L\big) + \boldsymbol{\Gamma}(2 X_H X_L)$ (with $X_H, X_L \in \mathbb{F}_{p^6}$). We found no method to go below 156 cycles on both designs.
- Multiplication by the line in the Miller loop: half the values of B are equal to 0, therefore the schoolbook multiplication costs 144 cycles. Interpolation techniques do not manage to take the advantage of the half size operand.
- Squaring during final exponentiation: during the hard part of FE, the operands to be squared are in $G_{\Phi_6}(\mathbb{F}_{p^2})$. We use the formulae given in [21] (Algorithm

A.1). On Design I, we need 84 cycles to implement it. This approach is much more efficient than interpolation techniques. [1]

- Multiplication during final exponentiation: Without counting pipeline bubbles, Karatsuba requires 278 and 254 cycles on Design I and Design II, respectively, while the schoolbook method requires 288 on both. Note that there are only 20 such multiplications in the whole pairing for BN_{126}, and the savings are really limited. Because of the moderate gain and the complication brought to the sequencer, we decided not to implement it.

6.3 $\mathbb{F}_p$ Inversion

In the final exponentiation of pairing, an $\mathbb{F}_{p^{12}}$ inversion is required. It can be done with only one inversion, 35 reductions and additional multiplications/additions in $\mathbb{F}_p$ [14, 23], but the remaining inversion in $\mathbb{F}_p$ is very expensive. Since comparison in RNS is difficult, inversion through exponentiation ($X^{-1} \equiv X^{p-2} \bmod p$) is used. For this operation, we use a simple square and multiply algorithm with least significant bit first. It is more memory consuming, but it allows to perform multiplications in parallel. On a pipelined datapath, LSB-first exponentiation is more efficient than the MSB-first method. In total, an inversion needs $\lfloor log(p) - 1 \rfloor$ squarings and many multiplications, but the cost of multiplications is hidden in the pipeline.

6.4 Higher Level Scheduling

Based on the operation over $\mathbb{F}_p$ and its extension fields, we are ready to implement the whole pairing computation. The following two points are important to efficiently utilize the datapath and to have limited memory usage:

- control of the number of local variables to limit the size of RAMs;
- control of the dependency between operations, to avoid pipeline bubbles.

For the first point, the step which requires the most live local variables is in the `hard-part` of the final exponentiation. We use the algorithm described by Scott et al. [43], which is to date the fastest way to compute the `hard-part` (Algorithm A.1). To keep its full power while limiting the number of local variables (with a classical register allocation technique) we slightly rearranged their original formulae. For the second point, excluding the $\mathbb{F}_p$ inversion where idle states cannot be avoided, the most constrained part is the $\mathbb{F}_{p^2}$ arithmetic in the Miller loop. Therefore we rearranged the projective coordinates addition and doubling formulae to emphasize the inherent parallelism. Formulas can be found in Algorithm 4 and 5 in appendix. On both architectures, we managed to eliminate almost all pipeline bubbles in the Miller loop. Idle states on the multiplier remain less than %1 of the time on both architectures in the pairing computation, excluding the $\mathbb{F}_p$ inversion.

[1] More recently, Karabina introduced a compressed form for elements in $\mathbb{F}_{p^{12}}$ which require less operations to be squared [3, 28]. Unfortunately, this method involves extra inversions so that it is not suitable for our designs.

7 Implementation Results and Analysis

7.1 Area

The prototype of the proposed pairing coprocessors were implemented on commercial FPGAs. Table 1 gives the logic utilization of both designs.

Design I. We synthesized the first design for $n = 8$ on three different FPGAs :
- EP2C35: A low cost (less than \$100) 65 nm node Altera FPGA. It is designed for industrial series production;
- EP2S30: A 65 nm node high end series of Altera. It is picked for the sake of comparison with the Xilinx Virtex-4 used in [15].
- EP3SE50: A 40 nm node high end FPGA. Note that it is the smallest FPGA of the series. BN_{192} ($n = 19$) is also implemented on this device.

The area consumption is given in ALM, the equivalence of the Xilinx Virtex-4 slice, and in LE for the Cyclone, which can be considered as a simple 4×4 LUT with carry chain and registers. We refer the reader to [1] for details.

We let the synthesizer decide how to implement multiplications (with speed constraints). Note that this choice lead to the maximal use of DSP blocks. We used also embedded RAMs: the RAM of each Rower is built the M4k or M9k (depending on the FPGA). We gave no instructions to the design suite for the implementation of the sequencer (containing 20 kB microcode), therefore they were placed in the big available memories. The same design provides support for the two curves defined in 2.2, with the only difference being the content of the RAM blocks (precomputed values and the sequencer).

Design II. Design II is implemented on a Xilinx Virtex-6 XC6VLX240T-2 FPGA, which embeds 25×18 DSP slices. As there are 8 Rowers and each Rower contains 4 DSP slices, the total number of DSPs is 32. Data RAMs used in each Rower are implemented with distributed memory blocks. The block RAMs (BRAMs) serve as the microcode sequencer. Thanks to the fine-tuned pipeline, the coprocessor can operate at 250MHz.

Table 1. Logic Utilization

	n	Device	Freq.	Multipliers	Logic Elements	Data Memory	Sequencer
Design I	8	Cyclone II	91 MHz	35 18-bit Mult.	14274 LE	32 M4k	35 M4k
		Stratix II	165 MHz	72 DSP18el	4227 ALMs	32 M4k	1 M512
		Stratix III	165 MHz	72 DSP18el	4233 ALMs	16 M9k	1 M144 +2 M9k
	19	Stratix III	131 MHz	171 DSP18el	9910 ALMs	38 M9k	1 M144 +2 M9k
Design II	8	Virtex-6	250 MHz	32 DSP48E1s	7032 Slices	-	45 18Kb BRAMs

7.2 Performance

Table 2 gives number of cycles used by the sub-functions used in optimal ate pairing on both Design I and Design II. Due to the carefully selected bases and dual mode multiplier, Design II achieved a lower cycle count. The improvement mainly comes from the speedup of RNS reduction: 19 cycles on Design I compared to 12 cycles on Design II when $n = 8$.

Table 2. Cycle count in one optimal pairing

	Curve	Mul./Red. in $\mathbb{F}_p$	2T and $g_{(T,T)}(P)$	T+Q and $g_{(T,Q)}(P)$	f^2	$f \cdot g$	Miller's Loop	Final Exp.	Total
Design I	BN_{126}	2 / 19	507	581	384	372	86,530	89,581	176,111
	BN_{128}						92,480	94,101	192,502
	BN_{192}	2 / 41	947	1153	648	636	401,565	388,284	789,849
Design II	BN_{126}	2 / 12	320	430	301	289	61,116	81,995	143,111

7.3 Comparison and Discussion

Table 3 lists the performance of software and hardware implementations reported in recent literature. Both Design I and II achieve a better performance than previous hardware implementations [16, 19, 27]. Due to the use of different platforms, a fair comparison is difficult. Nevertheless, compared with the design of [16] which also uses Virtex-6, Design II achieves a speedup of factor 2. The software implementations achieve very high performance. Although we did not break the software record, the speed of our design is already close to that of software.

Table 3. Performance comparison of software and hardware implementations of pairings

Design	Pairing	Security [bit]	Platform	Algorithm	Area	Freq. [MHz]	Cycle [$\times 10^3$]	Delay [ms]
Design I	optimal ate	126	Altera FPGA (Cyclone II)	RNS Montgomery	14274 LE 35 mult.	91	176	1.93
			Altera FPGA (Stratix III)		4233 A 72 DSPs	165	176	1.07
		192	Altera FPGA (Stratix III)		9910 A 171 DSPs	131	790	6.03
Design II	optimal ate	126	Xilinx FPGA (Virtex-6)	RNS Montgomery	7032 slices 32 DSPs	250	143	0.573
[16]	ate	128	Xilinx FPGA (Virtex-6)	Hybrid Montgomery	4014 slices 42 DSPs	210	336	1.60
	optimal ate						245	1.17
[19]	Tate	128	Xilinx FPGA (Virtex-4)	Blakley	52k Slices	50	1,730	34.6
	ate						1,207	24.2
	optimal ate						821	16.4
[27]	Tate	128	ASIC (130 nm)	Montgomery	97 kGates	338	11,627*	34.4
	ate						7,706*	22.8
	optimal ate						5,340*	15.8
[15]	Tate over $\mathbb{F}_{3^{5 \cdot 97}}$	128	Xilinx FPGA (Virtex-4)	-	4755 Slices 7 BRAMs	192	429	2.23
[2]	optimal Eta over $\mathbb{F}_{2^{367}}$	128	Xilinx Virtex-4	-	4518 Slices	220	774*	3.52
[23]	ate	128	64-bit Core2	Montgomery	-	2400	15,000	6.25
	optimal ate						10,000	4.17
[20]	ate	128	64-bit Core2	Montgomery		2400	14,429	6.01
[36]	optimal ate	128	Core2 Quad	Hybrid Mult.	-	2394	4,470	1.86
[8]	optimal ate	126	Core i7	Montgomery	-	2800	2,330	0.83
[3]	optimal ate	126	Phenom II	Montgomery	-	3000	1,562	0.52
[4]	η_T over $\mathbb{F}_{2^{1223}}$	128	Xeon	-	-	2000	3,020	1.51
[9]	η_T over $\mathbb{F}_{3^{509}}$	128	Core i7	-	-	2900	5,423	1.87
[2]	opt. Eta $\mathbb{F}_{2^{367}}$	128	Core i5	-	-	2530	2,440	0.96

* Estimated by the authors.

The speedup comes mainly from three improvements. First, RNS multiplication has lower complexity than traditional integer multiplications. For example, an 256-bit multiplication on RNS involves 16 33×33 multiplications, while a 256-bit integer multiplication requires 27 32×32 multiplications using Karatsuba's method. Second, the use of lazy reduction reduces the cost of multiplications in extension fields. Third, RNS representation is very parallelization-friendly. Indeed, it only involves relatively small numbers (no long carry propagation) and always uses data in the local memory (no inter-core communication overhead). The performance of both Design I and II has demonstrated the efficiency of RNS in pairing implementations, and confirms with actual implementations on FPGA the analysis of [14].

8 Conclusions

In this paper, we demonstrated that RNS together with lazy reduction is a really competitive approach for pairing computation in hardware. Thanks to the use of RNS, both datapath and memory can be nicely parallelized. The results show that both designs we proposed achieve higher speed than all previous designs in hardware. The fastest version computes an optimal ate pairing at 126-bit security level in 0.573 ms, which is 2 times faster than all previous hardware implementations. Moreover, we also reported the first hardware pairing implementation at 192-bit security level.

For future work, we would like to further optimize the pipeline structure to achieve higher speed, particularly for 192-bit optimal pairing. We would like implement using RNS a pairing processor that provides 256-bit security. For these levels, BN curves may become less competitive, and have to be compared to other approaches such like KSS curves[26]. Also, as shown in Section 5, carefully selected RNS bases can help to reduce the complexity of RNS base extension. However, it is not clear how much we can benefit from it when the number of channels is relatively large, and this is definitely part of the design exploration for the implementation of pairings at higher security levels.

References

1. Altera web site, `http://www.altera.com`
2. Aranha, D., Beuchat, J.-L., Detrey, J., Estibals, N.: Optimal eta pairing on supersingular genus-2 binary hyperelliptic curves. Cryptology ePrint Archive, Report 2010/559 (2010), `http://eprint.iacr.org/`
3. Aranha, D., Karabina, K., Longa, P., Gebotys, C.H., López, J.: Faster explicit formulas for computing pairings over ordinary curves. In: Paterson, K.G. (ed.) EUROCRYPT 2011. LNCS, vol. 6632, pp. 48–68. Springer, Heidelberg (2011)
4. Aranha, D., López, J., Hankerson, D.: High-speed parallel software implementation of the η_T pairing. In: Pieprzyk, J. (ed.) CT-RSA 2010. LNCS, vol. 5985, pp. 89–105. Springer, Heidelberg (2010)
5. Bajard, J.-C., Didier, L.-S., Kornerup, P.: An RNS Montgomery modular multiplication algorithm. IEEE Transactions on Computers 47(7), 766–776 (1998)

6. Barreto, P., Kim, H., Lynn, B., Scott, M.: Efficient algorithms for pairing-based cryptosystems. In: Yung, M. (ed.) CRYPTO 2002. LNCS, vol. 2442, pp. 354–369. Springer, Heidelberg (2002)
7. Barreto, P., Naehrig, M.: Pairing-friendly elliptic curves of prime order. In: Preneel, B., Tavares, S. (eds.) SAC 2005. LNCS, vol. 3897, pp. 319–331. Springer, Heidelberg (2006)
8. Beuchat, J.-L., González-Díaz, J., Mitsunari, S., Okamoto, E., Rodríguez-Henríquez, F., Teruya, T.: High-Speed Software Implementation of the Optimal Ate Pairing over Barreto–Naehrig Curves. In: Joye, M., Miyaji, A., Otsuka, A. (eds.) Pairing 2010. LNCS, vol. 6487, pp. 21–39. Springer, Heidelberg (2010)
9. Beuchat, J.-L., López-Trejo, E., Martínez-Ramos, L., Mitsunari, S., Rodríguez-Henríquez, F.: Multi-core Implementation of the Tate Pairing over Supersingular Elliptic Curves. In: Garay, J.A., Miyaji, A., Otsuka, A. (eds.) CANS 2009. LNCS, vol. 5888, pp. 413–432. Springer, Heidelberg (2009)
10. Boneh, D., Franklin, M.: Identity-Based Encryption from the Weil Pairing. In: Kilian, J. (ed.) CRYPTO 2001. LNCS, vol. 2139, pp. 213–229. Springer, Heidelberg (2001)
11. Boneh, D., Lynn, B., Shacham, H.: Short signatures from the Weil pairing. Journal of Cryptology 17(4), 297–319 (2004)
12. Cha, J.C., Cheon, J.H.: An Identity-Based Signature from Gap Diffie-Hellman Groups. In: Desmedt, Y.G. (ed.) PKC 2003. LNCS, vol. 2567, pp. 18–30. Springer, Heidelberg (2002)
13. Costello, C., Lange, T., Naehrig, M.: Faster pairing computations on curves with high-degree twists. In: Nguyen, P.Q., Pointcheval, D. (eds.) PKC 2010. LNCS, vol. 6056, pp. 224–242. Springer, Heidelberg (2010)
14. Duquesne, S.: RNS arithmetic in $\mathbb{F}_{p^k}$ and application to fast pairing computation. Cryptology ePrint Archive, Report 2010/555 (2010), `http://eprint.iacr.org/` to appear in Journal of Mathematical Cryptology
15. Estibals, N.: Compact hardware for computing the tate pairing over 128-bit-security supersingular curves. In: Joye, M., Miyaji, A., Otsuka, A. (eds.) Pairing 2010. LNCS, vol. 6487, pp. 397–416. Springer, Heidelberg (2010)
16. Fan, J., Vercauteren, F., Verbauwhede, I.: Efficient hardware implementation of $\mathbb{F}_p$-arithmetic for pairing-friendly curves. IEEE Transactions on Computers PP(99), 1 (2011)
17. Freeman, D., Scott, M., Teske, E.: A taxonomy of pairing-friendly elliptic curves. Journal of Cryptology 23, 224–280 (2010)
18. Frey, G., Rück, H.G.: A remark concerning m-divisibility and the discrete logarithm in the divisor class group of curves. Mathematics of Computation 62(206), 865–874 (1994)
19. Ghosh, S., Mukhopadhyay, D., Roychowdhury, D.: High speed flexible pairing cryptoprocessor on FPGA platform. In: Joye, M., Miyaji, A., Otsuka, A. (eds.) Pairing 2010. LNCS, vol. 6487, pp. 450–466. Springer, Heidelberg (2010)
20. Grabher, P., Großschädl, J., Page, D.: On software parallel implementation of cryptographic pairings. In: Avanzi, R.M., Keliher, L., Sica, F. (eds.) SAC 2008. LNCS, vol. 5381, pp. 35–50. Springer, Heidelberg (2009)
21. Granger, R., Scott, M.: Faster squaring in the cyclotomic subgroup of sixth degree extensions. In: Nguyen, P.Q., Pointcheval, D. (eds.) PKC 2010. LNCS, vol. 6056, pp. 209–223. Springer, Heidelberg (2010)
22. Guillermin, N.: A High Speed Coprocessor for Elliptic Curve Scalar Multiplications over $\mathbb{F}_p$. In: Mangard, S., Standaert, F.-X. (eds.) CHES 2010. LNCS, vol. 6225, pp. 48–64. Springer, Heidelberg (2010)

23. Hankerson, D., Menezes, A., Scott, M.: Software Implementation of Pairings. Cryptology and Information Security Series, vol. 2, pp. 188–206. IOS Press, Amsterdam (2009); M. Joye and G. Neven edition
24. Hess, F., Smart, N.P., Vercauteren, F.: The Eta pairing revisited. IEEE Transactions on Information Theory 52(10), 4595–4602 (2006)
25. Joux, A.: A one round protocol for tripartite Diffie-Hellman. Journal of Cryptology 17, 263–276 (2004)
26. Kachisa, E.J., Schaefer, E.F., Scott, M.: Constructing brezing-weng pairing-friendly elliptic curves using elements in the cyclotomic field. In: Galbraith, S.D., Paterson, K.G. (eds.) Pairing 2008. LNCS, vol. 5209, pp. 126–135. Springer, Heidelberg (2008)
27. Kammler, D., Zhang, D., Schwabe, P., Scharwaechter, H., Langenberg, M., Auras, D., Ascheid, G., Mathar, R.: Designing an ASIP for Cryptographic Pairings over Barreto-Naehrig Curves. In: Clavier, C., Gaj, K. (eds.) CHES 2009. LNCS, vol. 5747, pp. 254–271. Springer, Heidelberg (2009)
28. Karabina, K.: Squaring in cyclotomic subgroups. Cryptology ePrint Archive, Report 2010/542 (2010), `http://eprint.iacr.org/`
29. Kawamura, S., Koike, M., Sano, F., Shimbo, A.: Cox-rower architecture for fast parallel montgomery multiplication. In: Preneel, B. (ed.) EUROCRYPT 2000. LNCS, vol. 1807, pp. 523–538. Springer, Heidelberg (2000)
30. Koblitz, N.: Elliptic Curve Cryptosystem. Math. Comp. 48, 203–209 (1987)
31. Koblitz, N., Menezes, A.: Pairing-based cryptography at high security levels. In: Smart, N.P. (ed.) Cryptography and Coding 2005. LNCS, vol. 3796, pp. 13–36. Springer, Heidelberg (2005)
32. Lee, E., Lee, H.-S., Park, C.-M.: Efficient and generalized pairing computation on abelian varieties. IEEE Transactions on Information Theory 55(4), 1793–1803 (2009)
33. Menezes, A.J., Okamoto, T., Vanstone, S.A.: Reducing elliptic curve logarithms to logarithms in a finite field. IEEE Transactions on Information Theory 39(5), 1639–1646 (1993)
34. Miller, V.S.: Use of Elliptic Curves in Cryptography. In: Williams, H.C. (ed.) CRYPTO 1985. LNCS, vol. 218, pp. 417–426. Springer, Heidelberg (1986)
35. Miller, V.S.: The Weil pairing, and its efficient calculation. Journal of Cryptology 17, 235–261 (2004), doi:10.1007/s00145-004-0315-8
36. Naehrig, M., Niederhagen, R., Schwabe, P.: New software speed records for cryptographic pairings. In: Abdalla, M., Barreto, P.S.L.M. (eds.) LATINCRYPT 2010. LNCS, vol. 6212, pp. 109–123. Springer, Heidelberg (2010)
37. Nozaki, H., Motoyama, M., Shimbo, A., Kawamura, S.-i.: Implementation of RSA algorithm based on RNS montgomery multiplication. In: Koç, Ç.K., Naccache, D., Paar, C. (eds.) CHES 2001. LNCS, vol. 2162, pp. 364–376. Springer, Heidelberg (2001)
38. National Institute of Standard and technology. Key management (2007), `http://csrc.nist.gov/groups/ST/toolkit/key_management.html`
39. Pereira, G.C.C.F., Simplício, M.A., Naehrig, M., Barreto, P.S.L.M.: A family of implementation-friendly BN elliptic curves. Journal of Systems and Software (2011)
40. Posch, K., Posch, R.: Base extension using a convolution sum in residue number systems. Computing 50, 93–104 (1993)
41. Rivest, R.L., Shamir, A., Adleman, L.: A Method for Obtaining Digital Signatures and Public-Key Cryptosystems. Communications of the ACM 21(2), 120–126 (1978)

42. Scott, M.: Implementing cryptographic pairings. In: Pairing-Based Cryptography - Pairing 2007. LNCS, vol. 4575, pp. 117–196. Springer, Heidelberg (2007)
43. Scott, M., Benger, N., Charlemagne, M., Dominguez Perez, L., Kachisa, E.: On the Final Exponentiation for Calculating Pairings on Ordinary Elliptic Curves. In: Shacham, H., Waters, B. (eds.) Pairing 2009. LNCS, vol. 5671, pp. 78–88. Springer, Heidelberg (2009)
44. Vercauteren, F.: Optimal pairings. IEEE Transactions on Information Theory 56(1), 455–461 (2010)

A Appendix

A.1 Sub-routines for Optimal Ate Pairing

Algorithm 4. `dbl`, doubling step

Require: $T = (X_T\gamma^2, Y_T\gamma^3, Z_T) \in E(\mathbb{F}_{p^{12}})$ with X_T, Y_T and $Z_T \in \mathbb{F}_{p^2}$, $P = (x_P, y_P) \in E(\mathbb{F}_p)$.

Ensure: The point $2T$ and the evaluation in P of the equation of the tangent line in T to the curve up to multiplicative factors in $\mathbb{F}_{p^2}$.

1: $B \leftarrow Y_T^2, C \leftarrow 3Z_T^2, D \leftarrow 2X_TY_T$
2: $F \leftarrow B + 3\mathbf{i}C, G \leftarrow B - 3\mathbf{i}C, H \leftarrow 3C, t_3 \leftarrow B + \mathbf{i}C, A \leftarrow X_T^2, E \leftarrow 2Y_TZ_Y$
3: $X_{2T} \leftarrow DF, Y_{2T} \leftarrow G^2 + 4HC, Z_{2T} \leftarrow 4BE, t_0 \leftarrow Ey_P, t_1 \leftarrow -3Ax_P$
4: **return** $(X_{2T}\gamma^2, Y_{2T}\gamma^3, Z_{2T}), t_0 + t_1\gamma + t_3\gamma^3$

These algorithms are specific to the BN curve $y^2 = x^3 + 2$ and the tower of extensions given in Subsection 2.1. Jacobian coordinates are usually used for pairing computations [8, 23, 36] but projective coordinates are more interesting in our case [3, 13]. Using the degree 6 twist on the curve, the point Q can be written as $(x_Q\gamma^2, y_Q\gamma^3)$ with x_Q and $y_Q \in \mathbb{F}_{p^2}$. The "doubling" step of the Miller loop consists of two stages: the doubling of a temporary projective point $T = (X_T\gamma^2, Y_T\gamma^3, Z_T)$ with X_T, Y_T and $Z_T \in \mathbb{F}_{p^2}$ and the evaluation in P of the tangent line in T to the curve. This is given in Algorithm 4 where the classical formulae are rearranged in a way to highlight the reductions (every temporary result needs a reduction except F, G, H and t_3), and the inherent parallelism in the local variables (each line can be implemented in random order). This is important to avoid idle states in Cox-Rower. The total cost of this step is 4 multiplications, 5 squarings, 8 reductions in $\mathbb{F}_{p^2}$ and 2 modular multiplications of an element of $\mathbb{F}_{p^2}$ by an element of $\mathbb{F}_p$. Note that multiplications like $2X_TY_T$ can be transformed into squaring of $(X_T + Y_T)$ at the cost of some extra additions [3, 13], thus it is not interesting for our design. In the same way, the cost of the addition step given by Algorithm 5 is 11 multiplications, 2 squarings, 11 reductions in $\mathbb{F}_{p^2}$ and 2 modular multiplications of an element of $\mathbb{F}_{p^2}$ by an element of $\mathbb{F}_p$.

Algorithm 5. add, addition step

Require: $T = (X_T\gamma^2, Y_T\gamma^3, Z_T) \in E(\mathbb{F}_{p^{12}})$ with X_T, Y_T and $Z_T \in \mathbb{F}_{p^2}$, $Q = (x_Q\gamma^2, y_Q\gamma^3) \in E(\mathbb{F}_{p^{12}})$, $P = (x_P, y_P) \in E(\mathbb{F}_p)$.
Ensure: The point $T + Q$ and the evaluation in P of the equation of the line passing through T and Q up to multiplicative factors in $\mathbb{F}_{p^2}$.
$E \leftarrow x_Q Z_T - X_T, F \leftarrow y_Q Z_T - Y_T$
$E_2 \leftarrow E^2, F_2 \leftarrow F^2$
$A \leftarrow F_2 Z_T - 2X_T E_2 - EE_2, B \leftarrow X_T E_2, E_3 \leftarrow EE_2$
$X_{T+Q} \leftarrow AE, Z_{T+Q} \leftarrow Z_T E_3, t_3 \leftarrow Fx_Q - Ey_Q$
$Y_{T+Q} \leftarrow F(B - A) - y_Q E_3, t_0 \leftarrow Ey_P, t_1 \leftarrow -Fx_P$
return $(X_{T+Q}\gamma^2, Y_{T+Q}\gamma^3, Z_{T+Q}), t_0 + t_1\gamma + t_3\gamma^3$

Algorithm 6. hard-part, hard part of the final exponentiation according [43]

Require: $f \in \mathbb{F}_{p^{12}}$ of order $p^4 - p^2 + 1$, $x = |u|$.
Ensure: $f^{(p^4-p^2+1)/\ell}$ with p and ℓ as in 2.1.
{Computation of the y_i}
$y_0 \leftarrow f^p f^{p^2} f^{p^3}$, $y_1 \leftarrow f^x$, $y_3 \leftarrow y_1^x$, $y_5 \leftarrow y_3^x$, $y_4 \leftarrow y_5^p$, $y_6 \leftarrow y_4 y_5 \left(= f^{x^3}\left(f^{x^3}\right)^p\right)$
$y_5 \leftarrow y_3^p$, $y_2 \leftarrow y_5^{-1}$, $y_4 \leftarrow y_1 y_2 \left(= f^x / \left(f^{x^2}\right)^p\right)$
$y_2 \leftarrow y_5^p \left(= \left(f^{x^2}\right)^{p^2}\right)$, $y_5 \leftarrow y_3^{-1} \left(= 1/f^{x^2}\right)$, $y_3 \leftarrow y_1^p\,((m^x)^p)$, $y_1 \leftarrow f^{-1}$
{Multi-addition chain for computing $y_0.y_1^2.y_2^6.y_3^{12}.y_4^{18}.y_5^{30}.y_6^{36}$}
$t_0 \leftarrow y_6^2$, $t_0 \leftarrow t_0 y_4$, $t_0 \leftarrow t_0 y_5$, $t_1 \leftarrow y_3 y_5$, $t_1 \leftarrow t_1 t_0$, $t_0 \leftarrow t_0 y_2$, $t_1 \leftarrow t_1^2$
$t_1 \leftarrow t_1 t_0$, $t_1 \leftarrow t_1^2$, $t_0 \leftarrow t_1 y_1$, $t_1 \leftarrow t_1 y_0$, $t_0 \leftarrow t_0^2$, $t_0 \leftarrow t_0 t_1$
return t_0

Algorithm 7. Squaring during final exponentiation (hard-part)[21]

Require: $A = \sum_{i=0}^{5} a_i\gamma^i \in \mathbb{F}_{p^{12}}$ with $a_i \in \mathbb{F}_{p^2}, 0 \le i \le 5$
Ensure: A^2
$A_0 \leftarrow 3a_0^2 + 3(1+\mathbf{i})a_3^2 - 2a_0, A_1 \leftarrow 6(1+\mathbf{i})a_2a_5 + 2a_1$
$A_2 \leftarrow 3a_1^2 + 3(1+\mathbf{i})a_4^2 - 2a_2, A_3 \leftarrow 6a_0a_3 + 2a_3$
$A_4 \leftarrow 3a_2^2 + 3(1+\mathbf{i})a_5^2 - 2a_4, A_5 \leftarrow 6a_1a_4 + 2a_5$
return $\sum_{i=0}^{5} A_i\gamma^i$

A.2 RNS Parameter Selection

Since p should be around 254 bits, we set n to be 8 and the moduli are chosen near 2^{33}. We consider for $i, j \in [1,n]$ the following issues: (1) bitlength of $|B_i|_{c_j}$ and $|C_i|_{b_j}$; (2) Hamming weight of b_i and c_j. A simple (bounded) exhaustive search program returns the bases shown in Section 5.

We denote the bitlength of all $\tilde{B}_{i,j}$ in a length matrix, $L_{\tilde{B}}$, and the bitlength matrix of all $\tilde{C}_{i,j}$ as $L_{\tilde{C}}$. For the selected bases, we have the following $L_{\tilde{B}}$ and $L_{\tilde{C}}$.

$$L_{\tilde{B}} = \begin{pmatrix} 23 & 20 & 21 & 20 & 21 & 20 & 18 & 19 \\ 22 & 20 & 22 & 20 & 21 & 20 & 18 & 19 \\ 25 & 23 & 23 & 22 & 23 & 22 & 21 & 22 \\ 25 & 24 & 25 & 26 & 25 & 26 & 23 & 28 \\ 29 & 30 & 28 & 28 & 28 & 28 & 28 & 27 \\ 32 & 33 & 32 & 31 & 31 & 31 & 33 & 31 \\ 32 & 33 & 32 & 32 & 32 & 32 & 34 & 32 \\ 33 & 33 & 33 & 32 & 33 & 33 & 37 & 32 \end{pmatrix}, \; L_{\tilde{C}} = \begin{pmatrix} 24 & 23 & 23 & 20 & 21 & 20 & 20 & 19 \\ 25 & 25 & 26 & 24 & 26 & 25 & 24 & 24 \\ 26 & 27 & 25 & 24 & 24 & 23 & 23 & 23 \\ 30 & 31 & 30 & 31 & 29 & 29 & 29 & 28 \\ 28 & 28 & 27 & 27 & 26 & 26 & 26 & 25 \\ 30 & 30 & 30 & 30 & 29 & 28 & 28 & 28 \\ 27 & 27 & 27 & 26 & 28 & 29 & 29 & 31 \\ 30 & 30 & 30 & 33 & 29 & 29 & 29 & 29 \end{pmatrix}$$

After truncating the least significant zeros, we get the following bitlength, denoted as $L_{\tilde{B}'}$ and $L_{\tilde{C}'}$. Note that we applied the same truncation parameter for all the numbers in the same row. The number of zeros truncated is chosen as

$$min\{z(B_{1,j}), z(B_{2,j}), \cdots, z(B_{n,j})\}$$

where $z(B_{i,j})$ gives the number of zeros at the LSBs of $B_{i,j}$.

$$L_{\tilde{B}'} = \begin{pmatrix} 23 & 20 & 21 & 20 & 21 & 20 & 18 & 19 \\ 12 & 10 & 12 & 10 & 11 & 10 & 8 & 9 \\ 16 & 14 & 14 & 13 & 14 & 13 & 12 & 13 \\ 15 & 14 & 15 & 16 & 15 & 16 & 13 & 18 \\ 17 & 18 & 16 & 16 & 16 & 16 & 16 & 15 \\ 20 & 21 & 20 & 19 & 19 & 19 & 21 & 19 \\ 19 & 20 & 19 & 19 & 19 & 19 & 21 & 19 \\ 19 & 19 & 19 & 18 & 19 & 19 & 23 & 18 \end{pmatrix}, \; L_{\tilde{C}'} = \begin{pmatrix} 14 & 13 & 13 & 10 & 11 & 10 & 10 & 9 \\ 15 & 15 & 16 & 14 & 16 & 15 & 14 & 14 \\ 16 & 17 & 15 & 14 & 14 & 13 & 13 & 13 \\ 20 & 21 & 20 & 21 & 19 & 19 & 19 & 18 \\ 20 & 20 & 19 & 19 & 18 & 18 & 18 & 17 \\ 21 & 21 & 21 & 21 & 20 & 19 & 19 & 19 \\ 17 & 17 & 17 & 16 & 18 & 19 & 19 & 21 \\ 20 & 20 & 20 & 23 & 19 & 19 & 19 & 19 \end{pmatrix}$$

Now all of $\tilde{B}'_{i,j}$ and $\tilde{C}'_{i,j}$ are less than 25 bits, and they fit in one operand of an FPGA DSP slice, while the standard 34-bit operands do not. In fact, in the implementation we do not truncate more zeros as far as all elements in that row fit in 25 bits.

High Speed Cryptoprocessor for η_T Pairing on 128-bit Secure Supersingular Elliptic Curves over Characteristic Two Fields

Santosh Ghosh, Dipanwita Roychowdhury, and Abhijit Das

Computer Science and Engineering
Indian Institute of Technology Khaargpur
WB, India, 721302
{santosh,drc,abhij}@cse.iitkgp.ernet.in

Abstract. This paper presents an efficient architecture for computing cryptographic η_T pairing for providing 128-bit security. A cryptoprocessor is proposed for Miller's Algorithm with a new 1223-bit Karatsuba multiplier that exploits parallelism. To the best of our knowledge this is the first hardware implementation of 128-bit secure η_T pairing on supersingular elliptic curves over characteristic two fields. The design has been implemented on Xilinx FPGAs. The place-and-route results show that the proposed design takes only $190\mu s$ to complete an 128-bit secure η_T pairing on a Virtex-6 FPGA. The proposed cryptoprocessor achieves eight times speedup compared to the best known existing design. It also outperforms the previous designs with respect to *area* $\times$ *time* product.

Keywords: Pairing, Supersingular curves, characteristic two fields, FPGA, Karatsuba multiplier.

1 Introduction

Since 2000, pairing is used in cryptography for developing security schemes for various applications. It is well suited for identity based cryptography [8] which has gained lot of importance in recent times. As a natural consequence, implementations of pairings are also extremely important. The implementations should be cost effective, both in terms of time and space requirement. In practice, pairing could be implemented either as a software library executed on general purpose processors or as a dedicated cryptoprocessor. However, the later one is favored due to huge mathematical operations required for pairing computation [5]. This paper broadly addresses design techniques of a pairing cryptoprocessor for high security level.

Pairing for cryptographic applications are computed on elliptic or hyperelliptic curves defined over suitably large finite fields and having small embedding degree [19,13]. The security of a pairing depends on the underlying algebraic curves and respective field types. For example, 128-bit symmetric security could be achieved by computing η_T pairing [3,18] on a supersingular elliptic curve

B. Preneel and T. Takagi (Eds.): CHES 2011, LNCS 6917, pp. 442–458, 2011.

defined over $\mathbb{F}_{2^{1223}}$ and having embedding degree $k = 4$. As per NIST recommendation, 128-bit symmetric security is essential beyond 2030 [2]. Therefore, it is of importance to explore the efficient implementation techniques of 128-bit secure pairings on different platforms.

Hardware implementation of 128-bit secure pairings was introduced in 2009, individually by Kammler *et al.* [21] and Fan *et al.* [12]. Both of them described hardware implementation techniques for computing 128-bit secure pairings over Barreto-Naehrig curves (BN curves) [4]. These CMOS based designs take $15.8ms$ and $2.9ms$ for computing an optimal-ate pairing, respectively. Thereafter, designs in [10,14,1,9] are appeared in literature, which computes 128-bit secure pairings in $2.3ms$, $16.4ms$, $3.5ms$, and $1.07ms$ respectively. However, to the best of our knowledge there is no hardware implementation results available in the literature which computes 128-bit secure pairings below one ms time limit. High-speed software implementations reported in [6,7] compute 128-bit secure pairings in $0.832ms$ and $1.87ms$. The work proposed by Beuchat *et al.* [5] describes design architectures for η_T pairings on supersingular elliptic curves over characteristic two and three fields for a maximum of 105-bit and 109-bit security, respectively. However, to the best of the authors' knowledge no hardware architectures are available for computing η_T pairing on 128-bit secure supersingular elliptic curves over binary fields.

Contribution. This paper explores the hardware design techniques for η_T pairing on 128-bit secure supersingular elliptic curves over characteristic two fields. It first designs cost-effective and time-efficient hybrid architectures for Karatsuba multiplication over $\mathbb{F}_{2^{1223}}$ field, on which the respective supersingular elliptic curve is defined. The major contributions of the paper are highlighted here.

- The paper explores area-time tradeoff designs of hybrid Karatsuba multiplier over $\mathbb{F}_{2^{1223}}$ field.
- It further explores high speed architectures for computing η_T pairing on supersingular elliptic curves based on the proposed hybrid multiplier.
- It provides the first hardware implementation result of an 128-bit secure pairing on elliptic curves over characteristic two fields.
- The proposed design is the first one which computes an 128-bit secure pairing in less than one ms.

The proposed design of hybrid multiplier and parallelism techniques result in the high speed cryptoprocessor which achieves significant improvement on the performance of 128-bit secure η_T pairing on supersingular elliptic curves over small characteristic fields.

Organization of the Paper. Section 2 of the paper proposes design techniques of Karatsuba multipliers for $\mathbb{F}_{2^{1223}}$ field. Section 3 describes the proposed pairing cryptoprocessor. Results and comparisons are provided in Section 4. Finally, the paper is concluded in Section 5.

2 The $\mathbb{F}_{2^{1223}}$-Multiplier

Multiplication is the key operation of a pairing computation. The 128-bit secure η_T pairing could be computed on a supersingular elliptic curve defined over 1223-bit characteristic-two fields. Therefore, the multiplication in $\mathbb{F}_{2^{1223}}$ field is an essential operation in this context. Karatsuba multiplication [22] is one of the most efficient and popular techniques for fields like $\mathbb{F}_{q^m}$. This technique is based on divide-and-conquer algorithm, where a full m-bit multiplication is divided recursively into several m/k-bit multiplications with small $k \in \{2, 3\}$. It then accumulates the results of smaller multiplications for the generating final result. Karatsuba technique for $k = 2$ computes product $a \cdot b$ of two elements $a, b \in \mathbb{F}_{q^m}$ by the following way:

$$\begin{aligned} a \cdot b &= (a_1 x^{\lceil m/2 \rceil} + a_0)(b_1 x^{\lceil m/2 \rceil} + b_0) \\ &= a_1 b_1 x^m + [(a_1 + a_0)(b_1 + b_0) - a_1 b_1 - a_0 b_0] \, x^{\lceil m/2 \rceil} + a_0 b_0. \end{aligned} \quad (1)$$

Hence, an m-bit multiplication can be performed by three $m/2$-bit multiplications along with four m-bit and two $m/2$-bit addition/subtraction operations. Generalization of Karatsuba multiplication is provided in [29]. We refer to the reader [27,17] for getting idea about implementation techniques of Karatsuba multiplication. Efficient implementation of Karatsuba multiplication is challenging– mainly for larger field sizes like $m = 1223$. It is more challenging on resource-constrained environments like an FPGA platform where the number of logic cells are limited. We may follow several ways for making trade-off between the multiplication latency and hardware resources for developing a multiplier for $\mathbb{F}_{2^{1223}}$ field. Fig. 1 shows the decomposition of a 1223-bit operand for Karatsuba multiplication with $k = 2$. The operand is decomposed recursively up to their 19-bit or 20-bit levels as it gives the most optimum design [27].

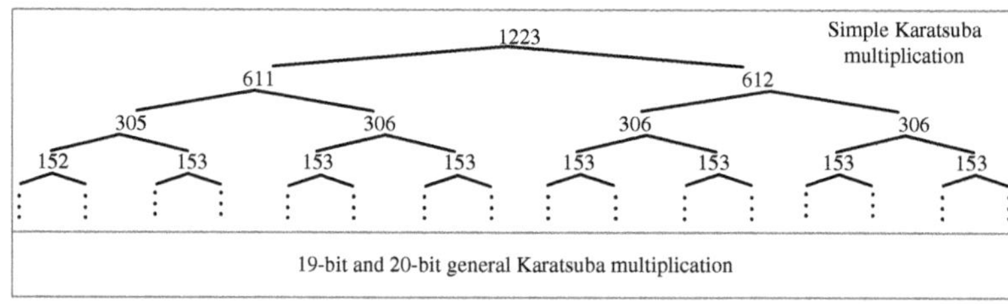

Fig. 1. The decomposition of an 1223-bit operand for Karatsuba multiplication

Fully Parallel Multiplier for $\mathbb{F}_{2^{1223}}$. A fully parallel Karatsuba multiplier can be designed for $\mathbb{F}_{2^{1223}}$ field by following the decomposition as shown in Fig. 1. After implementing it by Verilog (HDL) we synthesize the design by ISE tool for a Virtex-4 FPGA. The synthesis tool estimates 324342 LUTs for an 1223-bit fully parallel Karatsuba multiplier, which makes it infeasible to implement on a single Virtex-4 FPGA device.

Serial Use of 612-bit Parallel Multiplier. As an alternative to area-time tradeoff we take a fully parallel 612-bit Karatsuba multiplier on which three multiplications are performed in serial for computing multiplication in $\mathbb{F}_{2^{1223}}$. After synthesizing by ISE synthesis tool, it demands 95324 LUTs. Thus, it could be useful to implement a high throughput $\mathbb{F}_{2^{1223}}$ multiplier on a high-end single FPGA device. However, a pairing cryptoprocessor demands more circuits along with multipliers, which may not be put together on a single FPGA device.

2.1 Serial Use of 306-bit Parallel Multiplier

It is shown that the fully parallel multiplier as well as serial use of 612-bit parallel multiplier for $\mathbb{F}_{2^{1223}}$ are infeasible to implement a respective η_T pairing crypto-processor. Here we propose a serial use of 306-bit parallel Karatsuba multiplier for $\mathbb{F}_{2^{1223}}$ field. The current multiplier is based on a 306-bit fully parallel Karatsuba multiplier on which top two levels of Fig. 1 are performed in serial. The proposed architecture for computing 1223-bit multiplication based on this serial-parallel hybridization is shown in Fig. 2. The architecture follows the exact steps and nomenclatures of variables that are described in Algorithm 2, Appendix A.

The proposed architecture works as follows. During the initialization stage (Algorithm 2, step 1 to step 7) it breaks the operands a, and b into four parts by following two repeated Karatsuba decompositions. The smaller operands are generated by following way:

$$\begin{aligned} a \cdot b &= (a_1 x^{612} + a_0)(b_1 x^{612} + b_0) \\ &= a_1 b_1 x^{1222} + \left[(a_1 + a_0)(b_1 + b_0) - a_1 b_1 - a_0 b_0\right] x^{612} + a_0 b_0. \end{aligned}$$

The 1223-bit multiplication is performed by three 612-bit multiplications[1], $a_0 \cdot b_0$, $a_1 \cdot b_1$, and $(a_1 + a_0) \cdot (b_1 + b_0)$, which are further decomposed by following way.

$$\begin{aligned} a_0 \cdot b_0 &= (a_{01} x^{306} + a_{00})(b_{01} x^{306} + b_{00}) \\ &= a_{01} b_{01} x^{612} + \left[(a_{01} + a_{00})(b_{01} + b_{00}) - a_{01} b_{01} - a_{00} b_{00}\right] x^{306} + a_{00} b_{00} \\ &= a_{01} b_{01} x^{612} + \left[g_0 h_0 - a_{01} b_{01} - a_{00} b_{00}\right] x^{306} + a_{00} b_{00}, \end{aligned} \quad (2)$$

where, $g_0 = a_{01} + a_{00}$ and $h_0 = b_{01} + b_{00}$. Similarly, the second 612-bit multiplication is performed by following equation.

$$\begin{aligned} a_1 \cdot b_1 &= (a_{11} x^{306} + a_{10})(b_{11} x^{306} + b_{10}) \\ &= a_{11} b_{11} x^{611} + \left[(a_{11} + a_{10})(b_{11} + b_{10}) - a_{11} b_{11} - a_{10} b_{10}\right] x^{306} + a_{10} b_{10} \\ &= a_{11} b_{11} x^{611} + \left[g_1 h_1 - a_{11} b_{11} - a_{10} b_{10}\right] x^{306} + a_{10} b_{10}, \end{aligned} \quad (3)$$

where, $g_1 = a_{11} + a_{10}$ and $h_1 = b_{11} + b_{10}$. The third 612-bit multiplication is performed as:

[1] More accurately, two 612-bit multiplications and one 611-bit multiplication. For simplicity we say three 612-bit multiplications.

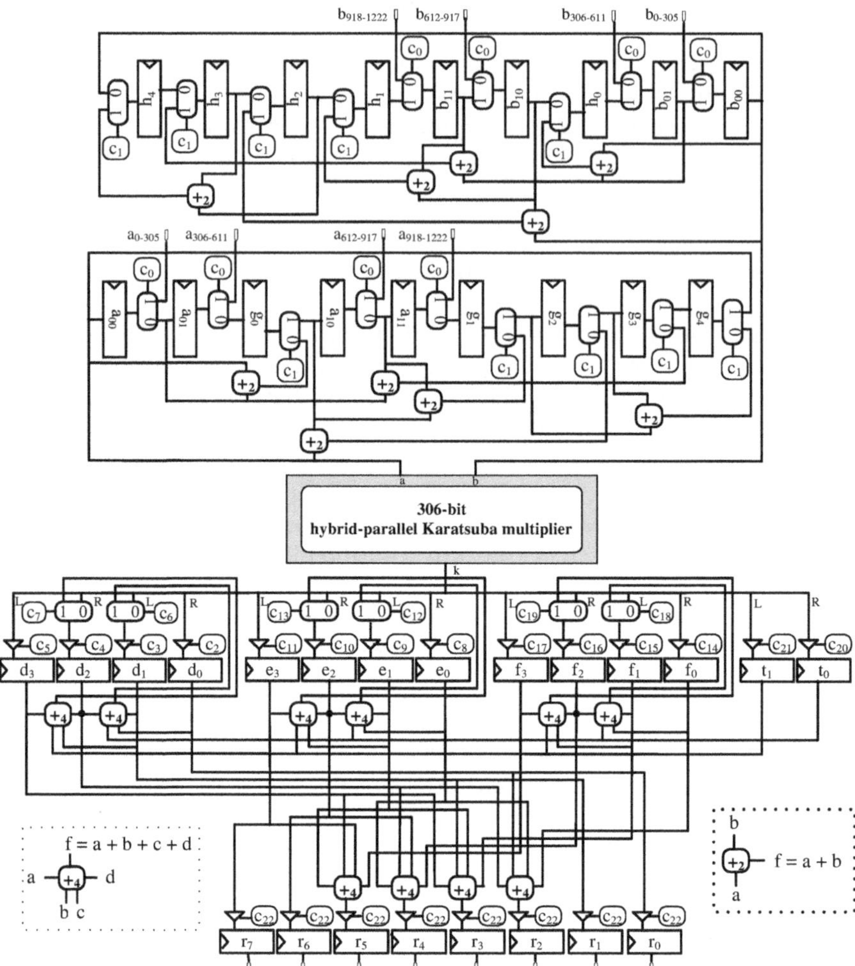

Fig. 2. The architecture of $\mathbb{F}_{2^{1222}}$ multiplier unit

$$
\begin{aligned}
g_2 &= a_{10} + a_{00}; \quad g_3 = a_{11} + a_{01} \\
h_2 &= b_{10} + b_{00}; \quad h_3 = b_{11} + b_{01} \\
(a_1 + a_0) \cdot (b_1 + b_0) &= (g_3 x^{306} + g_2)(h_3 x^{306} + h_2) \\
&= g_3 h_3 x^{612} + [(g_3 + g_2)(h_3 + h_2) - g_3 h_3 - g_2 h_2]\, x^{306} + g_2 h_2 \\
&= g_3 h_3 x^{612} + [g_4 h_4 - g_3 h_3 - g_2 h_2]\, x^{306} + g_2 h_2, \qquad (4)
\end{aligned}
$$

where, $g_4 = g_3 + g_2$ and $h_4 = h_3 + h_2$. Therefore, one 1223-bit multiplication is performed by nine 306-bit multiplications. In our proposed architecture (Fig. 2), the operands of these nine multiplications are stored into nine 306-bit parallel shift registers. These registers are automatically reloaded by synchronous shift operations so that the two correct operands of 306-bit parallel multiplier are available into a_{00} and b_{00} registers, respectively, at every clock. The 306-bit parallel Karatsuba multiplier takes only one clock cycle to compute one respective multiplication. The strategy of shift register is adopted for avoiding two complex 9-to-1 multiplexers to the multiplier input ports. The first three 309-bit multiplication results (Algorithm 2, step 8 to step 13) are combined to generate the intermediate result of first 612-bit multiplication $a_0 \cdot b_0$. The final result of $a_0 \cdot b_0$ as defined in Eq. 2 (Algorithm 2, step 14) is computed by means of two 306-bit 4-input parallel adders (4-input XORs in this case) and it is stored into the registers d_i, $0 \leq i \leq 3$. Similarly, the result of the second 612-bit multiplication as defined in Eq. 3 (Algorithm 2, step 15 to step 21) is stored into the registers e_i, $0 \leq i \leq 3$, and for the third one, Eq. 4, (Algorithm 2, step 22 to step 28) is stored into the registers f_i, $0 \leq i \leq 3$. Finally, in steps 29 to 31, the algorithm combines the final result of 1223-bit multiplication and stores into the registers r_i, $0 \leq i \leq 7$. The proposed architecture (Fig. 2) takes 10 clock cycles for completing one multiplication in the respective base field $\mathbb{F}_{2^{1223}}$.

Implementation Results on FPGA Platforms. The synthesis tool estimates 34325 LUTs on a Virtex-4 FPGA for implementing the proposed serial use of 306-bit parallel multiplier for $\mathbb{F}_{2^{1223}}$. In this paper, we are looking for a pairing cryptoprocessor on a medium-range FPGA device. The place-and-route results as summarized in Table 1 ensure that this multiplier is suitable for designing our target cryptoprocessor.

Table 1. Cost and time of 1223-bit multipliers on FPGA platforms

Multiplier type	FPGA family	LUTs	Frequency $[MHz]$	Serial use	Multiplication latency $[ns]$	$(A \cdot T)^{\S}$
Serial use of 306-bit parallel multiplier	Virtex-2	34 547	125	10	80.0	2.76
	Virtex-4	34 325	168	10	60.0	2.06
	Virtex-6	30 148	250	10	40.0	1.21
§ : $(A \cdot T)$ represents product of *area* in LUTs and *time* in milliseconds.						

However, designer may opt for *serial use of 153-bit parallel multiplier* with low resources. But, it requires 27 serial use, which slows down the multiplication. On a Virtex-4 FPGA one such multiplier takes 16231 LUTs and achieves maximum 185 MHz clock frequency. Therefore, this multiplier with lower resource requires $151ns$ for completing one 1223-bit multiplication which is 2.5 times slower than *serial use of 306-bit parallel multiplier*. The respective $A \cdot T$ value (2.46) of this design is 1.2 times higher than the same for the design with 306-bit parallel multiplier. Thus, *serial use of 306-bit parallel multiplier* provides the most optimized design with respect to the feasibility of implementation as well as *area* $\times$ *time* product.

3 The η_T Pairing Cryptoprocessor over $F_{2^{1223}}$

In this section, we present a high-speed cryptoprocessor for computing the η_T pairing over a large characteristic-two field $\mathbb{F}_{2^{1223}}$. The proposed architecture is depicted in Fig. 3. The pairing computation consists of two major operations – the non-reduced pairing (Miller's algorithm) and the final exponentiation. Beuchat *et al.* in [5] proposed two separate coprocessors on which these two tasks are pipelined. Two separate coprocessors in pipeline helps to reduce the computation time. But, at the same time it needs larger area. In case of a large field like $\mathbb{F}_{2^{1223}}$ it is important to take care of the overall area requirement for pairing computation as most of its applications demand area-constrained devices. It is observed that almost 50% datapath of both the above coprocessors are consumed by the base-field multipliers. This paper attempts to optimize the area of an η_T pairing cryptoprocessor. We propose here a common datapath for computing both the Miller's algorithm and the final exponentiation. Adequate parallelism is also applied in the datapath to achieve a high-speed cryptoprocessor. The supersingular elliptic curves, the representation of the fields, and the η_T pairing algorithm that are used in this paper are described in Appendix B.

The η_T pairing computation in characteristic-two field is described in [16]. We rewrite it, specifically for $\mathbb{F}_{2^{1223}}$, in Algorithm 1 with parenthesized indices in superscript in order to emphasize the intrinsic dependency as well as parallelism of the pairing computation. Two interdependent operations in the Miller's algorithm, namely, the computation of the $G^{(i)}$ (step 7 to step 10) and the sparse multiplication[2] $F^{(i-1)} \cdot G^{(i)}$ over $\mathbb{F}_{(2^{1223})^4}$ along with the computation of $x_2^{(i)}$, $y_2^{(i)}$ for next iteration (step 11 and step 12) are performed in serial, whereas we apply the parallelism within each of these two operations.

3.1 Computation of Miller's Loop

The proposed cryptoprocessor as shown in Fig. 3 first computes the non-reduced pairing based on Algorithm 1. It breaks this computation in three sub-parts as described here. We use the same nomenclature of Algorithm 1 for representing the intermediate results in the architecture.

Initialization. The registers $x_1^{(i)}$, $y_1^{(i)}$, $x_2^{(i-1)}$, $y_2^{(i-1)}$, and $s^{(i)}$ are initialized according to step 1 and step 2 of Algorithm 1 [Fig. 3]. During the initialization of $s^{(i)}$, the operation $x_1 + 1$ is performed simply by inverting the least significant bit of x_1 as $x_1 \in F_2[x]$. The variables $t_0^{(i)}$ and $t_1^{(i)}$ are initialized by two sets of 2-input XORs,[3] which perform $s^{(0)} + x_2^{(0)}$ and $y_1^{(0)} + y_2^{(0)}$, respectively. These two operations are performed on the fly. As defined in step 4, the initialization of

[2] An operand in $\mathbb{F}_{(2^{1223})^4}$ is sparse when some of its coefficients are trivial (i.e., either zero or one).

[3] The addition in $\mathbb{F}_2[x]$ is performed by simple bit-wise XOR. Therefore, addition and XOR are used with same meaning in this paper.

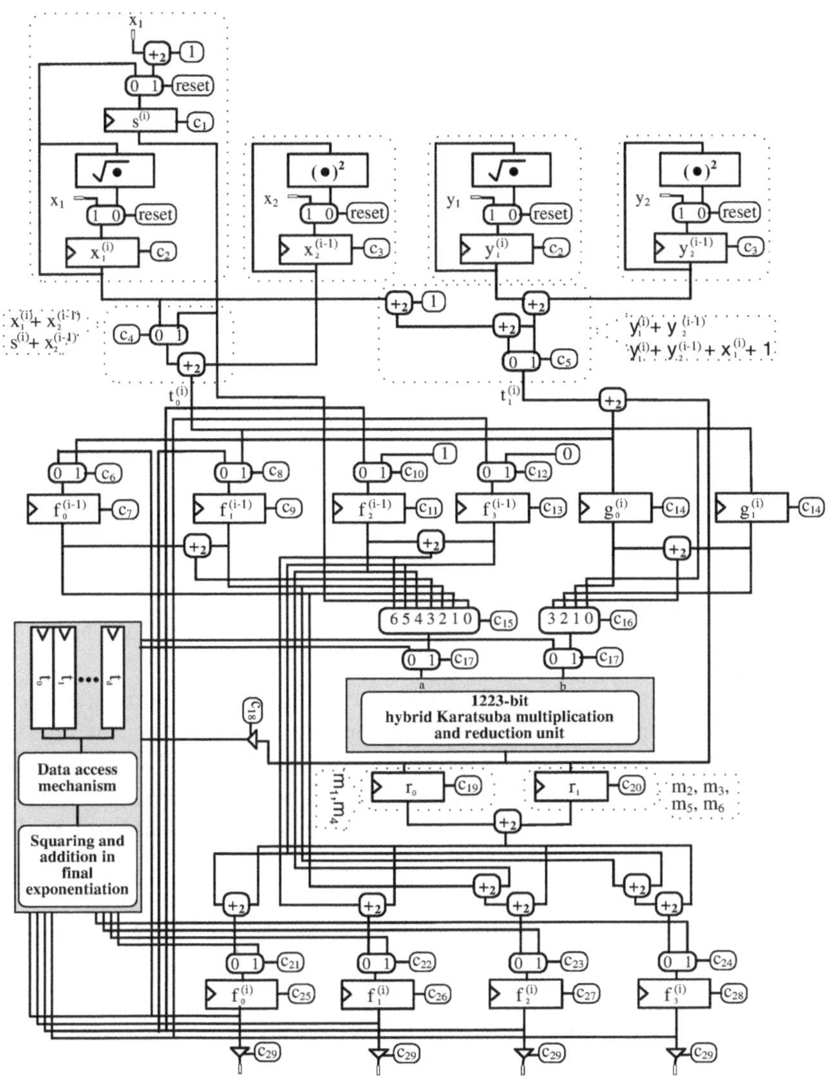

Fig. 3. The η_T pairing cryptoprocessor over $F_{2^{1223}}$

register $f_0^{(i-1)}$ is done by means of the output of a multiplication followed by a 2-input addition. Similarly, the initialization of $f_1^{(i-1)}$ requires a 2-input addition, whereas the same for $f_2^{(i-1)}$ and $f_3^{(i-1)}$ are trivial. In total, the initialization part of Miller's algorithm takes only 12 clock cycles in our proposed cryptoprocessor.

Computation of $G^{(i)}$. We represent $G^{(i)} \in \mathbb{F}_{(2^{1223})^4}$ in $\{1, u, v, uv\}$ basis. However, throughout Miller's loop $G^{(i)}$ contains sparse value which is represented as

---

Algorithm 1 . Computing the η_T pairing on $E/\mathbb{F}_{2^{1223}}$. Intermediate variables in uppercase belong to $\mathbb{F}_{(2^{1223})^4}$, whereas those in lowercase to $\mathbb{F}_{2^{1223}}$.

Input: $P(x_1, y_1)$ and $Q(x_2, y_2) \in E(\mathbb{F}_{2^{1223}})[r]$.
Output: $\eta_T(P, Q)$.

---

1. $x_1^{(0)} \leftarrow x_1$; $y_1^{(0)} \leftarrow y_1$; $x_2^{(0)} \leftarrow x_2$; $y_2^{(0)} \leftarrow y_2$;
2. $s^{(0)} \leftarrow x_1 + 1$;
3. $t_0^{(0)} \leftarrow s^{(0)} + x_2^{(0)}$; $t_1^{(0)} \leftarrow y_1^{(0)} + y_2^{(0)}$;
4. $f_0^{(0)} \leftarrow s^{(0)} \cdot t_0^{(0)} + t_1^{(0)}$; $f_1^{(0)} \leftarrow s^{(0)} + x_2^{(0)}$; $f_2^{(0)} \leftarrow 1$; $f_3^{(0)} \leftarrow 0$;
5. $F^{(0)} \leftarrow f_0^{(0)} + f_1^{(0)} u + f_2^{(0)} v + f_3^{(0)} uv$;
6. **for** *i from* 1 *to* 612 **do**
7. $s^{(i)} \leftarrow x_1^{(i-1)}$, $x_1^{(i)} \leftarrow \sqrt{x_1^{(i-1)}}$; $y_1^{(i)} \leftarrow \sqrt{y_1^{(i-1)}}$;
8. $t_0^{(i)} \leftarrow x_1^{(i)} + x_2^{(i-1)}$; $t_1^{(i)} \leftarrow y_1^{(i)} + y_2^{(i-1)} + x_1^{(i)} + 1$;
9. $g_0^{(i)} \leftarrow s^{(i)} \cdot t_0^{(i)} + t_1^{(i)}$; $g_1^{(i)} \leftarrow s^{(i)} + x_2^{(i-1)}$;
10. $G^{(i)} \leftarrow g_0^{(i)} + g_1^{(i)} u + v$;
11. $F^{(i)} \leftarrow F^{(i-1)} \cdot G^{(i)}$;
12. $x_2^{(i)} \leftarrow (x_2^{(i-1)})^2$; $y_2^{(i)} \leftarrow (y_2^{(i-1)})^2$;
13. **end for**
14. **return** $(F^{(612)})^{(2^{2446}-1)(2^{1223}-2^{612}+1)}$.

$g_0^{(i)} + g_1^{(i)} u + 1$. The computation of $g_0^{(i)}$ (Algorithm 1, step 9) is performed by means of one multiplication in $\mathbb{F}_{2^{1223}}$ followed by one 2-input addition. The operands of above multiplication $s^{(i)}$ and $t_0^{(i)}$ are generated on the fly after computing two square-root operations (in step 7) in parallel. Current cryptoprocessor computes the square-root operations inexpensively by means of simple shift and XOR operations. Let, $a = \sum a_i x^i \in \mathbb{F}_{2^{1223}}$, then $\sqrt{a} = \sum a_{2j} x^j + (x^{612} + x^{128}) \sum a_{2j+1} x^j$, which is computed in one clock. Therefore, in the proposed cryptoprocessor (Fig. 3) the control signal c_2 is activated only in that respective clock cycle during the execution of each iteration of Miller's algorithm. After computing $x_1^{(i)}$ and $y_1^{(i)}$ at the next clock the cryptoprocessor starts the multiplication $s^{(i)} \cdot t_0^{(i)}$. The multiplication in $\mathbb{F}_{2^{1223}}$ takes 10 clock cycles and immediately at the next clock the cryptoprocessor updates $g_0^{(i)}$ and $g_1^{(i)}$ registers. Therefore, in total the computation of $G^{(i)}$ takes 12 clock cycles by the proposed cryptoprocessor.

Sparse Multiplication over $\mathbb{F}_{(2^{1223})^4}$. The operation $F^{(i-1)} \cdot G^{(i)}$ in Algorithm 1, step 11 is identified as sparse multiplication in $\mathbb{F}_{(2^{1223})^4}$ as $G^{(i)}$ consists only two non-trivial coefficients. The computation of this sparse multiplication is much easier than a full multiplication in the above extension field. The computation procedure on our proposed cryptoprocessor is described in Table 2.

In the proposed cryptoprocessor multiplications $m_i, 1 \leq i \leq 6$, are performed in serial on a single $\mathbb{F}_{2^{1223}}$ multiplier core. The registers r_0 and r_1 (in Fig. 3) are alternatively used to hold the multiplication outputs. After completing m_1 and m_2 we start m_3 at the next clock when in parallel the value of $f_0^{(i)}$ is computed

Table 2. Computation of $F^{(i-1)} \cdot G^{(i)}$

m_1 : $r_0 \leftarrow f_0^{(i-1)} \cdot g_0^{(i)}$;	m_4 : $r_0 \leftarrow f_2^{(i-1)} \cdot g_2^{(i)}$;
m_2 : $r_1 \leftarrow f_1^{(i-1)} \cdot g_1^{(i)}$;	m_5 : $r_1 \leftarrow f_3^{(i-1)} \cdot g_3^{(i)}$;
$x_4^{(1)}$: $f_0^{(i)} \leftarrow (r_0 + r_1) + f_4^{(i-1)}$;	$x_4^{(3)}$: $f_2^{(i)} \leftarrow (r_0 + r_1) + (f_1^{(i-1)} + f_3^{(i-1)})$;
m_3 : $r_1 \leftarrow (f_0^{(i-1)} + f_1^{(i-1)}) \cdot (g_0^{(i)} + g_1^{(i)})$;	m_6 : $r_1 \leftarrow (f_2^{(i-1)} + f_3^{(i-1)}) \cdot (g_0^{(i)} + g_1^{(i)})$;
$x_4^{(2)}$: $f_1^{(i)} \leftarrow (r_0 + r_1) + (f_3^{(i-1)} + f_4^{(i-1)})$;	$x_4^{(4)}$: $f_1^{(i)} \leftarrow (r_0 + r_1) + (f_2^{(i-1)} + f_4^{(i-1)})$;

by two sets of 2-input XORs as defined by $x_4^{(1)}$ in Table 2. Similarly, we perform m_4 and $x_4^{(2)}$ in parallel and also do m_6 and $x_4^{(3)}$. Finally, after m_6 we execute $x_4^{(4)}$ for computing $f_3^{(i)}$ at the next clock cycle. Therefore, the computation of sparse multiplication $F^{(i-1)} \cdot G^{(i)}$ takes 61 clock cycles in the proposed cryptoprocessor.

Computation of $x_2^{(i)}$ and $y_2^{(i)}$. Squaring over $\mathbb{F}_2[x]$ is free. Let $a = \sum a_i x^i \in \mathbb{F}_{2^{1223}}$ then $a^2 = \sum a_i x^{2i}$. However, the reduction after squaring requires some XOR operations, which are performed in parallel in only one clock cycle. The computation of $x_2^{(i)}$ and $y_2^{(i)}$ are independent of the last step of sparse multiplication (step $x_4^{(4)}$ of Table 2). Therefore, they are computed in parallel with $x_4^{(4)}$ which does not take any additional time.

Computation Cost of Miller's Algorithm. One iteration of Miller's algorithm is performed by following three parts. The computation of $G^{(i)}$ which takes 12 clock cycles, the computation of $F^{(i-1)} \cdot G^{(i)}$ which takes 61 clock cycles, and the computation of $x_2^{(i)}, y_2^{(i)}$ which is free. Thus, in total, each iteration of Algorithm 1 takes 73 clock cycles, which incurs 44688 clock cycles for computing whole Miller's algorithm including initialization.

3.2 Computation of Final Exponentiation

The output $F^{(612)} \in \mathbb{F}_{(2^{1223})^4}$ of the Miller's algorithm is raised to the power $(2^{2446}-1)(2^{1223}-2^{612}+1)$. The 2^{1223}-th powering an element $G = g_0 + g_1 u + g_2 v + g_3 uv$ in $\mathbb{F}_{(2^{1223})^4}$ is easily computed by following equation.

$$G^{2^{1223}} = (g_0 + g_1 + g_2) + (g_1 + g_2 + g_3)u + (g_2 + g_3)v + g_3 uv, \tag{5}$$

which is computed by three additions (one 2-input and two 3-input additions). Thus, two clock cycles are taken for computing $(F^{(612)})^{2^{2446}}$ by the current cryptoprocessor. Further we perform one inversion followed by one multiplication in $\mathbb{F}_{(2^{1223})^4}$ for computing $(F^{(612)})^{2^{2446}-1}$.

The Inversion in $\mathbb{F}_{(2^{1223})^4}$. Let $G = g_0 + g_1 u + g_2 v + g_3 uv$ and $H = G^{-1} = h_0 + h_1 u + h_2 v + h_3 uv$ then $(g_0 + g_1 u + g_2 v + g_3 uv)(h_0 + h_1 u + h_2 v + h_3 uv) = 1$. This could follow the matrix representation :

$$\begin{bmatrix} g_0 & g_1 & g_3 & g_2+g_3 \\ g_1 & g_0+g_1 & g_2+g_3 & g_2 \\ g_2 & g_3 & g_0+g_2 & g_1+g_3 \\ g_3 & g_2+g_3 & g_1+g_3 & g_0+g_1+g_2+g_3 \end{bmatrix} \begin{bmatrix} h_0 \\ h_1 \\ h_2 \\ h_3 \end{bmatrix} = \begin{bmatrix} 1 \\ 0 \\ 0 \\ 0 \end{bmatrix}$$

From which the value of h_1, h_2, h_3, and h_4 could be solved by $(I + 36M + 8S + 57A)$, where I, M, S, A stand for inversion, multiplication, squaring, and addition in the base field $\mathbb{F}_{2^{1223}}$. Operation I is performed by Itoh-Tsujii algorithm [20], which requires $(14M + 1222S)$. Thus the cost for computing inversion in $\mathbb{F}_{(2^{1223})^4}$ is $(50M + 1230S + 57A)$. Thereafter, a multiplication in $\mathbb{F}_{(2^{1223})^4}$ is performed by $(16M + 22A)$ operations. Thus, in total, the computation of $(F^{(612)})^{2^{2446}-1}$ requires $(66M + 1230S + 82A)$ operations.

The Exponentiation by $(2^{1223} - 2^{612} + 1)$. The second part of the exponent $(2^{1223} - 2^{612} + 1)$ is raised to the power of $(F^{(612)})^{2^{2446}-1}$ by means of $(32M + 612S + 53A)$. This is possible as the inverse of $(F^{(612)})^{2^{2446}-1}$ in $\mathbb{F}_{(2^{1223})^4}$ is performed by computing $((F^{(612)})^{2^{2446}-1})^{2^{2446}}$, which is easy as shown in Eq. 5. Thus, major operations in the second part are two multiplications in $\mathbb{F}_{(2^{1223})^4}$.

Computation Costs of Final Exponentiation and η_T Pairing. The final exponentiation is performed by means of $(98M + 1842S + 135A)$ operations. In our proposed cryptoprocessor (Fig. 3) the multiplier unit is shared by both Miller's algorithm and the final exponentiation. The control signal c_{17} selects operands from one of these two operations. The squaring and additions of final exponentiation are performed separately from the Miller's algorithm. Some of squaring and additions are performed in parallel. The proposed cryptoprocessor computes final exponentiation in 2922 clock cycles, which is much less than the cycle count for computing Miller's algorithm. Total clock cycle count for computing an 128-bit secure η_T pairing is 47610 on our proposed architecture.

4 Results

The whole design has been done in Verilog (HDL). All results have been obtained from the place-and-route report of Xilinx ISE Design Suit. Table 3 shows the implementation results. The critical path of the design is formed in between the input and the output of the hybrid 306-bit Karatsuba multiplier (in Fig. 2). We produce the results for fair comparison, observing the performance of the proposed cryptoprocessor on different FPGA platforms. The Virtex-6 is the latest FPGA family of Xilinx, on which the proposed design runs at a maximum frequency of $250MHz$. In total, it uses 15167 logic slices including whole data path (for Miller's algorithm and for final exponentiation), the controller logic, and registers on the Virtex-6 FPGA, where it finishes computation of one 128-bit secure η_T pairing in $190\mu s$.

Table 3. Implementation results of the η_T pairing cryptoprocessor

Platform	Slice	LUT	Frequency $[MHz]$	Clock Cycles	Security $[bit]$	Times $[\mu s]$
Virtex-2	36534	69367	125			381
Virtex-4	35458	69367	168	47610	128	286
Virtex-6[‡]	15167	54681	250			190
‡ : One Virtex-6 slice consists of four LUTs and eight flip-flops.						

4.1 Comparison with Existing Designs

Two aspects of the proposed design are considered when it is compared with the existing designs. First, we compare it with the existing η_T pairing processors over characteristic-two fields as summarized in Table 4. We consider only the design results with maximum security level provided by the respective authors. To the best of the authors' knowledge no hardware implementation is available

Table 4. Hardware designs for the η_T pairing

Designs	Curve	Security $[bit]$	FPGA	Area $[Slices]$	Frequency $[MHz]$	Times $[\mu s]$
Shu *et al.* [28]	$E/\mathbb{F}_{2^{557}}$	96	xc4vlx200-10	37931	66	675.5
Beuchat *et al.* [5]	$E/\mathbb{F}_{2^{691}}$	105	xc4vlx200-11	78874	130	18.8
This work	$E/\mathbb{F}_{2^{1223}}$	128	xc4vlx200-11	35458	168	286.0
This work	$E/\mathbb{F}_{2^{1223}}$	128	xc6vlx130t-3	15167	250	190.0

for computing 128-bit secure η_T pairing on supersingular elliptic curves over characteristic-two fields. The existing designs in this respect are for a maximum of 105-bit secure design over $\mathbb{F}_{2^{691}}$ field, which is proposed by Beuchat *et al.* in [5]. The design proposed in [5] computes 105-bit secure η_T pairing and achieves a very good speed of $18.8\mu s$. However, compared to the respective design in [5] our design with higher security level demands much lesser, less than half, number of slices on the same FPGA family. As a result, the proposed design could be implemented on a medium-range Virtex-4 device, whereas the existing one's demand a high-range device in the same FPGA family. This makes our design more useful in resource-constrained identity-aware devices.

The second aspect of the design is considered on the fact of 128-bit secure pairing computation irrespective of underlying curve and field types. Table 5 summarizes the comparative studies of related designs. The proposed design is the first one which computes an 128-bit secure pairing in less than one millisecond ($190\mu s$ on a Virtex-6 FPGA) on a dedicated hardware.

All the existing designs except [1] are based on elliptic curves. The design of [1] computes optimal-eta pairing on 128-bit secure supersingular Genus-2 binary hyperelliptic curves. The compact design proposed in [10] computes η_T pairing on supersingular elliptic curves over $\mathbb{F}_{3^{m'}}$ fields. Due to its low area the

Table 5. Hardware designs for 128-bit secure pairings

Designs	**Curve**	**FPGA**	**Area**	**Freq.** $[MHz]$	**Times** $[\mu s]$	$A \cdot T$†
Duquesne et al. [9]§	$E/\mathbb{F}_{p256}$	Stratix III	4233 A‡	165	1070	-
Fan et al. [11]	$E/\mathbb{F}_{p256}$	xc6vlx240-3	4014 Slices, 42 DSP	210	1170	-
Kammler *et al.* [21]	$E/\mathbb{F}_{p256}$	130*nm* CMOS	97000 Gates	338	15800	-
Fan *et al.* [12]	$E/\mathbb{F}_{p256}$	130*nm* CMOS	183000 Gates	204	2900	-
Ghosh *et al.* [14]	$E/\mathbb{F}_{p256}$	xc4vlx200-12	52000 Slices	50	16400	852.8
Estibals [10]	$E/\mathbb{F}_{3^{5 \cdot 97}}$	xc4vlx200-11	4755 Slices	192	2227	10.6
Aranha *et al.* [1]	$Co/\mathbb{F}_{2^{367}}$	xc4vlx25-11	4518 Slices	220	3518	15.9
This work	$E/\mathbb{F}_{2^{1223}}$	xc4vlx200-11	35458 Slices	168	286	10.1
This work	$E/\mathbb{F}_{2^{1223}}$	xc6vlx130t-3	15167 Slices	250	190	2.9

† $A \cdot T$ represents product of *area* in slices and *time* in seconds.
§ It provides 126-bit security instead of 128-bit.
‡ It has 8 Rowers, each consisting of two 36x36 DSP blocks and one 9x9 multiplier.

Table 6. Software for 128-bit secure pairings

Reference	**Platform**	**Pairing**	**Curve**	**Frequency** $[MHz]$	**Times** $[ms]$
Beuchat et al. [7]	core i7 2.8GHz	modified Tate	$E/\mathbb{F}_{3^{509}}$	2800	1.87
			$E/\mathbb{F}_{2^{1223}}$	2800	3.08
Naehrig et al. [26]	core2 Q6600	optimal-ate	$E/\mathbb{F}_{p256}$	2394	1.86
Beuchat et al. [6]	core i7 2.8GHz	optimal-ate	$E/\mathbb{F}_{p256}$	2800	0.83
Hankerson et al. [16]	64-bit core2	optimal-ate	$E/\mathbb{F}_{p256}$	2400	6.25
		η_T	$E/\mathbb{F}_{2^{1223}}$	2400	16.25
		η_T	$E/\mathbb{F}_{3^{509}}$	2400	13.75
Grabher et al. [15]	64-bit core2	ate	$E/\mathbb{F}_{p256}$	2400	6.01

design of [10] is useful to resource constrained applications. It is analyzed in Section 5.2, [5] that the number of base-field multiplications required for computing an η_T pairing with high security level over $\mathbb{F}_{2^m}$ and $\mathbb{F}_{3^{m'}}$ are almost same. This is also true for other fields with 128-bit security. For example, the *optimal-ate* pairing on $E/\mathbb{F}_{p_{256}}$ reported in [9,11,12,14,21] requires 15093 multiplications in the base field [16]. On the other hand, the η_T pairing on $E/\mathbb{F}_{2^{1223}}$ requires 4566 multiplications in the base field, which is only 1/3 of *optimal-ate* pairing. Furthermore, the base field size of $\mathbb{F}_{2^{1223}}$ is 1223 bits which is 4.8 times longer than the size of $\mathbb{F}_{p256}$. Thus, the operation complexities for computing both the pairings are almost same. To sum up, the proposed design achieves a significant performance improvement for computing 128-bit secure pairings on hardware platforms. With respect to the $A \cdot T$ product too, the proposed design gives the best results compared to all existing designs. The software implementation results of 128-bit secure pairings computed over different elliptic curves are enlisted in Table 6. The most efficient software for computing 128-bit pairings on supersingular elliptic curves over $\mathbb{F}_{2^{1223}}$ is proposed in [7]. It takes $3.08ms$ on eight parallel cores of a *core i7 2.8GHz* processor.

5 Conclusion

In this paper we have proposed an area and time optimized hybrid Karatsuba multiplier for $\mathbb{F}_{2^{1223}}$. Sufficient parallelism has been employed in the architecture for which we have achieved a high-speed η_T pairing cryptoprocessor. A common datapath for both non-reduced pairing and final exponentiation has been shared which reduces the overall logic cells in its FPGA implementation. The proposed design achieves a significant improvement with respect to two aspects of the design. It computes η_T pairing in characteristic-two field with higher security (128:105) in half area. On the other hand, it achieves eight times speedup and also provides the best $area \times time$ product among existing designs for computing 128-bit secure pairings.

References

1. Aranha, D.F., Beuchat, J.L., Detrey, J., Estibals, N.: Optimal Eta pairing on supersingular genus-2 binary hyperelliptic curves. Cryptology ePrint Archive, Report 2010/559, `http://eprint.iacr.org/`
2. Barke, E., Barker, W., Burr, W., Polk, W., Smid, M.: Recommendation for key management part 1: General (revised). National Institute of Standards and Technology, NIST Special Publication 800-57 (2007)
3. Barreto, P.S.L.M., Galbraith, S.D., ÓhÉigeartaigh, C., Scott, M.: Efficient pairing computation on supersingular Abelian varieties. Designs, Codes and Cryptography 42, 239–271 (2007)
4. Barreto, P.S.L.M., Naehrig, M.: Pairing-friendly elliptic curves of prime order. In: Preneel, B., Tavares, S. (eds.) SAC 2005. LNCS, vol. 3897, pp. 319–331. Springer, Heidelberg (2006)
5. Beuchat, J.L., Detrey, J., Estibals, N., Okamoto, E., Henríguez, F.R.: Fast architectures for the η_T pairing over small-characteristic supersingular elliptic curves. IEEE Transactions on Computers 60(2) (2011)
6. Beuchat, J.-L., González-Díaz, J.E., Mitsunari, S., Okamoto, E., Rodríguez-Henríquez, F., Teruya, T.: High-speed software implementation of the optimal ate pairing over barreto–naehrig curves. In: Joye, M., Miyaji, A., Otsuka, A. (eds.) Pairing 2010. LNCS, vol. 6487, pp. 21–39. Springer, Heidelberg (2010)
7. Beuchat, J.L., Trejo, E.L., Ramos, L.M., Mitsunari, S., Henríquez, F.R.: Multi-core Implementation of the Tate Pairing over Supersingular Elliptic Curves. Cryptology ePrint Archive, Report 2009/276 (2009), `http://eprint.iacr.org/`
8. Boneh, D., Franklin, M.K.: Identity-Based Encryption from the Weil Pairing. In: Kilian, J. (ed.) CRYPTO 2001. LNCS, vol. 2139, pp. 213–229. Springer, Heidelberg (2001)
9. Duquesne, S., Guillermin, N.: A FPGA pairing implementation using the residue number system. Cryptology ePrint Archive, Report 2011/176 (2011), `http://eprint.iacr.org/`

10. Estibals, N.: Compact Hardware for Computing the Tate Pairing over 128-Bit-Security Supersingular Curves. In: Joye, M., Miyaji, A., Otsuka, A. (eds.) Pairing 2010. LNCS, vol. 6487, pp. 397–416. Springer, Heidelberg (2010)
11. Fan, J., Vercauteren, F., Verbauwhede, I.: Efficient Hardware Implementation of $\mathbb{F}_p$-arithmetic for Pairing-Friendly Curves. IEEE Trasaction on Computers (to appear, 2011)
12. Fan, J., Vercauteren, F., Verbauwhede, I.: Faster $\mathbb{F}_p$-Arithmetic for Cryptographic Pairings on Barreto-Naehrig Curves. In: Clavier, C., Gaj, K. (eds.) CHES 2009. LNCS, vol. 5747, pp. 240–253. Springer, Heidelberg (2009)
13. Galbraith, S.: Pairings. In: Blake, I.F., Seroussi, G., Smart, N.P. (eds.) Advances in Elliptic Curve Cryptography. London Mathematical Society Lecture Note Series, vol. ch. IX, Cambridge University Press, Cambridge (2005)
14. Ghosh, S., Mukhopadhyay, D., Roychowdhury, D.: High speed flexible pairing cryptoprocessor on FPGA platform. In: Joye, M., Miyaji, A., Otsuka, A. (eds.) Pairing 2010. LNCS, vol. 6487, pp. 450–466. Springer, Heidelberg (2010)
15. Grabher, P., Großschädl, J., Page, D.: On software parallel implementation of cryptographic pairings. In: Avanzi, R.M., Keliher, L., Sica, F. (eds.) SAC 2008. LNCS, vol. 5381, pp. 35–50. Springer, Heidelberg (2009)
16. Hankerson, D., Menezes, A., Scott, M.: Software implementation of pairings. Cryptology and Information Security Series, ch. 12, pp. 188–206. IOS Press, Amsterdam (2009)
17. Henríquez, F.R., Koç, Ç.K.: On fully parallel Karatsuba multipliers for $GF(2^m)$. In: International Conference on Computer Science and Technology CST 2003, pp. 405–410 (2003)
18. Hess, F., Smart, N.P., Vercauteren, F.: The Eta pairing revisited. IEEE Transactions on Information Theory 52(10), 4595–4602 (2006)
19. Hoffstein, J., Pipher, J., Silverman, J.H.: An introduction to mathematical cryptography. Springer, Heidelberg (2008)
20. Itoh, T., Tsujii, S.: A fast algorithm for computing multiplicative inverses in GF(2^m) using normal bases. Inf. Comput. 78(3), 171–177 (1988)
21. Kammler, D., Zhang, D., Schwabe, P., Scharwaechter, H., Langenberg, M., Auras, D., Ascheid, G., Mathar, R.: Designing an ASIP for Cryptographic Pairings over Barreto-Naehrig Curves. In: Clavier, C., Gaj, K. (eds.) CHES 2009. LNCS, vol. 5747, pp. 254–271. Springer, Heidelberg (2009)
22. Karatsuba, A., Ofman, Y.: Multiplication of Multidigit Numbers on Automata. Soviet Physics Doklady (English Translation) 7(7), 595–596 (1963)
23. Lee, E., Lee, H.S., Park, C.M.: Efficient and generalized pairing computation on abelian varieties. Cryptology ePrint Archive, Report 2009/040 (2009), `http://eprint.iacr.org/`
24. Lenstra, A.: Unbelievable security: Matching AES security using public key systems. In: Boyd, C. (ed.) ASIACRYPT 2001. LNCS, vol. 2248, pp. 67–86. Springer, Heidelberg (2001)
25. Miller, V.S.: The Weil pairing, and its efficient calculation. Journal of Cryptology 17, 235–261 (2004)
26. Naehrig, M., Niederhagen, R., Schwabe, P.: New software speed records for cryptographic pairings. Cryptology ePrint Archive, Report 2010/186, `http://eprint.iacr.org/`
27. Rebeiro, C., Mukhopadhyay, D.: High speed compact elliptic curve cryptoprocessor for FPGA platforms. In: Chowdhury, D.R., Rijmen, V., Das, A. (eds.) INDOCRYPT 2008. LNCS, vol. 5365, pp. 376–388. Springer, Heidelberg (2008)

28. Shu, C., Kwon, S., Gaj, K.: Reconfigurable computing approach for tate pairing cryptosystems over binary fields. IEEE Transactions on Computers 58(9), 1221–1237 (2009)
29. Weimerskirch, A., Paar, C.: Generalizations of the Karatsuba algorithm for efficient implementations. Cryptology ePrint Archive, Report 2006/224 (2006), `http://eprint.iacr.org/`

Appendix A

We describe 1223-bit multiplication based on *the serial use of 306-bit parallel multiplier* in Algorithm 2. The variable names of the algorithm are similar to the registers and intermediate results computed by the proposed 1223-bit multiplier as shown in Fig. 2.

Appendix B

The η_T Pairing on Supersingular Elliptic Curves over $\mathbb{F}_{2^{1223}}$. This paper considers the η_T pairing computed over characteristic two field $\mathbb{F}_{2^{1223}}$, which is represented as $\mathbb{F}_2[x]/(x^{1223} + x^{255} + 1)$ in the polynomial basis with irreducible polynomial $(x^{1223} + x^{255} + 1)$. The supersingular elliptic curve E over above field is defined as:

$$E/\mathbb{F}_{2^{1223}} : Y^2 + Y = X^3 + X, \tag{6}$$

which has embedding degree $k = 4$. It forms a large subgroup with prime order $r = (2^{1223} + 2^{612} + 1)/5$. The η_T pairing on $E/\mathbb{F}_{2^{1223}}$ attains 128-bit security level because Pollard's rho method for computing discrete logarithms in above order-r subgroup has running time at least 2^{128}, as do the index-calculus algorithms for computing discrete logarithms in the extension field $\mathbb{F}_{(2^{1223})^4}$. We refer the reader to [3,18,24] for more details about the computation techniques of η_T pairing and its respective security. We represent the extension field $\mathbb{F}_{(2^{1223})^4}$ using tower field extensions $\mathbb{F}_{(2^{1223})^2} = \mathbb{F}_{2^{1223}}[u]/(u^2+u+1)$ and $\mathbb{F}_{(2^{1223})^4} = \mathbb{F}_{(2^{1223})^2}[v]/(v^2+v+u)$, where a basis for $\mathbb{F}_{(2^{1223})^4}$ over $\mathbb{F}_{2^{1223}}$ is [1, u, v, uv].

Algorithm 2 . The 1223-bit multiplication based on Karatsuba technique†.

Input: $a = \sum_{i=0}^{1222} a_i x^i$ and $b = \sum_{i=0}^{1222} b_i x^i$.
Output: $a \cdot b$.

$a_{00} \leftarrow \sum_{i=0}^{305} a_i x^i$; $a_{01} \leftarrow \sum_{i=306}^{611} a_i x^i$;
$a_{10} \leftarrow \sum_{i=612}^{917} a_i x^i$; $a_{11} \leftarrow \sum_{i=918}^{1222} a_i x^i$;
$b_{00} \leftarrow \sum_{i=0}^{305} b_i x^i$; $b_{01} \leftarrow \sum_{i=306}^{611} b_i x^i$;
$b_{10} \leftarrow \sum_{i=612}^{917} b_i x^i$; $b_{11} \leftarrow \sum_{i=918}^{1222} b_i x^i$;
$g_0 \leftarrow a_{00} + a_{01}$; $g_1 \leftarrow a_{10} + a_{11}$; $g_2 \leftarrow a_{00} + a_{10}$; $g_3 \leftarrow a_{01} + a_{11}$;
$h_0 \leftarrow b_{00} + b_{01}$; $h_1 \leftarrow b_{10} + b_{11}$; $h_2 \leftarrow b_{00} + b_{10}$; $h_3 \leftarrow b_{01} + b_{11}$;
$g_4 \leftarrow g_2 + g_3$; $h_4 \leftarrow h_2 + h_3$;
$k \leftarrow a_{00} \cdot b_{00}$;
$d_1 \leftarrow k_L$; $d_0 \leftarrow k_R$;
$k \leftarrow a_{01} \cdot b_{01}$;
$d_3 \leftarrow k_L$; $d_2 \leftarrow k_R$;
$k \leftarrow g_0 \cdot h_0$;
$t_1 \leftarrow k_L$; $t_0 \leftarrow k_R$;
$d_1 \leftarrow d_1 + d_0 + d_2 + t_0$; $d_2 \leftarrow d_2 + d_1 + d_3 + t_1$;
$k \leftarrow a_{10} \cdot b_{10}$;
$e_1 \leftarrow k_L$; $e_0 \leftarrow k_R$;
$k \leftarrow a_{11} \cdot b_{11}$;
$e_3 \leftarrow k_L$; $e_2 \leftarrow k_R$;
$k \leftarrow g_1 \cdot h_1$;
$t_1 \leftarrow k_L$; $t_0 \leftarrow k_R$;
$e_1 \leftarrow e_1 + e_0 + e_2 + t_0$; $e_2 \leftarrow e_2 + e_1 + e_3 + t_1$;
$k \leftarrow g_2 \cdot h_2$;
$f_1 \leftarrow k_L$; $f_0 \leftarrow k_R$;
$k \leftarrow g_3 \cdot h_3$;
$f_3 \leftarrow k_L$; $f_2 \leftarrow k_R$;
$k \leftarrow g_4 \cdot h_4$;
$t_1 \leftarrow k_L$; $t_0 \leftarrow k_R$;
$f_1 \leftarrow f_1 + f_0 + f_2 + t_0$; $f_2 \leftarrow f_2 + f_1 + f_3 + t_1$;
$r_0 \leftarrow d_0$; $r_1 \leftarrow d_1$; $r_2 \leftarrow d_2 + d_0 + e_0 + f_0$;
$r_3 \leftarrow d_3 + d_1 + e_1 + f_1$; $r_4 \leftarrow e_0 + d_2 + e_2 + f_2$;
$r_5 \leftarrow e_1 + d_3 + e_3 + f_3$; $r_6 \leftarrow e_2$; $r_7 \leftarrow e_3$;
return $(r_7 \cdot x^{2142} + r_6 \cdot x^{1836} + r_5 \cdot x^{1530} + r_4 \cdot x^{1224} + r_3 \cdot x^{918} +$
$r_2 \cdot x^{612} + r_1 \cdot x^{306} + r_0)$.

† In the algorithm, k_R represents the least significant m bits of $(2m-1)$-bit result k, and k_L represents the most significant $m-1$ bits of k.

Fast Multi-precision Multiplication for Public-Key Cryptography on Embedded Microprocessors

Michael Hutter and Erich Wenger

Institute for Applied Information Processing and Communications (IAIK),
Graz University of Technology, Inffeldgasse 16a, 8010 Graz, Austria
{Michael.Hutter,Erich.Wenger}@iaik.tugraz.at

Abstract. Multi-precision multiplication is one of the most fundamental operations on microprocessors to allow public-key cryptography such as RSA and Elliptic Curve Cryptography (ECC). In this paper, we present a novel multiplication technique that increases the performance of multiplication by sophisticated caching of operands. Our method significantly reduces the number of needed *load* instructions which is usually one of the most expensive operation on modern processors. We evaluate our new technique on an 8-bit ATmega128 microcontroller and compare the result with existing solutions. Our implementation needs only 2,395 clock cycles for a 160-bit multiplication which outperforms related work by a factor of 10 % to 23 %. The number of required *load* instructions is reduced from 167 (needed for the best known hybrid multiplication) to only 80. Our implementation scales very well even for larger Integer sizes (required for RSA) and limited register sets. It further fully complies to existing multiply-accumulate instructions that are integrated in most of the available processors.

Keywords: Multi-precision Arithmetic, Microprocessors, Elliptic Curve Cryptography, RSA, Embedded Devices.

1 Introduction

Multiplication is one of the most important arithmetic operation in public-key cryptography. It engross most of the resources and execution time of modern microprocessors (up to 80 % for Elliptic Curve Cryptography (ECC) and RSA implementations [6]). In order to increase the performance of multiplication, most effort has been put by researchers and developers to reduce the number of instructions or minimize the amount of memory-access operations.

Common multiplication methods are the schoolbook or Comba [4] technique which are widely used in practice. They require at least $2n^2$ *load* instructions to process all operands and to calculate the necessary partial products. In 2004, Gura et al. [6] presented a new method that combines the advantages of these methods (hybrid multiplication). They reduced the number of *load* instructions

B. Preneel and T. Takagi (Eds.): CHES 2011, LNCS 6917, pp. 459–474, 2011.

to only $2\lceil n^2/d\rceil$ where the parameter d depends on the number of available registers of the underlying architecture. They reported a performance gain of about 25 % compared to the classical Comba multiplication. Their 160-bit implementation needs 3,106 clock cycles on an 8-bit ATmega128 microcontroller. Since then, several authors applied this method [7,12,14,15,17] and proposed various enhancements to further improve the performance. Most of the related work reported between 2,593 and 2,881 clock cycles on the same platform.

In this paper, we present a novel multiplication technique that reduces the number of needed *load* instructions to only $2n^2/e$ where $e > d$. We propose a new way to process the operands which allows efficiently caching of required operands. In order to evaluate the performance, we use the ATmega128 microcontroller and compare the results with related work. For a 160-bit multiplication, 2,395 clock cycles are necessary which is an improvement by a factor of 10 % compared to the best reported implementation of Scott et al. [14] (which need 2,651 clock cycles) and by a factor of about 23 % compared to the work of Gura et al. [6]. We further compare our solution with different Integer sizes (160, 192, 256, 512, 1,024, and 2,048) and register sizes (e = 2, 4, 8, 10, and 20). It shows that our solution needs about 15 % less clock cycles for any chosen Integer size. Our solution also scales very well for different register sizes without significant loss of performance. Besides this, the method fully complies with common architectures that support multiply-accumulate instructions using a (Comba-like) triple-register accumulator.

The paper is organized as follows. In Section 2, we describe related work on that topic and give performance numbers for different multiplication techniques. Section 3 describes different multi-precision multiplication techniques used in practice. We describe the operand scanning, product scanning, and the hybrid method and compare them with our solution. In Section 4, we present the results of our evaluations. We describe the ATmega128 architecture and give details about the implementation. Summary and conclusions are given in Section 5.

2 Related Work

In this section, we describe related work on multi-precision multiplication over prime fields. Most of the work given in literature make use of the hybrid-multiplication technique [6] which provides best performance on most microprocessors. This technique was first presented at CHES 2004 where the authors reported a speed improvement of up to 25 % compared to the classical Comba-multiplication technique [4] on 8-bit platforms. Their implementation requires 3,106 clock cycles for a 160-bit multiplication on an ATmega128 [1]. Several authors adopted the idea and applied the method for different devices and environments, e.g. sensor nodes. Wang et al. [18] and Ugus et al. [16] made use of this technique and implemented it on the MICAz motes which feature an ATmega128 microcontroller. Results for the same platform have been also reported by Liu et al. [11] and Szczechowiak et al. [15] in 2008 who provide software libraries (TinyECC and NanoECC) for various sensor-mote platforms. One of the first who improved the implementation of Gura has been due to Uhsadel et al. [17]. They have been able to reduce the number of needed clock cycles to only 2,881. Further improvements have been also reported by Scott et al. [14]. They introduced additional registers (so-called *carry catchers*) and could increase the performance to 2,651 clock cycles. Note that they fully unrolled the execution sequence to avoid additional clock cycles for loop instructions. Similar results have been also obtained by Kargl et al. [7] in 2008 which reported 2,593 clock cycles for an un-rolled 160-bit multiplication on the ATmega128.

In 2009, Lederer et al. [9] showed that the needed number of addition and move instructions can be reduced by simply rearranging the instructions during execution of the hybrid-multiplication method. Similar findings have been also reported recently by Liu et al. [12] who reported the fastest looped version of the hybrid multiplication needing 2,865 clock cycles in total.

3 Multi-precision Multiplication Techniques

In the following subsections, we describe common multiplication techniques that are often used in practice. We describe the operand scanning, product scanning, and hybrid multiplication method[1]. The methods differ in several ways how to process the operands and how many *load* and *store* instructions are necessary to perform the calculation. Most of these methods lack in the fact that they load the same operands not only once but several times throughout the algorithm which results in additional and unnecessary clock cycles. We present a new multiplication technique that improves existing solutions by efficiently reducing the *load* instructions through sophisticated caching of operands.

Throughout the paper, we use the following notation. Let a and b be two m-bit large Integers that can be written as multiple-word array structures $A = (A[n-1], \ldots, A[2], A[1], A[0])$ and $B = (B[n-1], \ldots, B[2], B[1], B[0])$. Further let W be the word size of the processor (*e.g.* 8, 16, 32, or 64 bits) and $n = \lceil m/W \rceil$ the number of needed words to represent the Integers a or b. We denote the result of the multiplication by $c = ab$ and represent it in a double-size word array $C = (C[2n-1], \ldots, C[2], C[1], C[0])$.

3.1 Operand-Scanning Method

Among the most simplest way to perform large Integer multiplication is the operand-scanning method (or often referred as *schoolbook* or *row-wise* multiplication method). The multiplication can be implemented using two nested loop operations. The outer loop loads the operand $A[i]$ at index $i = 0 \ldots n-1$ and keeps the value constant inside the inner loop of the algorithm. Within the inner loop, the multiplicand $B[j]$ is loaded word by word and multiplied with the operand $A[i]$. The partial product is then added to the intermediate result of the same column which is usually buffered in a register or stored in data memory.

[1] Note that we do not consider multiplications methods such as Karatsuba-Ofman or FFT in this paper since they are considered to require more resources and memory accesses on common microcontrollers than the given methods [8].

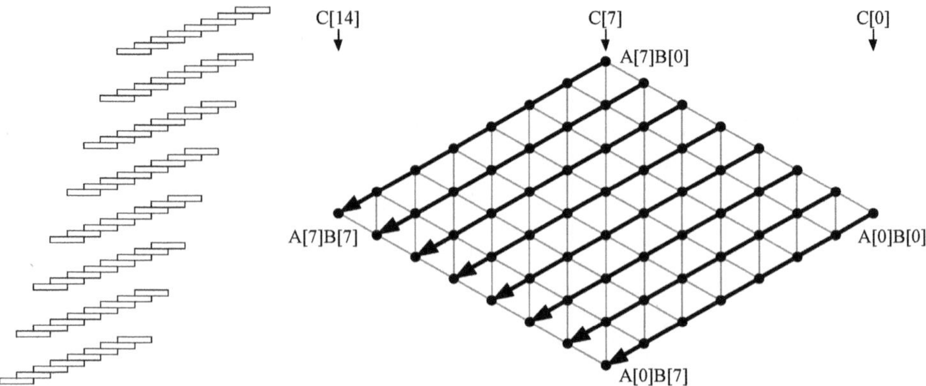

Fig. 1. Operand-scanning multiplication of 8-word large Integers a and b

Figure 1 shows the structure of the algorithm on the left side. The individual row levels can be clearly discerned. On the right side of the figure, all n^2 partial products are displayed in form of a rhombus. Each point in the rhombus represents a multiplication $A[i] \times B[j]$. The most right-sided corner of the rhombus starts with the lowest indices $i, j = 0$ and the most left-sided corner ends with the highest indices $i, j = n - 1$. By following all multiplications from the right to the lower-mid corner of the rhombus, it can be observed that the operand $A[i]$ keeps constant for any index $i \in [0, n)$. The same holds true for the operand $B[j]$ and $j \in [0, n)$ by following all multiplications from right to the upper-mid corner of the rhombus. Note that this is also valid for the left-handed side of the rhombus.

For the operand-scanning method, it can be seen that the partial products are calculated from the upper-right side to the lower-left side of the rhombus (we marked the processing of the partial products with a black arrow). In each row, n multiplications have to be performed. Furthermore, $2n$ *load* operations and n *store* operations are required to load the multiplicand and the intermediate result $C[i+j]$ and to store the result $C[i+j] \leftarrow C[i+j] + A[i] \times B[j]$. Thus, $3n^2 + 2n$ memory operations are necessary for the entire multi-precision multiplication. Note that this number decreases to $n^2 + 3n$ for architectures that can maintain the intermediate result in available working registers.

3.2 Product-Scanning Method

Another way to perform a multi-precision multiplication is the product-scanning method (also referred as *Comba* [4] or *column-wise* multiplication method). There, each partial product is processed in a column-wise approach. This has several advantages. First, since all operands of each column are multiplied and added consecutively (within a multiply-accumulate approach), a final word of the result is obtained for each column. Thus, no intermediate results have to be stored or loaded throughout the algorithm. In addition, the handling of carry propagation

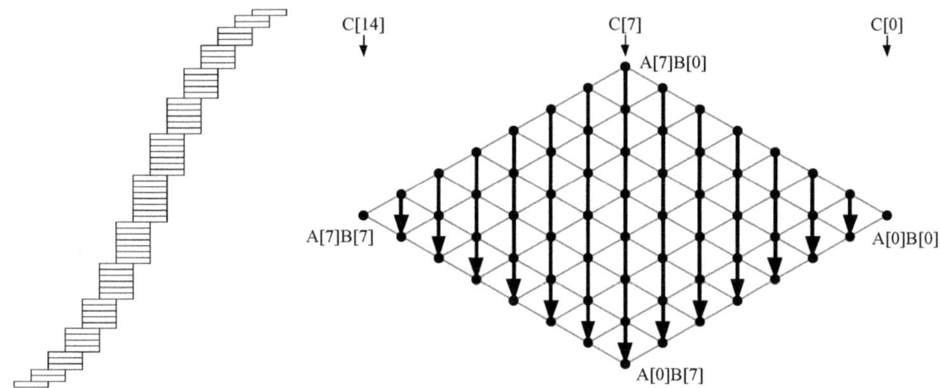

Fig. 2. Product-scanning multiplication of 8-word large Integers a and b

is very easy because the carry can be simply added to the result of the next column using a simple register-copy operation. Second, only five working registers are needed to perform the multiplication: two registers for the operand and multiplicand and three registers for accumulation[2]. This makes the method very suitable for low-resource devices with limited registers.

Figure 2 shows the structure of the product-scanning method. By having a look at the rhombus, it shows that by processing the partial products in a column-wise instead of a row-wise approach, only one *store* operation is needed to store the final word of the result. For the entire multi-precision operation, $2n^2$ *load* operations are necessary to load the operands $A[i]$ and $B[j]$ and $2n$ *store* operations are needed to store the result. Therefore, $2n^2+2n$ memory operations are needed.

3.3 Hybrid Method

The hybrid multiplication method [6] combines the advantages of the operand-scanning and product-scanning method. It can be implemented using two nested loop structures where the outer loop follows a product-scanning approach and the inner loop performs a multiplication according to the operand-scanning method.

The main idea is to minimize the number of *load* instructions within the inner loop. For this, the accumulator has to be increased to a size of $2d+1$ registers. The parameter d defines the number of rows within a processed block. Note that the hybrid multiplication is equals to the product-scanning method if parameter d is chosen as $d = 1$ and it is equal to the operand-scanning method if $d = n$.

Figure 3 shows the structure of the hybrid multiplication for $d = 4$. It shows that the partial products are processed in form of individual blocks (we marked the processing sequence of the blocks from 1 to 4). Within one block, all operands are processed row by row according to the operand-scanning approach. Note that

[2] We assume the allocation of three registers for the accumulator register whereas $2 + \lceil log_2(n)/W \rceil$ registers are actually needed to maintain the sum of partial products.

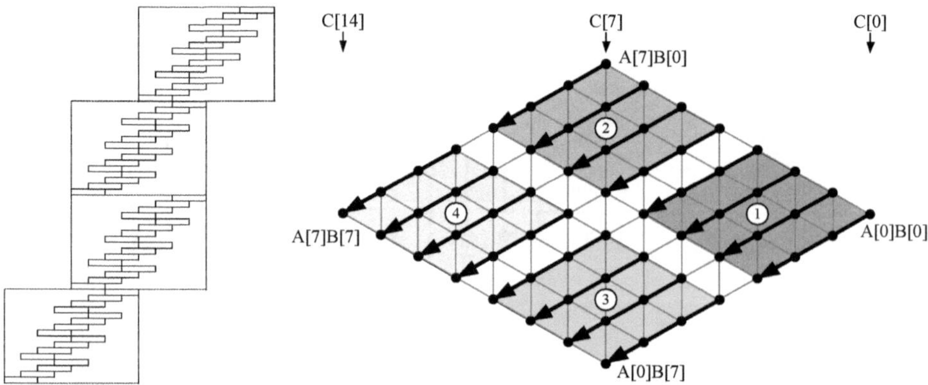

Fig. 3. Hybrid multiplication of 8-word large Integers a and b ($d = 4$)

these blocks use operands with a very limited range of indices. Thus, several *load* instructions can be saved in cases where enough working registers are available. However, the outer loop of the hybrid method processes the blocks in a column-wise approach. So between two consecutive blocks no operands can be shared and all operands have to be loaded from memory again. This becomes clear by having a look at the processing of Block 1-3. Block 2 and 3 do not share any operands that possess the same indices. Therefore, all operands that have already been loaded for Block 1 and that can be reused in Block 3 have to be loaded again after processing of Block 2 which requires additional and unnecessary *load* instructions. However, in total, the hybrid method needs $2\lceil n^2/d\rceil + 2n$ memory-access instructions which provides good performances on devices that feature a large register set.

3.4 Operand-Caching Method

We present a new method to perform multi-precision multiplication. The main idea is to reduce the number of memory accesses to a minimum by efficiently caching of operands. We show that by spending a certain amount of *store* operations, a significant amount of *load* instructions can be saved by reusing operands that have been already loaded in working registers.

The method basically follows the product-scanning approach but divides the calculation into several rows. In fact, the product-scanning method provides best performance if all needed operands can be maintained in working registers. In such a case, only $2n$ *load* instructions and $2n$ *store* instructions would be necessary. However, the product-scanning method becomes inefficient if not enough registers are available or if the Integer size is too large to cache a significant amount of operands. Hence, several *load* instructions are necessary to reload and overwrite the operands in registers.

In the light of this fact, we propose to separate the product-scanning method into individual rows $r = \lfloor n/e \rfloor$. The size e of each row is chosen in a way that all

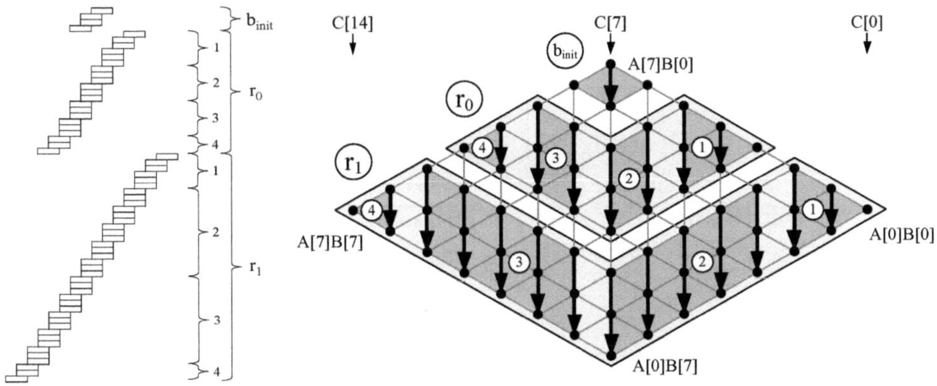

Fig. 4. Operand-caching multiplication of 8-word large Integers a and b ($e = 3$)

needed words of one operand can be cached in the available working registers. Figure 4 shows the structure of the proposed method for parameter $e = 3$. That means, 3 registers are reserved to store 3 words of operand a and 3 registers are reserved to store 3 words of operand b. Thus, we assume $f = 2e + 3 = 9$ available registers including a triple-word accumulator. The calculation is now separated into $r = \lfloor 8/3 \rfloor = 2$ rows, *i.e.* r_0 and r_1, and consists of one remaining block which we further denote as initialization block b_{init}. This block calculates the partial products which are not processed by the rows.

All rows are further separated into four parts. Part 1 and 4 use the classical product-scanning approach. Part 2 and 3 perform an efficient multiply-accumulate operation of already cached operands.

The algorithm starts with the calculation of b_{init} and processes the individual rows afterwards (starting from the the smallest to the largest row, *i.e.* from the top to the bottom of the rhombus). Furthermore, all partial products are generated from right to left. In the following, we describe the algorithm in a more detail.

Initialization Block b_{init}. This block (located in the upper-mid of the rhombus) performs the multiplication according to the classical product-scanning method. The Integer size of the b_{init} multiplication is $(n - re)$, *i.e.* $8 - 6 = 2$ in our example, which is by definition smaller than e. Because of that, all operands can be loaded and maintained within the available registers resulting in only $4(n - re)$ memory-access operations. Note that the calculation of b_{init} is only required if there exist remaining partial products, *i.e.* $n \bmod e \neq 0$. If $n \bmod e = 0$, the calculation of b_{init} is skipped. Furthermore, consider the special case when $n < e$ where only b_{init} has to be performed skipping the processing of rows (trivial case).

Processing of Rows. In the following, we describe the processing of each row $p = r - 1 \ldots 0$. Each row consists of four parts.

Part 1. This part starts with a product-scanning multiplication. All operands for that row are first loaded into registers, *i.e.* $A[i]$ with $i = pe \ldots e(p + 1) - 1$

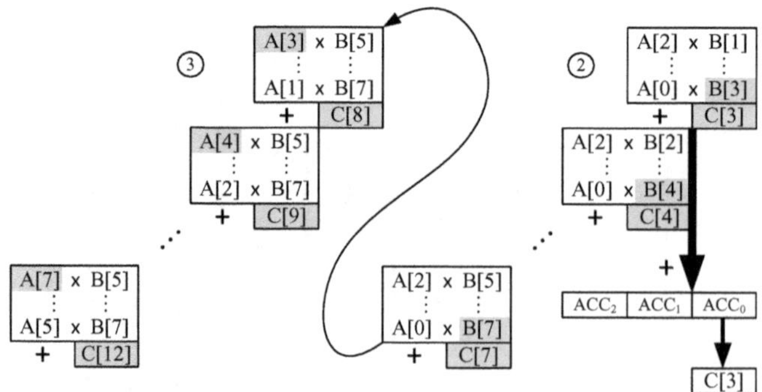

Fig. 5. Processing of Part 2 and 3 of the row r_1

and $B[j]$ with $j = 0 \ldots e-1$. The sum of all partial products $A[i] \times B[j]$ is then stored as intermediate result to the memory location $C[i]$ (same index range as $A[i]$). Therefore, $2e$ *load* instructions and e *store* instructions are needed.

Part 2. The second part, processes $n - e(p+1)$ columns using a multiply-accumulate approach. Since all operands of $A[i]$ were already loaded and used in Part 1, only one word $B[j]$ has to be loaded from one column to the next. The operands $A[i]$ are kept constant throughout the processing of Part 2. Next to the needed *load* instructions for $B[j]$, we have to load and update the intermediate result of Part 1 with the result obtained in Part 2. Thus, $2(n - e(p+1))$ *load* and $n - e(p+1)$ *store* instructions are required for that part.

Part 3. The third part performs the same operation as described in Part 2 except that the already loaded operands $B[j]$ are kept constant and that one word $A[i]$ is loaded for each column. Figure 5 shows the processing of Part 2 and 3 of row r_1 ($p = 0$). For each column, two *load* instructions are necessary (marked in grey). All other operands have been loaded and cached in previous parts. Operands which are not required for further processing are overwritten by new operands, *e.g.* $B[1] \ldots B[4]$ in Part 2 of our example.

Part 4. The last part calculates the remaining partial products. In contrast to Part 1, no *load* instructions are required since all operands have been already loaded in Part 3. Hence, only e memory-access operations are needed to store the remaining words of the (intermediate) result c.

Table 1 summaries the memory-access complexity of the initialization block and the individual parts of a row p. By summing up all *load* instructions, we get

$$2(n - re) + \sum_{p=0}^{r-1}(4n - 4pe - 2e) = 2n + 4rn - 2er^2 - 2er \leq \frac{2n^2}{e}. \qquad (1)$$

The total number of *store* operations can be evaluated by

$$2(n - re) + \sum_{p=0}^{r-1}(2n - 2pe) = 2n + 2rn - er^2 - er \leq \frac{n^2}{e} + n. \qquad (2)$$

Table 1. Memory-access complexity of b_{init} and each part of row $p = 0 \ldots r-1$

Component	Load Instr.	Store Instr.	Total
b_{init}	$2(n-re)$	$2(n-re)$	$4(n-re)$
Part 1	$2e$	e	$3e$
Part 2	$2(n-e(p+1))$	$n-e(p+1)$	$3(n-e(p+1))$
Part 3	$2(n-e(p+1))$	$n-e(p+1)$	$3(n-e(p+1))$
Part 4	0	e	e

Table 2 lists the complexity of different multi-precision multiplication techniques. It shows that the hybrid method needs $2\lceil\frac{n^2}{d}\rceil$ *load* instructions whereas the operand-caching technique needs about $\frac{2n^2}{e}$. Since the total number of available registers f equals to $2e+3$ for the operand-caching technique ($2e$ registers for the operand registers and three registers for the accumulator) and $3d+2$ for the hybrid method ($d+1$ registers for the operands and $2d+1$ registers for the accumulator), we obtain

$$2e+3 = 3d+2 \implies e = \frac{3d-1}{2} \quad \text{and} \quad e > d. \tag{3}$$

If we compare the total number of memory-access instructions for the hybrid and the operand-caching method and express both runtimes using f, we get

$$2\left\lceil\frac{3n^2}{f-2}\right\rceil + 2n > \frac{6n^2}{f-3} + n \tag{4}$$

Note that there are more parameters to consider. The number of additions of the operand-caching method is $3n^2$ and the number of additions of the hybrid method is $n^2(2+d/2)$ (upper bound). Also the pseudocode of Gura et al. [6] for the hybrid multiplication method is inefficient in the special case of $n \bmod d \neq 0$.

Table 2. Memory-access complexity of different multiplication techniques

Method	Load Instructions	Store Instructions	Memory Instructions
Operand Scanning	$2n^2+n$	n^2+n	$3n^2+2n$
Product Scanning [4]	$2n^2$	$2n$	$2n^2+2n$
Hybrid [6]	$2\lceil n^2/d\rceil$	$2n$	$2\lceil n^2/d\rceil+2n$
Operand Caching	$\mathbf{2n^2/e}$	$\mathbf{n^2/e+n}$	$\mathbf{3n^2/e+n}$

4 Results

We used the 8-bit ATmega128 microcontroller for evaluating the new multiplication technique. The ATmega128 is part of the megaAVR family from Atmel [1]. It has been widely used in embedded systems, automotive environments, and

Table 3. Unrolled instruction counts for a 160-bit multiplication on the ATmega128

Method	**Instruction**						**Clock**
	LD	ST	MUL	ADD	MOVW	Others	**Cycles**
Operand Scanning	820	440	400	1,600	2	464	5,427
Product Scanning	800	40	400	1,200	2	159	3,957
Hybrid (d=4)	200	40	400	1,250	202	109	2,904
Operand Caching (e=10)	**80**	**60**	**400**	**1,240**	**2**	**68**	**2,395**

sensor-node applications. The ATmega128 is based on a RISC architecture and provides 133 instructions [2]. The maximum operating frequency is 16 MHz. The device features 128 kB of flash memory and 4 kB of internal SRAM. There exist 32 8-bit general-purpose registers (R0 to R31). Three 16-bit registers can be used for memory addressing, i.e. R26:R27, R28:R29, and R30:R31 which are denoted as X, Y, and Z. Note that the processor also allows pre-decrement and post-increment functionalities that can be used for efficient addressing of operands. The ATmega128 further provides an hardware multiplier that performs an 8×8-bit multiplication within two clock cycles. The 16-bit result is stored in the registers R0 (lower word) and R1 (higher word).

We used register $R25$ to store a zero value. Furthermore, we reserved R23, R24, and R25 as accumulator register. Thus, 20 registers, *i.e.* R2...R21, can be used to store and cache the words of the operands ($e = 10$ registers for each operand a and b). All implementations have been done by using a self-written code generator that allows the generation of (looped & unrolled) assembly code.

In order to demonstrate the performance of our method, we implemented all multiplication techniques described in Section 3. For comparison reasons, we decided to implement a 160×160-bit multiplication as it has been done by most of the related work. Note that for RSA and ECC, larger Integer sizes are recommended in practice [10,13]. The Standards for Efficient Cryptography (SEC) already removed the recommended secp160r1 elliptic curve from their standard since SEC version 2 of 2010 [3].

Table 3 summarizes the instruction counts for the operand scanning, product scanning, hybrid, and operand-caching implementation. The operand-scanning and product-scanning methods have been implemented without using all the available registers (as it usually would be implemented). For hybrid multiplication, we applied $d = 4$ because it allows a better optimization regarding necessary addition operations compared to a multiplication with $d = 5$. The carry propagation problem has been solved by implementing a similar approach as proposed by Liu et al. [12]. Thus, 200 `MOVW` instructions have been necessary to handle the carry propagation accordingly. For a fair comparison, all methods have been optimized for speed and provide unrolled instruction sequences. Furthermore, we implemented all accumulators as ring buffers to reduce necessary `MOV` instructions. After each partial-product generation, the indices of the accumulator registers are shifted so that no `MOV` instructions are necessary to copy the carry.

Table 4. Comparison of multiplication methods for different Integer sizes

Size [bit]	Op. Scan.	Prod. Scan.	Hybrid Method	Operand Caching
160	5,427	3,957	2,904	2,395
192	7,759	5,613	4,144	3,469
256	13,671	9,789	7,284	6,123
512	53,959	38,013	28,644	24,317
1,024	214,407	149,757	113,604	96,933
2,048	854,791	594,429	452,484	387,195

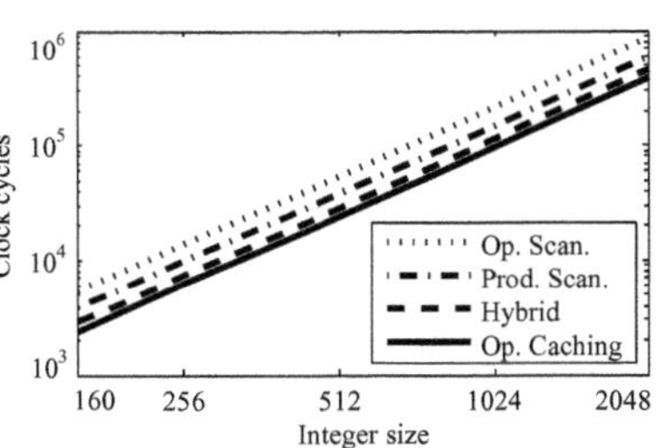

Fig. 6. Comparison chart

Best results have been obtained for the operand-caching technique. By trading additional 20 *store* instructions, up to 120 *load* instructions could be saved when we compare the result with the best reference values (hybrid implementation). Note that *load*, *store*, and *multiply* instructions on the ATmega128 are more expensive than other instructions since they require two clock cycles instead of only one. For operand-caching multiplication, almost the same amount of *load* and *store* instructions are required. In total 2,395 clock cycles are needed to perform the multiplication. Compared to the hybrid implementation, a speed improvement of about 18 % could be achieved.

We also compare the performance of the implemented multi-precision methods for different Integer sizes. Table 4 shows the result for Integer sizes from 160 up to 2,048 bits[3]. The operand-caching technique provides the best performance for any Integer size. It is therefore well suited for large Integer sizes such as it is in the case of RSA. In average, a speed improvement of about 15 % could be achieved compared to the hybrid method. Figure 6 shows the appropriate performance chart in a double logarithmic scale.

Table 5. Performance of operand-caching multiplication for different Integer sizes and available registers

Size	e=2	e=4	e=8	e=10	e=20
160	3,915	2,965	2,513	2,395	2,205
192	5,611	4,255	3,577	3,469	3,207
256	9,915	7,531	6,339	6,123	5,671
512	39,291	29,915	25,227	24,317	22,451
1,024	156,411	119,227	100,635	96,933	89,529
2,048	624,123	476,027	401,979	387,195	357,581

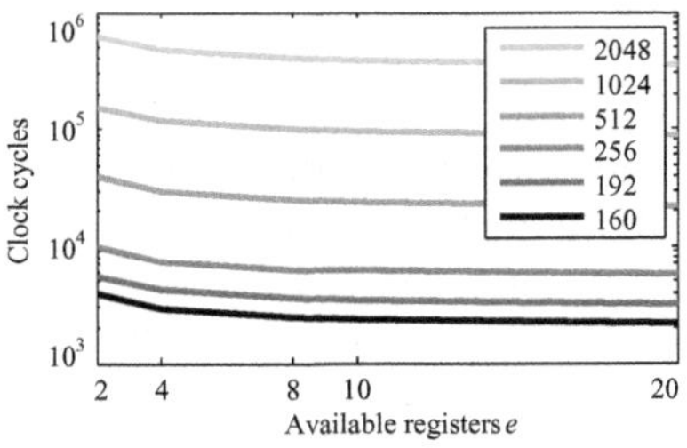

Fig. 7. Performance chart

[3] Note that due to a fully unrolled implementation such large Integer multiplications might be impractical due to the huge amount of code.

Table 6. Comparison with related work

Method	Instruction						Clock
	LD	ST	MUL	ADD	MOVW	Others	Cycles
Hybrid							
Gura et al. [6] (d=5)	167	40	400	1,360	355	197	3,106
Uhsadel et al. [17] (d=5)	238	40	400	986	355	184	2,881
Scott et al. [14] (d=4)[a]	200	40	400	1,263	70	38	2,651
Liu et al. [12] (d=4)	200	40	400	1,194	212	179	2,865
Operand Caching							
with looping[a,c] (e=9)	**92**	**66**	**400**	**1,252**	**41**	**276**	**2,685**
unrolled[b,c] (e=10)	**80**	**60**	**400**	**1,240**	**2**	**68**	**2,395**

[a] b_{init}, Part 1, and Part 4 unrolled. Part 2 and Part 3 looped.
[b] Fully unrolled implementation without overhead of loop instructions.
[c] w/o `PUSH/POP/CALL/RET`.

Table 5 and Figure 7 show the performance for different Integer sizes in relation to parameter e. The parameter e is defined by the number of available registers to store words of one operand, *i.e.* $e = \frac{f-3}{2}$, where $f = 2e + 3$ denotes the number of available registers in total (including the triple-size register for the accumulator). It shows that for $e > 10$ no significant improvement in speed is obtained. The performance decrease for smaller e and higher Integer sizes. However, if we compare our solution (160-bit multiplication with smallest parameter $e = 2 \rightarrow f = 7$ registers) with the product-scanning method (needing $f = 5$ registers), we obtain 3,915 clock cycles for the operand-caching method and 3,957 clock cycles for the product scanning method. It therefore provides a good performance even for a smaller set of available registers. For the special case $e = 20$, where all 20 words of one 160-bit operand can be maintained in registers (ideal case for product scanning), it shows that the number of clock cycles reaches nearly the optimum of 2,160 clock cycles, i.e. $4n = 80$ memory-access instructions, $n^2 = 400$ multiplications, and $3n^2 = 1,200$ additions.

We compare our result with related work in Table 6. For a fair comparison, we also implemented a operand-caching version that does not unroll the algorithm but includes additional loop instructions. It shows that the operand-caching method provides best performance. Compared to Gura et al. [6] 23 % less clock cycles are needed for a 160-bit multiplication. A 10 % improvement could be achieved compared to the best solution reported in literature [14]. Note that most of the related work need between 167 to 238 *load* instructions which mostly explains the higher amount of needed clock cycles.

5 Conclusions

We presented a novel multiplication technique for embedded microprocessors. The multiplication method reduces the number of necessary *load* instructions through sophisticated caching of operands. Our solution follows the product-scanning approach but divides the processing into several parts. This allows the scanning of sub-products where most of the operands are kept within the register-set throughout the algorithm.

In order to evaluate our solution, we implemented several multiplication techniques using different Integer sizes on the ATmega128 microcontroller. Using operand-caching multiplication, we require 2,395 clock cycles for a 160-bit multiplication. This result improves the best reported solution by a factor of 10 % [14]. Compared to the hybrid multiplication of Gura et al. [6], we achieved a speed up of 23 %. Our evaluation further showed that our solution scales very well for different Integer sizes used for ECC and RSA. We obtained an improvement of about 15 % for bit sizes between 256 and 2,048 bits compared to a reference implementation of the hybrid multiplication.

It is also worth to note that our multiplication method is perfectly suitable for processors that support multiply-accumulate (`MULACC`) instructions such as ARM or the dsPIC family of microcontrollers. It also fully complies to architectures which support instruction-set extensions for `MULACC` operations such as proposed by Großschädl and Savaş [5].

Acknowledgements. The work has been supported by the European Commission through the ICT program under contract ICT-2007-216646 (European Network of Excellence in Cryptology - ECRYPT II) and under contract ICT-SEC-2009-5-258754 (Tamper Resistant Sensor Node - TAMPRES).

References

1. Atmel Corporation. 8-bit AVR Microcontroller with 128K Bytes In-System Programmable Flash (August 2007), `http://www.atmel.com/dyn/resources/prod_documents/doc2467.pdf`
2. Atmel Corporation. 8-bit AVR Instruction Set (May 2008), `http://www.atmel.com/dyn/resources/prod_documents/doc0856.pdf`
3. Certicom Research. Standards for Efficient Cryptography, SEC 2: Recommended Elliptic Curve Domain Parameters, Version 2.0. (January 2010), `http://www.secg.org/`
4. Comba, P.: Exponentiation cryptosystems on the IBM PC. IBM Systems Journal 29(4), 526–538 (1990)
5. Großschädl, J., Savaş, E.: Instruction Set Extensions for Fast Arithmetic in Finite Fields GF(p) and GF(2^m). In: Joye, M., Quisquater, J.-J. (eds.) CHES 2004. LNCS, vol. 3156, pp. 133–147. Springer, Heidelberg (2004)
6. Gura, N., Patel, A., Wander, A., Eberle, H., Shantz, S.C.: Comparing Elliptic Curve Cryptography and RSA on 8-bit CPUs. In: Joye, M., Quisquater, J.-J. (eds.) CHES 2004. LNCS, vol. 3156, pp. 119–132. Springer, Heidelberg (2004)
7. Kargl, A., Pyka, S., Seuschek, H.: Fast Arithmetic on ATmega128 for Elliptic Curve Cryptography. Cryptology ePrint Archive Report 2008/442 (October 2008), `http://eprint.iacr.org/`
8. Koç, Ç.K.: High Speed RSA Implementation. Technical report, RSA Laboratories, RSA Data Security, Inc. 100 Marine Parkway, Suite 500 Redwood City (1994)

9. Lederer, C., Mader, R., Koschuch, M., Großschädl, J., Szekely, A., Tillich, S.: Energy-Efficient Implementation of ECDH Key Exchange for Wireless Sensor Networks. In: Markowitch, O., Bilas, A., Hoepman, J.-H., Mitchell, C.J., Quisquater, J.-J. (eds.) Information Security Theory and Practice. LNCS, vol. 5746, pp. 112–127. Springer, Heidelberg (2009)
10. Lenstra, A., Verheul, E.: Selecting Cryptographic Key Sizes. Journal of Cryptology 14(4), 255–293 (2001)
11. Liu, A., Ning, P.: TinyECC: A Configurable Library for Elliptic Curve Cryptography in Wireless Sensor Networks. In: International Conference on Information Processing in Sensor Networks - IPSN 2008, St. Louis, Missouri, USA, Mo, April 22-24, pp. 245–256 (2008)
12. Liu, Z., Großschädl, J., Kizhvatov, I.: Efficient and Side-Channel Resistant RSA Implementation for 8-bit AVR Microcontrollers. In: Workshop on the Security of the Internet of Things - SOCIOT 2010, 1st International Workshop, Tokyo, Japan, November 29. IEEE Computer Society, Los Alamitos (2010)
13. National Institute of Standards and Technology (NIST). SP800-57 Part 1: DRAFT Recommendation for Key Management: Part 1: General (May 2011), `http://csrc.nist.gov/publications/drafts/800-57/Draft_SP800-57-Part1-Rev3_May2011.pdf`
14. Scott, M., Szczechowiak, P.: Optimizing Multiprecision Multiplication for Public Key Cryptography. Cryptology ePrint Archive, Report 2007/299 (2007), `http://eprint.iacr.org/`
15. Szczechowiak, P., Oliveira, L.B., Scott, M., Collier, M., Dahab, R.: NanoECC: Testing the Limits of Elliptic Curve Cryptography in Sensor Networks. In: Verdone, R. (ed.) EWSN 2008. LNCS, vol. 4913, pp. 305–320. Springer, Heidelberg (2008)
16. Ugus, O., Hessler, A., Westhoff, D.: Performance of Additive Homomorphic EC-ElGamal Encryption for TinyPEDS. In: GI/ITG KuVS Fachgespräch Drahtlose Sensornetze, RWTH Aachen, UbiSec 2007 (July 2007)
17. Uhsadel, L., Poschmann, A., Paar, C.: Enabling Full-Size Public-Key Algorithms on 8-bit Sensor Nodes. In: 4th European Workshop on Security and Privacy in Ad-hoc and Sensor Networks, ESAS 2007, Cambridge, UK, July 2-3 (2007)
18. Wang, H., Li, Q.: Efficient Implementation of Public Key Cryptosystems on Mote Sensors (Short Paper). In: Ning, P., Qing, S., Li, N. (eds.) ICICS 2006. LNCS, vol. 4307, pp. 519–528. Springer, Heidelberg (2006)

A Algorithm for Operand-Caching Multiplication

The following pseudo code shows the algorithm for multi-precision multiplication using the operand-caching method. Variables that are located in data memory are denoted by M_x where x represents the name of the Integer a or b. The parameter e describes the number of locally usable registers $R_a[e-1,\ldots,0]$ and $R_b[e-1,\ldots,0]$. The triple-word accumulator is denoted by $ACC = (ACC_2, ACC_1, ACC_0)$.

```
Require: word size n, parameter e, n ≥ e, Integers a, b ∈
  [0, n), c ∈ [0, 2n).
Ensure: c = ab.
  r = ⌊n/e⌋.                                                   ⎫
  R_A[e − 1, . . . , 0] ← M_A[n − 1, . . . , re].              ⎪
  R_B[e − 1, . . . , 0] ← M_B[n − re − 1, . . . , 0].          ⎪
  ACC ← 0.                                                     ⎪
  for i = 0 to n − re − 1 do                                   ⎪
    for j = 0 to i do                                          ⎪
      ACC ← ACC + R_A[j] ∗ R_B[i − j].                         ⎪
    end for                                                    ⎪
    M_C[re + i] ← ACC_0.                                       ⎪
    (ACC_1, ACC_0) ← (ACC_2, ACC_1).                           ⎪
    ACC_2 ← 0.                                                 ⎬ b_init
  end for                                                      ⎪
  for i = 0 to n − re − 2 do                                   ⎪
    for j = i + 1 to n − re − 1 do                             ⎪
      ACC ← ACC + R_A[j] ∗ R_B[n − re − j + i].                ⎪
    end for                                                    ⎪
    M_C[n + i] ← ACC_0.                                        ⎪
    (ACC_1, ACC_0) ← (ACC_2, ACC_1).                           ⎪
    ACC_2 ← 0.                                                 ⎪
  end for                                                      ⎪
  M_C[2n − re − 1] ← ACC_0.                                    ⎪
  ACC_0 ← 0.                                                   ⎭
  for p = r − 1 to 0 do                                        } Row Loop:
    R_A[e − 1, . . . , 0] ← M_A[(p + 1)e − 1, . . . , pe].     ⎫
    R_B[e − 1, . . . , 0] ← M_B[e − 1, . . . , 0].             ⎪
    for i = 0 to e − 1 do                                      ⎪
      for j = 0 to i do                                        ⎪
        ACC ← ACC + R_A[j] ∗ R_B[i − j].                       ⎬ Part 1
      end for                                                  ⎪
      M_C[pe + i] ← ACC_0.                                     ⎪
      (ACC_1, ACC_0) ← (ACC_2, ACC_1).                         ⎪
      ACC_2 ← 0.                                               ⎪
    end for                                                    ⎭
    for i = 0 to n − (p + 1)e − 1 do                           ⎫
      R_B[e − 1, . . . , 0] ← M_B[e + i], R_B[e − 2, . . . , 1]. ⎪
      for j = 0 to e − 1 do                                    ⎪
        ACC ← ACC + R_A[j] ∗ R_B[e − 1 − j].                   ⎪
      end for                                                  ⎬ Part 2
      ACC ← ACC + M_C[(p + 1)e + i].                           ⎪
      M_C[(p + 1)e + i] ← ACC_0.                               ⎪
      (ACC_1, ACC_0) ← (ACC_2, ACC_1).                         ⎪
      ACC_2 ← 0.                                               ⎪
    end for                                                    ⎭
```

> **for** $i = 0$ **to** $n - (p+1)e - 1$ **do** — Part 3
> $R_A[e-1, \ldots, 0] \leftarrow M_A[(p+1)e+i], R_A[e-2, \ldots, 1]$.
> **for** $j = 0$ **to** $e - 1$ **do**
> $ACC \leftarrow ACC + R_A[j] * R_B[e-1-j]$.
> **end for**
> $ACC \leftarrow ACC + M_C[(n+i]$.
> $M_C[n+i] \leftarrow ACC_0$.
> $(ACC_1, ACC_0) \leftarrow (ACC_2, ACC_1)$.
> $ACC_2 \leftarrow 0$.
> **end for**
> **for** $i = 0$ **to** $e - 2$ **do** — Part 4
> **for** $j = i+1$ **to** $e - 1$ **do**
> $ACC \leftarrow ACC + R_A[j] * R_B[e-j+i]$.
> **end for**
> $M_C[2n-(p+1)e+i] \leftarrow ACC_0$.
> $(ACC_1, ACC_0) \leftarrow (ACC_2, ACC_1)$.
> $ACC_2 \leftarrow 0$.
> **end for**
> $M_C[2n-1-pe] \leftarrow ACC_0$.
> $ACC_0 \leftarrow 0$.
> **end for**
> **Return** c.

B Example: 160-Bit Operand-Caching Multiplication

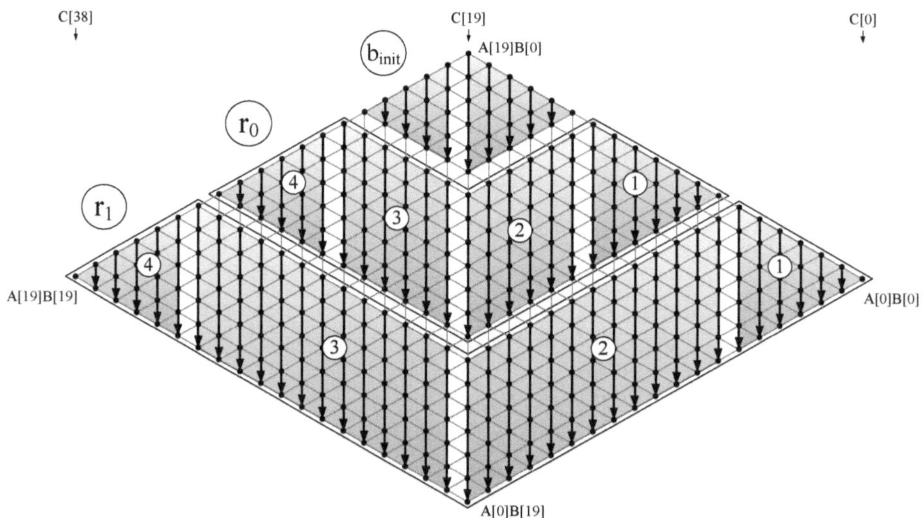

Fig. 8. Operand-caching multiplication for $n = 20$ and $e = 7$

Small Public Keys and Fast Verification for $\mathcal{M}$ultivariate $\mathcal{Q}$uadratic Public Key Systems

Albrecht Petzoldt[1], Enrico Thomae[2], Stanislav Bulygin[1], and Christopher Wolf[2]

[1] Technische Universität Darmstadt and
Center for Advanced Security Research Darmstadt (CASED)
apetzoldt@cdc.informatik.tu-darmstadt.de, Stanislav.Bulygin@cased.de
[2] Horst Görtz Institute for IT-security
Faculty of Mathematics
Ruhr-University of Bochum, 44780 Bochum, Germany
Enrico.Thomae@rub.de, Christopher.Wolf@rub.de,
chris@Christopher-Wolf.de

Abstract. Security of public key schemes in a post-quantum world is a challenging task—as both RSA and ECC will be broken then. In this paper, we show how post-quantum signature systems based on $\mathcal{M}$ultivariate $\mathcal{Q}$uadratic ($\mathcal{MQ}$) polynomials can be improved up by about 9/10, and 3/5, respectively, in terms of public key size and verification time. The exact figures are 88% and 59%. This is particularly important for small-scale devices with restricted energy, memory, or computational power. In addition, we provide evidence that this reduction does not affect security and that it is also optimal in terms of possible attacks. We do so by combining the previously unrelated concepts of reduced and equivalent keys. Our new scheme is based on the so-called *Unbalanced Oil and Vinegar* class of $\mathcal{MQ}$-schemes. We have derived our results mathematically and verified the speed-ups through a C++ implementation.

Keywords: Multivariate Quadratic Cryptography, Post-Quantum Cryptography, Implementation, Unbalanced Oil and Vinegar Signature Scheme.

1 Introduction

When finding an old sonnet of Shakespeare, we can usually determine its validity accurately by checking the wording, the ink, the paper, and so on. Similar techniques apply in disputes over last wills - or other documents of historical or financial interest. Even if they are several decades old, we can fairly certainly determine if they have been written by the person in question and sometimes even date them accurately.

For digital documents, this is a much more difficult task. They are electronically signed with the help of so-called *digital signature schemes*. The ones widely used today are Digital Signature Algorithms (DSAs) based on RSA and elliptic curve cryptography (ECDSA). Unfortunately, all these schemes are broken

B. Preneel and T. Takagi (Eds.): CHES 2011, LNCS 6917, pp. 475–490, 2011.

if large enough quantum computers will be built. The reason is the algorithm of Shor which breaks all cryptographic algorithms based on the difficulty of factoring and the discrete logarithm (DL) problem [16]. This covers DL over numbers, RSA, and ECDSA. Even if unlikely now, quantum computers may be available in the medium future and are hence a concern for long-term-validity of authentication data. We must be sure that a document signed today is not repudiated 50 years later. Likewise, we do not want a signature that is generated today to be forged in the future. So to guard security even in the presence of quantum computers, post-quantum cryptography is needed and has hence become a vital research area [1]. One possible solution in this context is the so-called *$\mathcal{M}$ultivariate $\mathcal{Q}$uadratic cryptography.* It is widely believed that it is secure against attacks with quantum computers.

In addition, $\mathcal{M}$ultivariate $\mathcal{Q}$uadratic (or $\mathcal{MQ}$ for short) signature schemes have nice properties in terms of speed of signature generation and verification which make them superior to DL, RSA and ECDSA. Note that ECDSA is the most efficient of the three. However, even when comparing to signature generation in $\mathcal{MQ}$ and ECC, the former are a factor of 2 - 50 faster on FPGA than the latter [3]. Similar results have been demonstrated for comparison with RSA and ECC in software [21], [4], [5]. One of the main reasons for this higher efficiency is the comparably small finite fields, *e.g.* $\mathbb{F}_{2^8}$ which allows for efficient hardware and software implementations. The other operations usually boil down to vector-matrix functions, which can be implemented efficiently, too. As an immediate consequence, we can use $\mathcal{MQ}$ schemes in restricted devices, *i.e.* with low energy or computational power.

Another point is the high flexibility of $\mathcal{MQ}$-schemes. This allows for the use of sparse polynomials in the *private key* as done in the TTS schemes of Yang, Chen, and Chen [21]. This leads both to a significant reduction of the time needed for signature generation, as well as for the size of the private key. Another way to reduce the private key is by choosing the coefficients of the private maps from smaller fields (*e.g.* $\mathbb{F}_{16}$ instead of $\mathbb{F}_{256}$), [4]. In addition, we want to mention the so-called *similar keys* which exploit linear relations between public and private key [9]. However, they are not applicable to schemes like UOV. Finally, one research direction deals with reducing the public key directly. In [13, 14] it is shown how to reduce the public key size of the UOV scheme by choosing public coefficients in a structured way, cf. Section 3.

1.1 Achievement

Combining two previously unrelated ideas, we deal with reducing the size of the public key. For $\mathcal{MQ}$ schemes like *Unbalanced Oil and Vinegar* (UOV - see below), typical choices of parameters lead to around 80 kByte for the public key. We use the approach of [14] to bring this size down to about 9 kB. We show that we can use this idea to reduce the verification time, too. By choosing them partially to be 0 or 1 only, verification time is reduced by up to 59%. This way, $\mathcal{MQ}$ verification can be performed in low-power, low-energy devices. For example for mobile devices, we can easily imagine a scenario where a server signs data which needs to be verified by a (comparably restricted) phone. As further contribution, we give arguments that this reduction in size does *not* affect security. This is due to an observation regarding equivalent keys of [19, 20].

In addition, our modification also works for restricting the choice of the coefficients. Using Turán graphs we demonstrate that this further reduction in size and verification time resists all know attacks.

1.2 Organization

The structure of this paper is as follows: After giving some introduction in Section 1, we continue with the background on $\mathcal{MQ}$-schemes and in particular the UOV in Section 2. In Section 3, we review the cyclic construction from [13]. This is followed by security considerations regarding cyclic keys in UOV in Section 4. Using these results, we outline our new constructions, its implementation, efficiency, and security implications in Section 5. The paper concludes with Section 6. Some background on Turán graphs and how this relates to our monomial ordering in Subsection 5.3 can be found in the full version of this paper [15].

2 $\mathcal{M}$ultivariate $\mathcal{Q}$uadratic Cryptography

The main idea behind $\mathcal{M}$ultivariate $\mathcal{Q}$uadratic cryptography is to choose a system $\mathcal{F}$ of m quadratic polynomials in n variables which can be easily inverted. Here $\mathcal{F}$ is called the *central map*. In addition, we need invertible affine maps S and T to hide the structure of the central map $\mathcal{F}$. The public key of the cryptosystem is now composed as $\mathcal{P} = T \circ \mathcal{F} \circ S$. For a secure $\mathcal{MQ}$-system, $\mathcal{P}$ must be difficult to invert. The private key consists of $(\mathcal{F}, T, S)$ and therefore allows efficient inversion of $\mathcal{P}$. More information on $\mathcal{M}$ultivariate $\mathcal{Q}$uadratic schemes can be found in [6, 18].

2.1 Notation

Solving non-linear systems of m equations in n variables over a finite field is a difficult problem in general. Restricting it to the seemingly easy case of quadratic equations is still difficult. Actually this problem is also known as $\mathcal{MQ}$-problem which is proven to be NP-hard in the worst-case [8], even over $\mathbb{F}_2$.
Let $\mathcal{P}$ be an $\mathcal{MQ}$ system of the form

$$\begin{aligned} p^{(1)}(x_1, \ldots, x_n) &= 0 \\ p^{(2)}(x_1, \ldots, x_n) &= 0 \\ &\vdots \\ p^{(m)}(x_1, \ldots, x_n) &= 0, \end{aligned} \tag{1}$$

with

$$p^{(k)}(x_1, \ldots, x_n) := \sum_{1 \le i \le j \le n} p_{ij}^{(k)} x_i x_j + \sum_{1 \le i \le n} p_i^{(k)} x_i + p_0^{(k)} \quad (k = 1, \ldots, m). \tag{2}$$

Let $\pi^{(k)}$ be the coefficient vector of $p^{(k)}(x_1,\ldots,x_n)$ w.r.t. graded lexicographic ordering of monomials, *i.e.*

$$\pi^{(k)} = (p_{11}^{(k)}, p_{12}^{(k)}, \ldots, p_{1n}^{(k)}, p_{22}^{(k)}, p_{23}^{(k)}, \ldots, p_{nn}^{(k)}, p_1^{(k)}, \ldots, p_n^{(k)}, p_0^{(k)}). \quad (3)$$

Let M_P be the corresponding coefficient matrix

$$M_P := \begin{pmatrix} \pi^{(1)} \\ \vdots \\ \pi^{(m)} \end{pmatrix}. \quad (4)$$

Note that the ordering of monomials (and thus coefficients) in the matrix M_P does not necessarily have to be graded lexicographic ordering. We may want to order monomials of the public key in a certain way. Therewith, the ordering of coefficients (columns of M_P) is then changed accordingly.

2.2 Unbalanced Oil and Vinegar

In this subsection we introduce the Oil and Vinegar Signature Scheme, which was proposed by J. Patarin in [12]. Let $\mathbb{F}_q$ be a finite field. Denote the number of oil variables by $o \in \mathbb{N}$, the number of vinegar variables by $v \in \mathbb{N}$ and set $n := o+v$. Let $V := \{1,\ldots,v\}$ and $O := \{v+1,\ldots,n\}$ denote the sets of indices of vinegar and oil variables. The private key $\mathcal{F} := (f^{(1)},\ldots,f^{(o)})$ is defined by

$$f^{(k)}(u_1,\ldots,u_n) := \sum_{i\in V, j\in O} f_{ij}^{(k)} u_i u_j + \sum_{i,j\in V, i\leq j} f_{ij}^{(k)} u_i u_j \quad (k = 1,\ldots,o). \quad (5)$$

Remark: We omit the linear part of $\mathcal{F}$ and the constant part of S, because it was shown in [11, 10] that it does not contribute to the security of UOV.

For the inversion of the map $\mathcal{F}$ it is important that the variables in $f^{(k)}$ are not completely mixed, *i.e.* oil variables are only multiplied by vinegar variables and never by oil variables. This construction leads to an efficient way to invert $\mathcal{F}$. If we assign arbitrary values to the vinegar variables we obtain a system of o linear equations in o variables. With high probability this system has a solution. If not we try again with a different choice for the vinegar variables $x_1,\ldots,x_v$. In the public key $\mathcal{P}$, the central map $\mathcal{F}$ is hidden by composing it with a linear map $S : \mathbb{F}_q^n \to \mathbb{F}_q^n$, *i.e.* $\mathcal{P} := \mathcal{F} \circ S$.

Note that, in opposite to other multivariate schemes, the second linear map T is not needed for the security of UOV. So it can be dropped.

Figure 1 shows the signature generation and verification process for UOV.

Signature Generation: To sign a document d, one uses a hash function $\mathcal{H} : \mathbb{F}_q^\star \to \mathbb{F}_q^m$ to compute the hash value $\mathbf{h} = \mathcal{H}(d) \in \mathbb{F}_q^m$. After that one computes first $\mathbf{u} := \mathcal{F}^{-1}(\mathbf{h})$ and then $\mathbf{x} := S^{-1}(\mathbf{u})$. The signature of the document d is

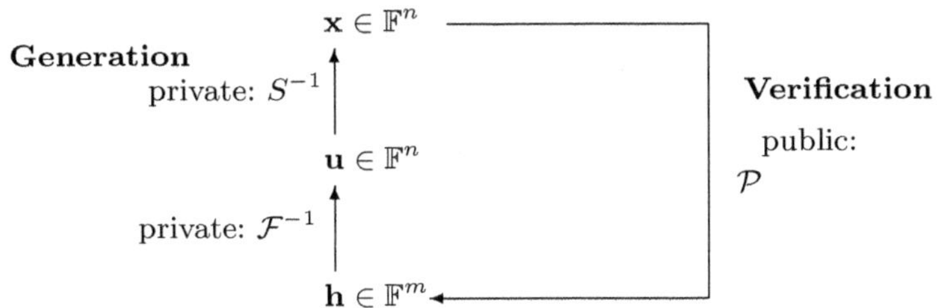

Fig. 1. Signature generation and verification for UOV

$\mathbf{x} \in \mathbb{F}_q^n$. In a slight abuse of notation we write $\mathcal{F}^{-1}(\mathbf{h})$ for finding one (of possibly many) pre-image of $\mathbf{h}$ under $\mathcal{F}$.

Signature Verification: To verify the authenticity of a signature, one computes the hash value $\mathbf{h}$ of the corresponding document and the value $\mathbf{h}' = \mathcal{P}(\mathbf{x})$. If $\mathbf{h} = \mathbf{h}'$ holds, the signature is accepted, otherwise rejected.

In his original paper [12], Patarin suggested to use $o = v$ (Balanced Oil and Vinegar - OV). After this scheme was broken by Kipnis and Shamir in [11], it was proposed in [10] to use $v \geq 2o$ (Unbalanced Oil and Vinegar (UOV)). UOV parameters $q = 2^8$, $(o, v) = (26, 52)$ give 80-bit security against the most efficient attacks currently known [2].

3 Reviewing Cyclic Keys

In this section we review the approach of [13] to create a UOV-based scheme with a partially cyclic public key. Remember that, in the case of the Unbalanced Oil and Vinegar signature scheme [10], the public key $\mathcal{P}$ is given as the concatenation of the central UOV-map $\mathcal{F}$ and a linear invertible map S, *i.e.* $\mathcal{P} = \mathcal{F} \circ S$.

In [13] it is observed, that this equation (after fixing the linear map S), leads to a linear relation between the coefficients of the quadratic monomials of $\mathcal{P}$ and $\mathcal{F}$ of the form

$$p_{ij}^{(k)} = \sum_{r=1}^{n} \sum_{s=r}^{n} \alpha_{ij}^{rs} \cdot f_{rs}^{(k)}, \tag{6}$$

where $p_{ij}^{(k)}$ and $f_{ij}^{(k)}$ are the coefficients of $x_i x_j$ in the k-th component of $\mathcal{P}$ and $\mathcal{F}$ respectively and the α_{ij}^{rs} are given as

$$\alpha_{ij}^{rs} = \begin{cases} s_{ri} \cdot s_{si} & (i = j) \\ s_{ri} \cdot s_{sj} + s_{rj} \cdot s_{si} & \text{otherwise} \end{cases}. \tag{7}$$

Here $s_{ij} \in \mathbb{F}_q$ denote the coefficients of the linear map S. Let $D := \frac{v \cdot (v+1)}{2} + ov$ be the number of non-zero quadratic terms in any component of $\mathcal{F}$ and $D' := \frac{n \cdot (n+1)}{2}$ be the number of quadratic terms in the public polynomials. Let

M_P and M_F be the coefficient matrices of $\mathcal{P}$ and $\mathcal{F}$ respectively (w.r.t. graded lexicographic ordering of monomials). The matrices M_P and M_F are divided into submatrices as shown in Figure 2. Note that, due to the absence of oil × oil terms in the central polynomials, we have a block of zeros on the right side of M_F.

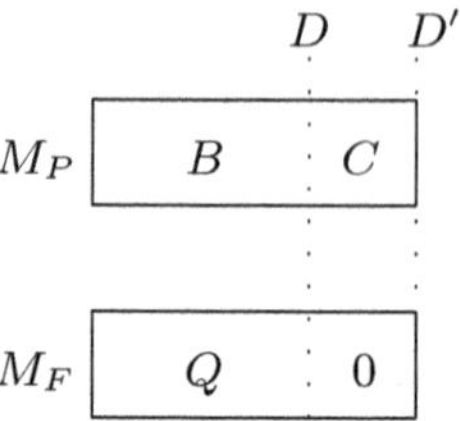

Fig. 2. Layout of the matrices M_P and M_F

Furthermore, the authors of [13] defined the so called transformation matrix $A_{UOV} \in \mathbb{F}_q^{D \times D}$ containing the coefficients α_{ij}^{rs} of equation (6), *i.e.* $A_{UOV} = \left(\alpha_{ij}^{rs}\right)$ for $1 \leq r \leq v, r \leq s \leq n$ for the rows and $1 \leq i \leq v, i \leq j \leq n$ for the columns.

$$A_{UOV} = \begin{pmatrix} \alpha_{11}^{11} & \alpha_{12}^{11} & \dots & \alpha_{vn}^{11} \\ \alpha_{11}^{12} & \alpha_{12}^{12} & \dots & \alpha_{vn}^{12} \\ \vdots & & & \vdots \\ \alpha_{11}^{vn} & \alpha_{12}^{vn} & \dots & \alpha_{vn}^{vn} \end{pmatrix}. \tag{8}$$

With this notation, equation (6) yields

$$B = Q \cdot A_{UOV}. \tag{9}$$

If matrix A_{UOV} is invertible, this equation has a solution for Q. Experiments indicate that this condition is fulfilled with high probability. By solving equation (9) for Q, the authors of [13] were able to insert a partially circulant matrix B into the UOV public key. By doing so, they reduced the public key size of the scheme by a factor of 6. After choosing matrix B, we can use Algorithm 1 to compute the corresponding key.

4 Security of UOV

Due to equivalent keys [19, 20] UOV contains a lot of redundancy. We show which part of the public key is important for security and which part can be chosen such that the public key gets as small as possible.

It is rather intuitive that the linear and constant part of the public key do not provide extra security because we can easily separate them from the quadratic part. This was previously exploited by Kipnis and Shamir in their cryptanalysis of (balanced) Oil and Vinegar [11]. But it is quite surprising that also a fraction of the quadratic part is not essential for security. This is implied by the observation of equivalent keys by Wolf and Preneel [19, 20].

Algorithm 1. Alternative Key Generation for UOV schemes

1: Choose an $o \times D$ matrix B (e.g. partially circulant or generated by an LRS).
2: Choose randomly a linear map $\mathcal{S}$ (represented by an $n \times n$-matrix S). If S is not invertible, choose again.
3: Compute for S the corresponding transformation matrix A_{UOV} (using equations (7) and (8)). If A_{UOV} is not invertible, go back to step 2.
4: Solve the linear system given by equation (9) to get the matrix Q and there with the coefficients of the central polynomials.
5: Compute the public key as $\mathcal{P} = \mathcal{F} \circ \mathcal{S}$.

Definition 1. *Let $(\mathcal{F}, S)$ and $(\mathcal{F}', S')$ be two UOV private keys. They are called equivalent if they result in the same UOV public key, i.e. $\mathcal{F} \circ S = \mathcal{F}' \circ S' =: \mathcal{P}$.*

The set of all private keys resulting in a given public key $\mathcal{P}$ is denoted by $EQ_{\mathcal{P}}$.

In order to produce equivalent keys we use the following transformation Ω on the variables $\mathbf{u}$ that preserves the structure of $\mathcal{F}$.

$$\Omega = \begin{pmatrix} \Omega^{(1)}_{v \times v} & 0 \\ \Omega^{(2)}_{o \times v} & \Omega^{(3)}_{o \times o} \end{pmatrix} \; \textit{resp.} \; \Omega = (\omega_{ij})_{i,j=1}^{n} \tag{10}$$

Let $\mathcal{F}(\mathbf{u}) := (f^{(1)}(\mathbf{u}), \ldots, f^{(o)}(\mathbf{u}))$ be a UOV central map (i.e. no quadratic cross terms in oil variables). Let $\mathbf{u} = \Omega \cdot \mathbf{u}'$. The vinegar variables $u_1, \ldots u_v$ are computed as sums of vinegar variables $u'_1, \ldots, u'_v$. Therefore we get (for $k = 1, \ldots, o$):

$$\begin{aligned} f^{(k)}(\mathbf{u}) &= \sum_{i,j \in V} f_{ij}^{(k)} u_i u_j + \sum_{i \in V, j \in O} f_{ij}^{(k)} u_i u_j \\ &= \sum_{i,j \in V} f_{ij}^{(k)} \left(\sum_{l \in V} \omega_{il} u'_l \right) \cdot \left(\sum_{m \in V} \omega_{jm} u'_m \right) \\ &+ \sum_{i \in V, j \in O} f_{ij}^{(k)} \left(\sum_{l \in V} \omega_{il} u'_l \right) \cdot \left(\sum_{m \in V} \omega_{jm} u'_m + \sum_{m \in O} \omega_{jm} u'_m \right) \\ &= \sum_{i,j \in V} \widetilde{f_{ij}^{(k)}} u'_i u'_j + \sum_{i \in V, j \in O} \widetilde{f_{ij}^{(k)}} u'_i u'_j \text{ for some } \widetilde{f_{ij}^{(k)}}(f_{ij}, \omega_{lm}). \end{aligned}$$

Due to $\mathcal{F} \circ S = \mathcal{F} \circ \Omega^{-1} \circ \Omega \circ S$ both $(\mathcal{F}, S)$ and $(\mathcal{F} \circ \Omega^{-1}, \Omega \cdot S)$ are equivalent private keys for the public key $\mathcal{P}$.

Lemma 1. *For every UOV public key $\mathcal{P}$ there exists a UOV private key $(\mathcal{F}, S) \in EQ_{\mathcal{P}}$ such that S has the form*

$$S = \begin{pmatrix} I & \widetilde{S} \\ 0 & I \end{pmatrix} \tag{11}$$

for some $(v \times o)$ matrix $\widetilde{S}$ (except for permutations of rows and columns).

Proof. Let Ω be of form (10) and $\mathcal{F}$ and

$$S = \begin{pmatrix} S^{(1)}_{v\times v} & S^{(2)}_{v\times o} \\ S^{(3)}_{o\times v} & S^{(4)}_{o\times o} \end{pmatrix}$$

an arbitrary private key. There exists a permutation of rows and columns such that $S^{(1)}$, $S^{(4)}$ and $I - S^{(3)}S^{(1)^{-1}}S^{(2)}S^{(4)^{-1}}$ are invertible[1]. Now we choose Ω such that $\Omega^{(1)}S^{(1)} = I$, $\Omega^{(2)}S^{(2)} + \Omega^{(3)}S^{(4)} = I$ and $\Omega^{(2)}S^{(1)} + \Omega^{(3)}S^{(3)} = 0$, i.e. $\Omega^{(1)} = S^{(1)^{-1}}$, $\Omega^{(3)} = S^{(4)^{-1}} \cdot (I - S^{(3)}S^{(1)^{-1}}S^{(2)}S^{(4)^{-1}})^{-1}$ and $\Omega^{(2)} = -\Omega^{(3)}S^{(3)}S^{(1)^{-1}}$. □

For the following, we assume w.l.o.g. that the linear map S has the form (11), since the further analysis is not changed by row and column permutations.
The next lemma shows that the vinegar × vinegar coefficients in the public key do not hide any information about the secret map.

Lemma 2. *In the case of S having the form* (11) *we get $f_{ij}^{(k)} = p_{ij}^{(k)}$ for $i, j \in \{1, \ldots, v\}$ and $k \in \{1, \ldots, o\}$.*

Proof. We consider for $k \in \{1, \ldots o\}$ the quadric forms of $p^{(k)}$ and $f^{(k)}$, i.e. $MP^{(k)} = S^T \cdot MF^{(k)} \cdot S$. For S having the form (11) this yields $MP_1^{(k)} = MF_1^{(k)}$ $(k = 1, \ldots, o)$. □

For a key recovery attack it is sufficient to find any of the equivalent keys. Thus an attacker can search for a private key, with S of the form (11). This means we can assume that the attacker actually knows all coefficients of squared vinegar variables in the private map. This does not effect the overall knowledge of the attacker. So the $p_{ij}^{(k)}$ with $i, j \in V$ in the public map do not hide any secret and thus we can choose them in an arbitrary way.

Proposition 1. *The first $\frac{v(v+1)}{2}$ coefficients of each public polynomial do not provide any security in the sense of key recovery attacks. Arbitrarily fixing these coefficients does not give advantage to an attacker who wants to recover the whole private key.*

By equation (9) we are even able to choose the first $\frac{v(v+1)}{2} + ov$ coefficients of each public polynomial of a special form and thus save memory. Proving the security of this construction is not as obvious as in the latter construction. We have to show that additionally fixing ov coefficients does not give advantage to an attacker in the sense of key recovery attacks

Usually, we fix the coefficients of the central polynomials $\mathcal{F}$ and the linear map S and then compute the public polynomials $\mathcal{P}$. However, for our construction we turn things around by *first* fixing parts of the public polynomials and *then* computing the central polynomials. Intuitively, this should be equally secure. We capture this in the following proposition.

[1] Our experiments showed, that these three conditions are fulfilled for 99.2 % of all UOV private keys without changing rows and columns.

Proposition 2. *Let the $o \times D$ matrix B be an MDS matrix (i.e. every choice of o columns leads to an invertible submatrix). Then, fixing the first $\frac{v(v+1)}{2} + ov$ coefficients of each public polynomial to the elements of B does not give the attacker any advantage in the sense of key recovery attacks.*

Proof. We start our proof with equation (9)

$$B = Q \cdot A_{UOV}.$$

For S having the form (11) we can write this as

$$(B_1|B_2) = (F_1|F_2) \cdot \begin{pmatrix} I & \Sigma \\ 0 & 1 \end{pmatrix} \tag{12}$$

with a $o \times o \cdot v$ matrix F_2 containing the coefficients f_{ij}, $(i \in \{1, \ldots, v\}$, $j \in \{v+1, \ldots, n\})$ and a $\frac{v \cdot (v+1)}{2} \times o \cdot v$ matrix Σ linear in the elements of S.
This leads to $F_1 = B_1$ and

$$B_2 = F_1 \cdot \Sigma + F_2 = B_1 \cdot \Sigma + F_2. \tag{13}$$

Equation (13) yields $o \cdot o \cdot v$ linear equations in the $(o+1) \cdot o \cdot v$ unknowns s_{ij} and $f_{ij}^{(k)}$. We can use the last $o \cdot v$ of these equations to eliminate the s_{ij} and get $(o-1) \cdot o \cdot v$ linear relations between the coefficients $f_{ij}^{(k)}$. If the map

$$S \mapsto B_2 - B_1 \cdot \Sigma(S) := F_2$$

is injective, there remain exactly $o \cdot v$ coefficients $f_{ij}^{(k)}$, which have to be guessed correctly to obtain a valid private key. The injectivity follows from the fact that all square submatrices of B_1 are invertible, which is the property we obtain by using an MDS matrix[2].

This is exactly the same situation we obtain for the standard UOV scheme. Therefore, fixing the matrix B to an MDS-matrix does not make key recovery attacks easier. □

We use this observation by proposing a variant of the UOV, which is provably as secure as the original scheme, but reduces the public key size by a huge factor.

In comparison to the case of Algorithm 1 the $(o \times D)$ matrix B is now fixed to an MDS matrix and system parameter of the algorithm. In the remainder of this paper, we refer to this scheme as Compressed UOV (see Algorithm 2).

5 The New Construction

We are now investigating, how much additional structure we can hide in B to speed up the verification process. We do this by choosing the elements of the matrix B from the ground field $\mathbb{F}_2$. In order to make sure that message recovery attacks are still difficult, we have to choose the ordering of monomials appropriately, as explained in Subsection 5.3.

[2] For large enough q, e.g. $q = 2^8$, the matrix B with coefficients chosen uniformly at random is MDS with high probability.

Algorithm 2. Key Generation for Compressed UOV

1: Choose randomly a linear map S (represented by an $n \times n$-matrix S). If S is not invertible, choose again.
2: Compute for S the corresponding transformation matrix A_{UOV} (using equations (7) and (8)). If A_{UOV} is not invertible, go back to step 1.
3: Solve the linear system given by equation (9) to get the matrix Q and therewith the quadratic coefficients of the central polynomials.
4: Compute the public key as $\mathcal{P} = \mathcal{F} \circ S$.

5.1 Message Recovery Attacks

Let $M_P = (B|C)$ with B an $(o \times D)$ matrix. After we claimed that fixing B does not give to an attacker advantage in the sense of key recovery attacks, we have to clarify how $B \in \mathbb{F}_2^{o \times D}$ can be chosen without decreasing security against message recovery attacks. Obviously $B = 0$ is a bad choice, as this would imply $C = 0$. We also have to assure that B has full rank, as otherwise C would also not have full rank. In general our goal is that solving our structured system using Gröbner bases is as difficult as solving a random instance over $\mathbb{F}_q$.

We now first introduce our choice of B. Afterwards we explain, why message recovery remains hard.

5.2 Choice of B

The first $(o \times o)$ block in B can be chosen to be the identity matrix $\mathrm{I}_{o \times o}$ as every attacker is able to reach this situation by Gaussian Elimination. Furthermore this ensures B to have full rank. The remaining part B_1 of B has to be chosen in such a way that there are no systematic dependencies between the elements of B, *i.e.* every m columns with $m \geq o$ have a big chance to have full rank. Otherwise we could produce large zero-blocks which would decrease the complexity of Gröbner bases algorithms.

We suggest to choose every element of the matrix B_1 uniformly at random from $\mathbb{F}_2$. Note that for such a B_1 the rank property above is fulfilled with overwhelming probability. Note that B is no longer part of the public key. Once B is constructed, it is fixed, and thus it is a part of the key generation algorithm.

5.3 Ordering of Monomials

In contrast to the method described in [13, 14], we need to choose a special monomial ordering for our construction. In order to understand why this monomial ordering is constructed, let us recall how direct (Gröbner) attacks on multivariate signature schemes work. In the message recovery attack, the attacker is facing the problem of solving the public UOV system $\mathcal{P}(\mathbf{x}) = \mathbf{h}$ directly. This system is defined over $\mathbb{F}_q$ and has o equations in $n = o + v$ variables. Such a system has on average $q^{n-o} = q^v$ solutions. Considering the values of v usually used (*e.g.* $v = 52$), such a system has a huge amount of solutions (for $q = 2^8$ and $v = 52$ it is 2^{416}). Gröbner bases methods have a great difficulty in solving such a system, since they have to describe a huge variety. Since the attacker is interested in only *one* solution for the signature forgery, recovering all solutions is unnecessary. By fixing values of any v variables in the public system, an attacker obtains a quadratic system in o variables and o equations. On average such a system has a unique solution. Solving this new system with Gröbner bases methods is much easier.

Going back to the matrix M_P we see that C (as defined in Section 3) contains coefficients of monomials $x_i x_j$ with $i, j \in \{v+1, \ldots, n\}$, since there we used the graded lexicographic ordering of monomials. Now if the attacker fixes values for the variables $x_{v+1}, \ldots, x_n$, the monomials represented by C will become constants. Therewith, the resulting quadratic system will have only quadratic terms over $\mathbb{F}_2$ coming from the matrix B. Clearly, Gröbner bases computations will be much easier then, since the attacker does not have to deal with $\mathbb{F}_{2^8}$ arithmetics that much. Thus we have to ensure that an attacker is not able to remove too many monomials with coefficients in $\mathbb{F}_{2^8}$ by assigning v variables to some values.

Note that we do not consider monomials of the form x_i^2. If such monomials remain after fixing v variables they do not force us to calculate in $\mathbb{F}_{2^8}$ as they are linear due to the Frobenius homomorphism. Note that for UOV schemes over fields with odd characteristic, it makes sense to consider such monomials.

Denote by $\overline{C}$ the set of monomials whose coefficients are contained in the matrix C. We can represent this set as a graph $G(V, E)$ with $V := \{x_1, \ldots, x_n\}$ being the vertices and $E := \{e(x_i, x_j) \mid x_i x_j \in \overline{C}\}$ being the edges. By construction we have $|E| = \frac{o(o+1)}{2}$. In the following our goal is to construct the graph G in such a way that the induced monomial ordering precludes an attacker from removing too many $\mathbb{F}_{2^8}$-terms (independent of the choice of variables he fixes). Note also that by "monomial ordering" we do not mean a monomial well-ordering as in the theory of Gröbner bases, but just some ordering of monomials w.r.t. which the columns of the coefficient matrix M_P are ordered. For the following we need two definitions.

Definition 2. *Let $G(V, E)$ be a graph. A subset $V' \subseteq V$ is called a k-*independent set, *if $|V'| = k$ and $\{e(v_i, v_j) : v_i, v_j \in V'\} \cap E = \emptyset$.*

Definition 3. *For a graph $G(V, E)$ the set $V' \subseteq V$ is called a k-*clique, *if $|V'| = k$ and all the vertices $v_i \in V'$ are pairwise connected,* i.e. *$\{e(v_i, v_j) : v_i, v_j \in V'\} \subseteq E$.*

We observe the following. If G contains an o-independent set, an attacker is able to fix v variables in such a way that all the monomials in $\overline{C}$ become constants.

So our task is to choose the edges of G in such a way, that G does not contain a k-independent set (for minimal k). For fixed k, the problem of finding a graph without k-independent set and minimal number of edges is solved by the complementary Turán graph [17].

So we start with $k = 1$ and construct the complementary Turán graph $\mathrm{CT}(n, 1)$. We then increase k until the number of edges in $\mathrm{CT}(n, k)$ will be less or equal to $\frac{o\cdot(o+1)}{2}$. If the number of edges in $\mathrm{CT}(n,k)$ is less than $\frac{o\cdot(o+1)}{2}$, we add arbitrarily edges until we reach the number of monomials in $\overline{C}$. By doing so we get a graph G with $\mathrm{CT}(n,k) \leq G$.

Example 1. In our case ($o = 26, v = 52$) we find a solution for $k = 8$ and thus it is assured that at least 30 monomials over $\mathbb{F}_{2^8}$ remain after fixing v variables. The best attack on this parameter set is called HybridF$_5$ [2] and uses fixing v and then guessing two variables before applying Faugères F_5 algorithm [7] to compute a Gröbner basis. But even if we fix/guess $v+2$ variables, there will remain at least 24 monomials over $\mathbb{F}_{2^8}$. So an attacker can not hope to transfer the system into a smaller field. More details to these experiments can be found in the full version of this paper [15].

Once we have constructed our graph G as above, it defines which monomials are in the set $\overline{C}$. Therefore, we can now define an induced ordering on quadratic monomials, such that monomials from $\overline{C}$ are smaller than those that are not from this set. For the monomials not being in $\overline{C}$ we define real squares (*i.e.* x_i^2 for $i = 1, \ldots, n$) to be bigger than other monomials. Once we defined an ordering of monomials, it is fixed and is a system parameter.

Let us investigate the effect of the new ordering on the construction of matrix A_{UOV}. In Section 3 the columns of the matrix A_{UOV} corresponded to the first D monomials w.r.t. graded lexicographic ordering. Now we have to choose the columns of the transformation matrix in such a way that its columns correspond to the first D monomials in the monomial ordering defined above. With respect to the graph G, if the i-th edge of the complementary graph $\overline{G}$ (which is actually a subgraph of the Turán graph $\mathrm{T}(n,k)$) connects the vertices v_{i_1} and v_{i_2}, we have

$$\widetilde{A_{UOV}} = \begin{pmatrix} \alpha_{11}^{11} & \alpha_{22}^{12} & \ldots & \alpha_{nn}^{11} & \widetilde{\alpha}_1^{11} & \widetilde{\alpha}_2^{11} & \ldots & \widetilde{\alpha}_{D-n}^{11} \\ \alpha_{11}^{12} & \alpha_{22}^{12} & \ldots & \alpha_{nn}^{12} & \widetilde{\alpha}_1^{12} & \widetilde{\alpha}_2^{12} & \ldots & \widetilde{\alpha}_{D-n}^{12} \\ \vdots & & & & & & & \vdots \\ \alpha_{11}^{vn} & \alpha_{22}^{vn} & \ldots & \alpha_{nn}^{vn} & \widetilde{\alpha}_1^{vn} & \widetilde{\alpha}_2^{vn} & \ldots & \widetilde{\alpha}_{D-n}^{vn} \end{pmatrix}. \tag{14}$$

Here, the coefficients α_{ii}^{rs} are given by equation (7) and the $\widetilde{\alpha}_i^{rs}$ are given by

$$\widetilde{\alpha}_i^{rs} = s_{rv_{i1}} \cdot s_{sv_{i2}} + s_{rv_{i2}} \cdot s_{sv_{i1}}. \tag{15}$$

With this notation we get (as in Section 3)

$$B = Q \cdot \widetilde{A_{UOV}}. \tag{16}$$

In the case of $\widetilde{A_{\mathrm{UOV}}}$ being invertible we can use equation (16) to compute the matrix Q and therewith the non zero coefficients of the central map $\mathcal{F}$.

In Algorithm 3 the matrix B chosen as shown in Subsection 5.2 with a fixed matrix $B_1 \in_R \mathbb{F}_2^{o\times(D-o)}$.

5.4 Efficiency of the Verification Process

During the verification process one has to evaluate for each public polynomial the equations

$$P_i(\mathbf{z}) = (z_1, \ldots, z_n) \cdot P_i \cdot (z_1, \ldots, z_n)^T, 1 \leq i \leq o,$$

Algorithm 3. Key Generation for 0/1 UOV

1: Choose randomly a linear map S (represented by an $n \times n$-matrix S). If S is not invertible, choose again.
2: Compute for S the corresponding transformation matrix $\widetilde{A_{\mathrm{UOV}}}$ (using equations (7), (15) and ((14)). If $\widetilde{A_{\mathrm{UOV}}}$ is not invertible, go back to step 1.
3: Solve the linear system given by equation (16) to get the matrix Q and therewith the quadratic coefficients of the central polynomials.
4: Compute the public key as $\mathcal{P} = \mathcal{F} \circ S$.

with $\mathbf{z} = (z_1, \ldots, z_n)$ being the signature of the message and P_i being the (upper triangular) matrix representing the i-th public polynomial.

To evaluate this equation for a randomly chosen P_i one needs $\frac{n\cdot(n+1)}{2} + n$ multiplications in $\mathbb{F}_{2^8}$ for each of the o polynomials, or $o \cdot \frac{n\cdot(n+3)}{2}$ $\mathbb{F}_{2^8}$-multiplications and the same number of additions for the whole key.

For our reduced version we can do better. We first compute $\mathbf{z} \cdot P_i$ $(i = 1, \ldots, o)$. For this we divide each of the matrices P_i into two parts $P_i^{(1)}$ and $P_i^{(2)}$, which we cover by for loops. For an element $a \in P_i^{(1)}$ we test if $a = 0$ or $a = 1$. In the first case, we don't have to do anything, if $a = 1$ we have to carry out one addition. Only for the elements from P_2 we have to perform one multiplication.

By doing so, we can reduce the number of multiplications needed during the verification process to $\frac{o\cdot(o+1)}{2} + n$. For the parameters $(o, v) = (26, 52)$, we get therefore a reduction of 85 %.

However, since we have to perform a number of other operations, for practical implementations we don't get such a hugh reduction in time.

5.5 Security of 0/1 UOV

Since we do not have MDS matrices over $\mathbb{F}_2$, we can not use Proposition 2 to prove the security of our scheme. Therefore we checked the security of our schemes against known attacks, including

1. Direct attacks
2. Rank attacks
3. UOV-Reconciliation attack
4. UOV attack

and found that these attacks cannot use the special structure of our public keys. The results of our experiments can be found in the full version of this paper [15].

5.6 Parameters and Implementation

In this section, we give our choice of parameters and show how they transfer to a practical C++ implementation. More concrete, based on our security considerations, we propose for our scheme the same parameters as for the standard

UOV scheme, namely field size $q = 2^8$, $(o, v) = (26, 52)$. Additionally, Table 1 gives one more conservative parameter set, namely $(q, o, v) = (2^8, 28, 56)$.
We implemented our scheme and the standard UOV in C++.

Key Generation: For the key generation we follow closely Algorithm 3. The linear system in step 3 of the algorithm is solved by inverting the matrix $\widetilde{A_{UOV}}$ and then computing the matrix product $B \cdot \widetilde{A_{UOV}}$. For both we use the M4RIE library for efficient linear algebra over finite fields and Travolta tables. By doing so, we can compute a key pair for 0/1 UOV(2^8,26,52) in roughly 27 sec on an Intel Dual Core 2 with 2.53 MHz.

Signature Generation: The signature generation process works as for the standard UOV Scheme. The running time of the signature generation process is about 0.69 ms.

Signature Verification: The signature verification process works as desribed in Subsection 5.4. For each parameter set listed in Table 2 we carried out 1,000,000 verification processes on the Intel machine as well as on an AMD Athlon XP 2400+ with 2.00 GHz. Table 2 shows the results.

Table 1. Proposed parameters for UOV schemes

Scheme(q, o, v)	system parameter (kB)	public key size (kB)	private key size (kB)	reduction of public key size (%)
UOV(2^8,26,52)	-	78.2	75.3	-
Compr.UOV(2^8,26,52)	69.3	8.9	75.3	88.6
0/1 UOV(2^8,26,52)	8.7	8.9	75.3	88.6
UOV(2^8,28,56)	-	97.6	93.4	-
Compr.UOV(2^8,28,56)	86.5	11.1	93.4	88.6
0/1 UOV(2^8,28,56)	10.8	11.1	93.4	88.6

Table 2. Running time of the verification process

	Intel Dual Core 2 2.53 GHz			AMD Athlon XP 2400+ 2.00 GHz		
(o, v)	UOV	0/1 UOV	reduction factor	UOV	0/1 UOV	reduction factor
(26,52)	0.49 ms	0.21 ms	57 %	0.68 ms	0.29 ms	57 %
(28,56)	0.54 ms	0.22 ms	58 %	0.74 ms	0.32 ms	58 %
(32,64)	0.75 ms	0.30 ms	59 %	1.03 ms	0.43 ms	58 %

6 Conclusion

In this paper, we have shown that $\mathcal{M}$ultivariate $\mathcal{Q}$uadratic public key schemes can benefit from much smaller public key sizes (cf. Table 1) without any degeneration of security. The overall idea requires some flexibility in the private key.

To our knowledge, only the two $\mathcal{MQ}$-schemes UOV and Rainbow have these. UOV was covered in this article. Rainbow has a more difficult internal structure, so we have to leave a concrete application of our improvement to Rainbow as an open question, which we plan to address.

The security arguments made use of the idea of equivalent keys. Hereby, each public key can be assigned many private keys. We have turned this idea around by considering transformations of the public key $\mathcal{P}$ instead and showed that an attacker does not gain from this specific structure.

As we can enforce a specific form on the public key $\mathcal{P}$, we can also use it to speed up public key operations, namely verification of signatures. As we see in Table 2, this reduces the overall time by about 59% or a markable factor of 2.4.

As the construction is very general, it can be used on other platforms (*e.g.* GPU, FPGA) as well. We actually expect similar gains in area reduction or speed there, too.

From a theoretical perspective, forcing a specific structure on the central polynomials $\mathcal{F}$ or the public polynomials $\mathcal{P}$ are equivalent: We can do either. Hence, for specific application domains it might be useful to find a certain trade-off. For example, we could reduce the computational workload on a server based on the maximal available memory on a smart card.

Acknowledgements. We thank Ishtiaq Shah for doing the implementation of our scheme. Furthermore we want to thank our financial supporters. The first author is supported by the Horst Görtz Foundation (HGS) within the project where the third author is the principal investigator. The third author is supported by the DFG grant BU 630/22-1. The second author was supported by the German Science Foundation (DFG) through an Emmy Noether grant where the forth author is principal investigator.

References

[1] Bernstein, D.J., Buchmann, J., Dahmen, E. (eds.): Post-Quantum Cryptography. Springer, Heidelberg (2009)

[2] Bettale, L., Faugére, J.-C., Perret, L.: Hybrid approach for solving multivariate systems over finite fields. Journal of Mathematical Cryptology 3, 177–197 (2009)

[3] Bogdanov, A., Eisenbarth, T., Rupp, A., Wolf, C.: Time-area optimized public-key engines: $\mathcal{MQ}$-cryptosystems as replacement for elliptic curves? In: Oswald, E., Rohatgi, P. (eds.) CHES 2008. LNCS, vol. 5154, pp. 45–61. Springer, Heidelberg (2008)

[4] Chen, A.I.-T., Chen, C.-H.O., Chen, M.-S., Cheng, C.-M., Yang, B.-Y.: Practical-sized instances of multivariate pKCs: Rainbow, TTS, and ℓIC-derivatives. In: Buchmann, J., Ding, J. (eds.) PQCrypto 2008. LNCS, vol. 5299, pp. 95–108. Springer, Heidelberg (2008)

[5] Chen, A.I.-T., Chen, M.-S., Chen, T.-R., Cheng, C.-M., Ding, J., Kuo, E.L.-H., Lee, F.Y.-S., Yang, B.-Y.: SSE implementation of multivariate pKCs on modern x86 cPUs. In: Clavier, C., Gaj, K. (eds.) CHES 2009. LNCS, vol. 5747, pp. 33–48. Springer, Heidelberg (2009)

[6] Ding, J., Gower, J.E., Schmidt, D.: Multivariate Public Key Cryptography. Cambridge University Press, Cambridge (2006)
[7] Faugére, J.-C.: A new efficient algorithm for computing Gröbner bases without reduction to zero (F5). In: ISSAC 2002, pp. 75–83. ACM Press, New York (2002)
[8] Garey, M.R., Johnson, D.S.: Computers and Intractability: A Guide to the Theory of NP-Completeness. W.H. Freeman and Company, New York (1979)
[9] Hu, Y., Wang, L., Chou, C., Lai, F.: Similar keys of multivariate quadratic public key cryptosystems. In: Desmedt, Y.G., Wang, H., Mu, Y., Li, Y. (eds.) CANS 2005. LNCS, vol. 3810, pp. 211–222. Springer, Heidelberg (2005)
[10] Kipnis, A., Patarin, J., Goubin, L.: Unbalanced oil and vinegar signature schemes. In: Stern, J. (ed.) EUROCRYPT 1999. LNCS, vol. 1592, pp. 206–222. Springer, Heidelberg (1999)
[11] Kipnis, A., Shamir, A.: Cryptanalysis of the oil & vinegar signature scheme. In: Krawczyk, H. (ed.) CRYPTO 1998. LNCS, vol. 1462, pp. 257–266. Springer, Heidelberg (1998)
[12] Patarin, J.: The oil and vinegar signature scheme. Presented at the Dagstuhl Workshop on Cryptography, transparencies (September 1997)
[13] Petzoldt, A., Bulygin, S., Buchmann, J.: A multivariate signature scheme with a partially cyclic public key. In: SCC 2010, pp. 229–235 (2010)
[14] Petzoldt, A., Bulygin, S., Buchmann, J.: Linear recurring sequences for the UOV key generation. In: Catalano, D., Fazio, N., Gennaro, R., Nicolosi, A. (eds.) PKC 2011. LNCS, vol. 6571, pp. 335–350. Springer, Heidelberg (2011)
[15] Petzoldt, A., Thomae, E., Bulygin, S., Wolf, C.: Small Public Keys and Fast Verification for Multivariate Quadratic Public Key Systems (full version), `http://eprint.iacr.org/2011/294`
[16] Shor, P.W.: Polynomial-time algorithms for prime factorization and discrete logarithms on a quantum computer. SIAM Journal on Computing 26(5), 1484–1509 (1997)
[17] Turán, P.: On an extremal problem in Graph Theory. Matematiko Fizicki Lapok 48, 436–452 (1941)
[18] Wolf, C., Preneel, B.: Taxonomy of public key schemes based on the problem of multivariate quadratic equations, `http://eprint.iacr.org/2005/077`
[19] Wolf, C., Preneel, B.: Superfluous keys in Multivariate Quadratic asymmetric systems. In: Vaudenay, S. (ed.) PKC 2005. LNCS, vol. 3386, pp. 275–287. Springer, Heidelberg (2005)
[20] Wolf, C., Preneel, B.: Equivalent keys in multivariate quadratic public key systems. Journal of Mathematical Cryptology (to appear, 2011)
[21] Yang, B.-Y., Chen, J.-M., Chen, Y.-H.: TTS: High-speed signatures on a low-cost smart card. In: Joye, M., Quisquater, J.-J. (eds.) CHES 2004. LNCS, vol. 3156, pp. 371–385. Springer, Heidelberg (2004)

Throughput vs. Area Trade-offs in High-Speed Architectures of Five Round 3 SHA-3 Candidates Implemented Using Xilinx and Altera FPGAs*

Ekawat Homsirikamol, Marcin Rogawski, and Kris Gaj

ECE Department, George Mason University, Fairfax, VA 22030, U.S.A.
{ehomsiri,mrogawsk,kgaj}@gmu.edu
http://cryptography.gmu.edu

Abstract. In this paper we present a comprehensive comparison of all Round 3 SHA-3 candidates and the current standard SHA-2 from the point of view of hardware performance in modern FPGAs. Each algorithm is implemented using multiple architectures based on the concepts of folding, unrolling, and pipelining. Trade-offs between speed and area are investigated, and the best architecture from the point of view of the throughput to area ratio is identified. Finally, all algorithms are ranked based on their overall performance, and the characteristic features of each algorithm important from the point of view of its implementation in hardware are identified.

Keywords: benchmarking, hash functions, SHA-3, hardware, FPGA.

1 Introduction

Performance in hardware has proven to be an important tie-breaker in the contests for new cryptographic standards. For example, in the AES contest [14], performance in FPGAs and ASICs has played a major role, because all five finalists have been judged to have adequate security, and their performance in hardware varied substantially.

In this paper, we focus on comparing hardware performance of the remaining five final candidates in the SHA-3 contest organized by NIST in the period from 2007 to 2012 [1]. The unique and novel feature of our approach is the investigation of multiple hardware architectures of each algorithm. Our goal is to analyze the entire performance space in terms of the throughput to area trade-offs, for all Round 3 SHA-3 candidates, as well as the current standard, SHA-2. This investigation is very important because the exact requirements on the speed and area of a hash function core depend on a particular application and very in

* This work has been supported in part by NIST through the Recovery Act Measurement Science and Engineering Research Grant Program, under contract no. 60NANB10D004.

B. Preneel and T. Takagi (Eds.): CHES 2011, LNCS 6917, pp. 491–506, 2011.

a wide range. Knowledge of alternative architectures may allow the developer to substantially reduce the relative area of a hash core in a system-on-chip, or move to a substantially less expensive part in case of a stand-alone implementation of a hash core in an FPGA.

We perform our investigation using four high-performance FPGA families from two major vendors: Virtex 5 and Virtex 6 from Xilinx, and Stratix III and Stratix IV from Altera. All algorithms have been implemented based on their updated Round 3 specifications, published in January 2011.

2 Previous Work

Previous results on comparison of Round 2 SHA-3 candidates in hardware are summarized in [2]. These results are classified into four major categories, based on the technology (FPGA vs. ASIC), and the optimization target (High-Speed vs. Low-Area). The previous results most relevant to the subject of this paper belong to the category of High-Speed Implementations in FPGAs. The most comprehensive results belonging to this category have been reported in [5][8][12][13]. All these papers include results for all 14 Round 2 candidates. Majority of published results concern 256-bit variants of the candidates, implemented using Xilinx Virtex 5 FPGAs. In [12], results for 256-bit and 512-bit variants of all algorithms, implemented using 10 FPGA families from Xilinx and Altera are discussed. Additionally, pipelined implementations of three Round 2 SHA-3 candidates have been investigated in [4].

Some of the most interesting low-area implementations of the SHA-3 candidates have been described in [6][7][15]. The most comprehensive studies of the ASIC implementations of the Round 2 SHA-3 candidates are presented in [10][11][16].

All results obtained based on the Round 2 specifications of SHA-3 candidates carry without any changes for Keccak and Skein. The specifications of BLAKE, Groestl, and JH have been tweaked at the start of Round 3, in January 2011, and at the time of writing, we are not aware of any published reports on the high-speed FPGA implementations of the Round 3 variants of these algorithms.

3 Performance Metrics

Three major performance metrics used in our study are throughput, area, and throughput to area ratio. Throughput is understood as the throughput for long messages, or cumulative throughput for a large number of small messages (where processing and input/output functions overlap in time). The resource utilization in FPGAs is a vector, with coordinates specific to the given FPGA family, e.g.

$$Resource\,Utilization_{Virtex\,5} = (\#CLB\_slices, \#BRAMs, \#DSPs) \quad (1)$$

$$Resource\,Utilization_{Stratix\,III} = (\#ALUTs, \#memory\_bits, \#DSPs). \quad (2)$$

In these formulas: $\#CLB\_slices$ is the number of Configurable Logic Block slices, $BRAM$ stands for Block RAM, DSP is a Digital Signal Processing unit, $\#ALUTs$ represents the number of Adaptive Look-Up Tables, and $\#memory\_bits$ is the number of bits placed in dedicated Altera FPGA memories. Taking into account that vectors cannot be easily compared to each other, we have decided to opt out of using any dedicated resources in the hash function implementations used for our comparison. Thus, all coordinates of our vectors, other than the first one have been forced (by choosing appropriate options of the synthesis and implementation tools) to be zero. This way, our resource utilization (further referred to as Area) is characterized using a single number, specific to the given family of FPGAs, namely $\#CLB\_slices$ for Xilinx Virtex 5 and Virtex 6, and $\#ALUTs$ in Stratix III and Stratix IV. We believe that the capability of using embedded resources should be treated as a measure of the algorithm flexibility, and should be investigated independently from this study.

4 Investigated Hardware Architectures

A starting point for our exploration of various architectures of hash functions is the basic iterative architecture, shown in Fig. 1a. The characteristic features of this architecture are as follows: a) datapath width = state size (denoted by s), b) one round is performed in a single clock cycle, c) only one message is processed at a time. The minimum block processing time is typically given by (3),

$$T_{block} = (r + f) \cdot T \tag{3}$$

where r is the number of rounds, f is the number of clock cycles required to finalize computations for a block (typically 0 or 1), and T is the minimum clock period. The corresponding throughput is given by (4),

$$Tp = b/T_{block} \tag{4}$$

where b is the size of a message block in bits. We denote the area of this architecture by $Area$. The basic iterative architecture is typically an architecture of choice for high-speed hardware implementations of SHA-1, SHA-2, and SHA-3 candidates.

If a round of a hash function has a symmetric structure, with two or more similar operations performed one after another, horizontal folding is possible. In Fig. 1b, horizontal folding by a factor of two is demonstrated. We will denote this architecture by /2(h). In this architecture, a half of a round is implemented as combinational logic, and the entire round is executed using two clock cycles. The datapath width stays the same as in the basic iterative architecture, and is equal to the state size, s. The block processing time is given by (5),

$$T_{block-/2(h)} = (2 \cdot r + f) \cdot T_{/2(h)} \tag{5}$$

where $T/2 < T_{/2(h)} < T$, ideally $T_{/2(h)} \approx T/2$, and $Area/2 < Area_{/2(h)} < Area$. As a result, the block processing time (and thus also throughput) stays

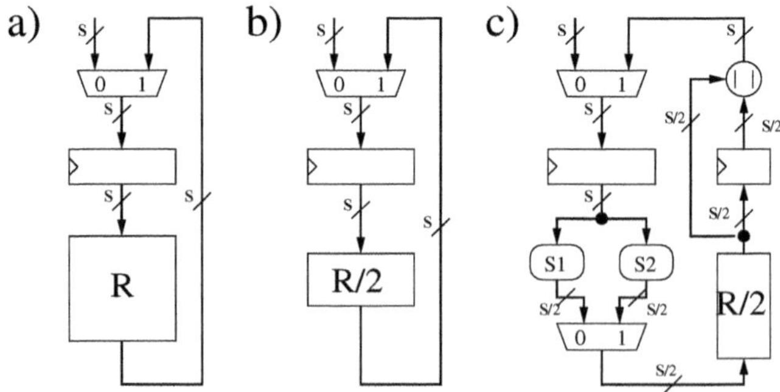

Fig. 1. Three hardware architectures of a hash function: a) basic iterative: x1, b) folded horizontally by a factor of 2: /2(h), c) folded vertically by a factor of 2: /2(v). R – round, S1, S2 – selection functions.

approximately the same, and area decreases. These dependencies lead to the overall increase of the Throughput to Area ratio. In general, folding by a factor of k might be possible, and the corresponding architecture will be denoted by /k(h).

Among the five finalists, the only candidate that can benefit substantially from horizontal folding is BLAKE. The round of BLAKE consists of two horizontal layers of identical G functions, separated only by a permutation. By implementing only one layer in combinational logic, horizontal folding by a factor of two can be easily achieved. Additionally, each G function has a very symmetric structure along the horizontal axis, and can be easily folded horizontally by a factor of 2. As a result a folding factor of 4, is achieved for the entire round. Other SHA-3 finalists do not demonstrate any similar symmetry.

In case horizontal folding is either not possible or does not achieve the required reduction in area, vertical folding may be attempted. In Fig. 1c, we demonstrate vertical folding by a factor of 2, which we denote by /2(v). In this architecture, the datapath width is reduced by a factor of two. As a result two clock cycles are required to complete a round. In the first clock cycle, only bits of the internal state affecting the first half of the round output are provided to the input of R/2. In the second clock cycle, the remaining bits of the internal state are processed. The first output is stored in an auxiliary register of the size of $s/2$ bits. This output is concatenated with the output from the second iteration to form a new internal state.

The clock period of this architecture is approximately equal to the clock period of the basic iterative architecture, $T_{/2(v)} \approx T$. As a result, the block processing time, increases approximately by a factor of two compared to the basic architecture, as shown in the equation below:

$$T_{block-/2(v)} = (2 \cdot r + f) \cdot T_{/2(v)} \approx (2r + f) \cdot T. \quad (6)$$

The area reduction is also smaller than in case of horizontal folding, because of the need for an extra $s/2$-bit register and multiplexer. As a result the throughput to area ratio is likely to go down. In general, vertical folding by a factor of k might be possible, and the corresponding architecture will be denoted by /k(v).

Out of five final SHA-3 candidates, BLAKE and Groestl are most suitable for vertical folding. JH can be folded, but the gain in area is not expected to be substantial, because the round of JH is very simple, and does not dominate the total area of the circuit. For Skein and Keccak, the internal round symmetry, necessary for implementation of vertical folding, is missing.

In order to increase the throughput of a hash function, different architectures must be applied. The three common approaches are unrolling, pipelining, and parallel processing. Unrolling is suitable for increasing throughput of a single long message. Pipelining and parallel processing increase the combined data throughput in case of processing multiple messages (e.g., multiple packets) at the same time.

In Fig. 2a, architecture with unrolling by a factor of two is demonstrated. We will denote this architecture by x2. The datapath width stays the same as in the basic iterative architecture. The combinational logic of a round is replicated, so now two rounds are performed per clock cycle. Since the total number of clock cycles is reduced approximately by a factor of two, and the clock period increases by a factor less than two (due to optimizations on the boundaries of two rounds, and the smaller relative contributions of the multiplexer delay, the register delay, and the register setup time), the total throughput increases. Unfortunately, at the same time, the area of the circuit is likely to increase by a factor close to the unrolling factor. As a result, in most cases, the throughput to area ratio decreases substantially compared to the basic iterative architecture. As such, architectures with unrolling are typically used only when throughput for single long messages is of the utmost concern, and area is abundant. Nevertheless, there are exceptions to this rule. Unrolling can improve the throughput to area ratio when rounds used by an algorithm in subsequent iterations are not the same. Among the five final SHA-3 finalists, this situation happens only for Skein.

In majority of practical applications of hash functions, the messages that are processed are relatively short (typically smaller than 1500 bytes), and multiple messages (packets) are available for processing by a hashing unit at the same time. For example, in the most widespread Internet security protocols, such as IPSec, SSL, and WLAN (802.11), the inputs to a hash unit are packets. The maximum size of a packet for Internet is limited by so called Maximum Transmission Unit (MTU). The typical size of MTU for Ethernet based networks is 1500 bytes. The Maximum Transmission Unit for the Internet IPv4 path is even smaller, and set at 576 bytes. As a result, in a typical internet node, up to 80% of packets processed have the size of 576 bytes or less, and 100% of packets have sizes equal or smaller than 1500 bytes. Such small sizes of packets mean that hundreds of packets could be easily buffered in the processing nodes, in the form of packet queues, without introducing any significant latency to the total packet travel time from the source to destination. In this paper, we will assume that the number of messages available in parallel is large (at least 10), and we will look at the combined throughput for all available streams of data.

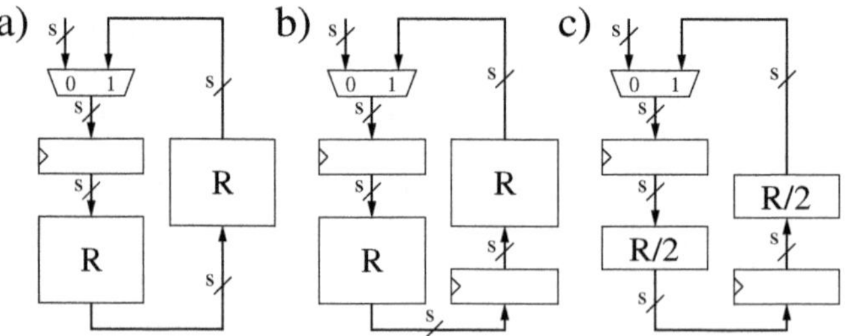

Fig. 2. Three hardware architectures of a hash function a) unrolled by a factor of 2: x2, b) unrolled by a factor of 2 with 2 pipeline stages: x2-PPL2, c) basic iterative with 2 pipeline stages: x1-PPL2

The easiest way to implement pipelining in hash functions is to first unroll, and then introduce pipeline registers between adjacent rounds. The simplest case is the architecture that is two times unrolled, and has two pipeline stages, as shown in Fig. 2b. We will denote this architecture as x2-PPL2. The clock period of this architecture is approximately equal to the clock period of the basic iterative architecture, T. Processing a single block takes the same number of clock cycles as in the basic iterative architecture. However, since two blocks belonging to two different messages are processed simultaneously, the combined throughput increases by a factor of two. The throughput to area ratio remains roughly the same, and may be either larger or smaller than in the basic iterative architecture, depending on a particular algorithm.

The more challenging way of using pipelining is to introduce pipeline registers inside of a hash function round. The improvement in throughput compared to the basic iterative architecture is than equal (either exactly or at least approximately) to the ratio of the new clock frequency to the original clock frequency. Since the critical path is reduced, the increase in throughput is guaranteed, but its level depends on how well the critical path has been divided by pipeline registers into shorter paths with approximately equal delays. At the same time, the area of the circuit increases by the area of pipeline registers, plus any logic required for simultaneous processing of multiple streams of data. The throughput to area ratio may increase, but the improvement is not guaranteed for all algorithms, and all FPGA families, and may be small or negative in case the basic iterative architecture operates already at the clock frequency close to the maximum clock frequency supported by the given FPGA family.

The formulas for the block processing time and the throughput of all aforementioned architectures are summarized in Table 1.

5 Design Methodology and Design Environment

Our designs for the basic, folded, and unrolled architectures use the interface and the communication protocol proposed in [8]. Our designs for the pipelined architectures, use the interface and surrounding logic shown in Fig. 3.

Table 1. Formulas for the time required to process a single message block, T_{block}, and the Throughput, Tp, for all investigated architectures. Notation: b – block size, r – number of rounds, f – number of clock cycles required to finalize computations for a block ($f = 0$ for Keccak and Groestl $(P + Q)$, $f = 1$ for all remaining algorithms), k – folding factor or unrolling factor, n – number of pipeline stages, T – clock period.

Architecture	Time required to process a single message block	Throughput
Basic iterative, x1	$T_{block} = (r + f) \cdot T$	$Tp = b/T_{block}$
Folded by a factor of k, /k	$T_{block} = (k \cdot r + f) \cdot T$	$Tp = b/T_{block}$
Unrolled by a factor of k, xk	$T_{block} = (r/k + f) \cdot T$	$Tp = b/T_{block}$
Basic iterative with n pipeline stages, x1-PPLn	$T_{block} = (n \cdot r + f) \cdot T$	$Tp = n \cdot b/T_{block}$
Folded by a factor of k with n pipeline stages, /k-PPLn	$T_{block} = (n \cdot k \cdot r + f) \cdot T$	$Tp = n \cdot b/T_{block}$
Unrolled by a factor of k with n pipeline stages, xk-PPLn	$T_{block} = (n \cdot r/k + f) \cdot T$	$Tp = n \cdot b/T_{block}$

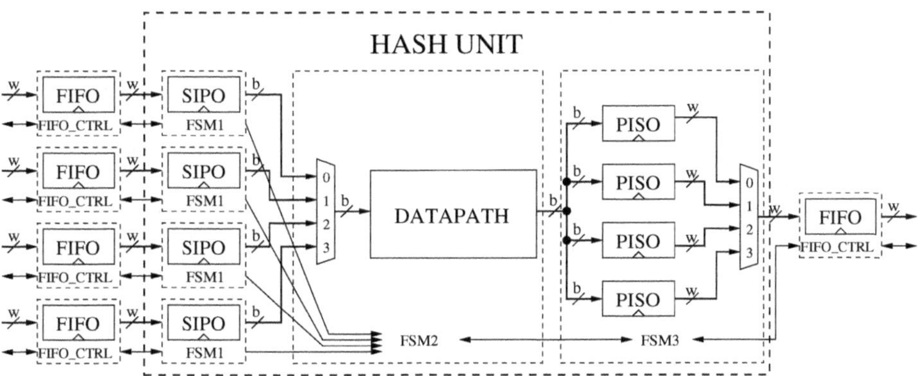

Fig. 3. Interface, high-level block diagram, and surrounding logic of the Hash Unit for the pipelined architecture with four pipeline stages. Notation: SIPO – Serial-In Parallel-Out unit, PISO – Parallel-In Serial-Out unit, w – input/output bus width, $w = 64$ for all investigated algorithms, except SHA-2-256, where $w = 32$.

Input FIFOs serve as packet queues. Each FIFO communicates with the corresponding Serial-In Parallel-Out (SIPO) unit and the associated Finite State Machine 1 (FSM1). FSM 1 is responsible for reading in the next block of data, using b/w clock cycles, possibly in parallel with the Datapath processing the previous block under the control of FSM2. Outputs corresponding to four independent packets are first stored in the corresponding Parallel-In Serial-Out Units, and then multiplexed to the output FIFO.

All architectures have been modeled in VHDL-93. All VHDL codes have been thoroughly verified using a universal testbench, capable of testing an arbitrary hash function core. A special padding script was developed in Perl in order to pad messages included in the Known Answer Test (KAT) files, distributed as a part of each candidates submission package.

For synthesis and implementation, we have used tools developed by FPGA vendors themselves:

- for Xilinx: Xilinx ISE Design Suite v. 12.4, including Xilinx XST,
- for Altera: Quartus II v. 10.1 Subscription Edition Software.

The generation of a large number of results was facilitated by an open source benchmarking environment, ATHENa (Automated Tool for Hardware EvaluatioN) [3][9]. The details of results and selected source codes are available at [3].

6 Results

The results of our implementations are summarized in Figs. 4-8, and in Tables 2 and 3. In Fig. 4, we present the detailed throughput vs. area graphs for all implemented architectures of the 256-bit variants of six investigated algorithms in Xilinx Virtex 5 FPGAs.

For BLAKE (see Fig. 4a), the two best architectures in terms of the throughput to area ratio are: /4(h)/4(v), i.e., architecture with horizontal folding by a factor of 4, combined with vertical folding by a factor of 4; and x1-PPL2, i.e., basic architecture with two pipeline stages. The good performance of the former of these two architectures is associated with the significant reduction of the complexity of the BLAKE PERMUTE function as a result of vertical folding by 4. The good performance of the latter is associated with the perfectly symmetric structure of the round, which makes it easy to divide the datapath into two well-balanced pipeline stages. The two less successful architectures include x1 and /2(h)-PPL4. These architectures are not included in our combined graphs shown in Figs. 5-8.

For Groestl (see Fig. 4b), we consider two major architectures: a) parallel architecture, denoted (P+Q), in which Groestl permutations P and Q are implemented using two independent units, working in parallel, and b) quasi-pipeline architecture, denoted (P/Q), in which, the same unit is used to implement both P and Q, and the computations belonging to these two permutations are interleaved [16]. The best architecture overall appears to be the parallel architecture (P+Q) in the basic version, with two pipeline stages, x1-PPL2. Vertical folding by 2 provides quite substantial reduction in area, but at the price of an even greater reduction in throughput. An attempt to pipeline Groestl using 7 pipeline stages (x1-PPL7), using logic-only implementation of S-boxes, appeared to be rather unsuccessful.

For JH (see Fig. 4c), we consider two major types of architectures: a) with round constants stored in memory, JH (MEM), and b) with round constants calculated on the fly, JH (OTF). Both approaches seem to result in a very similar performance for the basic iterative architectures, x1. Neither vertical folding nor pipelining seem to be efficient when applied directly to the basic architecture. Vertical folding, somewhat unexpectedly, increases area, and the

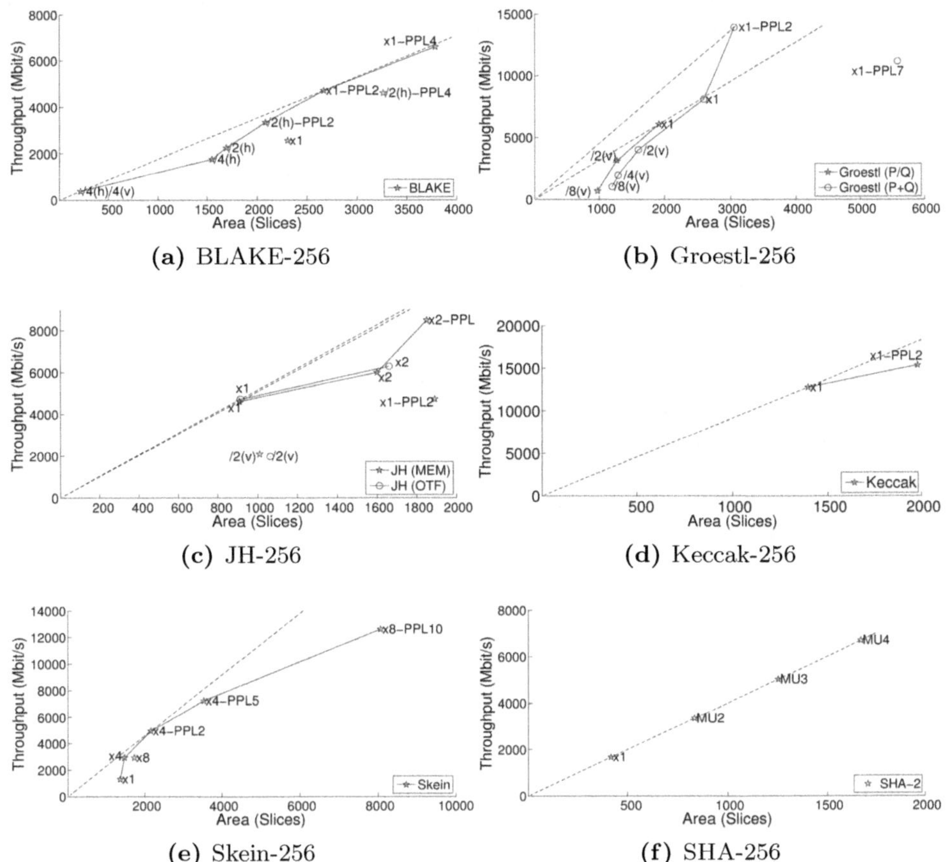

Fig. 4. Throughput vs. Area graphs for multiple architectures of a) BLAKE-256, b) Groestl-256, c) JH-256, d) Keccak-256, e) Skein-256, and f) SHA-256, implemented in Xilinx Virtex 5 FPGAs. Notation: x1 – basic iterative architecture, /k(h) – horizontally folded by a factor of k, /k(v) - folded vertically by a factor of k, xk – unrolled by a factor of k, PPLn – pipelined with n pipeline stages, $(P+Q)$ – parallel architecture of Groestl, P/Q – quasi-pipelined architecture of Groestl, MEM – architecture of JH with round constants stored in memory, OTF – architecture of JH with round constants calculated on the fly, MU – multi-unit architecture.

basic architecture with two pipeline stages does not improve throughput. Both undesired effects can be tracked back to the simplicity of the main round. Folding does not reduce area, because of extra registers and multiplexers introduced to a very simple round. Pipelining does not increase throughput, because a simple basic round is hard to divide into two well balanced pipeline stages. As a result, the basic iterative architecture remains most efficient in terms of the throughput to area ratio.

For Keccak (see Fig. 4d), neither horizontal nor vertical folding applies. Two pipeline stages increase throughput, but by a factor smaller than the increase in the circuit area.

For Skein (see Fig. 4e), the unrolled by 4 architecture, x4, appears to be significantly more efficient than the basic architecture, x1. At the same time, unrolling by 8 does not give any additional improvement. The best results are obtained by first unrolling basic architecture by a factor of four, and then pipelining the obtained circuit using two pipeline stages. Five pipeline stages have been attempted as well because of an extra addition executed every fourth round, but did not improve the overall throughput to area ratio.

For SHA-2 (see Fig. 4f), none of the discussed techniques applies. The implementation of this function is already small, so reducing area is not necessary. The best way to speed up this function is by using multiple independent units of SHA-2 working in parallel. We denote this architecture by MUn, where n denotes the number of hash units.

The combined graphs for the 256-bit variants and the 512-bit variants of all algorithms, implemented using Xilinx Virtex 5 FPGAs, are presented in Figs. 5 and 6. Individual dots placed in regular intervals on the dashed lines represent multi-unit architectures. Algorithms can be ranked first in terms of the throughput to area ratio of their best architecture (as identified above). This is because this architecture can be easily replicated, allowing for processing n streams of data in parallel. Both throughput and area will increase by a factor of n.

The secondary criterion is the area of the best architecture. The smaller the area, the denser is the graph representing possible locations of a given function on the throughput vs. area graph.

The results for the 256-bit variants of hash functions, shown in Fig. 5, indicate that the order of the SHA-3 candidates in terms of throughput, for implementations using 1500 or more CLB slices is: 1) Keccak, 2) JH, 3) Groestl, 4) Skein, and 5) BLAKE. Keccak and JH clearly outperform SHA-2, while Groestl becomes faster only with more than 3000 CLB slices. At the same time, only BLAKE and SHA-2 have implementations based on basic iterative architecture and/or folding, with area below 500 CLB slices.

The results for the 512-bit variants of hash functions, shown in Fig. 6, are quite similar, with the exception that JH performs almost equally well as Keccak (because of the decrease in the Keccak message block size from 1088 to 576 bits), SHA-2 is ranked third, Skein slightly outperforms Groestl (because of the increase in the number of rounds of Groestl from 10 to 14), and BLAKE is a distant sixth.

The performance for Altera devices, shown in Figs. 7 and 8 is somewhat different. For the 256-bit versions of the algorithms, Keccak is the only function that outperforms SHA-2 in terms of the throughput to area ratio. JH is the third in ranking, with two architectures offering the similar ratio as SHA-2. BLAKE, Groestl, and Skein are in tie with each, with Groestl being somewhat disadvantaged by approximately twice as large area of its most efficient architecture. For the 512-bit versions of the algorithms (see Fig. 8), Keccak and JH outperform

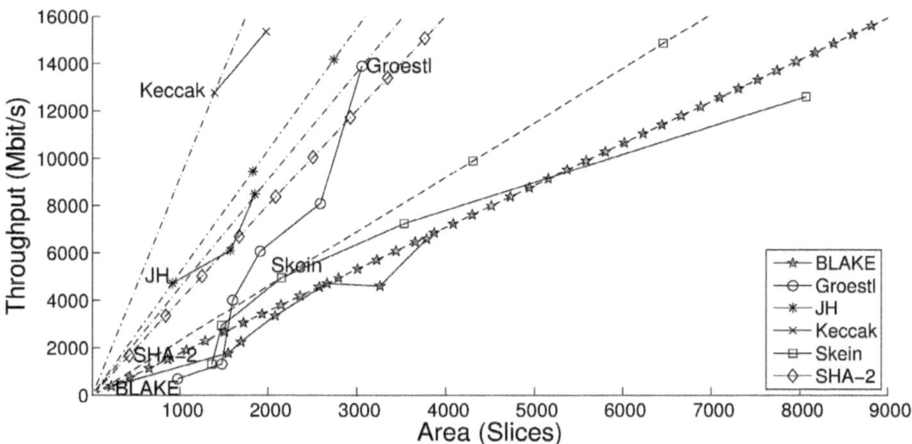

Fig. 5. Combined Throughput vs. Area graph for multiple hardware architectures of the 256-bit variants of BLAKE, Groestl, JH, Keccak, Skein, and SHA-2, implemented in Xilinx Virtex 5 FPGAs

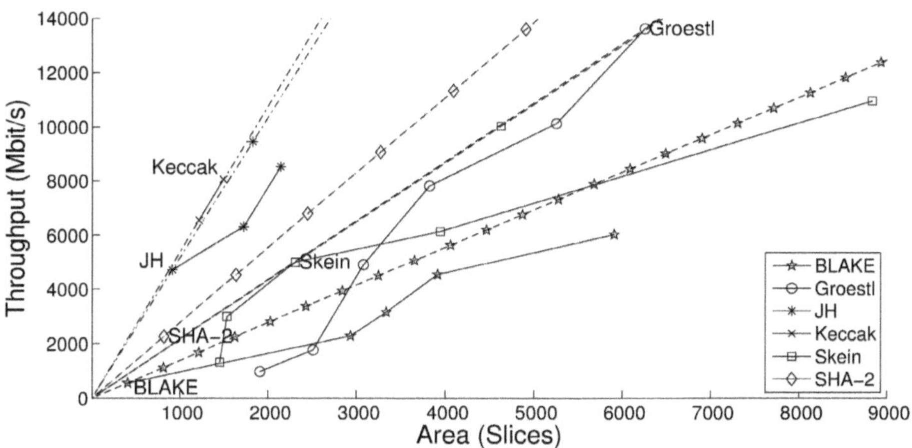

Fig. 6. Combined Throughput vs. Area graph for multiple hardware architectures of the 512-bit variants of BLAKE, Groestl, JH, Keccak, Skein, and SHA-2, implemented in Xilinx Virtex 5 FPGAs

SHA-2, Skein is in tie with SHA-2, Groestl and BLAKE fall significantly behind the current standard.

The numerical results for all our implementations are summarized in Tables 2 and 3. The best values of the throughput to area ratios and the best architectures for each hash function are listed in **bold** in these tables.

Additionally, we have also performed an initial study on the influence of padding units on the ranking of the candidates. Based on this study, the largest

Table 2. Results for the 256-bit variants of the Round 3 SHA-3 candidates and SHA-2, implemented using all investigated architectures and four FPGA families: Virtex 5 and Virtex 6 from Xilinx, and Stratix III and Stratix IV from Altera. Notation: Tp – throughput, A – area, Tp/A – Throughput to Area Ratio.

Arch	**Virtex 5**			**Virtex 6**			**Stratix III**			**Stratix IV**		
	Tp	**A**	**Tp/A**	**Tp**	**A**	**Tp/A**	**Tp**	**A**	**Tp/A**	**Tp**	**A**	**Tp/A**
BLAKE-256												
/4(h)/4(v)	381	215	1.77	412	181	2.28	370	915	0.40	378	915	0.41
/4(h)	1770	1547	1.14	1784	888	2.01	1708	3153	0.54	1747	3157	0.55
/2(h)	2253	1691	1.33	1956	1247	1.57	2151	3603	0.60	2302	3605	0.64
/2(h)-PPL2	3346	2083	1.61	3069	1792	1.71	3149	4571	0.69	3471	4570	0.76
x1	2561	2306	1.11	2388	1721	1.39	2195	4745	0.46	2305	4742	0.49
/2(h)-PPL4	4609	3261	1.41	5002	2516	1.99	4894	5080	0.96	5312	5049	1.05
x1-PPL2	4714	2666	**1.77**	5156	2206	2.34	4487	5420	0.83	4704	5431	0.87
x1-PPL4	6596	3784	1.74	7937	2616	**3.03**	7524	6273	**1.20**	8186	6278	**1.30**
Groestl-256 (P+Q)												
/8(v)	1042	1197	0.87	1161	980	1.18	1103	2716	0.41	1094	2736	0.40
/4(v)	1948	1287	1.51	2289	1134	2.02	2129	4079	0.52	2058	4093	0.50
/2(v)	4014	1598	2.51	4890	1560	3.13	4623	6130	0.75	4349	6073	0.72
x1	8081	2591	3.12	9340	2630	3.55	9608	11122	0.86	9122	11154	0.82
x1-PPL2	13894	3057	**4.55**	17084	3034	**5.63**	13793	11727	**1.18**	13749	11727	**1.17**
x1-PPL7	11167	5582	2.00	N/A	N/A	N/A	13964	14487	0.96	14392	14470	0.99
Groestl-256 (P/Q)												
/8(v)	691	973	0.71	808	813	0.99	812	2141	0.38	791	2141	0.37
/4(v)	1322	1477	0.89	1687	996	1.69	1401	3660	0.38	1378	3658	0.38
/2(v)	3136	1270	2.47	3301	1074	3.07	3198	4208	0.76	3209	4216	0.76
x1	6072	1912	3.18	4621	1737	2.66	6041	7498	0.81	5586	7287	0.77
JH-256 (MEM)												
/2(v)	2088	1010	2.07	2202	861	2.56	2104	3365	0.63	2066	3377	0.61
x1	4624	909	5.09	5700	847	**6.73**	5146	3207	**1.60**	4868	3209	**1.52**
x1-PPL2	6000	1600	3.75	7093	1328	5.34	6225	5607	1.11	6001	5574	1.08
x2	4728	1891	2.50	4986	1613	3.09	5314	4254	1.25	5378	4262	1.26
x2-PPL2	8487	1851	4.58	8846	1934	4.57	9816	6303	1.56	9522	6259	**1.52**
JH-256 (OTF)												
/2(v)	1981	1064	1.86	2219	915	2.42	2039	3464	0.59	2010	3469	0.58
x1	4725	914	**5.17**	5306	1039	5.11	5028	3380	1.49	4965	3383	1.47
x2	6297	1661	3.79	6983	1441	4.85	6141	6079	1.01	5620	6043	0.93
Keccak-256												
x1	12777	1395	**9.16**	11843	1165	10.17	12971	3909	3.32	13159	4129	3.19
x1-PPL2	15362	1980	7.76	16236	1446	**11.23**	19193	4955	**3.87**	18610	4953	**3.76**
x1-PPL4	12652	3849	3.29	13201	2785	4.74	16019	5391	2.97	17913	5402	3.32
Skein-256												
x1	1307	1364	0.96	1382	1127	1.23	1108	3538	0.31	1247	3539	0.35
x4	2937	1476	1.99	3523	1216	2.90	2455	3965	0.62	2621	3968	0.66
x8	2931	1728	1.70	3275	1510	2.17	3178	5586	0.57	3372	5493	0.61
x4-PPL2	4950	2154	**2.30**	5858	1860	**3.15**	4273	4421	0.97	4596	4423	1.04
x4-PPL5	7240	3532	2.05	7465	2839	2.63	6772	5920	**1.14**	7693	5935	**1.30**
x8-PPL10	12602	8065	1.56	N/A	N/A	N/A	12283	10994	1.12	11378	10996	1.03
SHA-256												
x1	1675	418	**4.01**	2273	286	**7.95**	1654	988	**1.67**	1744	988	**1.77**

Table 3. Results for the 512-bit variants of the Round 3 SHA-3 candidates and SHA-2, implemented using all investigated architectures and four FPGA families: Virtex 5 and Virtex 6 from Xilinx, and Stratix III and Stratix IV from Altera. Notation: Tp – throughput, A – area, Tp/A – Throughput to Area Ratio.

Arch	Virtex 5			Virtex 6			Stratix III			Stratix IV		
	Tp	**A**	**Tp/A**	**Tp**	**A**	**Tp/A**	**Tp**	**A**	**Tp/A**	**Tp**	**A**	**Tp/A**
BLAKE-512												
/4(h)/4(v)	563	406	**1.39**	612	324	**1.89**	485	1664	0.29	546	1675	0.33
/4(h)	2287	2935	0.78	2709	1936	1.40	2230	6137	0.36	2477	6161	0.40
/2(h)	3159	3337	0.95	3187	2628	1.21	2905	7127	0.41	3288	7128	0.46
/2(h)-PPL2	4544	3912	1.16	4821	3642	1.32	4033	8960	0.45	4780	8962	0.53
x1	3401	3984	0.85	3273	3823	0.86	2947	9251	0.32	3310	9268	0.36
/2(h)-PPL4	6035	5911	1.02	6948	4922	1.41	5535	9698	**0.57**	7521	9703	**0.78**
x1-PPL2	6405	5730	1.12	6426	4922	1.31	5549	10616	0.52	6222	10627	0.59
x1-PPL4	3825	7497	0.51	3607	6234	0.58	4952	12100	0.41	5594	12100	0.46
Groestl-512 (P+Q)												
/8(v)	1351	2249	0.60	1484	1837	0.81	1496	5312	0.28	1367	5303	0.26
/4(v)	2533	2263	1.12	2933	2237	1.31	2902	8031	0.36	2692	7945	0.34
/2(v)	4914	3079	1.60	6257	3154	1.98	5985	12295	0.49	5851	12216	0.48
x1	10124	5254	1.93	11566	5106	2.27	12393	21854	0.57	12164	21847	0.56
x1-PPL2	13628	6258	**2.18**	N/A	N/A	N/A	17050	22570	**0.76**	17196	22412	**0.77**
x1-PPL7	12669	11194	1.13	N/A	N/A	N/A	17635	29320	0.60	18395	28976	0.63
Groestl-512 (P/Q)												
/8(v)	984	1908	0.52	1037	1406	0.74	1052	4749	0.22	1010	4744	0.21
/4(v)	1783	2516	0.71	2145	1787	1.20	1855	6000	0.31	1945	6301	0.31
/2(v)	4139	2161	1.92	4442	2172	2.04	4550	7417	0.61	4352	7426	0.59
x1	7819	3821	2.05	8468	3658	**2.31**	8029	14445	0.56	7944	14461	0.55
JH-512 (MEM)												
/2(v)	2187	1102	1.98	2392	1002	2.39	2199	3621	0.61	2076	3620	0.57
x1	4624	909	5.09	5700	847	**6.73**	5146	3207	**1.60**	4868	3209	**1.52**
x1-PPL2	4610	1973	2.34	5007	1745	2.87	5208	4514	1.15	5261	4527	1.16
x2	5792	1506	3.85	7018	1405	5.00	6309	5769	1.09	6176	5806	1.06
x2-PPL2	8523	2148	3.97	8321	1895	4.39	9704	6335	1.53	9328	6307	1.48
JH-512 (OTF)												
/2(v)	2068	1041	1.99	2287	955	2.39	2099	3701	0.57	2134	3816	0.56
x1	4725	914	**5.17**	5306	1039	5.11	4912	3548	1.38	4860	3549	1.37
x2	6043	1554	3.89	6827	1515	4.51	6270	6320	0.99	5752	6261	0.92
Keccak-512												
x1	6556	1220	5.37	7225	1231	5.87	6859	3477	1.97	6805	3470	1.96
x1-PPL2	8056	1498	**5.38**	8853	1470	**6.02**	9836	4431	**2.22**	9490	4418	**2.15**
x1-PPL4	7095	3756	1.89	7202	2650	2.72	9440	5175	1.82	9328	5201	1.79
Skein-512												
x1	1325	1457	0.91	1356	1155	1.17	1082	3632	0.30	1229	3631	0.34
x4	2999	1537	1.95	3321	1258	2.64	2398	4074	0.59	2619	4093	0.64
x8	2810	1658	1.70	3113	1591	1.96	3111	5571	0.56	3296	5573	0.59
x4-PPL2	5013	2314	**2.17**	5649	1818	**3.11**	4236	4677	0.91	4607	4680	0.98
x4-PPL5	6141	3942	1.56	7664	3209	2.39	6378	6189	**1.03**	6977	6179	**1.13**
x8-PPL10	10973	8831	1.24	11982	7323	1.64	10839	11204	0.97	10945	11203	0.98
SHA-512												
x1	2267	818	**2.77**	3081	592	**5.20**	2146	2072	**1.04**	2399	2073	**1.16**

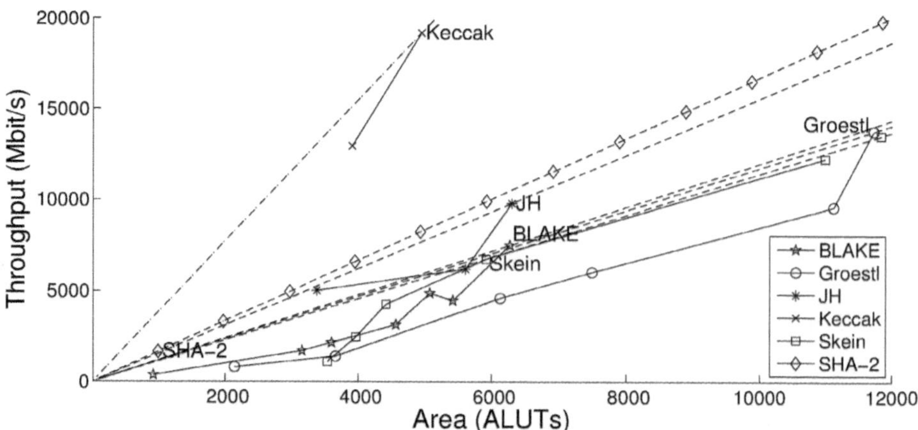

Fig. 7. Combined Throughput vs. Area graph for multiple hardware architectures of the 256-bit variants of BLAKE, Groestl, JH, Keccak, Skein, and SHA-2, implemented in Altera Stratix III FPGAs

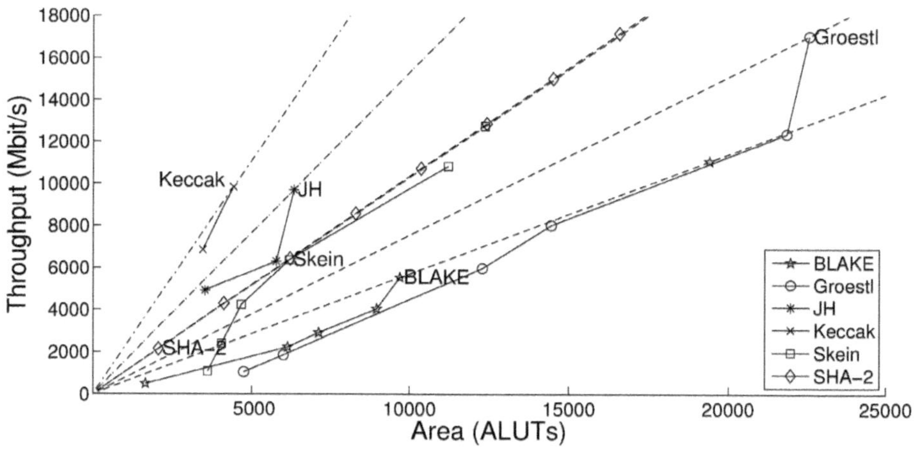

Fig. 8. Combined Throughput vs. Area graph for multiple hardware architectures of the 512-bit variants of BLAKE, Groestl, JH, Keccak, Skein, and SHA-2, implemented in Altera Stratix III FPGAs

decrease in the throughput to area ratio caused by adding a padding unit to the basic architecture of a SHA-3 candidate has not exceeded 16%. So small variations in this ratio are not likely to affect the overall ranking of the candidates.

7 Conclusions

In this paper, we have performed a systematic investigation of high-speed hardware architectures for the five final SHA-3 candidates. The investigated architectures were based on the concepts of the basic iterative architecture, horizontal folding, vertical folding, unrolling, pipelining, and parallel processing using multiple independent units. Each architecture was implemented using four high-performance FPGA families: Virtex 5 and Virtex 6 from Xilinx, and Stratix III and Stratix IV from Altera. Based on the obtained results, we have identified the most efficient hardware architecture for each of the investigated algorithm, based on the best throughput to area ratio.

In case of four out of five candidates (all except JH), the most efficient architecture appeared to be a pipelined architecture. The optimum number of pipeline stages was specific to the algorithm, and was equal to two for Keccak and Groestl, and four for BLAKE. The optimum pipelined architecture for Skein was the architecture with four rounds unrolled, and n pipeline stages, where the optimum value of n was equal to two for Xilinx high-performance FPGAs, and five for Altera high-performance FPGAs.

The results for all investigated functions, and the most successful architectures have been then summarized on the comprehensive throughput vs. area graphs. These graphs have revealed that Keccak is the only candidate that consistently outperforms SHA-2 for all considered FPGA families and two hash function variants (with 256-bit and 512-bit output). The only drawback of this function appears to be that it is not suitable for any kind of folding, and thus requires a quite substantial minimum area (in the range of 1400 CLB slices in Virtex 5) to be implemented in its basic iterative version.

JH performed better than SHA-2 in three out of four scenarios. It was outperformed by SHA-2 only for the 256-bit function variants implemented using Altera FPGAs. Interestingly, JH is most efficient in its basic iterative architecture, and is not suitable for either folding or inner-round pipelining.

Groestl was the only other candidate outperforming SHA-2 in at least one scenario, for the 256-bit variants implemented using Virtex 5. However this advantage was reached only for the relatively large area of about 3000 CLB slices. Although Groestl appeared to be very suitable for vertical folding, the very nature of this technique caused that the decrease in area was accompanied by the very significant decrease in speed.

Skein is the only finalist that can substantially benefit from unrolling. It is also the fastest for the pipelined versions of the 4x unrolled architecture, and is the only algorithm that can be pipelined up to 10 times. It performs particularly well compared to other algorithms for the 512-bit variants of hash functions implemented using Altera.

BLAKE is the algorithm with the highest flexibility, and the largest number of potential architectures. It can be easily folded horizontally and vertically by factors of two and four. It can also be easily pipelined even in the folded architectures. It is also the only algorithm that has a relatively efficient architecture that is smaller than the basic iterative architecture of SHA-2.

Our future work will include experimental testing of all developed high-speed architectures of the SHA-3 finalists, using high-performance FPGA boards based on Xilinx and Altera FPGAs, equipped with high-speed communication interface, such as PCI Express.

Acknowledgments. The authors would like to thank Ambarish Vyas for preliminary results regarding hash cores with padding units, and Rajesh Velegalati for extensive help with multiple ATHENa runs.

References

1. Cryptographic Hash Algorithm Competition, `http://csrc.nist.gov/groups/ST/hash/sha-3`
2. SHA-3 Hardware Implementations, `http://ehash.iaik.tugraz.at/wiki/SHA-3_Hardware_Implementations`
3. ATHENa Project Website, `http://cryptography.gmu.edu/athena`
4. Akin, A., Aysu, A., Ulusel, O.C., Savas, E.: Efficient Hardware Implementation of High Throughput SHA-3 Candidates Keccak, Luffa and Blue Midnight Wish for Single- and Multi-Message Hashing. In: 2nd SHA-3 Candidate Conf. (2010)
5. Baldwin, B., et al.: FPGA Implementations of the Round Two SHA-3 Candidates. In: 2nd SHA-3 Candidate Conf. (2010)
6. Beuchat, J.-L., Okamoto, E., Yamazaki, T.: A Compact FPGA Implementation of the SHA-3 Candidate ECHO. Cryptology ePrint Archive, Report 2010/364 (2010)
7. Detrey, J., Gaudry, P., Khalfallah, K.: A Low-Area yet Performant FPGA Implementation of Shabal. In: Biryukov, A., Gong, G., Stinson, D.R. (eds.) SAC 2010. LNCS, vol. 6544, pp. 99–113. Springer, Heidelberg (2011)
8. Gaj, K., Homsirikamol, E., Rogawski, M.: Fair and Comprehensive Methodology for Comparing Hardware Performance of Fourteen Round Two SHA-3 Candidates Using FPGAs. In: Mangard, S., Standaert, F.-X. (eds.) CHES 2010. LNCS, vol. 6225, pp. 264–278. Springer, Heidelberg (2010)
9. Gaj, K., Kaps, et al.: ATHENa – Automated Tool for Hardware EvaluatioN: Toward Fair and Comprehensive Benchmarking of Cryptographic Hardware using FPGAs. In: Proc. FPL 2010 (2010)
10. Guo, X., Huang, S., Nazhandali, L., Schaumont, P.: Fair and Comprehensive Performance Evaluation of 14 Second Round SHA-3 ASIC Implementations. In: 2nd SHA-3 Candidate Conf. (2010)
11. Henzen, L., Gendotti, P., Guillet, P., Pargaetzi, E., Zoller, M., Gürkaynak, F.K.: Developing a Hardware Evaluation Method for SHA-3 Candidates. In: Mangard, S., Standaert, F.-X. (eds.) CHES 2010. LNCS, vol. 6225, pp. 248–263. Springer, Heidelberg (2010)
12. Homsirikamol, E., Rogawski, M., Gaj, K.: Comparing Hardware Performance of Fourteen Round Two SHA-3 Candidates Using FPGAs. Cryptology ePrint Archive, Report 2010/445 (2010)
13. Matsuo, S., et al.: How Can We Conduct "Fair and Consistent" Hardware Evaluation for SHA-3 Candidate? In: 2nd SHA-3 Candidate Conf. (2010)
14. Nechvatal, J., et al.: Report on the Development of the Advanced Encryption Standard (AES), `http://csrc.nist.gov/archive/aes/round2/r2report.pdf`
15. Sklavos, N., Kitsos, P.: BLAKE HASH Function Family on FPGA: From the Fastest to the Smallest. In: Proc. ISVLSI 2010 (2010)
16. Tilich, S., et al.: High-speed Hardware Implementations of Blake, Blue Midnight Wish, Cubehash, ECHO, Fugue, Groestl, Hamsi, JH, Keccak, Luffa, Shabal, Shavite-3, SIMD, and Skein, Cryptology ePrint Archive, Report 2009/510 (2009)

Efficient Hashing Using the AES Instruction Set

Joppe W. Bos[1], Onur Özen[1], and Martijn Stam[2]

[1] Laboratory for Cryptologic Algorithms, EPFL, Station 14, CH-1015 Lausanne, Switzerland
{joppe.bos,onur.ozen}@epfl.ch
[2] Department of Computer Science, University of Bristol,
Merchant Venturers Building, Woodland Road, Bristol, BS8 1UB, United Kingdom
stam@cs.bris.ac.uk

Abstract. In this work, we provide a software benchmark for a large range of 256-bit blockcipher-based hash functions. We instantiate the underlying blockcipher with AES, which allows us to exploit the recent AES instruction set (AES-NI). Since AES itself only outputs 128 bits, we consider double-block-length constructions, as well as (single-block-length) constructions based on RIJNDAEL-256. Although we primarily target architectures supporting AES-NI, our framework has much broader applications by estimating the performance of these hash functions on any (micro-)architecture given AES-benchmark results. As far as we are aware, this is the first comprehensive performance comparison of multi-block-length hash functions in software.

1 Introduction

Historically, the most popular way of constructing a hash function is to iterate a compression function that itself is based on a blockcipher (this idea dates back to Rabin [49]). This approach has the practical advantage—especially on resource-constrained devices—that only a single primitive is needed to implement two functionalities (namely encrypting and hashing). Moreover, trust in the blockcipher can be conferred to the corresponding hash function. The wisdom of blockcipher-based hashing is still valid today. Indeed, the current cryptographic hash function standard SHA-2 and some of the SHA-3 candidates are, or can be regarded as, blockcipher-based designs. In the 1980s, several methods were proposed with an eye towards using the then-standard Data Encryption Standard (DES) as the underlying primitive [40,28,14]. At present, the contemporary Advanced Encryption Standard (AES [41]) is a more obvious choice instead.

A well-studied class of blockcipher-based hash functions are the PGV hash functions (after Preneel, Govaerts and Vandewalle [48]), encompassing Davies–Meyer (DM) and Matyas–Meyer–Oseas (MMO) as special cases. When based on a blockcipher operating on n-bit blocks with k-bit keys, these functions compress k bits per blockcipher call and they output an n-bit digest. The PGV hash functions are simple (low overhead) and are provably secure in the ideal-cipher model [10]. Yet they suffer from one major drawback: in order to achieve an acceptable level of collision resistance, one needs a primitive operating on more than 160 bits. This rules out most existing blockciphers, *including* AES (which operates on 128-bit blocks only).

As a remedy, *double-block-length* and more generally *multi-block-length* compression and hash functions were introduced. These are compression functions outputting

B. Preneel and T. Takagi (Eds.): CHES 2011, LNCS 6917, pp. 507–522, 2011.

an rn-bit digest (for an integer $r \geq 2$, $r = 2$ for the double-block-length case), even though they are based on a primitive operating only on n-bit blocks. The longer digest size opens up the possibility of collision resistance of 2^n time (primitive evaluations) even when using a relatively small primitive. Today, there is truly a wealth of suitable blockcipher-based constructions to choose from and Table 1 gives an overview of the constructions we consider. We do not consider *all* possibilities, for instance we omit versions of GRØSTL, JH or SPONGE based on RIJNDAEL-256. As can be seen, we instantiate the underlying blockcipher with either AES-128, AES-256 or RIJNDAEL-256. The latter option allows us to consider single-block-length constructions achieving a 256-bit digest (using an AES-related primitive).

Our choice of constructions includes several different design ideas and paradigms. For years, most cryptographic hash function designs revolved around the same principle [49,40,14]: the Merkle-Damgård paradigm. In this cascaded mode of operation, the main focus is to construct a secure and efficient compression function; these properties are then inherited by the overall hash function. Later constructions started to deviate from this paradigm, for instance by some form of strengthening [35,7] or by only targeting security in the iteration [10,6].

A more fundamental design shift occurred in the way the blockcipher itself is used. A blockcipher, operating on n bits with a k-bit key, can already be regarded as a compressing primitive itself. This facilitates the transformation into a proper compression function, but a disadvantage of using a blockcipher this way is that it requires frequent re-keying, which tends to be expensive (see Section 2 for details). For this reason, there have been substantial efforts in recent years to design permutation-based compression functions. Obviously, given a blockcipher one can construct a permutation by simply fixing the key (we focus on permutations with either $n = 128$ or 256 bits).

While the design and analysis of multi-block-length compression functions have garnered significant attention, the focus in the literature seems squarely at security evaluation and theoretical notions of efficiency (expressed as the ratio of message blocks compressed per blockcipher call). Although the latter is known to give only a coarse indication of real-life efficiency, actual performance benchmarks, in hard- or software, are normally left as future work. (A notable exception is the work by Bogdanov et al. [11], who provide hardware benchmarks for some multi-block-length compression functions in hardware using the lightweight blockcipher PRESENT as the underlying building block.)

Our Contribution. In this work we bring together the mainly theoretical world of compression function designs with the practical demand of fast implementations. Instantiating the blockcipher-based primitives with AES-128, AES-256, respectively RIJNDAEL-256 (and their fixed-key versions to build permutations), we obtain hash functions with a fixed 256-bit digest size. Apart from three constructions (LANE*, LUFFA* and KNUDSEN–PRENEEL) all constructions have known proofs of security in the ideal-cipher model (we refer to the full version of this work [12] for a more detailed discussion on the security of our target constructions). The former SHA-3 candidates LANE* and LUFFA* do not have security proofs, neither for collision resistance nor for preimage resistance. We include them in our benchmark (with different building blocks)

Table 1. A brief taxonomy of the schemes considered. The number of rounds N_r is 10 for AES-128 and 14 for AES-256 and RIJNDAEL-256.

Blockcipher (dimensions)	Variable-key Constructions	Fixed-key Constructions
AES-128 ($k = 128, n = 128$)	MDC-2, MJH, PEYRIN ET AL.(I)	LP362
AES-256 ($k = 256, n = 128$)	ABREAST-DM, HIROSE-DBL, KNUDSEN–PRENEEL, MJH-DOUBLE, QPB-DBL, PEYRIN ET AL.(II)	n.a.
RIJNDAEL-256 ($k = 256, n = 256$)	DAVIES–MEYER	LANE*, LUFFA*, LP231, SHRIMPTON–STAM

to illustrate their performance capabilities; KNUDSEN–PRENEEL is another exception where a good collision resistance lower bound is still an open problem.

To the best of our knowledge, this is the first overview of software implementations of the most studied and influential blockcipher- and permutation-based compression and hash functions. The target designs (see also Table 1) have been implemented and measured on an Intel Core i5 650 (3.20GHz) using C intrinsics to implement the various SS(S)E{2,3,4} and the recent AES instruction set (AES-NI) extensions [18,19]. Although measured on a single Intel architecture with AES-NI we expect the relative performance obtained to be representative for other Intel architecture families with AES-NI support as well. The Intel compiler version 12.0.0 and GNU Compiler Collection (gcc) version 4.4.3 were used for code compilation. For each design we performed specific optimizations to fully exploit AES-NI. The details are discussed in Section 3, with Table 3 providing a summary of our findings.

The Choice for AES. Our choice for AES (and RIJNDAEL-256) is a natural one: it is the official US and de facto world standard blockcipher. AES' prime position has led to a large body of research on AES, both on its security and implementation. Consequently, AES runs very fast in hard- and software, making AES an obvious choice from a performance perspective. The deal is sweetened further by the recent introduction of AES-NI. Indeed, as reported in [19], one can achieve significant speed using the new instruction set (e.g. up to 1.3 cycles/byte on a single core Intel Core i7-980X for AES-128 in parallel modes). To benefit from synergy with AES and AES-NI in particular, several SHA-3 candidates were instantiated by using some of AES' components as well (e.g. the AES round function), which was later demonstrated to indeed lead to fast hashing [2]. Our goal here is to investigate the potential of AES-NI for fast hashing even further by focusing on well-known blockcipher- and permutation-based (compression function) designs that can be instantiated with AES (or more generally RIJNDAEL).

From a security perspective, AES remains unbroken as a blockcipher in the standard setting. It has survived many years of cryptanalysis and a practical break of this cipher would have a significant impact on the cryptographic landscape. Nonetheless, our choice for AES will not be without detractors as a consequence of recent related-key attacks on AES [8,9]. The theoretical ramifications of a related-key attack to hash-function security are still unclear. Any serious related-key attack undermines the assumption that the blockcipher behaves 'ideally', but this need not lead to any deviant behaviour of the hash function itself (especially if its proof uses the weaker unforgeable-cipher model). Of course, in practice a related-key attack is often underpinned by some other (well-defined) weakness and exploiting this weakness directly (ignoring the derived related-key attack) might be more fruitful when attacking the hash function. For instance, Khovratovich [24, Corollary 2] states unambiguously that "AES-256 in the Davies–Meyer hashing mode leads to an insecure hash function" but later provides solace by remarking that it is not known how the techniques used against AES-256 in Davies–Meyer mode can be modified to attack double-block-length constructions (the focus of this paper).

As a final remark, the timings we obtain evidently depend strongly on the number of rounds used by AES. While one can argue that the number of rounds used should be fine-tuned for each of the hash functions (increasing or decreasing, depending on the perceived security margin), we believe that using AES as is will give the cleanest comparison (and any changes might be considered contentious).

2 Preliminaries

The Blockciphers AES-128, AES-256 and RIJNDAEL-256. AES is a member of the RIJNDAEL blockcipher suite [41]. It was standardized by the US National Institute of Standards and Technology (NIST) after a public competition similar to the one currently ongoing for SHA-3 [43]. AES operates on an internal state of 128 bits while supporting 128-, 192-, and 256-bit keys. The internal state is organized in a 4×4 array of 16 bytes, which is transformed by a round function N_r times. The number of rounds is $N_r = 10$ for the 128-bit key, $N_r = 12$ for the 192-bit key, and $N_r = 14$ for the 256-bit key variants. In order to encrypt, the internal state is initialized, then the first 128-bits of the key are XORed into the state, after which the state is modified $N_r - 1$ times according to the round function, followed by a slightly different final round (for the exact details see the AES specification [41]). The larger state variant of AES, RIJNDAEL-256, operates almost in the same way with a state size of 256 bits, a 256-bit key and $N_r = 14$ rounds.

Nine years after becoming the symmetric encryption standard, the only theoretical attack on the full AES is restricted to the related key scenario and even then applies only to the 192-bit [8] and 256-bit key versions [8,9]. So far no theoretical attacks on all rounds of AES-128 are known. More cryptanalytic success has been achieved by using the characteristics from the actual implementation of AES, e.g. cache attacks [60,3] can recover an AES key in only 65 milliseconds (Tromer et al. [60] give a more detailed survey of side-channel attacks against AES). However, side-channel analysis is far less of a concern for hash functions (except for MACs based on hash functions, such as HMAC) and we will blithely ignore the issue in this paper.

The AES Instruction Set (AES-NI). In the last decade, use of the *single instruction, multiple data* (SIMD) paradigm has become a general trend in computer architecture design. It enhances the speed of software implementations by offloading the computational work to special units which operate on larger data types, improving overall throughput. In 1999, Intel introduced the streaming SIMD extensions (SSE), a SIMD instruction set extension to the x86 architecture. One of the latest additions to these extensions is the AES instruction set [18,19] available in the 2010 Intel Core processor family based on the 32nm Intel micro-architecture named Westmere. This instruction set will also be supported by AMD in their next-generation CPU "Bulldozer". (Note that previously several instruction set extensions have been suggested towards improving the performance of AES [58,5,59].) AES-NI does not only increase the performance of AES (as well as any version of RIJNDAEL) but also runs in data-independent time and by avoiding the use of any table lookups the aforementioned cache attacks are avoided. This instruction set consists of six new instructions. At the same time, a new instruction for performing carry-less multiplication is released in the `CLMUL` instruction set extension. We can summarize the new instructions as follows [18,19,20]:

- `AESENC` and `AESDEC` perform a single round of encryption, resp. decryption.
- `AESENCLAST` and `AESDECLAST` perform the last round of encryption resp. decryption.
- `AESKEYGENASSIST` is used for generating the round keys used for encryption.
- `AESIMC` is used for converting the encryption round keys to a form usable for decryption using the Equivalent Inverse Cipher.
- `PCLMULQDQ` performs carry-less multiplication of two 64-bit operands to an 128-bit output.

Many of the constructions targeted in this paper require the computation of more than one call to a blockcipher (with or without a fixed-key). If these two or more calls can be run concurrently (while possibly sharing the key expansion), a performance gain can be expected as AES round instructions are pipelined and can be dispatched theoretically every 1-2 CPU clock cycles, provided that all data is available on time and there is no dependency between such subsequent calls [18]. Since the latency of a single round instruction is 5 cycles [17], running multiple independent blockciphers increases the overall throughput. The same reasoning holds when implementing a single RIJNDAEL-256 component. This sibling of AES works on an internal state of 256 bits and it is implemented using two data-independent calls to `AESENC`.

In the context of encryption, several performance results of AES exploiting AES-NI have been presented [19,20,37]. These works show that using AES-NI tends to give very fast implementations when multiple blockcipher calls can be made in parallel (incidentally, they also show that the optimal way to interleave the instructions is hard to pin down). However, they are of limited use to predict the runtimes of AES-based hash functions as re-keying tends to be far more frequent in the hashing scenario than in the encryption one. Indeed, for blockcipher-based compression functions considered in this paper, the key-scheduling needs to be performed for every compression function evaluation and that results in a significant overhead. For this reason, we start with a detailed performance overview of AES and RIJNDAEL-256 that takes re-keying into account. Table 2 contains performance details when running multiple key expansions, encryptions or a combination of the two. In order to conduct these experiments we created a code generator which, when given a number of x key expansions and y encryptions, tries different strategies to implement these functionalities. The performance numbers presented in Table 2 are an average over millions of runs. For comparison, we also included timings from Gueron's hand-crafted assembly code [19,20] as used in the Intel AES-NI sample library. (Note that, roughly speaking, our measure **1E** coincides

Table 2. Our experimental results on the encryption and key expansion routines for AES-128 (A128), AES-256 (A256) and RIJNDAEL-256 (R256). The entries show the results in cycles per operation together with the compiler, icc (i) or gcc (g), resulting in the fastest code. In the table **K** and **E** denote the key expansion and the encryption respectively. The upper part of the table shows the results of several independent key expansions and encryption operations that are called in parallel. In the lower part, **xKyE** denotes x independent key schedules followed by y independent encryptions. If $x = 1$ all encryptions use the same expanded key, if $x = y$ all encryptions use a different expanded key. For comparison, the performance details of the Intel AES-NI sample library on our platform are stated as well.

				Operation				
	1K	**2K**	**3K**	**4K**	**1E**	**2E**	**3E**	**4E**
A128	97.7 (g)	126.1 (g)	163.4 (g)	226.7 (i)	60.2 (i)	60.6 (i)	67.7 (i)	84.7 (i)
A256	125.5 (g)	147.2 (g)	202.6 (i)	287.2 (i)	82.0 (i)	83.0 (i)	93.6 (i)	113.9 (i)
R256	291.6 (g)	316.6 (g)	412.6 (g)	570.3 (i)	182.9 (i)	219.2 (g)	281.4 (i)	352.6 (g)
	1K1E	**2K2E**	**3K3E**	**4K4E**		**1K2E**	**1K3E**	**1K4E**
A128	107.4 (g)	149.2 (g)	200.0 (g)	269.9 (g)		120.1 (g)	135.3 (g)	137.8 (g)
A256	152.8 (g)	178.1 (g)	249.7 (g)	337.9 (g)		154.0 (g)	158.4 (g)	164.9 (g)
R256	285.3 (i)	407.5 (i)	620.5 (i)	867.3 (i)		312.0 (g)	373.3 (i)	463.7 (g)
				Intel AES-NI Sample Library				
	1K	**1E**	**4E**			**1K**	**1E**	**4E**
A128	98.8	62.1	79.6		A256	124.4	84.6	108.8

with AES run in a chaining mode such as CBC or CFB, whereas AES run in a parallel mode such as CTR or ECB is closer to the best time we get for **xE**, see Table 2 for the performance details).

Finite Field Arithmetic ($\mathbb{F}_{2^m}$ Full/Scalar Multiplication). Some of the compression function designs we consider require finite field multiplication, in particular in $\mathbb{F}_{2^{128}}$ and $\mathbb{F}_{2^{256}}$. There is some freedom in how to represent the fields—the security proofs for the hash functions are independent of this choice—so we opt for the usual representation of elements in $\mathbb{F}_{2^m}$ as polynomials over $\mathbb{F}_2$ reduced modulo an irreducible polynomial of degree m. We use $x^{128} + x^7 + x^2 + x + 1$ as irreducible polynomial for $m = 128$ and $x^{256} + x^{10} + x^5 + x^2 + 1$ for $m = 256$.

Multiplication in $\mathbb{F}_{2^{128}}$ is implemented using the code examples as described in [20] in the setting of implementing the Galois counter mode. This is realized by using the new instruction `PCLMULQDQ` to implement the multiplication; this instruction calculates the carry-less product of the two 64-bit input to an 128-bit output. Note that this instruction has a latency of 12 cycles and can be dispatched every 8 cycles [17]. Hence, compared to other SSE instructions, some of which can be dispatched in pairs of three every clock cycle, this instruction might not always be the optimal choice from a performance perspective. An example where the usage of the `PCLMULQDQ` instruction might not lead to a speed-up is in the case of (field) multiplication by x. This can be computed by shifting the input one position to the left (the polynomial multiplication by x) and performing a conditional `XOR` with the reduction polynomial (depending on the bit shifted out). Unfortunately, the SSE instruction set has no bit shift operation shifting the full 128-bit vector. Shifting the two 64-bit, four 32-bit or eight 16-bit in SIMD fashion is possible but the bits shifted out locally are lost. We outline a novel approach (with the SSE instruction in parentheses) to obtain the desired result in the setting of $\mathbb{F}_{2^{128}}$ where we exploit the fact that the second largest exponent of the reduction polynomial is < 32 (which also holds in the setting of $\mathbb{F}_{2^{256}}$). Given an input A we

1. swap the two 64 bit halves of A to t (PSHUFD),
2. create a mask m (either all ones or zeros in each 64-bit half) depending if bits 63 and 127 of t are set (PCMPGTQ),
3. use m to extract the correct 64-bit parts of a precomputed constant $[1, R]$ in t (PAND),
4. shift both 64-bit parts of A left by one bit and store this in s (PSLLQ),
5. perform the reduction plus restoring the local carry bit by combining s and t (PXOR).

Here R denotes the hexadecimal representation of the reduction polynomial, excluding the term with the highest exponent, stored in a 64-bit word. Note that this computation might be sped up, depending on the setting, in the following way. Replace step 1 by a byte shuffle (PSHUFD) which moves bits 63 and 127 to bit position 95 and 31 respectively and set the other 14 bytes to zero. The resulting vector, viewed as four 32-bit signed integers, contains two 32-bit words where only the sign bit may be set. Now step 2 can be replaced by using an arithmetic right shift of 31 positions (PSRAD) creating the mask by using the fact that this instruction shifts in the sign bit. In order to overcome this instruction set limitation (no 128-bit single-bit shift instruction) we tried if field multiplication by x^8 is faster. Now the input needs to be shifted eight bits, which can be performed using a single byte-shuffle instruction. The reduction, a subtraction by $i \cdot R$, where $0 \leq i < 2^8$, depends on the eight bits shifted out. Since the reduction polynomial is constant we can precompute the 256 multiples and use the shifted-out byte as in index for this look-up table. We found that, using our implementation of both approaches, the performance of both field multiplications, by x and x^8, are comparable with a slight advantage when multiplying by x.

3 Implementations of the Target Algorithms

Table 3 contains an overview of the benchmarks we obtained. The measurements have been carried out analogously to [19]; i.e. with the help of the time stamp counter which is read using the RDTSC instruction. The presented performance results are an average over thousands of times compressing a random 4KB message. In the sequel, we provide separate treatments for constructions based on a (variable-key) blockcipher versus a permutation (in which case we fix the key of the blockcipher). Due to space limitations, we refer to the original works for exact specifications of the various algorithms (references, including those relevant for security results, are given in Table 3; see also the full version [12] for a more detailed analysis and illustrated specifications).

Two of the designs considered are based on past SHA-3 candidates. For those, we instantiate the underlying permutation by (fixed-key) RIJNDAEL-256, rather than the originally submitted permutation. For compression functions supporting more than 256-bit output (e.g. KNUDSEN–PRENEEL and LUFFA*) an output transformation (after MD-iteration) can be used to reduce the final output to 256-bit, however we neither implemented nor timed this.

Table 3. The achieved speeds (in cycles per byte) using the AES-NI for the designs considered in this work. Also mentioned are the number of b bytes which are absorbed per compression function call and how many unique key scheduling calls are made (see Table 1 for the primitives employed). Predicted speed estimates are based on the results from Table 2. The last column provides additional references.

Algorithm	b	Key Scheduling	Predicted Speed Range	Achieved Speed	Security Reference
ABREAST-DM [28]	16	two	$11.1 + \epsilon$	11.21	[16,29,33]
DM [39]	32	one	$[6.8, 10.2]$	8.69	[48,10]
HIROSE-DBL [21]	16	one, shared	9.6	9.82	[21,27]
KNUDSEN–PRENEEL [26]	32	four	10.6	10.58	[44,46]
LANE* (Sec. 3)	64	fixed	11.7	11.71	[22]
LP231 [51,52]	32	fixed	$12.6 + \epsilon$	13.04	[51,52,30]
LP362 [51,52]	16	fixed	$11.8 + \epsilon$	12.09	[51,52,31]
LUFFA* (Sec. 3)	32	fixed	$8.8 + \epsilon$	10.22	[15]
MDC-2 [13]	16	two	$[9.3, 11.7] + \epsilon$	10.00	[57,25]
MJH [32]	16	one, shared	$6.6 + \epsilon$	7.45	[32]
MJH-DOUBLE [32]	32	one, shared	$4.1 + \epsilon$	4.82	[32]
QPB-DBL [55]	16	one	$9.5 + \epsilon$	14.12	[55]
PEYRIN ET AL.(I) [47]	16	three, shared	$[12.5, 16.3]$	15.09	[53]
PEYRIN ET AL.(II) [47]	32	three, shared	$[7.8, 10.7]$	8.75	[53]
SHRIMPTON–STAM [54]	32	fixed	12.6	12.39	[54]

3.1 Blockcipher-Based Constructions

Davies–Meyer (DM). Davies–Meyer (DM) [39] is a single-block-length compression function design. It is one of the most popular ways of creating a secure hash function using a blockcipher: many cryptographic hash functions, including MD5 [50] (for $n = 128$, $k = 512$) and SHA-256 [42] (for $n = 256$, $k = 512$), follow the DM design philosophy. DM is one of the most efficient PGV-type compression functions as it allows to run several key schedules independently in the MD-iteration. In our implementations, we exploit this feature; yet we also study other possible optimizations. Namely, these are the three flavors of DM that we have considered in our benchmark:

1. Standard iterative approach: compression function calls are made sequentially for each step in the MD-iteration. The compression function evaluation starts with the key schedule and continues with the encryption call. Independent key schedule and encryption rounds are interleaved to get more efficient results.
2. Partially pipelined: the encryption call of the current round and the key schedule of the next round are being processed concurrently.
3. Fully pipelined: j key schedules are called in parallel for some (integer) $j > 1$ followed by j iterative encryption calls . Several experiments were run for varying j and the best result is obtained for $j = 4$. Note that this approach allows to interleave the first encryption round calls with the key scheduling stage to hide latencies and obtain faster results.

Among the three approaches the fully pipelined version gives the best result and is the one reported in Table 3. We included a prediction of the performance of DM based on the vanilla timings of RIJNDAEL-256 provided in Section 2. Here the timing for **4K4E** serves as a lower bound, as it makes the encryption calls in parallel. The timing for **4K** plus four times **1E** serves as an upper bound for DM because the first encryption can be scheduled during the four key scheduling stages hiding the instruction dependencies in the encryption improving the overall throughput. (Similar strategies are used for the constructions discussed subsequently. If the predictions in Table 3 include an ϵ, this indicates that certain computations, for instance finite field multiplications, are not considered in the prediction.)

ABREAST-DM. ABREAST-DM and its sister design TANDEM-DM, both proposed in the early 90s [28], are two of the classical examples of double-block-length compression/hash function designs. We only consider ABREAST-DM instantiated with AES-256 for our benchmark. We expect that TANDEM-DM has a slightly worse performance compared to ABREAST-DM due to its sequential structure. In our implementations, we make extensive use of the parallelism inside the ABREAST-DM compression function by calling two key schedules in parallel followed by two concurrent encryption calls (where the 'follow' is on a fine-grained per AES-round basis). Hence, the prediction for ABREAST-DM is based on the performance numbers for AES-256 in the **2K2E** setting (see also [12] for the discussion on another alternative yet slower method to implement ABREAST-DM).

HIROSE-DBL. ABREAST-DM suffers from a performance drawback that, although run in parallel, the underlying blockciphers require separate key schedule routines. Hirose's construction [21] overcomes this problem by sharing the key scheduling for the two blockcipher calls. In our implementations, we apply the same approach as for ABREAST-DM to HIROSE-DBL and our results are in accordance with the predicted speed based on the **1K2E** setting for AES-256. Our timings also demonstrate that Hirose's scheme is indeed faster than ABREAST-DM.

MDC-2. MDC-2 [13] is one of the oldest double-block-length hash functions available and it has been specified in the ANSI X9.31 and ISO/IEC 10118-2 standards [1,23]. Although originally designed for use with DES, we consider the obvious generalization where one can use two calls to a single-key blockcipher (where $k = n$ with AES-128). Since MDC-2 is based on MMO, it is difficult to pipeline multiple MDC-2 compression function calls in the MD-iteration (as we did for DM). Yet, one can benefit from the parallelism naturally present within a single compression function evaluation by making the two blockcipher calls concurrently (corresponding to **2K2E**). This is indeed how we have achieved our best result, matching the predicted speed.

MJH. Recently, an alternative construction called MJH was proposed by Lee and Stam [32]. It is inspired by the compression function of JH [61] (one of the SHA-3 finalists). The main design rationale behind MJH is to reduce the number of key-schedules required in a single compression function evaluation—as in HIROSE-DBL—and call several key schedules in parallel in multiple iterations—as in DM. Obviously, this results in an efficient design. More interestingly, the security of the construction still holds once the message block (size) to the compression function is doubled (this is what we call MJH-DOUBLE). This leads to a significantly more efficient scheme, although the cost of key set-up increases. We investigate the performance of MJH in accordance with our optimizations on DM and HIROSE-DBL. Based on our results, we note that MJH-DOUBLE has achieved the best cycle count in our benchmark. We implemented different strategies when interleaving $1 \leq i \leq 8$ iterations of the compression function, the best results are obtained with $i = 2$. Hence, the predictions are based on the setting **2K2E+2E**, ignoring the cost of the finite field (scalar) multiplications.

KNUDSEN–PRENEEL. One of the classical examples of multi-block-length compression functions is provided by Knudsen and Preneel [26] who proposed several constructions with multiple blockcipher calls in parallel using the generator matrices of various linear error correcting codes. We consider one of their proposals, which is based on a $[4, 2, 3]$ linear code over $\mathbb{F}_{2^3}$, to show its performance capabilities with AES-NI. When based on AES-256, this gives rise to a $6n \rightarrow 4n$ bit compression function with security expected to be at least that of a $2n$-bit compression function. We base our exact specification on the later analysis by Özen et al. [44]. One of the nice features of this construction is that one can call four independent key schedules followed by four independent encryptions where one can interleave the rounds of both operations to hide latencies. This makes it much easier to give an accurate performance estimate since this scenario is exactly the **4K4E** case for AES-256.

PEYRIN ET AL.-DBL. All the designs considered so far follow a very similar approach: there exist linear pre- and post-processing functions that operate on the blocks of data, interacting with the underlying primitives. Based on this general model, Peyrin et al. [47] determined, under a very general attack-based approach (i.e. only considering time-complexity upper bounds), necessary conditions to have a secure compression function (where they used smaller ideal $2n \rightarrow n$ and $3n \rightarrow n$ bits compression functions as underlying primitives which are replaced by single-key, resp. double-key, blockciphers in DM mode in our framework). To investigate the performance using AES-NI, we consider their two concrete proposals: one uses five AES-128 calls and leads to a $3n \rightarrow 2n$ bit compression function, the other uses five AES-256 calls for a $4n \rightarrow 2n$ bit compression function. In our implementations, we make use of the high parallelism inside a single compression function evaluation by calling several shared key-schedules. In both scenarios the predicted time corresponds to **3K5E**, since among the five encryptions two keys are used twice. This case is not considered in Table 2 and we estimate the performance by considering the performance interval [**3K3E**, **3K3E+2E**] for AES-128 and AES-256 instead.

QPB-DBL. We finish this section with the interesting scenario of constructing a $2n$-bit digest while making only a single call to the blockcipher (theoretically, this would provide optimal efficiency). Lucks [36] provided the first construction of this type, although it is secure only in the iteration (see [45] for a detailed discussion of the security of Lucks' construction). The main practical overhead in Lucks' construction are the costly finite field multiplications that are bound to be performed sequentially. Later, Stam [55] gave another, more practical, construction in the public random function model using a quadratic-polynomial based design (hence the name QPB-DBL). This construction was generalized [56,34] to the ideal cipher model by replacing the random function with a double-length-key blockcipher running in DM mode. For our benchmarks, we use a slightly modified compression function, in that we shuffle the inputs

$$A' = \begin{pmatrix} 1\,0\,0\,0\,0 \\ 0\,1\,0\,0\,0 \\ 1\,2\,1\,1\,0 \\ 1\,1\,2\,4\,2 \end{pmatrix}, \quad \tilde{A} = \begin{pmatrix} 1\,0\,0\,0\,0 \\ 0\,1\,0\,0\,0 \\ 1\,1\,1\,1\,0 \\ 1\,0\,1\,0\,1 \end{pmatrix} \quad \text{and } A'' = \begin{pmatrix} 1\,0\,0\,0\,0\,0\,0\,0\,0 \\ 0\,1\,0\,0\,0\,0\,0\,0\,0 \\ 0\,0\,1\,0\,0\,0\,0\,0\,0 \\ 1\,0\,0\,1\,0\,0\,0\,0\,0 \\ 0\,1\,0\,0\,1\,0\,0\,0\,0 \\ 0\,0\,1\,0\,0\,1\,1\,0\,0 \\ 1\,1\,1\,1\,1\,1\,1\,0\,1 \\ 1\,2\,4\,1\,2\,4\,0\,1\,0 \end{pmatrix}$$

Fig. 1. The matrices used for LP231, LP362 and SS. The field elements denoted 1, 2 and 4 correspond to polynomials $1, x$ and x^2, respectively (see Section 2).

slightly. This allows us to benefit from increased parallelism in the iteration, without violating the security proof. As already argued, in the QPB-DBL compression function the main overhead consists of costly finite field multiplications (which we try to minimize by using the features of the new PCLMULQDQ instruction). Our tweak allows us to interleave the key-scheduling of round $i+1$ with the two (sequential) finite field multiplications of round i. The predicted performance of QPB-DBL is based on the **1K1E** setting for AES-256 and ignores the relatively high cost of the two (full) finite field multiplications.

3.2 Permutation-Based Constructions

Rogaway and Steinberger's LP and SHRIMPTON–STAM. Rogaway and Steinberger introduced a class of linearly-determined, permutation-based compression functions $\{0,1\}^{mn} \to \{0,1\}^{rn}$ making k calls to the different permutations π_i for $i \in \{1, \ldots, k\}$ (hence the notation LPmkr throughout). Let (x_i, y_i) denote the input-output pair corresponding to the permutation π_i. The main ingredient of Rogaway and Steinberger's LP design is a $(k+r) \times (k+m)$ matrix A over $\mathbb{F}_{2^n}$. This matrix determines the block-wise interaction between the inputs to the compression function (V, M), (x_i, y_i) pairs and the output Z of the compression function in the following way: for the row vector a_i (of A), the inputs to the underlying permutations are determined by the scalar product $x_i = a_i \cdot (V_1, \ldots, V_r, M_1, \ldots, M_{m-r}, y_1, \ldots, y_{i-1}, 0^{k-i})$ whereas the output Z (which is treated as a concatenation of r n-bit blocks Z_i) is computed by $Z_i = a_{k+i} \cdot (V_1, \ldots, V_r, M_1, \ldots, M_{m-r}, y_1, \ldots, y_k)$. There is considerable choice for the matrices A, as long as a certain independence criterion is satisfied [51,52]). For optimal performance, we use the matrices A' over $\mathbb{F}_{2^{256}}$ (suggested in [30]), and A'' over $\mathbb{F}_{2^{128}}$ (taken from [31]) as given in Fig. 1 for LP231, respectively LP362.

In independent work, Shrimpton and Stam [54] proved security for a compression function (SS) that can be regarded as an LP231 scheme (based on matrix $\tilde{A}$), even though their matrix does not satisfy the independence criterion imposed by Rogaway and Steinberger.

There are multiple ways how one can implement these constructions in practice. We chose to implement LP231 in three stages where we first run two permutations and a field multiplication by x in parallel, followed by one permutation and field multiplication by x and x^2 and finally the remaining multiplication by x^2. This corresponds to the setting **2E+1E** $+ \epsilon$ for RIJNDAEL-256 on which we base our performance prediction. After experimenting with different strategies we settled on the following regarding LP362. Again three stages are used where we do three, two and one permutation in parallel in every stage. The multiplications are calculated in the last two stages in order to hide the relatively high latencies of especially the single permutation. Hence, the predicted performance is based on the **3E+2E+1E**$+\epsilon$ setting for AES-128. The implementation of SS is straightforward corresponding to the setting **2E+1E** for RIJNDAEL-256, note that this is the only case where the actual construction (slightly) outperforms the predicted speed. This anomaly might be explained by the fact that SS has to load (store) the input (output) only once for both operations while in the performance benchmark setting this has to be done twice.

LANE*. LANE [22] is a permutation-based hash function design submitted to the SHA-3 competition by Indesteege (supported by the COSIC research group). For our purposes, we consider the LANE compression function with 256-bit digest size which is instantiated by eight calls to the fixed-key RIJNDAEL-256 and denoted by LANE*. Although some weaknesses have been exploited [38] for the original proposal, it is not immediate that the attacks carry over to LANE* as the attacks exploit weaknesses in the original permutations (in particular the relatively low number of rounds). In our implementations, we exploit the high parallelism inside a single compression function evaluation by running several permutation calls in parallel. Although possible, we did not investigate further pipelining options along the MD-iteration due to sufficient number of independent permutation calls in a single compression function evaluation. The predicted speed for LANE* is based on the setting of **6E+2E** for RIJNDAEL-256. Note that the original version of LANE, performs significantly faster (4.3 cycles per byte) on our platform due to the relatively light permutations given in the submitted version.

LUFFA*. LUFFA [15] is a second round permutation-based SHA-3 candidate designed by De Cannière and Watanabe which can possibly benefit from the AES-NI once the underlying permutations are modified accordingly. To this end, we instantiate the three underlying permutations of LUFFA-256 with fixed-key RIJNDAEL-256 and denote this version by LUFFA*. In the implementation of LUFFA*, we follow a standard approach: first the multiplications required in the message injection step are computed (see [15] for a description of how to implement these efficiently), followed by the computation of the three independent permutations. The predicted performance results (**3E**$+\epsilon$ using RIJNDAEL-256) is too optimistic, the ϵ incorporates the cost of the multiple polynomial multiplications. Note that our implementation is slightly faster then the original version of LUFFA (which runs at 10.49 cycles per byte) using the fastest implementation (called `SSSE3-PS-2`) submitted to eBASH [4].

4 Discussion and Conclusion

In this work, we presented the first comprehensive performance comparison of many multi-block-length hash functions (old and new alike) in software on a modern architecture supporting AES-NI. Our results are summarized in Table 3 in conjunction with speed predictions based on the vanilla AES timings from Table 2. Based on these results, we can draw the following conclusions:

1. Our major conclusion is that, when assuming that the underlying primitives behave ideally, one can obtain fast and provably secure blockcipher-based hash functions on soon to be mainstream architectures supporting AES-NI. Indeed, the algorithms studied provide reasonable collision and preimage resistance and require between 4 and 15 cycles per byte on our target platform, so in this sense almost *all* of them outperform SHA-256 while several of them are faster than SHA-512.[1] As discussed in the introductions, our results are obtained with the original number of rounds for AES and RIJNDAEL-256. Relative performance results follow by increasing or decreasing the number of rounds, depending on the security margin.
2. Among the blockcipher-based compression functions, DM is the fastest algorithm when optimal security (in terms of proven collision resistance lower bound) is desired. For practical security levels, MJH-DOUBLE significantly outperforms the others (including the permutation-based designs). Note that both constructions require only one key schedule call inside a single compression function evaluation.
3. In the permutation-based setting, the LUFFA* compression function is the fastest, but it is being outperformed by many blockcipher-based constructions. This is partly due to the higher number of primitive calls, but one can argue that our methodology (use AES as is) results in a relatively more conservative security margin for fixed-key constructions. Among the provably secure constructions LP362 performs the best, showing the possibility of achieving higher speed despite the increased number of primitive calls.

Finally, we remark that all the constructions we consider are generic in the sense that they can be instantiated with any secure blockcipher (or permutation, where relevant). Hence, it is well possible that one can achieve better performance with different blockciphers or permutations. In particular, any AES-inspired yet more efficient primitive, for instance a round-reduced version or a tweaked version with more secure and efficient key-scheduling, would result in a faster scheme on our target platform. We believe that our benchmark provides a valuable toolbox to see the relative performance figures for a majority of blockcipher- and permutation-based compression and hash functions.

Acknowledgements. This work was supported by the Swiss National Science Foundation under grant numbers 200020-132160, 200021-119776, and 200021-122162 and by the European Commission through the ICT programme under contract ICT-2007-216676 ECRYPT II. We gratefully acknowledge Çağdaş Çalık and Institute of Applied Mathematics at Middle East Technical University for granting us access to the Intel i5 with AES-NI to benchmark our programs and Thorsten Kleinjung for useful discussions on how to optimize the SSE field multiplication by x. We would like to thank the anonymous reviewers for their useful comments and suggestions.

[1] Compared SHA-256 and SHA-512 speeds (13.90 and 10.47 respectively) are based on the fastest publicly available implementation on eBACS [4] run on Intel Core i5 M 520 (2.4 GHz with AES-NI).

References

1. American National Standards Institute: Public key cryptography using reversible algorithms for the financial services industry. American National Standards Institute (1998)
2. Benadjila, R., Billet, O., Gueron, S., Robshaw, M.J.B.: The Intel AES instructions set and the SHA-3 candidates. In: Matsui, M. (ed.) ASIACRYPT 2009. LNCS, vol. 5912, pp. 162–178. Springer, Heidelberg (2009)
3. Bernstein, D.J.: Cache-timing attacks on AES (2005), http://cr.yp.to/papers.html#cachetiming
4. Bernstein, D.J., Lange, T. (eds.): eBACS: ECRYPT Benchmarking of Cryptographic Systems (2010), http://bench.cr.yp.to
5. Bertoni, G., Breveglieri, L., Farina, R., Regazzoni, F.: Speeding up AES by extending a 32 bit processor instruction set. In: Application-specific Systems, Architectures and Processors, pp. 275–282. IEEE Computer Society, Los Alamitos (2006)
6. Bertoni, G., Daemen, J., Peeters, M., Van Assche, G.: On the indifferentiability of the sponge construction. In: Smart, N.P. (ed.) EUROCRYPT 2008. LNCS, vol. 4965, pp. 181–197. Springer, Heidelberg (2008)
7. Biham, E., Dunkelman, O.: A framework for iterative hash functions – HAIFA. Presented at Second NIST Cryptographic Hash Workshop, Santa Barbara, USA (2006)
8. Biryukov, A., Khovratovich, D.: Related-key cryptanalysis of the full AES-192 and AES-256. In: Matsui, M. (ed.) ASIACRYPT 2009. LNCS, vol. 5912, pp. 1–18. Springer, Heidelberg (2009)
9. Biryukov, A., Khovratovich, D., Nikolić, I.: Distinguisher and related-key attack on the full AES-256. In: Halevi, S. (ed.) CRYPTO 2009. LNCS, vol. 5677, pp. 231–249. Springer, Heidelberg (2009)
10. Black, J., Rogaway, P., Shrimpton, T., Stam, M.: An analysis of the blockcipher-based hash functions from PGV. Journal of Cryptology 23(4), 519–545 (2010)
11. Bogdanov, A., Leander, G., Paar, C., Poschmann, A., Robshaw, M.J.B., Seurin, Y.: Hash functions and RFID tags: Mind the gap. In: Oswald, E., Rohatgi, P. (eds.) CHES 2008. LNCS, vol. 5154, pp. 283–299. Springer, Heidelberg (2008)
12. Bos, J.W., Özen, O., Stam, M.: Efficient hashing using the AES instruction set. Cryptology ePrint Archive, Report 2010/576 (2010)
13. Brachtl, B., Coppersmith, D., Hyden, M., Matyas Jr., S., Meyer, C., Oseas, J., Pilpel, S., Schilling, M.: Data authentication using modification detection codes based on a public one-way encryption function. U.S. Patent No 4,908,861 (1990)
14. Damgård, I.: A design principle for hash functions. In: Brassard, G. (ed.) CRYPTO 1989. LNCS, vol. 435, pp. 416–427. Springer, Heidelberg (1990)
15. De Cannière, C., Sato, H., Watanabe, D.: Hash function Luffa: Supporting document. Submission to NIST (Round 2) (2009), http://www.sdl.hitachi.co.jp/crypto/luffa/Luffa_v2_SupportingDocument_20090915.pdf
16. Fleischmann, E., Gorski, M., Lucks, S.: Security of cyclic double block length hash functions. In: Parker, M.G. (ed.) Cryptography and Coding 2009. LNCS, vol. 5921, pp. 153–175. Springer, Heidelberg (2009)
17. Fog, A.: Instruction tables, lists of instruction latencies, throughputs and microoperation breakdowns for Intel, AMD and VIA CPUs (2010), http://www.agner.org/optimize/
18. Gueron, S.: Intel's new AES instructions for enhanced performance and security. In: Dunkelman, O. (ed.) FSE 2009. LNCS, vol. 5665, pp. 51–66. Springer, Heidelberg (2009)
19. Gueron, S.: Intel advanced encryption standard (AES) instructions set. Tech. rep., Intel (2010), http://software.intel.com/file/24917

20. Gueron, S., Kounavis, M.E.: Intel carry-less multiplication instruction and its usage for computing the GCM mode. Tech. rep., Intel (2010), http://software.intel.com/file/24918
21. Hirose, S.: Some plausible constructions of double-block-length hash functions. In: Robshaw, M. (ed.) FSE 2006. LNCS, vol. 4047, pp. 210–225. Springer, Heidelberg (2006)
22. Indesteege, S.: The LANE hash function. Submission to NIST (2008), http://www.cosic.esat.kuleuven.be/publications/article-1181.pdf
23. International Organization for Standardization: ISO/IEC 10118-2: hash functions using an n-bit block cipher (2010)
24. Khovratovich, D.: New Approaches to the Cryptanalysis of Symmetric Primitives. Ph.D. thesis, University of Luxembourg (2010)
25. Knudsen, L.R., Mendel, F., Rechberger, C., Thomsen, S.S.: Cryptanalysis of MDC-2. In: Joux, A. (ed.) EUROCRYPT 2009. LNCS, vol. 5479, pp. 106–120. Springer, Heidelberg (2009)
26. Knudsen, L.R., Preneel, B.: Construction of secure and fast hash functions using nonbinary error-correcting codes. IEEE Transactions on Information Theory 48(9), 2524–2539 (2002)
27. Krause, M., Armknecht, F., Fleischmann, E.: Preimage resistance beyond the birthday barrier – the case of blockcipher based hashing. Cryptology ePrint Archive, Report 2010/519 (2010)
28. Lai, X., Massey, J.L.: Hash functions based on block ciphers. In: Rueppel, R. (ed.) EUROCRYPT 1992. LNCS, vol. 658, pp. 55–70. Springer, Heidelberg (1993)
29. Lee, J., Kwon, D.: The security of Abreast-DM in the ideal cipher model. Cryptology ePrint Archive, Report 2009/225 (2009)
30. Lee, J., Park, J.H.: Adaptive preimage resistance and permutation-based hash functions. Cryptology ePrint Archive, Report 2009/066 (2009)
31. Lee, J., Park, J.H.: Preimage resistance of LPmkr with $r = m - 1$. Information Processing Letters 110(14-15), 602–608 (2010)
32. Lee, J., Stam, M.: MJH: A faster alternative to MDC-2. In: Kiayias, A. (ed.) CT-RSA 2011. LNCS, vol. 6558, pp. 213–236. Springer, Heidelberg (2011)
33. Lee, J., Stam, M., Steinberger, J.: The collision security of Tandem-DM in the ideal cipher model. In: Rogaway, P. (ed.) CRYPTO 2011. LNCS, vol. 6841, pp. 561–568. Springer, Heidelberg (2011)
34. Lee, J., Steinberger, J.P.: Multi-property-preserving domain extension using polynomial-based modes of operation. In: Gilbert, H. (ed.) EUROCRYPT 2010. LNCS, vol. 6110, pp. 573–596. Springer, Heidelberg (2010)
35. Lucks, S.: A failure-friendly design principle for hash functions. In: Roy, B. (ed.) ASIACRYPT 2005. LNCS, vol. 3788, pp. 474–494. Springer, Heidelberg (2005)
36. Lucks, S.: A collision-resistant rate-1 double-block-length hash function. In: Symmetric Cryptography. No. 07021 in Dagstuhl Seminar Proceedings, Internationales Begegnungs- und Forschungszentrum für Informatik (IBFI), Schloss Dagstuhl, Germany (2007)
37. Manley, R., Magrath, P., Gregg, D.: Code generation for hardware accelerated AES. In: 21st IEEE International Conference on Application-specific Systems Architectures and Processors (ASAP), pp. 345–348 (2010)
38. Matusiewicz, K., Naya-Plasencia, M., Nikolic, I., Sasaki, Y., Schläffer, M.: Rebound attack on the full LANE compression function. In: Matsui, M. (ed.) ASIACRYPT 2009. LNCS, vol. 5912, pp. 106–125. Springer, Heidelberg (2009)
39. Menezes, A.J., van Oorschot, P.C., Vanstone, S.: CRC-Handbook of Applied Cryptography. CRC Press, Boca Raton (1996)
40. Merkle, R.C.: One way hash functions and DES. In: Brassard, G. (ed.) CRYPTO 1989. LNCS, vol. 435, pp. 428–446. Springer, Heidelberg (1990)
41. NIST: FIPS-197: Advanced encryption standard (AES) (2001), http://www.csrc.nist.gov/publications/fips/fips197/fips-197.pdf

42. NIST: Secure hash standard. FIPS 180-2, NIST (August 2002), http://www.itl.nist.gov/fipspubs/fip180-2.htm
43. NIST: Cryptographic hash algorithm competition (2008), http://csrc.nist.gov/groups/ST/hash/sha-3/index.html
44. Özen, O., Shrimpton, T., Stam, M.: Attacking the Knudsen-Preneel compression functions. In: Hong, S., Iwata, T. (eds.) FSE 2010. LNCS, vol. 6147, pp. 94–115. Springer, Heidelberg (2010)
45. Özen, O., Stam, M.: Another glance at double-length hashing. In: Parker, M. (ed.) Cryptography and Coding 2009. LNCS, vol. 5921, pp. 176–201. Springer, Heidelberg (2009)
46. Özen, O., Stam, M.: Collision attacks against the Knudsen-Preneel compression functions. In: Abe, M. (ed.) ASIACRYPT 2010. LNCS, vol. 6477, pp. 76–93. Springer, Heidelberg (2010)
47. Peyrin, T., Gilbert, H., Muller, F., Robshaw, M.J.B.: Combining compression functions and block cipher-based hash functions. In: Lai, X., Chen, K. (eds.) ASIACRYPT 2006. LNCS, vol. 4284, pp. 315–331. Springer, Heidelberg (2006)
48. Preneel, B., Govaerts, R., Vandewalle, J.: Hash functions based on block ciphers: A synthetic approach. In: Stinson, D. (ed.) CRYPTO 1993. LNCS, vol. 773, pp. 368–378. Springer, Heidelberg (1994)
49. Rabin, M.O.: Digitalized signatures. In: Foundations of Secure Computations, pp. 155–166. Academic Press, London (1978)
50. Rivest, R.: The MD5 message-digest algorithm, request for comments (RFC) 1320. Tech. rep., Internet Activities Board, Internet Privacy Task Force (1992)
51. Rogaway, P., Steinberger, J.: Security/efficiency tradeoffs for permutation-based hashing. In: Smart, N. (ed.) EUROCRYPT 2008. LNCS, vol. 4965, pp. 220–236. Springer, Heidelberg (2008)
52. Rogaway, P., Steinberger, J.P.: Constructing cryptographic hash functions from fixed-key blockciphers. In: Wagner, D. (ed.) CRYPTO 2008. LNCS, vol. 5157, pp. 433–450. Springer, Heidelberg (2008)
53. Seurin, Y., Peyrin, T.: Security analysis of constructions combining FIL random oracles. In: Biryukov, A. (ed.) FSE 2007. LNCS, vol. 4593, pp. 119–136. Springer, Heidelberg (2007)
54. Shrimpton, T., Stam, M.: Building a collision-resistant compression function from non-compressing primitives. In: Aceto, L., Damgård, I., Goldberg, L., Halldórsson, M., Ingólfsdóttir, A., Walukiewicz, I. (eds.) ICALP 2008, Part II. LNCS, vol. 5126, pp. 643–654. Springer, Heidelberg (2008)
55. Stam, M.: Better security/efficiency tradeoffs for compression functions. In: Wagner, D. (ed.) CRYPTO 2008. LNCS, vol. 5157, pp. 397–412. Springer, Heidelberg (2008)
56. Stam, M.: Blockcipher-based hashing revisited. In: Dunkelman, O. (ed.) FSE 2009. LNCS, vol. 5665, pp. 67–83. Springer, Heidelberg (2009)
57. Steinberger, J.P.: The collision intractability of MDC-2 in the ideal-cipher model. In: Naor, M. (ed.) EUROCRYPT 2007. LNCS, vol. 4515, pp. 34–51. Springer, Heidelberg (2007)
58. Tillich, S., Großschädl, J.: Instruction set extensions for efficient AES implementation on 32-bit processors. In: Goubin, L., Matsui, M. (eds.) CHES 2006. LNCS, vol. 4249, pp. 270–284. Springer, Heidelberg (2006)
59. Tillich, S., Herbst, C.: Boosting AES performance on a tiny processor core. In: Malkin, T. (ed.) CT-RSA 2008. LNCS, vol. 4964, pp. 170–186. Springer, Heidelberg (2008)
60. Tromer, E., Osvik, D.A., Shamir, A.: Efficient cache attacks on AES, and countermeasures. Journal of Cryptology 23, 37–71 (2010)
61. Wu, H.: The hash function JH. Submission to NIST (updated) (2009), http://icsd.i2r.a-star.edu.sg/staff/hongjun/jh/jh_round2.pdf

Author Index

GPSR Compliance

The European Union's (EU) General Product Safety Regulation (GPSR) is a set of rules that requires consumer products to be safe and our obligations to ensure this.

If you have any concerns about our products, you can contact us on ProductSafety@springernature.com

In case Publisher is established outside the EU, the EU authorized representative is:

Springer Nature Customer Service Center GmbH
Europaplatz 3
69115 Heidelberg, Germany

Batch number: 09473985

Printed by Printforce, the Netherlands